ENCYCLOPEDIA OF PHYSICS

EDITED BY

S. FLÜGGE

VOLUME XLIV

NUCLEAR INSTRUMENTATION I

CO-EDITOR

E. CREUTZ

WITH 225 FIGURES

SPRINGER-VERLAG

BERLIN · GÖTTINGEN · HEIDELBERG

1959

HANDBUCH DER PHYSIK

HERAUSGEGEBEN VON

S. FLÜGGE

BAND XLIV

INSTRUMENTELLE HILFSMITTEL
DER KERNPHYSIK I

MITHERAUSGEBER

E. CREUTZ

MIT 225 FIGUREN

SPRINGER-VERLAG

BERLIN · GÖTTINGEN · HEIDELBERG

1959

ISBN 978-3-642-45928-3   ISBN 978-3-642-45926-9 (eBook)
DOI 10.1007/978-3-642-45926-9

Inhaltsverzeichnis.

Anmerkung der Herausgeber.

Einige Beiträge zu diesem Bande wurden vor mehr als zwei Jahren geschrieben und die letzten Änderungen oder Zusätze zu ihnen bei der Korrektur fanden zwischen Dezember 1957 und Februar 1958 statt.

Editorial Note.

Several articles in this volume have been written more than two years ago and the last changes or additions to them in proof have been made between December 1957 and February 1958.

Kaskadengeneratoren.

Von

E. BALDINGER.

Mit 50 Figuren,

1. Einleitung. Im Jahre 1920 hat H. GREINACHER [G 20] in einer allgemeinen Betrachtung über die Gleichrichtung von Wechselspannungen Schaltungen angegeben, die zur Erzeugung hoher Gleichspannungen sehr geeignet sind.

Diese Arbeiten gehen auf die Entwicklung des Ionometers, eines Apparates zur Messung der Ionisation von Radium- und Röntgenstrahlen, zurück. Die notwendige Spannung von einigen 100 Volt sollte auf möglichst einfache Art aus dem Lichtnetz erzeugt werden. Die Lösung dieser Aufgabe wurde durch die sog. Greinacher-Schaltung (Fig. 4) und den Kaskadengenerator (Fig. 24) möglich. In diesem Zusammenhang muß daran erinnert werden, daß zur damaligen Zeit noch keine Elektronenröhren als Ventile zur Verfügung standen. Verwendet wurden „Graetzsche Zellen", die aus einer Aluminium- und einer Eisenelektrode in einer $NaHCO_3$-Lösung als Elektrolyt bestanden und die eine Sperrspannung von nur etwa 30 Volt aufwiesen.

Die von H. GREINACHER angegebenen Schaltungen blieben lange Zeit unbeachtet, bis 1930 J. D. COCKCROFT und E. T. S. WALTON mit ihren Versuchen zur künstlichen Erzeugung von Kernumwandlungen begannen [C 30] und im Jahre 1932 einen Kaskadengenerator von 800 kV Gleichspannung veröffentlichten [C 32]. 1933 berichtete A. BOUWERS[1] über einen Höchstspannungsgenerator, dem die gleiche Schaltung zugrunde lag. Aber schon 1919 hat M. SCHENKEL [S 19] eine ähnliche Vervielfacherschaltung (Fig. 44) beschrieben, die allerdings den Nachteil aufweist, daß die Kondensatoren mit einer von Stufe zu Stufe zunehmenden Spannung beansprucht werden. Ferner hat sich herausgestellt, daß J. SLEPIAN [S 28] in den Vereinigten Staaten 1928 ein Patent erteilt wurde, dem der gleiche Gedanke zugrunde liegt.

Mit der Arbeit von J. D. COCKCROFT und E. T. S. WALTON hat die Verwendung von Kaskadenschaltungen auf dem Gebiete der Kernphysik ihren Anfang genommen. Die heutige Bedeutung solcher Kaskadengeneratoren läßt sich an der Vielfalt der bekannten Anwendungen ermessen. Abgesehen von den bereits erwähnten Hochspannungsanlagen zur Beschleunigung von Protonen oder Deuteronen dienen solche Schaltungen zur Erzeugung der Hochspannung von Röntgenanlagen, Elektronenmikroskopen, Szintillationszählern, Zählrohren, Fernsehapparaten und Kathodenstrahloszillographen.

Es ist das Ziel dieses Artikels, die Möglichkeiten einer rechnerischen Behandlung derartiger Gleichrichterschaltungen aufzuzeigen und ihre allgemeinen Eigenschaften zu erläutern. Die Theorie des Kaskadengenerators läßt sich mit gewissen vereinfachenden Annahmen auf die Theorie des Einweggleichrichters zurückführen, mit der wir uns aus diesem Grunde zunächst beschäftigen wollen. Das Verhalten von einfachen Gleichrichterschaltungen wurde von verschiedenen Autoren theoretisch und experimentell untersucht. Wir werden hier eine

---

[1] A. BOUWERS: Vortrag American Congress of Radiology, Chicago 1933.

Darstellung bevorzugen, die sich für die Erweiterung auf Kaskadengeneratoren besonders eignet, um im Kapitel II eine typische Kaskadenschaltung in ihren Einzelheiten zu diskutieren und insbesondere zu zeigen, welche Einflüsse den Spannungsabfall bei Belastung verursachen und wie sich dieser Spannungsabfall in guter Näherung rechnerisch erfassen läßt. Der letzte Abschnitt dieses Kapitels behandelt verschiedene Varianten, und es wird kurz dargelegt, daß die am speziellen Modell entwickelte Theorie zwanglos auf solche abgeänderte Schaltungen erweitert werden kann. Es zeigt sich ferner, daß es oft günstig ist, eine verhältnismäßig hohe Frequenz der Wechselspannung zu verwenden. Heute werden nur noch selten Anlagen gebaut, die mit 50 oder 60 periodigem Wechselstrom arbeiten. In großen Anlagen benützt man Umformergruppen mit Frequenzen bis zu 10 kHz. In kleineren Anlagen empfiehlt sich die Verwendung von Röhrenoszillatoren mit Frequenzen zwischen 20 und 100 kHz. Abgesehen von der Oszillator- und Leistungsverstärkerstufe tritt dann als neues Problem die zweckmäßige Dimensionierung des Transformators auf, der normalerweise in der Nähe der sekundärseitigen Resonanzfrequenz betrieben wird. Über Oszillatoren und Leistungsverstärker besteht eine umfangreiche Fachliteratur. Infolgedessen beschränkt sich unsere Darstellung auf die Behandlung der passenden Hochspannungstransformatoren, die im Kapitel III zu finden ist.

I. Theoretische Behandlung einfacher Gleichrichterschaltungen.

a) Gleichrichter mit Kondensatoreingang.

2. Spannungsabfall und Spitzenstrom. Zur Berechnung eines Gleichrichters nach Fig. 1 wollen wir die Streuinduktivität des Transformators vernachlässigen und die Kennlinie der Gleichrichterröhre entsprechend Fig. 5 idealisieren, wobei R_v den Innenwiderstand der Röhre im leitenden Zustand bedeutet.

Mit R_t sei der auf die Sekundärseite reduzierte Widerstand des Transformators und mit R die Summe $(R_v + R_t)$ bezeichnet.

Unsere Rechnung ist somit auf der Annahme aufgebaut, daß sich das Ventil im Ersatzschema durch einen konstanten Widerstand R_v in Serie mit einem synchronisierten Schalter darstellen läßt. Die in Wirklichkeit nichtlineare Kennlinie der Gleichrichterröhre kann durch passende Wahl des Widerstandes R_v berücksichtigt werden (vgl. Ziff. 5). Dabei ist zu beachten, daß zur Bestimmung des Spitzenwertes, des Mittelwertes und des Effektivwertes des Ventilstromes für ein und dieselbe Schaltung verschiedene Widerstandswerte zugrunde gelegt werden müssen. Der Spitzenwert des Ventilwiderstandes $\hat{R}$ sei als Quotient vom Spitzenwert der Anodenspannung in Leitrichtung zum Spitzenwert des Ventilstromes definiert. Ganz entsprechend verstehen wir unter $\bar{R}$ den Quotienten aus der im leitenden Zustande auftretenden mittleren Anodenspannung zum Mittelwert des Stromes und unter $|R|$ das Verhältnis der Anodenverlustleistung zum Quadrat des Effektivwertes des Ventilstromes (vgl. Ziff. 5).

Wie O. H. Schade [S 43] gezeigt hat, bilden bei Hochvakuumdioden die so definierten Widerstände eine Proportion, die von der Belastung des Gleichrichters und von den Parametern der Schaltung weitgehend unabhängig ist. Mit einem Fehler, der 5% nicht überschreitet, gelten für Kennlinien der Form $i = \varkappa \cdot u^{\frac{3}{2}}$ folgende Beziehungen:

$$\bar{R} = 1{,}14\,\hat{R}, \quad |R| = 1{,}07\,\hat{R}. \tag{2.1}$$

Fig. 6 zeigt einige Kennlinien von Gleichrichterröhren. Bei bekanntem Spitzenstrom läßt sich aus diesen Kennlinien der Widerstand $\hat{R}$ bestimmen.

Um einen Gleichrichter zu dimensionieren, wird zunächst der Spitzenstrom geschätzt, mit dem daraus bestimmten $\hat{R}$ bzw. $\bar{R}$ der Gleichrichter berechnet und der zugehörige, berechnete Spitzenstrom mit der Schätzung verglichen. Wenn

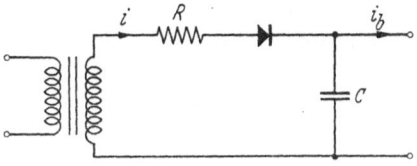

Fig. 1. Einweggleichrichter mit Kondensatoreingang.
$R = R_t + R_v.$

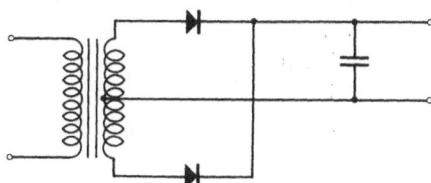

Fig. 2. Doppelweggleichrichter.

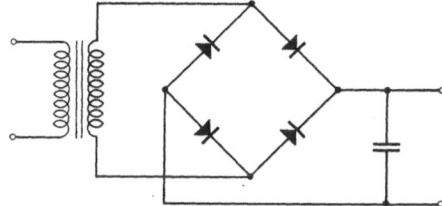

Fig. 3. Graetz-Schaltung.

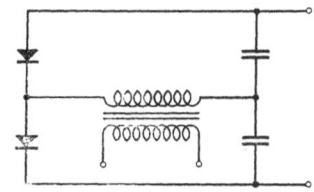

Fig. 4. Greinacher-Schaltung (Spannungsverdopplung).

notwendig, läßt sich die ursprüngliche Annahme solange verbessern, bis man Übereinstimmung erhält. Da zu $\bar{R}$ der Widerstand des Transformators in Serie liegt, wirkt sich in der Regel ein Fehler in der Schätzung nicht sehr stark aus. Insbesondere gilt diese Bemerkung für Kaskadengeneratoren, bei denen, wie später gezeigt wird, der Einfluß des Transformatorwiderstandes proportional zur Stufenzahl wächst und die Wirkung der Ventilwiderstände oft beträchtlich überwiegt.

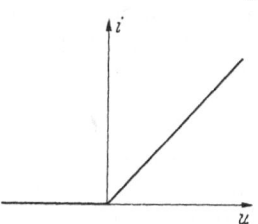

Fig. 5. Idealisierte Kennlinie der Gleichrichter. $i = (R_v)^{-1} \cdot u$ für $u > 0$; $i = 0$ für $u < 0$.

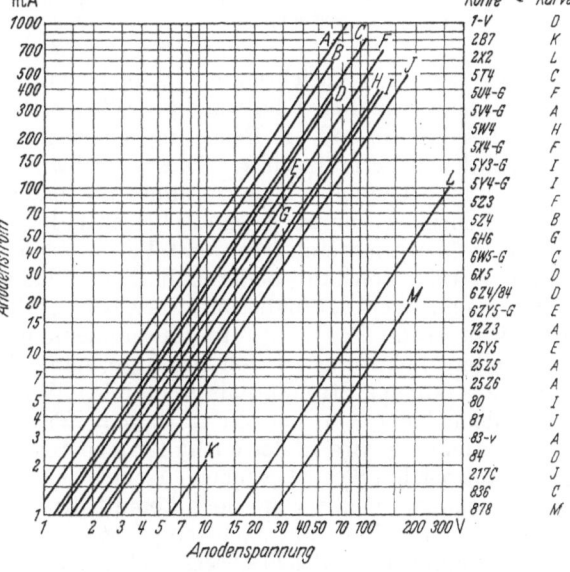

Fig. 6. Kennlinien von Gleichrichterröhren (Mittelwerte), welche das $U^{\frac{3}{2}}$-Gesetz befolgen, zur Bestimmung des Widerstandes $\hat{R}$. (Aus H. J. Reich [1], S. 576.)

Eine gasgefüllte Gleichrichterröhre läßt sich durch eine äquivalente Schaltung, bestehend aus einer elektromotorischen Gegenkraft von der Größe der Zündspannung, zu der ein kleiner Widerstand und ein synchronisierter Schalter in

Serie liegt, ersetzen. Für kleine Gleichrichterröhren hat dieser Widerstand einen Wert von etwa 4 Ohm und die elektromotorische Gegenkraft beträgt etwa 10 Volt.

Der zeitliche Verlauf der Ausgangsspannung eines Einweggleichrichters nach Fig. 1 ist in Fig. 7 dargestellt. Das Ventil leitet, wenn das Potential der Anode positiv gegenüber dem der Kathode ist, also von t_1 bis t_2 (Ladezeit). Während dieser Ladezeit läßt sich der Verlauf der Spannung u_c am Kondensator durch die folgende Gleichung beschreiben:

$$\left.\begin{aligned} U_0 \cos \omega t &= u_c + i\,R \\ &= U_g - \frac{\delta U}{2} + \int_{t_1}^{t} \frac{i - i_b}{C}\,dt + i\,R \quad (t_1 \leq t \leq t_2). \end{aligned}\right\} \quad (2.2)$$

Dabei bedeutet δU die Rippelspannung (vgl. Fig. 7) und U_g den Mittelwert zwischen der höchsten und der tiefsten Spannung des Kondensators. Die Gleichspannung $\bar{u}_c$, also der zeitliche Mittelwert von u_c, weicht in der Regel nur wenig von dem soeben definierten Wert U_g ab. Wir dürfen deshalb in sehr guter Näherung

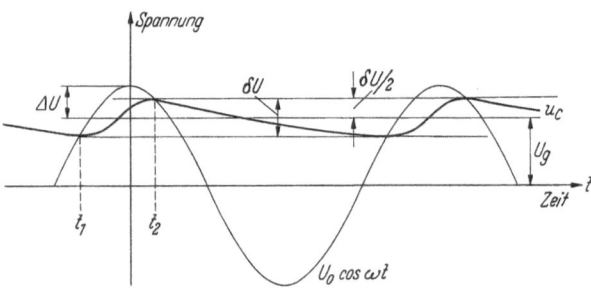

Fig. 7. Zeitlicher Verlauf der Ausgangsspannung des Einweggleichrichters nach Fig. 1.

$$\bar{u}_c \approx U_g \qquad (2.3)$$

setzen. In gewissen Fällen, wie z. B. linearer Entladung und linearer Aufladung des Kondensators, ist die Beziehung (2.3) naturgemäß exakt erfüllt.

Über den vom Gleichrichter abgegebenen Strom i_b sind von Fall zu Fall verschiedene Annahmen notwendig. Wird z. B. an den Ausgang in Fig. 1 ein Ohmscher Widerstand als Belastung angeschlossen, so variiert i_b entsprechend dem zeitlichen Verlauf von u_c. Dieses Problem wurde von O. H. Schade [S 43] vollständig behandelt. Eine zusammenfassende Darstellung seiner Ergebnisse ist im Lehrbuch von H. J. Reich [1] enthalten. In vielen Fällen ist indessen zwischen Gleichrichterausgang und Verbraucher ein Tiefpaßfilter geschaltet, so daß i_b zeitlich konstant bleibt und die Entladung des Kondensators in guter Näherung linear erfolgt (näherungsweise behandelt durch R. G. Mitchell [Mi 43], J. Kammerloher [2] u. a.). Im Zusammenhang mit Kaskadenschaltungen ist eine weitere Belastungsart von Interesse, nämlich der Fall, daß der Belastungsstrom aus kurzen Stromimpulsen besteht. Während der Ladezeit soll i_b verschwindend klein sein, und die in diesem Zeitintervall dem Kondensator C zugeführte Ladung wird in der nachfolgenden Sperrzeit des Ventils als kurzzeitiger Stromstoß nach außen abgegeben. Keiner der oben erwähnten Fälle trifft exakt beim Kaskadengenerator zu, doch kommt die dritte Annahme den tatsächlichen Verhältnissen weitaus am nächsten [B 56]. Es sei noch bemerkt, daß die Ausgangsspannung des Gleichrichters erst bei größerer Belastung, also bei merklichem Spannungsabfall, auf die Art der Annahme über den zeitlichen Verlauf von i_b empfindlich wird.

Mit Rücksicht auf die Kaskadenschaltungen bevorzugen wir in den folgenden Betrachtungen die dritte Annahme und setzen in (2.2)

$$i_b = 0 \quad \text{und} \quad i = \frac{U_0 \cos \omega t - u_c}{R}. \qquad (2.4)$$

Unter Verwendung der Abkürzungen

$$\frac{u_c}{U_0} = u_c^*, \quad \alpha = \omega R C, \quad \varphi = \omega t \tag{2.5}$$

erhält man aus (2.2) die Differentialgleichung

$$\frac{d u_c^*}{d \varphi} + \frac{u_c^*}{\alpha} = \frac{\cos \varphi}{\alpha}. \tag{2.6}$$

Ihre Lösung lautet:

$$u_c^* = \frac{\cos \varphi + \alpha \sin \varphi}{1 + \alpha^2} + \left(u_{c0}^* - \frac{1}{1 + \alpha^2} \right) e^{-\varphi/\alpha}, \tag{2.7}$$

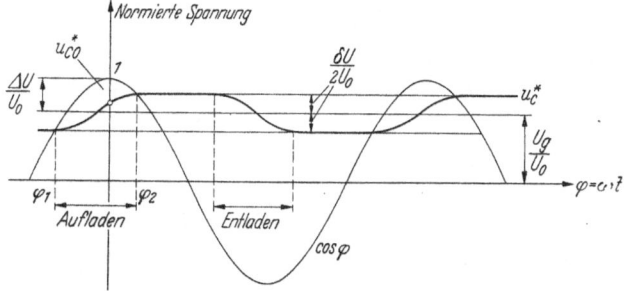

Fig. 8. Zur Berechnung des Spannungsabfalles eines Einweggleichrichters bei impulsmäßiger Belastung.

wobei u_{c0}^* die normierte Spannung des Kondensators für $\varphi = 0$ bedeutet (vgl. Fig. 8). Die beiden Phasenwinkel φ_1 und φ_2, die Anfang und Ende der Ladezeit kennzeichnen, ergeben sich aus der Bedingung

$$u_c^* = \cos \varphi, \tag{2.8}$$

die zu folgender Bestimmungsgleichung für φ_1 und φ_2 führt:

$$\frac{u_{c0}^* (1 + \alpha^2) - 1}{\alpha^2} = \left(\cos \varphi - \frac{\sin \varphi}{\alpha} \right) e^{\varphi/\alpha}. \tag{2.9}$$

Die Kenntnis von φ_1 und φ_2 gestattet für jede Belastung die minimale und die maximale Spannung des Kondensators und damit die Gleichspannung des Einweggleichrichters zu ermitteln. Das Ergebnis einer numerischen Rechnung ist in den Fig. 9 und 10 zusammengestellt, und in Fig. 11 ist ferner der Spitzenwert des Ventilstromes in Abhängigkeit von der Belastung mit $\alpha = \omega R C$ als Parameter aufgetragen.

Zur Berechnung des Spitzenwertes des Ventilstromes bestimmen wir aus Gl. (2.7) die Spannung u, die während der Aufladezeit über dem Ventil bzw. über dem Widerstand R in Fig. 1 liegt[1].

$$u^* = \cos \varphi - \frac{\cos \varphi + \alpha \sin \varphi}{1 + \alpha^2} - \left(u_{c0}^* - \frac{1}{1 + \alpha^2} \right) e^{-\varphi/\alpha}. \tag{2.10}$$

Der Phasenwinkel φ_m, bei dem die größte Spannung $\hat{u}$ und damit der Spitzenstrom $\hat{\imath}$ auftritt, erhält man aus der Bedingung $\partial u^*/\partial \varphi = 0$, welche die Bestimmungsgleichung

$$\frac{u_{c0}^* (1 + \alpha^2) - 1}{\alpha^2} = (\cos \varphi_m + \alpha \sin \varphi_m) e^{\varphi_m/\alpha} \tag{2.11}$$

liefert. $\varphi_m = f(u_{c0}^*, \alpha)$ kann dann graphisch oder numerisch ermittelt werden.

[1] Alle mit einem * bezeichneten Größen sind auf die Spannung U_0 normiert.

Durch Einsetzen von (2.11) in (2.10) erhält man die einfache Beziehung:

$$\sin \varphi_m = -\frac{\hat{u}^*}{\alpha} = -\frac{\hat{u}}{U_0 \alpha} = -\frac{\hat{\imath}}{\omega C U_0} \tag{2.12}$$

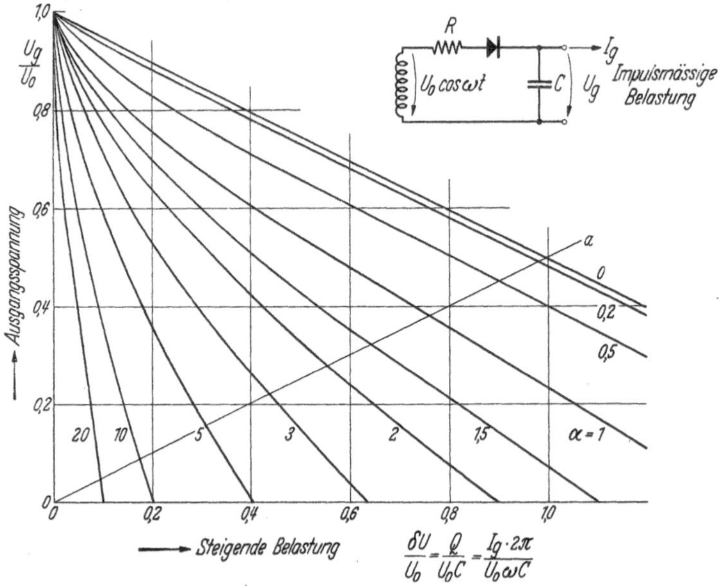

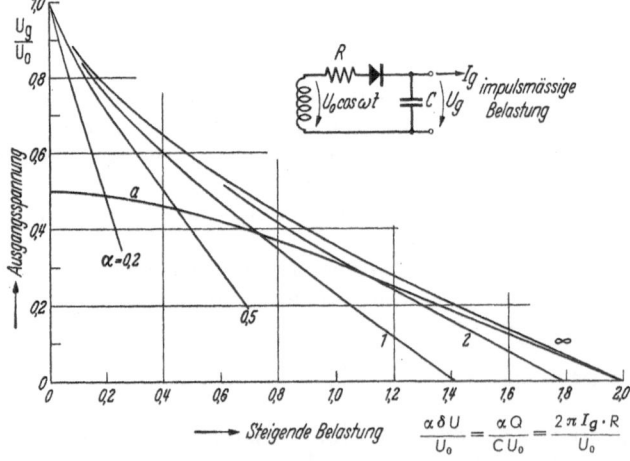

Fig. 9. Ausgangsspannung des Einweggleichrichters in Abhängigkeit von $\delta U/U_0$ bei impulsmäßiger Belastung mit $\alpha = \omega R C$ als Parameter. Die Gerade a verbindet Punkte, für die der Einschaltwinkel $\varphi_1 = -\pi/2$ beträgt und der Kondensator jedesmal auf die Spannung 0 entladen wird.

mit deren Hilfe sich der Spitzenstrom aus φ_m leicht bestimmen läßt. Es sei noch bemerkt, daß (2.12) nicht an die eingangs getroffene Voraussetzung einer

Fig. 10. Ausgangsspannung des Einweggleichrichters in Abhängigkeit von $\alpha\,\delta U/U_0$ bei impulsmäßiger Belastung mit $\alpha = \omega R C$ als Parameter. Die Kurve a entspricht der Geraden a in Fig. 9. Im Unterschied zu Fig. 9 zeigt Fig. 10 das Verhalten bei großen Werten des Parameters α.

ideal geknickten Ventil-Kennlinie gebunden ist, sondern ganz allgemein für beliebige Kennlinien des Ventils gilt, vorausgesetzt daß der Spannungsabfall am Ventil mit wachsendem Strom zunimmt. Um dies zu beweisen, greifen wir auf

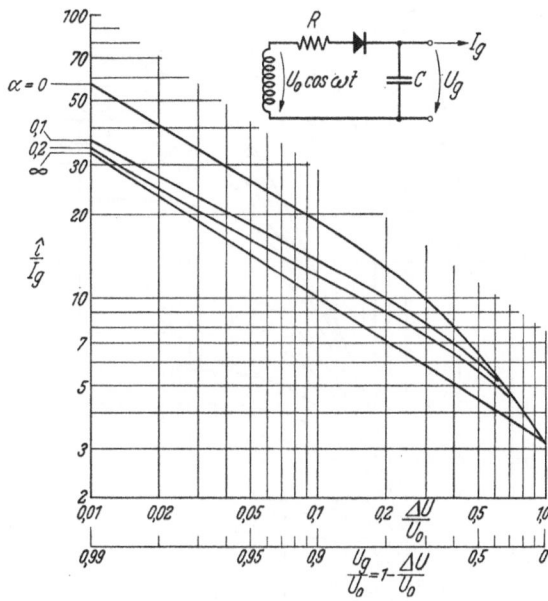

Fig. 11. Spitzenwert ($\hat{\imath}$) zu Mittelwert (I_g) des Ventilstromes in Funktion der Ausgangsspannung U_g/U_0 bei impulsmäßiger Belastung.

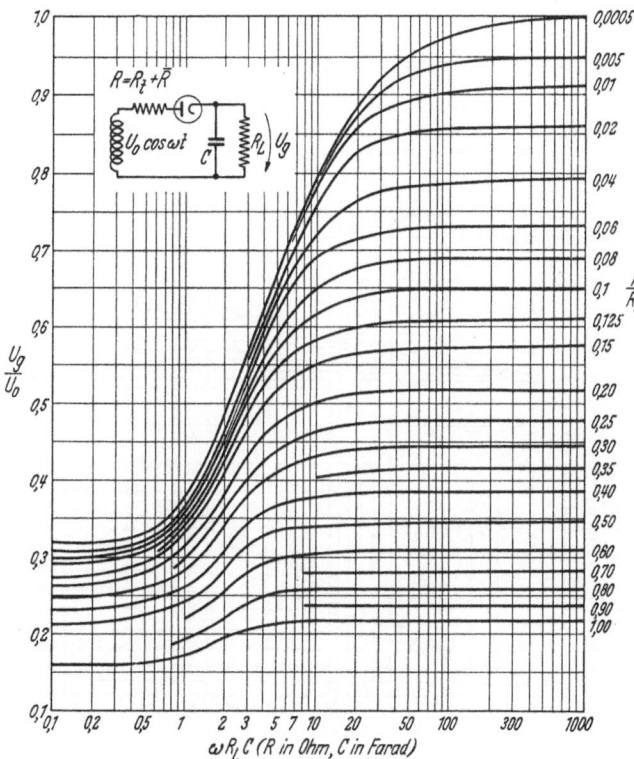

Fig. 12. Ausgangsspannung des Einweggleichrichters in Abhängigkeit von $\omega R_L C$ bei ohmscher Belastung. R_L Lastwiderstand. R_t Widerstand des Transformators. $\overline{R}$ Äquivalenter Ventilwiderstand nach Gl. (5.2). (Aus H. J. Reich [1], S. 577.)

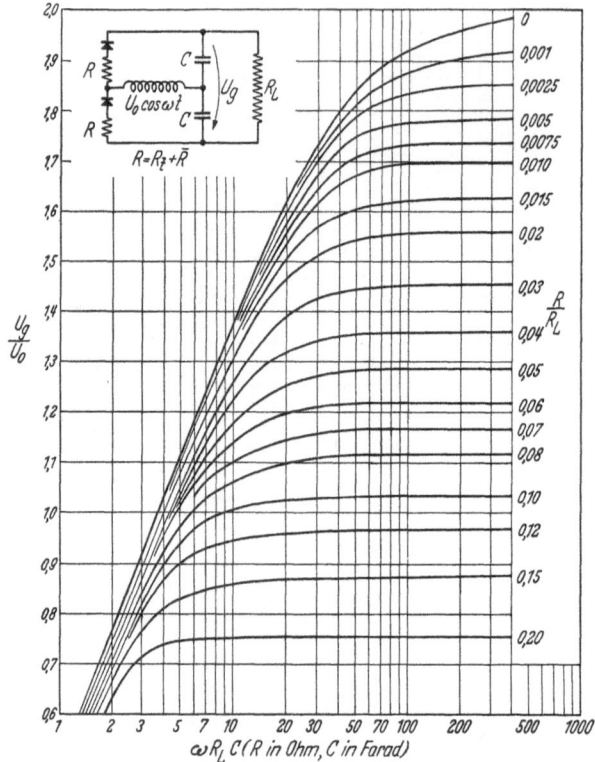

Fig. 13. Ausgangsspannung der Greinacher-Schaltung in Abhängigkeit von $\omega R_L C$ bei ohmscher Belastung. R_L Lastwiderstand. R_t Widerstand des Transformators. $\bar{R}$ Äquivalenter Ventilwiderstand nach Gl. (5.2).

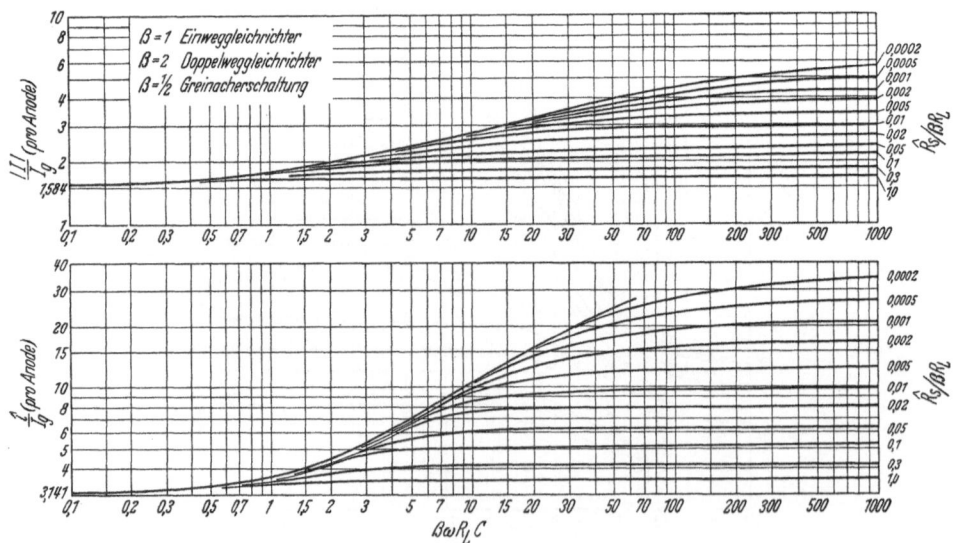

Fig. 14. Effektivwert $|I|$ und Spitzenwert $\hat{\imath}$ des Ventilstromes bezogen auf den im Mittel fließenden Gleichstrom bei ohmscher Last R_L in Abhängigkeit von $\beta \omega R_L C$. Nach O. H. Schade [S 43] gilt für $U^{\frac{3}{2}}$ Kennlinien des Ventils $\hat{R}_s = R_t + \hat{R}$
$R_t =$ Widerstand des Transformators. $\hat{R}$ nach Gl. (5.1). (Aus H. J. Reich [1], S. 580.)

Gl. (2.2) zurück, die mit den Abkürzungen $\omega t = \varphi$, $iR/U_0 = u^*$ und mit $i_b = 0$ wie folgt lautet:

$$\cos \varphi = \text{const} + \int_{\varphi_1}^{\varphi} \frac{i}{\omega C U_0} d\varphi + u^* \qquad (2.13)$$

und in dieser Form für beliebige Kennlinien gilt. Man beachte, daß bei impulsmäßiger Belastung des Gleichrichters $i_b = 0$ ist. Differenzieren wir diese Gleichung nach φ und berücksichtigen, daß zum Zeitpunkt des Spitzenstromes

$$\frac{\partial u^*}{\partial \varphi} = 0$$

wird, so folgt unmittelbar die Beziehung (2.12).

Wie bereits bemerkt, hängt der Spannungsabfall des Gleichrichters mit wachsendem Strom immer stärker vom Charakter der Belastung ab. Ein Vergleich von Fig. 9 und 10 mit Fig. 12 bestätigt diese Feststellung. In Fig. 12 sind die von O. H. SCHADE [S 43] ermittelten Kurven dargestellt, welche für einen am Ausgang mit einem Ohmschen Widerstand R_L belasteten Einweggleichrichter gelten.

Fig. 13 zeigt die analogen Kurven für die Greinacher-Schaltung (Spannungsverdopplung) und Fig. 14 die in beiden Anordnungen auftretenden Spitzenströme, sowie den für die Erwärmung maßgebenden Effektivwert des Ventilstromes.

Die Greinacher-Schaltung läßt sich übrigens in einfacher Weise auf den Einweggleichrichter reduzieren. Die entsprechenden Belastungskurven sind hier gesondert dargestellt, weil die Spannungsverdopplerschaltung in Hochspannungsanlagen bis zu 100 kV oft verwendet wird (z.B. zur Erzeugung der Beschleunigungsspannung von Elektronenmikroskopen).

3. Näherungsformel zur Berechnung des Spannungsabfalles bei impulsmäßiger Belastung. Die Diskussion einer Reihe von Fragen wird insbesondere beim Kaskadengenerator dadurch erschwert, daß die exakte Behandlung des Gleichrichters auf transzendente Gleichungen führt, deren Lösung zunächst nur numerisch möglich ist. Es besteht deshalb ein Bedürfnis nach einem analytischen Ausdruck für die in Fig. 9 und Fig. 10 dargestellten Ergebnisse. Tatsächlich läßt sich der Spannungsabfall ΔU eines Gleichrichters in sehr guter Näherung durch folgende halbempirische Formel darstellen:

$$\Delta U = U_0 - U_g \approx \sqrt[3]{\Delta U_C^3 + \Delta U_R^3} \; . \qquad (3.1)$$

Hier bedeutet ΔU_R den Spannungsabfall, welcher bei unendlich großem Kondensator infolge des Widerstandes R und ΔU_C den Spannungsabfall, der bei vernachlässigbar kleinem Widerstand infolge des endlichen Wertes der Kapazität C auftritt. Die beiden Größen ΔU_R und ΔU_C lassen sich in einfacher Weise berechnen und wir gelangen so zu einer expliziten Darstellung der Ausgangsspannung, die sich in vielen Fällen als brauchbar erweist. Das Zustandekommen der empirischen Näherungsformel (3.1) läßt sich wie folgt erläutern:

Es ist vernünftig zu verlangen, daß die Grenzfälle unendlich großer Kapazität C und verschwindenden Widerstandes R exakt durch (3.1) wiedergegeben werden. Die von MITCHELL[1] angegebenen Kurven legen nahe, eine Interpolation zwischen diesen Grenzfällen in der Form

$$\Delta U = \sqrt[n]{\Delta U_C^n + \Delta U_R^n} \qquad (3.1\,\text{a})$$

_____

[1] [Mi 43], besonders Fig. 5.

vorzunehmen. Eine numerische Auswertung von (3.1a) und ein Vergleich mit exakt bekannten ΔU zeigt, daß eine ausreichende Übereinstimmung mit $n = 3$ erzielt wird.

Bei impulsmäßiger Belastung ist ΔU_C exakt gleich der halben Rippelspannung:

$$\Delta U_C = \frac{\delta U}{2} . \tag{3.2}$$

Für ΔU_R läßt sich, wie im nächsten Abschnitt gezeigt wird, eine einfache Näherungsformel herleiten [vgl. Gl. (4.7) für $n = 1$]. Man erhält

$$\frac{Q}{C\,U_0} \approx \frac{4}{3}\frac{\sqrt{2}}{\alpha}\left(\frac{\Delta U_R}{U_0}\right)^{\frac{3}{2}}, \tag{3.3}$$

wobei Q die während der Ladezeit dem Kondensator C zugeführte Elektrizitätsmenge darstellt. Andererseits muß

$$Q = \delta U \cdot C \tag{3.4}$$

sein. Mit Gl. (3.4) und (3.3) bestimmt sich somit aus (3.1) der Spannungsabfall ΔU näherungsweise zu:

$$\Delta U = \frac{\delta U}{2}\sqrt[3]{1 + \left(\frac{3}{2}\alpha\right)^2\left(\frac{U_0}{\delta U}\right)} . \tag{3.5}$$

Ein Vergleich dieser einfachen Beziehung mit den in Fig. 9 und 10 dargestellten Ergebnissen einer strengen Rechnung zeigt eine so gute Übereinstimmung, daß wir bei der später erfolgenden Behandlung des Kaskadengenerators von der Beziehung (3.5) unbedenklich Gebrauch machen dürfen, ohne daß größere Fehler zu erwarten sind. Diese Fehler lassen sich übrigens mit Hilfe der Tabelle 1 leicht abschätzen.

b) Gleichrichter mit sehr großem Kondensator.

4. Die Gleichrichterkennlinie folge der allgemeinen Beziehung $i = \varkappa\,u^n$, $(u > 0)$. Wird der Wert des Kondensators C in Fig. 1 genügend groß gewählt, so darf seine Spannung als zeitlich konstant angesehen werden. Die Berechnung des Spannungsabfalles läßt sich unter dieser speziellen Annahme wesentlich vereinfachen. Es ist ohne weiteres ersichtlich, daß der zeitliche Verlauf des Laststromes i_b nun keinen Einfluß mehr auf das Ergebnis ausüben kann.

Zunächst wollen wir eine brauchbare Näherungsformel herleiten, welche für beliebige Gleichrichterkennlinien der Form

$$i = \varkappa\,u^n \quad \text{für} \quad u > 0, \quad i = 0 \quad \text{für} \quad u < 0 \tag{4.1}$$

gilt, und für die Exponenten $n = 1$, $n = 1,5$ und $n = 2$ diese Näherungsformel mit den Ergebnissen einer strengen Rechnung vergleichen. Wie die Angabe von P. Lorrain und Mitarbeitern zeigt [L 56], deren Ventile eine $u^{1,18}$ Kennlinie befolgen (vgl. auch [P 55]), ist eine Rechnung für beliebige Exponenten n von praktischem Interesse. Das Resultat dieser Untersuchung gestattet, die in Gl. (3.3) verwendete Beziehung abzuleiten.

Während einer Periode der Wechselspannung $U_0\cos\omega t$ wird dem Kondensator die Ladung Q

$$\left.\begin{aligned} Q &= \varkappa\int_{t_1}^{t_2}(U_0\cos\omega t - U_g)^n\,dt, \\[2mm] Q &= \frac{2\varkappa}{\omega}U_0^n\int_0^{\varphi_2}\left(\cos\varphi - \frac{U_g}{U_0}\right)^n d\varphi \end{aligned}\right\} \tag{4.2}$$

Tabelle 1. *Vergleich der Näherungsformel (3.5) mit den Ergebnissen einer exakten Rechnung für den impulsmäßig belasteten Einweggleichrichter.*

$\dfrac{\delta U}{U_0}$	$\dfrac{U_g}{U_0}$ exakt	$\dfrac{U_g}{U_0}$ (3.5)	$\dfrac{\delta U}{U_0}$	$\dfrac{U_g}{U_0}$ exakt	$\dfrac{U_g}{U_0}$ (3.5)
$\alpha = 0$			$\alpha = 1$		
0	1	1	0	1	1
0,2	0,9	0,9	0,062	0,891	0,89
0,6	0,7	0,7	0,204	0,759	0,77
1	0,5	0,5	0,429	0,590	0,60
1,4	0,3	0,3	0,756	0,378	0,40
2	0	0	1,414	0,000	0,03
$\alpha = 0,2$			$\alpha = 1,5$		
0	1	1	0	1	1
0,137	0,916	0,92	0,022	0,932	0,93
0,449	0,756	0,76	0,138	0,766	0,77
0,910	0,526	0,53	0,363	0,544	0,55
1,397	0,282	0,29	0,555	0,387	0,40
1,961	0,000	0,01	1,109	0,000	0,02
$\alpha = 0,5$			$\alpha = 5$		
0	1	1	0	1	1
0,178	0,849	0,86	0,008	0,925	0,92
0,458	0,686	0,70	0,055	0,724	0,72
0,901	0,451	0,47	0,100	0,590	0,59
1,125	0,335	0,36	0,233	0,287	0,27
1,789	0,000	0,02	0,392	0,000	−0,03

$\alpha \dfrac{\delta U}{U_0}$	$\dfrac{U_g}{U_0}$ exakt	$\dfrac{U_g}{U_0}$ (3.5)
$\alpha = \infty$		
0	1	1
0,060	0,900	0,90
0,315	0,700	0,70
0,685	0,500	0,49
1,148	0,300	0,28
2,000	0,000	−0,04

zugeführt. Die Phasenwinkel φ_1 und φ_2, die Anfang und Ende der Aufladung des Kondensators kennzeichnen, lassen sich aus

$$\cos \varphi = \frac{U_g}{U_0} \tag{4.3}$$

leicht berechnen. Entwickeln wir (4.3) nach Potenzen von φ und brechen die Reihe mit dem quadratischen Gliede ab, so folgt

$$\varphi_{1,2} = \mp \sqrt{\frac{2\Delta U}{U_0}}\,, \tag{4.4}$$

wobei ΔU den Spannungsabfall $U_0 - U_g$ bedeutet. Entwickelt man in (4.2) den Integranden entsprechend wie bei (4.3), so erhält man für die Ladung Q in erster Näherung

$$Q = \frac{2\varkappa U_0^n}{\omega} \int\limits_0^{\varphi_2} \left(\frac{\Delta U}{U_0} - \frac{\varphi^2}{2} \right)^n d\varphi. \tag{4.5}$$

Für gewisse Exponenten läßt sich (4.5) geschlossen integrieren; bei beliebigem n entwickeln wir den Integranden in eine Binominalreihe und erhalten

$$Q = \frac{2\varkappa U_0^n}{\omega} \int\limits_0^{\varphi_2} \left[\left(\frac{\Delta U}{U_0} \right)^n - n \left(\frac{\Delta U}{U_0} \right)^{n-1} \frac{\varphi^2}{2} + \frac{n(n-1)}{2!} \left(\frac{\Delta U}{U_0} \right)^{n-2} \left(\frac{\varphi^2}{2} \right)^2 \mp \cdots \right] d\varphi. \tag{4.6}$$

Integriert man gliedweise und setzt für die obere Grenze $\varphi_2 = + \sqrt{2 \frac{\Delta U}{U_0}}$ ein, so wird

$$Q = \frac{2\sqrt{2}}{\omega} \varkappa A_n U_0^n \left(\frac{\Delta U}{U_0}\right)^{n+\frac{1}{2}}, \tag{4.7}$$

wobei A_n durch die folgende Reihe gegeben ist

$$A_n = \left\{ 1 - \frac{n}{3} + \frac{n(n-1)}{1 \cdot 2 \cdot 5} - \frac{n(n-1)(n-2)}{1 \cdot 2 \cdot 3 \cdot 7} \pm \cdots \right\}. \tag{4.8}$$

Für einige wichtige Exponenten sind die Konstanten A_n in Tabelle 2 zusammengestellt.

Ein Vergleich der Näherungsformel (4.7) mit den Ergebnissen einer exakten Integration, wie sie z.B. im Lehrbuch von J. Kammerloher [2] für die Exponenten $n = 1$, $n = \frac{3}{2}$ und $n = 2$ durchgeführt ist (vgl. auch W. H. Aldous [A 36]), zeigt, daß die Abweichungen für alle normalerweise auftretenden Spannungsabfälle 5% kaum erreichen. Um den Vergleich zu erleichtern, schreiben wir die strenge Lösung der Gln. (4.2) und (4.3) in der Form:

Tabelle 2. *Die Konstante A_n aus Gl. (4.7)*.

n	Kennlinie	A_n
$\frac{1}{2}$	$i = \varkappa \sqrt{u}$	$\pi/4 = 0{,}785$
1	$i = \varkappa u$	$\frac{2}{3} = 0{,}667$
$\frac{3}{2}$	$i = \varkappa u^{\frac{3}{2}}$	$3\pi/16 = 0{,}588$
2	$i = \varkappa u^2$	$\frac{8}{15} = 0{,}533$

$$Q = \frac{2\sqrt{2}}{\omega} \varkappa U_0^n \left(\frac{\Delta U}{U_0}\right)^{n+\frac{1}{2}} B_n, \tag{4.9}$$

wobei B_n für jedes n schwach von $\Delta U/U_0$ abhängt. So läßt sich B_n für den speziellen Exponenten $n = \frac{3}{2}$ durch die folgende Reihe darstellen [2]:

$$B_{\frac{3}{2}} = A_{\frac{3}{2}} \left[1 + \frac{1}{12} \left(\frac{\Delta U}{2 U_0}\right) + \frac{3}{128} \left(\frac{\Delta U}{2 U_0}\right)^2 + \frac{5}{512} \left(\frac{\Delta U}{2 U_0}\right)^3 + \cdots \right]. \tag{4.10}$$

Die Zusammenstellung in Tabelle 3 zeigt das Ergebnis einer numerischen Rechnung, wobei für B_2 aus der Integration von (4.2) und der Definitionsgleichung (4.9)

$$B_2 = \frac{1}{\sqrt{2}} \left(\frac{1}{1 - U_g/U_0}\right)^{\frac{5}{2}} \left\{ \left(\frac{1}{2} + \left(\frac{U_g}{U_0}\right)^2\right) \arccos\left(\frac{U_g}{U_0}\right) - \frac{3}{2} \frac{U_g}{U_0} \sqrt{1 - \left(\frac{U_g}{U_0}\right)^2} \right\}$$

folgt. Die Näherung (4.7) stimmt befriedigend mit der strengen Rechnung überein, so daß Gl. (4.7) mit $n = 1$ in (3.3) verwendet werden darf. In Tabelle 3 ist

Tabelle 3. *Vergleich zwischen der näherungsweise bestimmten Größe A_n mit dem Ergebnis einer strengen Rechnung (B_n).*

U_g/U_0	$n=1$		$n=\frac{3}{2}$			$n=2$		
	A_1	B_1	$A_{\frac{3}{2}}$	$B_{\frac{3}{2}}$	$B_1/B_{\frac{3}{2}}$	A_2	B_2	B_1/B_2
1,0	0,667	0,667	0,588	0,588	1,13	0,533	0,533	1,25
0,8	0,667	0,673	0,588	0,593	1,13	0,533	0,537	1,25
0,6	0,667	0,681	0,588	0,598	1,14	0,533	0,541	1,26
0,4	0,667	0,687	0,588	0,604	1,14	0,533	0,545	1,26
0,2	0,667	0,697	0,588	0,610	1,14	0,533	0,549	1,27
0,0	0,667	0,707	0,588	0,616	1,15	0,533	0,555	1,27

ferner der Quotient B_1/B_n aufgeführt, der weitgehend von der Belastung unabhängig ist und somit eine gute Konstante darstellt, die für jeden Exponenten n einen charakteristischen Wert annimmt. Von diesem Resultat werden wir im nächsten Abschnitt Gebrauch machen.

5. Die Widerstände $\overline{R}$, $\widehat{R}$ und $|R|$ und ihre Bestimmung für Kennlinien der Form $i = \varkappa u^n$. Wie in der Einleitung erwähnt, läßt sich in Gleichrichtern, jedem Ventil, das eine Kennlinie der Form $i = \varkappa u^n$ für $u > 0$ und $i = 0$ für $u < 0$ aufweist, ein Ersatzventil mit einer ideal geknickten Kennlinie $i = \dfrac{u}{R}$, $u > 0$ so zuordnen, daß beide Anordnungen denselben Spannungsabfall zeigen. Dieses Vorgehen hat den großen Vorteil, daß sich eine gekrümmte Gleichrichterkennlinie eliminieren läßt, und der äquivalente Ersatzwiderstand zu weiteren in Serie zum Ventil liegenden Ohmschen Widerständen addiert werden darf.

Wir definieren $\widehat{R}$ als Spitzenwert der Ventilspannung in Leitrichtung dividiert durch den Spitzenwert des Ventilstromes.

$$\widehat{R} = \frac{\widehat{u}}{i}, \qquad (5.1)$$

ferner $\overline{R}$ als Quotienten aus mittlerer Ventilspannung in Leitrichtung zum mittleren Ventilstrom

$$\overline{R} = \frac{\int_{t_1}^{t_2} u \, dt}{\int_{t_1}^{t_2} i \, dt} \qquad (5.2)$$

und schließlich $|R|$ als das Verhältnis der Anodenverlustleistung zum Quadrat des Effektivwertes des Ventilstromes. Die so definierten Größen bilden eine Proportion, die weitgehend von den Einzelheiten der Schaltung und der Belastung unabhängig ist.

Der erste Schritt in unserer Betrachtung besteht im Nachweis, daß der äqui-

Fig. 15a u. b. Einweggleichrichter (a) und zugeordnete Anordnung (b) mit ideal geknickter Kennlinie des Ventils. Beide Gleichungen gelten während der Ladezeit des Kondensators, die zur Zeit t_1 beginne. Der Belastungsstrom bestehe aus kurzen Stromimpulsen der Ladung Q (Fig. 8).

valente Ersatzwiderstand R_{ae} in guter Näherung durch den oben definierten Widerstand $\overline{R}$ gegeben ist, während wir im zweiten Schritt den Quotienten $\overline{R}/\widehat{R}$ diskutieren.

Zunächst betrachten wir die Schaltung Fig. 15a, bei der die Kennlinie die Form $i = \varkappa u^n$ besitzt, und vergleichen sie mit der Schaltung Fig. 15b, deren Ventil definitionsgemäß der Kennlinie $i_{ae} = u_{ae}/R_{ae}$ folge. Die beiden Schaltungen bezeichnen wir als äquivalent, wenn:

1. Die Belastungsströme und die Transformatorspannung gleich sind.
2. Die Kapazitäten denselben Wert besitzen.
3. Am Ausgang gleiche Spannungen U_g erzeugt werden.

Aus diesen Voraussetzungen folgt, daß pro Periode der Wechselspannung in den beiden Anordnungen dieselbe Ladung transportiert wird

$$Q = \int_{t_1}^{t_2} i \, dt = \int_{t_1}^{t_2} i_{ae} \, dt. \qquad (5.3)$$

Gleiche Ladung Q und gleiche Kondensatoren C bedingen bei impulsmäßiger Belastung die gleiche Größe der Rippelspannung und somit:

$$D = D_{ae}.$$

Aus beiden in Fig. 15 angegebenen Gleichungen folgern wir weiter

$$\int\limits_{t_1}^{t} \frac{i}{C}\, dt + u = \int\limits_{t_1}^{t} \frac{i_{ae}}{C}\, dt + u_{ae}.$$

Diese Beziehung integrieren wir nach t in den Grenzen t_1 bis t_2 und erhalten:

$$\int\limits_{t_1}^{t_2} u\, dt = \int\limits_{t_1}^{t_2} u_{ae}\, dt - \frac{1}{C} \int\limits_{t_1}^{t_2} dt \left\{ \int\limits_{t_1}^{t} (i - i_{ae})\, dt \right\}. \tag{5.4}$$

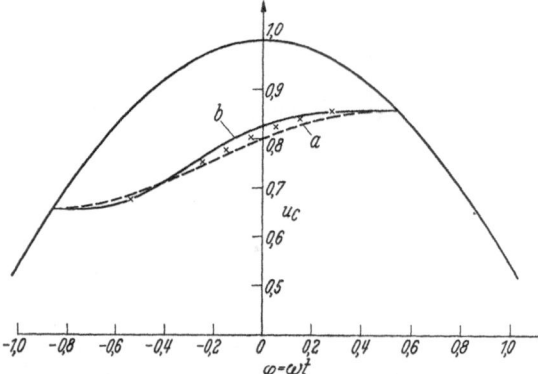

Fig. 16. Die Spannung u_c in Abhängigkeit von φ für $\alpha = 1$. $\alpha = \omega R_{ae} C \approx \omega \overline{R} C$. Die Kurve a gilt für die ideal geknickte Kennlinie des äquivalenten Gleichrichters (berechnet), die Kurve b für eine Ventilkennlinie der Form $i = \varkappa u^3$ und die Punkte für $i = \varkappa u^2$ (beide durch graphische Integration gewonnen). Aus den Kurven lassen sich $\hat{i}/\hat{i}_{ae}$ und $\overline{R}/\hat{R}$ mit einer Genauigkeit von etwa 5% bestimmen. Man erhält: $i = \varkappa u^2$: $\hat{i}/\hat{i}_{ae} \approx 1{,}25$, $\overline{R}/\hat{R} \approx 1{,}27$; $i = \varkappa u^3$: $\hat{i}/\hat{i}_{ae} \approx 1{,}37$, $\overline{R}/\hat{R} \approx 1{,}45$.

Gl. (5.4) liefert uns einen Zusammenhang zwischen u und u_{ae}, den wir zunächst für zwei Grenzfälle diskutieren.

1. $C \to \infty$ ($\alpha = \omega R C \to \infty$). Weil sowohl i als auch i_{ae} endlich bleiben, verschwindet der zweite Term der rechten Seite von (5.4).

2. C klein ($\alpha = \omega R C \to 0$) d.h. der Spannungsabfall ΔU besteht im wesentlichen aus der halben Rippelspannung $\delta U [\hat{u} \ll \delta U]$. In diesem Falle wird die Aufladegeschwindigkeit des Kondensators C durch die Wechselspannung direkt vorgeschrieben, und der Ventilstrom wird von der Kennlinie des Ventils weitgehend unabhängig, d.h. $i \to i_{ae}$. Der zweite Term der rechten Seite von (5.4) verschwindet ebenfalls.

Es läßt sich weiter zeigen, daß auch in allen zwischen 1. und 2. liegenden Fällen

$$\int\limits_{t_1}^{t_2} u\, dt \approx \int\limits_{t_1}^{t_2} u_{ae}\, dt \tag{5.5}$$

gilt (vgl. dazu Fig. 16). Aus der Definition von R_{ae} folgt unter Berücksichtigung von (5.5)

$$R_{ae} = \frac{u_{ae}}{i_{ae}} = \frac{\int\limits_{t_1}^{t_2} u_{ae}\, dt}{\int\limits_{t_1}^{t_2} i_{ae}\, dt} \approx \frac{\int\limits_{t_1}^{t_2} u\, dt}{\int\limits_{t_1}^{t_2} i\, dt};$$

d.h. aber

$$R_{ae} \approx \overline{R}. \tag{5.6}$$

Der äquivalente Ersatzwiderstand R_{ae} ist somit in guter Näherung gleich dem eingangs definierten mittleren Ventilwiderstand $\overline{R}$. Liegt zum Ventil in Fig. 15a ein konstanter Widerstand R_t in Serie, so folgt mit ähnlichen Argumenten für die Ersatzschaltung Fig. 15b

$$R_{ae} \approx R_t + \overline{R} = \overline{R}_s. \tag{5.6a}$$

Von besonderem Interesse ist der Quotient $\hat{R}/\overline{R}$, von dem wir noch zeigen müssen, daß er von der Belastung und vom Parameter α weitgehend unabhängig ist und für jeden Exponenten der Ventilkennlinie einen charakteristischen Wert annimmt.

1. Der Grenzfall des großen Kondensators ($\alpha \to \infty$) wurde in Ziff. 4 behandelt. Mit (5.3) erhält[1] man aus (4.9)

$$\varkappa \varDelta U^{n-1} \cdot B_n = \frac{1}{\overline{R}} B_1.$$

Man beachte, daß für $\alpha \to \infty$ $\hat{u} = \varDelta U$ wird. Im Zeitpunkt des Spitzenstromes folgt aus (4.1)

$$\hat{\imath} = \varkappa \varDelta U^n = \frac{1}{\hat{R}} \varDelta U,$$

und $\hat{R}/\overline{R}$ bestimmt sich streng zu

$$\frac{\hat{R}}{\overline{R}} = \frac{B_n}{B_1}. \tag{5.7}$$

Die in Tabelle 3 für $n = \frac{3}{2}$ und $n = 2$ berechneten Werte bestätigen die weitgehende Unabhängigkeit von der Belastung.

Zur Berechnung der Quotienten $\hat{R}/\overline{R}$ und $\hat{R}/|R|$ bei beliebigem Exponenten n dürfen wir deshalb den Fall des kurzgeschlossenen Gleichrichters ($U_g = 0$) betrachten. Dieser Spezialfall läßt sich einfach behandeln, indem wir von (4.2) ausgehen und beachten, daß wegen $U_g = 0$ aus (4.3) $\varphi_2 = \pi/2$ folgt. Wir erhalten für Kennlinien der Form $i = \varkappa u^n$

$$\frac{\hat{R}}{\overline{R}} = \int\limits_0^{\pi/2} \cos^n \varphi \, d\varphi = \frac{\sqrt{\pi}}{2} \cdot \frac{\varGamma\left(\frac{n+1}{2}\right)}{\varGamma\left(\frac{n}{2}+1\right)}. \tag{5.8 a}$$

Gehen wir andererseits von der Definition des Widerstandes $|R|$ aus, so ergibt sich weiter

$$\frac{\hat{R}}{|R|} = \frac{\int\limits_0^{\pi/2} \cos^{2n} \varphi \, d\varphi}{\int\limits_0^{\pi/2} \cos^{n+1} \varphi \, d\varphi} = \frac{\varGamma\left(\frac{n+3}{2}\right) \varGamma\left(n+\frac{1}{2}\right)}{\varGamma\left(\frac{n}{2}+1\right) \varGamma(n+1)}. \tag{5.8 b}$$

Die Ergebnisse sind in Tabelle 4 zusammengefaßt.

2. Der Grenzfall des kleinen Kondensators ($\alpha \to 0$) läßt sich ebenfalls leicht überblicken. Der Ventilstrom ist dann durch

$$i \approx \omega U_0 C \frac{d}{d\varphi} (\cos \varphi), \qquad \varphi_1 \leqq \varphi \leqq 0$$

gegeben. Aus der Definition von $\overline{R}$ und aus der Tatsache, daß der Maximalwert des Stromes beim Einschaltwinkel $\varphi_1 = \omega t_1$ auftritt, folgt dann

$$\frac{\overline{R}}{\hat{R}} = \frac{\sin \varphi_1}{1 - \cos \varphi_1} \frac{\int\limits_0^{\varphi_1} \sin^{1/n} \varphi \, d\varphi}{(\sin \varphi_1)^{1/n}}.$$

Entwickeln wir nach Potenzen von φ, so erhält man für nicht zu große[2] Winkel $|\varphi_1|$

$$\frac{\overline{R}}{\hat{R}} = \frac{2n}{1+n}. \tag{5.9}$$

---

[1] Es gilt $(\varkappa)_{n=1} = 1/\overline{R}$.
[2] Für $\varphi_1 = -\pi/2$ lautet das Ergebnis $\overline{R}/\hat{R} = \int\limits_0^{\pi/2} \sin^{1/n} \varphi \, d\varphi$ und man erhält etwas kleinere Zahlenwerte als nach Gl. (5.9).

ein Ergebnis, das mit den Zahlenwerten von Tabelle 4 noch befriedigend übereinstimmt und als Beweis für die weitgehende Unabhängigkeit von α angesehen werden darf. Dabei ist zu berücksichtigen, daß im Grenzfall (2) der Einfluß des Ventilwiderstandes auf den Spannungsabfall sehr klein ist, während im Grenzfall (1) die Kennlinie des Ventils den Spannungsabfall vollständig bestimmt. Aus diesem Grunde sind für die Dimensionierung die Zahlenwerte nach (5.8) maßgebend.

Als Ergänzung zu dieser Betrachtung zeigt Fig. 16 an einem Beispiel für $\alpha = 1$, daß diese Zahlenwerte in den zwischen (1) und (2) liegenden Fällen als recht gute Näherung angesehen werden dürfen.

Liegt zum Ventil in Fig. 15a ein konstanter Widerstand R_t in Serie, so ergibt diese Kombination einen Spitzenwiderstand $\hat{R}_s$ von $\hat{R}_s = R_t + \hat{R}$ und der Quotient $\overline{R}_s / \hat{R}_s$ wird

$$\frac{\overline{R}_s}{\hat{R}_s} = \frac{R_t/\overline{R} + 1}{R_t/\overline{R} + \hat{R}/\overline{R}},$$

wobei die Zahlenwerte von $\hat{R}/\overline{R}$ wiederum der Tabelle 4 entnommen werden dürfen. Damit läßt sich in einfacher Weise die Wirkung eines konstanten Widerstandes in Serie zu einem Ventil der Kennlinie $i = \varkappa u^n$ erfassen.

Tabelle 4. $\overline{R}/\hat{R}$ und $|R|/\hat{R}$ für verschiedene Exponenten der Kennlinien $i = \varkappa u^n$ und Gleichrichter mit Kondensatoreingang.

n	0,5	1	1,5	2	3
$\overline{R}/\hat{R}$	0,833	1	1,14	1,27	1,5
$|R|/\hat{R}$	0,873	1	1,08	1,13	1,20

In der Literatur wird öfters, ohne nähere Begründung, $\hat{R}$ als äquivalenter Ersatzwiderstand gewählt. Die Tabelle 4 zeigt, daß diese Näherung zwar nicht schlecht ist, sich aber wesentlich verfeinern läßt. Die Tabelle vermittelt ferner ein Bild über die Zunahme des Spitzenstromes mit wachsendem Exponenten n. Bei großem Kondensator $(\alpha \to \infty)$ gilt

$$\frac{\hat{i}}{\hat{i}_{ae}} = \frac{\overline{R}}{\hat{R}},$$

und der Spitzenstrom ist somit bei einer $u^{\frac{3}{2}}$ Kennlinie des Ventils um 14% und bei einer u^3 Kennlinie bereits um 50% größer als beim äquivalenten Gleichrichter mit ideal geknickter Kennlinie. Im Grenzfall des kleinen Kondensators $(\alpha \to 0)$ wird andererseits der Quotient $\hat{i}/\hat{i}_{ae} \approx 1$ (vgl. S. 14).

In allen zwischen den Grenzen $\alpha = 0$ und $\alpha = \infty$ liegenden Fällen gilt

$$1 \leq \frac{\hat{i}}{\hat{i}_{ae}} \leq \frac{\overline{R}}{\hat{R}},$$

eine Beziehung, die bei nicht zu großem Exponenten einen guten Überblick über den Verlauf des Spitzenstromes ermöglicht. Fig. 16 bestätigt diese Zusammenhänge.

Bezeichnen wir die Verluste im Ventil Fig. 15a mit W und diejenigen im Ventil Fig. 15b mit W_{ae}, so erhält man aus den in der Einleitung gegebenen Definitionen unter Berücksichtigung von (5.6)

$$\frac{W}{W_{ae}} = \frac{|I|^2 |R|}{|I_{ae}|^2 \overline{R}} = \lambda \frac{|R|}{\overline{R}},$$

wobei die $|I|$ die zugeordneten Effektivwerte der beiden Ventilströme bedeuten. Der Quotient λ geht für $\alpha \to 0$ gegen 1 und erreicht für $\alpha \to \infty$ die Werte

n	1	2	3
$\lambda_{\alpha \to \infty}$	$\equiv 1$	$\approx 1{,}2$	$\approx 1{,}40$

Aus den Zahlenwerten von $|R|/\overline{R}$ (Tabelle 4) und denjenigen von λ geht hervor, daß W/W_{ae} nur wenig mit dem Exponenten variiert. In allen praktisch vorkommenden Belastungsfällen darf deshalb

$$\frac{W}{W_{ae}} \approx 1$$

gesetzt werden. Für $1 \leq n \leq 3$ bleibt dabei der Fehler innerhalb etwa 10%. Es ist somit möglich, den Effektivwert des Ventilstromes und die Erwärmung des Ventils mit ausreichender Genauigkeit abzuschätzen.

6. Die Gleichrichterkennlinie folge dem Exponentialgesetz. Halbleiterdioden nach Fig. 19 folgen in guter Näherung einem Exponentialgesetz

$$i = I_s \left(e^{\frac{qu}{kT}\varrho} - 1 \right). \tag{6.1}$$

Dabei bedeutet I_s den Sättigungsstrom in Sperrichtung, q die Elementarladung, k die Boltzmann-Konstante und T die absolute Temperatur. ϱ ist ein dimensionsloser Faktor, dessen Wert zwischen $\frac{1}{2}$ und 1 liegt[1]. Für die folgenden Rechnungen wollen wir

$$\frac{kT}{q\varrho} = U_m, \qquad \frac{kT}{q} = 25{,}9\,\text{mV} \quad \text{für} \quad T = 300°\,\text{K} \tag{6.2}$$

als Abkürzung einführen. Mit einer Kennlinie nach Gl. (6.1) bestimmt sich beim Einweggleichrichter die pro Periode dem Kondensator zugeführte Ladung zu:

$$Q = \frac{1}{\omega} \int\limits_{-\pi}^{+\pi} i\,d\varphi = \frac{2I_s}{\omega} \left[e^{-U_g/U_m} \int\limits_{0}^{\pi} e^{U_0 \cos\varphi/U_m}\,d\varphi - \pi \right], \tag{6.3}$$

wobei wiederum ein genügend großer Kondensator C vorausgesetzt wurde. Das Integral in (6.3) ist die Bessel-Funktion nullter Ordnung und somit wird

$$Q = \frac{2\pi I_s}{\omega} \left[e^{-U_g/U_m}\,J_0(-jx) - 1 \right], \qquad x = \frac{U_0}{U_m}. \tag{6.4}$$

Die Reihenentwicklung von $J_0(-jx)$ für $x < 1$, bzw. die asymptotische Darstellung für $x > 1$ (halbkonvergente Reihen von HANKEL), führen auf die in J. KAMMERLOHER [2] angegebenen Reihen.

In Fig. 17 ist die Funktion $J_0(-jx) \cdot e^{-x} \cdot \sqrt{2\pi x}$ in Abhängigkeit von x aufgetragen, sie geht für $x > 10$ gegen eins:

$$h = J_0(-jx)\,e^{-x}\sqrt{2\pi x} \approx 1, \quad \text{für} \quad x > 10. \tag{6.5}$$

$x > 10$ bedeutet eine Amplitude der Wechselspannung von $> 0{,}26$ Volt, eine Bedingung, die bei Hochspannungsgleichrichtern wohl immer erfüllt ist, so daß sich (6.4) mit (6.5) zu

$$I_g = \frac{\omega}{2\pi}Q = I_s \left[\frac{e^{(U_0 - U_g)/U_m}}{\sqrt{2\pi U_0/U_m}} - 1 \right]. \tag{6.6}$$

vereinfacht.

---

[1] Vgl. z.B. M. B. PRINCE: Bell Syst. Techn. J. **35**, 661 (1956), insbesondere S. 669.

Es ist von Interesse, darauf hinzuweisen, daß ein Einweggleichrichter mit großem Kondensator und einer Kennlinie nach Gl. (6.1) bei Vernachlässigung des Ohmschen Widerstandes des Transformators einen bemerkenswert kleinen Spannungsabfall und einen entsprechend großen Spitzenstrom aufweist. Einerseits ist ein derart großer Spitzenstrom oft unzulässig, andererseits bedingt der

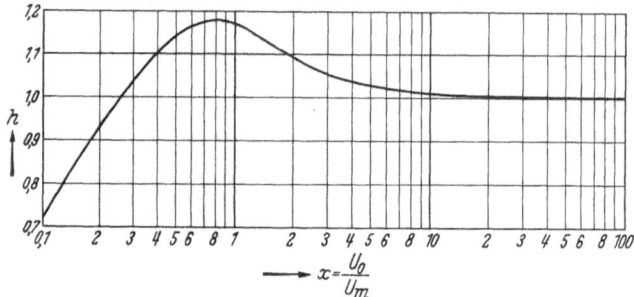

Fig. 17. Die Größe h [Gl. (6.5)] in Abhängigkeit von x. (Aus J. Kammerloher [2], S. 56.)

kleine Spannungsabfall einen stark überdimensionierten Transformator. Infolgedessen wird in vielen Anwendungen der Widerstand des Ventils nach Gl. (6.1)

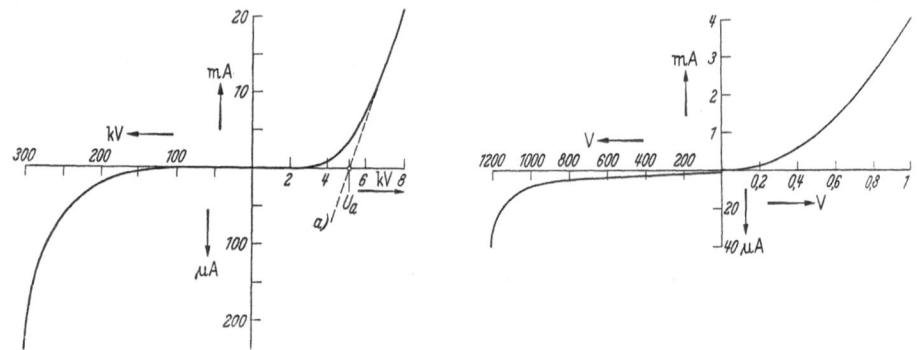

Fig. 18.   Fig. 19.

Fig. 18. Kennlinie einer Säule aus Selengleichrichtern, wie sie in Kaskadengeneratoren benützt wird. (Type: ss Hib 296 der Siemens-Werke, nach Messungen der Firma E. Haefely & Co., Basel.)

Fig. 19. Kennlinie einer Ge-Flächendiode. [W. J. Pietenpol: Phys. Rev. 82, 120 (1951).

neben dem Widerstand des Transformators bzw. der Impedanz der Streuinduktivität zu vernachlässigen sein. Unter welchen Bedingungen dies der Fall ist, kann durch Vergleich der Formeln (6.6) und (7.2) bzw. (10.2) leicht entschieden werden.

Germanium- bzw. Silizium-Flächendioden werden heute in Hochspannungsanlagen noch nicht verwendet. Prinzipielle Gründe, die gegen ihre Einführung sprechen, sind jedoch nicht bekannt. Als Vorteil dieser Dioden mögen ihre kleinen Abmessungen und ihre geringen Verluste erwähnt werden.

Selengleichrichter hingegen finden seit der Arbeit von W. R. Arnold [A 50] eine weitgehende Anwendung. Der Wegfall der Glühkathoden bedeutet eine wesentliche Vereinfachung des Aufbaus von Kaskadengeneratoren. Außerdem sind die praktischen Erfahrungen mit Selengleichrichtern in solchen Hochspannungsanlagen sehr zufriedenstellend. Fig. 18 zeigt die Kennlinie einer Säule aus vielen einzelnen Gleichrichterelementen. Zur Berechnung des Spannungsabfalles empfiehlt sich bei starker Aussteuerung die Approximation durch eine ideal geknickte Kennlinie (Gerade a in Fig. 18)[1]. Der Achsenabschnitt U_a bedeutet eine elektromotorische Gegenkraft im Sinne der Bemerkung auf S. 3, und U_a ist von der Amplitude der Wechselspannung abzuziehen (vgl. auch [Mü 54]). Der Sperrstrom läßt sich als zusätzliche äußere Belastung des Gleichrichters einführen. Bei kleiner Aussteuerung kann man die Kennlinie in Fig. 18 durch ein Exponentialgesetz darstellen.

Es ist noch darauf hinzuweisen, daß Selengleichrichter, die aus vielen einzelnen Gleichrichterelementen aufgebaut sind, bei Frequenzen über etwa 10 kHz Schwierigkeiten bereiten. Die Ursache liegt darin, daß bei genügend hohen Frequenzen die Spannungsverteilung in Sperrichtung längs der Gleichrichtersäule durch die Kapazitäten der einzelnen Elemente bestimmt wird. Große Streuung der Kapazitätswerte hat eine ungleichmäßige Spannungsverteilung zur Folge, die zu Durchschlägen Anlaß geben kann.

7. Einige weitere Beziehungen für Gleichrichter mit großem Eingangskondensator und ideal geknickter Kennlinie des Ventils. In diesem Abschnitt sind einige weitere Beziehungen zusammengestellt, die sich bei der Dimensionierung einfacher Gleichrichterschaltungen in manchen Fällen als nützlich erweisen und oft eine rasche und einfache Bestimmung aller für einen Gleichrichter wichtigen Größen gestatten. Dies ist besonders dann wahr, wenn der Einfluß des Transformatorwiderstandes denjenigen des Ventils in Leitrichtung überwiegt und deshalb die Annahme einer ideal geknickten Kennlinie ($i = u/R$ für $u > 0$ und $i = 0$ für $u < 0$) eine gute Approximation darstellt.

Die dem Kondensator C in Fig. 1 oder 2 jeweils zugeführte Ladung erhalten wir aus (4.7) mit $n = 1$ und $\varkappa = 1/R$ zu:

$$Q = \frac{2\sqrt{2}}{\omega R} \frac{2}{3} U_0 \left(\frac{\Delta U}{U_0}\right)^{\frac{3}{2}}. \tag{7.1}$$

Bei Einweggleichrichtung wird der Kondensator $\omega/2\pi$ mal und bei Doppelweggleichrichtung ω/π mal pro Sekunde geladen. Zwischen dem Spannungsabfall ΔU bei Belastung und dem Belastungsstrom I_g des Gleichrichters besteht daher die folgende einfache Beziehung:

$$I_g = \frac{2\beta\sqrt{2}\,U_0}{3\pi R} \left(\frac{\Delta U}{U_0}\right)^{\frac{3}{2}} \tag{7.2}$$

(Einweggleichrichter $\beta = 1$, Doppelweggleichrichter $\beta = 2$).

Das Verhältnis des Spitzenstromes des Ventils $\hat{i} = \Delta U/R$ zum Gleichstrom I_g bestimmt sich dann zu

$$\frac{\hat{i}}{I_g} = \frac{3\pi}{2\beta\sqrt{2}} \sqrt{\frac{U_0}{\Delta U}}. \tag{7.3}$$

Die Erwärmung der Wicklung des Transformators sowie diejenige der Anode der Gleichrichterröhre wird durch den Effektivwert des Ventilstromes

$$|I| = \left(\frac{1}{TR^2} \int_{t_1}^{t_2} (U_0 \cos \omega t - u_c)^2\, dt\right)^{\frac{1}{2}}, \qquad T = \frac{2\pi}{\beta\omega} \tag{7.4}$$

bedingt. Mit den bereits früher benützten Vereinfachungen (vgl. S. 11) folgt für $|I|$ der Ausdruck:

$$|I| \approx \left(\frac{8\beta\,\Delta U^2}{15\pi R^2} \sqrt{2\frac{\Delta U}{U_0}}\right)^{\frac{1}{2}}, \tag{7.5}$$

---

[1] Als nächste Approximation ist natürlich ein Ansatz der Form $i = \varkappa\,(u - U_a')^n$ für $(u - U_a') > 0$ und $i = 0$ für $(u - U_a') < 0$ möglich. U_a' bedeutet eine passend zu wählende Konstante.

der unter Berücksichtigung von (7.2) in die übersichtlichere Form

$$\frac{|I|}{I_g} = \left(\frac{6\pi}{5\beta}\right)\sqrt{\frac{U_0}{2\Delta U}}^{\frac{1}{2}} = \frac{1,63}{\sqrt{\beta}}\left(\frac{U_0}{\Delta U}\right)^{\frac{1}{4}} \tag{7.6}$$

gebracht werden kann.

Bei Doppelweggleichrichtung fließt durch jede Hälfte der Sekundärwicklung des Transformators ein Strom mit dem Effektivwert

$$|I|^* = \frac{|I|}{\sqrt{2}} \tag{7.7}$$

und durch die Primärwicklung der Magnetisierungsstrom sowie der auf die Primärseite reduzierte Belastungsstrom $|I|\ddot{u}$ ($\ddot{u}$ = Übersetzungsverhältnis des Transformators). Zum gewöhnlichen Einweggleichrichter ist insofern eine Bemerkung notwendig, als die Vormagnetisierung des Transformators durch den Gleichstrom I_g zu beachten ist. Diese Vormagnetisierung bewirkt bei zu hoher Sättigung des Eisens eine Vergrößerung des primären Magnetisierungsstromes. Infolge des Ohmschen Widerstandes der Primärseite und infolge der Streuinduktivität entsteht dadurch ein zusätzlicher Spannungsabfall, der bei Hochspannungsgleichrichtern beträchtliche Werte annehmen kann. In einem solchen Falle ist es günstig, im Eisenkern des Transformators einen Luftspalt vorzusehen, ähnlich wie dies bei vormagnetisierten Drosselspulen gebräuchlich ist. Im allgemeinen verdient jedoch die Greinacher-Schaltung (oder eine ähnliche Variante) den Vorzug, weil sich so die unnötige Gleichstrommagnetisierung des Transformators vermeiden läßt.

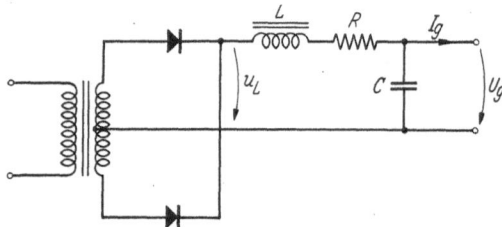

Fig. 20. Doppelweggleichrichter mit Drosselspuleneingang.

Für die Erwärmung der Sekundärseite des Transformators ist bei Einweggleichrichtung der Effektivwert $|I|$ nach Gl. (7.6) maßgebend (wobei $\beta = 1$ zu setzen ist), während auf der Primärseite einerseits der Strom $|I|_p$

$$|I|_p^2 = (|I|^2 - I_g^2)\,\ddot{u}^2 \tag{7.8}$$

und andererseits der durch die Vormagnetisierung vergrößerte Magnetisierungsstrom die Kupferverluste verursacht.

c) Gleichrichter mit Drosselspuleneingang.

Drosselspuleneingang findet in der Regel nur bei Doppelweggleichrichtung (Fig. 2 und 3) Verwendung. Wir beschäftigen uns hier mit diesem Problem, weil wir so zu einer Abschätzung des Einflusses der Streuinduktivität gelangen, und entscheiden können, unter welchen Umständen die Vernachlässigung der Streuinduktivität beim Gleichrichter mit Kondensatoreingang berechtigt ist.

8. Große Belastung. Bei genügend großer Belastung des Gleichrichters wird die Eingangsspannung u_L des Filters durch den Transformator aufgeprägt, wie dies in Fig. 21 dargestellt ist. Die Entwicklung von u_L in eine Fourier-Reihe ergibt:

mit

$$\left.\begin{aligned} u_L &= U_0\,|\cos\omega t| = \frac{2U_0}{\pi} + \sum_{n=1}^{\infty} a_n \cos(2n\,\omega t) \\ a_n &= \frac{-4U_0(-1)^n}{\pi(4n^2-1)}. \end{aligned}\right\} \tag{8.1}$$

Am Kondensator C entsteht eine Gleichspannung vom Betrage:

$$U_g \approx \frac{2\,U_0}{\pi} - I_g R. \tag{8.2}$$

Dabei ist unter R die Summe der Ohmschen Widerstände von Drosselspule, Gleichrichterröhre und Transformator zu verstehen. Bei einer Ventilkennlinie der Form $i = \varkappa u^n$ ist als äquivalenter Widerstand $\widehat{R}$ einzusetzen. Dies folgt aus der Tatsache, daß der Ventilstrom in guter Näherung einen rechteckförmigen Verlauf zeigt, so daß $|R|/\widehat{R} \approx 1$ und $\overline{R}/\widehat{R} \approx 1$ wird.

Die bisherigen Betrachtungen gelten unter der Voraussetzung, daß der Strom in der Gleichrichterröhre und somit in der Drossel nie verschwindet. Da der Strom im Ventil sein Vorzeichen nicht umkehren kann, muß der Gleichstrom größer als die Amplitude des überlagerten Wechselstromes sein. Diese ist im wesentlichen durch die Amplitude der ersten Harmonischen bestimmt. Im Falle eines großen Kondensators erhält man so folgende Bedingungsgleichung:

$$I_g \geq \frac{4\,U_0}{3\pi\,2\omega L}. \tag{8.3}[1]$$

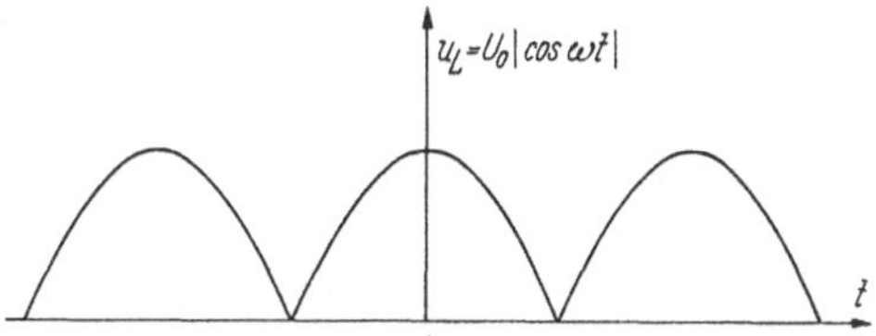

Fig. 21. Zeitlicher Verlauf der Spannung u_L in Fig. 20.

Die Amplitude der ersten Harmonischen der Störspannung U_{st} am Kondensator C hat den Wert:

$$U_{st} \approx \frac{U_0}{3\pi\omega^2 L C}. \tag{8.4}$$

Die Amplitude der zweiten Harmonischen ist bereits 20mal geringer und kann vernachlässigt werden.

Durch Beziehung (8.3) ist für jeden Belastungsfall der Mindestwert der Induktivität vorgeschrieben. Siebdrosseln enthalten einen Eisenkern, und ihre Induktivität L ist von der Vormagnetisierung, d.h. vom Gleichstrom I_g abhängig. Kann der Belastungsstrom eines Gleichrichters zwischen einem Mindestwert und einem Höchstwert schwanken, so muß dafür gesorgt werden, daß bei jeder Belastung die Beziehung (8.3) erfüllt ist, da sonst der Gleichrichter einen meistens unerwünschten Verlauf der Belastungskennlinie aufweisen wird. Es ist aber möglich, die Drossel derart zu dimensionieren, daß ihre Induktivität L über einen großen Bereich etwa umgekehrt proportional zum Gleichstrom I_g variiert. Die Bedingung (8.3) läßt sich dann mit einem minimalen Aufwand realisieren (Schwingdrossel).

9. Kleine Belastung. Im folgenden Abschnitt wollen wir die Verhältnisse diskutieren, die eintreten, wenn die Bedingung (8.3) nicht erfüllt ist. Der Kondensator C sei so groß gewählt, daß seine Spannung als zeitlich konstant betrachtet werden darf. Unter dieser Voraussetzung ist der zeitliche Verlauf der Eingangsspannung u_L des Filters für zwei verschiedene Belastungsfälle in Fig. 22 aufgetragen. Im Belastungsfalle nach Fig. 22a wird im folgenden Abschnitt die Gleichspannung des Gleichrichters unter Vernachlässigung der Ohmschen Widerstände von Drossel, Röhre und Transformator berechnet. Zwischen dem

---

[1] Gl. (8.3) wurde zuerst auf empirischem Wege von F. S. Dellenbaugh und R. S. Quinby [QST **16** (1932) Februar, 14] in der Form $L \geq \dfrac{U_g}{I_g \cdot 1000}$ gültig für $\omega = 377 \text{ sec}^{-1}$ $(f = 60 \text{ Hz})$, gefunden. Zum Vergleich ist für U_g der Wert $2U_0/\pi$ einzusetzen.

Momentanwert des Stromes i durch die Drossel und der Spannung über der Drossel besteht die Beziehung

$$L \frac{di}{dt} = u_L - U_g. \tag{9.1}$$

Daraus folgt:

$$L Q = \int_{t_1}^{t_2} dt \int_{t_1}^{t} [U_0 \cos \omega t - U_g] \, dt, \tag{9.2}$$

wobei Q die Ladung bedeutet, welche dem Kondensator C pro Halbwelle zugeführt wird. Die Integration von (9.2) führt auf

$$\frac{\omega^2 L Q}{U_0} = \cos \omega t_1 - \cos \omega t_2 - [\omega t_2 - \omega t_1] \left[\sin \omega t_1 + \frac{U_g}{2 U_0} (\omega t_2 - \omega t_1) \right]. \tag{9.3}$$

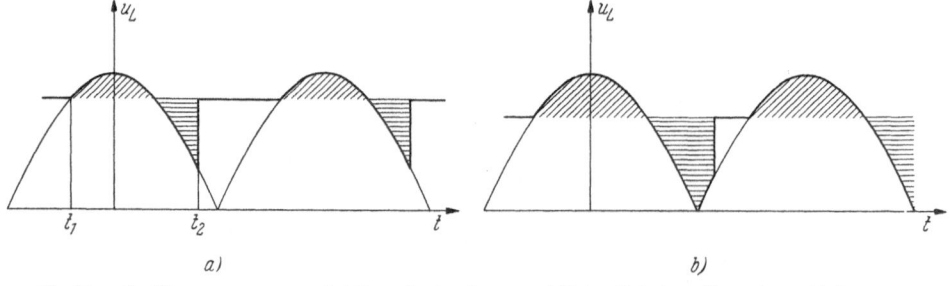

Fig. 22a u. b. Eingangsspannung u_L bei Drosselspuleneingang und kleiner Belastung für zwei verschiedene Belastungsfälle (a) und (b).

Da der Kondensator ω/π mal pro Sekunde geladen wird, fließt der Gleichstrom

$$I_g = \frac{\omega}{\pi} Q. \tag{9.4}$$

Beginn und Ende der Ladezeit des Kondensators sind nach Fig. 22a durch t_1 und t_2 bestimmt. Für t_1 gilt:

$$U_0 \cos \omega t_1 = U_g. \tag{9.5}$$

Der Strom durch die Gleichrichterröhre und somit durch die Drossel kann seine Richtung nie umkehren, d.h. die in Fig. 22 schraffierten Flächen müssen gleich groß sein, was die weitere Beziehung

$$\frac{U_0}{\omega} (\sin \omega t_2 - \sin \omega t_1) - U_g (t_2 - t_1) = 0 \tag{9.6}$$

zur Folge hat.

Aus (9.5) und (9.6) lassen sich die Zeiten t_1 und t_2 leicht näherungsweise berechnen:

$$t_1 \approx -\frac{1}{\omega} \sqrt{\frac{2 \Delta U}{U_0}}, \qquad t_2 \approx -2 t_1, \tag{9.7}$$

wobei $\Delta U = U_0 - U_g$ die Spannungsdifferenz zwischen Leerlauf und Belastung des Gleichrichters darstellt. Entwickelt man (9.3) in eine Potenzreihe, bricht mit $(\omega t)^4$ ab und berücksichtigt ferner (9.4) und (9.7), so erhält man eine einfache Beziehung zwischen Spannungsabfall und Gleichstrom I_g,

$$I_g \approx \frac{9 U_0}{2 \pi \omega L} \left(\frac{\Delta U}{U_0} \right)^2. \tag{9.8}$$

Die Gl. (9.8) setzt voraus, daß $\omega t_2 < \pi/2$ bleibt. Die exakte Auflösung von (9.3), (9.5) und (9.6) liefert für $\omega t_2 = \pi/2$ einen Spannungsabfall von $0{,}29\,U_0$, wobei der Gleichstrom durch

$$\left(\frac{I_g\,\pi\,\omega\,L}{U_0}\right)_{\omega t_2 = \pi/2} \approx 0{,}36 \tag{9.9}$$

bestimmt wird.

Mit der Näherungsbeziehung (9.8) und mit (8.2) läßt sich der Spannungsabfall eines Gleichrichters mit Drosselspuleneingang über den gesamten Belastungsbereich leicht überblicken (vgl. Fig. 23).

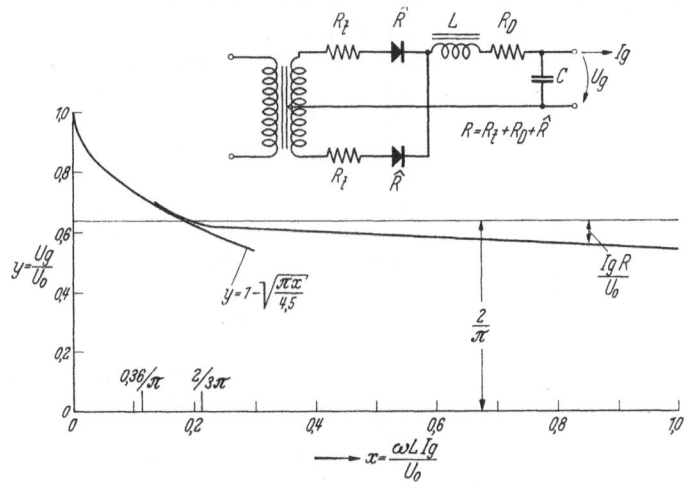

Fig. 23. Belastungskennlinie eines Gleichrichters mit Drosselspuleneingang. $y = 1 - \sqrt{\pi\,x/4,5}$: Näherung nach Gl. (9.8).

10. Vernachlässigung der Streuinduktivität beim Einweggleichrichter mit Kondensatoreingang. Um zu einer Abschätzung zu gelangen, unter welchen Bedingungen die Streuinduktivität des Transformators beim Einweggleichrichter mit Kondensatoreingang (Fig. 1) vernachlässigt werden darf, vergleichen wir die Spannungsabfälle nach (7.1) und (9.8). In beiden Fällen wird ein sehr großer Kondensator C vorausgesetzt. Nach (7.1) beträgt die während der Ladezeit dem Kondensator in Fig. 1 zugeführte Ladung Q_R

$$Q_R = \frac{4\sqrt{2}\,U_0}{3\,R\,\omega}\left(\frac{\Delta U}{U_0}\right)^{\frac{3}{2}} \tag{10.1}$$

und nach (9.8) und (9.4) die dem Kondensator in Fig. 20 zugeführte Ladung Q_L

$$Q_L = \frac{9\,U_0}{2\,\omega^2 L}\left(\frac{\Delta U}{U_0}\right)^2. \tag{10.2}$$

Damit die Streuinduktivität bei Kondensatoreingang vernachlässigt werden darf, muß $Q_R \gg Q_L$ sein. Dies führt zu folgender Ungleichung:

$$\frac{\omega L}{R} \ll 2{,}4\,\sqrt{\frac{\Delta U}{U_0}}. \tag{10.3}$$

Unsere Bedingung stimmt mit Messungen von R. G. MITCHELL [Mi 43] überein, der experimentell gefunden hat, daß für

$$\frac{\omega L}{R} \lesssim 0{,}1 \text{ bis } 0{,}2 \tag{10.4}$$

der Einfluß der Streuinduktivität vernachlässigbar ist. Zum Vergleich der beiden Bedingungen ist zu beachten, daß in der Arbeit von R. G. Mitchell $\sqrt{\dfrac{\Delta U}{U_0}}$ etwa zwischen 0,3 und 0,7 liegt. Unter Berücksichtigung dieser experimentellen Ergebnisse läßt sich die Ungleichung (10.3) etwas verfeinern. Es genügt, daß die linke Seite etwa 10mal kleiner ist als die rechte, und wir erhalten damit aus (10.3)

$$\frac{\omega L}{R} \leq 0,2 \sqrt{\frac{\Delta U}{U_0}} . \tag{10.5}$$

II. Kaskadengeneratoren.

Der Kaskadengenerator (Fig. 24) zeichnet sich dadurch aus, daß die abgegebene Gleichspannung ein Vielfaches der Wechselspannung des Transformators

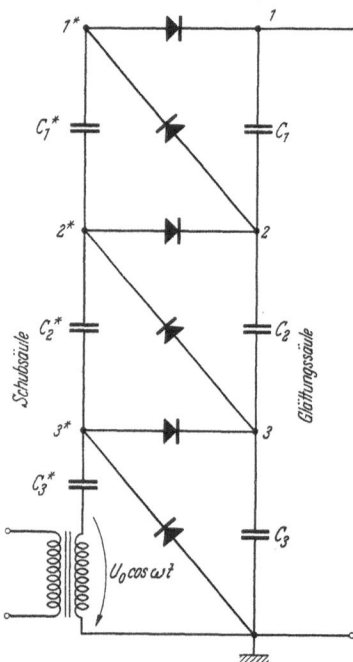

sowie der Sperrspannung der Ventile beträgt und die Kondensatoren nur mit einem Bruchteil der zu erzeugenden Gleichspannung belastet sind. Es ist ferner möglich, an den einzelnen Stufen der Kaskade Teilspannungen abzugreifen, die z.B. zur Steuerung der Potentialverteilung eines Beschleunigungsrohres zweckmäßig verwendet werden können. Im Vergleich zum Van de Graaff-Generator gestattet der Kaskadengenerator die Erzeugung größerer Gleichströme und benötigt keine mechanisch bewegten Teile, was als weiterer Vorteil angesehen werden darf. Als Nachteil sind die Schwierigkeiten zu erwähnen, die sich einer sehr guten Stabilisierung hoher Gleichspannungen entgegenstellen, ein Problem, das beim Van de Graaff-Generator leichter zu beherrschen ist.

Fig. 24. Dreistufiger Kaskadengenerator.

Die höchste Spannung, welche sich mit Kaskadengeneratoren erreichen läßt, kann nicht mit Bestimmtheit angegeben werden. Diese obere Schranke wird mehr durch die Eigenschaften der zur Zeit im Handel erhältlichen Schaltelemente bestimmt, als durch die Eigenschaften der Schaltung selber. Vier Millionen Volt Gleichspannung erscheinen heute durchaus möglich, und eine solche Anlage ist gegenwärtig am Physikalischen Institut der Universität Basel im Bau [H 55].

Die Anwendung des Kaskadengenerators ist jedoch nicht auf die Erzeugung höchster Gleichspannungen beschränkt. Auch Spannungen von nur einigen tausend Volt lassen sich mit Vorteil in dieser Art erreichen. Als Beispiel sei auf die in Fig. 45 dargestellte Anlage sowie auf die Ausführungen in Kapitel III (S. 47) hingewiesen.

Um das Verständnis der rechnerischen Behandlung solcher Schaltungen zu erleichtern, ist es zweckmäßig, die Wirkungsweise eines Kaskadengenerators und die wesentlichen Ursachen des Spannungsabfalles kurz zu erläutern.

a) Übersicht über die Problemstellung.

11. α) *Der Kaskadengenerator im Leerlauf.* Zunächst betrachten wir den Kaskadengenerator im Leerlauf, und setzen eine sinusförmige Transformatorspannung voraus.

Die Ventile leiten nur während einer sehr kurzen Zeit, nämlich dann, wenn das Potential der Anode positiv gegenüber der Kathode ist. Fig. 24 zeigt eine dreistufige Anordnung. Unter der Annahme idealer Ventile und unter Vernachlässigung der Streukapazitäten stellen sich an den Knotenpunkten 3, 2, 1 und 3*, 2*, 1* die in Fig. 25 wiedergegebenen Spannungen ein. Der hier dargestellte, idealisierte Generator liefert eine Leerlaufspannung von $6U_0$, wobei mit U_0 die Amplitude der Transformatorspannung $U_0 \cos \omega t$ bezeichnet ist.

β) *Der Kaskadengenerator bei Belastung.* Wird der Kaskadengenerator durch einen Widerstand oder durch ein Beschleunigungsrohr belastet, so sinkt natur-

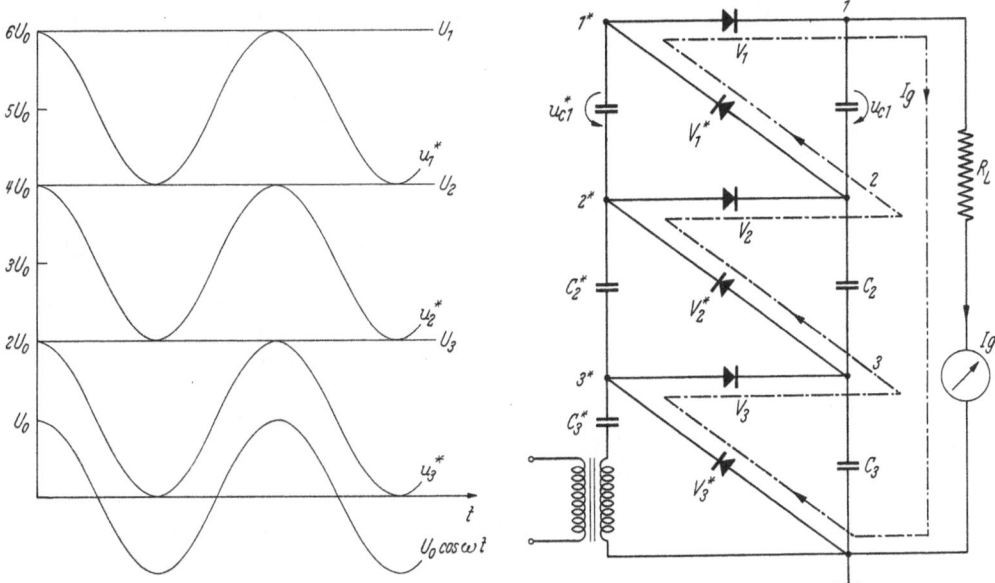

Fig. 25. Leerlaufspannungen beim Generator in Fig. 24.   Fig. 26. Der belastete Kaskadengenerator.

gemäß die Ausgangsspannung, und zwar umso stärker, je größer der Belastungsstrom I_g ist. Unter I_g wollen wir den vom Generator gelieferten, arithmetischen Mittelwert des Stromes, also den abgegebenen Gleichstrom verstehen. Während einer Periode $1/f$ der Wechselspannung wird der Glättungssäule somit die Ladung

$$Q = \frac{I_g}{f} \qquad (11.1)$$

entzogen. Falls ein stationärer Zustand bestehen soll, muß diese Ladung periodisch wieder zugeführt werden. Dies geschieht dadurch, daß während einer Halbwelle der Wechselspannung die Ladung Q von den Punkten 3* nach 3, bzw. 2* nach 2 und 1* nach 1 fließt, während in der andern Halbwelle die Ladung Q von Erde nach 3* bzw. von 3 nach 2* und von 2 nach 1* transportiert wird.

Zur Vereinfachung der rechnerischen Behandlung wollen wir annehmen, daß alle Ventile V_3^* bis V_1^* im gleichen Zeitpunkt t_1 zu leiten und im gleichen Zeitpunkt t_2 zu sperren beginnen, also gleichzeitig schalten. Auch die Ventile V_3 bis V_1 sollen jeweils gleichzeitig öffnen und schließen (Fig. 26). Eine einfache Überlegung zeigt, daß diese Annahme nur näherungsweise richtig sein kann. Wegen des wachsenden Spannungsabfalles in den beiden Kondensatorsäulen muß nämlich das Ventil in der höher gelegenen Stufe immer früher einschalten als das entsprechende Ventil der nächst niedrigeren Stufe. Infolgedessen steht dem oberen Ventil für den Stromdurchgang etwas mehr Zeit zur Verfügung. Die Berücksichtigung dieses Effektes, der allerdings erst bei großem Spannungsabfall von Bedeutung werden kann, macht die theoretische Behandlung der Schaltung sehr unübersichtlich. Für alle folgenden Betrachtungen werden wir daher ausdrücklich gleiche Schaltzeiten voraussetzen. Unter dieser Annahme

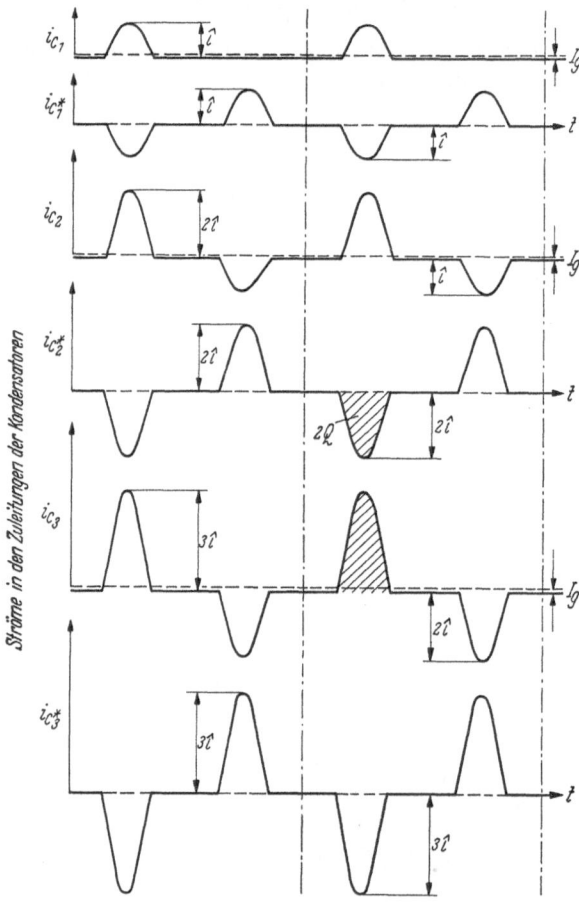

und der weiteren Voraussetzung, daß alle Ventile gleichartig sind, weisen dann alle Ventilströme denselben zeitlichen Verlauf auf. Den Fig. 27 und 28 liegen diese Voraussetzungen zugrunde.

γ) Spannungsabfall infolge der Streukapazitäten. Schon von H. Mehlhorn [*Me 43*] wurde darauf hingewiesen, daß die unvermeidlichen Streukapazitäten auch im Leerlauf einen zusätzlichen Spannungsabfall sowie eine Rippelspannung bewirken. Durch diese Kapazitäten fließen besonders bei hoher Frequenz der Wechselspannung beträchtliche Ströme, die den Transformator belasten und einen Spannungsabfall längs der Schubsäule zur Folge haben.

Eine tiefe Frequenz ist aus technischen Gründen unerwünscht. Nur bei hohen Frequenzen können die Kondensatoren der beiden Säulen genügend klein gehalten werden. Man ist bei der Erzeugung sehr hoher Spannungen gezwungen, den durch die Streukapazitäten

Fig. 27. Die Ströme in den Kondensatorzuleitungen eines dreistufigen Kaskadengenerators nach Fig. 26.

hervorgerufenen unerwünschten Spannungsabfall sowie die kapazitive Belastung des Transformators durch Drosseln zu reduzieren.

Zusammenfassend können wir feststellen, daß drei Einflüsse den Spannungsabfall eines Kaskadengenerators wesentlich bestimmen:

1. Die Größe der Kapazitäten der Schub- und der Glättungssäule.
2. Die Widerstände des Transformators und der Ventile.
3. Die Wechselströme, die durch die unvermeidlichen Streukapazitäten fließen.

Im folgenden wollen wir diese Einflüsse einzeln untersuchen und zunächst den Spannungsabfall berechnen, der durch das Umladen der Kondensatoren zustande kommt (Ziff. 12), um anschließend eine allgemeinere Theorie zu entwickeln, in der die Widerstände mitberücksichtigt werden (Ziff. 13). Diese

Theorie erlaubt uns, die Grenzen der Formel von A. BOUWERS und A. KUNTKE, Gl. (12.5), anzugeben. In Ziff. 14 schließlich wird der Einfluß der Streukapazitäten eingehender diskutiert.

b) Der belastete Kaskadengenerator.

12. Spannungsabfall ΔU und Rippelspannung δU verursacht durch das Umladen der Kondensatoren. Die der Ausgangsspannung U_g überlagerte Rippelspannung δU läßt sich, wie J. D. COCKCROFT und E. T. S. WALTON [C 32] gezeigt haben, relativ einfach berechnen. Wir betrachten hierzu den Zeitpunkt, in dem die Glättungssäule aufgeladen wird. Wie Fig. 29 zeigt, wird in diesem Zeitintervall jedem Knotenpunkt die Ladung Q zugeführt. Die Ausgangsspannung an der Glättungssäule erhöht sich infolgedessen um den Wert

$$\delta U = Q \sum_{n=1}^{N} \frac{n}{C_n}. \qquad (12.1)$$

δU stellt die Rippelspannung dar, welche sich der Gleichspannung U_g überlagert. Die während des Aufladens der Glättungssäule durch I_g entzogene Ladung $\varepsilon_2 Q$ wurde in (12.1) vernachlässigt. Berücksichtigt man diesen Effekt, so lautet die Formel:

$$\left.\delta U = Q \left[\sum_{n=1}^{N} \frac{n}{C_n} - \varepsilon_2 \sum_{n=1}^{N} \frac{1}{C_n} \right]. \right\} \qquad (12.2)$$

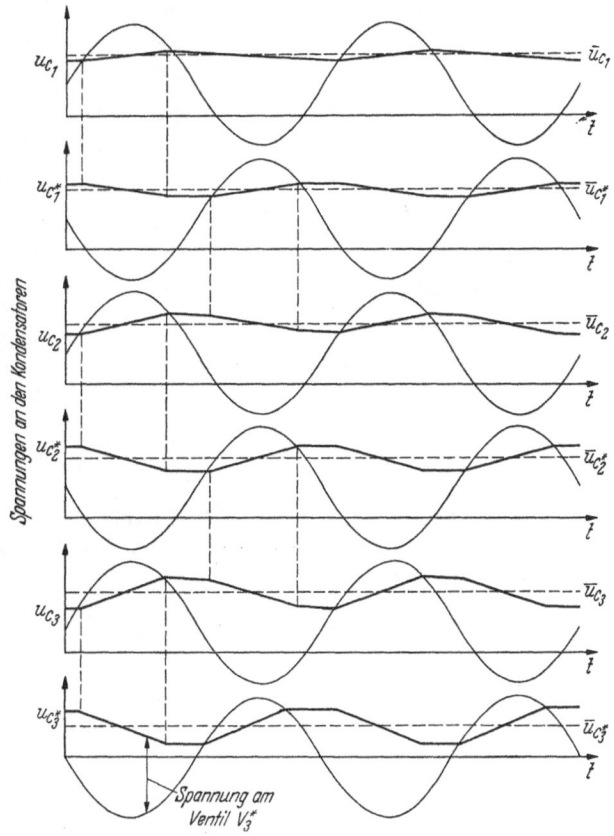

Fig. 28. Die Spannungen über den Kondensatoren der Kaskade Fig. 26 [*Mi 45*].

Der zweite Term wird in den meisten Fällen nur eine unwesentliche Korrektur bedeuten.

Außer der Rippel-Spannung δU interessiert uns der Spannungsabfall ΔU, d.h. der Unterschied zwischen der theoretischen Leerlaufspannung und der Klemmenspannung bei Stromentnahme. Unter der Annahme, daß die Widerstände des Transformators und der Ventile vernachlässigbar klein sind, berechnen wir diesen Spannungsabfall für eine beliebige Stufenzahl N.

Bei vernachlässigbar kleinen Widerständen wird der unterste Kondensator C_N^* der Schubsäule auf die Amplitude der Wechselspannung aufgeladen. Bereits der Kondensator C_N der Glättungssäule erreicht jedoch nicht mehr $2 U_0$, denn während des Aufladens dieses Kondensators wird die Schubsäule gleichzeitig entladen. Die höchste Spannung an C_N tritt am Ende seiner Ladezeit auf. In diesem Augenblick ist die Spannung des Punktes N gleich der Spannung des Punktes N^*.

Durch das Aufladen der Glättungssäule wird dem Kondensator C_N^* die Ladung NQ entzogen. Infolgedessen sinkt seine Spannung um

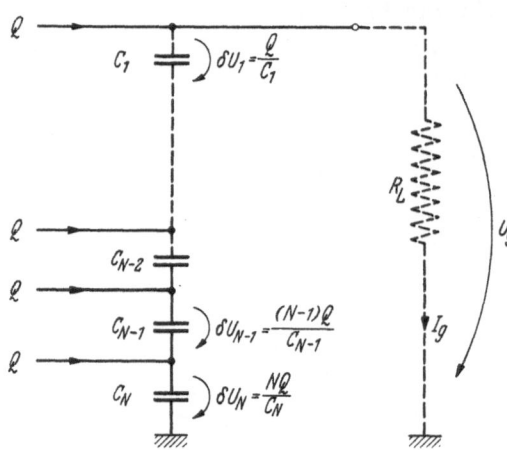

$$\Delta U_N^* = \frac{QN}{C_N^*}.$$

Die Spannung am Kondensator C_N erreicht somit einen Höchstwert von

$$2U_0 - \frac{QN}{C_N^*}.$$

Als Folge des Belastungsstromes I_g wird C_N bis zum nächsten Aufladen der Schubsäule die Ladung $(1 - \varepsilon_1)Q$ und ferner während der Aufladezeit selbst die Ladung $(N-1)Q$ entzogen (vgl. Fig. 30). Am Ende dieser Aufladezeit erreicht die Spannung an C_N den Wert $2U_0$ vermindert um

Fig. 29. Glättungssäule bestehend aus N Kondensatoren, im Zeitpunkt der Aufladung.

$$\Delta U_N = NQ\left(\frac{1}{C_N^*} + \frac{1}{C_N}\right) - \frac{\varepsilon_1 Q}{C_N}.$$

Daraus bestimmt sich der Höchstwert der Spannung am Punkte $(N-1)^*$. Unter Berücksichtigung der nachfolgenden Ladungsentnahme wird somit die Größe ΔU_{N-1}^* gegeben durch

$$\Delta U_{N-1}^* = QN\left(\frac{1}{C_N^*} + \frac{1}{C_N}\right) - \frac{\varepsilon_1 Q}{C_N} + \frac{QN}{C_N^*} + \frac{Q(N-1)}{C_{N-1}^*}.$$

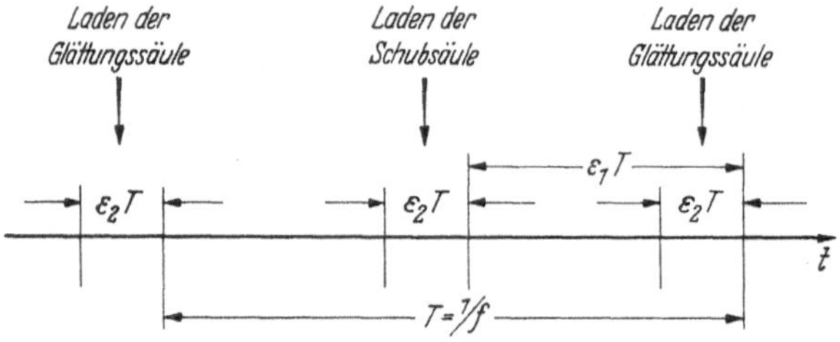

Fig. 30. Zur Berechnung des Spannungsabfalles. f Frequenz der Transformatorspannung.

$Q(N-1)/C_{N-1}^*$ ist dabei der Spannungsabfall über dem Kondensator C_{N-1}^*, der durch das Aufladen der Glättungssäule entsteht, während QN/C_N^* den Spannungsabfall über dem Kondensator C_N^* darstellt, der sich additiv überlagert.

Entsprechend erhält man weiter

$$\Delta U_{N-1} = Q\left[2N\left(\frac{1}{C_N^*} + \frac{1}{C_N}\right) + (N-1)\left(\frac{1}{C_{N-1}^*} + \frac{1}{C_{N-1}}\right) - \varepsilon_1\left(\frac{2}{C_N} + \frac{1}{C_{N-1}}\right)\right].$$

Verfolgen wir in dieser Weise den Spannungsabfall längs der Kaskade bis zur Ausgangsklemme des Generators, so erhält man schließlich den gesamten Spannungsabfall ΔU, der sich als Summe

$$\Delta U = Q \sum_{n=1}^{N} \left[n^2 \left(\frac{1}{C_n^*} + \frac{1}{C_n} \right) - \frac{\varepsilon_1 n}{C_n} + \frac{\varepsilon_1 - \varepsilon_2}{C_n} \right] \tag{12.3}$$

darstellen läßt.

Der zusätzliche Term $\sum_{n=1}^{N} \dfrac{\varepsilon_1 - \varepsilon_2}{C_n}$ berücksichtigt die Entladung der Glättungssäule während der Zeit $(\varepsilon_1 - \varepsilon_2)\,T$ durch den Belastungsstrom I_g (vgl. Fig. 30).
Die Ausgangsspannung der Kaskade variiert somit zwischen den Werten:

$$\left. \begin{aligned} U_{\max} &= 2N U_0 - \Delta U + \delta U, \\ U_{\min} &= 2N U_0 - \Delta U, \end{aligned} \right\} \tag{12.4}$$

und die abgegebene Gleichspannung bestimmt sich zu:

$$U_g = 2N U_0 - \Delta U_g = 2N U_0 - \Delta U + \frac{\delta U}{2}. \tag{12.5}$$

In der vorliegenden Form wurde diese Gleichung zuerst von H. MEHLHORN [Me 43]
hergeleitet. Sie geht mit $\varepsilon_1 = \varepsilon_2 = 0$ in die ältere Formel von A. BOUWERS und
A. KUNTKE [Bo 37] über. Es sei noch bemerkt, daß in allen praktischen Fällen
$\varepsilon_1 = \frac{1}{2}$ gesetzt und der Term mit ε_2 vernachlässigt werden darf ($\varepsilon_2 \approx 0$).

Spezialfall gleich großer Kondensatoren. Unter der Annahme, daß sämtliche
Kondensatoren denselben Wert C besitzen, lassen sich die Summen in (12.2)
und (12.3) leicht berechnen. Man erhält

$$\delta U = \frac{Q}{C} \sum_{n=1}^{N} (n - \varepsilon_2) = \frac{Q}{C} \left[\frac{N^2 + N}{2} - \varepsilon_2 N \right], \tag{12.6}$$

$$\left. \begin{aligned} \Delta U_g &= \frac{Q}{C} \sum_{n=1}^{N} \left[2n^2 - \left(\varepsilon_1 + \frac{1}{2} \right) n + \left(\varepsilon_1 - \frac{\varepsilon_2}{2} \right) \right] \\ &= \frac{Q}{C} \left[\frac{2}{3} N^3 + N^2 + \frac{N}{3} - \left(\varepsilon_1 + \frac{1}{2} \right) \left(\frac{N^2 + N}{2} \right) + N \left(\varepsilon_1 - \frac{\varepsilon_2}{2} \right) \right]. \end{aligned} \right\} \tag{12.7}$$

Die Größe ΔU_g ist in (12.5) definiert. Für die weitere Diskussion wollen wir der
Einfachheit halber $\varepsilon_1 = \frac{1}{2}$ und $\varepsilon_2 = 0$ annehmen und mit

$$\Delta U_g = \frac{Q}{C} \left[\frac{2}{3} N^3 + \frac{N^2}{2} + \frac{N}{3} \right] \tag{12.8}$$

rechnen.

Man erkennt, daß die untersten Kondensatoren den Hauptteil der Welligkeit
und den größten Teil des Spannungsabfalles bewirken. Es wäre denkbar, die
Kapazität der Kondensatoren nach unten hin gestaffelt zunehmen zu lassen.
Dies ist jedoch nur in einem sehr beschränkten Umfang möglich, da sonst bei
plötzlichen Kurzschlüssen an einzelnen Kondensatoren unzulässig hohe Spannungen auftreten können. Hingegen ist es zweckmäßig, der Kapazität des ersten
Kondensators der Schubsäule den doppelten Wert der übrigen Kondensatoren
zu erteilen. Da an ihm nur die halbe Gleichspannung liegt, ist das Volumen
dieses Kondensators selbst dann noch kleiner als dasjenige der übrigen.

Formel (12.8) zeigt ferner, daß die Ausgangsspannung einer Kaskade nicht
beliebig vergrößert werden kann. Mit wachsender Stufenzahl N nimmt der
Spannungsabfall rasch zu. Für jeden vorgegebenen Belastungsstrom I_g existiert

deshalb eine günstigste Stufenzahl. Eine weitere Erhöhung von N würde zu einer kleineren Ausgangsspannung führen. Man erhält N_{opt} aus Gl. (12.5) und (12.8) zu:

$$N_{\mathrm{opt}} = -\frac{1}{4} + \sqrt{\frac{CU_0}{Q} - \frac{5}{48}} \approx \sqrt{\frac{CU_0}{Q}} = \sqrt{\frac{CU_0 f}{I_g}}. \tag{12.9}$$

So ergibt sich z.B. für einen Generator mit $U_0 = 110$ kV, $f = 200$ Hz, $C = 0,02$ μF bei einem Belastungsstrom $I_g = 4$ mA, N_{opt} zu 10 Stufen. Bei einer Leerlaufspannung von 2,2 MV wird mit der vorgesehenen Belastung eine Gleichspannung von $U_g = 1,5$ MV erzielt. Wir müssen wohl kaum besonders betonen, daß in der Praxis eine kleinere Stufenzahl zu wählen ist. So würden z.B. 8 Stufen immer noch $U_g = 1,38$ MV (Leerlauf 1,76 MV) liefern und die beiden weiteren Stufen (4 Kondensatoren und 4 Ventile) bedeuten im Verhältnis zum Aufwand nur einen sehr bescheidenen Gewinn.

13. Allgemeine Behandlung des belasteten Kaskadengenerators ohne Streukapazitäten.

Wie R. G. Mitchell [Mi 45] gezeigt hat, läßt sich durch die Einführung passender Ersatzgrößen die Theorie des Kaskadengleichrichters näherungsweise auf die im Kapitel I dargestellte Theorie des Einweggleichrichters zurückführen. So wird es möglich, außer dem Einfluß der Kondensatoren auch den der Widerstände des Transformators und der Ventile zu berücksichtigen.

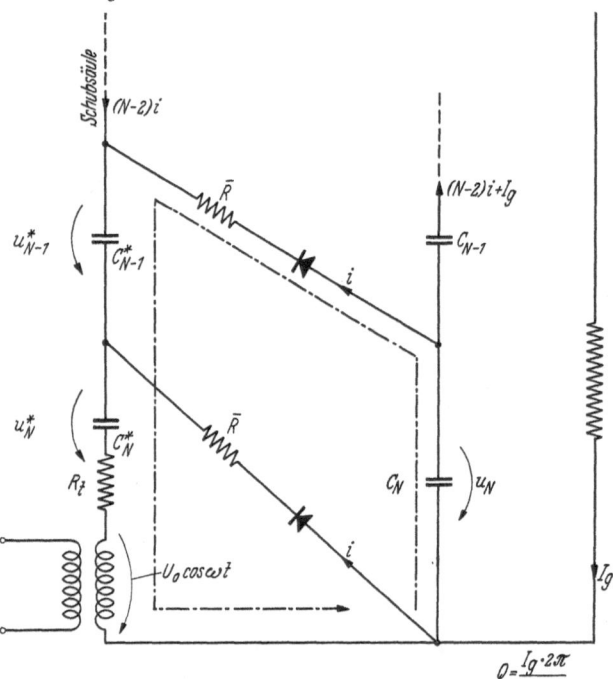

Fig. 31. Aufladung der Schubsäule eines Generators aus N Stufen. Es bedeuten: R_l: Auf die Sekundärseite reduzierter Widerstand des Transformators $R_l = R_s + \ddot{u}^2 \cdot R_p$. $\ddot{u}$: Übersetzungsverhältnis des Transformators. R_p: Widerstand der Primärseite des Transformators. $\bar{R}$: Äquivalenter Widerstand der Ventile in Leitrichtung. $U_0 \cos \omega t$: EMK des Transformators.

Wir bezeichnen als $\nu_n U_0$ bzw. $\nu_n^* U_0$ die Gleichspannungen über den Ventilen V_n bzw. V_n^* (vgl. auch Fig. 26 bis Fig. 28). Die Gleichspannung U_g am Ausgang der Kaskade läßt sich somit durch die Beziehung

$$U_g = U_0 \sum_{n=1}^{N} (\nu_n^* + \nu_n) \tag{13.1}$$

darstellen. Es wird sich zeigen, daß sich ν_n bzw. ν_n^* als normierte Gleichspannungen in zugeordneten Ersatzgleichrichtern auffassen lassen.

Die Aufgabe der folgenden Rechnung besteht also darin, die Größen ν_n bzw. ν_n^* zu berechnen. Hierzu müssen wir das Aufladen der Schubsäule und dasjenige der Glättungssäule gesondert betrachten. Der Einfachheit halber soll die Streuinduktivität des Transformators vernachlässigt werden. Eine Abschätzung ihres Einflusses ist in Ziff. 10 enthalten. Im weiteren wird der Sperrstrom der Ventile nicht berücksichtigt[1].

Schubsäule. Um zu zeigen, wie sich eine Stufe der Schubsäule auf einen Einweggleichrichter zurückführen läßt, sei als Beispiel die Gleichung für das Aufladen des Kondensators C_{N-1}^* betrachtet. Die Maschengleichung für den in Fig. 31 eingezeichneten Weg lautet:

$$
\left.
\begin{aligned}
U_0 \cos \omega t = & -U_{gN} - \frac{\delta U_N}{2} + \int_{t_1}^t \frac{(N-1)\,i + I_g}{C_N}\,dt + i\,\overline{R} + \\
& + U_{gN-1}^* - \frac{\delta U_{N-1}^*}{2} + \int_{t_1}^t \frac{i\,(N-1)}{C_{N-1}^*}\,dt + \\
& + U_{gN}^* - \frac{\delta U_N^*}{2} + \int_{t_1}^t \frac{N\,i}{C_N^*}\,dt + N\,i\,R_i .
\end{aligned}
\right\} \tag{13.2}
$$

Bei der Aufstellung von Gl. (13.2) ist zu beachten, daß zur Zeit t_1 der Kondensator C_N auf die maximale Spannung aufgeladen ist, während über C_N^* und C_{N-1}^* die minimale Gleichspannung liegt. Die Vorzeichen von $\delta U_n/2$ sind daher entsprechend zu wählen. U_{gn} bedeutet die Gleichspannung über dem entsprechenden Kondensator, $\delta U_n/2$ die Amplitude der dieser Gleichspannung überlagerten Rippelspannung (vgl. Fig. 28) und t_1 den Zeitpunkt, in welchem die eingezeichneten Ventile zu leiten beginnen. Es wurde vorausgesetzt, daß alle in Betracht fallenden Ventile zur gleichen Zeit einschalten bzw. ausschalten. Zwar bietet es keine besonderen Schwierigkeiten, (13.2) auf die in Wirklichkeit auftretenden, voneinander verschiedenen Einschaltzeiten zu erweitern, hingegen wird die Behandlung des Kaskadengleichrichters dann recht verwickelt. Zur weiteren Diskussion von (13.2) ist es zweckmäßig, folgende Abkürzungen einzuführen

$$
\left.
\begin{aligned}
& i_{eN-1} = (N-1)\,i, \quad \delta U_{eN-1}^* = \delta U_N^* + \delta U_{N-1}^* + \delta U_N = \frac{Q_{eN-1}}{C_{eN-1}^*}, \\
& R_{eN-1} = \frac{N R_i + \overline{R}}{N-1}, \quad U_0\,\nu_{N-1}^* = U_{gN}^* + U_{g\,N-1}^* - U_{gN}, \\
& \frac{1}{C_{eN-1}^*} = \frac{N}{N-1}\frac{1}{C_N^*} + \frac{1}{C_{N-1}^*} + \frac{1}{C_N}, \quad Q_{eN-1} = (N-1)\,Q \\
& \hspace{9cm} = (N-1)\int_{t_1}^{t_2} i\,dt .
\end{aligned}
\right\} \tag{13.3}
$$

Wir erhalten so die einfachere und übersichtlichere Darstellung

$$
\left.
\begin{aligned}
U_0 \cos \omega t = & \; i_{eN-1} R_{eN-1} + \nu_{N-1}^* U_0 - \frac{\delta U_{eN-1}^*}{2} + \\
& + \int_{t_1}^t \left(\frac{i_{eN-1}}{C_{eN-1}^*} + \frac{I_g}{C_N} \right) dt .
\end{aligned}
\right\} \tag{13.4}
$$

Der Term mit I_g/C_N ist normalerweise klein und soll im folgenden vernachlässigt werden. Dies bedeutet wie früher $\varepsilon_2 = 0$ und entspricht einer rein impulsmäßigen Belastung im Sinne von Fig. 8. Ein Vergleich von (13.4) mit der Gl. (2.2) des Einweggleichrichters zeigt, daß beide dieselbe Form besitzen. Die Einführung der Ersatzgrößen $i_{e,n}$, $C_{e,n}$, $R_{e,n}$, $\delta U_{e,n}$, ν_n erlaubt uns, die in Kapitel I

---

[1] Der Einfluß des Sperrstroms ist nur bei Halbleiter-Gleichrichtern von Bedeutung und kann gegebenenfalls leicht abgeschätzt werden. Die Berücksichtigung des Sperrstromes bedeutet eine zusätzliche Belastung.

hergeleiteten Beziehungen [Gl. (3.1) u. f.] und Kurven (Fig. 9 bis 11) über impuls-mäßige Belastung zu benützen.

Für eine beliebige Stufe n der Schubsäule nehmen die Ersatzgrößen folgende Werte an:

$$\left.\begin{aligned}
&i_{en} = n\,i, \qquad Q_{en} = n\,Q, \qquad \delta U^*_{en} = \frac{Q_{en}}{C^*_{en}}, \\
&R_{en} = \frac{N R_t + \overline{R}}{n}, \\
&\frac{1}{C^*_{en}} = \frac{1}{n}\left\{\frac{N}{C^*_N} + (N-1)\left(\frac{1}{C_N} + \frac{1}{C^*_{N-1}}\right) + \cdots + n\left(\frac{1}{C_{n+1}} + \frac{1}{C^*_n}\right)\right\}, \\
&\qquad = \frac{1}{n}\left\{\frac{N}{C^*_N} + \sum_{\lambda=1}^{N-n}(N-\lambda)\left(\frac{1}{C_{N-\lambda+1}} + \frac{1}{C^*_{N-\lambda}}\right)\right\}.
\end{aligned}\right\} \tag{13.5}$$

$1/C^*_{en}$ wird durch eine Reihe mit $(N-n+1)$ Gliedern dargestellt.

Im Spezialfall gleich großer Kondensatoren wird:

$$\frac{1}{C^*_{en}} = \frac{1}{n\,C}\left[N^2 - n^2 + n\right]. \tag{13.6}$$

Die gleichen Überlegungen lassen sich auf das Aufladen der *Glättungssäule* an-wenden. Die entsprechende Maschengleichung führt unter denselben Verein-fachungen für eine beliebige Stufe n zu folgenden Ersatzgrößen:

$$\left.\begin{aligned}
&i_{en} = n\,i, \qquad Q_{en} = n\,Q, \qquad \delta U_{en} = \frac{Q_{en}}{C_{en}}, \\
&R_{en} = \frac{N R_t + \overline{R}}{n}, \\
&\frac{1}{C_{en}} = \frac{1}{n}\sum_{\lambda=0}^{N-n}(N-\lambda)\left(\frac{1}{C^*_{N-\lambda}} + \frac{1}{C_{N-\lambda}}\right).
\end{aligned}\right\} \tag{13.7}$$

Falls alle Kondensatoren denselben Wert besitzen, reduziert sich $1/C_{en}$ auf

$$\frac{1}{C_{en}} = \frac{1}{n\,C}\left[N^2 - n^2 + n + N\right]. \tag{13.8}$$

Spannungsabfall bei verschwindend kleinen Widerständen: Es ist von Interesse zu zeigen, daß Formel (12.5) als Spezialfall in den soeben dargestellten Ergeb-nissen enthalten ist. Sind die Widerstände des Transformators R_t und der Ven-tile $\overline{R}$ verschwindend klein, so wird der Spannungsabfall pro Kondensator der Kaskade $= \delta U_{en}/2$ bzw. $\delta U^*_{en}/2$, wobei δU_{en} bzw. δU^*_{en} die als Ersatzgrößen eingeführten Rippelspannungen bedeutet [Gl. (13.7) bzw. (13.5)]. Nun ist zu beachten, daß

$$\left.\begin{aligned}
\delta U_{en} = \frac{Q_{en}}{C_{en}} = \frac{n\,Q}{C_{en}}, \\
\delta U^*_{en} = \frac{Q_{en}}{C^*_{en}} = \frac{n\,Q}{C^*_{en}}
\end{aligned}\right\} \tag{13.9}$$

ist. Infolgedessen erhalten wir für die Ausgangsspannung der belasteten Kaskade

$$U_g = 2N U_0 - \sum_{n=1}^{N}\left(\frac{n\,Q}{2 C^*_{en}} + \frac{n\,Q}{2 C_{en}}\right). \tag{13.10}$$

Man kann sich leicht davon überzeugen, daß diese Gleichung mit Formel (12.5) übereinstimmt, falls in (12.2) und (12.3) $\varepsilon_1 = \frac{1}{2}$ und $\varepsilon_2 = 0$ eingesetzt wird. $\varepsilon_2 = 0$

ist eine Folge der Vernachlässigung des Terms mit I_g in Gl. (13.4) und in der entsprechenden Gleichung der Glättungssäule.

Spannungsabfall bei sehr großen Kondensatoren. Bei genügend großen Kondensatoren ist der Spannungsabfall ausschließlich in der Wirkung der Widerstände R_t (Transformator) und $\overline{R}$ (Ventil) begründet[1]. Im Kapitel I wurde dieser Spezialfall ausführlich behandelt und wir dürfen die dort abgeleiteten Formeln, Gl. (7.2), übernehmen. Pro Kondensator bzw. Ventil tritt ein Spannungsabfall von

$$\Delta U_R = \Delta U_n = \Delta U_n^* = U_0 \frac{1}{2} (3\pi)^{\frac{2}{3}} \left(\frac{NR_t + \overline{R}}{U_0} I_g \right)^{\frac{2}{3}} \tag{13.11}$$

auf. Da alle ΔU_R denselben Wert besitzen, erhält man für die Ausgangsspannung der Kaskade:

$$U_g = 2N U_0 \left[1 - \frac{1}{2} (3\pi)^{\frac{2}{3}} \left(\frac{NR_t + \overline{R}}{U_0} I_g \right)^{\frac{2}{3}} \right]. \tag{13.12}$$

Gl. (13.12) zeigt, daß selbst mit unendlich großen Kondensatoren in der Schub- und in der Glättungssäule eine bestimmte Stufenzahl nicht überschritten werden soll. Die optimale Stufenzahl, welche bei vorgegebener Belastung I_g die größte Ausgangsspannung liefert, bestimmt sich mit der Annahme $NR_t \gg \overline{R}$ zu

$$N_{\text{opt}} \approx \left(\frac{6}{5} \right)^{\frac{3}{2}} \frac{1}{3\pi} \frac{U_0}{R_t I_g}. \tag{13.13}$$

Nach Gl. (12.9) würde bei beliebig großen Kondensatoren (bzw. genügend hoher Frequenz) die Stufenzahl des Kaskadengenerators nicht begrenzt sein. Da der Widerstand R_t des Transformators relativ niedrige Werte annehmen kann, führt (13.13) zu bemerkenswert großen N_{opt}. Mit $R_t = 50\,\text{k}\Omega$, $U_0 = 80\,\text{kV}$, $I_g = 6{,}75\,\text{mA}$ (vgl. Fig. 39) erhält man beispielsweise $N_{\text{opt}} \approx 33$. Bei derart hohen Stufenzahlen ist indessen der Einfluß der Streukapazitäten der bestimmende Faktor.

Geltungsbereich der Formel (12.5) *nach A. Bouwers und A. Kuntke.* In Kapitel I wurde am Beispiel des Einweggleichrichters [Gl. (3.1)] gezeigt, daß die folgende Beziehung näherungsweise gilt:

$$\Delta U = \sqrt[3]{\Delta U_C^3 + \Delta U_R^3}. \tag{13.14}$$

Dabei bedeutet ΔU_C den Spannungsabfall, der bei vernachlässigbar kleinen Widerständen ($R_t = \overline{R} = 0$) auftritt und ΔU_R den Spannungsabfall, welcher bei unendlich großem Kondensator als Folge der Widerstände zu beobachten ist. Mit Hilfe dieser Näherungsgleichung und den in diesem Abschnitt berechneten Ersatzgrößen läßt sich nun eine Grenze angeben, innerhalb der die Bouwerssche Formel (12.5) verwendet werden darf. Da der größere der beiden Spannungsabfälle sehr stark überwiegt, ist diese Grenze ziemlich scharf.

Betrachten wir einen beliebigen äquivalenten Gleichrichter der Kaskade. Der kritische Wert der Ersatzkapazität C_{kr} wird durch

$$\Delta U_R = \Delta U_C \tag{13.15}$$

bestimmt:

$$C_{\text{kr}} = \left(\frac{\sqrt{2}}{1{,}5\,R_{en}\omega} \right)^{\frac{3}{2}} \left(\frac{Q_{en}}{2U_0} \right)^{\frac{1}{2}}. \tag{13.16}$$

---

[1] Dabei wird die Streuinduktivität des Transformators vernachlässigt. Es ist möglich, daß der Einfluß der Streuinduktivität denjenigen der Widerstände überwiegt. In diesem Falle kann unter Berücksichtigung von (10.2) der Spannungsabfall in ähnlicher Weise berechnet werden.

Für $C_{en} < C_{kr}$ darf der Einfluß der Widerstände und für $C_{en} > C_{kr}$ derjenige der Kondensatoren vernachlässigt werden. Insbesondere gilt die Gl. (12.5) dann, wenn die Ersatzkapazität für sämtliche Kondensatoren der Kaskade kleiner als C_{kr} ist. In der Praxis liegen die Verhältnisse oft so, daß C_{en} für die untersten

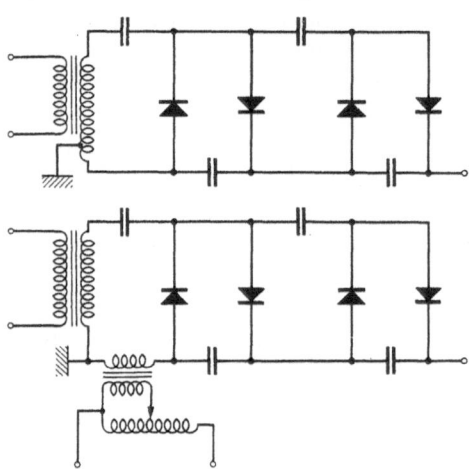

Stufen größere und für die oberen Stufen kleinere Werte als C_{kr} annimmt. Die Ursache liegt darin, daß C_{en} von unten nach oben sukzessive und ziemlich rasch abnimmt. Die Beziehung (13.16) erlaubt zusammen mit (13.9) und (13.11) eine verhältnismäßig einfache und bequeme Abschätzung des Spannungsabfalles in einer belasteten Kaskade.

14. Spannungsabfall durch Streukapazitäten.

Von den in der Einleitung erwähnten Ursachen des Spannungsabfalles eines Kaskadengenerators sind zwei bereits besprochen worden. Die dritte Ursache, nämlich die Wirkung der Streukapazitäten, erfordert eine gesonderte Betrachtung.

Fig. 32. Schaltungen zur Kompensation der Rippel-Spannung.

Durch die Anordnung und den Aufbau der Kondensatoren, Ventile und Abschirmungen entstehen schädliche Kapazitäten, deren Einfluß in den wenigsten Fällen vernachlässigt werden darf. Als Folge dieser Streukapazitäten treten

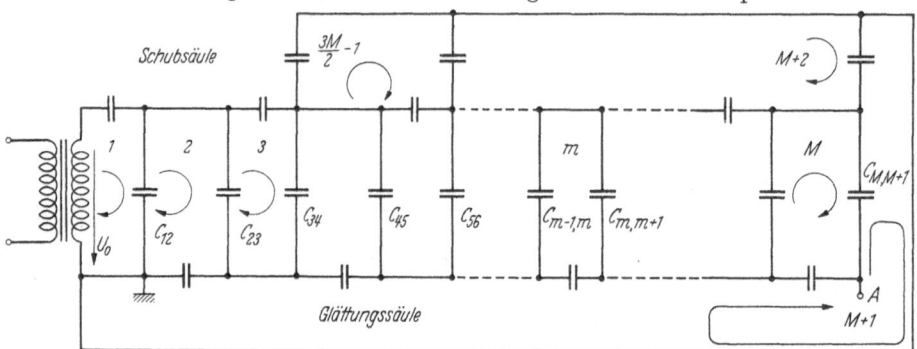

Fig. 33. Kaskadengenerator mit Streukapazitäten. Die Ventile sind nicht eingezeichnet. Sie liegen parallel zu den Kapazitäten C_{12} bis $C_{M,\,M+1}$. V ist die Gleichspannung zwischen A und Erde.

unerwünschte Ströme auf, die einerseits eine niedrigere Klemmenspannung (auch bei unbelastetem Generator) zur Folge haben und andererseits der Gleichspannung eine zusätzliche Welligkeit überlagern, deren Größe beträchtlich sein kann. Außerdem stellen diese Ströme oft die wesentliche Belastung des Transformators dar. Es ist möglich, die erwähnte Welligkeit der Gleichspannung zu kompensieren. Solche Kompensationsschaltungen sind in Fig. 32 dargestellt. Selbstverständlich gestatten die gleichen Schaltungen auch die Kompensation der Grundwelle der Rippelspannung, welche infolge der äußeren Belastung des Kaskadengenerators entsteht.

In Fig. 33 sind alle nennenswerten Streukapazitäten eingezeichnet. Eine allgemeine Behandlung des belasteten Generators unter Berücksichtigung sämtlicher Kapazitäten wäre recht verwickelt, so daß wir zunächst die Verhältnisse im Leerlauf diskutieren.

Leerlaufspannung. Um die Leerlaufspannung V zu erhalten, müssen wir die Scheitelwerte der Wechselspannungen, die sich über den einzelnen Ventilen ausbilden, addieren, oder was dasselbe bedeutet, die Amplituden der Spannungen, die sich an den Kondensatoren C_{12} bis $C_{M,M+1}$ einstellen, zusammenzählen. Die kapazitiven Ströme durch die Glättungssäule erzeugen ferner eine sinusförmige Wechselspannung, die sich der Gleichspannung überlagert. Da es sich im Leerlauf um eine kapazitive Unterteilung handelt, ist es offensichtlich, daß Spannungsabfall und Rippelspannung bei konstanter Amplitude der Klemmspannung $U_0 \cos \omega t$ des Transformators unabhängig von der Frequenz sind.

Die Maschengleichungen für das in Fig. 33 dargestellte Schema lauten[1]:

$$
\left.
\begin{aligned}
U_0 &= \frac{I_1}{j\omega C_{11}} - \frac{I_2}{j\omega C_{12}}, \\
0 &= -\frac{I_1}{j\omega C_{12}} + \frac{I_2}{j\omega C_{22}} - \frac{I_3}{j\omega C_{23}}, \\
0 &= \qquad -\frac{I_2}{j\omega C_{23}} + \frac{I_3}{j\omega C_{33}} - \frac{I_4}{j\omega C_{34}}. \\
\cdots & \cdots \cdots \cdots \cdots \cdots \cdots
\end{aligned}
\right\}
\begin{array}{c} \dfrac{3M}{2} - 1 \\[4pt] \text{Gleichungen} \end{array}
\qquad (14.1)
$$

Wie üblich bedeutet $1/j\omega C_{11}$ die Impedanz der ersten Masche usw. Aus diesem Gleichungssystem berechnen sich die Ströme

$$
I_m = \frac{\Delta_{1m}}{\Delta} U_0 \qquad (m = 1, 2, 3 \cdots), \qquad (14.2)
$$

wobei Δ_{1m} den Kofaktor[2] der Determinante Δ zum Glied 1, m darstellt. Daraus bestimmt sich die Hochspannung V im Leerlauf zu:

$$
V = \sum_{m=1}^{M} \left| \frac{I_m - I_{m+1}}{j\omega C_{m,m+1}} \right|. \qquad (14.3)
$$

Bei vielen Stufen und einer beliebigen Verteilung der Kapazitäten führt (14.3) zu umständlichen numerischen Rechnungen. Der Überblick über den quantitativen Einfluß der Streukapazitäten geht verloren. Es besteht deshalb ein Bedürfnis nach einer übersichtlicheren Behandlung, die eine rasche Abschätzung des Spannungsabfalles gestattet. Nach E. EVERHART und P. LORRAIN [E 53] ist dies möglich, wenn der Kaskadengenerator als einfache, aus gleichen Gliedern zusammengesetzte, elektrische Leitung aufgefaßt werden kann (vgl. Fig. 34).

Diese Bedingungen erfüllen insbesondere die in Fig. 36 und Fig. 41 dargestellten symmetrischen Anordnungen. Die erwähnten symmetrischen Schaltungen besitzen ferner den großen praktischen Vorteil, daß an der Hochspannungsklemme im Leerlauf die Wechselspannung verschwindet und deshalb der Hochspannung V keine Rippelspannung überlagert ist.

Vereinfachte Behandlung mit Hilfe der Theorie elektrischer Leitungen [E 53]. Den folgenden Betrachtungen sei das vereinfachte Schema Fig. 34 zugrunde gelegt, welches für die symmetrischen Anordnungen bei gleichen Kondensatoren der einzelnen Glieder streng richtig ist, während es für die Schaltung Fig. 38 nur

---

[1] Die Größen U und I bedeuten hier und in allen folgenden Rechnungen nicht die Effektivwerte, sondern die Amplituden.
[2] Wir halten uns an die Nomenklatur von E. A. GUILLEMIN: The Mathematics of Circuit Analysis. New York: John Wiley Inc. 1951.

eine Näherung darstellt; C bedeutet die zu jedem Ventil parallel liegende Schalt-kapazität (Querkapazität). Jede Seriekapazität wird als das b^2-fache der Quer-kapazität angesetzt. Die Endkapazitäten $2b^2C$ sind so gewählt, daß sich die Leitung aus gleichen Gliedern zusammensetzt. Der Wert $2b^2C$ des Kondensators am linken Ende entspricht der Tatsache, daß in der Praxis der erste Schub-kondensator den doppelten Wert der übrigen Kondensatoren besitzt. Der Kon-densator am rechten Ende ist zunächst ohne Bedeutung, denn die Leitung arbeitet im Leerlauf. Spannung und Strom in der m-ten Masche werden beschrieben durch:

$$i(m, t) = A\, e^{j\omega t - \gamma m} + B\, e^{j\omega t + \gamma m} = I_m\, e^{j\omega t} \tag{14.4}$$

und

$$u(m, t) = z_w\, (A\, e^{j\omega t - \gamma m} - B\, e^{j\omega t + \gamma m}) = U_m\, e^{j\omega t}. \tag{14.5}$$

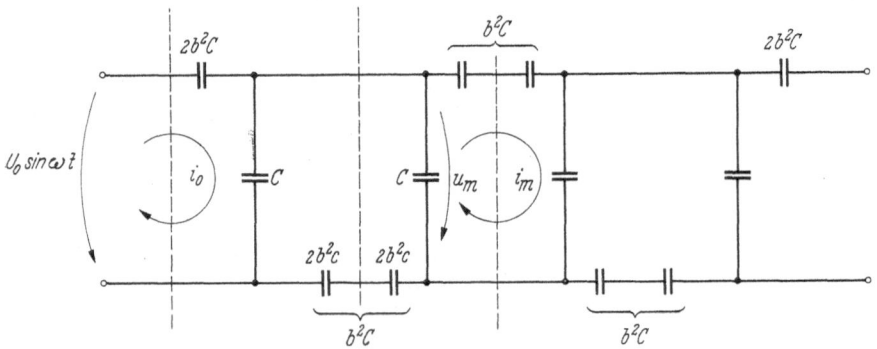

Fig. 34. Kaskadengenerator im Leerlauf, als Leitung mit gleichen Gliedern dargestellt. Die Ventile sind nicht eingezeichnet.

Die unterste Stufe wird durch $m = 0$, die oberste durch $m = M$ gekennzeichnet.

Die Fortpflanzungskonstante γ ist reell und hat den Wert

$$\operatorname{Cos}\gamma = 1 + \frac{1}{2b^2}. \tag{14.6}$$

In praktischen Fällen liegt b^2 in der Größenordnung von etwa 100, so daß in guter Näherung gilt:

$$\gamma \approx \frac{1}{b}. \tag{14.7}$$

In der gleichen Näherung erhält man für den Wellenwiderstand z_w

$$z_w \approx \frac{1}{j\omega b C}. \tag{14.8}$$

Unter Benützung von (14.4), (14.5) und (14.8) bestimmt sich die Impedanz in der m-ten Masche zu

$$z_m = \frac{1}{j\omega b C}\, \frac{1 - (B/A)\, e^{2\gamma m}}{1 + (B/A)\, e^{2\gamma m}}. \tag{14.9}$$

Da die Leitung am Ende offen ist (M-te Masche) wird $z_M = \infty$. Für das Verhältnis B/A folgt daraus

$$\frac{B}{A} = -\, e^{-2\gamma M}. \tag{14.10}$$

und für die Impedanz der m-ten Masche erhält man

$$z_m = \frac{1}{j\omega b C}\, \operatorname{Cot}\left(\frac{M - m}{b}\right). \tag{14.9a}$$

Um die Leerlaufspannung V zu erhalten, müssen wir die Amplituden der Spannungen an den Ventilen zusammenzählen. Es gilt

$$U_m = (I_{m-1} - I_m)\, \frac{1}{j\,\omega\,C}\,. \tag{14.11}$$

Da sämtliche Ströme die gleiche Phasenlage besitzen, wird

$$V = \sum_{m=1}^{M} |U_m| = \sum_{m=1}^{M} \left| \frac{I_{m-1} - I_m}{j\,\omega\,C} \right| = \left| \frac{I_0 - I_M}{j\,\omega\,C} \right|. \tag{14.12}$$

Mit Gl. (14.4) folgt

$$V = \frac{1}{\omega\,C}\left(A + B - A\,\mathrm{e}^{-\gamma M} - B\,\mathrm{e}^{\gamma M}\right), \tag{14.13}$$

ein Ergebnis, das sich mit Gl. (14.10) zu

$$V = \frac{A}{\omega\,C}\left(1 - \mathrm{e}^{-2\gamma M}\right) \tag{14.14}$$

vereinfacht. Diese Spannung ist mit der idealen Leerlaufspannung

$$V_0 = M\,U_0 = 2N\,U_0 \tag{14.15}$$

der Kaskade zu vergleichen[1].

Aus (14.5) folgt

$$U_0 = |z_w(A - B)| \tag{14.16}$$

und mit (14.8) und (14.10)

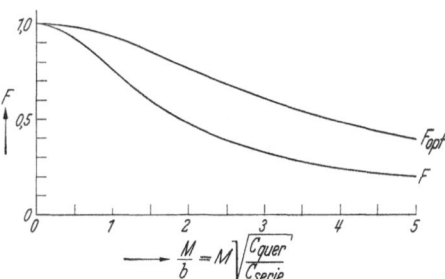

Fig. 35. Verhältnis F der wirklichen zur idealen Leerlaufspannung des Kaskadengenerators in Abhängigkeit von M/b. Kurve F nach Gl. (14.18) und Kurve F_{opt} nach Gl. (14.20).

$$U_0 = \frac{A}{\omega\,b\,C}\left(1 + \mathrm{e}^{-2\gamma M}\right). \tag{14.17}$$

Damit bestimmt sich das Verhältnis F der Gleichspannung V der Kaskade zur idealen Leerlaufspannung V_0 zu:

$$F = \frac{V}{V_0} = \frac{b\left(1 - \mathrm{e}^{-2\gamma M}\right)}{M\left(1 + \mathrm{e}^{-2\gamma M}\right)} = \frac{b}{M}\,\mathrm{Tan}\left(\frac{M}{b}\right). \tag{14.18}[1]$$

Die Funktion F ist in Fig. 35 dargestellt und vermittelt uns ein gutes Bild über den Spannungsabfall als Folge der Streukapazitäten und der Stufenzahl des Generators.

Wie zu erwarten war, ist F unabhängig von der Frequenz der Wechselspannung, ferner müssen die Seriekapazitäten groß gegenüber den Querkapazitäten (Streukapazitäten) sein, damit F keine unerwünscht niedrigen Werte annimmt. Der Vollständigkeit halber soll noch die Eingangsimpedanz z_0 erwähnt werden. Diese müssen wir kennen, um die Belastung des Transformators und der Kondensatoren abzuschätzen. Dabei braucht nur der unterste Kondensator der Kaskade berücksichtigt zu werden, welcher den größten Wechselstrom zu übertragen hat. Aus (14.9a) folgt mit $m = 0$

$$z_0 = \frac{\mathrm{Cot}\,(M/b)}{j\,\omega\,b\,C}\,. \tag{14.19}$$

Es ist möglich, die nachteilige Wirkung der Streukapazität C_q zu verringern, indem z.B. zu jedem Ventil eine, mit C_q auf Resonanz abgestimmte Spule parallel

---

[1] Wegen der Numerierung in Fig. 34 bedeutet M die Zahl der Ventile, es gilt daher $M = 2N$, wobei N wie früher die Stufenzahl des Generators bedeutet.

geschaltet wird, wobei ein weiterer Kondensator zur Sperrung der Gleichspannung erforderlich ist. Dieser zusätzliche Kondensator fällt bei der symmetrischen Schaltung Fig. 41a weg, weil nun die Spulen zwischen entsprechenden Punkten der beiden Schubsäulen angeordnet sein können. Damit nicht unzulässige Verluste auftreten, muß selbstverständlich die Spulengüte Q genügend groß sein. Messungen, die in Basel durchgeführt wurden, zeigen, daß Spulen für hohe Span-

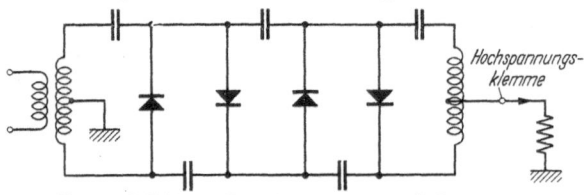

Fig. 36. Modifizierter Kaskadengenerator nach E. EVERHART und P. LORRAIN [E 53].

nungen mit $L = 10$ Henry und einem $Q = 120$ bei 10 kHz durchaus möglich sind [B 57], und diese Lösung ist für den im Bau begriffenen 4 MV-Generator des Basler Physikalischen Institutes vorgesehen.

Eine andere bemerkenswerte Methode [L 56] besteht in der Verwendung einer einzigen Spule L zwischen den beiden Hochspannungsklemmen der beiden Kondensatorsäulen in Fig. 36. Mit einer solchen Spule ist es möglich, die Wechselspannung über dem letzten Ventil so zu erhöhen, daß sie über die

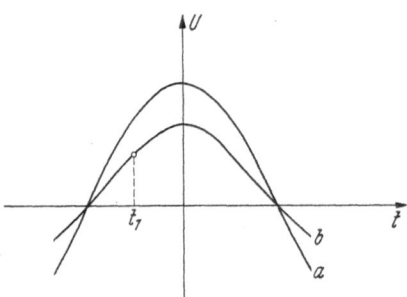

Fig. 37. Wechselspannung über dem Ventil. a im Leerlauf ohne Streukapazitäten, b im Leerlauf mit Streukapazitäten.

Transformatorspannung $U_0 \cos \omega t$ steigt. Nach E. EVERHART und P. LORRAIN [E 53] erhält man optimale Verhältnisse, wenn die Wechselspannungen der Ventile am Eingang und am Ausgang der Kaskade genau gleich groß sind.

Die Rechnung führt dann zu

$$L_{\text{opt}} = \frac{1}{\omega^2 b C} \operatorname{Cot}\left(\frac{M}{2b}\right),$$

mit einem Faktor F_{opt} von

$$F_{\text{opt}} = \left(\frac{2b}{M}\right) \operatorname{Tan}\left(\frac{M}{2b}\right). \qquad (14.20)$$

F_{opt} ist in Fig. 35 eingetragen.

Der belastete Kaskadengenerator mit Streukapazitäten. Wie wir bereits erwähnten, ist eine allgemeine Behandlung des belasteten Kaskadengenerators mit Berücksichtigung der Streukapazitäten in einfacher und übersichtlicher Weise nicht bekannt, und wahrscheinlich auch nicht möglich. Hingegen läßt sich eine Näherungsbeziehung angeben, die in den meisten Fällen ausreichend sein dürfte.

Wir betrachten zunächst Fig. 37. Bis zum Zeitpunkt t_1, in dem das Ventil zu leiten beginnt, verläuft die Spannung nach Kurve b. Sobald das Ventil leitet, kann hingegen der Einfluß der Streukapazität C_q vernachlässigt werden, denn die Zeitkonstante gebildet aus dem Widerstand $\bar{R}$ des Ventils in Leitrichtung und der Streukapazität, darf wohl immer als sehr klein verglichen mit einer Periode der Wechselspannung betrachtet werden

$$\bar{R} C_q \ll 1/f.$$

Im Grenzfall $N R_t \gg \bar{R}$ und bei genügend großen Längskapazitäten gilt dann streng

$$U_g' = U_g \cdot F. \qquad (14.21)$$

Dabei bedeutet U_g die in Ziffer 13 berechnete Ausgangsspannung unter Vernachlässigung der Streukapazitäten und U_g' die Ausgangsspannung unter Berücksichtigung der Streukapazitäten. F stellt den in (14.18) definierten Faktor, also das Verhältnis der wirklichen zur idealen Leerlaufspannung des Generators dar.

Es erscheint nun vernünftig (14.21) als Näherung auch dann zu verwenden, wenn für die Längskapazitäten und die Ventilwiderstände die oben angegebenen Beschränkungen nicht erfüllt sind. Um so mehr, als mit wachsender Belastung die Wechselspannung an der Schubsäule die Tendenz hat, einen mehr rechteckförmigen Verlauf anzunehmen, so daß die Amplituden kleiner werden. Damit werden aber auch die Ströme durch die Streukapazitäten bzw. die durch sie verursachte Verminderung der Gleichspannung geringer.

15. Zusammenstellung der Berechnungsgrundlagen eines Kaskadengenerators und Vergleich mit Meßergebnissen. Um zu einer raschen und übersichtlichen

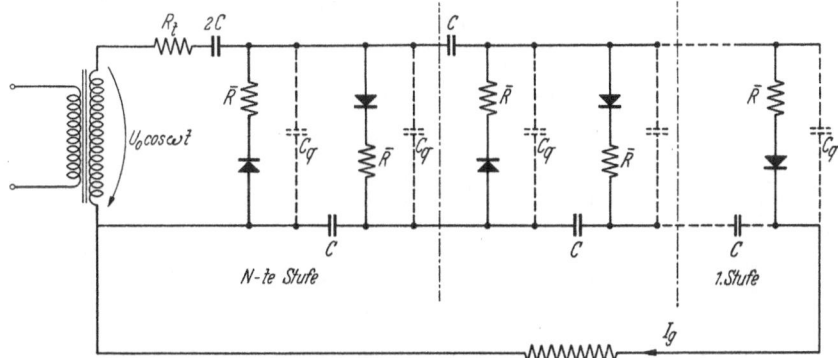

Fig. 38. Kaskadengenerator, Gln. (15.1) bis (15.5).

Berechnung des Kaskadengenerators nach Fig. 38 zu gelangen, sollen die bisherigen Ergebnisse kurz zusammengefaßt werden. Wir beschränken uns auf die in der Praxis meistens verwendete Anordnung, bei welcher der unterste Kondensator der Schubsäule den doppelten Wert aller übrigen besitzt. Es bedeuten:

I_g Gleichstrom mit dem der Generator belastet ist;

f Frequenz der Wechselspannung;

$Q = I_g/f$ Ladung die pro Periode der Wechselspannung nach außen abgegeben wird;

U_0 Amplitude der Transformatorspannung;

R_t auf die Sekundärseite reduzierter Widerstand des Transformators;

$\overline{R}$ äquivalenter Widerstand der Ventile in Leitrichtung (im Sinne von Abschnitt 5).

N Gesamtzahl der Stufen der Kaskade;

n laufende Nummer der Stufen (oberste Stufe $n = 1$, unterste Stufe $n = N$);

F Verhältnis der Leerlaufspannung des Kaskadengenerators (Streukapazitäten inbegriffen) zur Spannung $2 N U_0$.

Für den in Fig. 38 dargestellten Generator ergibt die Reduktion auf den Einweggleichrichter die folgenden Ersatzgrößen:

a) Glättungssäule:

$$\left.\begin{aligned}
\delta U_{en} &= \frac{nQ}{C_{en}} = \frac{Q}{C}\left(N^2 + \frac{N}{2} - n^2 + n\right), \\
\alpha_{en} &= \omega R_{en} C_{en} = \frac{\omega C(N R_t + \overline{R})}{\left(N^2 + \dfrac{N}{2} - n^2 + n\right)}.
\end{aligned}\right\} \qquad (15.1)$$

Nach Gl. (3.1) bzw. (3.5) führt dies zu einem Spannungsabfall ΔU_G an den Kondensatoren der Glättungssäule von

$$\Delta U_G = \sum_{n=1}^{N} \sqrt[3]{\left(\frac{\delta U_{en}}{2}\right)^3 + (\Delta U_R)^3} = \sum_{n=1}^{N} \frac{\delta U_{en}}{2} \sqrt[3]{1 + \frac{9\,\alpha_{en}^2\,U_0}{4\,\delta U_{en}}} \,. \tag{15.2}$$

b) Schubsäule

$$\delta U_{en}^* = \frac{n\,Q}{C_{en}^*} = \frac{Q}{C}\left(N^2 - \frac{N}{2} - n^2 + n\right),$$

$$\alpha_{en}^* = \frac{\omega\,C\,(N\,R_t + \overline{R})}{N^2 - \frac{N}{2} - n^2 + n}\,,$$

$$\Delta U_S = \sum_{n=1}^{N} \sqrt[3]{\left(\frac{\delta U_{en}^*}{2}\right)^3 + (\Delta U_R)^3} = \sum_{n=1}^{N} \frac{\delta U_{en}^*}{2} \sqrt[3]{1 + \frac{9\,\alpha_{en}^{*2}\,U_0}{4\,\delta U_{en}^*}} \,. \tag{15.3}$$

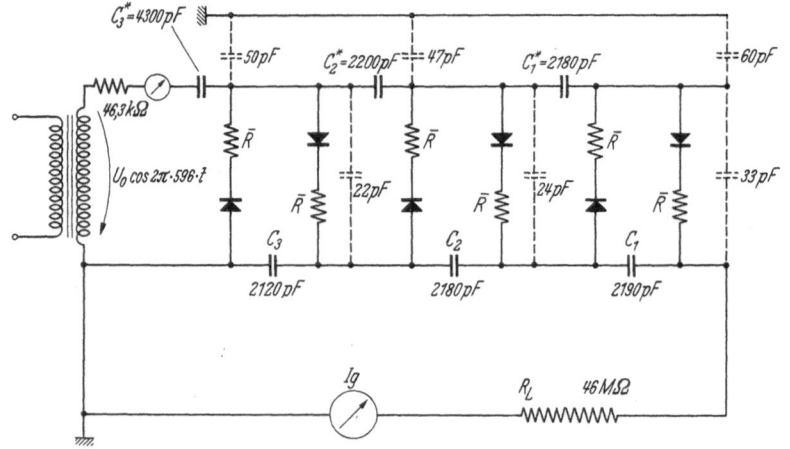

Fig. 39. Dreistufiger Kaskadengenerator nach H. Mehlhorn mit den gemessenen Zahlenwerten der Schaltelemente. $\overline{R} = 20$ kΩ.

Damit bestimmt sich der Spannungsabfall ΔU der Kaskade zu

$$\Delta U = \Delta U_G + \Delta U_S \,. \tag{15.4}$$

Wird ferner der Einfluß der Streukapazitäten C_q berücksichtigt, so erhält man für die resultierende Ausgangsspannung U_g' des Kaskadengenerators den Ausdruck

$$U_g' = (2N\,U_0 - \Delta U)\,F\,, \tag{15.5}$$

wobei F die in Fig. 35 gezeichnete Funktion darstellt[1].

Die Formeln (15.2) bis (15.5) gestatten in den meisten Fällen eine zuverlässige Dimensionierung von Kaskadengeneratoren.

Vergleich mit Meßergebnissen: Um ein Maß für die Güte der von uns betrachteten Näherungen zu erhalten, wenden wir unsere Rechnung auf die von Mehlhorn [Me 43] beschriebene und sehr sorgfältig ausgemessene Anlage an. Im Schema Fig. 39 sind die Zahlenwerte der Schaltelemente dieses 3stufigen Generators eingetragen. Unsere Näherungsformeln ergeben die in Tabelle 5 zusammengestellten Werte. Man beachte, daß beim untersten Kondensator der Schubsäule

---

[1] Mit den Bezeichnungen in Fig. 38 wird $M/b = 2N\sqrt{C_q/C}$.

der Spannungsabfall praktisch nur durch die Widerstände von Transformator und Ventil bestimmt wird, während in den folgenden Stufen der Einfluß der Widerstände rasch verschwindet. Dies stimmt mit der früher erwähnten Tatsache überein, daß die wirksame Kapazität C_{en} von unten nach oben stark abnimmt (vgl. S. 34).

Tabelle 5. *Zur Berechnung des Spannungsabfalles des Generators*[1] *Fig. 39 bei einem Belastungsstrom* $I_g = 6{,}75\ mA$ *und einer Transformatorspannung* $U_0 = 80\ kV$.

C	C_{en} bzw. C_{en}^* pF	ΔU_R kV	$\Delta U_C = \dfrac{\delta U_{en}}{2}$ kV	ΔU_n bzw. ΔU_n^* kV
C_3^*	4300	10	3,92	10,20
C_3	1420	10	11,88	13,90
C_2^*	784	10	14,35	15,84
C_2	508	10	22,13	22,80
C_1^*	288	10	19,48	20,35
C_1	206	10	27,20	27,65

$$\sum_n \Delta U_n = \Delta U = 110{,}74$$

Einfluß der Streukapazitäten: Die in Fig. 39 eingezeichneten Streukapazitäten wurden von H. MEHLHORN rechnerisch abgeschätzt und durch Messungen kontrolliert. Mit den eingetragenen Zahlenwerten bestimmt sich der Faktor F nach (14.3) zu

$$F = 0{,}85\ .$$

Damit erhalten wir für die Ausgangsspannung U_g' der belasteten Kaskade unter Berücksichtigung der Streukapazitäten schließlich

$$U_g' = (480 - 110{,}74) \cdot 0{,}85 = 314\ \text{kV}.$$

Der an der gebauten Anlage experimentell gefundene Wert von 315 kV weicht nur unwesentlich und innerhalb der Fehlergrenzen der Messung von diesem Ergebnis ab.

Die vereinfachte Theorie von E. EVERHART und P. LORRAIN stellt für das Schema der Fig. 39 eine relativ grobe Näherung dar. Trotzdem liefert (14.18) eine brauchbare Abschätzung für F, vorausgesetzt, die Querkapazitäten in Fig. 34 werden so gewählt, daß der kapazitive Belastungsstrom des Transformators in beiden Fällen den gleichen Wert annimmt. H. MEHLHORN hat den Effektivwert dieses Stromes zu 46 mA gemessen. Bestimmen wir daraus die Eingangsimpedanz $z_0 = \dfrac{8{,}0 \cdot 10^4}{\sqrt{2}\,46 \cdot 10^{-3}} = 1{,}23 \cdot 10^6\ \Omega$, so ergibt (14.19) mit einer mittleren Seriekapazität von $\overline{C_s} = 2170$ pF eine wirksame Schaltkapazität von $C_q = 44{,}7$ pF. (14.18) liefert dann den annähernd richtigen Wert von $F = 0{,}81$.

Kennlinie bei konstantem Lastwiderstand. In Fig. 40 ist der von H. MEHLHORN gemessene Laststrom I_g in Abhängigkeit von der Transformatorspannung U_0 aufgetragen, wobei der Hochspannungsgenerator mit einem konstanten Widerstand R_L belastet wurde. Es sei noch bemerkt, daß der Sprühstrom der Anlage vernachlässigt werden darf, beträgt er doch bei 400 kV erst 0,1 mA. Fig. 40

---

[1] Bei der Berechnung der Tabelle sind wir nicht von den Ausdrücken (15.1) bis (15.3), sondern von den allgemeineren auf S. 32 ausgegangen, welche die Berücksichtigung der gemessenen Werte der Kondensatoren gestatten. ΔU_R s. Gl. (13.11); $\Delta U_C = \dfrac{\delta U_{en}}{2}$ bzw. $\dfrac{\delta U_{en}^*}{2}$ s. Gl. (13.9); $\Delta U_n = \sqrt[3]{\Delta U_R^3 + \Delta U_C^3}$.

zeigt einen weitgehend linearen Zusammenhang zwischen I_g und U_0. Infolgedessen muß auch ein linearer Zusammenhang zwischen der Gleichspannung des Generators und U_0 bestehen. Dieses Ergebnis stimmt durchaus mit unserer Theorie überein. Für einen konstanten Lastwiderstand R_L wird

$$I_g = \frac{U_0}{R_L} = Q \cdot f.$$

Setzen wir diesen Ausdruck in Formel (15.4) zur Berechnung des Spannungsabfalles ein, so folgt, daß $\Delta U/U_0$ nur eine Funktion von U_g/U_0 ist

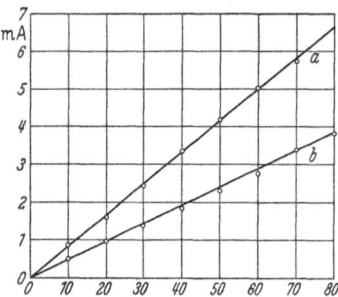

Fig. 40. Gleichstrom I_g in Abhängigkeit von der Transformatorspannung U_0 bei einer Belastung des Generators Fig. 39 mit $R_L \approx 46$ MΩ (a) und $R_L \approx 91$ MΩ (b).

$$\frac{\Delta U}{U_0} = f\left(\frac{U_g}{U_0}\right). \qquad (15.6)$$

Bei konstantem Lastwiderstand R_L ist infolgedessen U_g proportional der Transformatorspannung U_0. Der Faktor F, mit dem diese Spannung zu multiplizieren ist, um die wirkliche Ausgangsspannung U_g' zu erhalten, ändert an diesem linearen Zusammenhang nichts. Allerdings muß zusätzlich verlangt werden, daß die wirksamen Widerstände R_{en} unabhängig vom Laststrom I_g sind. Nach (13.5) gilt

$$R_{en} = \frac{N R_t + \overline{R}}{n}.$$

Da normalerweise der Widerstand $\overline{R}$ des Ventils in Leitrichtung klein ist gegenüber $N R_t$ ($R_t =$ Widerstand des Transformators), ist die oben erwähnte Forderung meistens erfüllt (vgl. z.B. Zahlenwerte in Fig. 39).

Es sei noch bemerkt, daß der lineare Zusammenhang nach (15.6) Modellversuche gestattet und erlaubt, die geplante Hochspannungsanlage an einer Schaltung mit stark reduzierter Spannung auszumessen.

c) Varianten des Kaskadengenerators.

Zur Verbesserung der Eigenschaften des Kaskadengenerators werden in der Literatur eine Reihe von Maßnahmen diskutiert. Sie bezwecken einerseits die Verminderung der Welligkeit und andererseits die Herabsetzung des Spannungsabfalles. Insbesondere ist in der bisher behandelten Anordnung die Rippelspannung zu erwähnen, welche als Folge der unvermeidlichen Streukapazitäten auftritt. Es ist zwar möglich, die Grundwelle dieser Rippelspannung sowohl im Leerlauf als auch bei Belastung zu kompensieren (vgl. Fig. 32). Oft ist es jedoch störend, daß die notwendige Kompensationsspannung mit der Belastung variiert.

Man hat ferner versucht die Rippelspannung dadurch zu verringern, daß an den Hochspannungsausgang eine besondere Filterkondensatorsäule über einen hochohmigen Widerstand angeschlossen wird [T 52]. Diese Maßnahme führt aber bei Generatoren für höchste Spannungen, die mit Vorteil in einen Drucktank eingebaut werden, zu großen konstruktiven Schwierigkeiten.

Ein weiterer Nachteil des üblichen Kaskadengenerators besteht darin, daß — mindestens bei kleinen Kondensatoren — der Spannungsabfall annähernd mit der dritten Potenz der Stufenzahl steigt.

Im folgenden sollen die Verbesserungen diskutiert werden, die sich durch passende Änderungen im Aufbau der Kaskadenschaltung erzielen lassen.

16. Symmetrische Schaltungen. Die Varianten a und b in Fig. 41 sind weitgehend äquivalent. In beiden Schaltungen muß der Eingangstransformator die doppelte Sekundärspannung erzeugen, aber nur für die Hälfte dieser Spannung isoliert sein. Bei der Schaltung nach Fig. 41a sind doppelt so viele Ventile notwendig wie in Fig. 41b. Beide Varianten a und b benötigen eine größere Zahl von Kondensatoren als die einfache Kaskade in Fig. 24.

Die Rippelspannung infolge der Streukapazitäten wird aus Symmetriegründen kompensiert, und diejenige infolge der äußeren Belastung wesentlich reduziert. In der bisher behandelten Anordnung wirkt die

Fig. 41 a u. b. Symmetrische Anordnung des Kaskadengenerators. G Glättungssäule, S Schubsäulen. a Schaltung nach W. HEILPERN [H 55]. b Schaltung nach A. H. B. WALKER und L. H. PETER [W 48], [W 49].

Glättungssäule zugleich als Schubsäule. Sie muß, ebenso wie die Schubsäule, den Ladungstransport nach oben besorgen und zugleich Ladung nach außen abgeben. In Fig. 41 dagegen übernehmen die beiden Schubsäulen den Ladungstransport allein, während die Glättungssäule nur Ladung nach außen liefert. Die Funktionen des Ladungsnachschubes und der Glättung sind in den beiden symmetrischen Schaltungen scharf getrennt. Die Rippelspannung läßt sich deshalb besonders einfach berechnen, denn die Glättungssäule wirkt für die Gesamtspannung wie ein einziger Kondensator. Besitzen alle Kondensatoren der Glättungssäule den gleichen Wert C, so beträgt ihre gesamte wirksame Kapazität C/N, womit sich die Rippelspannung zu

$$\delta U = Q \left[\sum_{n=1}^{N} \left(\frac{1}{C_n} - \varepsilon_2 \right) \right] = \frac{Q}{\beta} \frac{N}{C} (1 - \varepsilon_2), \qquad Q = \frac{I_g}{\beta f}, \left. \right\} \quad (16.1)$$

(Fig. 41 a: $\beta = 2$,   Fig. 41 b: $\beta = 1$)

ergibt. $\varepsilon_2 T = \varepsilon_2 / f$ bedeutet wie früher die Aufladezeit der Glättungssäule.

Bei Vernachlässigung der Ohmschen Widerstände des Transformators und der Ventile erhält man den Spannungsabfall infolge der Umladung der Kondensatoren für die Schaltung von Fig. 41a zu:

$$\varDelta U = Q \sum_{n=1}^{N} \frac{n^2}{C_n^*} = \frac{Q}{C^*}\left[\frac{N^3}{3} + \frac{N^2}{2} + \frac{N}{6}\right]. \tag{16.2}$$

Der rechte Teil der Gl. (16.2) gilt unter der Voraussetzung gleicher Werte C^* der Kapazitäten in den beiden Schubsäulen. Die Ausgangsspannung der Kaskade variiert zwischen

$$\left.\begin{aligned} U_{\max} &= 2N\,U_0 - \varDelta U, \\ U_{\min} &= 2N\,U_0 - \varDelta U - \delta U, \end{aligned}\right\} \tag{16.3}$$

und die abgegebene Gleichspannung hat den Wert

$$U_g = 2N\,U_0 - \varDelta U - \frac{\delta U}{2}. \tag{16.4}$$

Dieser Ausdruck ist mit (12.3) bis (12.5) der gewöhnlichen Kaskade zu vergleichen und wir stellen fest, daß der Spannungsabfall bei etwa gleichem Volumen der Kondensatoren 2- bis 4mal kleinere Werte annimmt.

Falls man in unserer symmetrischen Schaltung Fig. 41a die eine Hälfte der Gleichrichter entfernt und so zur reduzierten Schaltung Fig. 41b übergeht, nimmt der Spannungsabfall nur unwesentlich zu, und zwar wird die Zunahme mit wachsender Stufenzahl kleiner. Diese Bemerkung gilt sowohl für den Anteil des Spannungsabfalles infolge der endlichen Werte der Kondensatoren, als auch für den Anteil, der durch die Widerstände des Transformators und der Ventile bedingt ist. Es ist ferner leicht einzusehen, daß die Ventile in Fig. 41b, verglichen mit der vollständigen Schaltung Fig. 41a, durch doppelt so große Ströme beansprucht werden.

Unter die symmetrischen Schaltungen ist ferner die bereits früher erwähnte Variante Fig. 36 zu zählen [E 53], in welcher in eleganter Art und Weise eine weitgehende Kompensation des nachteiligen Einflusses der Streukapazitäten erreicht wird. In dieser Anordnung übernehmen die beiden Kondensatorsäulen gleichzeitig die Funktionen der Glättung und des Ladungsnachschubs, so daß eine Unterteilung in Schub- und Glättungssäule nicht mehr sinnvoll ist[1].

17. Allgemeine Behandlung der belasteten Kaskade nach Fig. 41a. In Abschnitt 13 wurde gezeigt, wie sich die Theorie des Kaskadengenerators in guter Näherung auf diejenige des Einweggleichrichters zurückführen läßt. Die gleichen Überlegungen lassen sich auch auf die symmetrischen Schaltungen in Fig. 41 anwenden. Nehmen wir an, daß alle Ventilströme denselben zeitlichen Verlauf aufweisen und daß die einander entsprechenden Ventile gleichzeitig schalten, so dürfen die Verbindungen zwischen den Punkten N bis 2 und der Glättungssäule aufgetrennt werden (Fig. 41a) und zwar ohne die Wirkungsweise der Anordnung irgendwie zu beeinflussen.

Zur Berechnung des Spannungsabfalles der Anordnung Fig. 41a betrachten wir zunächst die unterste Stufe der Schubsäule. Es gilt:

$$U_{gN} = \nu_N U_0. \tag{17.1}$$

Der äquivalente Einweggleichrichter besitzt die Ersatzgrößen:

$$\left.\begin{aligned} C_{eN}^* &= C_N^*, \qquad R_{en} = \frac{N R_t + \overline{R}}{N}, \\ Q_{en} &= N Q = N \frac{I_g}{2f}. \end{aligned}\right\} \tag{17.2}$$

---

[1] Zur Kompensation der Streukapazität vgl. die Diskussion auf S. 38.

R_t bedeutet den Widerstand einer Hälfte der Sekundärwicklung des Transformators. Für alle übrigen Stufen $n \leq N - 1$ der Schubsäule gilt

$$U_{gn} = v_n 2 U_0, \tag{17.3}$$

mit den Ersatzgrößen

$$\left. \begin{aligned} \frac{1}{C_{en}^*} &= \frac{2}{n} \left[\left(\sum_{\lambda=0}^{N-n} (N - \lambda)\, \frac{1}{C_{N-\lambda}^*} \right) - \frac{n}{2 C_n^*} \right], \\ R_{en} &= \frac{2}{n} (N R_t + \overline{R}), \qquad Q_{en} = n Q = n \frac{I_g}{2f}. \end{aligned} \right\} \tag{17.4}$$

Zum Schluß ist noch der Spannungsabfall, der beim Aufladen der Glättungssäule auftritt, zu berücksichtigen. Unter C_t wollen wir die Gesamtkapazität dieser Kondensatorsäule verstehen

$$\frac{1}{C_t} = \sum_{n=1}^{N} \frac{1}{C_n}, \tag{17.5}$$

und wir erhalten für den obersten äquivalenten Gleichrichter $(n = 1)$

$$\left. \begin{aligned} \frac{1}{C_{e1}^*} &= \sum_{\lambda=0}^{N-1} \frac{N - \lambda}{C_{N-\lambda}^*} + \frac{1}{C_t}, \\ R_{e1} &= N R_t + \overline{R}, \qquad Q_{e1} = Q. \end{aligned} \right\} \tag{17.6}$$

Der gesamte Spannungsabfall des Kaskadengenerators läßt sich nun als Summe der Spannungsabfälle aller soeben erwähnten äquivalenten Einweggleichrichter darstellen. Es sei noch bemerkt, daß diese Summe unter Vernachlässigung der Widerstände des Transformators und der Ventile in unsere frühere Beziehung (16.4) übergeht. Beim Vergleich ist zu beachten, daß (17.6) die Vereinfachung $\varepsilon_2 = 0$ zugrunde liegt.

Der Einfluß der Streukapazitäten läßt sich in ähnlicher Weise wie früher berücksichtigen. Falls die beiden untersten Kondensatoren C_N^* der beiden Schubsäulen den doppelten Wert der übrigen C^* besitzen, bestimmt sich das Verhältnis F der wirklichen zur idealen Leerlaufspannung ähnlich wie früher zu

$$F = \frac{b}{M} \operatorname{Tan} \left(\frac{M}{b} \right).$$

Es bedeuten:

$b^2 = C_s/C_q$

$C_s =$ Kapazität der Kondensatoren der Schubsäulen, unterster Kondensator $= 2 C_s$,

$C_q/2 =$ Streukapazität parallel zu den Ventilen (vgl. Fig. 41a),

$M =$ Anzahl der Stufen, d.h. $\frac{1}{4}$ der Gesamtzahl der Ventile in Fig. 41a.

Zum Schluß sei auf eine weitere Variante (R. G. MITCHELL [Mi 45]) hingewiesen, bei welcher der Transformator in der Mitte der Kaskade angeordnet ist (Fig. 42) und die sich empfiehlt, falls der Mittelpunkt der Hochspannungsanlage geerdet werden darf. Die Rippelspannung zwischen den Punkten A und B in Fig. 42 ist stark reduziert, denn beim Aufladen einer Hälfte der Glättungssäule wird gleichzeitig die andere Hälfte in die Schubsäule entladen. Infolgedessen verursacht nur der äußere Belastungsstrom I_g eine Rippelspannung, während sich

die übrigen Beiträge gegenseitig kompensieren. Zur Berechnung des Spannungsabfalles kann jede Hälfte für sich betrachtet werden. Daraus geht unmittelbar
hervor, daß der Spannungsabfall, verglichen mit einer gewöhnlichen Kaskade
der vollen Stufenzahl, wesentlich niedriger ist.

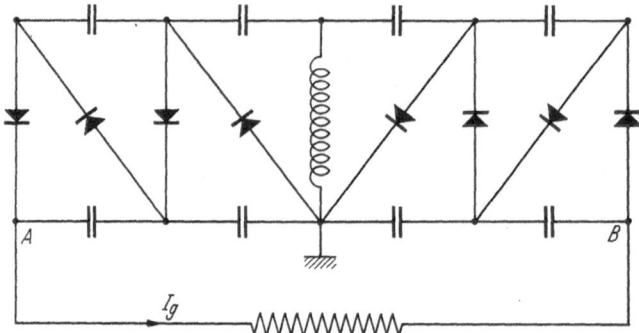

Fig. 42. Kaskadengenerator mit in der Mitte angeordnetem Transformator.

18. Greinacher-Schaltungen in Kaskadengeneratoren. Eine Hochspannungsanlage für etwa 1 MV, die aus 4 in Serie geschalteten Spannungsverdopplern
(Greinacher-Schaltungen) besteht, wur-
de 1940 von T. Bjerge u. a. [B 40] beschrieben.

In Fig. 43 ist das Schema dieser Anordnung dargestellt. Die Hochspannungstransformatoren $A_1 - A_4$ von je
125 kV Spitzenspannung werden über
die Isoliertransformatoren $B_1 - B_4$ gespeist. Es ist von Interesse zu bemerken,

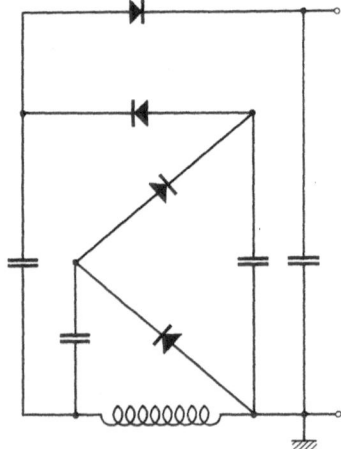

Fig. 43.   Fig. 44.

Fig. 43. 1 MV Generator bestehend aus vier Greinacher-Schaltungen (I bis IV). A_n Hochspannungstransformatoren.
B_n Isoliertransformatoren [B 40]. (Aus J. D. Craggs and J. M. Meek [6], S. 34.)
Fig. 44. Kaskadengenerator nach M. Schenkel [S 19].

daß diese Anlage durch Entfernen der Ventile und durch entsprechende Ände-
rung der Verbindungen zur Erzeugung hoher Wechselspannungen benützt werden

kann. Ferner läßt sie sich zu einem Stoßgenerator für 2 MV Impulse (Marx-Schaltung) umbauen.

Die Greinacher-Schaltung kann auch als erste Stufe einer gewöhnlichen Kaskade verwendet werden. Wie H. MEHLHORN [Me 43] gezeigt hat, erlaubt dies eine — allerdings nicht sehr große — Reduktion des Spannungsabfalles.

Abschließend zeigt Fig. 44 die Anordnung von M. SCHENKEL, die wohl die älteste Kaskadenschaltung darstellt. Sie besitzt, wie bereits erwähnt, den Nachteil, daß die Kondensatoren für die Gesamtspannung der betreffenden Stufe dimensioniert werden müssen.

III. Erzeugung der Hochspannung mit Hilfe eines Röhrenoszillators.

19. Einleitung. Wie schon O. H. SCHADE [Sc 43] bemerkte, werden konventionelle Gleichrichter, die mit Netzfrequenz arbeiten, infolge der Abmessungen des Hoch-

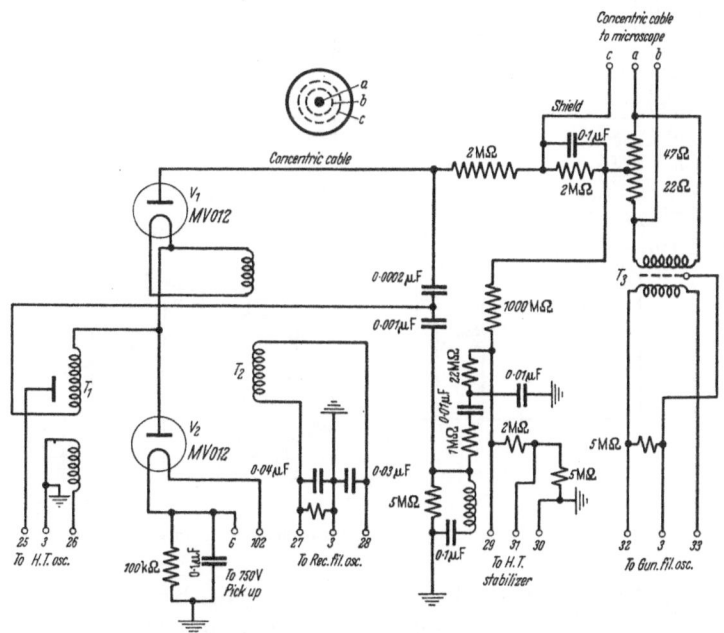

Fig. 45. Schaltschema der Hochspannungsanlage von 100 kV zur Speisung eines Elektronenmikroskopes [H 50].
(Aus J. D. CRAGGS and J. M. MEEK [6], S. 11.)

spannungstransformators und der übrigen Bauteile relativ groß und unhandlich. Für Spannungen von einigen kV bis zu einigen 100 kV empfiehlt sich deshalb in manchen Fällen die Verwendung eines Röhrenoszillators, dessen Frequenz immer so gewählt werden kann, daß der Transformator und die Kondensatoren verhältnismäßig kleine Abmessungen erhalten. Solche Anlagen sind heute zur Erzeugung der Beschleunigungsspannung von Kathodenstrahlröhren und Elektronenmikroskopen allgemein gebräuchlich. Ein weiterer wesentlicher Vorteil dieser Hochspannungsgleichrichter besteht darin, daß die Ausgangsspannung mit einfachen Mitteln stabilisiert werden kann. So haben z. B. M. E. HAINE u. a. [H 50] eine Anordnung beschrieben, die wahlweise 25, 50, 75 kV (bei je 1 mA Belastung) sowie 100 kV (bei 0,4 mA Belastung) liefert und deren Spannung elektronisch konstant gehalten wird. Die Schaltungen des Hochspannungsgenerators, des Oszillators und der Stabilisierung sind in den Fig. 45 und 46 dargestellt. Drei

Netzanschlußgeräte speisen drei Röhrenoszillatoren von 30, 50 und 50 kHz, welche zur Erzeugung der Hochspannung, zur Heizung der Ventile und zur Heizung der Kathode des Elektronenmikroskopes benötigt werden. Der Hochspannungsgleichrichter in Greinacher-Schaltung ist in einen Öltank eingebaut. Die Stabilisierung erfolgt durch negative Rückkopplung. Die zu stabilisierende Spannung wird an einem hochohmigen Spannungsteiler abgegriffen, verstärkt und dem Schirmgitter der Oszillatorröhre zugeführt. Als Vergleichsspannung für diese Regulierung dienen Trockenbatterien von 180 Volt.

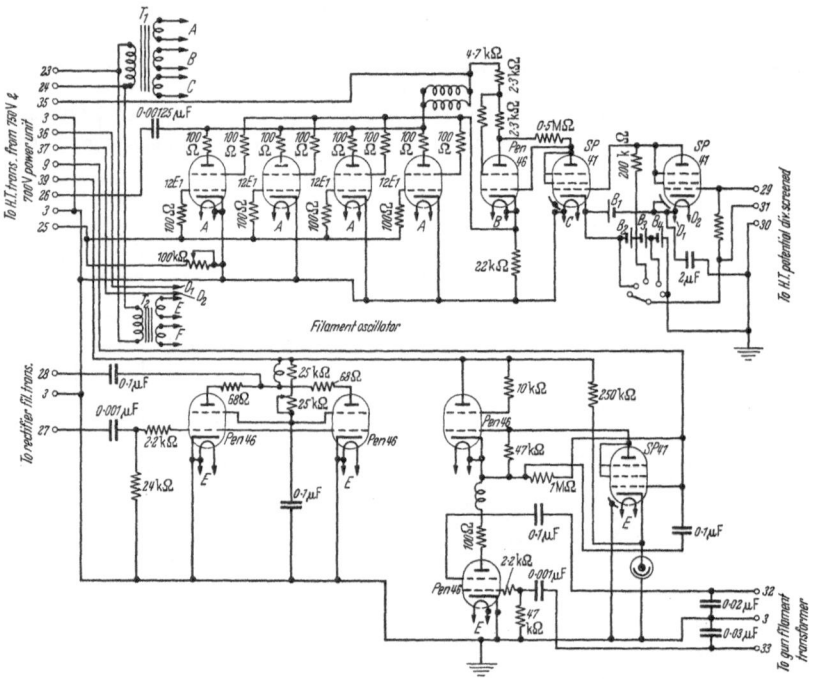

Fig. 46. Schaltschema der Oszillatoren des Hochspannungsgenerators in Fig. 45 [H 50].

Die gleichen Argumente, die bei den soeben beschriebenen Gleichrichtern für höhere Frequenz der Speisespannung sprechen, treffen auch auf Kaskadengeneratoren zu. Kaskadengleichrichter mit Röhrenoszillatoren werden in jüngster Zeit selbst für Spannungen von 500 kV gebaut. Eine solche Anlage ist in Fig. 47 schematisch dargestellt. Die Verwendung eines Drucktankes und eine Oszillatorfrequenz von etwa 30 kHz ermöglichen es, eine sehr kompakte Anlage zu bauen, die durch ihren geringen Raumbedarf auffällt [L 56].

Die theoretischen Grundlagen zur Berechnung der Gleichrichterschaltungen sind im Kap. II bereits ausführlich dargelegt worden. In den meisten Fällen wird die hohe Frequenz der Wechselspannung zur Folge haben, daß der Einfluß der endlichen Werte der Kondensatoren auf den Spannungsabfall vernachlässigbar ist und lediglich die Widerstände der Ventile und vor allem der Einfluß der Schaltkapazitäten berücksichtigt werden müssen. Hingegen tritt ein neues Problem auf, nämlich die zweckmäßige Dimensionierung des Transformators, der normalerweise in der Nähe der sekundärseitigen Resonanzfrequenz betrieben wird. Beim Einbau der Anlage in einen Drucktank kann es vorteilhaft sein, eine Linkverbindung zu benützen. In den folgenden Abschnitten wird deshalb der

Transformator mit und ohne Linkverbindung diskutiert. Die vorliegende Darstellung stützt sich auf Arbeiten, die an der Universität von Montréal in Canada durchgeführt wurden [*L 56*].

20. Der Eingangstransformator ohne Linkverbindung. Der Rechnung liegt das Schaltbild Fig. 48 zugrunde. Primärseitig sei der Transformator von einem Endverstärker im *C*-Betrieb gespeist.

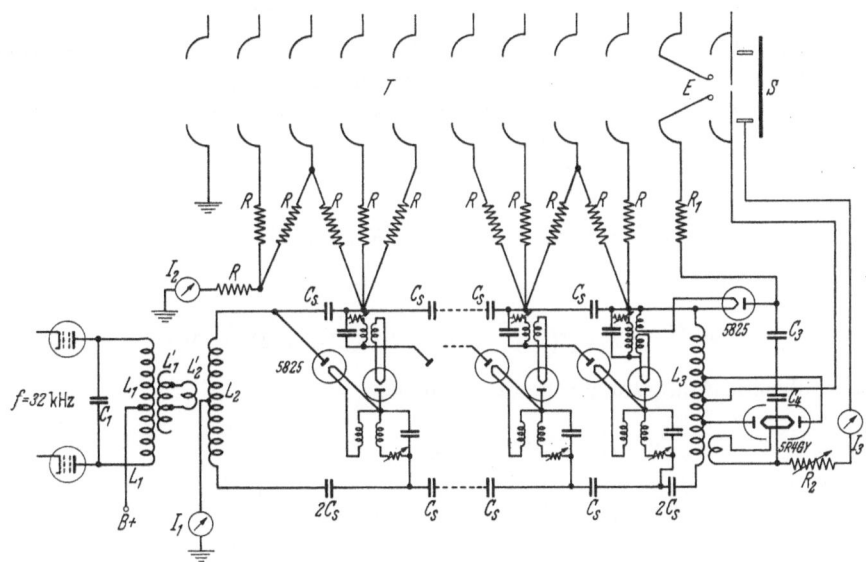

Fig. 47. Schema des 500 kV Generators mit Beschleunigungsrohr und Ionenquelle nach P. LORRAIN und Mitarbeitern [*L 56*]. Der Anlage liegt Fig. 36 zugrunde. Zur Heizung der Ventile werden abgestimmte Autotransformatoren verwendet, in denen der Wechselstrom durch die Streukapazitäten der Ventile ausgenützt wird. Die Spule L_3 ist nach Gl. (14.20) dimensioniert. Abgesehen von der Verkleinerung des Spannungsabfalles gestattet sie, die notwendige Hochfrequenzleistung abzunehmen, welche zur Erzeugung der Hilfsspannungen (Ionenquelle) erforderlich ist.

R_L bedeutet den äußeren Belastungswiderstand des Hochspannungsgleichrichters und R'_L den auf den Eingang der Gleichrichterkaskade bezogenen Belastungswiderstand. Die Transformation von R_L auf R'_L hat so zu erfolgen, daß beide dieselbe Leistung verbrauchen. Näherungsweise gilt

$$R'_L \approx R_L/8N^2. \qquad (20.1)$$

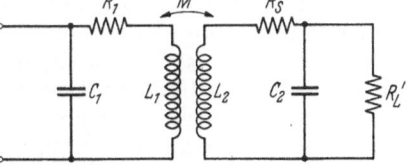

Fig. 48. Prinzipschema des Transformators.

Dabei wurde der Spannungsabfall des Kaskadengenerators vernachlässigt und ferner nicht berücksichtigt, daß der Belastungsstrom des Transformators aus kurzzeitigen Stromstößen besteht und schließlich weitere Verluste in den Ventilen auftreten. Die erste Vernachlässigung ergibt einen etwas zu kleinen und die zweite Vereinfachung einen etwas zu großen Wert für R'_L, so daß sich beide Einflüsse zum Teil kompensieren. Für die nachfolgenden Untersuchungen ist es zweckmäßig, die Parallelschaltung von R'_L und C_2 durch eine äquivalente Serieschaltung von C_2 und einem Widerstand vom Werte $(\omega^2 C_2^2 R'_L)^{-1}$ zu ersetzen. Diese Transformation gilt unter der zulässigen Annahme

$$(\omega C_2 R'_L)^2 \gg 1. \qquad (20.2)$$

Damit bestimmen sich die Ströme I_1 und I_2, welche in die Induktivitäten L_1 und L_2 fließen, zu:

$$I_1 = \frac{U_1}{(R_1 + j\omega L_1) + \dfrac{\omega^2 M^2}{R_2 + jX}}, \tag{20.3}$$

$$I_2 = \frac{j\omega M U_1}{(R_1 + j\omega L_1)(R_2 + jX) + \omega^2 M^2}. \tag{20.4}$$

Es bedeuten R_1 den Seriewiderstand der Spule L_1, R_2 die Summe aus dem Widerstand R_s der Spule L_2 und dem äquivalenten Lastwiderstand $(\omega^2 C_2^2 R_L')^{-1}$. Die Reaktanz X sei eine Abkürzung für den Term $(\omega L_2 - 1/\omega C_2)$. Unter Berücksichtigung der Annahme (20.2) bestimmt sich die Spannungsübersetzung g des belasteten Transformators zu

$$g = \frac{U_2}{U_1} \approx \frac{I_2}{U_1 j\omega C_2} = \frac{M}{C_2[(R_2 + jX)(R_1 + j\omega L_1) + \omega^2 M^2]}. \tag{20.5}$$

Der Betrag von g wird

$$|g| = \frac{M}{R_1 R_2 C_2[(1 - Q_1 Q_2 \Omega + K^2/K_c^2)^2 + (Q_2 \Omega + Q_1)^2]^{\frac{1}{2}}}, \tag{20.6}$$

wobei folgende Abkürzungen benützt wurden:

$$\left.\begin{array}{cc}
Q_1 = \dfrac{\omega L_1}{R_1}, & \omega_2^2 = \dfrac{1}{L_2 C_2}, \\[2mm]
Q_2 = \dfrac{\omega L_2}{R_2}, & \Omega = 1 - \dfrac{\omega_2^2}{\omega^2}, \\[2mm]
K^2 = \dfrac{M^2}{L_1 L_2}, & K_c^2 = \dfrac{1}{Q_1 Q_2}.
\end{array}\right\} \tag{20.7}$$

K stellt den Kopplungskoeffizienten und K_c den speziellen Wert der kritischen Kopplung der beiden Spulen dar. Q_1 und Q_2 sind die Spulengüten, wobei in Q_2 die an die Kaskade abgegebene Leistung inbegriffen ist. Die Spannungsübersetzung g des belasteten Transformators erreicht bei $\Omega \approx 0$, also in der Nähe der Resonanzfrequenz der Sekundärseite, ein Maximum. Die Gl. (20.5) zeigt ferner, daß eine Vergrößerung von R_2 (d.h. kleineres R_L) die Sekundärspannung erniedrigt. Wie zu erwarten ist, sinkt also mit wachsender Belastung des Kaskadengenerators die Ausgangsspannung des Transformators. Oft ist es wünschenswert, daß diese Spannung auf Lastschwankungen möglichst unempfindlich ist. Dies läßt sich mit einer Betriebsfrequenz erreichen, die nach oben oder unten genügend von der Resonanzfrequenz abweicht. Selbstverständlich ist mit dieser Maßnahme eine Reduktion der Sekundärspannung verbunden.

Um die Abhängigkeit der transformierten Spannung vom Lastwiderstand (der in R_2 enthalten ist) besser zu überblicken, ist es zweckmäßig, Gl. (20.6) umzuformen und die Größe Q_2 zu eliminieren:

$$\frac{M^2}{|g|^2 C_2^2 R_1^2} = (R_2^2 + X^2)(1 + Q_1^2) + Q_1 \omega L_2 K^2 [2R_2 + \omega L_2 Q_1 (K^2 - 2\Omega)]. \tag{20.8}$$

Um den Einfluß der Belastung klein zu halten, muß zunächst

$$X^2 \gg R_2^2 \tag{20.9}$$

sein. Es bleibt nun zu zeigen, daß mit der Bedingung (20.9) auch der zweite Summand der rechten Seite von (20.8) unabhängig von R_2 wird. Der fragliche

Term lautet $2Q_1\omega L_2K^2R_2$ für den wir $2Q_1Q_2R_2^2K^2$ setzen können. Dieser Ausdruck ist sicher nicht größer als der bereits vernachlässigbare Anteil $R_2^2Q_1^2$ des ersten Summanden. Beachtet man nämlich, daß $Q_2<Q_1$ gilt und K in der Größenordnung von 0,5 liegt, so folgt sofort:

$$Q_1^2R_2^2 > Q_1Q_2R_2^2\,2K^2. \tag{20.10}$$

Die Bedingung (20.9) ist somit ausreichend, um (20.8) als unabhängig von R_2 betrachten zu können.

Unter dem Wirkungsgrad η des Transformators wird, wie üblich, das Verhältnis von abgegebener zu primärseitig aufgenommener Leistung verstanden. Zur Berechnung von η ist es zweckmäßig, zunächst die Energieübertragung von der Primärspule auf die Sekundärspule und anschließend diejenige von der Sekundärspule auf den Lastwiderstand R_L' zu betrachten. Den ersten Faktor erhält man

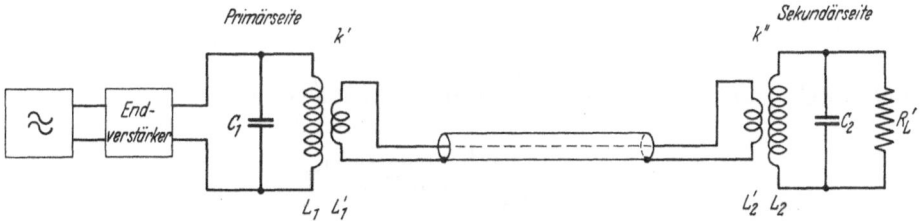

Fig. 49. Eingangstransformator mit Linkverbindung. R_L' berücksichtigt die Belastung des Hochspannungstransformators durch den Kaskadengenerator.

durch Berücksichtigung der Widerstände R_1 und R_r, während der zweite Faktor durch $\dfrac{R_2-R_s}{R_2}$ gegeben ist. Der Wirkungsgrad η bestimmt sich dann zu

$$\eta = \frac{R_r}{R_r+R_1}\;\frac{1}{1+\omega^2C_2^2R_L'R_s}.$$

R_1+R_r bedeutet den Realteil der von der Primärseite gesehenen Impedanz des belasteten Transformators [vgl. (21.12)]. Eliminiert man R_r, so wird

$$\eta = \frac{K^2/K_c^2}{(K^2/K_c^2 + 1 + Q_2^2\Omega^2)(1+\omega^2C_2^2R_L'R_s)}. \tag{20.11}$$

21. Der Eingangstransformator mit Linkverbindung. Bei der Verwendung eines Drucktankes empfiehlt es sich, den Ausgang des Endverstärkers und den Hochspannungstransformator über eine sog. Linkverbindung zu koppeln, wie dies in Fig. 49 schematisch dargestellt ist.

Derartige Linkverbindungen bieten in unserem Problem zwei wichtige Vorteile. Einerseits läßt sich der eigentliche Kaskadengenerator vom Wechselstromgenerator ohne hochspannungsführende Kabel und Verbindungen trennen. Andererseits erlauben Linkverbindungen eine konstruktiv einfache Lösung des Problems, die Speiseleitungen durch die Wand des Drucktankes hindurchzuführen. Auf der Übertragungsleitung ist die Spannung stark heruntertransformiert, so daß die Kapazität der kurzen Leitung in der folgenden Rechnung vernachlässigt werden darf. Eine niedrige Spannung bedingt aber einen großen Strom, der so beträchtliche Werte annehmen kann, daß spezielle Kabel verwendet werden müssen.

Wir wählen wie üblich $L_1'=L_2'=L'$ und $R_1'=R_2'=R'$, wobei R_1' und R_2' die Seriewiderstände der Spulen L_1' und L_2' darstellen. Der Strom auf der Primärseite berechnet sich zu:

$$I_{1L} = \frac{U_1}{z_1 + \dfrac{\omega^2 M'^2}{2z' + \dfrac{\omega^2 M''^2}{z_2}}}. \tag{21.1}$$

Es bedeuten:

$$\left. \begin{array}{ll} z_1 = R_1 + j\omega L_1, & z_2 = R_2 + jX, \\ z' = R' + j\omega L', & \end{array} \right\} \tag{21.2}$$

ferner sind M' bzw. M'' die gegenseitigen Induktivitäten zwischen L' und L_1 bzw. L' und L_2.

Der Strom I' in der Übertragungsleitung wird

$$I' = \frac{j\omega M' I_1}{2z' + \dfrac{\omega^2 M''^2}{z_2}}, \tag{21.3}$$

und mit $Q' = \omega L'/R'$ der Strom I_{2L} auf der Sekundärseite

$$I_{2L} = \frac{-\omega K' K'' \sqrt{L_1 L_2}\, U_1}{z_1 z_2 \left(\dfrac{2}{Q'} + 2j + \dfrac{\omega K''^2 L_2}{z_2} + \dfrac{\omega K'^2 L_1}{z_1} \right)}. \tag{21.4}$$

Mit den Abkürzungen

$$Z_2^2 = R_2^2 + X^2, \qquad P = \frac{\omega K''^2 L_2}{Z_2^2} \tag{21.5}$$

erhält man für den Betrag $|I_{2L}|$ den Ausdruck

$$|I_{2L}|^2 = \frac{K'^2 K''^2 L_2 U_1^2}{L_1 Z_2^2 \left[\left(\dfrac{2}{Q'} + PR_2 + \dfrac{K'^2}{Q_1} \right)^2 + (2 - PX - K'^2)^2 \right]}. \tag{21.6}$$

In Gl. (21.6) ist insofern eine Näherung enthalten, als $Q_1^2 \gg 1$ vorausgesetzt wurde, eine Annahme, die wohl immer zulässig ist.

Die Spannungsübersetzung g_L wird

$$g_L = \frac{I_{2L}}{U_1 j\omega C_2} \tag{21.7}$$

und zeigt in der Nähe der Resonanzfrequenz $\omega_2 = \dfrac{1}{L_2 C_2}$ der Sekundärseite ein Maximum.

Auch hier können wir verlangen, daß die Sekundärspannung durch die variable Belastung des Kaskadengenerators nicht stark beeinflußt werde, was wiederum eine von der Resonanzfrequenz abweichende Betriebsfrequenz bedingt. Zur Diskussion dieser Frage betrachten wir die folgende Gleichung:

$$\frac{K'^2 K''^2 L_2}{|g_L|^2 \omega^2 C_2^2 L_1} = Z_2^2 \left[\left(\frac{2}{Q'} + PR_2 + \frac{K'^2}{Q_1} \right)^2 + (2 - PX - K'^2)^2 \right]. \tag{21.8}$$

Unabhängigkeit von der Belastung, also von R_2, wird wie früher mit

$$X^2 \gg R_2^2 \tag{21.9}$$

erreicht. Die Größe P in Gl. (21.5) bestimmt sich dann näherungsweise zu

$$P \approx \frac{K''^2}{\omega L_2 \Omega^2}, \tag{21.10}$$

und der Klammerausdruck auf der rechten Seite von (21.8) läßt sich in der vereinfachten Form

$$\left(\frac{2}{Q'} + \frac{K'^2}{Q_1} + \frac{K''^2 R_2}{\omega L_2 \Omega^2}\right)^2 + \left(2 - \frac{K''^2}{\Omega} - K'^2\right)^2$$

darstellen. Solange $Q_2 \gtrless 5$ gilt, ist dieser Ausdruck von R_2 praktisch unabhängig. Um dies einzusehen, vernachlässigen wir die beiden ersten Terme der ersten Klammer und beachten, daß Ω, K' und K'' von der Größe $\sim 0{,}5$ sind. In dem in Tabelle 6 dargestellten Beispiel variiert Q_2 zwischen 5 (für 1 mA Laststrom) und 200 (im Leerlauf der Kaskade), womit die soeben erwähnte Bedingung erfüllt ist.

Der Wirkungsgrad η_L des Transformators mit Linkverbindung bestimmt sich zu

$$\eta_L = \frac{1}{\left(1 + \dfrac{R_1}{R_r}\right)} \cdot \frac{1}{\left(1 + \dfrac{\omega^3 L_2 C_2^2 R_L'}{Q_2'}\right)}, \tag{21.11}$$

wobei Q_2' die Spulengüte der Sekundärseite im Leerlauf des Kaskadengenerators bedeutet ($R_L \to \infty$) und R_r den auf die Primärseite transformierten Gesamtwiderstand darstellt:

$$z_r = R_r + j X_r = \frac{\omega L_1 K'^2 \left(\dfrac{2}{Q'} + P R_2\right) - j \omega L_1 K'^2 (2 - P X)}{\left(\dfrac{2}{Q'} + P R_2\right)^2 + (2 - P X)^2}. \tag{21.12}$$

Der erste Faktor der rechten Seite von (21.11) liefert den auf die Sekundärseite übertragenen Bruchteil der Eingangsenergie, und der zweite Faktor bestimmt den Teil dieser Energie, der an den Kaskadengenerator abgegeben wird. Der Ausdruck $C_2^2 R_L'/Q_2'$ ist weitgehend frequenzunabhängig. Für einen gegebenen Wert von L_2 ändert sich deshalb der zweite Faktor in Gl. (21.11) stark mit ω. Um einen guten Wirkungsgrad zu erreichen, darf also ω nicht zu hoch sein. Der erste Faktor in (21.11) ist eine Funktion des Widerstandes R_r, also des Realteils der Gl. (21.12). Unter Verwendung von (21.10) wird

$$R_r = \frac{\omega L_1 K'^2 \left(\dfrac{2}{Q'} + \dfrac{K''^2}{Q_2 \Omega^2}\right)}{\left(\dfrac{2}{Q'} + \dfrac{K''^2}{Q_2 \Omega^2}\right)^2 + \left(2 - \dfrac{K''^2}{\Omega}\right)^2}. \tag{21.13}$$

Die Spulengüten können als frequenzunabhängig angesehen werden. Da ω in der Nähe der Resonanzfrequenz der Sekundärseite liegt, hat Ω einen Wert von etwa 0,5. Für ein gegebenes L_1 ist somit R_r angenähert proportional der Kreisfrequenz ω.

Eine wesentliche Rolle für die Dimensionierung spielt die Erwärmung der Spulen L_1 und L_2. Der Verlust W_2 in L_2 wird durch

$$\left. \begin{aligned} W_2 &= I_{2L}^2 R_s = U_2^2 (\omega C_2)^2 R_s^2, \\ R_s &= \text{Verlustwiderstand der Spule } L_2 \end{aligned} \right\} \tag{21.14}$$

gegeben und ist bei konstanter Sekundärspannung unabhängig von der Belastung des Kaskadengenerators[1].

---

[1] Für U und I sind die Effektivwerte einzusetzen.

Die Verluste W_1 in der Primärspule L_1 bestimmen sich zu

$$W_1 = \frac{U_1^2 R_1}{(R_1 + R_r)^2 + (\omega L_1 + X_r)^2} \, . \tag{21.15}$$

In der Anlage von P. LORRAIN und Mitarbeitern gilt $(R_1 + R_r)^2 \ll (\omega L_1 + X_r)^2$. X_r selbst ist praktisch unabhängig von R_L (Lastwiderstand der Kaskade).

Mit zunehmender Last sinkt jedoch die Spannungsübersetzung g_L, so daß die Primärspannung erhöht werden muß, um U_2 konstant zu halten. Deshalb zeigt W_1 mit wachsender Last eine leichte Zunahme.

Die Dimensionierung des Transformators erfordert die Bestimmung der Betriebsfrequenz f und der Induktivitäten L_1 und L_2. Die von P. LORRAIN u.a. [L 56] beschriebene Anlage möge als Beispiel dienen. Für L_2 wurde eine in Öl getauchte Spule der Spellmann Television Co. mit $L_2 = 0{,}23$ Henry verwendet. Die Spulenkapazität von 11 pF und die wirksame Eingangskapazität des Kaskadengenerators von 65 pF ergeben ein C_2 von 76 pF und eine Resonanzfrequenz von 38,2 kHz. Die Betriebsfrequenz soll in der Nähe der Resonanzfrequenz liegen, so daß einerseits eine vernünftige Spannungsübersetzung g erreicht wird und andererseits die Bedingung (21.9) erfüllt ist.

Die Betriebsfrequenz kann zunächst höher oder tiefer als 38,2 kHz gewählt werden. Die tiefere Frequenz ist jedoch vorzuziehen, denn mit ihr erzielt man einen besseren Wirkungsgrad. Die Wahl von L_1 ist in verschiedener Hinsicht begrenzt. Kann der zu verwendende Litzendraht mit einen Strom von beispielsweise 13 A belastet werden, so muß

$$\frac{U_1}{\omega L_1} \approx I_1 < 13 \,\text{A}$$

Tabelle 6. *Zahlenwerte eines 12-stufigen Kaskadengenerators nach P. Lorrain und Mitarbeitern.* $f = 32$ kHz, Ausgangsspannung $= 500$ kV, $F = 0{,}81$ [L 56].

Abgegebener Gleichstrom		0	0,5 mA
R_r		1,08 Ω	4,98 Ω
X_r		68,4 Ω	68,0 Ω
g		4,04	3,90
η_L		0	0,592
U_2		18,2 kV	18,2 kV
U_1	(Effektivwerte)	4,50 kV	4,66 kV
I_1		7,7 A	7,98 A
I_2		0,278 A	0,278 A
W_1		180 Watt	192 Watt
W_2		19,4 Watt	19,4 Watt

sein. Der größte Wert von U_1 ist andererseits durch die Verstärkerröhren vorgeschrieben. Mit $U_1 = 6{,}7$ kV Effektivwert und $f \sim 30$ kHz ergibt sich in unserem Beispiel für L_1 ein Wert von etwa 2,7 mH. Der Anlage von P. LORRAIN und Mitarbeiter liegen folgende Zahlenwerte zugrunde:

$$L_1 = 2{,}57 \,\text{mH} \qquad K' = 0{,}585 \qquad Q_1 = 171$$
$$R_1 = 3{,}22 \,\Omega \qquad K'' = 0{,}484 \qquad Q' \sim 60$$
$$f_2 = 38{,}2 \,\text{kHz} \qquad \text{Betriebsfrequenz} \qquad f = 32{,}1 \,\text{kHz}.$$

Die Tabelle 6 vermittelt einige Daten, welche für den 12stufigen Kaskadengenerator in Fig. 47 mit einer Ausgangsspannung von 500 kV berechnet wurden.

Der Faktor F (vgl. Fig. 35) besitzt einen Wert von $F_{\text{opt}} = 0{,}81$. Über den Spannungsabfall infolge des Widerstandes der Ventilröhren liegen keine Angaben vor; er wurde offenbar bei der Berechnung der Zahlenwerte in Tabelle 6 vernachlässigt.

Es sei noch bemerkt, daß die Resonanzfrequenz der Sekundärseite und damit die Spannungsübersetzung g des Transformators mit Hilfe eines kleinen Kondensators parallel zu L_2 auf einfache Art justiert werden kann.

22. Kaskadengeneratoren für kleine Spannungen. Wie bereits erwähnt, eignen sich Kaskadengeneratoren auch zur Erzeugung relativ niedriger Spannungen von einigen kV, wie sie etwa zum Betrieb eines Proportionalzählrohres oder zur Speisung des Multipliers eines Szintillationszählers[1] benützt werden. Konventionelle, mit Netzfrequenz arbeitende Gleichrichter sind verhältnismäßig unhandlich und teuer. Man ist deshalb seit längerer Zeit zur Verwendung von Schaltungen mit Röhrenoszillatoren übergegangen[2]. Die höhere Frequenz (10 kHz bis 100 kHz) gestattet, kleinere Siebkondensatoren zu benützen und vereinfacht die Konstruktion des Hochspannungstransformators.

In den älteren Schaltungen, wie sie beispielsweise im Buche von ELMORE und SANDS[1] beschrieben sind, werden Einweggleichrichter und Greinacher-Schaltungen bevorzugt. Der Nachteil eines Einweggleichrichters soll an folgendem Beispiel erläutert werden: Angenommen, die Amplitude der Oszillatorspannung betrage 500 Volt und es soll eine Hochspannung von 5000 Volt erzeugt werden, so erfordert dies ein Übersetzungsverhältnis des Transformators von etwa 1:10. Infolge des großen Übersetzungsverhältnisses üben bei hoher Oszillatorfrequenz kleine Änderungen der Schaltkapazitäten von 1—2 pF auf der Hochspannungsseite des Transformators einen wesentlichen Einfluß aus, und die Verdrahtung sowie der Abgleich solcher Schaltungen ist aus diesem Grunde oft mit Schwierigkeiten verbunden. Um den Einfluß der Schaltkapazitäten zu verringern, empfiehlt sich die Verwendung eines Kaskadengenerators. Eine Wechselspannung von etwa 700 Volt ist durchaus genügend. Entsprechend der niedrigeren Wechselspannung werden durch diese Maßnahme auch die Wechselströme durch die Schaltkapazitäten wesentlich kleiner. Als Gleichrichterelemente lassen sich mit Vorteil kommerzielle Selengleichrichter benützen. Die Kaskadenschaltung gestattet ferner, die Potentialverteilung an der Gleichrichtersäule und an den Kondensatorsäulen in einfacher Weise zu steuern und erlaubt deshalb die Verwendung billiger und leicht erhältlicher Bauelemente.

Es ist von Interesse, darauf hinzuweisen, daß Kaskadenschaltungen vor etwa 40 Jahren entworfen wurden, mit der Absicht, Spannungen von nur einigen hundert Volt zu erzeugen [G 20], [G 16]. Im Zusammenhang mit Transistoren tritt die gleiche Problemstellung heute erneut auf. Apparate mit Transistoren benützen im allgemeinen Speisespannungen zwischen 4 und 20 Volt. Falls nun in einem solchen Gerät eine höhere Hilfspannung von z. B. 500 Volt erforderlich ist, wird es in den meisten Fällen angebracht sein, diese höhere Spannung mit Hilfe eines Transistor-Oszillators und eines Kaskadengenerators zu erzeugen, wobei als Gleichrichterelemente mit Vorteil kleine Flächendioden benützt werden können. Dieses Vorgehen ist insbesonders bei Batteriegeräten zu empfehlen, da sich so eine besondere Batterie höherer Spannung vermeiden läßt.

IV. Schlußbemerkungen.

Wie wir gesehen haben, ist es bei Kaskadengeneratoren vorteilhaft, eine größere Zahl von Stufen zu verwenden. Die nützliche Stufenzahl ist indessen begrenzt. Wird der Generator mit Netzfrequenz betrieben, so bestimmt vor allem die Größe der Kondensatoren und die Zahl der Stufen den Spannungsabfall bei Belastung. Das Volumen der notwendigen Kapazitäten läßt sich durch die Wahl einer höheren Frequenz der Wechselspannung beträchtlich verkleinern und die zulässige Stufen-

---

[1] Vgl. z. B. [Bo 41] und R. PATZELT, Referat an der Tagung der österr. Physik. Ges., Graz, 14.—16. 10. 1957 [Physik. Verh. **8**, 253 (1957)].

[2] W. C. ELMORE and M. SANDS: Electronics, Experimental Techniques. New York: McGraw-Hill Book Co. 1949.

zahl kann somit erhöht werden. Dabei ist zu beachten, daß mit wachsender
Frequenz die Wechselströme durch die Streukapazitäten immer größere Werte
annehmen. Die Streukapazitäten machen sich durch eine Verkleinerung der
Hochspannung, selbst des unbelasteten Generators, bemerkbar. Dieser Effekt
wird um so größer, je höher die Stufenzahl und je kleiner das Verhältnis von
Seriekapazität zu Streukapazität pro Stufe wird.

Eine weitere unerwünschte Wirkung der Streukapazitäten, zumindestens in
der ursprünglichen Kaskadenschaltung nach Fig. 24, besteht darin, daß schon
beim unbelasteten Generator eine störende Rippelspannung auftritt, die beträchtliche Werte annehmen kann. Diese Störung läßt sich durch Speisung mit einer
gegen Erde symmetrischen Wechselspannung beheben (vgl. Fig. 36). Ferner ist es

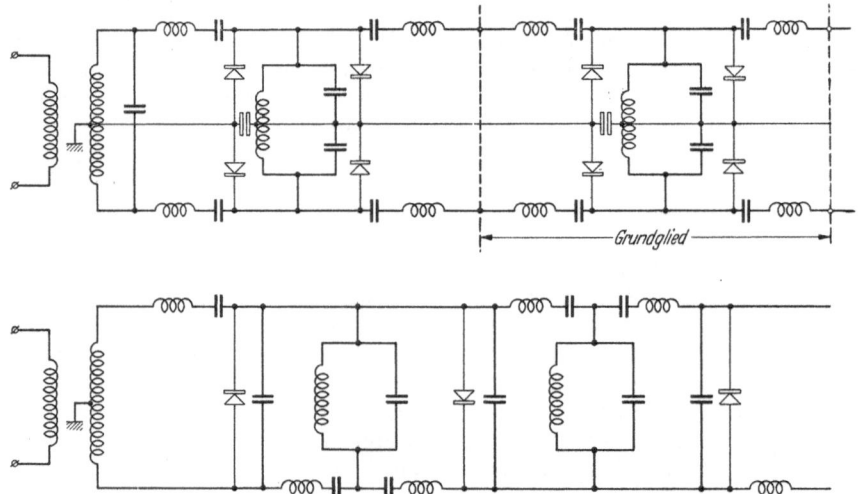

Fig. 50. Kaskadengeneratoren mit Bandfilterstruktur. Die zum Filter gehörenden Teile sind stark ausgezogen. Die
nur gleichstrommäßig bedingten Schaltelemente sind dünn eingetragen.

möglich, den durch die Streukapazitäten verursachten Spannungsabfall mit
geeigneten Maßnahmen (vgl. Diskussion auf S. 38) stark zu reduzieren. Die
konsequente Weiterführung dieses Gedankens führt dazu, den einzelnen Stufen
des Kaskadengenerators die Struktur eines Bandpaßfilters zu erteilen, wobei
die Frequenz der Wechselspannung etwa in der Mitte des Durchlaßbereiches
der so erhaltenen Filterkette liegen soll. Zwei Beispiele dieser Art, denen einfache
Bandfilterstrukturen zugrunde liegen, sind in Fig. 50 dargestellt. Die Schaltkapazitäten sowie die Induktivitäten der Seriekondensatoren und der Verdrahtung bilden nun Bestandteile der Filter und sind infolgedessen nicht mehr störend.
Die Stufenzahl kann durch diese Maßnahme weiter erhöht werden, wobei die
Zahl der nützlichen Stufen durch die Verluste der einzelnen Schwingungskreise
bestimmt wird. Es ist durchaus möglich, Spulen für hohe Spannungen, wie sie
in solchen Schaltungen notwendig sind, mit Q-Werten zwischen 100 und 200
und Eigenkapazitäten von nur einigen pF zu bauen ($B\,57$). Die Theorie solcher
Kaskadengeneratoren läßt sich in enger Anlehnung an die in der Nachrichtentech-
nik übliche Behandlung von Filtern aufbauen, wobei die Verluste der Spulen und
der Kondensatoren mit Reihenentwicklungen näherungsweise berücksichtigt werden können[1]. Die Leerlaufspannung erhält man dann als Summe der Amplituden

---

[1] Vgl. z.B. R. Feldtkeller: Einführung in die Siebschaltungstheorie. Leipzig: S. Hirzel
1939 (s. Kap. VII, S. 151).

der über den Ventilen liegenden Wechselspannungen bzw. als Summe der Spannungen, die sich an den entsprechenden, im Schema eingezeichneten Schaltkapazitäten ausbilden. Der Spannungsabfall bei Belastung läßt sich dadurch erfassen, daß in jeder Stufe ein äquivalenter, von der Belastung abhängiger Verlustwiderstand eingeführt wird. Es sollte in der hier skizzierten Art und Weise möglich sein, Kaskadengeneratoren zu bauen, deren Ausgangsspannung höher als 5 Millionen Volt liegt. Abschließend können wir somit feststellen, daß sich im gegenwärtigen Zeitpunkt keine obere Grenze der mit Kaskadengeneratoren möglichen Hochspannungen angeben läßt.

Literatur.

[A 34] ALLIBONE, T. E., A. BEETLESTONE and G. S. INNES: D. C. Generator using Continously Evacuated Rectifiers. Brit. J. Radiol. **7**, 83—96 (1934).

[A 36] ALDOUS, W. H.: The Characteristics of Thermionic Rectifiers. Wireless Engr. **13**, 576—580 (1936).

[A 39] ALLIBONE, T. E., F. E. BANCROFT and G. S. INNES: The St. Bartholomew's Hospital X-ray tube for one million volts. J. Instn. Electr. Engrs. **85**, 657—680 (1939).

[A 50] ARNOLD, W. R.: A 500 kilovolt linear accelerator using Selenium rectifiers. Rev. Sci. Instrum. **21**, 796—799 (1950).

[B 32] BRENZINGER, M.: Innere Vorgänge in Ventil-Kondensator-Schaltungen. Arch. Elektrotechn. **26**, 99—100 (1932).

[B 34] BELL, G. E.: Electrical Characteristics of Constant H. V. Generators for X-Ray Work. Part I. Theory of Half-Wave Generators. Brit. J. Radiol. **7**, 654—669 (1934).

[B 40] BJERGE, T., K. J. BROSTRØM, J. KOCH and T. LAURITSEN: A high tension apparatus for nuclear research. Kgl. danske Vid. Selsk. **18**, No. 1 (1940).

[B 47] BURCHAM, W. E.: The one million volt accelerator equipment of the Cavendish Laboratory Cambridge: Nature, Lond. **160**, 316—317 (1947).

[B 56] BALDINGER, E.: Zur Berechnung des Spannungsabfalles von Kaskadengeneratoren. Helv. phys. Acta **29**, 452—455 (1956).

[B 57] BALDINGER, E., u. W. HEILPERN: Kompensations-Drosselspulen hoher Güte für Kaskadengeneratoren. Helv. phys. Acta **30**, 282—287 (1957).

[Be 47] BENETT, W. H.: A Cold Cathode Rectifier. J. appl. Phys. **18**, 479—482 (1947).

[Bo 34] BOUWERS, A.: Design of X-ray apparatus. Radiology **22**, 163—173 (1934).

[Bo 37] BOUWERS, A., u. A. KUNTKE: Ein Generator für 3 Millionen Volt Gleichspannung. Z. techn. Phys. **18**, 209—219 (1937).

[Bo 41] BOUWERS, A., u. F. A. HEYN: Ein einfaches Gerät zur Elektronenzählung. Philips techn. Rdsch. **6**, 74—79 (1941).

[C 30] COCKCROFT, J. D., and E. T. S. WALTON: Experiments with high velocity positive ions. Proc. Roy. Soc. Lond., Ser. A **129**, 477—489 (1930).

[C 32] COCKCROFT, J. D., and E. T. S. WALTON: Experiments with high velocity positive ions—(I): Further developments in the method of obtaining high velocity positive ions. Proc. Roy. Soc. Lond., Ser. A **136**, 619—630 (1932).

[C 37] CHARLTON, E. E., G. HOTALING, W. F. WESTENDORP and L. E. DEMPSTER: An oil-immersed X-ray outfit for 500000 volts and an oil-immersed multi section X-ray tube. Radiology **29**, 329—340 (1937).

[C 40] CHARLTON, E. E., and H. S. HUBBARD: 1400000 Volt constant potential X-ray equipment. Gen. Electr. Rev. **43**, 272—279 (1940).

[C 42] COCKING, W. T.: Voltage Multiplying Rectifiers. Wireless World **48**, 60—61 (1942).

[C 55] CORK, B.: Proton Linear Accelerator Injector for the Bevatron. Rev. Sci. Instrum. **26**, 210—219 (1955).

[D 10] DELON, J.: Ersatz des Wechselstromes durch Gleichstrom bei der Spannungsprüfung elektrischer Leitungen. Elektrotechn. Z. [ETZ] **33**, 1179—1180 (1912). Übersetzt aus Bull. Soc. Int. Electr. **10** (1910).

[D 52] DAVENPORT, P. A.: A radio-frequency power supply for high voltage accelerators. Brit. J. appl. Phys. **3**, 124—126 (1952).

[E 53] EVERHART, E., and P. LORRAIN: The Cockcroft-Walton multiplying circuit. Rev. Sci. Instrum. **24**, 221—226 (1953).

[F 41] FROMMER, J. C.: The determination of operating data and allowable ratings of vacuum tube rectifiers. Proc. Inst. Radio Engrs. **29**, 481—485 (1941).

[G 14] GREINACHER, H.: Über einen Gleichrichter zur Erzeugung konstanter Gleichspannung. Elektrotechn. u. Maschinenbau [EuM] **1914**, Nr. 23. — Verh. dtsch. phys. Ges. **16**, 320—326 (1914).

[G 16] GREINACHER, H.: Eine neue Hochspannungsbatterie. Phys. Z. **17**, 343—346 (1916). — Bull. schweiz. elektrotechn. Ver. **9**, 85—88 (1918).

[G 20] GREINACHER, H.: Erzeugung einer Gleichspannung vom vielfachen Betrage einer Wechselspannung ohne Transformator. Bull. schweiz. elektrotechn. Ver. **11**, 59—66 (1920).

[G 21] GREINACHER, H.: Über eine neue Methode, Wechselstrom mittels elektrischer Ventile und Kondensatoren in hochgespannten Gleichstrom zu verwandeln. Z. Physik **4**, 195—205 (1921).

[G 36] GRADSTEIN, S.: Eine moderne Hochspannungsanlage. Philips techn. Rdsch. **1**, 6—10 (1936).

[G 40] GRAY, L. H., J. READ and J. G. WYATT: Neutron Generator for Biological Research. Brit. J. Radiol. **13**, 82—94 (1940).

[G 42] GREINACHER, H.: Über den Spannungsaufbau im Kaskadengenerator. Helv. phys. Acta **15**, 518—522 (1942).

[G 43] GREINACHER, H.: Der Kaskadengenerator als stabilisierte Spannungsquelle. Helv. phys. Acta **16**, 265—270 (1943).

[G 47] GERBER, TH.: Über den Spannungsaufbau im Kaskadengenerator und in ähnlichen Spannungsvervielfachern. Bull. schweiz. elektrotechn. Ver. **38**, 700—715 (1947).

[G 53] GREINACHER, H.: Zur Geschichte der Gleichrichterschaltungen. Schweiz. Arch. angew. Wiss. Techn. **18**, 1—4 (1953).

[H 50] HAINE, M. E., R. S. PAGE and R. G. GARFITT: The three stage electron microscope with stereographic dark field and electron diffraction capabilities. J. appl. Phys. **21**, 173—182 (1950).

[H 55] HEILPERN, W.: Kaskadengenerator zur Partikelbeschleunigung auf 4 MeV. Helv. phys. Acta **28**, 485—491 (1955).

[I 39] IMHOF, A.: Ein neuer Gleichspannungsgenerator für 3 Millionen Volt. Helv. phys. Acta **12**, 285—288 (1939).

[I 45] IMHOF, A.: Ein neuer Gleichrichter für hohe Spannungen. Bull. schweiz. elektrotechn. Ver. **36**, 333—337 (1945).

[J 42] JAGGI, M.: Untersuchung des Aufladevorganges im Kaskadengenerator zur Erzeugung hochgespannten Gleichstroms aus Wechselstrom. Helv. phys. Acta **15**, 339—341 (1942). — Bull. schweiz. elektrotechn. Ver. **34**, 386—399 (1943).

[K 37] KUNTKE, A.: Ein Generator zur Erzeugung sehr hoher Gleichspannungen. Philips techn. Rdsch. **2**, 161—165 (1937).

[L 49] LORRAIN, P.: On the design of a radiofrequency Cockcroft-Walton accelerator. Rev. Sci. Instrum. **20**, 216—217 (1949).

[L 56] LORRAIN, P., R. BÉIQUE, P. GILMORE, P. GIRARD, A. BRETON and P. PICHÉ: A High Frequency Cockcroft-Walton Accelerator. Institut de Physique, Université de Montréal. Canada, Févr. 1956. [Vervielfältigter Bericht.] Vgl. auch Canad. J. Phys. **35**, 299—312 (1957).

[M 37] MOULLIN, E. B.: The external characteristic of a diode rectifier. J. Instn Electr. Engrs. **80**, 553—563 (1937).

[M 39] MÜLLER, H.: Schaltungen zur Erzeugung hochgespannten Gleichstromes für Versuche. Arch. techn. Messen (1938) Z 43-1; (1939) Z 43-3.

[M 47] MAUTNER, R. S., and O. H. SCHADE: Television high voltage R-F supplies. RCA-Rev. **8**, 43—81 (1947).

[*Me 38*] MEHLHORN, H.: Hochspannungsanlage für 3 Millionen Volt konstante Gleichspannung. Siemens-Z. **18**, 417—422 (1938).

[*Me 43*] MEHLHORN, H.: Über die Greinacher-Ventilvervielfachungsschaltung und ihre Verwendung zur Erzeugung hoher konstanter Gleichspannung. Wiss. Veröff. Siemens-Werk **21**, 141—186 (1943).

[*Mi 43*] MITCHELL, R. G.: Vacuum rectifiers working with condenser input. Wireless Engr. **20**, 414—425 (1943).

[*Mi 45*] MITCHELL, R. G.: Cascade Generators. Wireless Engr. **22**, 474—483 (1945).

[*Mü 54*] MÜLLER-LÜBECK, K.: Gleichrichter mit Ladekondensator und Hochvakuumröhren oder Selengleichrichtern. Arch. Elektrotechn. **41**, 181—195 (1954).

[*N 40*] NITSCHKE, A.: Eine besonders leistungsstarke Röntgen-Therapieanlage für 1,2 MV. Elektrotechn. Z. [ETZ] **61**, 441—444 (1940).

[*P 47*] PLOOS VAN AMSTEL, J. J. A.: Kleine Selenventile. Philips techn. Rdsch. **9**, 267—276 (1947/48).

[*P 55*] PECK, R. A.: Characteristics of a high frequency Cockcroft-Walton voltage source Rev. Sci. Instrum. **26**, 441—444 (1955). — PECK, R. A., and H. P. EUBANK: High Current Cockcroft-Walton Accelerator for Neutron Production. Rev. Sci. Instrum. **26**, 444—448 (1955).

[*Pa 50*] PANZER, S.: Erzeugung hoher Gleichspannung mittels Hochfrequenz. Optik **7**, 290 (1950).

[*Pa 51*] PANZER, S.: Spannungsanlage für das AEG-ZEISS-Elektronenmikroskop. AEG-Mitt. **46**, Nr. 7/8, 10 (1951).

[*Pa 53*] PANZER, S.: HF-Hochspannungsanlage für die Mikroskopie mit Zwischenbeschleuniger. Optik **10**, 107 (1953).

[*Pa 55*] PANZER, S.: Eine Hochspannungsanlage bis 150 kV mit HF-betriebenen Trockengleichrichtern für elektronenoptische Zwecke. Phys. Verh. **6**, 20 (1955).

[*R 36*] ROBERTS, N. H.: The diode as half wave, full wave and doubling rectifier. Wireless Engr. **13**, 351—362, 423—430 (1936).

[*R 41*] REYVAL, J.: Générateurs de courant continu à haute tension utilisant des redresseurs à cuivre — oxyde de cuivre. Rev. gén. Electr. **50**, 294—299 (1941).

[*S 19*] SCHENKEL, M.: Eine neue Schaltung für die Erzeugung hoher Gleichspannungen. Elektrotechn. Z. [ETZ] **40**, 333—334 (1919).

[*S 28*] SLEPIAN, J.: High voltage direct current system. U.S. Patent 1666473 (1928).

[*S 32*] STARKE, H., u. R. SCHROEDER: Die Reihenschaltung von Gleichrichterventilen für die Erzeugung sehr hoher Gleichspannungen. Arch. Elektrotechn. **26**, 301—305 (1932).

[*S 35*] SLOAN, D. H.: A radio-frequency high-voltage generator. Phys. Rev. **47**, 62—71 (1935).

[*S 43*] SCHADE, O. H.: Analysis of rectifier operation. Proc. Inst. Radio Engrs. **31**, 341 to 361 (1943).

[*Sc 43*] SCHADE, O. H.: Radio-frequency-operated high-voltage supplies for cathode ray tubes. Proc. Inst. Radio Engrs. **31**, 158—163 (1943).

[*Sc 47*] SCHADE, O. H., u. L. A. MAUTNER: Television High-Voltage R-f-Supplies. R.C.A. Rev. **8**, 43 (1947).

[*Si 48*] SIEZEN, G. J., u. F. KERKHOF: Ein Fernsehempfänger mit Bildprojektion, III. Die Apparatur zur Erzeugung der Anodenspannung von 25 kV. Philips techn. Rdsch. **10**, 157—166 (1948). — SIEZEN, G. J., and F. KERKHOF: Pulse-Type High Voltage Supply. Proc. Inst. Radio Engrs. **36**, 401—407 (1948).

[*St 53*] STROJNIK, I.: Generation of High and Very Stable d.c. Voltages Using Audio Frequencies and Resonance. Elektroteh. Vestn. **1953**, Nr. 5, 131.

[*T 52*] TITTERTON, E. W.: 1,2 MeV Accelerator at the Australian National University. Nucleonics **10**, No. 5, 28—29 (May 1952).

[*V 01*] VILLARD, M. P.: Transformateur à haut voltage. J. de Phys., Sér. III **10**, 28—32 (1901).

[*V 38*]   Vanoni, E. P.: La progettazione dei circuiti moltiplicatori ad altissima tensione. L'Elettrotecnica **25**, 766—771 (1938).

[*V 39*]   Vanoni, E. P.: Ein 1000 kV-Generator zur Erzeugung von radioaktiven Substanzen. Strahlentherapie **65**, 304—314 (1939).

[*V 46*]   Van der Tuuk, J. H.: Hochspannungsgleichrichter für Röntgendiagnostik. Philips techn. Rdsch. **8**, 199—205 (1946).

[*V 47*]   Van Dorsten, A. C., W. J. Oosterkamp u. J. B. Le Poole: Ein experimentelles Elektronenmikroskop für 400 Kilovolt. Philips techn. Rdsch. **9**, 193—202 (1947).

[*V 48*]   Van Dorsten, A. C.: Die Stabilisierung der Beschleunigungsspannung für ein Elektronenmikroskop. Philips techn. Rdsch. **10**, 137—143 (1948).

[*W 41*]   Waidelich, D. L.: The full-wave voltage-doubling rectifier circuit. Proc. Inst. Radio Engrs. **29**, 554—558 (1941).

[*W 42*]   Waidelich, D. L., and C. H. Gleason: The half-wave voltage-doubling circuit. Proc. Inst. Radio Engrs. **30**, 535—541 (1942).

[*W 48*]   Walker, A. H. B.: Television E. H. T. Supply. Wireless World **54**, 169—173 (1948).

[*W 49*]   Walker, A. H. B., and L. H. Peter: Improvements relating to apparatus for the supply of high voltage uni-directional current from a relatively low voltage alternating current source. British Patent 645630 (1949).

[*Z 06*]   Zimmermann, C. I.: Electric converter. U.S. Patent 1099960 (1906).

Bücher.

[*1*]   Reich, H. J.: Theory and Applications of Electron Tubes. New York: McGraw-Hill Book Comp. 1944.

[*2*]   Kammerloher, J.: Hochfrequenztechnik, Teil III. Leipzig: C. F. Winter 1949.

[*3*]   Gray, T. S.: Applied Electronics. New York: John Wiley & Sons Inc. 1954.

[*4*]   Roth, A.: Hochspannungstechnik. Wien: Springer 1938.

[*5*]   Bouwers, A.: Elektrische Höchstspannungen. Berlin: Springer 1939.

[*6*]   Craggs, J. D., and J. M. Meek: High Voltage Laboratory Technique. London: Butterworth's Scientific Publ. 1954.

Bezeichnungen.

Einheiten: Volt, Amp. und sec.

a_n Koeffizienten in Gl. (8.1).

A Konstante in Gl. (14.4).

A_n Konstante in Gl. (4.7) (Einweggleichrichter).

b $b^2 = \dfrac{C\,\text{quer}}{C\,\text{serie}}$ beim Kaskadengenerator (Fig. 34).

B Konstante in Gl. (14.4).

B_n Funktion in Gl. (4.9) (Einweggleichrichter).

C Kapazität

C_1 bzw. C_2 Kapazität parallel zur Primär- bzw. Sekundärseite des Transformators (Fig. 48 und 49).

C_n bzw C_n^* Kapazitäten der Glättungssäule bzw. der Schubsäule eines Kaskadengenerators (Fig. 24).

C_{en} bzw. C_{en}^* Gl. (13.5) und (13.7). Auf die äquivalenten Einweggleichrichter reduzierte Kapazitäten des Kaskadengenerators.

D Konstante in Abschnitt 5 (vgl. Fig. 15).

e   Basis der natürlichen Logarithmen $= 2{,}718 \ldots$

f Frequenz.

F Verhältnis der wirklichen zur idealen Leerlaufspannung eines Kaskadengenerators (Fig. 35) $F = V/V'$.

g bzw. g_L Spannungsübersetzung des Transformators ohne bzw. mit Linkverbindung (Fig. 48 und 49).

h Funktion in Gl. (6.5) (Einweggleichrichter).

i zeitlich veränderlicher Strom.

i_{en} Gl. (13.5); auf die äquivalenten Einweggleichrichter reduzierte Ventilströme des Kaskadengenerators.

$\hat{i}$ Spitzenwert des Ventilstromes

I zeitunabhängiger Wert eines Stromes. In Kap. III Effektivwerte: I_1 bzw. I_{1L} Primärstrom des Transformators ohne bzw. mit Linkverbindung. I_2 bzw. I_{2L} Sekundärstrom ohne bzw. mit Linkverbindung.

I_g Gleichstrom.

I_s Gl. (6.1) Sättigungsstrom von Halbleiterdioden in Sperrichtung.

$|I|$ Effektivwert des Ventilstromes.

j $\sqrt{-1}$.

$J_0(z)$ Bessel-Funktion nullter Ordnung mit dem Argument z. Vgl. W. Magnus u. F. Oberhettinger: Formeln und Sätze für die speziellen Funktionen der mathematischen Physik, Springer Verlag (1948), E. Jahnke u. F. Emde: Tafeln höherer Funktionen. Teubner Verlag (1948), bzw. die amerikanische Ausgabe: Dover Publication (1945).

k Boltzmann-Konstante.

K Kopplungskoeffizient zweier Spulen.

$$K^2 = \frac{M^2}{L_1 L_2}, \qquad K_c^2 = \frac{1}{Q_1 Q_2}.$$

K' Kopplungskoeffizient zwischen Primärseite und Linkverbindung.

K'' Kopplungskoeffizient zwischen Linkverbindung und Sekundärseite.

L Induktivität.
Kap. III: L_1 Induktivität der Primär-, L_2 der Sekundärseite des Transformators, L' Induktivität einer Wicklung der Linkverbindung.

n Exponent der Ventilkennlinie $i = \varkappa u^n$, Abschnitt 2.

n laufende Nummer der Stufen einer Kaskade (oberste Stufe $n = 1$) Gl. (12.2).

N Gesamtstufenzahl einer Kaskade (12.2).

m Abschnitt 14 laufende Nummer der Maschen bei Kaskadengeneratoren.

M Abschnitt 14 Gesamtzahl der Ventile.

P Gl. (21.5).

q Elementarladung $= 1{,}6 \cdot 10^{-19}$ Coulomb.

Q Ladung, welche dem Kondensator eines Gleichrichters pro Periode zugeführt wird

$$Q = \frac{I_g}{f}.$$

Q Spulengüte $= \dfrac{\omega L}{R}$.

Kap. III: Q_1 und Q_2 Spulengüten der Primär- und der Sekundärseite, Q' Spulengüte der Linkverbindung. In Q_2 ist die an die Kaskade abgegebene Leistung inbegriffen. Q_2' Spulengüte der Sekundärseite bei unbelasteter Kaskade.

R Widerstand.

In Kap. III: R_1 bzw. R_2 Widerstand der Primär- bzw. der Sekundärseite, R' Widerstand einer Wicklung der Linkverbindung. R_r auf die Primärseite reduzierter Gesamtwiderstand Gl. (21.12), R_s Widerstand der Sekundärspule allein.

R_L Lastwiderstand eines Gleichrichters.

R_L' auf den Eingang der Kaskade reduzierter R_L Gl. (20.1).

$\hat{R}$ Spitzenwert des Ventilwiderstandes (Abschnitt 5) $\hat{R}_S = R_t + \hat{R}$.

$\bar{R}$ mittlerer Ventilwiderstand in Leitrichtung (Abschnitt 5).

$|R|$ Effektivwert des Ventilwiderstandes in Leitrichtung (Abschnitt 5).

$R_{ae} \cong \bar{R}$ Widerstand des äquivalenten Gleichrichters mit ideal geknickter Ventilkennlinie (Fig. 15).

R_t Auf die Sekundärseite reduzierter Widerstand des Transformators (Kap. I und II).

R_{en} Gl. (13.5). Auf die äquivalenten Einweggleichrichter reduzierte Widerstände beim Kaskadengenerator.

t Zeit.

t_1 bzw. t_2 Einschalt- bzw. Ausschaltzeit der Ventile.

$T = 1/f$ Zeitdauer einer Periode der Wechselspannung.

T Temperatur in Grad Kelvin.

u zeitlich veränderliche Spannung.

u_c Spannung am Kondensator (Fig. 7).

u_L Eingangsspannung bei Drosselspuleneingang (Fig. 20).

$u^* = u/U_0$ (Abschnitt 2), normierte Spannung.

$\hat{u}$ Spitzenwert der Ventilspannung in Leitrichtung.

U zeitunabhängiger Wert einer Spannung.

Kap. III: Effektivwerte U_1 Primär- und U_2 Sekundärspannung des Transformators.

U_0 Amplitude der Wechselspannung des Transformators.

U_g Gleichspannung.

δU Rippelspannung (z. B. Fig. 7 und 28).

δU_{en} bzw. δU_{en}^* auf die äquivalenten Einweggleichrichter reduzierte Rippelspannungen des Kaskadengenerators.

ΔU Spannungsabfall eines Gleichrichters bei Belastung (Fig. 7).

ΔU_R Spannungsabfall eines Gleichrichters bei großem Kondensator infolge der Widerstände des Transformators und der Ventile (Gl. 3.1).

ΔU_C Spannungsabfall eines Gleichrichters infolge des endlichen Wertes des Kondensators bei vernachlässigbar kleinen Widerständen der Ventile und des Transformators (Gl. 3.1).

$\ddot{u}$ Übersetzungsverhältnis eines Transformators.

V Leerlaufspannung des Kaskadengenerators.

V' ideale Leerlaufspannung des Kaskadengenerators.

W Verluste der Ventile (Abschnitt 5), der Transformatorenspulen (Kap. III).

x Variable in Gl. (6.4), $x = U_0/U_m$.

X Reaktanz $X = \omega L_2 - \dfrac{1}{\omega C_2}$ (Kap. III).

z Impedanz.

z_w Wellenwiderstand.

Z Betrag der Impedanz z.

α Parameter in Gleichrichterschaltungen $\alpha = \omega R C$ [z. B. Gl. (2.5)].

α_e bzw. α_e^* Gl. (15.1) und (15.3). Auf die äquivalenten Einweggleichrichter reduzierter Parameter α bei Kaskadengeneratoren.

β Parameter in Gleichrichterschaltungen.
$\beta = 1$ Einweggleichrichter, $\beta = 2$ Doppelweggleichrichter.

γ Fortpflanzungskonstante einer Welle.

$\varkappa$ Konstante der Ventilkennlinie $i = \varkappa u^n$ (Abschnitt 2).

λ Verhältnis von Strömen (Abschnitt 5).

ν bzw. ν^* Hilfsgröße bei der Reduktion des Kaskadengenerators auf äquivalente Einweggleichrichter [Gl. (13.1)].

π $3,141 \ldots$

$\omega = 2\pi f.$

$\omega_2 = \dfrac{1}{\sqrt{L_2 C_2}}$ Resonanzfrequenz der Sekundärseite des Transformators (Kap. III).

$\Omega = 1 - \dfrac{\omega_2^2}{\omega^2}$ (Kap. III).

$\varphi = \omega t$; φ_1 und φ_2 Einschalt- und Auschaltwinkel der Ventile, φ_m Phasenwinkel zur Zeit des Spitzenwertes des Ventilstromes.

η bzw. η_L Wirkungsgrad des Transformators ohne bzw. mit Linkverbindung.

Van de Graaff Generators.

By

R. G. HERB.

With 24 Figures.

Introduction.

The dramatic development of ROBERT VAN DE GRAAFF which was first reported in 1931 received immediate and wide recognition. His machine was very simple in principle and in its early form it was very simple to construct. It appeared at a time when a pressing need was being felt for a convenient machine to accelerate charged particles, and its characteristics appeared to be ideally suited to this purpose.

In VAN DE GRAAFF's first workable model a silk belt $2\frac{1}{2}$ inches wide was carried by two pulleys, one at ground potential which was powered by an electric motor, and the second inside a metal sphere of 24-inch diameter. The sphere was supported by glass rods.

Charge from a row of needle points held at a positive potential of about 10 kv was sprayed as corona current onto the belt surface just after it left the grounded pulley. Charge was removed from the belt by a second row of needle points well within the sphere before the belt reached the inner pulley. Here electric fields are only negligibly affected by charge on the outer surface of the sphere and charge passes freely from the belt to this outer surface of the sphere regardless of the potential of the sphere. If voltage is not adequately limited by current drain the sphere will rise in potential to where insulation fails, a spark occurs and the sphere is discharged.

The charging arrangement devised by VAN DE GRAAFF has been followed quite closely on most of the large number of electrostatic generators constructed up to this time. Especially for application to problems of nuclear physics the charging method has continued to satisfy requirements. Limitations which have restricted the utility of these machines have been imposed principally by electrical failure across insulating gas, along solid support members or inside the evacuated accelerating tube. This latter mode of failure has been especially troublesome in modern machines. Probably some ninety percent or more of the van de Graaff generators now in operation are limited in voltage by electrical discharge in the accelerating tube.

In early applications of VAN DE GRAAFF's development the simple form of his first model was rather closely followed. Much valuable work in nuclear physics was accomplished by machines of this simple form operating in air at atmospheric pressure. The first important departure from this simple early model was the utilization of high pressure gas for insulation instead of atmospheric air. This innovation which is due to BARTON, MUELLER and VAN ATTA working at Princeton and HERB, PARKINSON and KERST at Wisconsin was essential to wide and continued application.

The second contribution which was essential for widespread application of this machine was the development by the Wisconsin group of provisions to shape the electric field so as to achieve desirable characteristics such as a uniform gradient along insulators, the belt, and the accelerating tube, and to make practical the subdivision of insulators with potential control at each subdivision.

From the simple principles and simple form of the early machines a great amount of development work has been put forth to arrive at the rather intricate form of the successful modern machine insulated by high pressure gas. Contributions have come from many research and development groups throughout the world. The machines, although simple in principle, must satisfy many detailed complex requirements if they are to operate satisfactorily at high voltage. One small part incorrectly formed or incorrectly placed might lower the usable voltage by 50%.

To carry these machines from the 4.5 million volt range achieved at Wisconsin in 1940 to the 8 or more million achieved in the 1950's at M.I.T. and Los Alamos has required much ingenuity and a very great amount of painstaking work. McKibben and his group at Los Alamos and Turner in his work at Berkeley and later at Brookhaven made important contributions; but a great proportion of the developments came from Trump, van de Graaff and collaborators at M.I.T. and from the staff of the High Voltage Engineering Company.

Publication in this field has been very limited. The phenomena which cause failure are so poorly understood that the design of a machine must be largely determined by how a previous machine worked. Much of the material for this article must be a description of the design and limitations of certain machines. These machines must be ones with which the author was intimately associated. Other machines and the techniques of other groups cannot be treated with the thoroughness that would be desired.

A. Early developments.

1. Voltage limitations in air at atmospheric pressure. Prior to the 1930's the commonly accepted method for measurement of high voltage was by sparkover across sphere gaps. The dielectric strength of air at normal temperature and pressure had been determined to be approximately 30 kv/cm. A great amount of work had been done on the dielectric strength of air as a function of pressure and gap length and Paschen's law was established. Results of dielectric strength measurements in the pressure region above ten atmospheres showed considerable variation. Measurements at atmospheric pressure showed good agreement even with voltages of many hundreds of kilovolts.

Many of the early van de Graaff generators utilized spherical terminals which were relatively isolated from other conductors. The gradient at the surface of a sphere is given by

$$G = \frac{\Phi}{R} \tag{1.1}$$

where with the potential of the sphere Φ in volts and the sphere radius R in cm the gradient G is in volts/cm. In air at atmospheric pressure G should go to a maximum value of 30 kv/cm. Thus an ideal isolated sphere of 1 m radius should sustain a maximum of 3 million volts. In practice the spherical terminal cannot be isolated; the gradient near its surface is disturbed by supports, by openings for an accelerating tube and for a charging belt.

Steady running voltages achieved with van de Graaff generators operating in atmospheric air are approximately one-third of the values expected with ideal spheres. The cause for this severe limitation is not understood. High voltage terminals for many of these generators were well formed and much effort was spent to give them a smooth finish. Openings for accelerating tube and belt were provided with inwardly curved lips. Supports ran through openings with curved edges, to the inside, and were clamped in the field free region.

Sparks from the terminal of these machines were not confined to the region around openings. Many sparks came from smooth spherical portions. They appeared to be distributed in a relatively random fashion over the surface.

Fig. 1. Open air generator constructed by TUVE, HAFSTAD and DAHL in 1933 to 1934 at the Carnegie Institution of Washingto n. This generator operated up to 1.2 million volts. Large cylinder extending from the terminal encloses the resistance voltmeter.

Painstaking efforts to improve performance were frequently unrewarding. A terminal heavily covered with dust after prolonged operation might respond to a thorough cleaning and polishing by holding less voltage. Usually the maximum voltage rose somewhat with time during uninterrupted operation. A certain amount of sparking appeared to condition the terminal.

Open air machines suffered from other handicaps which restricted their application. They required relatively great amounts of space. High humidity was troublesome, not because of spark-over, since the dielectric strength of air is approximately independent of water vapor content, but because of leakage across insulators. Usually the charging belt was most sensitive to relative humidity.

In humid air it might refuse to carry charge or sparks along its surface might restrict terminal voltage.

Voltages obtained were disappointing but even the severe handicap of this limitation far from nullified the value of the method. A large number of these machines operating at atmospheric pressure were constructed for a wider variety of applications and several were used with great success for work in nuclear physics.

The 1.2 MV machine constructed in 1934 by Tuve, Hafstad, and Dahl (Fig. 1) at the Department of Terrestrial Magnetism of the Carnegie Institution of Washington[1] has an enviable record for output of data and in length of service it is excelled by few, if any, accelerators of any type. With a high voltage electrode radius of 1 meter its maximum voltage of 1.2 MV is slightly more than one-third of the theoretical maximum for an ideal sphere of this radius. Voltage tests from a sphere of $\frac{1}{2}$ meter radius at Tuve's laboratory gave a maximum usable value of 0.6 MV which corresponds to the same factor. The large M.I.T. machines[2] with spheres having 2.25 meter radii operated at a peak potential of about 2.7 MV giving a factor of 2.8 between usable voltage and the theoretical value.

2. Introduction of high pressure gas insulation. The relatively low usable gradients obtained across large air gaps at atmospheric pressure furnished the major impetus for the development of machines operating in high pressure gas. This innovation offered as its chief advantage a substantial reduction in size to offset a considerable loss in accessibility. Results of spark gap test data available in the literature showed considerable disagreement as to the variation of the dielectric strength of air with pressure. In general, however, spark-over voltages obtained in test work increased approximately linearly with pressure up to about ten atmospheres with a slow falling off of the curve at higher pressure. Voltages used in this early test work were usually below 100 kv and gap lengths consequently were very small.

The first machine built at Wisconsin[3] is shown in Fig. 2.

Sparking across the air gap from the high potential terminal to the tank wall limited voltage at atmospheric pressure to about 190 kv. The voltage for spark-over across the air gap rose somewhat less than linearly with pressure up to a pressure of about four atmospheres, absolute, which was the maximum the tank could withstand. In some tests in 1934 the voltage was found to go higher when a small percentage of CCl_4 was added to the air[4].

During the test work at the higher pressures and with CCl_4 vapor spark-over along the charging belt and along the surface of the textolite support tube frequently limited voltage. Further increase of pressure or addition of more CCl_4 then gave no improvement, and frequently led to failure at lower voltage. When an accelerating tube was added to this machine the upper voltage limit was set by discharge in this tube. Steady operation for nuclear bombardment data never went beyond 400 kv for the best accelerating tube.

Performance of this generator was in many respects gratifying and as a 0.4 Mev accelerator this simple design might even now be considered acceptable. As a model for a higher voltage machine it was clearly not satisfactory. The charging belt, the support tube and the accelerating tube were in a non-uniform field and

[1] M. A. Tuve, L. R. Hafstad and O. Dahl: Phys. Rev. **48**, 315 (1935).
[2] L. C. van Atta, D. L. Northrup, R. J. van de Graaff and C. M. van Atta: Rev. Sci. Instrum. **12**, 534 (1941).
[3] R. G. Herb, D. B. Parkinson and D. W. Kerst: Rev. Sci. Instrum. **6**, 261 (1935).
[4] M. T. Rodine and R. G. Herb: Phys. Rev. **51**, 508 (1937).

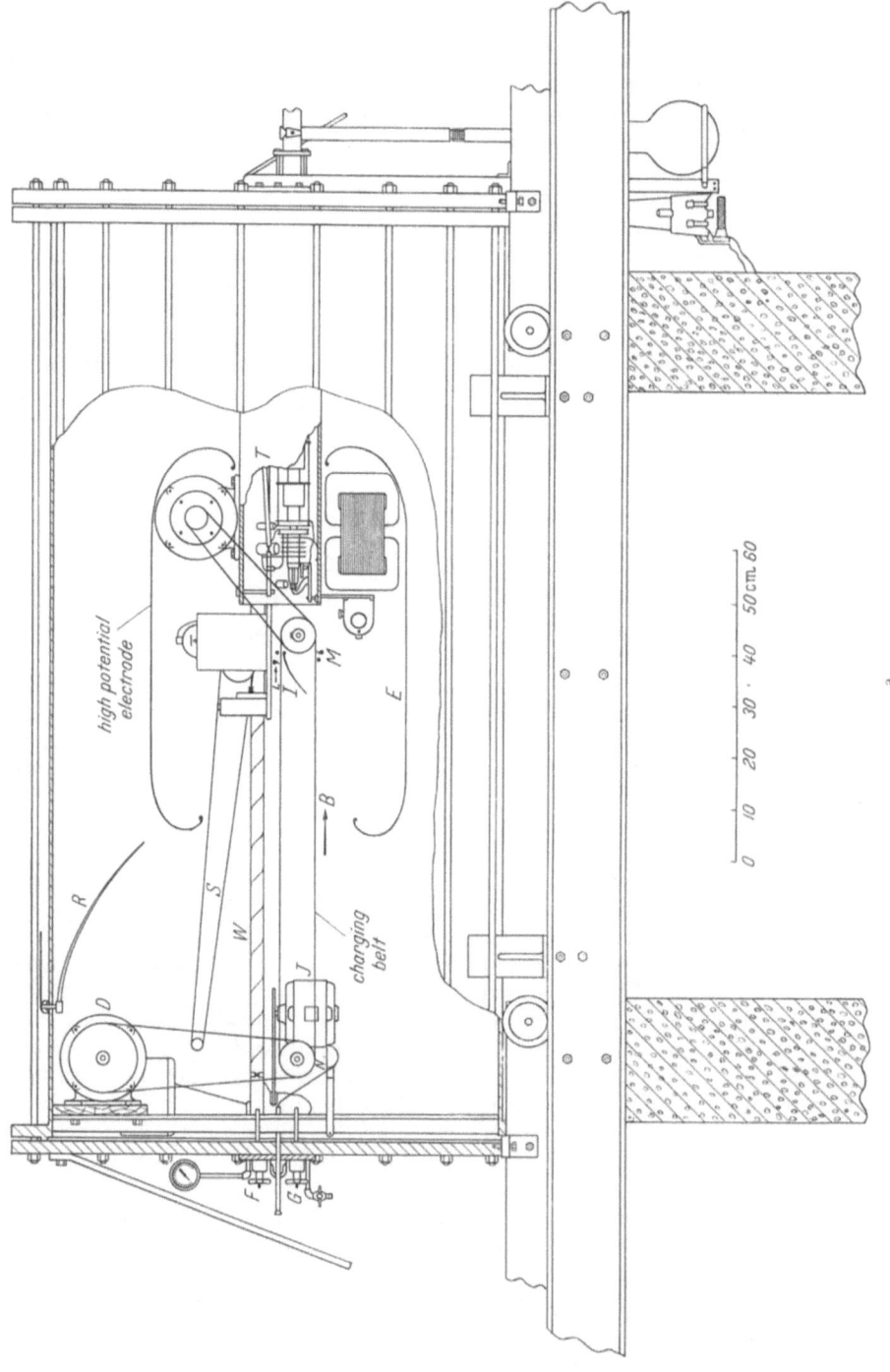

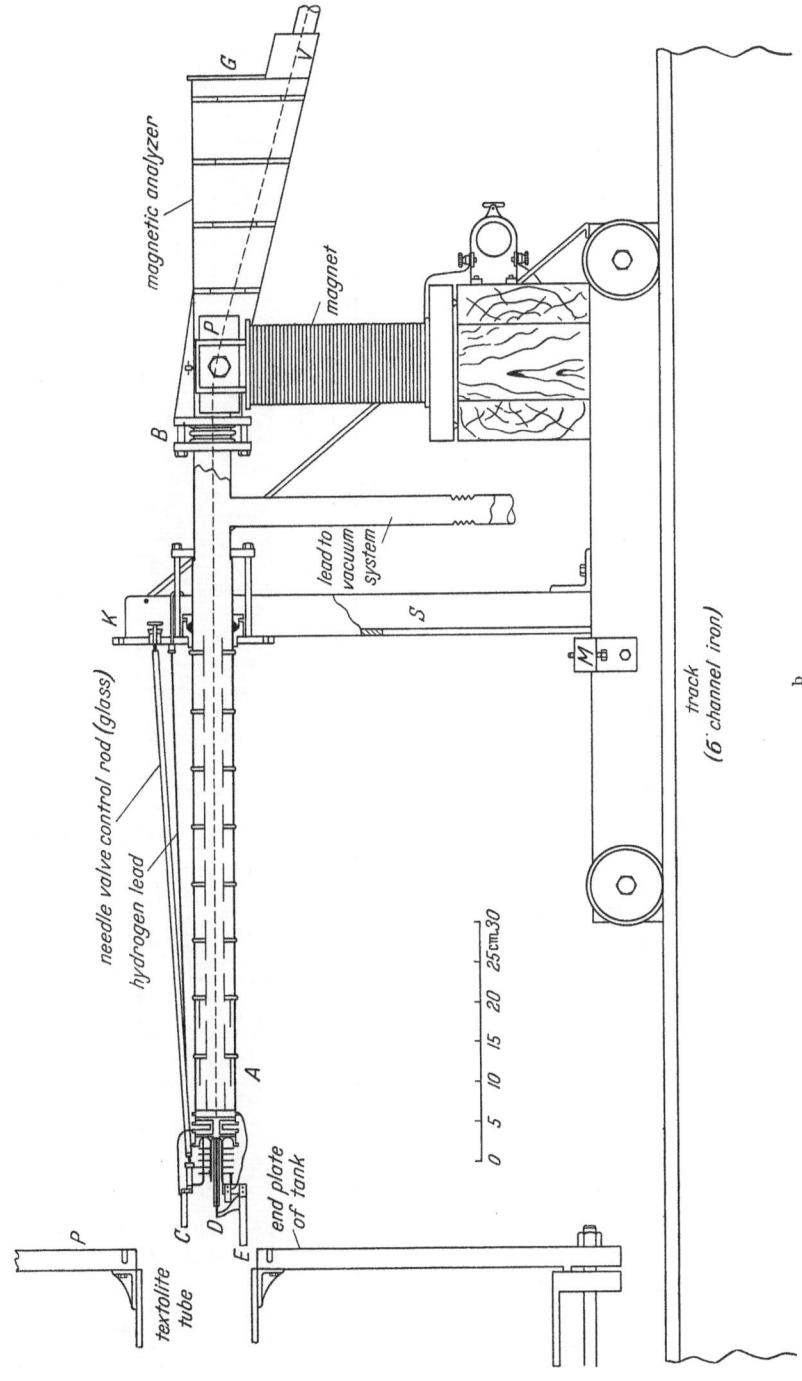

Fig. 2 a and b. (a) First generator constructed at Wisconsin with high pressure gas insulation. Operated up to about 700 kv without accelerating tube. (b) Accelerating tube for first generator. Consisted of eight glass cylinders joined by wax to metal electrodes. Operated up to 400 kv.

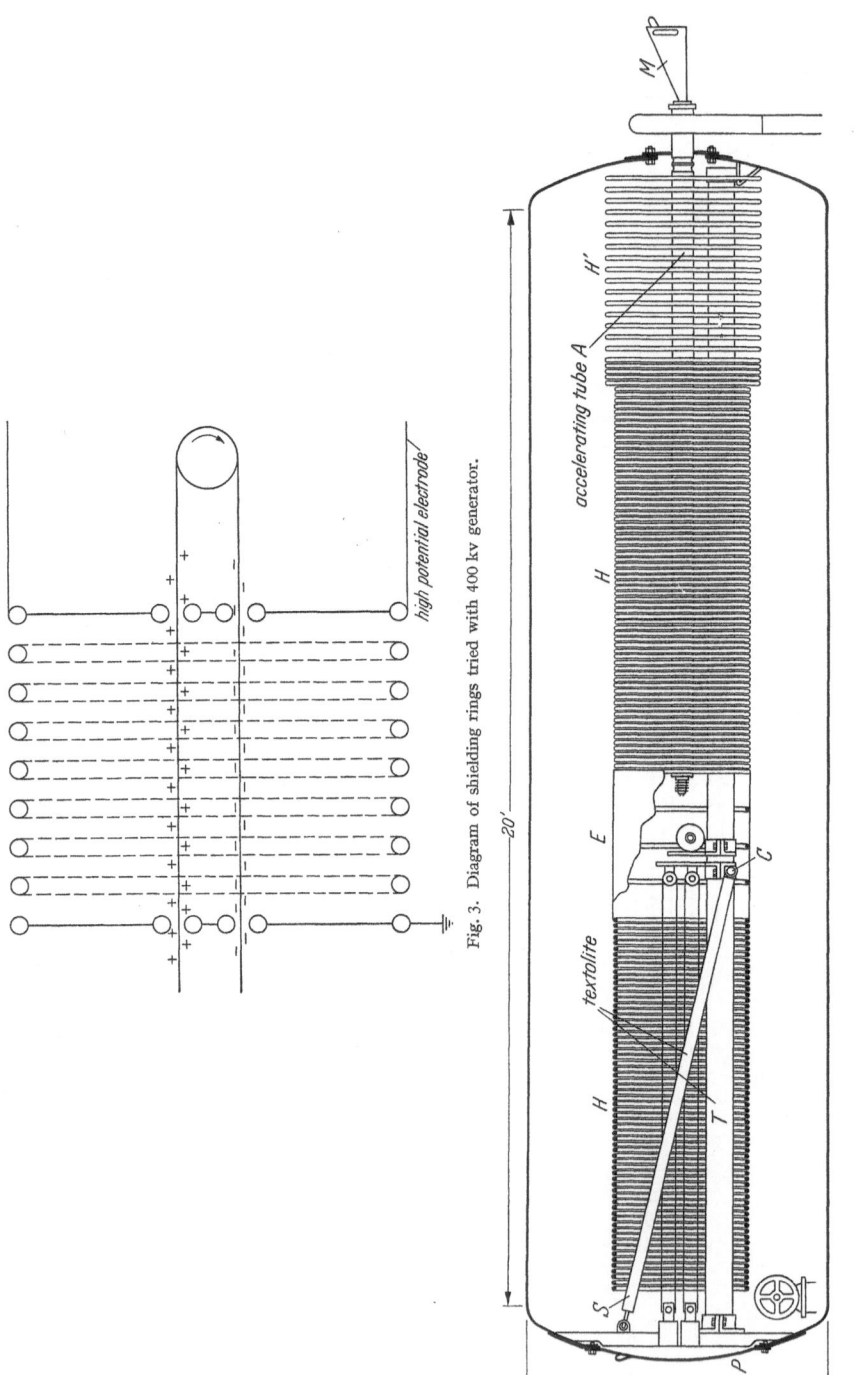

Fig. 3. Diagram of shielding rings tried with 400 kv generator.

Fig. 4. Second generator constructed at Wisconsin with high pressure gas insulation. Shielding rings and potential distribution system tried with the earlier generator were successfully used in this machine. Operated up to 2.5 MV.

were therefore used inefficiently. The generator did not appear to be well proportioned since the upper voltage limit for the accelerating tube was approximately one-half of the limiting values for other modes of failure.

An accelerating tube length greater in proportion to the other dimensions of the machine appeared to be called for. The usefulness of higher pressure or gases of higher dielectric strength appeared to depend on finding a solution to the problem of surface flash-over along solid insulators.

3. Electric field control. The high potential terminal of the generator shown in Fig. 2 was replaced by an electrode arrangement shown in Fig. 3. Corona current across point to plane gaps maintained a uniform gradient along the hoop system. If conducting planes are provided as shown in Fig. 3 at each end of the hoop system the field inside is uniform and parallel to the axis (neglecting the perturbing effect of a charged belt). Thin plates or wires perpendicular to the axis of the enclosure do not disturb the field. The interior of this enclosure is a suitable environment for a charging belt, an insulating support, and an accelerating tube.

Much data in the literature showed that the flash-over voltage across solid insulators does not increase linearly with insulator length even when the insulator is in a uniform field. Failure appears to be due to the establishment of a nonuniform gradient along the surface due to non-uniform creepage of charge. Subdivision of insulators appeared to be desirable to minimize this effect. Subdivision is easily accomplished for an insulator inside the hoop system of Fig. 3.

This first field control system functioned satisfactorily. It was not an important advantage for the small machine of Fig. 2 but it made possible the next advance, the 2.4-Mev machine[1] constructed at Wisconsin in 1936/37 where a uniform field was provided along supports, the charging belt and the accelerating tube. This machine which is shown in Fig. 4 will not be described in detail. Its performance helped to determine the design of later machines and the performance of the more modern machines will be described in some detail.

B. Electrode arrangements.

4. Cylindrical geometry. High potential electrodes with cylindrical sections and with hemispherical ends are commonly used in generators insulated by high pressure gas. A brief review of the field distribution between electrodes of these simple shapes may be helpful.

The field between two co-axial cylinders of infinite length is given by

$$E(r) = \frac{U}{r} \frac{1}{\log \frac{r_2}{r_1}} \tag{4.1}$$

where r_1 and r_2 are the radii of the inner and outer cylinder respectively and where U is the potential difference between the two cylinders. The field has a maximum value

$$E(r_1) = \frac{U}{r_1} \frac{1}{\log \frac{r_2}{r_1}} \tag{4.2}$$

at the surface of the inner cylinder. The value of $E(r_1)$ is plotted in Fig. 5 as a function of r_1/r_2 with r_2 held constant, as in the case where the choice of tank

---

[1] R. G. Herb, D. B. Parkinson and D. W. Kerst: Phys. Rev. **51**, 75 (1937).

diameter has been made and the most suitable diameter for the high potential electrode is being sought. The minimum in the curve may be obtained by setting

$$\frac{dE(r_1)}{dr_1} = 0 \tag{4.3}$$

giving $r_2/r_1 = e = 2.72$. If the optimum value is chosen for r_1 the field at r_1 is then given by U/r_1. For $U = 1$ MV and $r_1 = 40$ cm we have $E(r_1) = 25\,000$ volt/cm.

In a generator having a cylindrical high potential electrode enclosed in a cylindrical tank the optimum ratio of electrode to tank diameter would thus appear to be $1/2.7$. Machines of this type for positive ion work have in general utilized inner electrode diameters greater than this optimum because of space requirements in the high potential electrode, and the shape of the curve of Fig. 5 indicates considerable latitude in choice of this ratio without great effect on maximum voltage.

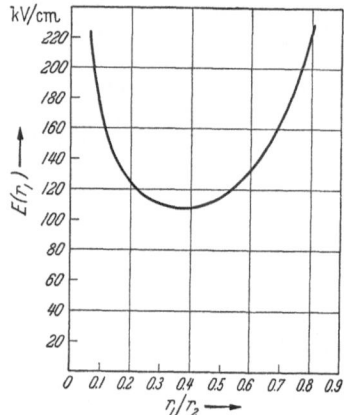

Fig. 5. Concentric cylinders of radii r_1 and r_2 with inner cylinder, radius r_1, at a potential of 4 MV. Value of r_2 fixed at 100 cm and value of r_1 varied. Values of the electric field at the surface of the inner cylinder are plotted as a function of r_1/r_2.

5. Multiple concentric cylindrical electrodes. Suppose a generator is to be built in a cylindrical tank six feet in diameter. A reasonable choice for the diameter of the high potential terminal might be three feet and the gradient at the cylindrical surface of the high potential electrode is then 1.45 times the average gradient. By subdivision of the gap by additional electrodes the maximum gradient can be brought closer to the average gradient. If in the six-foot tank we subdivide with one intermediate electrode the ratio of maximum to average field across the gap can be reduced from 1.45 to 1.2. By addition of a second intermediate electrode this ratio can be reduced to 1.14 but the 6% gain may not justify the added complexity. Radii $r_1, r_2, r_3, \ldots$ of successive shells should be chosen so that

$$\frac{r_1}{r_2} = \frac{r_2}{r_3} \cdots \tag{5.1}$$

in order to realize the best over-all ratio of maximum to average field.

If two co-axial cylindrical electrodes of radius r_1 and r_2 are capped by concentric hemispheres the field E_s at the surface of the inner sphere will be

$$E_s = \frac{U}{r_1} \frac{r_2}{(r_2 - r_1)} \tag{5.2}$$

and if $r_2/r_1 = 2$ this field weill be 1.38 times the field (E_c) over the surface of the inner cylinder. Ir $r_2/r_1 = 1.2$ then $E_s/E_c = 1.09$. Usually in practice to relieve the stress over the spherical surfaces the hemispherical caps are not made concentric. They are pushed apart to increase the gap over the major portion of the spherical surfaces.

In these cases the gradient will still be high at the junction between cylinder and sphere. Since sparking does not appear to be concentrated in these regions the problem does not appear to be of great practical importance.

6. The column enclosure. The term column will be used to denote the region where a uniform gradient is established by the use of metal hoops and a current drain for potential distribution.

To avoid accelerating tube failures column lengths must be made great compared to the distance across the high pressure gap. For best utilization of space provided for the accelerating tube it should be in a uniform gradient. The most ideal arrangement would appear to be a cylindrical enclosure of material with a suitable resistivity, to maintain, through leakage current, a uniform potential drop from the high potential electrode to ground. Frequently the equally attractive suggestion is made that an enclosing cylinder of insulating material be surfaced with a paint of the proper resistivity. Some exploration of these possibilities has been made but they do not look promising. Requirements on stability of resistivity over extreme ranges of gradients and under the intense localized power dissipated in a spark are very severe.

Metal shield rings properly spaced and graded in potential as tried with the first Wisconsin machine have proved satisfactory and have been used in all later machines.

To guide the choice of cross sectional diameter for the rings and ring spacing, accurate determination of the fields by plotting methods or by analysis would have been desirable so that these parameters could have been chosen to minimize fields. Because of lack of time and the urgent needs for other data which appeared to be more crucial accurate field determinations were not made. Rings for the Wisconsin 2.4-Mev machine had a cross sectional diameter of $\frac{3}{4}''$ and were spaced $1\frac{1}{4}''$ apart center to center.

These parameters were chosen after very approximate calculations of field strengths with generous allowance for inaccuracies in ring to ring spacing and in potential distribution, and with much stress on mechanical problems of fabrication and maintenance.

An analysis of the electric fields in a van de Graaff generator has been made by Boag[1]. He shows that to minimize fields at the surface of the shield rings the cross-sectional shape of the rings should be elliptic with the large radius portion of the ellipse subjected to the relatively high radial field. A generator with shield rings of elliptic cross section was designed and built by D. R. Chick and D. P. R. Petrie[2] (see Fig. 20).

7. Establishment of column gradient. Generators developed at Wisconsin have utilized corona current across point to plate gaps for distribution of potential down the column. The method is simple and has been widely used. Points are made negative and on some machines a number of points are operated in parallel. On many machines the point to plate separations can be adjusted from outside the generator so the current drain can be set to an optimum value at any generator voltage. In a machine built at Berkeley as an injector for the 32-Mev linear accelerator corona points and plates were enclosed in a textolite tube. Pressure in this tube could be controlled independently of pressure in the generator tank. The system functions as variable resistors. This innovation which is due to Dr. C. M. Turner did not work out satisfactorily because of mechanical difficulties. A recently constructed machine at Wisconsin utilizes corona gaps in a ceramic tube. Test work is not yet conclusive but the method shows considerable promise.

Each electrode of the current drain system is connected electrically to an electrode of the accelerating tube, to a ring around each support member, and to the hoop which surrounds the structure. All of these conductors should lie

[1] J. W. Boag: Proc. Inst. Electr. Engrs., Lond., Monograph No. 63, May, 1953 (**100**, Part IV, p. 63).

[2] D. R. Chick and D. P. R. Petrie: Proc. Inst. Electr. Engrs., Lond. **103**, Part B, No. 8 (1956).

in a plane perpendicular to the axis of the generator. Wires can be used to connect these electrodes providing the wires lie in the plane defined by the electrodes.

Ends of the column should be terminated by conducting planes with all openings such as those required for the belt having rounded edges. Conducting planes with rounded edges for all openings should also be provided where clamps short out part of the column. Otherwise field distortion can result in high field concentration at the edges or corners of a conductor such as a clamp.

Generators developed at M.I.T. have utilized resistors for current drain down the column. These generators have commonly operated at pressures too high for proper functioning of a corona gap. A large number of small resistors are connected in series across each subdivision of the column. The resistors are commonly arranged in a lucite cartridge. They are not adjustable and as voltage decreases current drain must decrease, but they have been dependable and the impossibility of adjustment has not normally proved to be a serious handicap.

C. Charging method.

8. Arrangement of belts. The charging belt as originally devised by VAN DE GRAAFF is still almost universally used. Belts in common use are of woven cotton with or without impregnation with materials such as rubber.

A charged belt which is relatively distant from surrounding objects is a good approximation, electrically, to an isolated plane sheet of charge where the field extending equally in both directions due to a charge density σ is $2\pi\sigma$. If we assume that a field of 20 kv/cm perpendicular to the belt surface is permissible we have for the charge density on the belt: $\sigma \approx 10 \frac{\text{esu}}{\text{cm}^2} = 3.3 \times 10^{-9}$ coulomb/cm².

Assume that a belt 20″ wide is driven by a pulley 5″ in diameter at 60 r.p.s. Surface displacement per sec is then 1.2×10^5 cm²/sec and a charging current of approximately 400 microamperes is obtained. If the down run of the belt carries an equal negative charge the total charging current is 800 microamperes. This is approximately what can be easily realized in high pressure gas.

A maximum charge density of 3.3×10^{-9} coulomb/cm² appears very low when it is observed that the field is then only 20 kv/cm while the belt is in high pressure gas which should have a dielectric strength ten to twenty times higher. At least part of the discrepancy can be explained as arising from factors which give instantaneous local fields far above the average. Some of these factors are: (1) belt roughness, (2) edge effects, (3) uneven charge distribution, and (4) improper positioning of the belt between conducting surfaces or between a conducting surface and an adjacent belt surface which carries a charge, (5) vectorial addition of the gradient from the belt and the gradient along the column.

The support system which extends from a ground plane to the high potential electrode is arranged to give a field which is as uniform as possible inside the column, and field lines which as closely as possible are parallel to the axis of the column. A charged belt surface inside the column distorts the field. Difficulties may develop if measures are not taken to minimize field distortion. In the schematic drawings of Fig. 6 the nature of these field distortions is illustrated. Fig. 6a shows a column containing a single charged belt surface. The equipotential surface S which without the charged belt would be in the plane of ring R is distorted to the shape illustrated.

If we assume that $L = 30$ cm and that the field from the surface of the belt is 20 kv/cm then the potential of point P on the belt, in the plane of ring R, is 600 kv higher than the potential of ring R.

In Fig. 6b conducting bars B connected to rings R are placed close to the belt so that the belt runs through a comparatively narrow slot. The equipotential surface S is now distorted from the plane of ring R only in the region between the bars B. The field at the surface of the belt is to a first approximation the same as in the arrangement of Fig. 6a. Actual fields at the belt surface will probably be higher in the second arrangement, first because of field concentrations

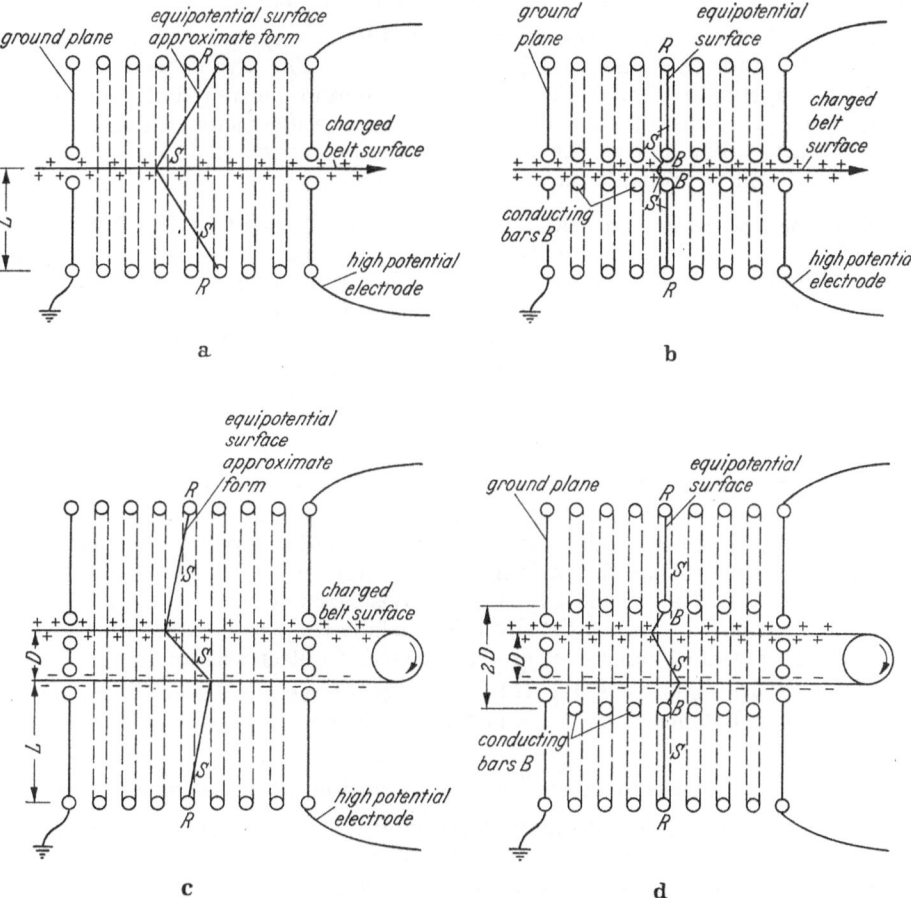

Fig. 6 a—d. Showing the form of equipotential surfaces within the shielding rings for several different belt arrangements.

introduced by the bars and, second because centering problems are in this case more severe. Potential differences between a point on the belt and neighboring bars will be small since distances are small. The successful utilization of this arangement is probably good evidence that maximum gradients which can be maintained in gases are higher across narrow gaps than across wide gaps. This arrangement lends itself readily to utilization of both runs of a belt. Each run is enclosed by spacer bars and they are completely independent.

In several machines the arrangement of Fig. 6c has been utilized. Distortion of equipotential surfaces is small if incoming charge density is equal to outgoing charge density of the opposite sign. The equipotential surface S passes through the plane of ring R midway between the runs of the belt.

If L is large compared to $D/2$ the field from the belt is mostly toward the opposite run and in magnitude it approaches $4\pi\sigma$.

Another variation is shown in Fig. 6d. Bars B are placed so as to equalize fields on both sides of each run when outgoing charge density is equal and opposite to ingoing density. Here the equipotential surfaces are not greatly disturbed and the field from the belt is $2\pi\sigma$. The arrangement is satisfactory electrically and mechanically and it offers convenience for belt replacement.

For low charging current the arrangement of Fig. 6c appears to be satisfactory. It is simple and is convenient for belt replacement. The arrangement of Fig. 6d should be superior to that of Fig. 6c for high charging current but comparative performance data are difficult to obtain. In this arrangement also belt replacement is simple. For high charging currents the arrangement of Fig. 6b is commonly used and is probably superior to the others.

9. Mechanical forces on belts. In the belt arrangement of Fig. 6c the two runs of the belt will be attracted toward each other like the two plates of a condenser with a force $f = 2\pi\sigma^2$ dynes per cm². The distance L of Fig. 6 is assumed to be large compared to $D/2$. Taking $\sigma = 10$ esu/cm² gives $f \approx 600$ dynes/cm² $\approx$ 1.2 lb per ft². Each run of the belt takes the form of a catenary (neglecting effects due to motion) and the deflection midway between the pulleys is given by

$$x = \frac{f l^2}{8 T}, \tag{9.1}$$

neglecting gravitational forces, where l is center to center distance between pulleys and T is belt tension. With a 10-foot belt run under a tension of 150 lbs per foot of width we have $x = 1''$ which may be troublesome for a belt separation of 4''. In most machines using this arrangement charge densities are usually below 3 esu per cm². Belting commonly used at Wisconsin weighs 0.2 lb per ft² and the machine is horizontal. Ordinarily this machine is operated with low charging currents and gravitational forces are comparable to electrical forces.

Belt runs arranged as in Fig. 6d experience no electrical force if charge on the two runs is balanced and if the belt runs are accurately located as shown in the figure.

If a belt run moves a distance x off the equilibrium points, a force $f = 4\pi\sigma^2 \dfrac{x}{D}$ dynes/cm² tends to cause it to move farther off. Forces are smaller than in the arrangement of Fig. 6c by the factor x/D and good belt stability should be expected for ordinary pulley diameters and belt lengths.

In a narrow slot as shown in Fig. 6b the same formula would apply but D is now very small and belt stability cannot be expected. In machines built in recent years at M.I.T. and by the High Voltage Engineering Company belts are confined to narrow slots and guide rods of insulating material must be used to keep the belt centered

10. Charging and discharging the belt. Arrangements currently used for charging and discharging belts do not differ greatly from methods originally proposed by van de Graaff. Design details of charging mechanisms are not critical for operation in air at atmospheric pressure. As gas pressure increases, problems become more difficult. The arrangement used at Wisconsin as shown in Fig. 7 works satisfactorily in air at pressures up to 100 lb/in². It also works if a partial pressure of 3 lb/in² of freon (CCl_2F_2) is added to the high pressure air. Under this condition the voltage required for needles is between 40 and 50 kv.

As the needles are shortened and made blunter and as the needle bar is made larger in diameter, the field at the needle tips decreases. If the needle array of Fig. 7 is moved back to where the belt is in contact with the pulley, the spray voltage required decreases but there may be difficulty with needle sparking if an unimpregnated belt is used. If the belt is well impregnated with rubber or a similar compound, charge may be sprayed onto the belt where it is in contact with the pulley.

The problem of charge removal from the belt is somewhat more difficult than the problem of charging. As a charged belt passes over a conducting pulley little charge is removed. Corona current is required. Suppose the high potential terminal (Fig. 8a) is at a potential V_s of $+1$ MV. As a point p on the belt moves into the enclosure its potential will rise above 1 MV by an amount $2\pi\sigma d$ where d is a constant which depends on the distance of point p from conductors or other surfaces at the terminal potential. If the pulley is connected to the terminal and is at position a the belt surface cannot rise high above terminal potential because it is never far from surfaces at terminal potential. Little charge can be removed. Position b is much better. The row of needle

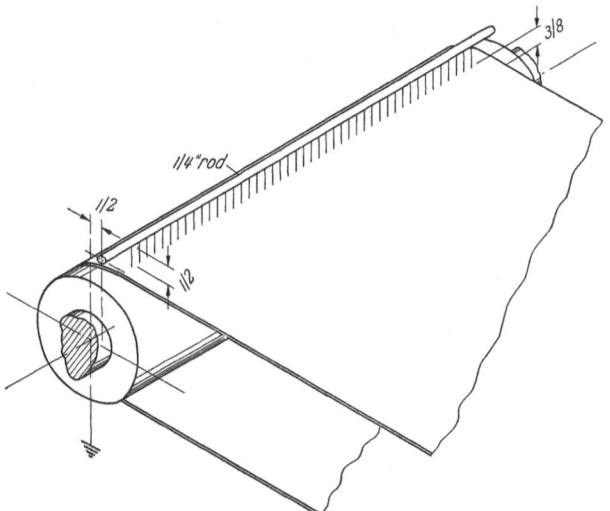

Fig. 7. Showing a charging system that has been successfully used in air at pressures up to about 8 atmospheres, and in air at 8 atmospheres with moderate amounts of CCl$_2$F$_2$ added. Needles commonly used are No. 9 sharp. With unimpregnated cotton belts the needles should be located about one-half inch forward from the position where the belt leaves the pulley. A distance of three-eighths inch from needle point to belt is satisfactory.

points should be approximately midway between the front plane of the terminal and the pulley. Relatively long needles and a needle support bar of modest proportions are desirable.

If a charge of opposite sign is desired on the down run of the belt the induction system of Fig. 8b can be used. The plane midway between the two runs of the belt is at terminal potential if ingoing and outgoing charge densities are equal. This situation depresses the potential on the incoming belt surface. The pulley should be back from the front plane of the terminal by one and one-half to two pulley diameters.

The pulley is now used as an inductor. If no charge is drained from the pulley it will rise in potential until corona current to needle array I ceases. Negative charge will be sprayed from needles S to neutralize positive charge and leave excess negative charge. In generators constructed at Wisconsin the geometry has been such that the negative charge on the outgoing run of the belt is considerably greater in magnitude than incoming positive charge unless some charge is drained from the pulley. A needle array C, controllable in position from outside the machine serves to drain charge from the pulley and by suitable adjustment of the distance of the needle points C from a plate connected to the pulley the outgoing negative charge can be made equal in magnitude to the incoming positive charge.

Belt charging by means of corona current from needle points becomes difficult in high pressure gas. POLLOCK and COOPER[1] showed that in air the voltage for corona onset from a positive needle point becomes equal to the spark-over voltage at a pressure of 13 atmospheres. At higher pressures, according to these results, no corona current is obtainable. These limitations were not observed for negative points.

TRUMP's group at M.I.T. have successfully operated machines with gas mixtures of 80% N_2 and 20% CO_2 at pressures up to 400 lb/in². To charge and discharge the belt they have used fine metal gauze with the sheared edge lightly rubbing against the belt surface. Sharp needles closely spaced with their tips very close to the belt have also been used by the group at M.I.T. and by the High Voltage Engineering Company.

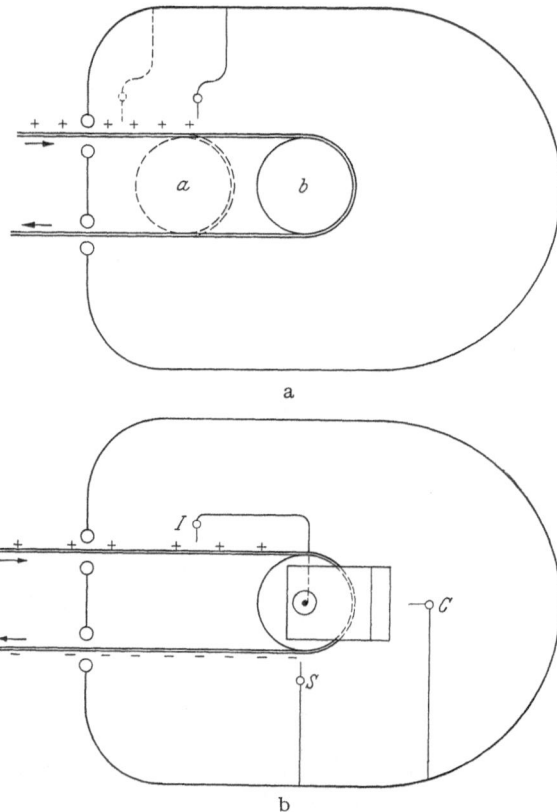

a

b

Fig. 8a and b. (a) Showing location of pulley in high potential electrode. With pulley in position b charge removal is more easily accomplished than with pulley in position a. (b) An induction system that works satisfactorily to remove positive charge from incoming run of belt and to provide negative charge on outgoing run of belt.

11. Other charging methods. At M.I.T. and at the High Voltage Engineering Company a method called "Conduction, Induction Charging" has been used successfully. A thin inner layer of the belt is made of material of suitable resistivity. Field from an inducing electrode causes charge to flow to the outer surface of this layer. As it moves through the column there is some current flow along the belt due to the column gradient. This can be held to a negligible value by a suitable choice of the resistivity of the layer.

At Wisconsin the belt of a small test generator was equipped with small metallic conductors which were charged and discharged by an induction plate as the conductors passed over the belt pulley (Fig. 9). The conductors are ordinary paper staples with the relatively uneven crimped ends on the outer surface of the belt. The method proved very satisfactory for a test generator where pressures up to 250 lb/in² were used and where tests were run with gases of high dielectric strength. Corona from the needle point system failed at the higher pressures. Charging currents with the paper staple conductors were independent of pressure and type of gas for a given voltage on the inductor plate. As pressure was increased the charge carrying capacity of the belt increased and the inductor plate could be operated at higher voltage.

---

[1] H. C. POLLOCK and F. S. COOPER: Phys. Rev. **56**, 170 (1939).

Staples were used with a variety of cotton belts, some impregnated with rubber and others untreated. With a belt 4″ wide running at a speed of 30 feet per sec the maximum charging current obtained was 50 microamperes. Ingoing and outgoing charge densities were balanced and outgoing negative charge was supplied by an inductor system. The density of staples in this case was approximately twelve per square inch. A readily obtainable charging current with most of the belts was about half the value given above.

A large generator recently constructed at Wisconsin for nuclear physics work is equipped with a stapled belt and performance is satisfactory. Belt charging

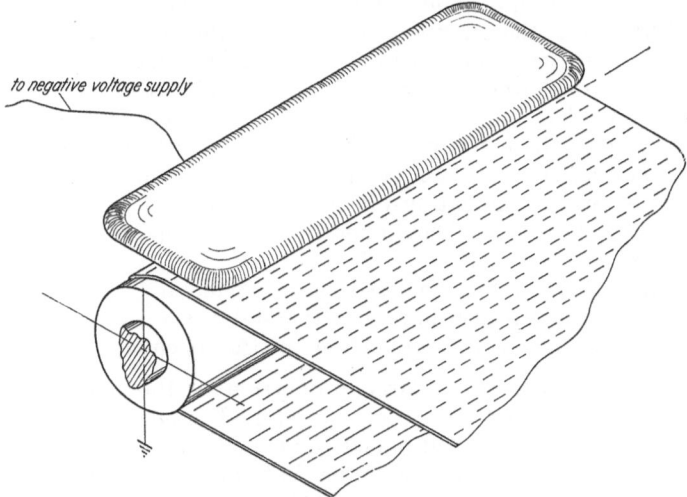

to negative voltage supply

Fig. 9. Belt equipped with paper stables charged by induction. Paper staples when crimped are one-half inch long. Densities from three to twelve per square inch of belt surface have been used. The crimped ends of the staples face outward so that the long bases of the staples touch the pulley. The induction plate should be made of metal approximately three-eighths inch thick with edges and corners smoothly rounded. The surface of this plate is commonly three-eighths inch from the belt surface. To minimize noise staples should be arranged in a complex pattern.

and charge removal are more easily controlled under a variety of conditions than in the case of corona charging. The High Voltage Engineering Company has adopted this method for machines where voltage stability is of importance. They have obtained good operating life but have not as yet realized the high charging currents with the stapled belts which they can achieve with their normal charging system.

A stapled belt in which staples have a simple pattern with many staples in each row across the belt makes a loud, high-pitched noise when it is operated over steel pulleys. In starting and stopping, the noise resembles a siren. A belt with a complex pattern such as shown in Fig. 9 makes much less noise. For recent stapled belts at Wisconsin Bostich type SBD-19¼″ staples have been used. These have sharp ends and appear to damage the belt fabric less during attachment than the blunt end type. About 6 staples are commonly used per square inch of belt surface.

D. Support structures.

12. Vertical machines. Problems of structural support are relatively straightforward in a vertical machine. A variety of insulating materials have been successfully employed. Disks of lucite, glass, or porcelain are commonly used between flat metal plates. The structure if properly proportioned can be stacked with no bonding and can have satisfactory stability. Good quality porcelain has a dielectric strength of about 300 kv/inch which should be adequate for column gradients now used. TRUMP's group at M.I.T. have had difficulty with puncture through porcelain after prolonged operation at high gradients. Violent surges during spark-over may have been responsible for the puncture of porcelain.

Some glasses and lucite are considerably superior to porcelain in dielectric strength. Lucite insulators have withstood up to one million volts per inch.

Most commonly, insulating supports fail due to flash-over along the insulator surface. Lucite appears to be less subject to surface flash-over then any material commonly used. Insulators of porcelain and glass must be carefully designed and installed to avoid this limitation. Most commonly the discharge appears to be initiated at the edge of the ceramic where it meets the metal. Here any roughness leaves voids and in these voids fields may be higher than average fields in the insulator by a factor G where G is the dielectric constant of the ceramic.

TRUMP's group has largely overcome this limitation by use of cements to bond the ceramic to the metal plate. The cement flows out beyond the edge of the ceramic to form a smooth rounded fillet. TRUMP's group found that corrugated surfaces are helpful in eliminating effects due to imperfections in the junction between metal and porcelain. They found that where imperfections are eliminated corrugations are not necessary.

The bonding of insulators to the separating metal plates with cements of good mechanical strength also improves the stability of a vertical column and permits greater freedom in the proportions of the machine.

13. Horizontal machines. A large proportion of the horizontal generators constructed prior to about 1953 are supported by Textolite tubing which consists of paper bonded with shellac and baked. This material is not ordinarily charred by sparks. It may be punctured or frayed but will continue to hold voltage. It has good mechanical strength and is machinable but generally the bonding is not good enough for satisfactory threading. It must be clamped to other structural members. A number of high strength plastics bonded with synthetic resin have excellent electrical properties but fail after prolonged operation in a generator because they are charred by sparks and become conducting.

The shellac bonded Textolite loses its excellent insulating properties if it is exposed to high humidity air for a prolonged period. At room temperature it dries slowly and according to some reports a drying period of a month or more may be required if it has been well saturated with water.

A number of machines have been built in which four Textolite tubes are clamped into a structure as shown schematically in Fig. 10a. Some helpful information in regard to the mechanical stress problem may be gained from the diagrams of Figs. 10b—c, and 10d where only two structural members are shown and where the load is assumed to be concentrated at the end of the structure.

The clamps and the metal cross members connecting these clamps are assumed to be very rigid in comparison with the Textolite tubing. Elongation of the upper Textolite members and compression of the lower members are neglected.

Further subdivision of the structure by clamps and cross members beyond that shown in Fig. 10d may not be profitable. Each such clamping framework is heavy and adds to the tension T. In the usual 4-Mev machine the clamps and connecting plates must be 4 to 6 inches wide to give sufficient rigidity. No voltage gradient can be imposed over this region and thus the effective column length is reduced.

14. Electrical subdivision of long insulators. Support structures utilizing Textolite tubing have given dependable performance at gradients of 300 kv per foot and in at least one generator 500 kv per foot has been held without trouble. A potential distribution system as shown in Fig. 11 is employed. The metal rings must be accurately made. They should clamp firmly leaving no gaps between metal and Textolite. They should be smooth and junctions J should meet accurately. Poorly made rings can easily lower maximum voltage by a factor of two over that obtainable from a well-designed machine.

Textolite tubing in several machines has been equipped with an inner clamping ring in the same plane as each outer ring. Inner and outer rings must be electrically connected. This method is excellent but is relatively difficult and expensive. It is effective only if very well made. If poorly made, it might easily be less effective than a better made system of rings on the outer surface only.

Where a support tube enters a clamp the end of the clamp should be rounded with a radius comparable to the cross-sectional radius of a ring.

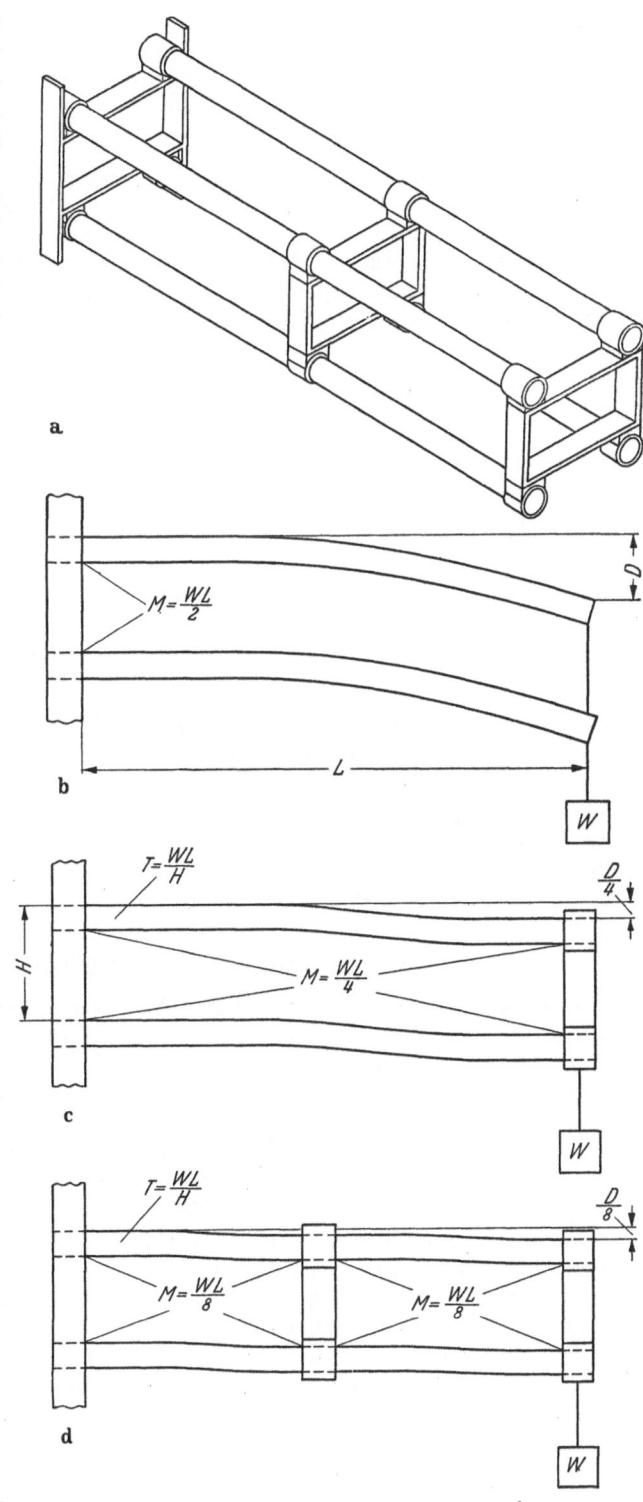

Fig. 10 a—d. In Fig. 10a a horizontal support structure is shown in which four insulating tubes are held by a system of metal clamps. In Figs. 10 b—d the function of the clamping system is illustrated. M is bending moment. W represents the load which in this idealized case is assumed to be concentrated at the end of the structure. T is tension and D is deflection. Clamps rigidly connected between upper and lower members serve to reduce deflection and stress in the insulating tubes.

Textolite support tubes in several machines have broken off where they enter metal clamps. The bending moment is at a maximum in these regions. If the clamp presses the Textolite very tightly at the end of the clamp stress is very heavily concentrated at this position. The clamp should be tapered so that radial pressure on the Textolite increases gradually with distance into the clamp.

15. Horizontal machines with ceramic insulators. In recent years a horizontal machine has been built at Wisconsin and others have been built by the High Voltage Engineering Company utilizing ceramic insulators. The Wisconsin machine utilizes porcelain cylinders bonded to metal by silver solder (eutectic mixture of silver and copper). Titanium hydride is applied to the porcelain surface to make the solder adhere.

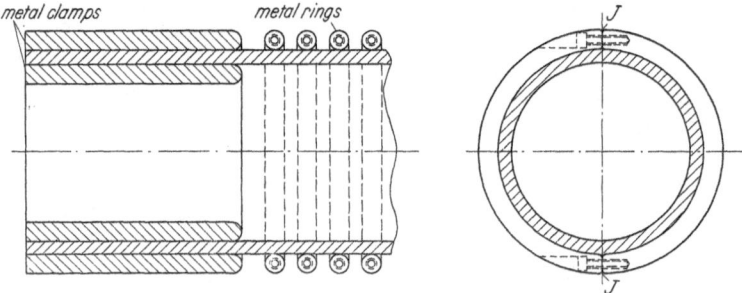

Fig. 11. Showing rings for distribution of potential along an insulating tube.

The structure is similar to that of Fig. 10d but more extensive subdivision was used to reduce bending moments.

Horizontal machines built by the High Voltage Engineering Company utilize glass insulators bonded to metal plates by high strength cements. Bending moments are taken up approximately equally by the metal plates which separate each set of four glass insulators from neighboring sets and bending moment stresses are thus made relatively small.

E. The accelerating tube.

16. Early tubes. Electrical discharge between metal electrodes in a vacuum or along the walls of the enclosing tube is a problem of long standing. Insulation of more than a few hundred kilovolts across a single vacuum gap is very difficult and in early work with canal rays and with X-rays subdivision of the evacuated tube was found to be helpful[1]. Experience gained over the past decades at many laboratories has not explained the nature of the discharge but certain empirical information has accumulated which serves to guide tube development. Tube performance has thereby slowly improved.

Factors which appear to be of importance in tube design can be most easily explained through a description of tubes with known performance characteristics. Fig. 12 shows an accelerating tube constructed by J. L. McKibben in 1938 and Fig. 13 shows a tube constructed by C. M. Turner and R. G. Herb which was completed in 1940. The tube of Fig. 12 was 5 ft long and held 2.5 MV satisfactorily. At the time of this writing it is still in operation after 18 years of use. No parts have been replaced. The tube of Fig. 13 holds up to 4.5 MV and has operated

[1] W. Wien: Ann. der Phys. **8**, 244 (1902). — F. Hoffmann: Ann. der Phys. **77**, 302 (1925). — W. D. Coolidge: J. Franklin Inst. **202**, 693 (1926).

for 16 years. The porcelain cylinder in the plane of a defective joint in a support tube has punctured three times. There has been no other deterioration.

Considerations entering into the design of these tubes are as follows:

1. The inner surfaces of the ceramic insulators must be shielded from the beam. An insulating surface might flash over if it were subjected to bombardment by scattered or defocussed ions or electrons. Conversely the beam must be shielded from insulating surfaces. These surfaces may become asymmetrically charged by ion or electron bombardment or by charge creepage. The ion beam might then be deflected off axis.

Ion beams of many generators move with time and with generator voltage with respect to the tube axis. This behavior is very troublesome when a target of small area is being studied and when accurate measurements are being attempted.

For these reasons the metal electrodes are telescoping.

2. The inner edge of the ceramic cylinder where it rests against the metal disk was recognized in much earlier tubes as a source of trouble.

Small voids where metal and ceramic meet will give rise to very high fields. Electrode surfaces are formed so that fields in these critical regions are reduced.

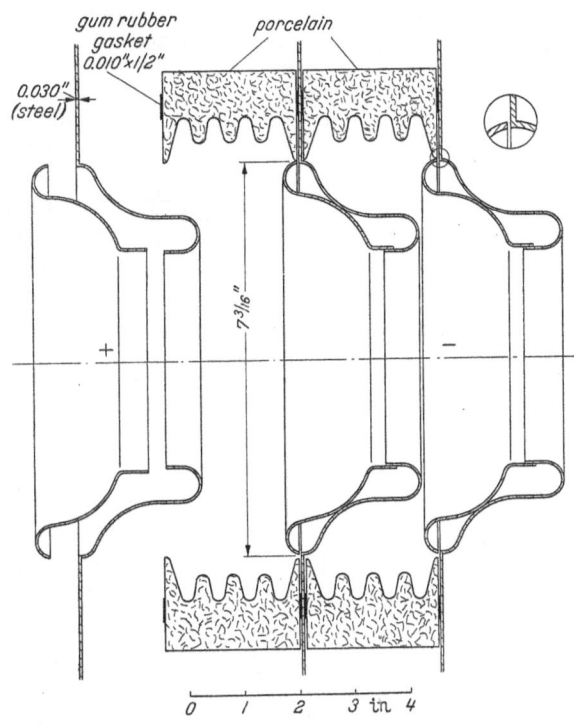

Fig. 12. Showing construction of accelerating tube utilized for 2.8-MV generator constructed by McKibben. The tube is five feet long. Generator constructed at Wisconsin and used since 1943 at the Los Alamos Scientific Laboratory.

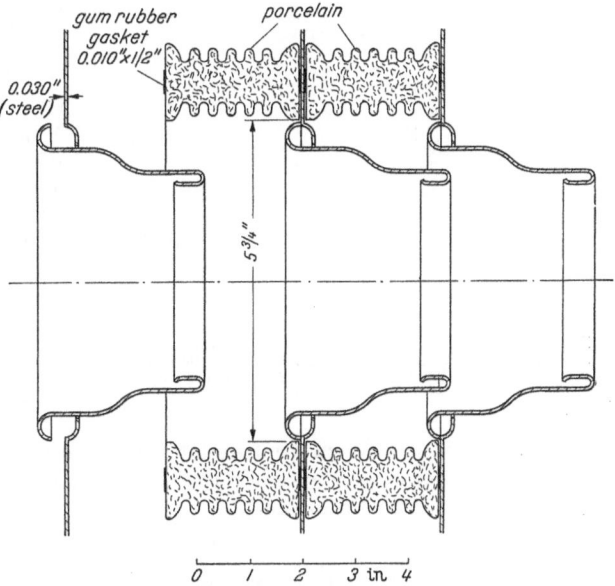

Fig. 13. Showing construction of accelerating tube utilized for the 4.5-MV generator of Fig. 19.

6*

3. The ceramic cylinders are corrugated to reduce the danger of failure due to discharge along the ceramic surfaces.

17. Recent tubes. Many modern tubes utilize straight-walled ceramic cylinders. The use of cements to bond metal to ceramic with a rounded fillet of cement extending out from the edge, has apparently eliminated the requirement for corrugations. This technique which was developed by TRUMP and his group at M.I.T. has also eliminated the need for the doughnut-shaped metal surfaces extending out from the metal-to-ceramic junction.

Many tubes have been constructed with the simple geometry of Fig. 14a. Cement bonds metal to ceramic and a fillet at the edges of the joints eliminates the need for corrugated ceramic or doughnut-shaped metal shields.

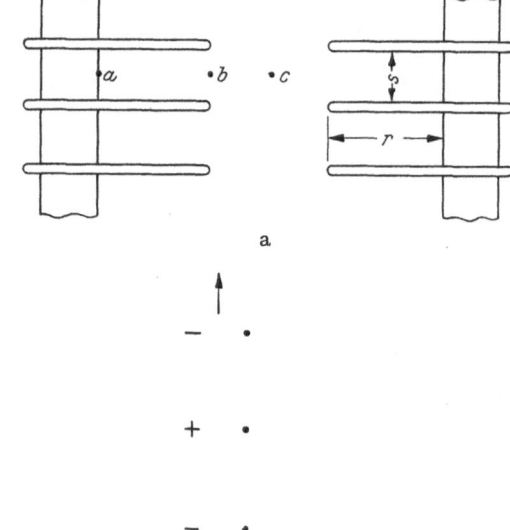

The ion beam will be adequately shielded from asymmetric charges on the porcelain wall if the value of r/s is sufficiently great. Approximate evaluations of the shielding provided by flat disks were made by I. MICHAEL and J. L. POWELL (private communication). They assumed that the geometry of Fig. 14b would provide a useful approximation to the problem. To obtain the potential at point b due to a unit charge at point a where a and b are midway between two grounded conducting planes separated by unit distance they computed contributions due to images in the conducting planes and obtained a series

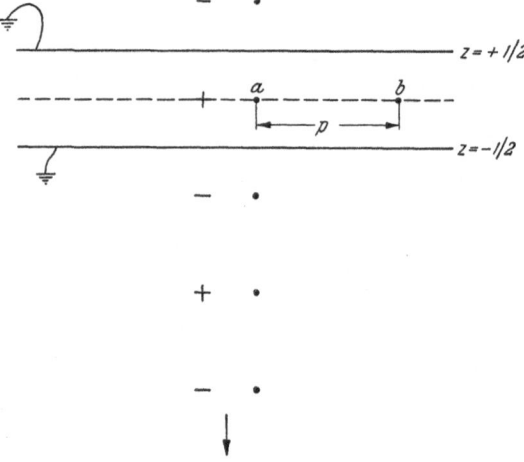

Fig. 14a and b. (a) Showing schematically construction of accelerating tube with flat disk electrodes. (b) Image method to compute field due to charge at position a.

$$\varphi = \frac{1}{\varrho} - 2\left\{\frac{1}{\sqrt{\varrho^2+1}} - \frac{1}{\sqrt{\varrho^2+2^2}} + \frac{1}{\sqrt{\varrho^2+3^2}} + \cdots\right\}. \tag{17.1}$$

This series was differentiated and terms were computed and added for several values of ϱ. For large values of ϱ the series converges rather slowly. J. L. POWELL then used the transformation

$$\frac{1}{\varrho} = \int_0^\infty e^{-\varrho t}\, dt \quad \text{and} \quad \frac{1}{\sqrt{\varrho^2+a^2}} = \int_0^\infty e^{-\varrho t} J_0(at)\, dt \tag{17.2}$$

to obtain the series

$$\varphi = 4 \{K_0(\pi \varrho) + K_0(3 \pi \varrho) \ldots\} \tag{17.3}$$

and

$$E = \frac{-d\varphi}{d\varrho} = 4\pi \{K_1(\pi \varrho) + 3 K_1(3 \pi \varrho) \ldots\} \tag{17.4}$$

where K_i is a Bessel function of the second kind[1]. For $\varrho \gg 1$ the first term is dominant and its asymptotic value gives

$$E \approx \frac{4\pi}{\sqrt{2\varrho}} e^{-\pi \varrho}. \tag{17.5}$$

In applying this result to tube sections with flat disk electrodes as shown in Fig. 14a the finite size of the disks, the presence of porcelain and effects of the tube aperture are neglected. The field at b due to unit charge at a is then

$$E = \frac{k}{r^2}$$

where

$$k = \frac{4\pi}{\sqrt{2}} \left(\frac{r}{s}\right)^{\frac{3}{2}} e^{-\frac{\pi r}{s}} \tag{17.6}$$

is the shielding factor. The additional approximation is then made that this shielding factor holds for the field at position c due to a charge at a. From the curve of Fig. 15 an r/s value of 1 appears to be relatively ineffective. A value of two on the other hand gives considerable shielding.

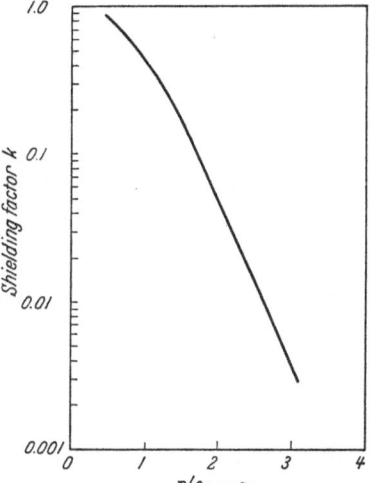

Fig. 15. Shielding provided by flat metal plates.

How far an ion is moved off axis due to a charge on the porcelain will depend on the total radial impulse I_r that is felt as the ion passes down the tube. The path length over which it is exposed to the coulomb field due to a charge at a will be decreased by the flat disk electrodes. As an ion passes through point a the force should rapidly fall to zero. Thus we should expect I_r to be decreased more by the electrodes than as given by the shielding constant k.

Shielding is most critical near the ion source where ions are moving slowly. A tube recently constructed at Wisconsin has an r/s value of 2.1 near the ion source. Beam movements with respect to the axis of the tube are small. A ratio of 2.5 or 3 should be completely adequate.

The insulating sections of most tubes built in recent years are shorter than those of Figs. 12 and 13. Insulator lengths of $\frac{1}{2}$ to 1 inch have been commonly used. As insulator lengths are decreased the shielding problem is made easier but the tube components become more expensive and fabrication work increases. There appears to be no convincing evidence for improvement in the voltage a tube will withstand as the length of insulating cylinders is decreased below 1 inch.

18. Tube materials. Porcelain, glass and Mykroy (mica flakes bonded with glass) have been used for insulators. Porcelain has performed well in a great many machines, with gradients up to 0.5 MV per foot. McKibben has successfully used Mykroy insulators for an accelerating tube. Trump's group at M.I.T. have found glass to be preferable and most of the tubes built by the High Voltage Engineering Company have employed glass.

---

[1] For definition see either J. Meixner's contribution to Vol. I of this Encyclopedia, or G. N. Watson: A treatise on the theory of Bessel functions. Cambridge 1944.

The severity of damage to accelerating tubes due to voltage surges can be greatly lessened by the proper employment of sphere gaps distributed around the accelerating tube. The M.I.T. group have found that three or four sets of sphere gaps are needed for each tube section.

The metal electrodes of Figs. 12 and 13 are made of mild steel. They were buffed to a smooth finish. McKIBBEN and TRUMP in recent years have obtained best results with aluminum.

19. Nature of the discharge in accelerating tubes. Two general types of failure have been observed. One type is characterized by violent discharges which in some cases may drop generator voltage to zero, with little tube activity and low X-ray background for periods in between discharges. The other type is characterized by an apparently steady current through the tube. In a positive voltage accelerator some positive ions are observed to come to the ground end when discharge is taking place. They are distributed throughout the tube aperture. Electrons pass up the tube and generate intense X-radiation. This type of discharge is referred to as electron loading. It frequently sets in at a rather sharply defined voltage and increases very rapidly in intensity as voltage is increased. A tube with an electron loading threshold at 2 MV might have an electron loading tube current of several hundred microamperes at 2.5 MV.

20. Modes of discharge. α) J. G. TRUMP and R. J. VAN DE GRAAFF[1] have investigated the possibility of a discharge building up through the mechanism of (a) positive ion impact toward the cathode end of the tube producing electrons which are accelerated toward the anode end; (b) electron impact toward the anode end of the tube producing positive ions which travel toward the cathode.

These multiplicative processes do not appear to be adequate to build up a discharge. The positive ion yield due to electron bombardment was found to be very low.

β) J. L. McKIBBEN[2] has assumed that negative ions play a prominent role. He has found evidence that some of the negatively charged particles are negative ions and has assumed that many are hydrogen negative ions.

γ) JOHN BLEWETT[3] has assumed that the Malter effect is responsible for at least part of the electron current. In this effect a thin insulating film becomes positively charged with respect to the underlying metal, pulls relatively large bursts of electrons from the metal, and ejects them into the evacuated space.

δ) C. M. TURNER[4] has investigated electron photon multiplicative processes and has shown that they can be of importance.

ε) McKIBBEN has assumed that surges can generate electrical oscillations between electrodes so that instantaneous electrode to electrode voltages can be very high.

ζ) W. DYKE has proposed that with gradients at which tube breakdown occurs field emission may be important. He has shown that points can build up on a metal surface to where a disruptive discharge occurs.

η) L. CRANBERG[5] has found that charged particles torn from electrodes at one end of a tube can become sufficiently energetic to heat and melt and even vaporize a small region of an electrode at the opposite end of the tube. Droplets formed this way can be accelerated to where they in turn can heat small regions where they hit.

[1] J. G. TRUMP and R. J. VAN DE GRAAFF: J. Appl. Phys. **18**, 327 (1947).
[2] J. L. McKIBBEN and K. BOYER: Phys. Rev. **82**, 315A (1951).
[3] J. P. BLEWETT: Phys. Rev. **81**, 305A (1951).
[4] C. M. TURNER: Bull. Amer. Phys. Soc. **1**, 134 (1956).
[5] L. CRANBERG: J. Appl. Phys. **23**, 518 (1952).

21. Performance characteristics. A few of the more striking characteristics of tube behavior observed experimentally are as follows:

α) *The long tube effect.* Tests on single tube sections frequently yield surprisingly good results. A tube section 1″ long may hold one hundred kilovolts or

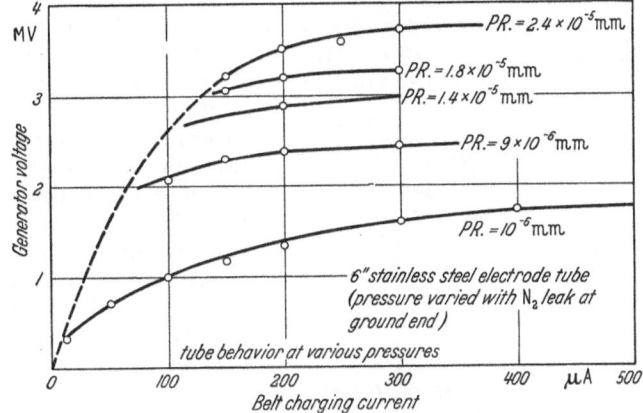

Fig. 16. Electron loading currents obtained in the tube of the generator which serves as the injector for the Brookhaven Cosmotron (unpublished data of C. M. TURNER).

more. When a number n of these sections, all identical to the one tested, are built up to form a tube the total voltage they will withstand is usually far below n times the voltage held by a single section. FORTESCUE[1] made extensive meas-

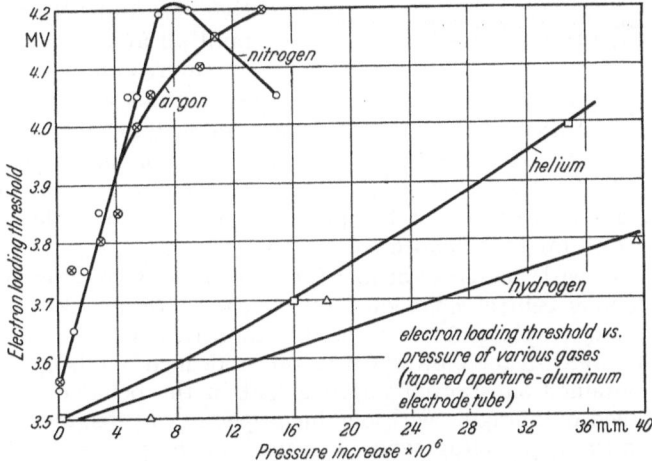

Fig. 17. Electron loading thresholds in the tube of the generator which serves as the injector for the Brookhaven Cosmotron (unpublished data of C. M. TURNER).

urements of maximum voltage held as a function of tube length. In certain extreme cases the total voltage went to a maximum and then decreased as tube length was further increased. CRANBERG examined experimental data available in the literature and found that the voltage a tube will withstand is approximately proportional to the square root of the tube length.

β) *Pressure effect.* The performance of many tubes that exhibit electron loading is improved as tube pressure is increased. Addition of any gas increases thresh-

---

[1] R. L. FORTESCUE: Progr. in Nucl. Phys. **1**, 21 (1950).

old voltage and at a given voltage above threshold X-ray production decreases as gases are added. A typical tube with its vacuum system will pump down to a pressure of 1 to 5×10^{-6} mm Hg. Addition of air to where the pressure is 2×10^{-5} mm Hg will substantially decrease electron loading and the ion beam is not appreciably spread out by scattering. Further increase of pressure gives further voltage improvement but as the pressure approaches 1×10^{-4} mm Hg the beam becomes spread out and with further pressure increase an ordinary gaseous discharge sets in which drops the voltage to a low value. Fig. 16 shows electron loading results obtained by C. M. Turner as he varied voltage and tube pressure with the generator which serves as the injector for the Brookhaven cosmotron. Figs. 17 and 18 show results of further tests where the electron loading threshold was studied. The threshold voltage appears to be a simple function of the mass of gas introduced into the accelerating tube. Turner suggests that the gas may influence electron loading by suppression (through some unknown mechanism) of thin film field emission.

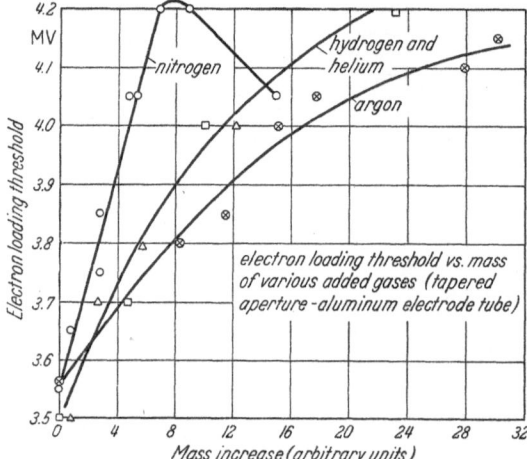

Fig. 18. Electron loading thresholds in the tube of the Cosmotron injector (unpublished data of C. M. Turner).

McKibben has suggested that addition of gas may destroy negative ions and help to destroy the multiplicative chain in which these ions were assumed to play a role.

γ) *Tube aperture.* The comparative performance of several tubes of similar structure strongly indicates that as tube aperture increases discharge difficulties become more severe. This might be expected for any of the suggested discharge mechanisms. As tube aperture is made smaller pumping speed decreases and difficulties might be expected if the pumping speed were made very low. An optimum aperture might be expected.

Several tubes with apertures of four inches or more were later equipped with diaphragms having central holes of 1 to 2 inches. Results were disappointing when only a few diaphragms were used. An occasional constriction did not appear to interfere with the multiplicative process which builds up to a discharge. McKibben[1] obtained substantial improvement in the behavior of a 20 ft tube when he installed diaphragms with one inch openings at distances of 2, 4, 8, 12 and 16 ft from the high voltage end. The field on the positive side of each diaphragm was reversed to prevent the escape of negative particles.

δ) *Contaminants in tube.* Tubes in many machines improve with use and deteriorate when they are not used for a prolonged period. On occasion a generator has been shut down for a week end and generator cooling water was not cut off. The tube was thus cooler than other parts of the vacuum system for a prolonged period. Tube behavior was then much below normal for a considerable period.

Vapors from pumps, from gaskets and from cements appear to play a part in tube failure. A generator has been constructed at Wisconsin in which all vapors are eliminated from the tube. An Evaporion pump is used; all organic

[1] J. L. McKibben: Bull. Amer. Phys. Soc. 1, 61 (1956).

materials are eliminated and the complete system can be baked out. Initial results have been disappointing and electrode modifications are now being tried.

22. Suggestions for possible improvement. Frequently the utilization of magnetic fields is suggested with the expectation that ions and especially electrons taking part in the multiplicative process might be deflected and captured by electrodes before gaining high energy. TURNER, FORTESCUE, and High Voltage Engineering personnel have performed a variety of experiments with magnetic fields. Results have in general been disappointing.

C. M. TURNER's suggestion of eleetron-photon multiplication and his calculations of photon yields show that electrodes of material with low atomic number might be helpful.

A tube with very small central apertures might be equipped with larger pumping holes toward the outer part of the electrodes. These holes could be arranged so they do not line up. Multiplicative processes might thus be largely suppressed. A tube of this design is now under construction at Wisconsin.

F. The ion source and associated equipment.

23. Ion source development[1]. Power and space available for an ion source are somewhat limited in an electrostatic generator and relatively few controls should be used. Long periods of trouble free operation are very important since repairs are difficult to make. Low gas consumption is desirable. The accelerating tube is subjected to a high gradient and a low energy ion beam entering the tube is very strongly converged by a lens action to where it may be overfocussed. Current drain down the accelerating tube as provided by corona gaps or resistors is usually limited to a few hundred microamperes. If an intense beam of ions is injected into the tube and if a substantial fraction of these ions diverge to where they strike tube electrodes the potential gradient along the tube may be thrown so far off uniformity that the generator will fail to function properly. These restrictions have been so severe that most van de Graaff generators built prior to the late 1940's were limited to beam currents of a few microamperes. Ion sources of the same type that yielded milliamperes with Cockcroft-Walton-generators were reduced to microamperes when used with van de Graaff generators.

Capillary arc sources of the type developed by TUVE, DAHL and HAFSTAD[2] and the Zinn type arc[3] were extensively used up to 1950, and are still in limited use. Both of these sources utilized a high pressure low voltage arc in a restricted region and ions were pulled from this arc by a probe having a canal one to two centimeters long and one to two millimeters in diameter. The monatomic beam from these sources was usually about equal in intensity to the diatomic plus triatomic beam. Careful measurements showed the monatomic beam to consist of three components. One component of highest energy originated from monatomic ions from the arc. A second component, perhaps a kilovolt lower in energy, originated as diatomic ions in the arc and these diatomic ions were broken apart while passing through the probe canal. A third component, still lower in energy, came from the triatomic ions pulled from the arc which were dissociated in the capillary.

---

[1] Ion sources have been treated in detail by D. KAMKE in Vol. XXXII of this Encyclopedia.

[2] M. A. TUVE, O. DAHL and L. R. HAFSTAD: Phys. Rev. **48**, 241 (1935).

[3] W. H. ZINN: Phys. Rev. **52**, 655 (1937).

With these sources a few electrostatic generators delivered monatomic ion currents above 10 microamperes. McKIBBEN with a Zinn type source achieved monatomic ion currents up to 40 and 50 microamperes in 1939.

Two sources developed in the late 1940's have improved the general situation in regard to ion currents. At Berkeley the P.I.G. (Phillips Ionization Gauge) ion source was developed by Gow and FOSTER[1]. It operates in an axial magnetic field and requires no filament. An oxidized aluminum or beryllium cathode is essential. For pulsed currents this ion source has been very successful in electrostatic generators. Pulse currents of several milliamperes are readily obtained. For steady currents it has not been outstanding.

During the late 1940's the radio frequency ion source[2-5] was developed to where it surpassed other types for steady current use in van de Graaff generators. The electrodeless discharge is confined by a pyrex envelope to maximize the ratio of monatomic ion yield to diatomic and triatomic yields. Hydrogen atoms have a lower combination coefficient on pyrex than on other common materials. Metals, which have recombination coefficients near 100%, are held to a minimum in the envelope. A large proportion of the ions from a clean pyrex *RF* source are monatomic. Gas consumption is low. Currents of 50 microamperes or higher are readily obtained. In general, ion currents from electrostatic generators constructed after about 1950 are limited by the current handling capabilities of the generator.

24. Focussing. The electric field in typical accelerating tubes is so nearly uniform that focussing effects are very small once the ions have entered. The entrance to the tube is a strong lens and its strength is relatively insensitive to the detailed shape of the entrance. McKIBBEN[6] has shown that the focal length of this lens is given approximately by $F = 4\varphi/g$ where φ is the energy of entering ions, g is tube gradient and F is focal length. The ion beam is to be brought to a focus on a target that is so distant that we can assume that the beam is to be made parallel. Thus F gives the distance from the source to the end of the tube, or if there is a cross-over between ion source and tube, F gives the distance between cross-over and tube entrance.

A more accurate and complete analysis of the focussing problem is given by ELKIND[7].

25. Gas supply. Gas flow to the ion source from containers in the high potential terminal is usually regulated by a needle valve or a palladium tube. Where helium is to be used the palladium tube cannot be used. At many laboratories both methods have been tried and neither has proved to be trouble-free. The flow through a palladium tube should be from outside to inside to avoid a build-up of impurities which collect and decrease hydrogen flow. Even with this precaution the palladium tube can become "poisoned" if unpurified hydrogen is used. Use of highly purified hydrogen has at times led to another difficulty; hydrogen may flow at room temperature and thus drain the supply when the generator is not operating.

---

[1] J. D. Gow and F. S. FOSTER jr.: Rev. Sci. Instrum. **24**, 606 (1953).

[2] J. G. RUTHERGLEN and J. F. I. COLE: Nature, Lond. **160**, 545 (1947).

[3] A. J. BAYLY and A. G. WARD: Canad. J. Res. A **26**, 69 (1948).

[4] P. THONEMAN, J. MOFFATT, O. ROAF and J. SANDERS: Proc. Phys. Soc. Lond. **61**, 483 (1948).

[5] C. D. MOAK, H. REESE jr. and W. M. GOOD: Nucleonics **44**, 18 (1951).

[6] J. L. McKIBBEN: LADC Report No. 604, 1949.

[7] M. M. ELKIND: Rev. Sci. Instrum. **24**, 129 (1953).

The hydrogen supply tank can be outside the generator and hydrogen can be fed to a needle valve or palladium tube regulator through a saran tube. This method was developed at Wisconsin and has been extensively used for hydrogen. It was not successful for helium. With helium in the saran tube the tube breaks down electrically if the generator is operated at high voltage. The saran tube may be broken into many charred pieces. A number of experimenters have found that a mixture of about 90% helium and 10% hydrogen yields good supplies of both kinds of ions and the ions desired are selected by a magnetic analyzer.

Until recently needle valves were usually made by the nuclear physicist. Good commercial valves are now available.

26. Power supply and controls. To minimize size and weight of components four hundred cycle power is commonly used in the high voltage terminal. Self-excited generators utilizing brushes are not satisfactory. A generator which functions completely dependably in normal air at atmospheric pressure will usually fail to start in the dry high pressure atmosphere of the generator. In early Wisconsin electrostatic generators dry cells, controlled by a switch were used to start the generator. Another trouble was rapid brush wear. The only satisfactory solution is a permanent magnetic generator. It can be driven by V-belt from the upper charging belt pulley, or better still, it may be built into the upper pulley as is done by the High Voltage Engineering Company.

For control of apparatus in the high potential terminal early generators were equipped with strings or rods which passed from the grounded end up to the high voltage terminal. In most cases they were operated by knobs at the end of the tank. Meters in the high voltage terminal could be read through a window. Remote operation of controls is now common. Selsyns are commonly used to transmit motion from the control console to the interior of the tank. From here to the terminal the string or control rod is still frequently used.

Control and data transmission are achieved by light beams in a number of generators. These systems are still relatively complex and maintenance problems nullify some of their advantages.

27. Helium ions. A large proportion of helium ions from an ion source are singly charged. BRADFORD, KIKUCHI and BENNETT[1] report a 0.05 microampere beam of doubly ionized helium ions from a modified Zinn type arc. This current is one-half percent of the total ion current. An ion source that has been used with hydrogen continues to yield hydrogen ions for a long period of time. Hydrogen diatomic ions can be separated from He^{++} only with an analyzer with good resolution. An ion source that has never been used with hydrogen may be used for He^{++} ions with an analyzer of only moderate resolution.

At Wisconsin[2] a 0.6 microampere beam of He^{++} ions was obtained in the following manner. A small belt charging generator was constructed in the high voltage terminal of the machine of Fig. 19. He^+ ions from an ion source were accelerated to approximately 400 kev. They then passed through a capillary $\frac{3}{16}''$ in diameter and $8''$ long. Oxygen was admitted to the center of this capillary and flowed out at both ends. Approximately 30% of the emergent ions were doubly charged. With this arrangement the final energy of He^{++} ions is sufficiently different from the energy of any hydrogen ions that separation is not difficult.

[1] C. E. BRADFORD, T. KIKUCHI and M. E. BENNETT: Phys. Rev. **93**, 931 (1954).
[2] JOHN W. BITTNER: Rev. Sci. Instrum. **25**, 1058 (1954).

G. Voltage measurement and control.

28. Early methods. The generating voltmeter was usually found to be unreliable for measurement of the voltage of open air generators. The field of view of the generating voltmeter commonly included not only the high voltage terminal but also insulating supports or part of a wall. The potential of non-conducting surfaces was not a well defined function of the terminal voltage.

At the Carnegie Institution, Tuve's group built a resistance voltmeter consisting of 1000 ten-megohm resistors, built up into an assembly which resembled the high voltage column of a modern generator. It was calibrated in sections and with care was dependable to a few percent.

In pressurized tanks, generating voltmeters could be made linear and reproducible to about 1%. These instruments consist of sectors of disks which are covered and uncovered by grounded rotating sectors. Current from early models used at Wisconsin was rectified by a commutator and measured with a galvanometer.

Modern machines still utilize the generating voltmeter. Usually the A.C. output is measured with a commercial A.C. vacuum tube voltmeter.

For precise work the ion beam energy is now measured by magnetic or electrostatic deflection of the ion beam and the generating voltmeter becomes a supplementary device, useful for test work and for putting the generator into operation.

29. The electrostatic analyzer. Focussing relationships for the cylindrical electrostatic analyzer were investigated by Hughes and Rojansky, by Herzog and others. These instruments are discussed by Bainbridge in Vol. II of the book, "Experimental Nuclear Physics" edited by Segrè, and in "Nuclear Physics" by Green.

An object slit of distance l' from the entrance to the deflection plates of an analyzer with a geometric mean radius r is imaged at a distance l'' beyond the exit of the deflecting gap, where

$$(l' - g)(l'' - g) = f^2,$$

$$g = \frac{r}{\sqrt{2}} \cot \sqrt{2}\,\Phi,$$

$$f = \frac{r}{\sqrt{2}} \cosec \sqrt{2}\,\Phi,$$

Φ is angular deflection.

The maximum practical angular deflection is $\pi/\sqrt{2}$ where with $l' = 0$ we have $l'' = 0$. As Φ is decreased the object slit or the image slit or both must be moved outward. If $l' = l''$ and with $\Phi = \frac{\pi}{2\sqrt{2}} \approx 63.5°$ we have $l' = l'' = \frac{r}{\sqrt{2}}$. Large values of l' or l'' are inconvenient because of space requirements.

If slits are not used at the conjugate positions the calibration of the analyzer may change if the ion beam from the electrostatic generator shifts in position or direction in the accelerating tube or if the analyzer shifts on its supports.

As an example of this difficulty we may choose $\Phi \approx 63.5°$ where with $l' = \infty$ we have $l'' = 0$. We may have image slits close to the exits and if the beam from the generator is parallel resolution may be excellent with no object slit, or we may consider the ion source as the object slit. Now, however, if the direction of the beam changes as it passes down the accelerating tube, the image will shift and the calibration of the analyzer will change.

For an analyzer with a deflection angle of $\pi/\sqrt{2}$ and a radius r the velocity dispersion is given by $Y = 2r\dfrac{\Delta v}{v}$ where Y is the distance at the image between two particles which originate at the same position in the object plane with the same direction but with velocities of v and $v + \Delta v$ respectively. Energy resolution is given by $R = r/w$ where w is slit width for the symmetric case and R is the reciprocal of the fractional energy change for which the image is displaced by an amount equal to its width.

Neglecting relativistic corrections generator voltage is given by $V = \frac{1}{2}Gr$ where G is the gradient in the deflecting gap. With a plate separation of 5 mm the maximum voltage difference that can be held without difficulty between large deflecting plates is about 50 kv. With a radius of 100 cm the maximum ion energy for which the analyzer is usable is 5 Mev for singly charged ions.

30. The magnetic analyzer.

Focussing relationships for magnetic analyzers where entrance and exit pole edges are perpendicular to the beam axis, are similar to the relationships for the electrostatic analyzers. For a wedge-shaped magnet having an angle Φ between the two faces we have $l = r\cos(\Phi/2)$ for the symmetric case where l is the distance between either slit and the nearest pole edge, r, is the radius of curvature of the particles in the magnetic field and Φ is angular deflection. For an angular deflection of 180° an object in the entrance plane is imaged in the exit plane and for an angle of 90° a parallel beam (object at infinity) is brought to a focus in the exit plane. The velocity disperion of a magnetic analyzer of deflecting angle π is given by the same expression as the velocity dispersion of an electrostatic analyzer with a deflecting angle of $\pi/\sqrt{2}$ and thus a magnetic analyzer will give approximately the same energy resolution as an electrostatic analyzer having the same physical dimensions and slit separations.

31. Control mechanisms.

Voltage control in early machines was achieved by observation of generating voltmeter readings and manual adjustment of current to the charging belt. Time lag was so great that variations could not easily be held below one or two percent.

Ashby and Hanson[1] at Wisconsin developed the corona triode. In utilizing this device for voltage control a needle or a group of fine wires are projected a short distance through a hole in the wall of the tank centrally located along the axis of the tank with respect to the high voltage terminal. Corona current from this needle to the high voltage terminal is a sensitive function of the bias voltage between needle and tank wall. With good geometry a ten kilovolt swing in needle potential can change the corona current drain from the terminal by 50 microamperes or more.

Usually a simple circuit is used in which the corona needle is connected to the plate of a high voltage triode such as the $4E27/8001$. The load is then the plate power supply. An error signal applied to the grid results in a corrective change in current drain from the high voltage electrode.

Response time is governed by the mobilities of ions as they cross the gap from the tank wall to the terminal. The time periods for ions to cross this gap are of the order of 10 milliseconds. Error signals of much faster response are advantageous. A change in corona current produces an immediate change in generator load current. As in the case of an ionization chamber, an ion cloud as it moves across the gap constitutes a current. The corona control system works well only for a generator operating at a positive potential; the needle is then negative.

---

[1] R. M. Ashby and A. O. Hanson: Rev. Sci. Instrum. **13**, 128 (1942).

McKibben developed the capacitor shell for voltage control. The tank is equipped with a metal liner relatively close to the tank wall but insulated from it. A voltage change V_1 on this liner will produce an instantaneous voltage change V_2 on the high voltage terminal. The ratio V_2/V_1 will depend on the solid angle subtended by the capacitor shell at the terminal. At the University of Notre Dame Miller and Waldman[1] used the capacitor shell system for a generator which is used for electron acceleration. Results were excellent.

Bonner's group[2] sent an electron stream up the accelerating tube from an electron gun at the grounded end. In this case response time is practically instantaneous and control is very simple. The X-rays generated would probably be a serious disadvantage for a machine in the 4 MV range.

King and Devons[3] utilize a soft X-ray tube which irradiates the gas in the gap between a high voltage terminal and the tank wall. They obtain good results when a small percentage of freon (CCl_2F_2) is present in the gas; the absorption coefficient is then adequate.

32. Voltage measurement and control systems. For the first control system used at Wisconsin the ion beam was separated into its components by a magnetic analyzer. With the monatomic beam directed at a target the diatomic beam was deflected by a small electrostatic analyzer. A pair of amplifiers gave no signal when they received equal current. Off balance current was amplified and fed to a circuit which controlled the voltage on the corona needles for the charging belt. This system was better than manual control but the time constant of the charging belt was too great to permit high accuracy.

At Bohr's lavoratory better results were achieved by utilizing a signal from the generating voltmeter to turn off apparatus for detection of reaction products when the voltage was not accurately at the desired value.

When error signals from a small electrostatic analyzer were fed to the Ashby-Hanson corona triode, results were excellent[4]. This system is very effective and has been widely used. Error signals from electrostatic or magnetic analyzers of even relatively low resolution when fed to the corona triode give very good voltage stability for short time periods.

The small electrostatic analyzer originally constructed at Wisconsin had an angular deflection of only 15°. Slits could not be put at object and image positions because distances would have been too great. Long term drifts caused much trouble and frequent calibration was required for precise work.

A 90° cylindrical electrostatic analyzer with a radius of 40 inches was constructed at Wisconsin in 1948. The object slit was located 30 inches from the entrance and the image was then at the exit. The image slits were located just beyond the exit plane. With an object slit separation of 1.2 mm and an image slit separation of 0.8 mm, the beam energy spread at half maximum current is approximately 0.1%. The monatomic ion beam was passed through the analyzer to the target. With this arrangement, ions which were not at the proper energy could not get through to the target. A shift in the ion beam as it came through the accelerator tube could decrease target current to zero, but could cause no error.

[1] W. C. Miller, B. Waldman, J. C. Noyes and J. E. van Hoomissen: Phys. Rev. **77**, 758 (1950).

[2] W. E. Bennett, T. W. Bonner, C. E. Mandeville and B. E. Watt: Phys. Rev. **70**, 882 (1946). — W. E. Bennett, T. W. Bonner, H. T. Richards and B. E. Watt: Phys. Rev. **71**, 11 (1947).

[3] M. B. King and S. Devons: Rev. Sci. Instrum. **25**, 933 (1954).

[4] J. L. McKibben, G. H. Frisch and J. M. Hush: Phys. Rev. **70**, 117 (1946).

When the error signal for the corona triode was taken from the image slits of this analyzer results were very good. A usable current was attained through to the target with the slits set to permit an energy spread of as little as 0.02%. The target current was in general satisfactorily steady, indicating that fluctuations in generator voltage were held within the limits of energy spread permitted by the analyzer. Short period variation in generator voltage were measured by means of a small capacitor plate mounted inside the cylindrical wall of the generator and connected through a resistor to ground. From observation of an oscilloscope connected across the resistor, short term variations in generator voltage appeared to be limited to approximately 0.02% when the analyzer was set to permit an energy spread of this magnitude.

Cylindrical analyzers of 1 meter radius have been constructed at a number of laboratories and have worked out well for voltage measurement and control. At the Argonne National Laboratory an electrostatic analyzer has been constructed with a radius of 1 meter with deflecting plates ground to spherical rather than cylindrical surfaces. At the Naval Research Laboratory, Washington, D.C., a cylindrical electrostatic analyzer has been constructed with a radius of 2 meters. Both of these instruments perform very well.

With an electrostatic analyzer the ratio of the electric field at any arbitrary point in the gap or in the fringing field region to the field at any other chosen point is completely independent of applied voltage. Determining by calibration the ion energy for one value of deflecting voltage is therefore adequate since the ion energy has a constant ratio to deflecting voltage.

An electrostatic analyzer is an energy measuring device. All ions from the generator with the same charge will undergo the same deflection. To separate the different ions that come from a typical ion source, the beam must first pass through a magnetic analyzer of low resolution. This requirement may be considered as a disadvantage of the electrostatic analyzer.

For energies above about 5 MV for singly charged ions, the electrostatic analyzer becomes somewhat difficult. The analyzer must be large. Until high electric fields can be readily attained across vacuum-insulated gaps, electrostatic analyzers will be relatively impractical in the energy region above about 8 or 10 Mev.

For equal deflection radii, deflection angles, and slit separations the energy resolution of magnetic analyzers is approximately the same as that of electrostatic analyzers. If a proton magnetic resonance device is used for magnetic field measurement high accuracy may be achieved. The momentum of ions may be determined by calibration for one value H_1 of field at a chosen position in the gap. However the ratio of ion momentum to field at this position is not in general independent of field strength. Calibrations may be made for a number of values of field strength with ion momentum at intermediate values then obtained by interpolation. Errors due to hysteresis effects may be largely avoided if in approaching a given field value the magnet is cycled each time in a similar manner.

H. Performance characteristics and limitations of an operational generator.

33. Pressure tank. The generator shown in Fig. 19 was completed in 1940 and has been operating almost continually since that time[1]. Its performance characteristics and especially its limitations are now well understood. A discussion of these characteristics and limitations may, therefore, be of value.

---

[1] R. G. HERB, C. M. TURNER, C. M. HUDSON and R. E. WARREN: Phys. Rev. **58**, 579 (1940).

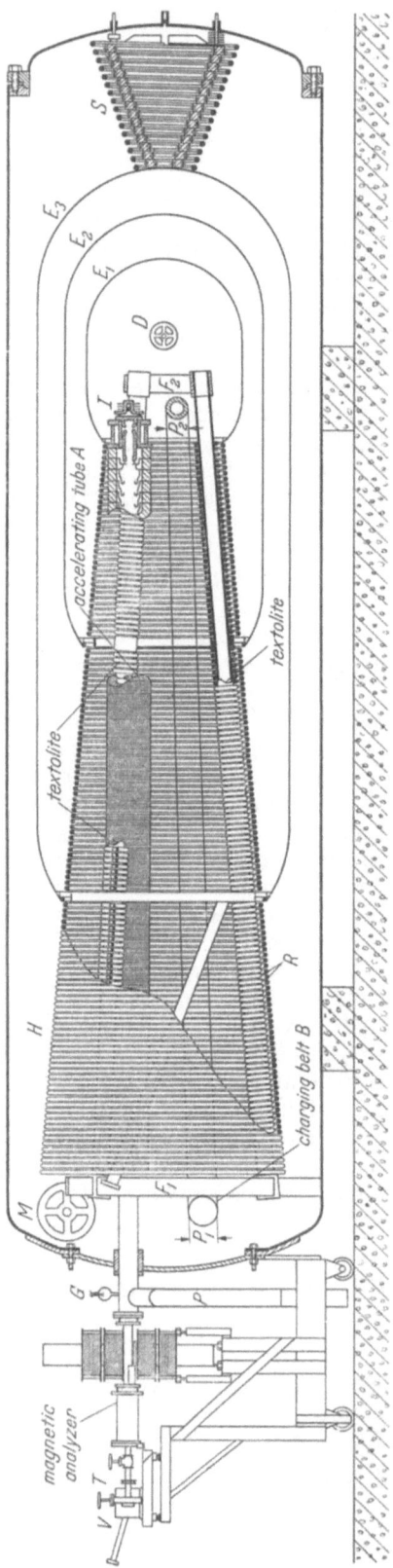

Fig. 19. Generator completed at Wisconsin in 1940. Operates up to 4.5 MV.

The enclosing tank for this machine was designed for a working pressure of 100 lb/square inch, gauge. It had been used for the 2.4-MV generator shown in Fig. 4 and was modified for the later machine. The wall thickness of the tank was not sufficient for safe evacuation. The general rule appears to be that the safe external pressure is about one-tenth of the internal working pressure. Without the possibility of evacuation transfer of a gas to a storage tank is somewhat impractical and to avoid undue expense, gases used in this machine have been restricted to air with a small percentage of freon (CCl_2F_2).

This tank is also disadvantageous in regard to serviceability. The tank cannot be rolled back from the enclosed apparatus and in making repairs work must be done in very restricted quarters.

34. Insulating gas. Rather surprisingly in view of difficulties in later machines the accelerating tube of this machine has rarely limited voltage. Several times after trouble with the vacuum system discharges in the tube gave difficulty. In each case after a period of time the tube recovered. Voltage limitation has been due usually to direct spark-over between concentric electrodes, flash-over along insulating supports, and flash-over along the charging belt. In early test work the voltage limit of about 4.5 MV was usually set by sparking along the charging belt.

As the pressure of ordinary air is increased in the tank up to a pressure of 100 lb per square inch the spark-over voltage increases from a value of about 0.60 MV at atmospheric pressure to a value of about 3.3 MV when the pressure is 100 lb per square inch. To operate at a higher voltage freon must be added. Ten pounds (weight) of freon added to the air gives a very substantial increase in spark-over voltage. The maximum operating voltage continues to increase as further freon is added up to about 25 lb (weight). This corresponds to a partial pressure of freon of approximately 3 lb per square inch. Additional freon usually has little effect on maximum voltage. If the generator is in a good state of repair and is reasonably clean it is usable at potentials up to about 4.5 MV.

If all parts of the machine are in good condition sparking at its uppermost limit will be between high voltage terminals. However, at this voltage the insulating supports and the charging belt appear to be close to the maximum voltage they will stand and in some cases there is failure along the belt or along an insulator.

The use of high pressure air in a generator has a number of disadvantages. It constitutes a serious fire hazard. At pressures used in this machine fires can be avoided by care in operation. If air pressure is increased further as has been done in a number of machines, the fire hazard becomes a very difficult problem. Other gases are much to be preferred. Oxygen in high pressure air also causes deterioration of organic materials. Ozone is formed by corona current and sparking, and this compound appears to be chiefly responsible for deterioration of charging belts and the V belts that are used for power transmission.

35. Control and information system. In this machine variation of voltages at the ion source and other adjustments in the high potential region are accomplished by means of strings which serve as belts to transmit rotary motion at the ground end of the tank to rotary motion of an instrument such as a Variac in the high voltage terminal. The control strings with their pulley systems have probably caused more time off for repairs than any other generator mechanism.

Silk fishline has proved to be more satisfactory than any other string. Nylon is satisfactory electrically but stretches so much that good control is difficult to achieve. Many of the difficulties with the system were purely mechanical and were due to inadequate care in design and construction of the apparatus. Other difficulties were due to electrical failure. A spark passing along the control string could cause a break. This difficulty could have been avoided if a better location has been provided for the strings.

Information in regard to ion source operation is obtained by visual observation of meters in the high voltage terminal. Observation ports for this purpose are located in the end of the tank near the accelerating tube entrance. This arrangement is not satisfactory when X-ray intensity or neutron intensity is high. Adherence to modern standards of safety is difficult unless a system is provided whereby controls may be operated from a remote control station. Provision for remote reading of critical ion source variables is also helpful.

36. Concentric electrodes. After a few years of generator operation the use of the largest of the concentric electrodes as shown in Fig. 19 was discontinued. Removal and reinstallation of this electrode took a sufficient amount of time and effort that the modest voltage gain which it appeared to give was not sufficiently rewarding. Some measurements were made with the intermediate terminal removed. This appeared to cause a substantial reduction of usable voltage. Part of this reduction may have been due to inadequate smoothing of the enclosing hoop system where this terminal had been attached. Regardless of the explanation, the use of this intermediate terminal is being continued.

37. Support column failure. This machine operated rather continuously for twelve years before any weakness or inadequacy of the support system became evident. The two upper Textolite tubes then snapped off just at the ends of the clamps at the grounded end of the generator. Failure occurred when the machine was not in operation. This failure may be some evidence for deterioration of

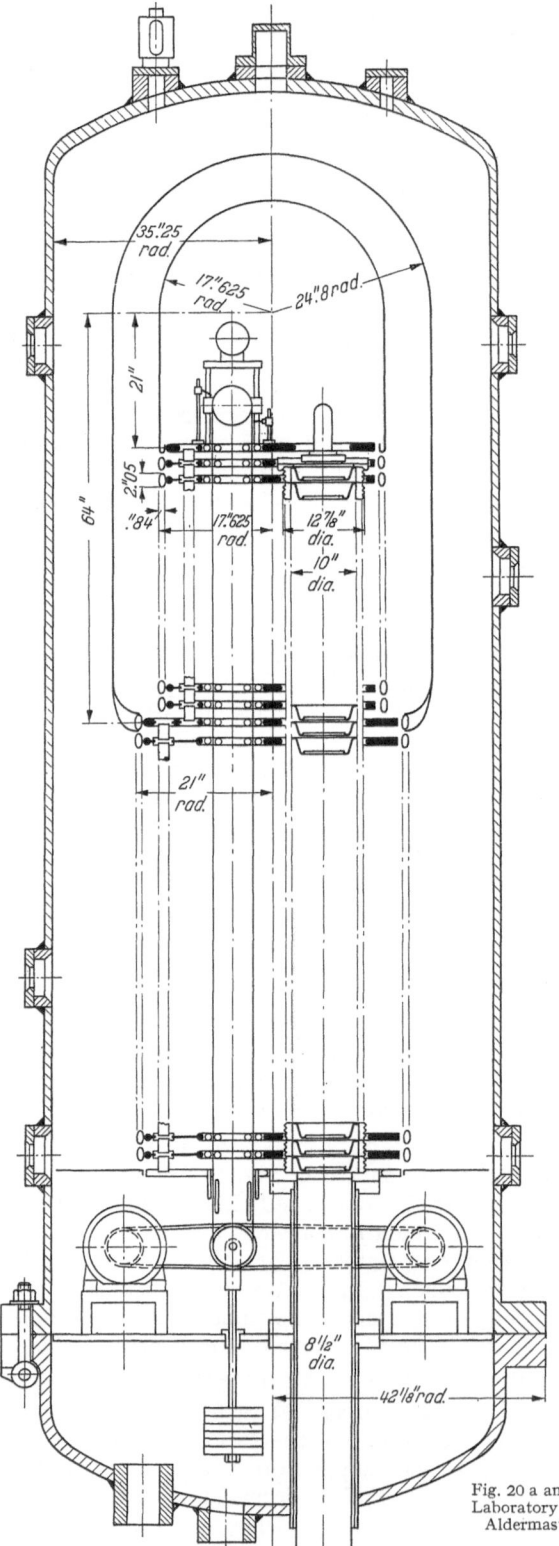

35.25 rad.

17.625 rad.

24.8 rad.

21"

2.05"

.84"

17.625 rad.

12⅞ dia.

10" dia.

64"

21" rad.

8½" dia.

42⅛"rad.

a

to vacuum

this material with age. These two members were replaced and approximately two years later the same failure reoccurred. Again these members were replaced but to guard against reoccurrence of this failure the column at the position where the intermediate shell is connected is supported by a sheet of lucite which rests directly on the tank wall. This additional support takes much load from the upper Textolite members and has not appeared to serve as a limitation in maximum generator voltage.

I. General discussion.

38. Performance extrapolations. One of the most general rules that appears to characterize performance of electrostatic generators is failure to follow what would appear to be reasonable extrapolations. We have the long tube effects where reasonable extrapolations from a single accelerating tube section or from a few sections fails so completely that the value of tests with single sections will be somewhat questionable until the processes involved in tube failure are better understood. Little can be done to check performance of a modified electrode design, or insulator design short of constructing and testing the complete tube. This is expensive and time-consuming and to a great extent has been responsible for the slow progress in accelerating tube performance.

Extrapolations from a short to a long insulator fail

Fig. 20 a and b. Generator constructed at the Research Laboratory of the Associated Electrical Industries, Ltd., Aldermaston, England. Shield rings have elliptic cross sections.

also although here with subdivision and with great care in all details of the design, extrapolations can be made to hold rather well as insulator length increases.

In regard to insulating materials, extrapolations cannot be safely made. The material to be used may be checked in tests to gradients higher than those to be held in the machine. Test results may look completely satisfactory, yet the insulator may fail after a relatively short period of use in a high voltage generator. A machine is relatively impracticable if occasional sparking can cause serious damage. When a spark occurs, insulators may be subjected to very high instantaneous gradients. If they char or puncture after repeated surges, the generator may become inoperative. Thus, the only adequate test for a new insulator material is to use it for a full-sized machine. Failure of insulators may require rebuilding the generator.

Fig. 20 b.

Wave fronts can be very steep in spark-over along an insulator. The discharge cannot be expected to follow a slightly longer path if a shorter path is available. Sphere gaps must be in line along an insulator if they are to give good protection. Sphere gaps on one side of a tubular insulator may not protect the other side.

Great caution must be exercised in the use of data on dielectric strengths of gases. and solid insulators to predict the voltage of a generator of new design. Perhaps the only rule that is well borne out is that the voltage of a generator differing appreciably from previous generators cannot be safely predicted. A faithful copy of a generator that is operating can be safely predicted in regard to performance.

39. Performance as a function of gas pressure and type of gas. Satisfactory generator performance in air at atmospheric pressure can be obtained with very relaxed design and construction standards. Gradients are relatively low and insulators are lightly stressed. As gas pressure is increased, good design and construction standards must be more rigorously followed if the increased pressure is to be beneficial.

At an absolute pressure of about 8 atmospheres the generator shown in Fig. 4 appeared to be approaching its maximum voltage. The slope of the voltage versus pressure curve at 8 atmospheres had decreased to about 0.3 of its value at one atmosphere. Some more recent machines give better behavior as shown by Figs. 21 and 22.

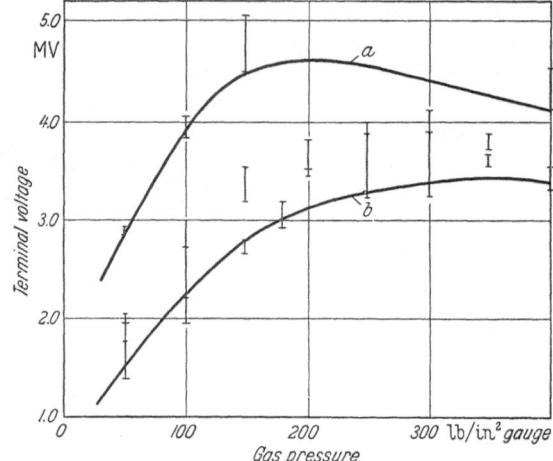

Fig. 21. Data taken with generator of Fig. 20 showing terminal voltage as a function of gas pressure for negative polarity, with intershield. Curve (a) nitrogen with 10% freon. Curve (b) nitrogen. Temperature 40° C.

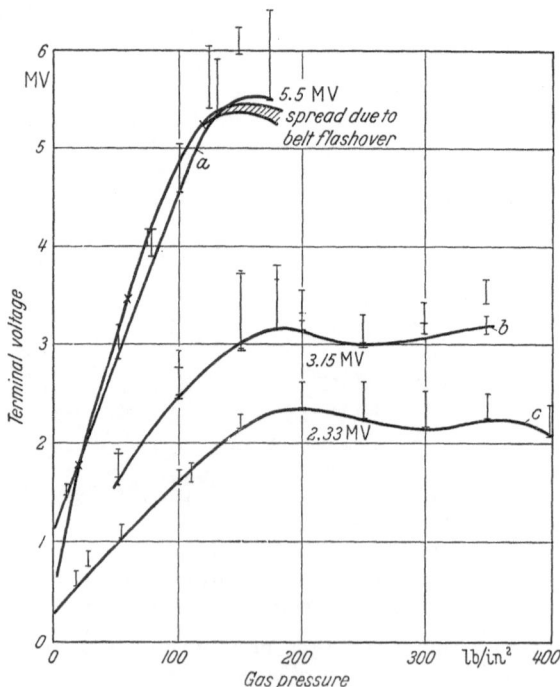

Fig. 22. Data taken with generator of Fig. 20 showing terminal voltage as a function of gas pressure for positive polarity. (a) Nitrogen with 10% freon. (b) Nitrogen. (c) Nitrogen with no intershield.

Departure from linearity in the voltage versus pressure curve may be partially due to surface roughness and to electrodes with small radii of curvature. Poorly shaped edges of electrodes may cause trouble even if these electrodes are not in

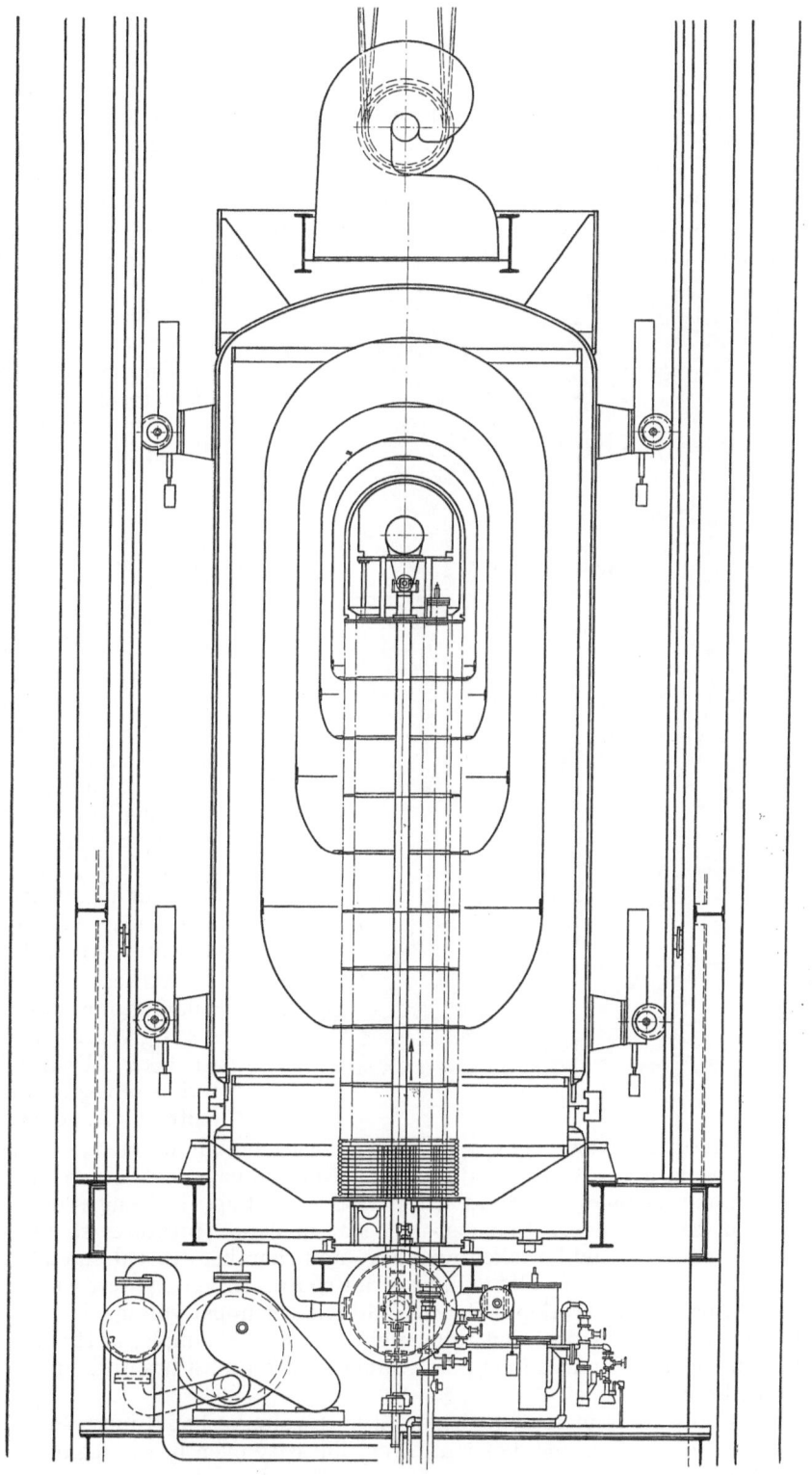

a

Fig. 23 a—c. Large generator constructed at Los Alamos. Maximum voltage without accelerating tube 14 MV. Maximum operating voltage with accelerating tube approximately 8 MV in 1956. Insulating column 20 feet high made up of ten sections each 2 feet high. Fig. 23 b shows the inner column which encloses the charging belt and accelerating tube. This inner column is surrounded by the outer column shown in Fig. 23 c. The outer column consists of Mykroy rings separated by metal rings with the joints bonded by cement. The tank and the outer column can be hoisted as a unit from the inner column without release of pressure from regions exterior to the outer column.

the most intense field. GOLDMAN and VUL[1] observed that sparking potentials from a positive point to a plane rise to a maximum and then decrease as gas pressure is increased. The gas pressure at which this maximum occurs is a function of temperature and varies greatly from one gas to another. In nitrogen at 13° C the maximum appeared at a pressure of about 9 atmospheres.

In some cases an apparent maximum in the voltage versus pressure curve may be due to flash-over along insulators. Once there is voltage limitation by insulator flash-over an increase in gas pressure may cause failure at lower voltage.

The dielectric strength of nitrogen is somewhat less than that of air. The dielectric strengths of freon and sulfurhexafluoride are approximately equal and are about three times that of air. At 21° C the vapor pressure of CCl_2F_2 is 85 psia and that of SF_6 is 330 psia.

Sulfurhexafluoride is a relatively inert and well-behaved compound although some products formed by ionization and sparking are poisonous. Small percentages of either CCl_2F_2 or SF_6 when mixed with air or nitrogen give a very marked increase in dielectric strength.

A discussion of the insulating properties of gases is given by TRUMP in the book, "Dielectric Materials and Applications", edited by A. VON HIPPLE.

Fig. 23 b.

40. Performance tests. For a machine of new design which has been constructed and is under test, the difficulties which appear and which determine the maximum voltage are found as spark-over voltage is being studied as a function of gas pressures. Some sparking at moderate pressures should do no damage. Voltage should be taken up slowly and charging current should be held to low values. The upper voltage limit is not sharp. In voltage test work, a uniform criterion should be adopted for determining the upper voltage limit in terms of the time between sparks. This time might be set, for example, at three minutes with the understanding that the upper voltage limit for most practical applications would be lower.

As gas pressure is increased, the spark-over voltage should increase although most generators show some departure from linearity even in the region of a few atmospheres.

---

[1] I. M. GOLDMAN and B. M. VUL: Techn. Physics USSR. **1**, 497 (1935).

Generators are commonly provided with viewing ports. For a generator of a new design, careful observation of sparking and corona can be very instructive. Even with careful observation, however, the cause of failure may be difficult to establish. A spark directly from the high voltage terminal to the tank wall may cause flash-over along an insulator as hoops discharge, or along a row of sphere gaps protecting an insulator. The surge, due to insulator flash-over, may trigger a spark from the terminal or from a hoop to the tank wall. Failure in the accelerator tube may be incorrectly diagnosed as failure along an insulator.

41. Accelerators utilizing negative ions. WILLARD H. BENNETT[1] first suggested the use of hydrogen negative ions to be accelerated to a positive terminal where they were to be stripped of electrons and then accelerated to a target at ground potential. ALVAREZ first attempted to use this method. Information on the formation of negative ions in hydrogen was obtained by WHITTIER[2] at Chalk River and practical ion sources for negative ions were developed at Los Alamos[3] and Wisconsin[4]. VAN DE GRAAFF has observed that negative ions might be generated in a terminal at a high negative potential. They may then be accelerated to another terminal at a high positive potential. Here they

Fig. 23 c.

may be stripped and then be accelerated to a terminal at a high negative potential where a target is located. If these terminals are all at a voltage V, hydrogen ions could arrive at the target with an energy of $4\,Ve$. Still higher energy multiplication might be realized with negative ions of elements of higher atomic number. At Wisconsin, a source giving 25 microamperes of hydrogen negative ions yielded in addition 25 microamperes of oxygen negative ions when the hydrogen admitted to the source was not dried.

According to calculations of HOLCIEN and MIDTDAL[5] helium negative ions can exist and a helium ion negative ion source may be a possibility. Experimental results are inconclusive[6,7].

---

[1] W. H. BENNETT: U.S. Patent 2, 206, 558.
[2] A. C. WHITTIER: Canad. J. Phys. **32**, 275 (1954).
[3] J. A. PHILLIPS and J. L. TUCK: Rev. Sci. Instrum. **27**, 97 (1956).
[4] J. A. WEINMAN and J. R. CAMERON: Rev. Sci. Instrum. **27**, 288 (1956).
[5] E. HOLCIEN and J. MIDTDAL: Proc. Phys. Soc. Lond., A **68**, 815 (1955).
[6] J. W. HIBY: Ann. Phys., Lpz. **34**, 473 (1939).
[7] P. G. KRUGER: Phys. Rev. **36**, 853 (1930).

Fig. 24. The 5.5-MV generator manufactured by the High Voltage Engineering Company, Cambridge, Mass. In 1956 these generators were in operation at the Oak Ridge National Laboratory, at the Rice Institute in Houston, Texas, and at Columbia University, New York.

Generators utilizing negative ions are now being developed at the High Voltage Engineering Company. A small machine is also being developed at Wisconsin. In addition to multiplication of ion energies, these machines appear to offer the advantage of less complex apparatus inside the pressurized tank.

Cyclotrons and Synchrocyclotrons.

By

BERNARD L. COHEN.

With 17 Figures.

1. Introduction. A cyclotron is an accelerator in which charged particles are constrained to move in quasi-circular orbits by a magnetic field, and accelerated by a radiofrequency electric field. It differs from a synchrotron in that ions are introduced with zero energy. The magnetic field extends throughout the area bounded by the outermost orbit, and does not vary with time.

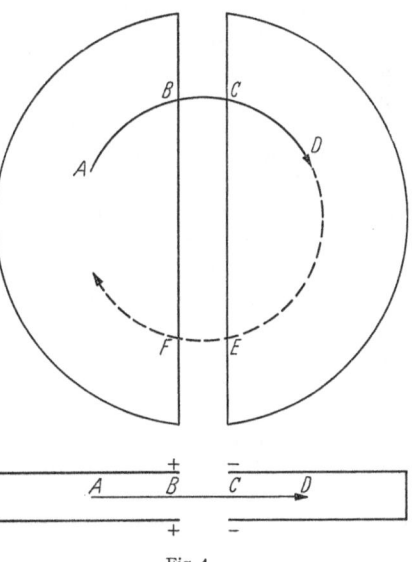

Fig. 1.

Although other methods have been considered, the accelerating electric field in all cyclotrons that have been constructed or proposed to date is applied by means of a "dee system". Dees are hollow, D-shaped electrodes as illustrated in Fig. 1. The radiofrequency voltage is applied to the two dees in opposite phase, and with a frequency approximately equal to the frequency with which the ions circulate under the action of the magnetic field. Consider a positively charged particle starting at A in Fig. 1. In moving from A to B, it is inside a dee so that it experiences no electric field; between B and C it is acted on by the electric field between the dees so that if the phase of the electric voltage is such that the left dee is positive with respect to the right dee, it will be accelerated. In going from C to D to E, the ion is again inside a dee and thus in a field-free region. It reaches E one-half cycle in its circulation after passing B. Since the electric frequency matches the circulating frequency, it too will have advanced one-half cycle, so that if the left dee was positive when the ion was at B, the right dee will now be positive and the electric field will again be accelerating as the ion passes from E to F. Beyond F it is again in a field-free region and returns eventually to a position and direction of motion similar to the starting condition at A, whence the motion is repeated. On each revolution the ion is accelerated to a higher energy, so that its radius of curvature in the magnetic field is increased. Thus the orbit radius expands in time until the ion reaches full energy at the outermost orbit. It may then be intercepted by an internal target, or deflected out of the magnetic field to an external target.

The system shown in Fig. 1 is a two-dee system. Instead of applying the radiofrequency voltage to the two dees in opposite phase, one dee could be grounded and the voltage applied only to the other dee. Only that part of the grounded dee near the accelerating gap need by retained, so that the remainder of this dee—called the "dummy dee"—is not used. This is known as a single dee system. Systems with three dees (called "triants") have been constructed for special purposes, and in principle, any number of dees may be used so long as the R.F. voltage is applied to them in the proper relative phase.

In order that the cyclotron action may proceed as described, it is necessary that:

(1) the phase of the R.F. voltage be such as to accelerate rather than decelerate (at least on an average). This phase problem is discussed in Part A II.

(2) the orbits be stable. This involves both stability against motion in a radial direction (i.e., across the magnetic field lines) other than the radial expansion due to the energy gain, and stability against motion in the vertical direction (i.e., parallel to the magnetic field lines). Radial instability reduces the usable area of the magnetic field, and thus reduces the energy obtainable, while vertical instability results in loss of ions by striking the dees. The problem of orbital stability is discussed in Part A I.

In these theoretical discussions, three classes of cyclotrons will be considered— the standard cyclotron, the synchrocyclotron, and the azimuthally varying field (AVF) cyclotron. The standard cyclotron uses a relatively uniform magnetic field and a constant frequency electric voltage. However, the small radial *de*crease in the magnetic field that is necessary (cf. Sect. 3) to achieve vertical focusing (i.e., orbital stability in the vertical direction) is inconsistent with the radial *in*crease that would be required to compensate the relativistic mass increase and thus keep the frequency with which the ions circulate constant. Thus the latter frequency must change monotonically with respect to the frequency of the electric voltage; this leads to a monotonic phase shift, so that the number of revolutions that an ion may make while the phase of the electric voltage is accelerating, is limited. This limits the energy that may be obtained from a standard cyclotron. The synchrocyclotron and AVF cyclotron represent methods of circumventing this difficulty.

In the synchrocyclotron, the frequency of the electric voltage applied to the dees is varied with time in order to maintain resonance with the circulation frequency of the ions. The phase problem is thus completely different from that in the standard cyclotron, although the problem of orbital stability is quite similar.

In the AVF cyclotron, the radial variations in the magnetic field are tailored to keep the frequency with which the ions circulate constant, so that with a constant electric frequency applied to the dees, there is no phase shifting and hence no phase problem. The vertical defocusing resulting from the radial increase in the magnetic field is compensated by the focusing action of the azimuthally varying field. However, the radially increasing and azimuthally varying field introduces complications into the orbital stability problem.

The various orbital stability and phase problems of the three types of cyclotrons are discussed in detail in Part A. Part B is then devoted to a discussion of the individual components of cyclotrons. The *physical* problems involved in the design of magnets, radiofrequency systems, ion sources, vacuum systems, beam deflection systems, targets, shielding, and controls and interlocks are described. In these sections, no effort is made to go into engineering details, beyond giving the reader a casual appreciation of them. Also, no consideration is given to the AVF cyclotron, since none of this type has reached the advanced design stage.

In Part C, several miscellaneous subjects are considered. Problems involved in operation are described, the status of cyclotrons is discussed, and some of the characteristics of individual cyclotrons are presented.

A. Theory.

I. Orbital stability.

a) Magnetic effects in standard cyclotrons and synchrocyclotrons.

2. Equations of motion and their solution for a uniform field. The force on a particle carrying an electric charge e and moving with velocity $\boldsymbol{v}$ in a magnetic field $\boldsymbol{H}$ is given by

$$\boldsymbol{F} = \frac{e}{c}\,(\boldsymbol{v} \times \boldsymbol{H}). \tag{2.1}$$

Setting this force equal to mass times acceleration gives the well known Lorentz equations of motion; introducing cylindrical coordinates (r, ϑ, z), they are

$$\left. \begin{aligned}
m\,\ddot{r} - m\,r\,\dot{\vartheta}^2 &= \frac{e}{c}\,(\dot{z}\,H_\vartheta - r\,\dot{\vartheta}\,H_z)\,, \\
m\,r\,\ddot{\vartheta} + 2\,m\,\dot{r}\,\dot{\vartheta} &= \frac{e}{c}\,(\dot{r}\,H_z - \dot{z}\,H_r)\,, \\
m\,\ddot{z} &= \frac{e}{c}\,(r\,\dot{\vartheta}\,H_r - \dot{r}\,H_\vartheta)
\end{aligned} \right\} \tag{2.2}$$

where a dot over a symbol indicates differentiation with respect to time, m is the mass of the particle, and H_r, H_ϑ, H_z, are the components of the vector H.

In the simplest situation of a uniform magnetic field in the z-direction, $H_r = H_\vartheta = 0$ and (2.2) becomes, after rearrangement of terms,

$$\left. \begin{aligned}
\ddot{r} &= r\,\dot{\vartheta}\,(\dot{\vartheta} - \omega)\,, \\
\ddot{\vartheta} &= -\frac{\dot{r}}{r}\,(2\dot{\vartheta} - \omega)\,, \\
\ddot{z} &= 0
\end{aligned} \right\} \tag{2.3}$$

where the symbol ω has been introduced for

$$\omega = \frac{e\,H}{m\,c}. \tag{2.4}$$

For a particle which initially has a velocity v, it is convenient to choose the origin of coordinates along a line normal to its direction of motion and at a distance r_0 given by v/ω. In this coordinate system, the initial conditions for the solution of (2.3) are

$$\dot{r} = 0, \quad \dot{\vartheta} = \omega. \tag{2.5}$$

From (2.3), then $\ddot{r}$ and $\ddot{\vartheta}$ are initially zero so that $\dot{r}$ and $\dot{\vartheta}$ do not change from (2.5). But since they do not, $\ddot{r}$ and $\ddot{\vartheta}$ remain zero, etc., so that the solutions of (2.3) are

$$\left. \begin{aligned}
r &= r_0 = \frac{v}{\omega}\,, \\
\vartheta &= \omega t + \vartheta_0\,, \\
z &= \dot{z}_0 t + z_0
\end{aligned} \right\} \tag{2.6}$$

where the subscript zero denotes values at $t = 0$. Thus the motion in the (r, ϑ) plane (generally called the "horizontal plane") is in a circular orbit traversed with angular frequency ω given by (2.4), and the motion in the z (vertical) direction is uniform. Since the distance available for particle motion in the vertical direction is quite small, uniform motion in the z-direction cannot be tolerated so that some sort of vertical focusing must be provided to keep the particle near the median plane of the magnetic field.

The first of (2.6) and (2.4) give

$$mv = \frac{eHr}{c} \tag{2.7}$$

as the relationship between momentum, magnetic field and radius of curvature. For low energies where relativistic effects can be neglected, the kinetic energy E is $\frac{1}{2}mv^2$, whence

$$E = \frac{1}{2m}\left(\frac{eHr}{c}\right)^2 \tag{2.8}$$

or

$$Hr = \frac{c}{e}\sqrt{2mE}. \tag{2.9}$$

In the relativistic region, the relationship between kinetic energy and momentum, with (2.7), gives

$$E = \sqrt{(eHr)^2 + (m_0 c^2)^2} - m_0 c^2 \tag{2.10}$$

or, solving for Hr,

$$Hr = \frac{1}{e}\sqrt{E(E + 2m_0 c^2)}. \tag{2.11}$$

Eqs. (2.8) to (2.11) give the total Hr necessary to contain an ion of a given energy, or conversely, the energy of an ion with a given radius of curvature in a given magnetic field.

3. Equations of orbital stability. In general, the magnetic field will not be perfect, so that H is a function of r and ϑ, and r deviates slightly from r_0. Assuming the imperfections to be small, H_z may be expanded as

$$H_z = \overline{H}_z(r_0)\left[1 + \sum_j h_j \cos(j\vartheta + \vartheta_{0j})\right] + \frac{\partial H_z}{\partial r}[r - r_0] \tag{3.1}$$

where $\overline{H}$ represents H averaged over ϑ, and h_j and ϑ_{0j} are the amplitudes and phases respectively of the terms in a Fourier expansion of the azimuthal variations in the magnetic field. By symmetry, the magnetic field lines pass through the median plane perpendicularly, so that on the median plane, about which we are expanding, $H_z = H$ and $H_r = H_\vartheta = 0$. For small z, $\overline{H}_r$ may therefore be approximated by the first non-zero term in the Taylor expansion

$$\overline{H}_r = \left(\frac{\partial \overline{H}_r}{\partial z}\right)_{z=0} z. \tag{3.2}$$

From the fact that H is irrotational, curl $H = 0$ whence $\left(\frac{\partial \overline{H}_r}{\partial z}\right)_{z=0} = \left(\frac{\partial \overline{H}_z}{\partial r}\right)_{z=0} = \frac{\partial \overline{H}}{\partial r}$, so that (3.2) becomes

$$\overline{H}_r = \frac{\partial \overline{H}}{\partial r} z. \tag{3.3}$$

Since there is no mechanism for transferring energy into or out of the system, the velocity remains constant and equal to $\overline{\omega} r_0$ where $\overline{\omega}$ is related to $\overline{H}$ by (2.4). Neglecting $\dot{r}^2$ and $\dot{z}^2$ relative to $(r\dot{\vartheta})^2$,

$$r\dot{\vartheta} = v = \overline{\omega} r_0 \tag{3.4}$$

and since the average value of r is r_0, ϑ in (3.1) is well approximated by $\overline{\omega} t$. Using (2.4), (3.1) to (3.4) in (2.2), and neglecting $\dot{r} H_\vartheta$ and $\dot{z} H_\vartheta$ as products of two small terms, gives

$$\left. \begin{aligned} \ddot{r} - \frac{\overline{\omega}^2 r_0^2}{r} &= - \overline{\omega}^2 r_0 \left[1 + \sum_j h_j \cos(j\overline{\omega} t + \vartheta_{0j})\right] - \overline{\omega}^2 \frac{r_0}{\overline{H}} \frac{\partial \overline{H}}{\partial r} [r - r_0], \\ \ddot{z} &= - \frac{\overline{\omega}^2 r_0^2}{\overline{H}} \frac{\partial \overline{H}}{\partial r}. \end{aligned} \right\} \tag{3.5}$$

These equations are simplified if new variables, ξ and η, are introduced as

$$\xi = \frac{r - r_0}{r_0}, \qquad \eta = \frac{z}{r_0}. \tag{3.6}$$

It is also convenient to define the field index, n, by

$$n = - \frac{r}{\overline{H}} \frac{\partial \overline{H}}{\partial r}. \tag{3.7}$$

By using (3.6) and (3.7) in (3.5) and approximating $\dfrac{1}{1+\xi}$ by $1 - \xi \ldots$ they become

$$\left. \begin{aligned} \ddot{\xi} + (1 - n) \overline{\omega}^2 \xi &= \sum_{j=1}^{\infty} h_j \cos(j\overline{\omega} t + \vartheta_{0j}), \\ \ddot{\eta} + n \overline{\omega}^2 \eta &= 0. \end{aligned} \right\} \tag{3.8}$$

These are the equations of radial and vertical oscillations; from (3.8) the frequencies of radial and vertical oscillation, ω_r and ω_z respectively, are

$$\left. \begin{aligned} \omega_r &= (1 - n)^{\frac{1}{2}} \overline{\omega}, \\ \omega_z &= n^{\frac{1}{2}} \overline{\omega}. \end{aligned} \right\} \tag{3.9}$$

These frequencies were first obtained by KERST and SERBER [15].

The second of (3.8) in conjunction with (3.7) indicates that the magnetic force gives a vertical focusing if the field decreases with increasing radius as mentioned in Sect. 1. The magnitude of this force increases linearly with increasing radius, so that once it becomes predominant, the amplitude of the vertical oscillations decreases as the particle is accelerated (unless the rate of field fall-off is decreased). This force and its effects will be discussed further in Sect. 13.

4. Instability due to first harmonic. From the first of (3.8) it is apparent that when n is small, there is an instability induced by the first harmonic in the magnetic field. The solution, for $n = 0$ and for the origin of ϑ chosen to set $\vartheta_{01} = 0$ is

$$\xi = \frac{h_1}{2} \overline{\omega} t \sin \overline{\omega} t + \text{periodic terms}. \tag{4.1}$$

This corresponds to a motion of the orbit center through a distance per revolution, $\Delta r / \Delta \nu$, given by

$$\frac{\Delta r}{\Delta \nu} = \pi r h_1. \tag{4.2}$$

This effect may be easily understood if we consider a magnetic field such as that in Fig. 2 which is larger in the upper half than in the lower half by the fraction 2Δ. From (2.7) an increase in magnetic field causes the radius of curvature of an ion of given velocity and charge to decrease in the same ratio. An ion starting at A moves in a semicircle with center at 0 and radius $r_0(1 - \Delta)$. At B, the radius of curvature becomes $r_0(1 + \Delta)$, so that the center is displaced to the left to $0'$, a distance of $2r_0\Delta$. At C, the radius again changes to $r_0(1 - \Delta)$, so that the center is again displaced to the left a distance $2r_0\varDelta$, to $0''$. Thus, in one turn, the orbit center precessed to the left a distance $4r_0\varDelta$. This could have been calculated from (4.2) by observing that a Fourier analysis of a square wave of amplitude $\varDelta$ gives a first harmonic of amplitude $4\varDelta/\pi$.

A radial oscillation with a frequency ω_r, equal to the frequency with which the ions circulate, $\overline{\omega}$, is equivalent to motion in a circular orbit with the center displaced from the origin of coordinates by a distance equal to the amplitude of the radial oscillation. If the frequency of the radial oscillation is slightly lower, the angle at which the maximum radial displacement occurs shifts back by an angle equal to 2π times the fractional difference between the two frequencies. This is equivalent as above to motion in a circular orbit the center of which precesses about the origin of coordinates (in this case, the center of symmetry of the magnetic field) with a frequency, ω_p, equal to the difference between $\overline{\omega}$ and ω_r. By applying this to the first of (3.8) with the term in n taken into account,

$$\omega_p = \overline{\omega}\left[1 - (1-n)^{\frac{1}{2}}\right]. \quad (4.3)$$

When $\omega_p t$ has increased by about unity, the direction of precession has changed, so that the motion of the orbit center given by (4.2) continues in the same general direction for about $\dfrac{1}{2\pi}(\overline{\omega}/\omega_p)$ revolutions. The total displacement of the orbit center, from (4.2), is thus about

$$\varDelta r_{\text{total}} = \frac{\overline{h}_1}{2}\frac{\overline{\omega}}{\omega_p}r \quad (4.4)$$

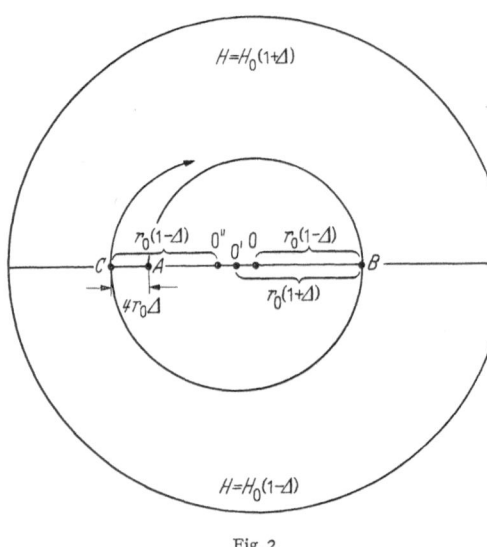

Fig. 2.

where $\overline{h}_1$ is the average value of h_1.

In most cyclotrons, $\overline{\omega}/\omega_p$ is typically of the order of 100, so that random errors in the magnetic field as large as one-part in a thousand cause relatively large precessions. It is thus profitable to expend a considerable effort to shim out the first harmonic in cyclotron magnetic fields. On the other hand, it is sometimes useful to introduce first harmonics at large radii in order to precess the beam in a desired direction, or to increase the separation between successive orbits as an aid to deflection.

5. Effect of rapid field fall-off at magnet edge. Near the outside edge of the magnet, the magnetic field fall-off becomes quite rapid, so that the frequencies of the radial and vertical oscillations, (3.9), deviate considerably from their values in uniform fields. It is therefore necessary to treat them in greater detail, taking into account the non-linear terms in the Lorentz equation. For the sake of brevity, we neglect azimuthal variations in the magnetic field, although the generalization to include them is straightforward [35] following the previous procedure. We expand the magnetic field as a polynomial series in ξ and η as

$$\left.\begin{array}{l}\dfrac{H_r}{\overline{H}} = a_{10}\xi + a_{01}\eta + \dfrac{1}{2}a_{20}\xi^2 + a_{11}\xi\eta + \dfrac{1}{2}a_{02}\eta^2 + \cdots, \\[2mm] \dfrac{H_z}{\overline{H}} = 1 + b_{10}\xi + b_{01}\eta + \dfrac{1}{2}b_{20}\xi^2 + b_{11}\xi\eta + \dfrac{1}{2}b_{02}\eta^2 + \cdots.\end{array}\right\} \quad (5.1)$$

From the conditions curl $H = 0$, and div $H = 0$

$$\frac{\partial H_r}{\partial \eta} = \frac{\partial H_z}{\partial \xi}, \qquad \frac{\partial H_r}{\partial \xi} = -\frac{\partial H_z}{\partial \eta}$$

whence

$$a_{01} = b_{10}, \quad a_{11} = b_{20}, \quad a_{02} = b_{11}, \quad a_{10} = b_{01}, \quad a_{20} = b_{11}, \quad a_{11} = b_{02}. \tag{2.5}$$

From the symmetry of H_z about the median plane

$$b_{01} = b_{11} = 0. \tag{5.3}$$

Since (5.1) is a Taylor series,

$$
\left.
\begin{aligned}
b_{10} &= \frac{1}{\overline{H}} \left(\frac{\partial \overline{H}}{\partial \xi} \right) = \frac{r}{\overline{H}} \frac{\partial \overline{H}}{\partial r} = -n, \\
b_{20} &= \frac{1}{\overline{H}} \left(\frac{\partial^2 \overline{H}}{\partial \xi^2} \right) = \frac{r}{\overline{H}} \frac{\partial^2 \overline{H}}{\partial r^2} = -n'
\end{aligned}
\right\} \tag{5.4}
$$

where n' is defined by the last expression. Inserting (5.2), (5.3) and (5.4) in (5.1), it becomes

$$
\left.
\begin{aligned}
\frac{H_r}{\overline{H}} &= -n\eta - n'\xi\eta + \cdots, \\
\frac{H_z}{\overline{H}} &= 1 - n\xi + \frac{1}{2} n'(\eta^2 - \xi^2) + \cdots.
\end{aligned}
\right\} \tag{5.5}
$$

Using (5.5) and (2.2), and proceeding as in the solution of (3.5) gives finally

$$
\left.
\begin{aligned}
\ddot{\xi} + \overline{\omega}^2 (1 - n)\,\xi &= \tfrac{1}{2} n' \eta^2, \\
\ddot{\eta} + \overline{\omega}^2 n\eta &= -n'\xi\eta.
\end{aligned}
\right\} \tag{5.6}
$$

The terms on the right of (5.6) give a coupling between the radial and vertical oscillations. To investigate this further, (5.6) may be solved by perturbation methods; the unperturbed solutions are

$$\xi = c_1 \cos (\omega_r t + c_2),$$
$$\eta = c_3 \cos (\omega_z t + c_4).$$

With these inserted in the right side of (5.6), it becomes

$$
\left.
\begin{aligned}
\ddot{\xi} + \omega_r^2 \xi &= \tfrac{1}{4} n' c_3^2 \cos (2\omega_z t + 2c_2) + \cdots, \\
\ddot{\eta} + \omega_z^2 \eta &= -\tfrac{1}{2} n' c_1 c_3 \cos \left[(\omega_r - \omega_z)\, t + c_2 - c_4 \right] + \cdots.
\end{aligned}
\right\} \tag{5.7}
$$

It is apparent that the solutions of (5.7) have secular terms if $2\omega_z = \omega_r$. From (3.9), this occurs when

$$n = 0.2. \tag{5.8}$$

Actually, the secular amplitude increase cannot continue since the total energy in the two oscillations must be conserved. A complete description of the motion would be very complex, but it consists essentially of a periodic transfer of the total energy between the radial and vertical oscillations. Since the energy stored in radial oscillations is relatively large, the transfer of this energy results, at times, in very large vertical oscillations causing most of the beam to be lost to the dees. This effect is especially serious in synchrocyclotrons since the ions make many revolutions within a small radial interval; it is therefore customary synchrocyclotron practice not to attempt acceleration beyond $n = 0.2$, but rather to postpone that point to as large a radius as possible.

In addition to the strong coupling at $n = 0.2$, there are weaker couplings at all radii where ω_r/ω_z is the ratio of small integers. For example, if the symmetry of H_z about the median plane is not perfect, b_{01} is not zero and terms linear in ξ and η appear on the right side of (5.6). By methods similar to those used above, it is seen that a strong coupling then occurs at $\omega_r = \omega_z$, which, from (3.9) occurs at $n = 0.5$.

The maximum radius at which the magnetic field can contain the ions is where Hr reaches its maximum; i.e., when $\frac{d}{dr}(Hr) = 0$. Solution of the latter gives $n = 1$. Beyond the radius where $n = 1$, the ions spiral outward from the machine.

b) Magnetic effects in the azimuthally varying field cyclotron[1].

6. Equations of orbital stability in the AVF cyclotron. In this section we consider the motion of an ion in a magnetic field of the form

$$H = \bar{H}\left[1 + f \cos p\,(\vartheta + \delta)\right]. \tag{6.1}$$

Here f is a measure of the magnitude of the azimuthal variations, p is the number of identical azimuthal sectors of which the field is composed (i.e., the field possesses p-fold azimuthal symmetry), and δ is defined by

$$r\,\frac{\partial \delta}{\partial r} = \tan \gamma \tag{6.2}$$

where γ is the angle between the axes of symmetry of the magnet sectors, and radial vectors; it is often referred to as the "*swirl*". The solution of the equations of motion for this field is very lengthy and complex, so that many simplifying assumptions will be necessary. It will be assumed throughout that p is large so that terms of the form $p^2 \pm 1$ will be approximated by p^2. In addition, it will be assumed that deviations from circular motion are small so that most of the simplifying assumptions of the derivation of (3.8) will be used here. On the other hand, no restriction will be assumed for $\partial\delta/\partial r$, as it may be made quite large.

By expanding H_r as in (3.2) and proceeding similarly for H_ϑ, the Lorentz equations (2.2) may be written

$$\left. \begin{aligned} \ddot{r} - r\,\dot{\vartheta}^2 &= -\frac{e}{mc}\,r\,\dot{\vartheta}H\,, \\ \ddot{z} &= \frac{e}{mc}\left(-\dot{r}\,\frac{z}{r}\,\frac{\partial H}{\partial \vartheta} + r\,\dot{\vartheta}\,z\,\frac{\partial H}{\partial r}\right). \end{aligned} \right\} \tag{6.3}$$

It is convenient here to use ϑ as the independent variable. Since (3.4) is valid in the approximation used here,

$$\frac{d}{dt} = \dot{\vartheta}\,\frac{d}{d\vartheta} = \frac{v}{r_0}\,\frac{d}{d\vartheta}$$

whence (6.3) becomes

$$\left. \begin{aligned} \frac{d^2 r}{d\vartheta^2} - \frac{r_0^2}{r} &= -r_0\,\frac{H}{\bar{H}}\,, \\ \frac{d^2 z}{d\vartheta^2} &= -\frac{\bar{\omega}z}{\bar{H}}\left[\frac{1}{v}\,\frac{\partial r}{\partial \vartheta}\,\frac{\partial H}{\partial \vartheta} - \frac{r_0^2}{v}\,\frac{\partial H}{\partial r}\right]. \end{aligned} \right\} \tag{6.4}$$

---

[1] The author is greatly indebted to T. A. WELTON for extensive discussions which greatly facilitated the preparation of this section.

We define here the equation of the stable orbit for the velocity v, as

$$r = q(\vartheta). \tag{6.5}$$

From the first of (6.4), it may be seen that if we adopt

$$q = \bar{q}\left[1 + \frac{f}{p^2}\cos p(\vartheta + \delta)\right] \tag{6.6}$$

and neglect all but lowest order terms, the correct expression for H from (6.1) is obtained. In the z-equation, all functions of r may be replaced by their values on the equilibrium orbit; making this substitution and using (3.6), the second of (6.4) becomes

$$\frac{d^2\eta}{d\vartheta^2} = -\frac{\bar{\omega}\,\eta}{\bar{H}}\left[\frac{1}{v}\frac{\partial q}{\partial \vartheta}\frac{\partial H}{\partial \vartheta} - \frac{\bar{q}^2}{v}\frac{\partial H}{\partial q}\right]. \tag{6.7}$$

Evaluating the terms in (6.7) from (6.1) and (6.6),

$$\begin{aligned}
\frac{\partial H}{\partial q} &= \frac{\partial H}{\partial \bar{q}}\frac{\partial \bar{q}}{\partial q} = \left[-\bar{H}p f \sin p(\vartheta + \delta)\frac{\partial \delta}{\partial \bar{q}} + \frac{\partial \bar{H}}{\partial \bar{q}}\right]\left[1 - \bar{q}\frac{f}{p}\sin p(\vartheta + \delta)\frac{\partial \delta}{\partial \bar{q}}\right], \\
\frac{\partial H}{\partial \vartheta} &= -\bar{H}p f \sin p(\vartheta + \delta), \\
\frac{\partial q}{\partial \vartheta} &= -\bar{q}\frac{f}{p}\sin p(\vartheta + \delta).
\end{aligned} \right\} \tag{6.8}$$

Inserting these in (6.7), replacing periodic terms with their values averaged over ϑ (where this is not zero), and using (6.2), gives

$$\frac{d^2\eta}{d\vartheta^2} + \eta\left[-\frac{\bar{q}}{\bar{H}}\frac{\partial \bar{H}}{\partial q} + \frac{1}{2}f^2 - p f \tan \gamma \sin p(\vartheta + \delta) + \frac{1}{2}f^2\tan^2\gamma\right] = 0. \tag{6.9}$$

The equation for ξ may be obtained in a somewhat similar fashion. In terms of q and $\bar{q}$, ξ is here defined as

$$\xi = \frac{r - q}{\bar{q}}.$$

Evaluating the terms in the first of (6.4)

$$\begin{aligned}
\frac{d^2r}{d\vartheta^2} &= \frac{d}{d\vartheta}\left(\frac{dr}{d\vartheta}\right) = \frac{d}{d\vartheta}\left(\frac{dq}{d\vartheta} + \bar{q}\frac{d\xi}{d\vartheta}\right) = -f\cos p(\vartheta + \delta) + \bar{q}\frac{d^2\xi}{d\vartheta^2}, \\
\frac{1}{r} &= \frac{1}{\bar{q}\left[1 + \frac{f}{p_2}\cos p(\vartheta + \delta) + \xi\right]} = \frac{1}{\bar{q}}\left[1 - \frac{f}{p_2}\cos p(\vartheta + \delta) - \xi\right], \\
H &= \bar{H}(\bar{q}) + \frac{\partial H}{\partial q}(r - \bar{q}) = \bar{H}(\bar{q}) + \frac{\partial H}{\partial q}\bar{q}\left[\frac{f}{p^2}\cos p(\vartheta + \delta) + \xi\right].
\end{aligned} \right\} \tag{6.10}$$

Inserting these into the first of Eqs. (6.4) and proceeding as was done in obtaining (6.9) gives finally

$$\frac{d^2\xi}{d\vartheta^2} + \xi\left[1 + \frac{\bar{q}}{\bar{H}}\frac{\partial \bar{H}}{\partial q} + p f \tan \gamma \cos p(\vartheta + \delta) - \frac{1}{2}f^2\tan^2\gamma\right] = 0. \tag{6.11}$$

(6.9) and (6.11) are the equations of orbital stability in the AVF cyclotron. It may be noted that for $p = 0$, $\bar{q} = r$, $\vartheta = \bar{\omega}t$, these are identical with the equations derived in Sect. 3 and 5 for the standard cyclotron and synchrocyclotron. For most purposes, the effect of the non-constant term in the coefficients of ξ and η

may be averaged over to give

$$\frac{d^2\xi}{d\vartheta^2} + v_r^2\xi = 0,$$ (6.12a)

$$\frac{d^2\eta}{d\vartheta^2} + v_z^2\eta = 0$$ (6.12b)

where, as shown in Appendix A,

$$v_r^2 = 1 + \frac{\bar{q}}{\bar{H}}\frac{\partial\bar{H}}{\partial\bar{q}},$$ (6.13a)

$$v_z^2 = -\frac{\bar{q}}{\bar{H}}\frac{\partial\bar{H}}{\partial\bar{q}} + \frac{1}{2}f^2 + f^2\tan^2\gamma.$$ (6.13b)

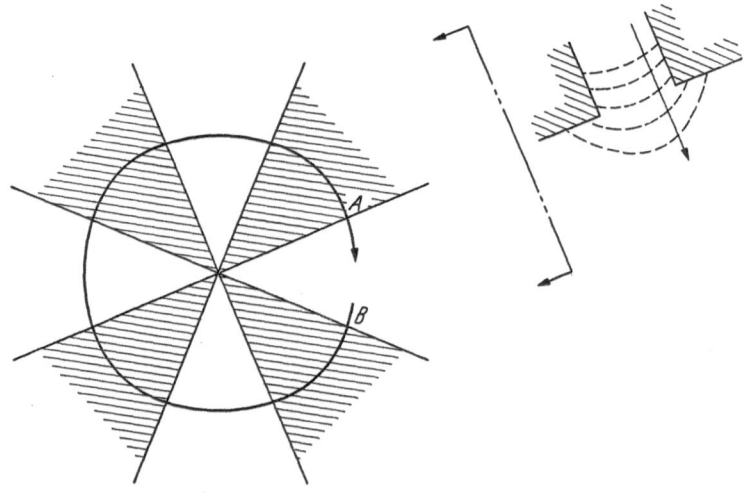

Fig. 3.

Eqs. (6.12b) and (6.13b) give the net focusing force in an AVF cyclotron. If it is desired to use a fixed frequency, the resonance condition (2.4) requires that the average magnetic field increase with radius at the same rate as the mass increases due to relativistic effects. This gives

$$\bar{H} = H_0\left(1 - \frac{\bar{\omega}^2 r^2}{c^2}\right)^{-\frac{1}{2}}.$$ (6.14)

Thus the magnetic field *increases* with radius, so that the first term in (6.13b) gives defocusing.

The second term of (6.13b) arises directly from the $\dot{r}H_\vartheta$ term in the LORENTZ equation for z; it was first recognized by THOMAS [36] and proposed as a basis for a high energy fixed frequency cyclotron. This "Thomas focusing" may be understood by considering a magnetic field with four-fold symmetry as shown in Fig. 3. The orbit deviates from a circle in that its curvature is greater in a strong field region than in a weak field region. From (6.3), the force in the z direction is

$$F_z = -\frac{e}{c}\dot{r}\frac{\partial H}{\partial\vartheta}\frac{z}{r}.$$ (6.15)

At an interface such as A, $\dot{r}$ is negative and $\partial H/\partial\vartheta$ is negative so that the force is restoring. At an interface such as B, both $\dot{r}$ and $\partial H/\partial\vartheta$ are positive, so that again the force is focusing. Since these two interfaces are representative of all the others, the force is at all times focusing.

The third term in (6.13 b) is important when large amounts of "swirl" are used. It was discovered and developed by the MURA group[1] as a form of alternate gradient focusing as illustrated in Fig. 4, although actually half of its magnitude arises from the Thomas effect. By making γ large, the magnitude of the field variations required to achieve a given focusing force may be greatly reduced.

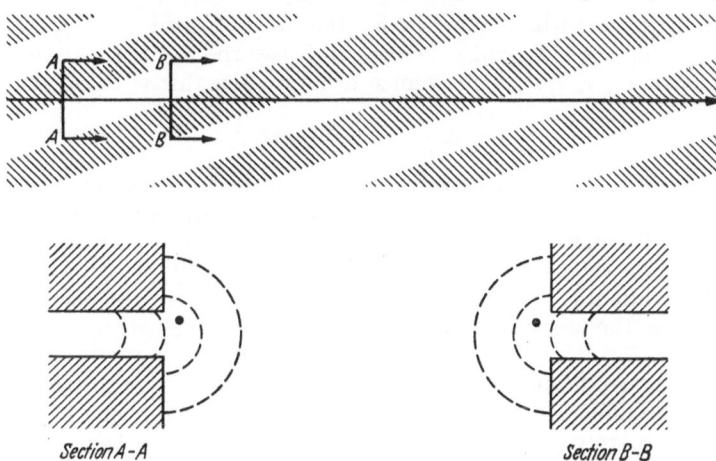

Section A-A   Section B-B

Fig. 4. In sections $A-A$ and $B-B$, path of particle is into paper.

7. Resonances of the stability equations. Eqs. (6.12a) and (6.13a) describe the radial oscillations in an AVF cyclotron. As was found in (3.8), a more complete treatment would give

$$\frac{d^2\xi}{d\vartheta^2} + v_r^2 \xi = \sum_{j=1}^{\infty} h_j \cos(j\vartheta + \vartheta_{0j}) \tag{7.1}$$

where the h_j are the coefficients of a Fourier analysis of the azimuthal variations in the magnetic field; they are large for $j = p$, $2p$, etc. It is clear from (7.1) that when

$$v_r = j \quad (j \text{ integer}) \tag{7.2}$$

a secular increase in the amplitude of the radial oscillations will occur. These are called the "integral resonances"[2].

The first integral resonance encountered is for $v_r = 1$ at the beginning of the motion. This is just the precession discussed in detail in Sect. 4, which causes the orbit center to move in accordance with (4.2); this is equivalent to a linear increase with time in the amplitude of the radial oscillations.

From (6.13), (6.14), and the relativistic expressions for mass and energy,

$$v_r = 1 + \frac{E}{m_0 c^2}. \tag{7.3}$$

Thus, the second integral imperfection resonance occurs when the kinetic energy is approximately equal to the rest energy. It may be expected that some difficulty will be encountered in accelerating beyond that energy.

[1] K. R. SYMON, L. W. JONES, D. W. KERST and K. M. TERWILLIGER: Phys. Rev. **98**, 1153 (1955).

[2] Difficulties from resonances were first pointed out by D. M. DENNISON and T. H. BERLIN: Phys. Rev. **69**, 542 (1946); **70**, 58, 764 (1946).

Even if the field at the equilibrium orbit is perfect so that all h_j are zero (except, of course, for $j = p$), trouble arises from imperfection in the field gradient. This has essentially the effect of causing ν_r^2 to vary. By making a Fourier analysis of this variation and moving the correction terms to the right side of (6.12a), that equation becomes

$$\frac{d^2 \xi}{d\vartheta^2} + \nu_r^2 \xi = \xi \sum_i b_i \cos (i\vartheta + \vartheta_{0i}). \qquad (7.4)$$

By solving with perturbation methods, the unperturbed ξ is $c_1 \cos(\nu_r \vartheta + c_2)$. Inserting this for ξ in the right side of (7.4) causes the right side to include terms with frequencies $\pm \nu_r \pm i$. A resonance occurs when this is equal to ν_r; that is, when

$$\nu_r = \frac{i}{2} \quad (i \text{ integer}). \qquad (7.5)$$

These are known as the "half-integral resonances". They are ordinarily less important than the integral resonances; however, they are quite important for the case $\nu_r = p/2$ which is called the "*essential* half-integral resonance", and since this occurs before the essential integral resonance, $\nu_r = p$, it will limit the maximum energy attainable in all AVF cyclotrons. From (7.3), the energy attainable before reaching the essential half integral resonance is about $(p/2 - 1)$ times the rest energy.

When non-linear terms are considered, resonances similar to the integral and half-integral resonances occur when ν_r is one-third integral, one-fourth integral, etc., with successively less importance. It has been shown that resonances[1] of higher fractional order than one-fourth cannot destroy the beam completely. If the one-fourth integral essential resonance cannot be passed, the maximum energy attainable is $(p/4 - 1)$ times the rest energy. This would probably be the energy limit of a cyclotron with a high degree of "swirl".

All of the considerations which have been applied to ν_r apply equally to ν_z. However, there is nothing which requires ν_z to become large in the way that (7.3) imposes that necessity on ν_r, so that ν_z may be kept as small as necessary to avoid dangerous resonances.

Considering the theory most generally, since the Lorentz equation is non-linear, equations such as (7.4) may include all terms of the form

$$\xi^{\varkappa'} \eta^{\varkappa''} \cos \varkappa \vartheta$$

where $\varkappa$, $\varkappa'$, and $\varkappa''$ are integers. Of especial interest are the terms which lead to equations of the form

$$\left. \begin{array}{l} \dfrac{d^2 \xi}{d\vartheta^2} + \nu_r^2 \xi = \sum_\varkappa \eta^{\varkappa''} \cos \varkappa \vartheta, \\[2mm] \dfrac{d^2 \eta}{d\vartheta^2} + \nu_z^2 \eta = \sum_\varkappa \xi^{\varkappa'} \cos \varkappa \vartheta. \end{array} \right\} \qquad (7.6)$$

Solving these by perturbation, $\xi = c_1 \cos(\nu_r \vartheta + c_2)$ and $\eta = c_3 \cos(\nu_z \vartheta + c_4)$ are inserted in the right sides. These equations then have secular solutions when

$$\left. \begin{array}{l} \pm \varkappa'' \nu_z \pm \varkappa = \nu_r, \\ \pm \varkappa' \nu_r \pm \varkappa = \nu_z \end{array} \right\}$$

or, combining these two, resonances occur when

$$\pm \varkappa_1 \nu_r \pm \varkappa_2 \nu_z = \varkappa_3 \quad (\varkappa_1, \varkappa_2, \varkappa_3 \text{ integer}). \qquad (7.7)$$

---

[1] P. A. STURROCK: AERE X/R 1771.

These are the so-called *"sum resonances"* and *"difference resonances"*. They are especially important because they couple the radial and vertical oscillations, feeding energy from one to the other. The beam blow-up at $n = 0.2$ in synchrocyclotrons discussed in Sect. 5 is actually a special case of this.

The simplest and perhaps most important resonances of this type are when

$$\nu_r \pm \nu_z = \varkappa_3. \tag{7.8}$$

A sum resonance of this type occurs somewhat before, and a difference resonance occurs somewhat after each integral imperfection resonance. This introduces a serious difficulty since integral resonances induce large radial oscillations the energy of which is fed into vertical oscillations at the succeeding difference resonance, causing part of the beam to strike the dees.

The effects of the various resonances and the limitations they impose on the accuracy required of the magnetic field are being studied by several groups using computers and, in some cases, electron model cyclotrons.

If the radiofrequency accelerating voltage is frequency modulated, the magnetic field need not increase with radius as fast as (6.14) so that ν_r need not increase with energy as rapidly as (7.3); thus the resonances maybe postponed to higher energy. This, of course, results in low intensity, pulsed beams such as those obtained from synchrocyclotrons.

c) Electric effects.

8. Electric vertical focusing. In considering effects of the radiofrequency electric field, it is convenient to use rectangular coordinates in the horizontal plane. These will be chosen such that $x = 0$ is the line half-way between the two dees and parallel to their flat edges. Thus, the electric field has essentially no y-component. The origin of coordinates is taken at the center of the cyclotron.

The Lorentz equations of motion in this coordinate system, neglecting non-uniformities in the magnetic field and assuming the vertical component of the velocity to be small, are

$$\left. \begin{aligned} m\ddot{x} &= -e\frac{\partial V}{\partial x} + \frac{e}{c}\dot{y}H, \\ m\ddot{y} &= -\frac{e}{c}\dot{x}H, \\ m\ddot{z} &= -e\frac{\partial V}{\partial z}. \end{aligned} \right\} \tag{8.1}$$

The time variation of the dee voltage is given by

$$V_t(x, z, t) = V(x, z) \cos(\omega t + \Theta) \tag{8.2}$$

where ω is the frequency of the electric voltage which, for purposes of this section, is taken equal to the resonant frequency (2.4), and Θ is the phase of the electric voltage at the time the ion crosses the gap. The equipotential lines of the electric field between the dees are as shown in Fig. 5 a[1,2]. The electric field in the x-direction, $-\frac{\partial V}{\partial x}$, is as shown in Fig. 5 b; it causes the ions to accelerate in the x-direction, gaining, as they cross the complete gap, an energy $2V_0 \cos\Theta$ where V_0 is the maximum dee-to-ground voltage in a two dee cyclotron or half this voltage in a single dee cyclotron. Fig. 5 c shows $-\frac{\partial V}{\partial z}$, the electric field in the z-direction; it indicates that the vertical force is focusing in the first half

---

[1] R. R. WILSON: Phys. Rev. **53**, 410 (1938).

[2] R. L. MURRAY and L. T. RATNER: J. Appl. Phys. **24**, 67 (1953).

gap, and defocusing in the second half gap, so that to first order, the net force is zero. The principal non-zero net force is due to the fact that the field is changing as the ion traverses it, in accordance with (8.2). If the field is increasing—that is, if Θ is negative—the force in the second half gap is larger so that the net effect is focusing.

In order to treat the problem mathematically [31], we consider the ion to move through the field shown in Fig. 5a with constant velocity v, and along a path of constant z. Since $\omega t \ll 1$ throughout the region where the electric field is present, we may use the small angle approximation

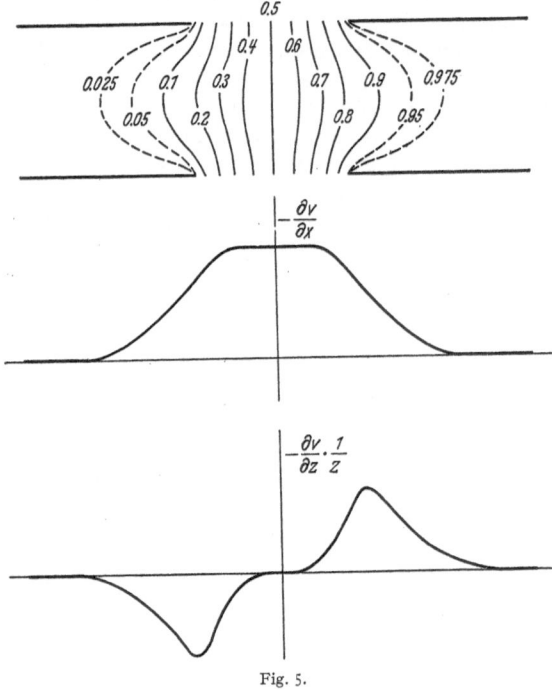

Fig. 5.

$$\cos(\omega t + \Theta) = \cos\Theta - \omega t \sin\Theta.$$

Putting (8.2) in the third of (8.1) then gives

$$\left.\begin{aligned}\frac{d^2 z}{dt^2} = -\frac{e}{m}\frac{\partial V}{\partial z}\times \\ \times(\cos\Theta - \omega t \sin\Theta).\end{aligned}\right\} \quad (8.3)$$

By transforming the independent variable from t to x, this becomes

$$\left.\begin{aligned}\frac{d^2 z}{dx^2} = -\frac{e}{2E}\frac{\partial V}{\partial z}\times \\ \times\left(\cos\Theta - \omega\frac{x}{v}\sin\Theta\right)\end{aligned}\right\} \quad (8.4)$$

where

$$E = \tfrac{1}{2}mv^2 \quad (8.5)$$

the kinetic energy.

Integrating both sides from $x = -\infty$ to $x = \infty$, gives the change in direction of the ions on a complete gap traversal, $\Delta(dz/dx)$, as

$$\Delta\left(\frac{dz}{dx}\right) = \int\limits_{-\infty}^{\infty}\frac{d^2 z}{dx^2}\,dx = -\frac{e}{2E}\left[\cos\Theta\int\limits_{-\infty}^{\infty} - \frac{\partial V}{\partial z}\,dx + \frac{\omega}{v}\sin\Theta\int\limits_{-\infty}^{\infty}x\,\frac{\partial V}{\partial z}\,dx\right]. \quad (8.6)$$

Since $\partial V/\partial z$ is antisymmetric (cf. Fig. 5c), the first integral vanishes. The second integral may be integrated by parts, as

$$\int\limits_{-\infty}^{\infty}x\,\frac{\partial V}{\partial z}\,dx = x\int\limits_{-\infty}^{\infty}\frac{\partial V}{\partial z}\,dx - \int\limits_{-\infty}^{\infty}dx\int\limits_{-\infty}^{x}\frac{\partial V}{\partial z}\,dx. \quad (8.7)$$

The first term is zero as discussed above. The second may be evaluated by noting that

$$\left.\begin{aligned}&\frac{\partial}{\partial z}\left[\int\limits_{-\infty}^{\infty}dx\int\limits_{-\infty}^{x}\frac{\partial V}{\partial z}\,dx\right] \\ &= \int\limits_{-\infty}^{\infty}dx\int\limits_{-\infty}^{x}\frac{\partial^2 V}{\partial z^2}\,dx = -\int\limits_{-\infty}^{\infty}dx\int\limits_{-\infty}^{x}\frac{\partial^2 V}{\partial x^2}\,dx = -\int\limits_{-\infty}^{\infty}\frac{\partial V}{\partial x}\,dx = -2V_0\end{aligned}\right\} \quad (8.8)$$

where the third expression was obtained from the second by use of LAPLACE's equation

$$\frac{\partial^2 V}{\partial x^2} + \frac{\partial^2 V}{\partial z^2} = 0. \tag{8.9}$$

Multiplying both sides of (8.8) by dz and integrating from $z = 0$ (where $\partial V/\partial z = 0$) to z, gives

$$\int_{-\infty}^{\infty} dx \int_{-\infty}^{x} \frac{\partial V}{\partial z} \, dx = -2 V_0 z. \tag{8.10}$$

Inserting (8.10) in (8.7) and the result in (8.6), we obtain

$$\Delta \left(\frac{dz}{dx} \right) = - \frac{e \, V_0 \sin \Theta}{E} \frac{\omega}{v} z. \tag{8.11}$$

This gives the electric focusing effect of the time variation of the field during the gap crossing as discussed above; actually, it is the first term of an expansion in powers of $(V_0/E)^{\frac{1}{2}}$. The higher order terms have been investigated in great detail [7], but they are generally very much smaller than (8.11) throughout the region where the expansion coverges rapidly enough to be useful. (8.11) becomes approximately valid when $E \approx 10 V_0$, which is after about three revolutions. At smaller radii, the problem must be treated numerically. The few calculations that have been made [7] indicate that the net effect of vertical forces in that region is quite small.

From Eq. (8.11) it is seen that the vertical forces are focusing when the phase is positive, and defocusing when the phase is negative (i.e., when the ion crosses the gap before the electric field reaches its maximum). Unfortunately, other considerations require that the phase be negative in the early part of the motion in standard cyclotrons. The net vertical force is thus defocusing causing a large fraction of the beam to be lost. This problem will be discussed in greater detail in Sect. 13.

Since the distance around a complete turn is $2\pi r$, and two gaps are passed per turn, (8.11) may be written, using (3.4)

$$\Delta \left(\frac{dz}{dv} \right) = - \frac{2\pi e \, V_0 \sin \Theta}{E} z \tag{8.12}$$

where $\Delta (dz/dv)$ is the change per turn of the rate of increase of the z-coordinate per turn. If the magnetic force from (3.8) is included, the final focusing formula becomes

$$\Delta \left(\frac{dz}{dv} \right) = \left(- \frac{2\pi e \, V_0 \sin \Theta}{E} - 2\pi^2 \frac{r}{H} \frac{dH}{dr} \right) z. \tag{8.13}$$

It may be noted from (8.13) that the electric force decreases as $1/E$, while the magnetic force increases as r, or equivalently, as $E^{\frac{1}{2}}$; thus, the magnetic force becomes predominent at large radii.

9. Electrically induced radial oscillations. In order to investigate the motion in the (x, y) plane, we seek to solve (8.1) with the initial conditions

$$\left\{ \begin{array}{ll} x = 0, & y = r \\ \dot{x} = \omega r, & \dot{y} = 0 \end{array} \right\} \quad \text{at} \quad t = 0. \tag{9.1}$$

Integrating the second of (8.1) from $t = 0$ to t gives, with use of (2.4),

$$\dot{y} = -\omega x. \tag{9.2}$$

Substituting this into the first of (8.1), it becomes

$$\ddot{x} + \omega^2 x = -\frac{e}{m}\frac{\partial V}{\partial x}. \tag{9.3}$$

The electric field, $-\partial V/\partial x$, may be expanded as

$$-\frac{\partial V}{\partial x} = \sum_{j=1}^{\infty} e_j \cos j\omega t \tag{9.4}$$

where e_j are the coefficients in a Fourier expansion of the electric field *experienced by the ion*. The principal accelerating electric field is experienced by the ion twice per revolution and thus contributes to the term $j = 2$. Its coefficient may be determined by multiplying both sides of (9.4) by $2\cos 2\omega t$ and integrating from 0 to 2π as

$$e_2 = \frac{2}{\pi}\int_0^{2\pi} -\frac{\partial V}{\partial x}\cos 2\omega t\, d\omega t. \tag{9.5}$$

The electric force is non-zero only where $\cos 2\omega t \approx 1$, and $\omega t \approx x/r$, whence (9.5) becomes

$$e_2 = \frac{4V_0}{\pi r}. \tag{9.6}$$

Since the voltage in most standard cyclotrons is applied from one end of the dees, the voltage at the two gap crossings is not the same, so that there is also a *first* harmonic in the electric field experienced by the ion. By approximating the voltage as

$$V(x, y, t) = V(x, t)\,[1 + \alpha_1 y + \alpha_2 y^2] \tag{9.7}$$

and proceeding as above,

$$e_1 = \frac{4\alpha_1}{\pi} V_0 \cos \Theta. \tag{9.8}$$

By inserting (9.6) and (9.8) in (9.5), and the result in (9.3)

$$\ddot{x} + \omega^2 x = \frac{4e\,V_0 \cos \Theta}{\pi m}\left[\alpha_1 \cos \omega t + \frac{1}{r}\cos 2\omega t\right]. \tag{9.9}$$

The solution of (9.9) is

$$x = r \sin \omega t + \frac{2}{\pi}\frac{e\,V_0 \cos \Theta}{m\omega^2}\left[\alpha_1 \omega t \sin \omega t + \frac{2}{3r}\cos 2\omega t\right]. \tag{9.10}$$

The first term on the right is the regular cyclotron circular motion due to the magnetic field. The third term leads to a radial oscillation with twice the frequency of revolution. Its amplitude is $\dfrac{2}{3\pi}\dfrac{e\,V_0 \cos \Theta}{E} r$ which is no more than a few percent of the orbit radius after about three revolutions, and damps down rapidly with increasing radius.

The second term, by the method used to obtain (4.2), leads to an orbital instability; i.e., it causes the orbit center to precess in the x-direction at a rate per turn, $\Delta x/\Delta \nu$

$$\frac{\Delta x}{\Delta \nu} = 2\alpha_1 r^2 \frac{e\,V_0 \cos \Theta}{E}. \tag{9.11}$$

Experience indicates that in a typical standard cyclotron, the value of α_1 is generally between $\dfrac{1}{10R}$ and $\dfrac{1}{50R}$ where R is the radius of the dees. It decreases as the distance from the "spider" to the dees is increased. By adopting the larger value of α_1, and defining the maximum energy, E_m, as

$$E_m = \tfrac{1}{2} m \omega^2 R^2 \qquad (9.12)$$

(9.11) becomes

$$\frac{\Delta x}{\Delta v} = \frac{R}{5} \frac{e V_0 \cos \Theta}{E_m}. \qquad (9.13)$$

As discussed in Sect. 4, this precession continues in the same direction for about $\dfrac{1}{2\pi} \dfrac{\omega_r}{\omega_p}$ revolutions, so that it may be an important contributor to radial oscillations in some cases. In synchrocyclotrons, the voltage is not applied from one end of the dee, so that $\alpha_1 = 0$ and this problem is avoided.

II. The phase problem.

a) Initial motion region.

10. Phase grouping. In the initial motion region—that is, the part of the motion from emission of the ion from the ion source until an appreciable fraction of its path lies inside the dees—it is appropriate to approximate the electric field, $-\dfrac{\partial V}{\partial x}$, by

$$-\frac{\partial V(x,t)}{\partial x} = \frac{V_0}{d} \left[1 - A \left(\frac{x}{d}\right)^2 + B \left(\frac{x}{d}\right)^4 + \cdots \right] \cos (\omega_e t + \Theta_0) \qquad (10.1)$$

where, as used previously, V_0 is the dee-to-ground voltage in a two dee cyclotron, and half of that quantity in a single dee machine, d is a distance parameter which, it turns out [7], is very nearly equal to the distance between the dees, $A, B, \ldots$ are parameters depending on the dee geometry, ω_e is the frequency of the electric voltage applied to the dees, and Θ_0 is the phase of the electric voltage when the ion under consideration is emitted from the ion source at $t = 0$. The initial motion problem has been studied [7] for an actual cyclotron electric field as given by (10.1), but for purposes of this review, we will work in the so called "uniform field" approximation obtained by assuming $A = B = \cdots = 0$ in (10.1), and the differences between this and an actual cyclotron will be quoted from Ref. [7]. The same procedure will be followed with regard to the early part of the motion in the outer region in which some of the characteristics of the initial motion region are still in evidence but damp out more or less rapidly. In addition, we will assume in this section that $\omega_e = \omega$. With one minor exception which will be pointed out in the discussion, the effect of the difference between these quantities is the same in this region as in the outer regions of the cyclotron and will therefore be discussed in later sections.

The Lorentz equations of motion for an ion in this region are given by (8.1). If we adopt as initial conditions at $t = 0$, $x = \dot{x} = y = \dot{y} = 0$, the solutions, obtained by the methods of Sect. 9, are

$$\left. \begin{aligned} x &= D \{- \sin \Theta_0 \sin \omega t + \omega t \sin (\omega t + \Theta_0)\,, \\ y - 2 D \sin \Theta_0 &= D \{- \sin (\omega t + \Theta_0) + \omega t \cos (\omega t + \Theta_0) - \sin \Theta_0 \cos \omega t\} \end{aligned} \right\} \quad (10.2)$$

where

$$D = \frac{e V_0}{2 m \omega^2 d} \qquad (10.3)$$

a length which recurs frequently in the theory. For the Oak Ridge cyclotron which is used as an example in Ref. [7], $D = 0.54$ inches. It is somewhat smaller for most other standard cyclotrons, and very much smaller for synchrocyclotrons. The initial motion region extends until x becomes as large as d which, from (10.2), occurs when $\omega t \approx d/D$. By this time, the second terms in (10.2) are predominant, so that

$$r = (x^2 + y^2)^{\frac{1}{2}} \approx D\omega t. \tag{10.4}$$

At times when $(\omega t + \Theta_0)$ is an integral multiple of π, the initial conditions (9.1) are approximately attained; thus the initial motion region blends smoothly into the outer regions of the cyclotrons discussed in other sections. This result for the "uniform field" cyclotron is also valid to the same approximation in an actual cyclotron.

Perhaps the most important property of the initial motion is the "phase grouping" which was first reported by BOHM and FOLDY [4]. The phase of the electric voltage with respect to an ion, Θ, is given by arc sin x/r at times when the electric field is at a maximum. From (10.2), this occurs when

$$(\omega t + \Theta_0) = \nu \pi \quad (\nu \text{ integer}). \tag{10.5}$$

Using (10.5) in (10.2) gives

$$\Theta \approx \frac{x}{r} = -\frac{D}{r}\sin^2 \Theta_0. \tag{10.6}$$

This treatment is valid until $r = d$, so that the phases in a "uniform field" cyclotron are given by

$$\Theta \approx -\frac{D}{d}\sin^2 \Theta_0 \tag{10.7}$$

which corresponds to a grouping of the phases of all ions to within a very few degrees of the phase of the electric voltage. The minus sign in (10.7) indicates that the ions have passed through the line $x = 0$ shortly *before* the electric field reaches its maximum.

For an actual cyclotron, a grouping equation of the form $d\Theta/d\nu = -k\Theta$ may be derived, and k is found to decrease continuously but quite rapidly in the region $r > d$. The net result is essentially the same as given by (10.7). If the difference between ω and ω_e is taken into account, it turns out that the grouping is with respect to a frequency half way between ω and ω_e.

In order to increase ion source output, many cyclotrons incorporate protrusions from the dee to increase the electric field at the ion source slit. These are usually known as ion source "feelers", and their principal effect is to change the initial conditions for $\dot{x}$ at $t = 0$ to the value given by

$$\tfrac{1}{2}m\dot{x}^2 = eV_0 \cos \Theta_0. \tag{10.8}$$

Using this changes the solution of (8.1) to

$$x = 2(D d \cos \Theta_0)^{\frac{1}{2}} \sin \omega t + \text{terms from (10.2)}. \tag{10.9}$$

The last time (10.5) is satisfied in the initial motion region,

$$\Theta \approx -\frac{x}{d} = -2\left(\frac{D}{d}\cos \Theta_0\right)^{\frac{1}{2}} \sin \Theta_0. \tag{10.10}$$

The product $(\cos \Theta_0)^{\frac{1}{2}} \sin \Theta_0$ reaches a maximum value of 0.67, so that the phase grouping is considerably spread. Actually (10.8) rather overemphasizes the effect of ion source feelers, so that the actual phases are probably intermediate between (10.7) and (10.10).

One very practical effect of the phase grouping is that the output beam of a cyclotron is very sharply pulsed with a frequency equal to the electric frequency applied to the dees, and with a pulse duration of only a very few percent of a cycle. This pulsed nature of the beam has been employed to advantage in time-of-flight experiments[1]; it also introduces a complication into the use of coincidence techniques.

In addition to the phase grouping, an ion in an actual cyclotron undergoes a small shift toward positive phases arising from the non-linearity of the equations of the motion (8.1) when the higher terms in (10.1) are taken into account. This positive shift is proportional to D, and for the Oak Ridge cyclotron which has a relatively large D, it amounts to only one degree per turn for $r < d$, and falls off quite rapidly in the region $r > d$. For purposes of this article, this effect will be neglected except insofar as it affects results quoted from Ref. [7].

b) Phase problem in the standard cyclotron.

11. Phase history of an ion. In a standard cyclotron, the frequency with which the ions circulate, as given by (2.4), decreases as the ion energy increases due to (a) the decrease of the magnetic field necessary to obtain vertical focusing as discussed in Sect. 3, and (b) the relativistic increase in mass, which may be expressed as

$$m = m_0 \left(1 + \frac{E}{m_0 c^2}\right) \tag{11.1}$$

where m_0 is the rest mass. Thus, if the electric frequency is equal to the circulating frequency at the center of the cyclotron, the phase history will be as shown by curve A of Fig. 6. The initial phase is taken as $0°$ in view of the results of Sect. 10. A not untypical cyclotron (18 Mev deuterons) might have an average magnetic field 1.25% less than the field at the center and an average relativistic mass increase of 0.5% (these averages must be taken with respect to r^2 since the ion spends approximately equal times at each energy rather than at each radius); the average phase shift would then be 1.75% or 6.3° per turn. Since the total allowable phase shift in this case is 90°, the ions may make a total of only 14 turns. Assuming $\overline{\cos \Theta} = 0.7$, to ob-

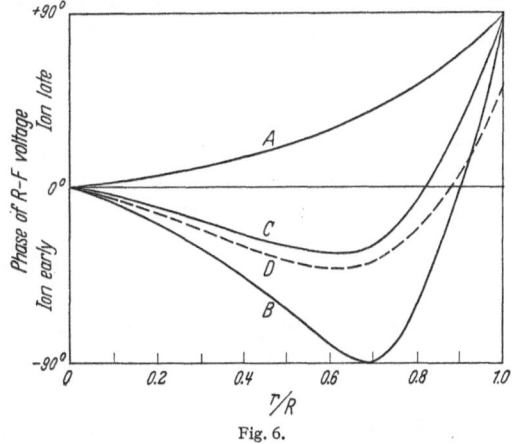

Fig. 6.

tain 18 Mev deuterons would require a dee-to-dee voltage of over 900 kv. It is immediately obvious that the same energy can be obtained with much smaller dee voltages if the electric frequency is somewhat lower than the resonant frequency at the center. In the extreme case, this leads to a phase history shown by curve B of Fig. 5. The total allowable phase shift due to magnetic field fall-off and relativistic mass increase is 90° plus the amount of negative phase shifting due to the low electric frequency, so that the required dee voltage is reduced by a large factor. This case is also unsatisfactory, however, since

---

[1] S. Bloom: Phys. Rev. **98**, 233 (1955).

according to (8.11) it leads to very strong electrostatic defocusing. Moreover, in allowing the phase to reach nearly $-90°$, many turns—and therefore much phase shifting—occur without appreciable increase in energy. In practice, cyclotrons generally operate with sufficient dee voltage to give a phase history as shown by curve C. The minimum phase is perhaps $-20°$, so that the energy gain remains near maximum. The electrostatic defocusing, while somewhat serious, is not catastrophic. On the other hand, appreciable increases in beam current can always be obtained by increasing the dee voltage and thus reducing the electrostatic defocusing.

It may be noted that curves A, B, and C of Fig. 5 all terminate at $+90°$. From the above discussion, it is apparent that this will give maximum current for a given dee voltage, since it keeps the region and magnitude of electrostatic defocusing to a minimum. On the other hand, the increase in radius per turn at the outside decreases to zero, which is an unsatisfactory condition for many purposes (e.g. deflection, spreading beam over a target, etc). Where a sacrifice in current to remedy this situation is warranted, the electric frequency may be slightly reduced from the value giving maximum circulating beam. This leads to a phase history as shown by curve D, where the voltage is the same as for curve C.

12. The phase equation. The qualitative considerations above can be expressed quantitatively as follows: The phase shift per turn, $d\Theta/d\nu$, is given by

$$\frac{d\Theta}{d\nu} = 2\pi \frac{\left(\omega_e - \frac{eH}{mc}\right)}{\omega} \tag{12.1}$$

where ω_e is the frequency of the R.F. voltage applied to the dees. By expanding H, m, and ω about their values at the center of the cyclotron, H_0, m_0, and ω_0, and retaining only lowest order terms, (12.1) may be written

$$\frac{d\Theta}{d\nu} = 2\pi \left(\frac{\omega_e - \omega_0}{\omega_0} + \frac{H_0 - H(r)}{H_0} + \frac{E(r)}{m_0 c^2}\right). \tag{12.2}$$

The first term on the right is a constant of the motion which may be adjusted to obtain the maximum current. The last two terms are functions of the radius r.

The energy gain per turn, $dE/d\nu$, is

$$\frac{dE}{d\nu} = 4eV_0 \cos \Theta \tag{12.3}$$

and the relationship between energy and radius is, to good approximation

$$E = \tfrac{1}{2} m\omega^2 r^2. \tag{12.4}$$

Eqs. (12.2) to (12.4) are three functional relations between the four variables Θ, ν, E, and r, and thus may be solved for the functional relationship between Θ and r in terms of their initial values. From the results of Sect. 10, we may use as the initial conditions $\Theta = 0$ at $r = 0$; strictly speaking, this cannot be used since (12.2) to (12.4) are not valid at small radii. However, in Ref. [7], this problem was investigated in detail, and its was found that no large error results if the solution is carried out under the assumption that those equations are valid at all radii. We therefore proceed with the solution under that assumption. By use of (12.2) to (12.4), the mathematical identity

$$\frac{d\sin\Theta}{dE} = \cos\Theta \, \frac{d\Theta/d\nu}{dE/d\nu} \tag{12.5}$$

may be written

$$\frac{d\sin\Theta}{d(r^2)} = \frac{\pi m \omega^2}{4 e V_0}\left(\frac{\omega_e - \omega_0}{\omega_0} + \frac{H_0 - H(r)}{H_0} + \frac{m\omega^2 r^2}{2 m_0 c^2}\right). \tag{12.6}$$

By integrating (12.6) and using the initial condition $\Theta = 0$ at $r = 0$,

$$\sin\Theta = \frac{\pi m \omega^2}{4 e V_0}\left\{\frac{\omega_e - \omega_0}{\omega_0} r^2 + \frac{1}{H_0}\int_0^r (H_0 - H)\,d(r^2) + \frac{m\omega^2 r^4}{4 m_0 c^2}\right\}. \tag{12.7}$$

This is the equation for the curves in Fig. 6. It gives the phase as a function of r, and the cyclotron parameters. According to the previous discussion, when the operation is such as to give a maximum circulating beam, Θ must be 90° at the outermost orbit radius, R. Inserting this into (12.7), we may solve for the electric frequency as

$$\frac{\omega_e - \omega_0}{\omega_0} = \frac{2 e V_0}{\pi E_m} - \frac{1}{H_0 r^2}\int_0^{R^2} (H_0 - H)\,d(r^2) - \frac{E_m}{2 m_0 c^2} \tag{12.8}$$

where E_m is the energy at radius R.

13. Applications of the phase equation. In order to illustrate the use of these equations, let us consider two example cyclotrons, one designed to give 18 Mev deuterons, and the other designed for 22 Mev protons. Let us assume the magnetic field falls of linearly with radius, and decreases a total of 1.8%. Then according to (12.8), if the deuteron machine has a maximum dee-to-dee voltage, of 300 kv, the electric frequency should be 1.2% less than the resonant frequency at the center. For the 22 Mev proton machine, assuming it to have a 400 kv maximum dee-to-dee voltage, the electric frequency should be about 1.8% less.

The radius at which the phase reaches its maximum negative value may be found by setting the right side of (12.6) equal to zero. This occurs in the deuteron cyclotron at $r = 0.52 R$, and in the proton cyclotron at $r = 0.58 R$. The maximum magnitude of this negative phase may then be found by substituting these results into (12.7). For the deuteron case this is 16°, and for the proton case it is 28°.

The dee-to-dee voltages required in order to avoid passing through a region of electrostatic defocusing may be obtained by setting the right side of (12.8) equal to zero. This gives 1050 kv for the deuteron case (in agreement with the rough calculations at the beginning of this section) and 1700 kv for the proton case. These voltages are far beyond the reach of present technology, so we see that a large loss of beam due to electrostatic defocusing is almost unavoidable, and certainly takes place in all existing cyclotrons.

The radius at which the electric defocusing is just counteracted by magnetic focusing, r_c, is determined by equating the two terms of (8.13) which gives

$$\frac{e V_0 \sin\Theta(r_c)}{E(r_c)} = \frac{r_c}{H}\left(\frac{dH}{dr}\right)_{r=r_0}. \tag{13.1}$$

This may be solved numerically using (12.7). It occurs for the deuteron cyclotron at $r_c = 0.50 R$ and for the proton cyclotron at $r_c = 0.61 R$. At this radius, the second time derivative of the vertical co-ordinate becomes negative, and the curvature, for the first time is toward the median plane. It is seen from (8.13) that the magnetic force becomes predominant very rapidly (the ratio between the magnetic and electric forces increases as r^3) so that not far beyond $r = r_c$, the vertical height reaches its maximum. Contrary to what has been believed until recently, the vertical motion up to this radius consists of a more or less *exponential* divergence from the median plane, with a very large fraction of the ions striking the dees. Because of the exponential nature of this divergence, the largest number strike the dees in the neighborhood of the radius r_c. Fig. 7 shows the vertical history of a proton in the ORNL 86 Inch Cyclotron [7] which starts from rest at the center of the cyclotron at the time the electric voltage is maximum (i.e. $\Theta_0 = 0°$). Ions starting at non-zero phase diverge much more than this; the maximum divergence of an ion with $\Theta_0 = \pm 30°$ is twice as large, and for ions with $\Theta_0 = \pm 60°$, it is about eight times as large as in the case shown in Fig. 7.

At radii larger than r_c, the curvature is always toward the median plane so that the vertical motion is oscillatory. The amplitude of these oscillations, Z, is, according to the WKB approximation, inversely proportional to the fourth root of the focusing force. Since the

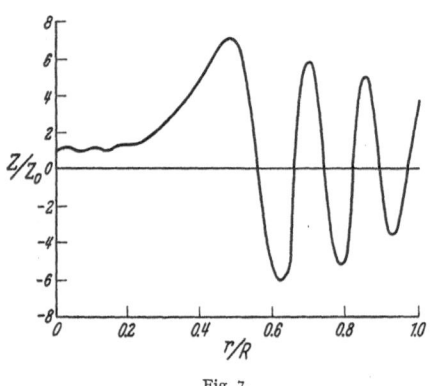

Fig. 7.

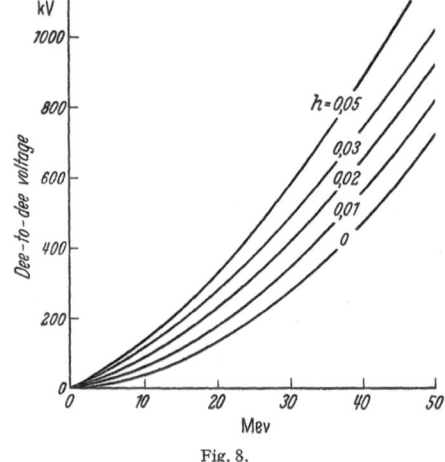

Fig. 8.

electric force rapidly becomes negligible, this is approximately

$$Z \propto \left(\frac{r}{H} \frac{dH}{dr} \right)^{-\frac{1}{4}}. \tag{13.2}$$

In order to obtain the maximum high energy circulating beam, the magnetic field should be tailored to keep Z in (13.2) constant at radii larger than r_c; this corresponds to a magnetic field fall off varying as $1/r$. If the fall off is any less rapid than this, more ions will be lost to the dees; if it is any more rapid, additional phase shifting will occur, so that (12.7) will give larger negative values of Θ and more ions will be lost due to electrostatic defocusing.

If the magnetic field is tailored in this way, the beam will fill the dees at the outside radius. This situation is favorable for the problem of cooling a target placed in the circulating beam, but it is unfavorable for deflection and for many other purposes. If a beam of small vertical height is desired, it can be achieved by increasing the field fall-off at large radii; this, of course, results in a decrease of circulating beam.

The minimum voltage with which a given energy particle may be obtained can be calculated by setting the minimum value of Θ equal to $-90°$. To do this, we first compute the relation between ω_e and r for the minimum value of Θ by setting the right side of (12.6) equal to zero; we then substitute this into (12.7) with $\sin \Theta$ set equal to minus one, and solve simultaneously with (12.8). The procedure is not unduly complicated when carried out numerically for any given field shape and maximum energy; however, for simplicity we will consider here only the case of a linear field fall off; that is, we will assume

$$\frac{H_0 - H}{H_0} = h\,\frac{r}{R} \tag{13.3}$$

where h is constant. The minimum voltage required to obtain a given maximum energy, as obtained from the calculation outlined above is

$$\left.\begin{aligned} V_0 &= 0.000\,14\,\frac{E_m^2}{e\,m_0\,c^2}\left(1 + \frac{4}{5}\,h\,\frac{m_0\,c^2}{E_m}\right), \quad \frac{4}{5}\,h \ll \frac{E_m}{m_0\,c^2}, \\ V_0 &= \frac{0.107\,h}{e}\,E_m\left(1 + \frac{5}{4}\,\frac{E_m}{h\,m_0\,c^2}\right), \quad\quad \frac{4}{5}\,h \gg \frac{E_m}{m_0\,c^2}. \end{aligned}\right\} \tag{13.4}$$

For the example cases we have been considering, the threshold dee-to-dee voltage is 110 kv for the deuteron machine and 220 kv for the proton machine.

Fig. 8 shows a plot of (13.4) for protons. The two regions represented by the two equations were found to join smoothly, so the curves should be valid for all energies and values of h. Fig. 8 may also be considered as a plot of maximum energies attainable with a given dee voltage.

The threshold voltages considered here are, of course, far too low to be practical. They would give ions with phase histories like curve B of Fig. 6, so that electrostatic defocusing would all but eliminate the beam. Since the voltages used in the example cyclotrons we have been considering are more or less in line with what experience has shown to be adequate, practical operating voltages should probably lead to a maximum negative phase no greater than 30°.

c) Phase problem in the synchrocyclotron.

14. Qualitative considerations. Since synchrocyclotrons accelerate ions to velocities comparable to the velocity of light, it is necessary to use the relativistic definition of kinetic energy, E, as

$$E = m c^2 - m_0 c^2 \tag{14.1}$$

where the relativistic mass, m, is given by

$$m = \frac{m_0}{\sqrt{1 - v^2/c^2}}. \tag{14.2}$$

The angular frequency with which the ions revolve may then be obtained from (2.4) by use of (14.1) as

$$\omega = \frac{e\,H\,c}{E + m_0\,c^2}. \tag{14.3}$$

The basic principle of a synchrocyclotron is that the electric frequency decreases with time so as to be equal to (14.3) at all times for at least some ions (these will be referred to as "synchronous" ions). For these ions, then

$$\omega_e = \frac{e\,H\,c}{E_s + m_0\,c^2} \tag{14.4}$$

where ω_e is the frequency of the electric voltage applied to dee. Since the magnetic field must decrease with increasing radius to insure magnetic focusing, H is a function of r and thus of E_s. The rate at which ω_e should change with time is determined by the rate at which the energy increases with time due to electric acceleration across the dee gap. This acceleration gives an energy increase per turn

$$\frac{2\pi}{\omega_e}\,\frac{dE_s}{dt} = e\,V_1 \sin\varphi_s \tag{14.5}$$

where V_1 is twice the maximum electric voltage across the dee gap (and thus $e V_1$ is the maximum possible energy gain per turn) and φ_s is the designed phase of the synchronous ions. In synchrocyclotron theory, it is customary to define "phase" as the angular position of the ions at the instant when the electric voltage is zero. This phase is greater than the phase Θ, used in Sects. 8 to 13 by

$$\varphi = 90° - \Theta. \tag{14.6}$$

One important purpose of the phase theory is to determine the optimum value of φ_s which, combined with (14.5) and (14.4), determines the rate at which the applied frequency should be decreased.

Since ions in a typical synchrocyclotron make many tens of thousands of revolutions, it is important that their motion be characterized by "phase stability"; that is, if their phase and energy at any time are different from φ_s and E_s, forces come into effect which tend to restore them toward their synchronous values. This will be the case only if φ_s is between 0° and 90° and the deviations are not large. For example, consider the case where φ is larger than φ_s. The energy gain will be larger than (14.5), so that the ion will soon have a higher energy than E_s. Its angular velocity thus decreases in accordance with (14.1) to (14.3) and becomes less than the electric frequency (14.4); this in turn causes φ to decrease, although $E - E_s$ continues to increase. When φ becomes less than φ_s, the rate of energy gain becomes less than the synchronous value, so that $E - E_s$ begins to decrease. This continues until the energy has decreased below the synchronous value, which causes the angular velocity of the ions to begin increasing relative to ω_e, and thus causes φ to begin increasing. Thus the phase and energy history of an ion consists of oscillations of $\varphi - \varphi_s$ and $E - E_s$ about zero, with the energy oscillations lagging the phase oscillations by 90°. For φ_s greater than 90°, a positive $\varphi - \varphi_s$ leads to less energy gain which leads to an increase in $\varphi - \varphi_s$, and thus, eventually, to an exponential divergence of both φ and E from their synchronous values. For $\varphi_s < 0°$, the ions do not gain energy, so that this case is of no practical interest.

15. Phase oscillations. In a quantitative analysis of these problems, the complexity is reduced if the variation of the magnetic field is neglected, so that this procedure will be adopted here. A complete treatment of the problem including the magnetic field fall-off is given by BOHM and FOLDY [4]. Their results are, in all cases, qualitatively and almost quantitatively the same as those obtained here. For our treatment, we introduce the angular variable of the ion in the cyclotron, α, which increases with angular velocity ω. The rate of phase shift, $d\varphi/d\alpha$, may thus be written

$$\frac{d\varphi}{d\alpha} = 1 - \frac{\omega_e}{\omega} \tag{15.1}$$

or, from (14.3) and (14.4),

$$\frac{d\varphi}{d\alpha} = 1 - \frac{\varepsilon}{\varepsilon_s} \tag{15.2}$$

where

$$\varepsilon = E + m_0 c^2 \tag{15.3}$$

the total energy. From the definition of α, it may be noted that,

$$\frac{d}{d\alpha} = \frac{dt}{d\alpha}\frac{d}{dt} = \frac{1}{\omega}\frac{d}{dt} = \frac{1}{\omega_e}\left(1 - \frac{d\varphi}{d\alpha}\right)\frac{d}{dt} \tag{15.4}$$

where the last was obtained using (15.1). By multiplying both sides of (15.2) by ε_s^2, and taking $d/d\alpha$, it becomes, with the use of (15.4) and reapplication of (15.2)

$$\frac{d}{d\alpha}\left(\varepsilon_s^2 \frac{d\varphi}{d\alpha}\right) = \frac{1}{\omega_e}\left(1 - \frac{d\varphi}{d\alpha}\right)\frac{d\varepsilon_s}{dt} \cdot \varepsilon_s\left(1 + \frac{d\varphi}{d\alpha}\right) - \varepsilon_s \frac{d\varepsilon}{d\alpha}. \tag{15.5}$$

Assuming that the phase and energy deviations from the synchronous values are small, $d\varphi/d\alpha$ is small from (15.2) so that its square may be neglected. Also, the energy gain per turn, $2\pi\, d\varepsilon/d\alpha$, is

$$2\pi \frac{d\varepsilon}{d\alpha} = eV_1 \cos\varphi. \tag{15.6}$$

Thus, (15.5) becomes, with the further use of (14.5) and (15.3)

$$\frac{d}{d\alpha}\left(\varepsilon_s^2 \frac{d\varphi}{d\alpha}\right) = \varepsilon_s \frac{eV_1}{2\pi}(\sin\varphi_s - \sin\varphi). \tag{15.7}$$

This is the equation of phase oscillation. If α is replaced by t, it is identical with the equation of motion of a pendulum acted upon by a constant torque. Some of the properties of the motion can best be visualized by referring to the pendulum model.

The equilibrium position is at $\varphi = \varphi_s$ since, from (15.7) there is no force acting at that point. The solution of (15.7) is simplified considerably if we consider only small oscillations, and define a new variable, β, as

$$\beta = \varphi - \varphi_s. \tag{15.8}$$

In view of the fact that β is small, we may use,

$$\sin\varphi = \sin(\beta + \varphi_s) = \sin\varphi_s + \beta\cos\varphi_s. \tag{15.9}$$

By using this in (15.7), it becomes

$$\frac{d}{d\alpha}\left(\varepsilon_s^2 \frac{d\beta}{d\alpha}\right) = -\beta\,\varepsilon_s \frac{eV_1}{2\pi}\cos\varphi_s. \tag{15.10}$$

From (15.4) and by again invoking the smallness of β, this may also be written

$$\frac{d}{dt}\left(\varepsilon_s^2 \frac{d\beta}{dt}\right) = -\beta\,\omega_e^2\,\varepsilon_s \frac{eV_1}{2\pi}\cos\varphi_s. \tag{15.11}$$

(15.11) is the familiar equation of simple harmonic motion as applied, for example, to a pendulum; the solution is

$$\beta = A_1 \sin\sqrt{\frac{k}{I}}\,(t - t_1) \tag{15.12}$$

where

$$\left.\begin{aligned} k &= \omega_e^2\,\varepsilon_s \frac{eV}{2\pi}\cos\varphi_s, \\ I &= \varepsilon_s^2 \end{aligned}\right\} \tag{15.13}$$

the restoring force and moment of inertia, respectively, of the equivalent pendulum, and A_1 and t_1 are integration constants determined by initial conditions. From (15.12) and (15.13), the frequency of the phase oscillations, ω_φ, is

$$\omega_\varphi = \sqrt{\frac{eV_1\cos\varphi_s}{2\pi\,\varepsilon_s}}\,\omega_e. \tag{15.14}$$

A beam modulation frequency given by (15.14) has been observed in the Berkeley 184 inch synchrocyclotron [33].

Since the explicit variation of ε_s and ω_e with time is small, the adiabatic theorem of periodic motion[1] states that the action integral, J, must be approximately

---

[1] Cf., e.g., R. B. Lindsay: Physical Mechanics, p. 276. New York: D. van Nostrand Co. 1933.

a constant of the motion; invoking the definition of J, this requires

$$J = \oint p_\beta d\beta = \text{const} \tag{15.15}$$

where p_β is the "angular momentum" of the β-variable and $\oint$ indicates that the integration is to be carried out over one complete cycle of the motion. By invoking the pendulum analogy,

$$p_\beta = I \frac{d\beta}{dt} = I A_1 \omega_\varphi \cos \omega_\varphi t ,$$

$$d\varphi = \frac{d\varphi}{d\omega_\varphi t} d\omega_\varphi t = A_1 \cos \omega_\varphi t \, d\omega_\varphi t$$

whence, from (15.15)

$$I A_1^2 \omega_\varphi = \text{const}$$

or, by using (15.13) and (15.14)

$$A_1 \propto \varepsilon_s^{-\frac{3}{4}} (e V \cos \varphi_s)^{-\frac{1}{4}} . \tag{15.16}$$

Thus the amplitude of the phase oscillations decreases as the energy increases, so that the oscillations are "damped". This introduces an important simplification into the phase stability problem, since it indicates that if the ion survives the first phase oscillation, it will survive all subsequent phase oscillations and be accelerated to high energy.

By application of (15.12) to (15.2), it is seen that the difference between the instantaneous and synchronous energies undergo oscillations with the same frequency as the phase oscillations, but lagging 90° behind them.

16. Phase stability. The next problem to be considered is that of phase stability. This requires an investigation of large oscillations, so that the approximations following (15.8) are not valid. From (15.7), the quantity $d^2\varphi/d\alpha^2$ changes sign when $\varphi = \pi - \varphi_s$. For larger values of φ than this, the "phase acceleration" is no longer restoring, so that this is the limit of stability. The equation of motion for the largest stable oscillation is thus obtained by integrating (15.7) and evaluating the integration constant from the condition

$$\frac{d\varphi}{d\alpha} = 0 \quad \text{when} \quad \varphi = \pi - \varphi_s . \tag{16.1}$$

Proceeding with the integration, we neglect the variation of ε_s with time and use the identity

$$\frac{d^2\varphi}{d\alpha^2} = \frac{d\varphi}{d\alpha} \frac{d}{d\varphi} \left(\frac{d\varphi}{d\alpha} \right)$$

which transforms (15.7) to

$$\frac{d\varphi}{d\alpha} d\left(\frac{d\varphi}{d\alpha} \right) = \frac{e V}{2\pi \varepsilon_s} (\sin \varphi_s - \sin \varphi) \, d\varphi . \tag{16.2}$$

Integrating both sides gives

$$\left(\frac{d\varphi}{d\alpha} \right)^2 = \frac{e V}{\pi \varepsilon_s} \left[\varphi \sin \varphi_s + \cos \varphi + c_1 \right] \tag{16.3}$$

where C_1 is the integration constant. Evaluating C_1 from (16.1), and substituting from (15.2) and (15.3), leads to

$$(E_s - E)^2 = \frac{e V \varepsilon_s}{\pi} \left[\varphi \sin \varphi_s + \cos \varphi + \cos \varphi_s - (\pi - \varphi_s) \sin \varphi_s \right] . \tag{16.4}$$

This gives the relationship of the energy and phase for the largest stable phase oscillation. If, for a given deviation of energy from the synchronous energy, the phase is larger than the value from (16.4), the ion will be lost; and conversely the ion will also be lost, if for a given phase, the energy deviation is larger than the value from (16.4). Plots of Eq. (16.4) are given by BOHM and FOLDY [4].

The most important application of (16.4) is the determination of the conditions for acceptance of ions emitted from the ion source into a stable phase. In Sect. 10, it was found that, to a very good approximation, the initial phase of all ions is $\varphi = \pi/2$. The condition for acceptance into a stable phase is thus determined by substituting $\varphi = \pi/2$ at $E = 0$ into (16.4). By using (15.3), this gives

$$\left.\begin{aligned} E_{s_0} &= \pm \sqrt{\frac{eV m_0 c^2}{\pi}} \times \\ &\times \sqrt{\left(\varphi_s - \frac{\pi}{2}\right)\sin\varphi_s + \cos\varphi_s} \, . \end{aligned}\right\} \quad (16.5)$$

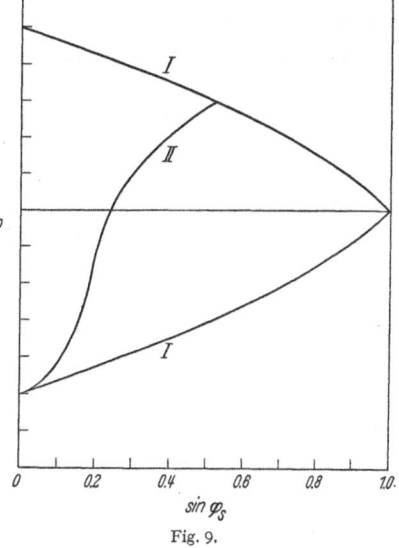

Fig. 9.

All ions emitted from the ion source when the synchronous energy is within the limits given by (16.5)—that is, when the electric frequency is within the limits obtained by substituting (16.5) into (14.4)—are accepted. All others are lost.

The maximum possible efficiency of a synchrocyclotron is obtained if the synchronous energy is varied from $-E_s$, to the final energy, E_m. Since, in accordance with (14.5), the synchronous energy increases with time at a rate inversely proportional to ω_e which, from (14.4) and (15.3) is inversely proportional to ε, the maximum possible efficiency for capture into stable phases, $\mathscr{E}$, is

$$\mathscr{E} = \frac{2E_{s_0}}{E_m + E_{s_0}}\frac{\varepsilon_0}{\bar\varepsilon} \approx \frac{\sqrt{eV_1 m_0 c^2}}{E_m\left(1 + \frac{E_m}{2m_0 c^2}\right)}\sqrt{\frac{4}{\pi}\left[\left(\varphi_s - \frac{\pi}{2}\right)\sin\varphi_s + \cos\varphi_s\right]} \quad (16.6)$$

where $\bar\varepsilon$, the average value of ε during the acceleration, is taken as $m_0 c^2 + E_m/2$, and $\varepsilon_0 = m_0 c^2$.

The synchronous phase φ_s should be chosen to give the maximum efficiency for this capture. From (16.6), this occurs when E_{s_0} is maximized. Fig. 9 shows a plot of E_{s_0} vs. φ_s from (16.5). From it we see that E_{s_0} is a maximum when φ_s approaches zero.

However, one other factor must be considered in the choice of φ_s—namely, the possibility that, in the course of the first energy oscillation, the ion will return to the origin and thus be lost. For φ_s near $\pi/2$, only very small phase oscillations are permissible within the limits of phase stability, so that the phase never becomes negative, deceleration never occurs, and hence return to the origin is not a problem. For φ_s near zero, however, deceleration is experienced for a long period of time. In fact, since the energy oscillation is about its synchronous value and the latter is near zero, the oscillation will surely lead to negative energies and thus to loss of ions by return to the origin *unless* the synchronous energy increases rapidly enough to prevent this from happening. For $\varphi_s = 0$, the synchronous energy does not increase, so that all ions are lost.

For small values of φ_s, ions which start when E_s is large and positive require a relatively long time to reach the minimum of their energy oscillation; E_s thus has a longer time in which to increase, so that these ions will not be lost by return to the ion source. Conversely, ions which start when E_s is negative will be lost for small φ_s.

This problem has been treated in detail by BOHM and FOLDY [4], and their result is shown by curve II of Fig. 9. They found that the smallest value of φ_s for which no phase stable ions return to the origin is about 30°. As seen from Fig. 9, the spread of E_{s_0} leading to accepted ions is greatest at that phase; it is therefore the most favorable synchronous phase, and synchrocylotrons are customarily designed for it.

17. Numerical illustration. The results of Sects. 15 and 16 would perhaps be more meaningful if illustrated with a typical numerical example. For this purpose, we choose a synchrocyclotron which accelerates protons to 400 Mev. We assume a magnetic field of 20000 gauss, a dee-to-ground voltage of 10 kv and a synchronous phase of 30°. From (2.11), the outermost (400 Mev) radius is 62.8 inches; at 100 Mev, the radius is 28.7 inches. According to (2.7), the velocities at 100 and 400 Mev are 0.43 and 0.71 times the velocity of light, and the masses are greater than the rest mass by 11% and 43% respectively. The electric frequency may be found from (14.4); at 0, 100, and 400 Mev, it is 30.7, 27.8, and 21.5 mc/sec. These correspond to elapsed times per revolution of 32.6, 36.0, and 46.5 millimicroseconds respectively. From its definition following (14.5), $eV_1 = 0.02$ Mev, so that $eV_1 \sin \varphi_s = 0.01$ Mev. The energy gain is thus at a rate of 100 revolutions per Mev, or a total of 40000 revolutions during the acceleration. This requires a time of 1580 microseconds, so that the maximum possible repetition rate is about 630 per second.

From (15.14), a single phase (or energy) oscillation requires 580 revolutions of the ion at zero energy; at 100 and 400 Mev, it takes 610 and 690 revolutions respectively. According to (15.16), the amplitude of the phase oscillation is reduced at the latter two energies to 93 and 77% respectively of the amplitude at zero energy.

In accordance with (16.5), ions will be admitted to phase stable orbits if they are emitted from the ion source while the synchronous energy is between -1.43 and $+1.43$ Mev; that is while the electric frequency is between 30.747 and 30.653 Mc/sec. From (14.5), this covers a time interval of 8.6 microseconds. Since, according to curve II of Fig. 9, no phase stable ions return to the origin when $\varphi_s = 30°$, all of the above ions will be accelerated to the full 400 Mev unless they are lost by striking the dees. Neglecting that source of loss which is small and occurs only during the first few turns, the maximum possible efficiency, as calculated from (16.6), is 0.58%. This could also have been obtained as the ratio of the acceptance time, 8.6 microseconds to the total acceleration time, 1580 microseconds. The actual efficiency is the acceptance time times the repetition rate. If the latter were 60 cycles per second, the efficiency would thus be 0.052%. For a given ion source output and neglecting ions lost to the dees, the current in this synchrocyclotron is less than the current in a standard fixed frequency cyclotron by that factor.

B. Cyclotron components.

a) The magnetic field.

18. Poles and pole faces. An electromagnet is a simple device, well known and much used in many fields of science and engineering. However, due to the fact that it is a very major cost item in a cyclotron, its design and construction requires extensive consideration. In this and the two succeeding sections, the three large components of the magnet—the poles, the exciting coils, and the yoke—will be discussed in turn. These are shown for identification purposes in Fig. 10.

The magnetic field of a cyclotron must extend over an area sufficient to confine the ions up to their maximum energy; this requires a radius as given by (2.9) or (2.11). The diameter to gap ratio is usually between 5 and 10. A ratio of about six is typical for small standard cyclotrons, while nine is more typical for large synchrocyclotrons. Throughout the usable portion, the field must be relatively uniform azimuthally and must decrease monotonically with increasing radius. In view of the discussion in Sect. 4, it is advisable to keep the first harmonic of the azimuthal variations to less than 0.1 % at all radii. It is customary to make the field fall-off approximately linear with radius, with a total decrease of about 2% in standard cyclotrons and about 5% in synchrocyclotrons. In the former, this is a compromise between the vertical focusing requirement and the strain imposed on the phase problem as discussed in Sect. 11 to 13; in synchrocyclotrons, the phase difficulty is completely overcome by the frequency modulation, so that the amount of decrease is limited only by the

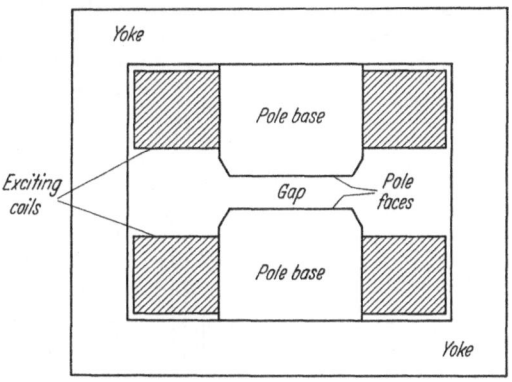

Fig. 10.

fact that it decreases the maximum energy obtainable with a given magnet. Since the first derivative of the magnetic field must be continuous everywhere, the linear fall-off cannot be extended to the center. In order to extend it as far toward the center as possible, the polefaces are sometimes designed with spikes at the center protruding into the gap.

The aim of magnet design is to obtain the desired energy, or what is equivalent, the desired HR, where R is the radius of the usable portion of the field, for a minimum cost. The cost of a magnet is proportional to its volume, so that it varies approximately as R^3, whereas the dependence of cost on magnetic field strength is roughly linear. Thus, for a given HR, there is every advantage in using very large magnetic fields and correspondingly small R. This argument maintains validity up to the point where saturation in the iron becomes a serious problem, so that cyclotron magnets customarily operate at fields between 15 000 and 22 000 gauss. Where it is planned to operate at different fields, as for example in a variable energy standard cyclotron, very high fields are not desirable since field shapes are quite sensitive to the magnitude of the field when iron is near saturation.

The cost of the magnet is essentially established once the flux through the pole base, φ_B, is determined. If all of this flux were useful, (i.e., if the "flux efficiency" were unity) it would be proportional to HR^2, so that the HR available from a given φ_B (and thus from a given cost) would be proportional to $1/R$. This indicates that there would be an advantage in tapering the poles so that the area of the pole face is smaller than that of the base. In practice, a tapered pole is not as favorable as in this idealized case, but at moderate fields, it is usually of some advantage. At very high fields, tapered poles are advantageous because the magnetic flux through the pole faces is limited by the saturation of the iron in the pole base. Using the tapered poles alleviates this problem and thus allows the attainment of higher fields in the gap and consequently of higher energy. At somewhat lower fields where the iron is not so completely saturated, the relative advantage of tapered poles is not easily understandable, so that the only reliable information is that obtained from model tests.

It is well known that flat, nearly parallel polefaces produce magnetic fields that begin to fall off appreciably at a distance of about one-half the magnetic gap inside of the outside edge. This immediately reduces the usable portion of the field to 80 to 90% of the pole face radius. An obvious method of improving this situation is to decrease the gap near the outside by making raised edges on the pole face.

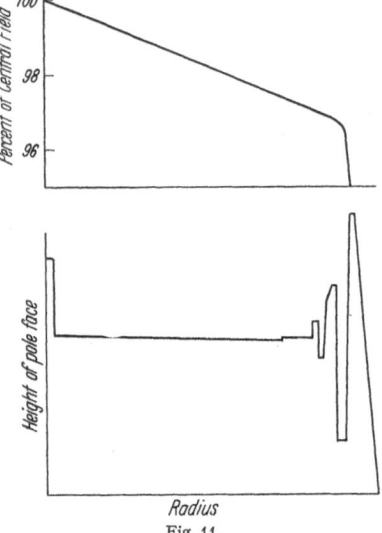

Fig. 11.

Probably the most complete theoretical and experimental analysis of cyclotron magnets was that carried out by Foss [9] and used in connection with the design of the magnet for the Carnegie Tech Synchrocyclotron. Since the magnetic field in the gap is irrotational, it must be derivable from a potential which satisfies Laplace's equation, and since the permeability of iron is very high, the polefaces are approximately equipotential surfaces. The potential problem can thus be solved by the well known Schwartz transformation method to give the field for any poleface configuration. Foss used this method in conjunction with experimental model studies to develop a very efficient pole face edge. Using a magnet with a nine to one diameter to gap ratio, he found that a flat pole face has a useful radius of only about 87% of the pole face radius. By using a raised edge, this could be increased to 92%. With a groove just inside this edge, it could be increased to 96%. The final design he obtained for the poleface is shown in Fig. 11 along with a plot of the magnetic field on the median plane. The usable radius obtained was 96.5% of the poleface radius.

Foss' studies were confined to very high fields where saturation of the iron in the pole base is an important limitation. He thus found that tapered poles were of considerable advantage, allowing about 10% higher energy for the same amount of iron and copper. In the design of the magnet for the Rochester Synchrocyclotron, Sewell[1] found that, at somewhat lower magnetic fields, tapered poles were of no advantage.

Since the thickness of the vacuum tank is an appreciable fraction of the magnet gap, it is customary to make the pole-tips part of the vacuum tank. In standard cyclotrons, they are usually made as "lids" for the sides of the vacuum tank, to which they are attached by gasketing or welding. Small gaps are usually left between the back of the lids and the remainder of the poles to allow insertion of "shims" to make small corrections in the field shape; the recent tendency, however, has been not to depend on this type of shimming. In the large synchrocyclotrons, the pole tips are attached directly to the poles, and the sides of the vacuum tank are gasketed to it.

---

[1] D. Sewell: Private communication to E. D. Hudson.

19. Exciting coils. The purpose of the magnet exciting coils is to apply a certain number of ampere-turns of magnetomotive force to the magnetic circuit. Assuming that the reluctance is made up entirely of the air gap, the poles are not tapered, and the flux efficiency is unity, the required number of ampere turns, NI, is

$$NI = 2.02\,Hg \qquad (19.1)$$

where H is the field in gauss and g is the gap in inches. In practice, between 1.5 and 3 times that many ampere-turns are usually needed. It is difficult to determine this accurately without model tests, or experience with a very similar magnet configuration.

Once the required number of ampere-turns has been determined, the current times the length of conductor, l, is essentially fixed; or

$$Il = C_0 \qquad (19.2)$$

where C_0 is a constant. The electric power, P, used in exciting the coils is

$$P = I^2 \varrho \frac{l}{A_c} \qquad (19.3)$$

where ϱ is the resistivity of the conductor and A_c is its cross sectional area. Combining (19.2) and (19.3),

$$P = C_0 \varrho \frac{I}{A_c}. \qquad (19.4)$$

The weight of conductor required, W_C, is proportional to its volume, or

$$W_C \propto l\,A_c. \qquad (19.5)$$

Using (19.2), this becomes

$$W_C \propto \frac{A_c}{I}. \qquad (19.6)$$

In addition, the weight of the yoke, W_y, is quite dependent on the volume of the coil, V_C; introducing the coil form factor, f_c, as the fraction of the cross section of the coil which is made up of conductor, we have

$$W_y \gtrsim V_c \propto \frac{W_c}{f_c} \propto \frac{1}{f_c}\frac{A_c}{I}. \qquad (19.7)$$

From (19.4), (19.6), and (19.7), we see that the current density in the conductor, I/A_c, is the determining factor in both the power costs and the material costs. High current densities save material costs, but lead to higher power costs, and vice versa. The compromise that is reached is determined by the relative costs of materials and power at the location of the machine. Current densities of one to two thousand amps per square inch are generally found most economical.

From (19.4) it is clear that the conductor should be of low resistivity material. Copper is therefore used most frequently, although aluminum is sometimes used. In achieving a given number of ampere turns with a given power expenditure, aluminum is somewhat cheaper than copper, but the bulkiness of the aluminum coil and the additional iron this necessitates for the yoke are generally considered to more than compensate the extra expense of copper.

From (19.7) it is clear that the size of the coil and therefore the cost of the iron yoke is minimized if the form factor is high. For this reason, conductor of rectangular rather than circular cross section is generally used in cyclotron magnets. Typical form factors are about 70%.

Once the number of ampere turns and the current density have been determined, there still remains the question of whether to achieve them with many turns of small conductor or with fewer turns of large cross section conductor. In many cases, the choice is determined by the availability of motor generator sets. If V_R and I_R are the rated voltage and current of the D.C. supply, the resistance of the coil, R_C, must be

$$R_C = \frac{V_R}{I_R}.\tag{19.8}$$

However, R_C is proportional to l/A_C. Since I/A_C and Il have been determined, $A_C \propto I \propto 1/l$. Thus

$$R_C \propto l^2 \propto \frac{1}{A_C^2}.\tag{19.9}$$

By combining (19.8) and (19.9), A_C is determined by the rated voltage and current of the motor generator set. For coils of the size used in cyclotrons, the most commonly available sets are suited to conductors of something less than one square inch cross sectional area, although supplies for much larger conductors are available and have been used. The principal advantages of large cross section conductors is that they allow better form factors and more efficient cooling. The principal disadvantage is that the coils are more difficult to fabricate.

Current densities of a few thousand amps per square inch cause heating of a few watts per cubic inch so that measures must be taken to cool the coils. The three methods most commonly used, in order of increasing efficiency, are:

1. The coils are wound as thin "pancakes" and water cooled plates are placed between them with a thin layer of insulating material between the coils and plates.

2. An insulating oil is made to pass around and between the coils. The oil is then cooled by water in a heat exchanger.

3. Cooling water passes though a hole in the center of the conductor, or through a tube soldered to it.

All three methods are good enough for the current densities used in most cyclotrons, although the first is somewhat marginal and leads to some temperature rise and a consequent increase in resistivity. The third is the most efficient; it has been used for current densities up to 50000 amp per square inch. Both the first and third methods can give trouble due to condensation of water on the conductor when the humidity is high. The oil cooling method is probably the most widely used. Its principal disadvantage is in the fire hazard; when used, it should be accompanied by adequate fire alarm and extinguishing facilities.

In addition to the three methods of cooling listed above, a few cyclotrons use only air cooling for the exciting coils. This requires a relatively large amount of copper, and appreciably increases the size of the magnet.

20. Yoke. The purpose of the magnet yoke is to complete the magnetic circuit with a low reluctance path, and to serve as a structural support. It is generally made of a low carbon steel rather than of iron since the former has almost equally favorable magnetic properties and is considerably cheaper. The cross sectional area of the yoke should be about equal to the cross sectional area of the poles, although steel in the yoke may be saved at the expense of additional ampere turns, so that the balance between the two should be determined by the economics of the situation. In considering a magnet yoke with two flux paths such as that shown in Fig. 10, the cross sectional areas of the two paths should be added. Since a magnet yoke requires very large masses of steel, relatively large savings may be achieved by seemingly minor considerations. For example, the average circumference is usually minimized by making it wide and thin, and outside corners which have little value as a flux path are often left off.

Because of the very large amounts of energy stored in the gap of the magnet, there are very large forces tending to pull the pole faces together. The magnitude expressed in pounds of force is 0.576 times the field in kilogauss times the pole face area in square inches. Typically, this force is several hundred thousand pounds. In addition to the difficulty introduced by this very large strain, the tolerable deflection without seriously affecting the field in the gap is very small. In some cyclotrons, this load is supported entirely by the yoke, although it is more usual to use spacers of non-magnetic material to keep the gap spacing correct and help support the load.

Magnet yokes are frequently constructed of laminated plates. This increases the original cost, but greatly facilitates the problems of transportation from the fabrication plant and installation at the site.

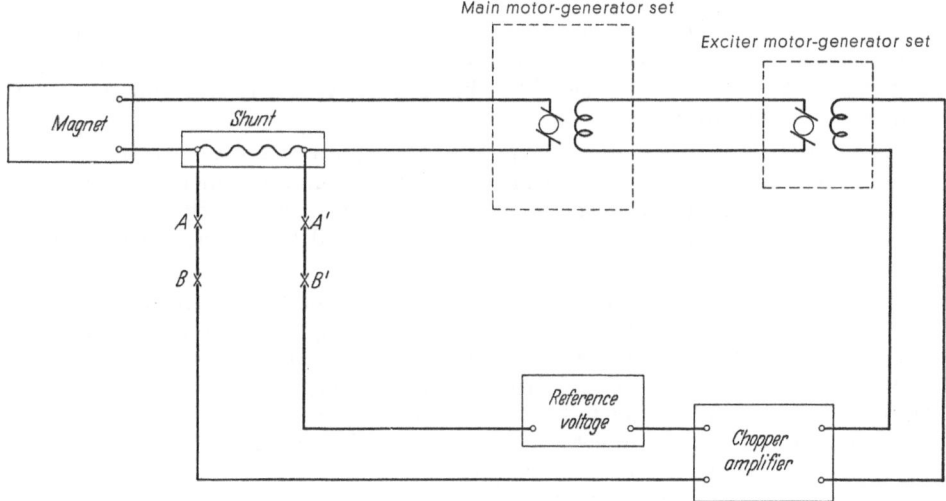

Fig. 12.

21. Miscellaneous magnetic field problems. $\alpha)$ *Current regulators.* The purpose of the current regulator is to smooth out small changes in the magnetic field occurring in time intervals short compared to those associated with variations in room temperature and other important external influences. Changes occurring over longer periods of time are usually compensated by manual regulation on the part of the operator.

Since the generator which energizes the magnetic field must be large and high powered, whereas the error signal from slight changes in the magnetic field is quite small, it is usually necessary to introduce an exciter generator whose field is affected by the error signal and which in turn excites the field of the main generator. A typical system is shown schematically in Fig. 12. The current through the magnet also passes through a series shunt, so that any change in this current results in a change in the voltage across the shunt. The voltage across the shunt is compared with a reference voltage (usually from a voltage divider across a battery), and the difference between the two is amplified and applied to the field of the exciter. This changes the exciter output voltage which changes the field of the main generator, and thus causes the current through the magnet to change in the direction which compensates the original change. A measure of the sensitivity of the device is the *"loop-gain"* which is the voltage multiplication around the loop. For example, consider the wires to be broken between A and B and between A' and B' in Fig. 12, and a one millivolt potential difference to be applied between B and B'. The current through the magnet will then change so as to produce a voltage between A and A' equal to the loop gain times one millivolt.

If T_1, T_2, and T_3 are the time constants of the magnet, the main generator field, and the exciter generator field respectively, it can be shown[1] that the maximum loop gain which does not lead to oscillation is

$$\text{Maximum Loop Gain} = (T_1 + T_2 + T_3)\left(\frac{1}{T_1} + \frac{1}{T_2} + \frac{1}{T_3}\right) - 1. \qquad (21.1)$$

Typical numerical values are $T_1 = 20$ sec, $T_2 = 4$ sec, $T_3 = 0.03$ sec. The maximum loop gain from (21.1) is then about 800.

The magnetic field may be regulated manually by changing the reference voltage. Long time stability could be achieved by taking pains to prevent the reference voltage from changing, but the frequency of the dee voltage is not sufficiently steady to warrant this.

As an alternative to taking the signal from a shunt in series with the magnet as in Fig. 12, some cyclotrons take the error signal from a fluxmeter which measures the magnetic field directly by nuclear magnetic resonance techniques. Instruments are commercially available which are equipped to provide such an error signal.

β) Model magnets. Since the magnetic field produced by a given configuration of steel and current carrying conductors is extremely difficult to calculate especially when saturation effects are included, it is usually necessary to employ model tests in the design of cyclotron magnets. The important quantity to be retained the same as in the full scale magnet is the magnetic field, since this determines the saturation effects. Linear dimensions are scaled down by a constant factor, S, which reduces the weight of material by the factor S^3; even for S as small as five to ten, the cost of the model thus becomes relatively insignificant.

Since the gap is reduced by a factor S, the number of ampere turns is also reduced by that factor. However, the cross sectional area of the coil is decreased by S^2, so that the current density is *increased* over that in the full scale model by a factor S. This introduces a very difficult cooling problem. The usual solution is to pulse the magnet with the required current long enough to make a measurement, and then allow time for cooling.

The most important information to be obtained from a model is the flux efficiency, since this determines the number of ampere turns necessary. This can be determined by a relatively rough model with S as large as 15 or 20. Models are also used to dermine pole shapes, yoke sizes and shapes, etc. Where more elaborate information is desired, smaller scale factors must be used.

γ) Shimming procedures. Once the magnetic field has been constructed, additional adjustment, known as "shimming" is always necessary to obtain the desired field shape. In order to determine what is necessary, accurate measurements must be made. This has usually been done by search coil methods, although with the advent of commercial nuclear fluxmeters, it has often been found simpler and more accurate to make absolute magnetic field measurements at each point.

---

[1] K. G. MACLEISH: Private communication via E. D. HUDSON.

Once the field pattern has been determined, it may be corrected to the desired shape by grinding iron off the pole faces in regions where the field is too high, and by adding thin sheets of iron (known as "shims") where the field is too low. These are attached by small screws, or clamped under non-magnetic materials. As a first guess, the amount of correction to be made is generally determined by assuming that the field strength varies inversely as the gap between the polefaces. This is usually not sufficiently accurate, so that additional measurements, and subsequent changes must be made. By this means, the field is tailored to give the desired rate of decrease with increasing radius, and first harmonics in the azimuthal variations at each radius are reduced as much as possible.

In addition, there is difficulty in some cyclotrons with the location of the magnetic median plane. Since the magnetic focusing is with respect to this plane, the ions will follow it closely; if it intersects the dees, the ions will be lost by striking the dees. In early cyclotrons, this was corrected by observing the beam (actually, the fluorescence caused by the ionization it produces). through a window and adding or removing shims from various locations to keep it from striking the dees. After the discovery of serious radiation hazards from exposure to fast neutrons, this process was abandoned. One good method for determining the location of the median plane depends on the fact that a weightless, current-carrying conductor is in equilibrium on the median plane of a magnetic field; when not on the median plane, it is repelled from it. Thus, by use of a loop of thin wire, the position of the median plane may be determined. It may be corrected, where necessary, by removing iron from one pole face and adding an equal amount to the other.

b) Radiofrequency systems.

22. General considerations. The purpose of the radiofrequency system is to excite the dees to high voltages with the proper frequency. Substituting the parameters of typical cyclotrons into (2.4), one finds frequencies of about 15 Mc/sec which is a frequency commonly encountered in laboratory type communication problems. The power requirements, on the other hand, are much more difficult to satisfy.

Consider first the situation in a typical standard cyclotron. Using the familiar formula for parallel plate condensers as applied to the dee-liner (or dee-poleface) system, one finds dee-to-dee capacitances of about 200 $\mu\mu$f. The capacitive reactance is thus about 50 ohm, so that in operating at 300 kv dee-to-dee voltage, the R.F. current is about 6000 amp. Due to skin effect[1], the R.F. current is confined within about 2×10^{-3} cm of the surface, so that, taking the length of the current path to be about three times its width, the resistance is found to be about 0.03 ohm. Thus, the I^2R losses are of the order of 100 kW. In addition, if the machine achieves a circulating current of 1000 μa of 15 Mev ions, 15 kW of power is being fed into the beam. As discussed in Sect. 13, an equal or greater amount of power is used to accelerate ions which are lost to the dees in the region of electrostatic defocusing. In synchrocyclotrons, the power requirements are somewhat less because of the lower dee voltage and the smaller beams of accelerated ions, but even here, the available R.F. power must be in the ten kilowatt range.

The radio frequency system in a cyclotron consists basically of three parts, the oscillator, the dee system, and the coupling lines between them. In Sects. 23 and 24, these will be discussed in turn for a standard cyclotron. In Sect. 25,

---

[1] Cf., for example, W. R. SMYTHE: Static and Dynamic Electricity, p. 391. New York: McGraw-Hill 1939.

a particular standard cyclotron R.F. system, namely that of the Oak Ridge 86-Inch Cyclotron, will be described in greater detail, and in Sect. 26, the R.F. systems of synchrocyclotrons will be discussed.

23. Oscillators. For purposes of understanding oscillator circuits, the dee and coupling system may be considered as a load consisting of an inductance, L, a capacitance, C, and a resistance, R, in series, with L and C chosen such that

$$\omega = \frac{1}{\sqrt{LC}} \tag{23.1}$$

where ω is approximately equal to the value from (2.4), and with $\omega L/R$, the " Q " of the circuit, very large (generally 1000 or larger). Using this simplification, we

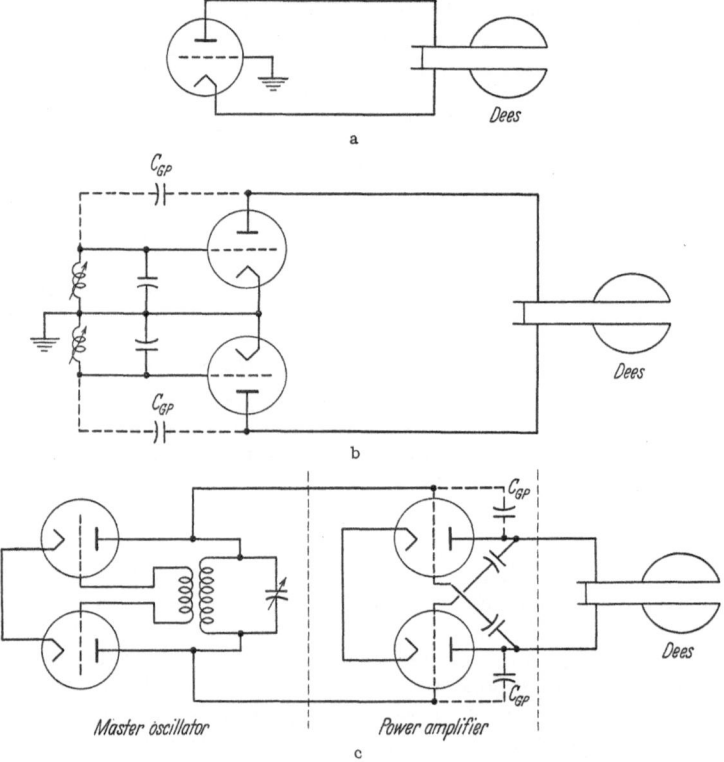

Fig. 13a—c. (a) Self-excited, grounded grid. (b) Self-excited, push-pull. (c) Master oscillator-power amplifier.

will consider the three types of oscillators most frequently used in cyclotrons. These are commonly referred to as the self-excited grounded grid, the self-excited push-pull, and the master-oscillator power-amplifier (MOPA) systems.

Fig. 13a shows, schematically, a self-excited, grounded-grid oscillator. It consists, essentially, of a high-power triode (or several triodes in parallel) with the plate circuit coupled to the dee system, the grid at R.F. ground, and the filament circuit receiving a feedback voltage from the plate load (the dee system). The feedback voltage is generally about 10 to 25% of the voltage supplied to the dees. The grounding of the grid suppresses spurious oscillations arising from the grid to plate capacitance. The oscillation frequency is determined directly by the plate load (the dee system) and is given by (23.1). Power is supplied to the dees only during the half cycle when the plate is positive, with the power during the other half cycle arising from stored energy.

In the grounded grid oscillator, there is an inherent unbalance arising from the fact that power is being fed to one dee, while the other dee is coupled only to a feedback loop. This difficulty is overcome in the self-excited push-pull system shown schematically in Fig. 13 b. Here there is complete symmetry between the two dees, and power is being supplied by one of the two triodes at all times. This suppresses spurious oscillations so that it is not necessary to ground the grid (note that the filament is grounded in Fig. 13 b). It is generally necessary to incorporate tuned grid circuits in this system, and care must be taken to ascertain that the oscillation frequency is as given by (23.1) and is not appreciably sensitive to small discrepancies between the resonant frequency of the grid circuits and that frequency. The feedback is by way of the grid to plate capacitances of the tubes. This circuit contains two power tubes and in general requires more complicated circuitry than the grounded grid oscillator.

One important difficulty in all types of self-excited oscillators is an electron-loading problem known as "multipactoring". Electrons accelerated across the gap between the dee and liner will knock out secondary electrons. At some relatively low voltage ($\sim$10 volt), the time required for these electrons to be-accelerated across the gap will be just equal to half the period of the radio frequency oscillations. In this condition, the number of electrons will be multiplied on each successive R.F. oscillation, with the result that the load on the oscillator becomes very large and thus prevents the voltage from building up. This "multipactoring" action takes place after each cessation of oscillation, as for example, after a sparking, and is therefore a constant nuisance. Several methods of overcoming this difficulty have been suggested, but the two most commonly used are applying a D.C. bias to the dees, and driving the system through the multipactoring region with a booster oscillator, called a "pre-exciter".

In the first, a D.C. voltage of perhaps 1000 volt is applied between the dees and the liner. This requires insulating the dees, but relatively simple techniques for accomplishing this are available. If the bias is such that the dees are negative, electrons coming from the liner will be repelled and thus prevented from reaching the dees, so that the multipactoring is quenched. A positive dee bias would lead to oscillations of the electrons near the ends of the dees, similar to the situation in a Phillips ion gauge. This is a serious loading effect, so that positive dee bias is not used.

In the pre-exciter method, an auxiliary oscillator is coupled to the dees and drives them to a voltage sufficiently high that multipactoring is not a problem. In order to prevent the main oscillator from feeding energy into the booster system once the former takes control, the booster oscillator is operated at a harmonic frequency.

While both the dee bias and booster oscillator method have usually proven quite satisfactory in eliminating the multipactoring problem, the early difficulties led some cyclotrons to abandon the self-excited oscillator and adopt a system in which the oscillator *drives* the dee system instead of including it as part of the oscillating system. This is known as the master-oscillator power-amplifier (MOPA) system, an example of which is shown schematically in Fig. 13 c.

The frequency is basically controlled by a low power master oscillator. Its output controls the grid of the power amplifier tube which then supplies the power to the dees. Care must be taken to prevent the power amplifier circuit from acting as an oscillator due to feedback through grid-to-plate capacitance. This is accomplished by suitably compensating that capacitance with an inductance, or with a capacitive feedback from a point which is in opposite phase (e.g., the plate of the other tube in a push-pull amplifier as in Fig. 13c). One important difficulty with the MOPA system arises from the fact that thermal or mechanical stress changes in the dee system cause small changes in its resonant frequency. Because of its very high Q, the dee cavity is extremely difficult to drive with a frequency even slightly out of resonance. Some mechanism is therefore necessary to keep the resonant frequency of the dee system equal to the master oscillator frequency. Quite often coupling through the circuit components is sufficient. In some cases, a dee trimmer condenser must be continually adjusted to hold the dee system resonance frequency at the proper value. This may be done either manually by the operator or by a servomechanism.

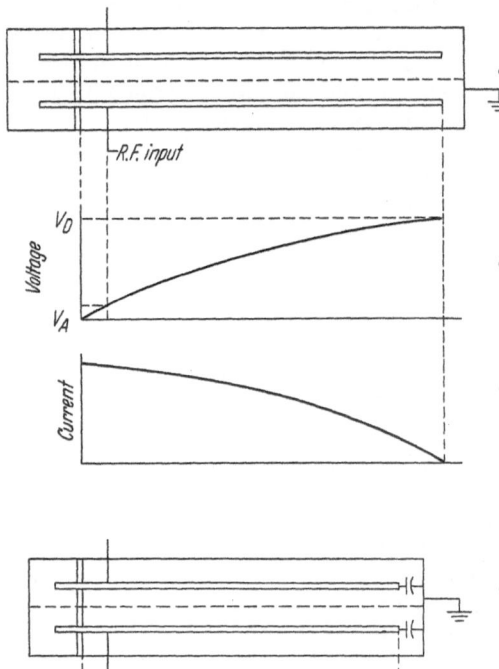

Fig. 14 a—e.

In general, it is very difficult to assess the relative advantages of various types of oscillators. Availability of parts and power supplies, and personal tastes and experiences of design engineers are usually the determining factors in the choice.

24. Dee and coupling systems. The dee system consists basically of the two dees and their dee-stems, a solid bar shorting the dee stems to each other and to ground known as the "*spider*", and the liner or whatever else serves as the grounded return path for the radiofrequency current (in low power machines, it may be the polefaces). Electrically, the dees may be considered as quarter wave transmission lines short circuited at one end (by the "spider"), and open circuited at the other (cf. Fig. 14a). As is well known, this leads to a standing wave which results in a sinusoidal distribution of voltage having a maximum at the open end and being zero at the shorted end as shown in Fig. 14b. Thus, by applying the R.F. voltage at a point near the spider, an amplification in voltage can be achieved, as shown by the ratio V_D/V_A in Fig. 14b. In principle, very large amplifications can be obtained, but higher dee voltages lead to higher R.F. currents and therefore to higher I^2R losses and greater power consumption. Thus the dee voltage is determined essentially by the power available from the oscillator. Amplification factors of about 10 are used in most cyclotrons.

The current is approximately 90° out of phase with the voltage so that its distribution is as shown in Fig. 14c. Since it is at its maximum at the spider, it is especially important that the contact between the dee stems and spider have very low resistance. This introduces a considerable complication in variable energy cyclotrons where it is necessary to move the spider in order to vary the resonant frequency of the system.

Due to the large dee-to-ground capacitance, the dees themselves are effectively not a portion of the quarter wave transmission line, but may be considered as a capacitance attached to the end of the dee stems; the transmission line effects thus neglected may be compensated by making this equivalent capacitance about 15% larger than the actual capacitance. A shunt capacitance, X_c, causes a phase shift equal to that of a transmission line length, l_c, given by

$$\cot \frac{2\pi l_c}{\lambda} = \frac{X_c}{Z_0} \tag{24.1}$$

where λ is the wave length and Z_0 is the characteristic impedance of the line. For typical values as $X_c = 100$ ohm, $Z_0 = 50$ ohm, the phase shift would be 27 electrical degrees, or $0.075 \, \lambda$. The distance from the spider to the end of the dee stems should thus be only $0.175 \, \lambda$ rather than the full quarter wavelength, $0.25 \, \lambda$. This effect is useful in suppressing harmonics of the principal frequency since X_c is frequency dependent. On the other hand, it reduces the voltage amplification as shown in Fig. 14d and e, and leads to other difficulties caused by deviations from the simple quarter wave transmission line. It is therefore customary to keep the dee capacitance as small as possible. In many cyclotrons, this is done by reducing the vertical height of the dees (and thus increasing the dee to ground gap) at the outside. This does not result in ions striking the dees because the vertical height of the beam is reduced at large radii by magnetic focusing (cf. Sects. 3 and 13).

The most common coupling between the power triode and the dee stem is by a half wave transmission line. In this case, it is relatively convenient for the feedback line to be of approximately equal length and thus provide a total of 360° phase shift between the grid and plate, which puts these voltages in proper phase. The half wavelength line also has the rather considerable advantage of reflecting load variations in the same sense as they occur. Thus, a lowered impedance in the dee system is reflected as lowered impedance in the plate load, so that the tube supplies more power which is just what is necessary to maintain the dee voltage.

In some cyclotrons, the oscillator is mounted directly adjacent to the dee stems which avoids the necessity of a transmission line. However, it has been found that transmission lines provide a valuable energy storage service which makes for steady operation of the oscillator.

Part of the length of the transmission line is replaced by capacitance to ground as discussed in connection with (24.1). Some of this arises from grid to plate capacity and from the capacity in the bushing through which the line enters the vacuum tank. However, additional capacity is sometimes added to give extra voltage amplification between the plate and the point of application to the dees (or to give negative amplification in the feedback line from the dees to the oscillator) by the principle shown in Fig. 14e.

The coupling between the transmission line and the dees (or between the oscillator and dees when no transmission line is used) may be direct, inductive, or capacitive. Probably the most widely used method is inductive coupling (sometimes called "loop coupling") as it may serve the additional purpose of matching the impedance of the dee system to that of the oscillator output tube.

25. Details of a particular R.F. system. The Oak Ridge 86-Inch Cyclotron accelerates protons to 22 Mev in a magnetic field of 8700 gauss which corresponds to a frequency of 13.4 mc/sec. The corresponding wave length is 76 feet. The maximum dee-to-dee voltage is between 400 and 500 kv. This is a relatively high voltage and thus requires a relatively high power, so that a detailed description of its radiofrequency system will show most of the problems that arise in cyclotron R.F. systems.

The circuit diagram for this system is shown in Fig. 15; it employs a grounded-grid, self-excited oscillator like that shown schematically in Fig. 13 a. The vacuum tube is a Federal-134 which is rated for 250 kW output power. The grid is grounded R.F.-wise through the 7000 μμf capacitor (impedance ≈ 20 ohm), and receives

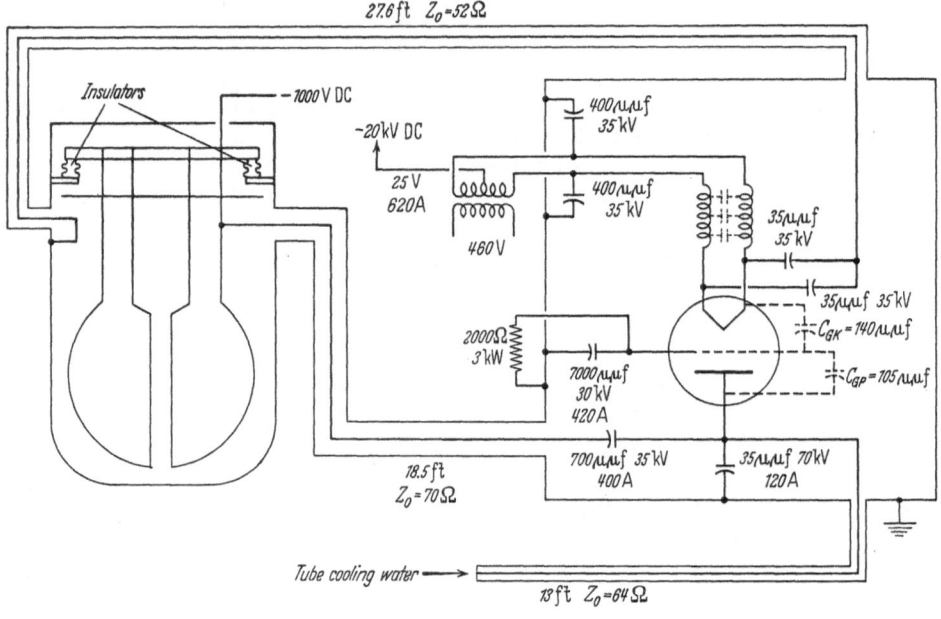

Fig. 15.

its D.C. bias of about 2000 volt by virtue of a 1 amp drain through the 2000 ohm grid leak resistor. The plate voltage of up to 20 kv is achieved by applying a negative high voltage to the center-tap of the filament transformer. The plate is at D.C. ground which facilitates its water cooling, and also allows the plate line to be direct-coupled to the dees. The plate line is a half wavelength when corrected for the various capacitances to ground by use of (24.1). It is directly coupled to one dee stem about 10 inches from the spider. The feedback voltage is taken by inductive coupling from the opposite dee stem and is applied to the filament through a half wavelength transmission line.

The dees are insulated from ground and given a D.C. bias of −1000 volt to overcome multipactoring. The oscillator will not operate without this bias, but once it is applied, no difficulty is experienced. The dee bias necessitates its D.C. insulation from the tube plate which is accomplished by the 700 μμf series condenser in the plate line. The filament feedback loop is isolated from the D.C. bias voltage by the two parallel 35 μμf condensers in series with the line. In order to protect the power supplies from the R.F. feedback voltage, the filament leads are wound as R.F. chokes and placed in close proximity to each other to provide bypass capacitance. Additional protection is afforded by the 400 μμf condensers to ground and the choking action of the transformer.

Water cooling is applied to the plate and to the plate line as far as the entrance to the vacuum system (the cooling is not so important near the ends of the line since the standing wave on the half wave length line gives maximum voltage at the ends and maximum current—and thus maximum I^2R heating—near the center) through a quarter wave length line shorted at the end. The 35 $\mu\mu$f condenser to ground is used to adjust the length of both this line and the plate to dee line to a quarter and a half wave length respectively. The relatively large part of these lengths replaced by capacitance are quite effective in suppressing harmonics because of the frequency dependence of the capacitative reactance.

The water cooling line bypasses the 700 $\mu\mu$f condenser with a Teflon jumper tube. The dees are cooled by passing a large amount of cooling water through tubes soldered to them; these tubes are isolated from ground by rubber hose connections. The water is demineralized to reduce the D.C. bias drain through it; this drain is only about 15 milliamp. The metal to glass joint at the entrance of the grid line to the tube is air cooled.

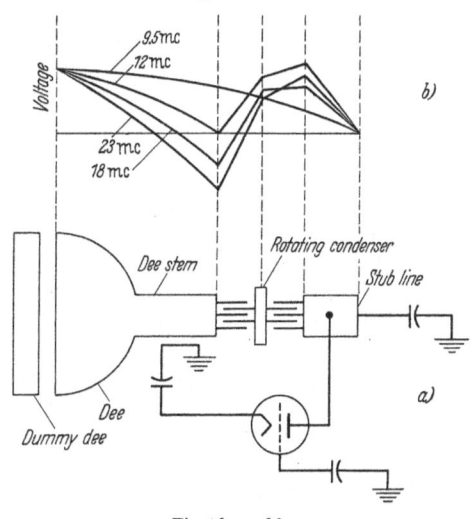

Fig. 16 a and b.

Most resistors and condensers shown are made up of several components in series or parallel to achieve necessary current carrying capacities. Non-water cooled, high voltage capacitors are commercially available with current ratings only up to 120 amp, so that the series capacitor in the plate line, for example, must consist of a number of condensers in parallel to achieve the rated current capacity of 420 amp. Water cooled condensers are available with current ratings up to 500 amp, but these are generally avoided as an unnecessary complication.

26. R.F. systems for synchrocyclotrons. Most large synchrocyclotrons accelerate protons to energies approaching half of their rest energy, so that the resonant frequency changes by a factor of about 1.5 during the acceleration. In addition, several machines are designed to accelerate deuterons and alpha particles. The non-relativistic resonant frequency for these is lower by a factor of two than that for protons, and an additional difference in frequency is needed to compensate their increase in mass with energy. The most convenient method of varying the frequency is by mechanically varying the capacitance of a large condenser in the dee system. This problem is greatly simplified if a single dee system is used, with the R.F. power fed into the single dee through a transmission line at a point on the periphery far from the dee gap.

Perhaps the most advanced design is the "three-quarter wave" system of McKenzie[1], shown schematically in Fig. 16a. In it, the end of the dee stem is coupled to a "stub line", the far end of which is grounded, by a rotary condenser. The total length of the dee stem, condenser, and stub line is about $\frac{3}{4}$ wavelength for the highest frequency used, whence the name. As the capacitance of the rotary condenser is varied, the system changes frequency and resonates in various modes as shown in Fig. 16b. The low frequency limit is reached when it resonates

[1] K. R. McKenzie: Rev. Sci. Instrum. **22**, 302 (1951).

as a quarter wave line, thus allowing a frequency swing approaching a factor of three.

The oscillator used for synchrocyclotrons is the self-excited grounded grid circuit described in Sect. 23 and 25. (It may be noted that the other two oscillators described in Sect. 23 include tuned components outside of the dee system and so are not adaptable for frequency modulation.) In order to suppress resonant modes in which there is no voltage across the rotary condenser, the plate and filament feedback lines enter the system on opposite sides of it. In Fig. 16a, the transmission line from the plate is directly coupled to the stub line while the filament line is inductively coupled to the dee system. The direct coupling of the plate line avoids voltage losses; this is important since the large voltage amplifications obtained in standard cyclotrons cannot be achieved here. The ratio of the voltage at the dee gap to the voltage at the coupling point in the stub line is not independent of frequency as is evident from Fig. 16b, nor is the ratio of the latter to the voltage at the oscillator plate. However, by proper choice of the length of the transmission line, these two effects may be made to approximately compensate, so that the dee voltage is kept roughly constant during the frequency swing.

Where the entire frequency swing is not used in the acceleration of any one particle, the oscillator may be pulsed off during the portion of the cycle that is not being used. This prevents undesirable resonance modes from becoming established, and is also advantageous from power economy and other standpoints. It is accomplished by applying a negative pulse to the oscillator grid, with the timing suitably controlled from the rotary condenser.

Several other R.F. systems have been used for various machines. In one of these, the dee system is a quarter wave line short circuited to ground at the end of the dee stem as in the standard cyclotron. The rotary condenser must then be near the dee gap and thus in the strong magnetic field where eddy current heating of the rotating metal parts is a problem. Some machines employ a half wave length line consisting of two stubs attached to opposite sides of the dee and short circuited to ground at the far ends. The dee is effectively at the center of the transmission line where the voltage of the standing wave is a maximum. The frequency is changed by changing the dee-to-ground capacity with a rotary condenser attached to the dee. Here again the rotary condenser must be in a magnetic field, and the amount of frequency variation obtainable is not as large as in the three-quarter wave system.

The engineering of rotary condensers involves many difficult problems. To achieve large frequency swings they must be relatively large (maximum capacitances of thousands of micromicrofarads are common), and since most machines use pulse rates between 50 and 500 per second, they must rotate at high speeds. R.F. and eddy current losses are quite appreciable, so that water cooling is usually necessary. In order to alleviate some of these problems, an alternative method has been developed at Berkeley employing the mechanical oscillations of a large, electrically driven tuning fork to vary the capacitance.

Since neither the varying condenser nor any point on the dees is always at R.F. ground potential, the entire system must be supported mechanically by ceramics to insulate it from ground. This also allows a D.C. bias to be applied to the dee system to eliminate multipactoring difficulties.

c) Ion sources.

27. Principles of ion source operation. The ion source is perhaps the least understood of all components of a cyclotron; this is most unfortunate since improvements in ion sources have in the past and will probably continues in the future to be one of the most fruitful approaches to improvement of cyclotron performance. Perhaps the most advanced design is that developed by LIVING-STON and JONES [25] shown schematically in Fig. 17. When the filament is heated, it emits electrons and these are accelerated by the 100 volt potential. However, low energy electrons in a strong magnetic field are constrained to move along the direction of the magnetic field lines. For example, if we assume that 50 volts of their kinetic energy is directed perpendicular to the field, application of (2.9) indicates that their motion in the plane transverse to the field follows circular orbits with a radius of about 0.001 inches. At first the electrons strike the anode, but eventually the anode becomes negatively charged and repels them causing them to oscillate back and forth between the filament and anode. Simultaneously, the gas is introduced, so that the electrons cause ionization, producing an assemblage of positive ions and electrons in thermal equilibrium. This assemblage, called an arc "plasma", is most dense in the cross hatched region of Fig. 17. As a result of local electric fields and collisions, the electrons drift slowly across the magnetic field lines and eventually drain to ground, mostly by way of the collimator. The positive ions move in directions transverse to the magnetic field much more easily than electrons—from (2.9) the radius of curvature for particles of a given energy is proportional to the square root of the mass—and many of them diffuse out of the exit slit. If they strike the filament or anode, or the chamber walls, they are neutralized, but they still have a high probability of drifting back into the plasma and being re-ionized. Thus, if the distance from the dense part of the plasma to the exit slit is of the order of the radius of curvature of the positive ions ($\sim \frac{1}{64}''$), the ratio of ions to neutral atoms drifting out of the exit slit is relatively large.

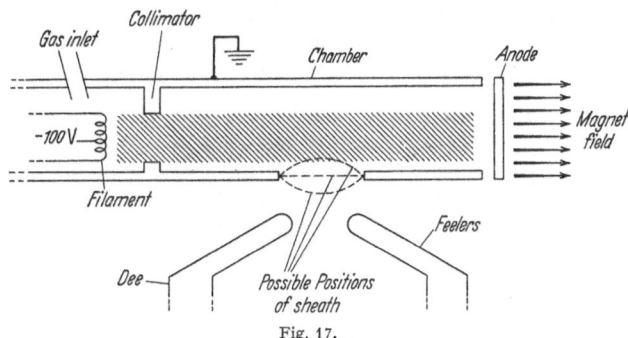

Fig. 17.

The current of ions drifting out of the ion source is limited by space charge considerations, which are, in turn, controlled by the accelerating voltage encountered by the ions as they leave the plasma; according to CHILD's law[1], the space charge limited current, I_{sc}, is proportional to

$$I_{sc} \propto \frac{V^{\frac{3}{2}}}{d_g^2} \tag{27.1}$$

where V is the accelerating voltage and d_g is the distance across the gap between the point where it is applied and the plasma boundary. (The plasma boundary is called the "arc sheath".) Since the accelerating voltage for a cyclotron ion source is applied from the dee opposite the exit slit, the ion source output can be considerably increased by attaching protrusions to this dee, called "feelers", to bring a point at dee potential close to the sheath and thus decrease d_g. This also has a generally favorable effect on the phase problem in a standard cyclotron as discussed in Sect. 13.

---

[1] Actually, CHILD's law has very poor quantitative validity in the usual operating region of cyclotron arcs.

Because of the very high mobility of electrons, no general electric fields can exist within the plasma, and the plasma as a whole assumes the potential of the point to which the electrons drain; this is ground potential in Fig. 17. In the region of the exit slit, the ground plane therefore follows the sheath. The accelerating field between the ion source and feelers is determined by its boundary conditions, which include the shape of the ground plane and therefore the shape of the sheath in the vicinity of the exit slit. Distortions of the sheath can therefore cause the electric field experienced by the ions to have a large vertical component with the result that they will ultimately strike the dees.

The shape of the sheath in the vicinity of the exit slit is determined by the balance between the magnitude of the arc current and the strength of the external electric field created by the feelers. If the latter is too small (large) for a given arc current, the sheath will bow outward (inward) as shown by the dashed lines in Fig. 17. Unfortunately, this balance cannot be maintained through an appreciable fraction of an R.F. cycle in a cyclotron because the electric field is varying with time. Some relatively poor compromise is therefore necessary.

28. Ion source design and operation. Most of the details of ion source operation follow more or less directly from the principles discussed in the previous section, although empirical experience has been a most important guide in some cases. For example, it has been found that cyclotron output is somewhat sensitive to slight rotations and displacements of the source relative to the feelers. (This is probably due to sheath distortion effects.) It is thus quite profitable to provide means for making slight positioning adjustments from the operating console. The spacing between the ion source and feelers is limited by voltage breakdown considerations. Typically, it is about $\frac{1}{8}$ inch for 10 to 15 kv dee-to-ground voltage, and $\frac{1}{2}$ inch for 200 kv dee-to-ground voltage. Each of these conditions leads to about the same output.

As the gas flow rate is increased from zero, the ion source output rises rapidly to a saturation point; this occurs at from a fraction to a few cubic centimeters per minute depending on the arc conditions. Excessive gas flow leads to vacuum difficulties and sparking in the region of the ion source. Increasing the arc voltage improves the efficiency of ionization and increases the average energy and thus the mobility across the magnetic field of the positive ions; it therefore improves ion source output up to the point where it causes voltage breakdown between the filament and the colimator. Increasing the arc current (usually by raising the filament current) improves the efficiency of ionization and thus increases the output within the bounds of the sheath distortion problem discussed above. In practice, the largest high energy cyclotron beams are obtained with relatively high arc currents, but not necessarily with the highest available. Consideration must also be given to the fact that very high arc currents shorten the life of a filament and other parts of the ion source. At the present state of development, ion sources with D.C. accelerating voltages are capable of producing about 500 milliamp of ion current per square centimeter of exit slit area. Exit slit areas cannot be too large lest the sheath be unduly disturbed; areas of about one square centimeter are usually used. The sinusoidal variation of the accelerating voltage leads to a very considerable loss relative to a D.C. accelerating voltage equal to its maximum. The voltage is in the wrong direction in one-half of the cycle, and too low to give appreciable output in a large fraction of the other half; the efficiency relative to the D.C. case is about 18%, so that maximum ion source output currents are about 100 milliamp. A large fraction of this is lost in the first few turns due to sheath distortion defocusing as discussed above. The maximum currents that have been obtained after several turns in a cyclotron are about 30 milliamp.

Several variations of the source shown in Fig. 17 are common. The anode is frequently not insulated from the chamber, or grounded separately, or omitted entirely (in which case the poleface acts as the anode). In these cases, the electrons do not oscillate, but drain steadily to the anode, so that the ionization efficiency of each electron is less. Several ion sources of more primitive design than those discussed here are still in use in some cyclotrons. Some of these are described in a review article by M. S. LIVINGSTON [20].

In the past, most ion sources have been made of copper or copper alloys; since they are heated rather severely, they must be water cooled at the base. Recently, carbon has been used to eliminate the necessity of cooling. Filaments are made of relatively thick (0.070 to 0.170 inches) wires of tungsten or tantalum, and heated with high currents (usually hundreds of amperes) of either D.C. or low frequency (100 to 200 kilocycles) R.F. power. The use of R.F. heating avoids strains induced by magnetic effects; these are quite appreciable on thin wire filaments. R.F. heating is unnecessary when thicker wires are used. Because of the high temperature, the filament support and sometimes the lead-in lines must be water cooled. In most cyclotrons, filaments burn out and must be replaced after about a hundred hours of operation. It is therefore imperative that means be provided for changing them without breaking the cyclotron vacuum.

d) Vacuum systems.

29. Design requirements. The vacuum requirements in a standard cyclotron are usually determined by the dee voltage breakdown problem; in synchrocyclotrons, where the voltage is lower but the distance traversed by the ions is very great, the limitation is set by scattering of the beam by the gas at low energies where the Rutherford cross section is large. In both cases, vacuums of about 10^{-5} mm of mercury are required. Fortunately vacuums in this range are attainable by methods well known from other fields of science and technology. The required pumping equipment, consisting of large diffusion pumps backed by small booster diffusion pumps which are in turn backed by large mechanical pumps, is available commercially at relatively moderate cost. In fact, the economics of the situation makes it profitable to use pumping systems large enough to tolerate small leaks. This saves leak hunting time and thus essentially buys relatively costly cyclotron operating time at the expense of relatively cheap initial pumping system cost. The usual rule is to provide a diffusion pump speed about equal to the volume of the system per second.

The diffusion pump must be elaborately baffled from the rest of the system to prevent an excessive amount of oil from getting into the cyclotron. These baffles are frequently refrigerated to very low temperatures (about $-40°$ C) with a freon system. In addition, cold traps may be provided to remove condensable vapors. These are sometimes cooled by liquid nitrogen, so that filling them at frequent intervals is an important part of cyclotron operation. A large valve is necessary to isolate the diffusion pumps from the main part of the vacuum system. Means are usually provided for closing it rapidly to protect the diffusion pump oil in case of a sudden leak. Where silicone diffusion pump oils are used, this is not necessary.

The mechanical pumps pass the same amount of gas as the diffusion pumps but at about a thousand times higher pressure, so that their speed may be slower by nearly that factor. However, the same mechanical pumps are usually used to pump the system down from atmospheric pressure to the pressure where the diffusion pumps become effective (about 100 microns of mercury), so that somewhat higher speeds are desirable. Typically, it is about equal to the volume of the system per minute.

Starting from atmospheric pressure, the system can usually be pumped down to operating vacuum in about one hour. If the system has been open to the atmosphere for an appreciable length of time, or if parts of system have been welded, additional time sometimes extending to several days is required for outgassing. This is especially true in high dee voltage machines where appreciable sparking occurs, since this constantly releases gas. If an appreciable amount of water has gotten into the system, it freezes during pumpdown and thus hinders operation for many hours.

When in operation, the ion source gas feed usually determines the pressure of the system. For example, a gas feed rate of 1 cc/minute corresponds to a leak rate of 1200 liter per second at a pressure of 10^{-5} mm Hg. With the ion source gas feed off, pressures as low as 2×10^{-6} mm Hg are sometimes obtained.

The vacuum tank includes the volume occupied by the dees, the pumping header, and in synchrocyclotrons, the rotating condenser. The magnet polefaces are generally part of the system, being attached to the remainder of the tank by gasketing or welding. The rest of the tank is usually made of non-magnetic steels using welded construction. Vacuum locks are provided for changing the ion source filament, or in some cases, the whole ion source, and for inserting and removing targets to be bombarded by the circulating beam. These are usually evacuated by a separate mechanical pump.

30. Vacuum techniques. In the early days of cyclotron development, gasket leaks were frequently a problem, but designs are now well standardized and very little difficulty is encountered. Gaskets have been developed to obtain various types of motion through the vacuum tank so as to make positioning adjustments of the ion source, deflector, dees, etc., and to allow rotation of targets for cooling. For many purposes where rapid motion through the seal is not necessary, an ordinary "O" ring is sufficient. The best known of the more elaborate gaskets is the Wilson seal which consists of two rubber gaskets with holes slightly smaller than the shaft which passes through them, and with the space between them evacuated by a mechanical pump. The gaskets are toed outward so that atmospheric pressure aids in making the seal; vacuum grease is liberally applied. For many purposes, only a single rubber gasket is sufficient.

Standard type vacuum indicating instruments are used at various points in the system. These include McLEOD, thermocouple, and Pirani gages for pressures down to about one micron, and triode and Phillips ion gages for the high vacuum indications. The Phillips ion gage is especially useful since it is quite rugged. It is frequently used to operate interlocks which protect the diffusion pump and oscillator, and to activate alarm systems. A large cyclotron may have as many as twenty separate vacuum gages of various types at various parts of the system.

A mass spectrometer leak detector is the principal instrument for vacuum leak hunting. During the construction phases, leaks are commonly found in welds and solder joints, and even in bad castings. These must be repaired or replaced. In the past, it has been common practice to stop leaks by painting with glyptal, but with improved techniques now available for both leak detection and repair, this is now not usually considered necessary or satisfactory.

e) Beam deflection systems.

31. Beam deflection in standard cyclotrons. Due to the strain on the phase problem caused by a large magnetic field fall-off as discussed in Sect. 13, the outermost circulating orbits in a standard cyclotron are usually well inside of the largest orbit that can be contained by the magnetic field. In order to extract a beam from the field, it is therefore necessary to achieve a separation between successive orbits of about one to three inches. In some cyclotrons, this is not enough to remove the beam completely, but it at least allows the field to be weakened sufficiently to achieve further separation without unduly influencing the field in the region of circulating orbits.

Two mechanisms which affect the separation between successive orbits have been discussed, namely the energy increase due to electric acceleration across the dee gap, and the precession of the orbit center (radial oscillations) due to the first harmonic in the magnetic (and electric) field experienced by the ion. Assuming the energy to be proportional to the square of the radius, an energy change ΔE causes a radius change Δr given by

$$\frac{\Delta E}{E} = 2 \frac{\Delta r}{r}. \tag{31.1}$$

Since the energy change per turn is $4V_0 \cos \Theta$ the separation between successive orbits, s, is

$$s = \frac{2V_0 \cos \Theta}{E} r. \tag{31.2}$$

Near the maximum energy, (31.2) seldom gives a separation larger than $\frac{1}{4}$ inch.

The maximum separation caused by procession of the orbit center is, from (4.2)

$$s = \pi r h_1 \tag{31.3}$$

where h_1 is the first harmonic in the magnetic field. The difficulty with achieving large orbit separation by this method is that h_1 must vary smoothly with radius, so that there is a large separation between many successive orbits and consequently inefficient use of the magnetic field.

Some other mechanism is therefore necessary to achieve the required separation of one inch or more and the method universally used is to pass the ions through a strong electric field. This requires an electrically conducting sheet, called the "*septum*", between the next to last and last circulating orbits and application of an electric potential between it and another plate, called the "*deflector*", a short distance outside of it. Ions which pass between these plates experience an electric field, $\mathscr{E}_n$, normal to their direction of motion. The force acting on an ion in this direction, F_n, is

$$F_n = e \mathscr{E}_n - \frac{e}{c} v H. \tag{31.4}$$

The effect, therefore, is equivalent to weakening the magnetic field by an amount

$$\Delta H = \frac{c}{v} \mathscr{E}_n. \tag{31.5}$$

The momentum of the ion, and therefore Hr, is not changed so that this results in an increase of the radius of curvature, Δr, given by

$$(H - \Delta H)(r + \Delta r) = Hr$$

whence

$$\Delta r = r \frac{\Delta H}{H - \Delta H}. \tag{31.6}$$

If the electric field is applied through an angle ϑ_e, the maximum orbit separation achieved is, from geometric arguments,

$$s = 2\,\varDelta r \sin \frac{\vartheta_e}{2}. \tag{31.7}$$

This maximum separation occurs at an angle of $\frac{\pi}{2} + \frac{\vartheta_e}{2}$ beyond the entrance to the deflection channel.

To illustrate the magnitudes involved, consider a cyclotron with 15 Mev deuterons ($v/c = 0.125$) and a radius of curvature of 20 inches, (i.e., $H = 15\,000$ gauss). For an electric field of 60 kv/cm (200 statvolt/cm), $\varDelta H = 1600$ gauss from (31.5), and from (31.6), $\varDelta r = 1.8$ inches. If the field extends through an angle of 90°, the maximum orbit separation, as obtained from (31.7), is 2.5 inches. It occurs 135° beyond the beginning of the channel, or 45° beyond the end of it.

The principal difficulty with an electrostatic deflector is that a large part of the beam strikes the septum. This loss can be minimized by maximizing the orbit separation at the entrance to the channel. In accordance with (31.2), this requires reducing the phase as far below 90° as possible; as discussed in Sect. 11 and Fig. 6, this reduces the circulating beam so that some compromise is necessary. A final phase of about 60 to 70° is perhaps typical. It may be noted that at this phase, if the septum is attached to the dee with its center about 60 to 70° before the gap, and the deflector is grounded, the R.F. voltage appearing between the septum and deflector is the maximum dee-to-ground voltage and is in the proper direction for deflection. Some cyclotrons use the R.F. voltage in this way to aid in deflection although most of these use it in conjunction with a D.C. potential. However, it is most common to isolate the septum from the dees. This increases the D.C. voltage required, but avoids difficulties from such things as R.F. discharge initiated by D.C. sparks between the dee and deflector, insulation for and by-passing the R.F. picked up by the deflector, and variations in the septum-deflector gap by mechanical and thermal distortion of the dees. Furthermore, independent control of the deflector voltage provides an additional parameter that may be varied to maximize the deflected beam.

Another method for increasing the orbit separation at the entrance to the deflector channel is by proper use of a small first harmonic in the magnetic field, as shown by (31.3). This may be done in shimming, or more conveniently, by current carrying coils wound half way around the magnet poles with the return path for the current through the poles[1]. Two such coils 90° apart allow complete adjustment of the first harmonics.

The current striking the septum is also minimized by keeping it as thin as possible. Very thin septums tend to warp excessively, so that it is common to use thicknesses of about 0.025 inches and file the leading edge thin to reduce the thickness at the entrance.

Because of the very high currents striking it, the septum is usually made of tungsten or tantalum sheet, mounted on water cooled supports. The heating of the septum (it usually runs "red hot") is frequently a limitation on the intensities of deflected beams. However, a successful septum has recently been made[2] of small diameter, thin walled aluminum tubes carrying high velocity water. It is capable of dissipating the heat from several milliamps of beam. Due to the variation of angle of entrance and the energy inhomogeneity, the beam widens radially as it passes through the deflection channel, so that the gap between the

---

[1] This method was developed by E. D. HUDSON, R. S. LORD and M. B. MARSHALL: Private Communication.

[2] E. D. HUDSON, R. S. LORD and M. B. MARSHALL: Private Communication.

septum and deflector is increased with the distance from the entrance. At the entrance, the gap is kept as small as possible consistent with the maximum value of orbit separation, s, in order to minimize the D.C. voltage required to obtain a given deflecting electric field. Methods of mechanically varying the position of the deflector from the control console must be provided.

Once the beam has passed through the deflection channel, it is sometimes necessary to provide additional deflection by weakening the magnetic field. This may be done by passing it between iron bars placed near the median plane; where greater weakening is required a large current carrying coil may be used to "buck out" the main magnetic field. In either case, care must be taken to ascertain that the disturbance is not large at the location of the outermost circulating orbit. This generally requires a considerable shimming effort.

After passing out of the magnetic field, the beam is usually focused by wedge shaped magnets [3], [5], [34] or quadrupole strong focusing magnets[1] (or a combination of both) and transmitted through the shield wall to an experimental area. When considered as an "optical" lens system, the beam effectively diverges from a point in the deflector channel, usually near the entrance. Considerable success has been achieved in transmitting deflected beams over large distances and focusing it into small areas.

32. Beam deflection in synchrocyclotrons. The problem of beam deflection in synchrocyclotrons is considerably more involved than in standard cyclotrons; the achievement of separation between successive orbits is much more difficult since the contribution from (31.2) is very small (~ 0.1 mm), and the effectiveness of electrostatic deflection as given by (31.5) is considerably reduced due to the high velocities. The problem of obtaining orbit separation by radial oscillations has been so troublesome that alternative methods of inherently low efficiency have been employed at some installations. One of these is to scatter the beam by passing it through a thin foil (~ 0.1 mm) of high atomic number material. This introduces a solid angle loss of about two orders of magnitude. Another low efficiency method consists of allowing the acceleration to continue to the radius where $n = 1$ so that the beam is no longer contained by the magnetic field (cf. Sect. 5). Before reaching this radius nearly all of the beam is lost due to the resonances in the coupling between radial and vertical oscillations at various radii ($n = 0.2, 0.5$, etc.). In only a few machines has any observable beam been obtained at $n = 1$.

The more straightforward attack on the problem of achieving separation between successive orbits is by inducing radial oscillations. The most successful method has been the "regenerative peeler" proposed by TUCK and TENG [37] and developed by CREWE and LE COUTEUR [8], [18]. It consists of a narrow ($\sim 5°$) sector where the magnetic field is weakened, called the "peeler", followed somewhat later in the ion orbit by a similar sector where the field is strengthened, called the "regenerator". In both cases, the weakening and strengthening increase linearly with radius. Upon reaching the radius where these begin, an ion passing through the peeler is deflected outward, starting a radial oscillation the maximum of which occurs 90° beyond. The regenerator deflects the particle inward starting it on a radial oscillation which has a maximum 270° beyond. If the regenerator were 90° beyond the peeler, the radial oscillations induced by the two would be 180° out of phase and would tend to cancel. However, the ion passed through the regenerator at a larger radius since it is at the maximum of the radial oscillation induced by the peeler, and thus the size of the magnetic disturbance is

---

[1] J. B. BLEWETT: Phys. Rev. **88**, 1197 (1952).

larger whence the amplitude of the oscillation is greater. At the maximum of this oscillation the ion passes through the peeler again at a still large radius thus inducing an even larger radial oscillation of opposite phase. Thus the phase of the radial oscillation reverses on each successive pass through the peeler and regenerator, but on each reversal, the amplitude becomes larger. The overall growth of amplitude is exponential, so that after a number of traversals, the separation between successive orbits becomes quite large.

The principal complication that must be considered is the effect on the vertical oscillations. In accordance with (5.6), the coupling between radial and vertical oscillations is, in lowest order, proportional to $\frac{d^2 H}{d r^2}$. Thus, the effect can be minimized if the magnetic field changes in the peeler and regenerator are made linear with radius. LE COUTEUR [18] investigated the complete problem in detail, and found that the ratio of vertical to radial amplitudes can be decreased by decreasing the angular separation between the peeler and regenerator from 90°. This, however, weakens the orbital separation obtainable with a given field disturbance. The compromise that seems most practical is an angular displacement of 60°. With this displacement and with magnetic field gradients of 1000 gauss per inch and sector extensions of 0.1 radians, a radial separation of 1.5 inches between successive orbits can be obtained after about 20 revolutions in typical synchrocyclotrons. Another alternative, which has recently been carried out successfully, is to omit the peeler completely, with its effect being obtained by the general radial decrease of the magnetic field.

Once an orbital separation has been achieved the beam usually must be conducted out of the magnetic field through a channel in which the field is weakened. The channel is made by inserting iron on either side, and the disturbance this creates in the region occupied by the circulating orbits must be eliminated by shimming. With an orbit separation of 1.5 inches, it has been found possible to do this successfully with a long and patient effort [8].

f) Targets.

33. Cyclotron targets. The most difficult cyclotron target problem is encountered when high intensity bombardments are made in standard cyclotrons. In order to achieve the largest available currents, it is necessary to avoid deflection losses (usually 75% or more) by intercepting the circulating beam with an internal target. Unfortunately, this introduces a very difficult cooling problem. In a typical case a beam of one milliamp of 15 Mev ions, which corresponds to 15 kW of heat production when stopped, might be available, and this beam is generally concentrated in an area of only a small fraction of a square inch. Thus, if a target were inserted so as to intercept it at normal incidence, a cooling system capable of dissipating several tens of kilowatts per square inch would be necessary. This is far beyond the reach of presently available techniques.

Two methods have been employed for alleviating this problem: rotating the target, and arranging the geometry so that the beam strikes the target at grazing incidence. As an example of the former, if the beam is $\frac{1}{4}$ inch in vertical height and strikes a 2-inch diameter cup-shaped target at a point on the periphery, the area over which the beam is spread is increased by a factor of 25 upon rotating the target. As an example of grazing incidence, a beam which at normal incidence has a radial width of $\frac{1}{4}$ inch has been spread over a target length of 8 inches, which represents a decrease in heat production per unit area by a factor of 32. The grazing incidence method requires careful measurements of the beam shape and painstaking adjustments of the target angle. However, grazing incidence targets may be more efficiently cooled because the complication of introducing water through a rotating seal is avoided.

A well designed water cooling system with a water velocity of about 50 feet per second is capable of dissipating about 3 to 5 kW of heat per square inch of heated surface area. To achieve this, it is necessary to use a large number of small cross section channels to direct the flow, and increase the cooling surface area, and to take special precautions to eliminate imperfections which might induce cavitation in the water and thus reduce the cooling efficiency. A large capacity, high pressure pumping system is required, and large cross section entrance and exit lines are necessary to reduce pressure drops between the pump and the target. The type of cooling used in most rotating targets falls considerably short of these specifications.

High intensity bombardments are usually made for production of radioactive isotopes or to give high fluxes of fast neutrons. In cyclotrons which are constructed for these applications, there is considerable advantage in reducing the magnetic field fall-off at large radii. This reduces the magnetic focusing and thus spreads the beam in the vertical direction, and has additional advantages discussed in Sect. 13. Internal cyclotron beams of up to 75 kW have been successfully received on well designed targets[1], and there is every reason to believe that considerably higher powers can be handled.

In most high current applications, it is desired to bombard special materials. When the material is metallic, it usually can be soldered or brazed to the target[1] without excessive loss of heat dissipation efficiency. Many non-metallic compounds may also be brazed[2], or pressed on to a roughened surface[3], or deposited by evaporation from solution, or enclosed in a capsule surrounded by cooling water[4]. Several other techniques have been developed for specialized purposes. When the material to be bombarded cannot be tolerated in the cyclotron vacuum, irradiations are sometimes made in the deflected beam close to the cyclotron. The beam is passed through a thin window into a special target area which is either evacuated or filled with helium. It is generally of low enough intensity and large enough cross sectional area that cooling is not a problem. The window is usually made of duraluminum about 0.001 inch thick; this is quite strong and causes relatively negligible energy loss. Bombardments with currents of about 100 microamps are commonly carried out in this arrangement.

Targets used in low intensity bombardments are of many types, and few generalizations can be made about them. They are generally determined by the nature of the experiment.

The current striking an internal target may be determined by connecting it to ground through a current meter. Care must be taken to eliminate (by inductive chokes, etc.) R.F. pick-up which may be rectified and thus give a false reading. Considerable trouble is also experienced with electron emission from the target, especially when the latter is hot. A more reliable method of measuring internal target current is by measuring the temperature rise in the cooling water; the water flow rate must, of course, be measured and well regulated.

In experiments with external beams, the R.F. pick-up problem is not present, and the electron emission problem may be eliminated by collecting the current in a Faraday cup, so that purely electrical measurements can be made very accurately. For most applications, it is most important to know the integrated

[1] J. A. MARTIN: Private Communication.
[2] Cf., for example, E. CREUTZ: Rev. Sci. Instrum. **24**, 330 (1953).
[3] F. N. D. KURIE: Phys. Rev. **55**, 241 (1939).
[4] J. A. MARTIN: Private Communication.

current. This may be measured by allowing the charge collected from the beam to accumulate on a condenser. The voltage across the condenser thus gives a measure of the integrated current. When this voltage reaches a convenient value, it can be made to "fire" a thyratron which discharges it. The voltage build-up then begins anew. Provision is usually made for automatically counting the number of discharges.

g) Shielding and radiation hazards.

34. Shielding. High energy particles bombarding almost any form of matter induce nuclear reactions which, with high probability, lead to the prompt emission of neutrons and gamma rays, and leave radioactive residual nuclei which, in turn, decay with the emission of beta and gamma rays. All of these radiations represent health hazards which must be carefully guarded against.

The probability for a particle in the high energy beam to induce a nuclear reaction, P_r, may be calculated with relatively good accuracy as

$$P_r = \int_0^{E_B} \sigma_r \frac{d\mathcal{R}}{dE} dE \qquad (34.1)$$

where σ_r is the reaction cross section, which at low energies is available from tables[1] and at high energies is just the geometric cross section $\pi r_0^2 A^{\frac{2}{3}}$ (where r_0 is approximately 1.4×10^{-13} cm and A is the atomic weight of the target material. For deuterons, alpha particles etc., the radius $r_0 A^{\frac{1}{3}}$ should be increased to include the size of the bombarding particles), $\mathcal{R}$ is the range which is available from range-energy curves in the literature[2], [$\mathcal{R}$ in (34.1) has units of atoms per cm² whereas range curves are usually given in terms of grams per cm². The latter is obtainable from the former by multiplying by AVOGADRO's number divided by the atomic weight), and the integration is carried from the incident energy of the bombarding particle, E_B, down to zero. For protons and deuterons bombarding a light element, P_r is roughly 10^{-3} at 8 Mev, 10^{-2} at 20 Mev, 10^{-1} at 90 Mev, and $\frac{1}{2}$ at 400 Mev. For a target of a heavy element, it is about 10^{-6} at 8 Mev, 10^{-4} at 15 Mev, 10^{-2} at 35 Mev, 10^{-1} at 70 Mev, and $\frac{1}{2}$ at 300 Mev.

The number of neutrons per incident particle is P_r times the number of neutrons per nuclear reaction. For energies up to about 50 Mev, compound nucleus reactions predominate so that one neutron is emitted for every 10 to 15 Mev of incident energy. At higher energies, a new reaction mechanism becomes predominant in which a large part of the incident energy is transferred to one emitted particle so that the number of neutrons per reaction increases less rapidly. In the region of the highest energy synchrocyclotrons it is generally between eight and fifteen.

To cite numerical examples, a one microampere beam (6×10^{12} particles per second) of 400 Mev protons on almost any target produces roughly $6 \times 10^{12} \times \frac{1}{2} \times 10 = 3 \times 10^{13}$ neutrons per second. A 1000 microampere beam of 20 Mev protons on beryllium produces $6 \times 10^{15} \times 0.01 \times 1.5 = 1 \times 10^{14}$ neutrons per second. The same current of 8 Mev protons yields $6 \times 10^{15} \times 10^{-3} \times 1 = 6 \times 10^{12}$ neutrons per second. At a distance of 10 meters, the fluxes in these three cases would be 3×10^6, 1×10^7, and 6×10^5 neutrons per cm²-sec respectively as compared with a generally accepted human tolerance flux of about 30 neutrons per cm²-sec.

In nuclear reactions at any energy, the residual nucleus continues to emit particles until this is no longer energetically possible. The remaining excitation

---

[1] J. M. BLATT and V. F. WEISSKOPF: Theoretical Nuclear Physics. New York: John Wiley & Sons 1951.
[2] Cf., for example, W. A. ARON, B. G. HOFFMAN and F. C. WILLIAMS: AECU-663.

energy, which may vary from zero to about ten Mev, is emitted in the form of gamma rays; an average of about three gammas is required to release this energy. Thus, in the above examples, the prompt gamma ray emission amounts to about 1×10^{13}, 2×10^{14}, and 2×10^{13} gammas per second respectively. At a distance of ten meters, these would give dosage rates of about 3, 60, and 6 roentgens per hour respectively as compared to the tolerance rate of 0.003 roentgens per hour.

In order to shield against this prompt neutron and gamma radiation, a shield wall is constructed around the cyclotron. As judged from the above examples, it should be of sufficient thickness to reduce the fast neutron flux by a factor of about 10^5 and the gamma ray flux by a factor of about 10^4. For economic reasons, shield walls are generally made of concrete or of water tanks.

In the region below about 50 Mev, the neutron energies are rapidly degraded by inelastic scattering which results in neutrons being reemitted with low energies and isotropic angular distributions. At energies of a few Mev or less (such energies are commonly emitted in the original reactions, and in any case, are rapidly reached by inelastic scattering), the elastic scattering angular distributions are also isotropic. Thus, the neutron attenuation length in most materials—that is, the distance within which the neutron flux is reduced by $1/e$—is essentially the mean free path for a scattering event. In ordinary concrete, this is about 11 cm. High density concretes with shorter attenuation lengths are also available, although at additional cost. For example, the attenuation lengths in barytes and iron aggragate concretes are about 8 cm and 6.5 cm respectively. Where water is used, advantage is taken of the energy degradation (averaging 50%) in elastic collisions between neutrons and protons, and of the large number of nuclei per unit volume; the attenuation length is about 10 cm. The gamma ray attenuation lengths are about 15 cm in normal concrete, about 6.5 cm in the heavy concretes, and about 25 cm in water. The required attenuations for both neutrons and gamma rays are thus realized if shield walls of from 3 to 6 feet in thickness surround the cyclotron on all sides.

Where neutron energies in the hundreds of Mev are involved as in the large synchrocyclotrons, the shielding problem becomes much more complex. The neutrons are emitted at small angles to the direction of the incident particle inducing the reaction. Thus, almost all of the high energy neutrons arrive at the shield wall near the median plane of the cyclotron. Furthermore, the reactions they induce in the shielding material will, with high probability, release neutrons of high energy moving in the same direction. Thus, the effectiveness of scattering is considerably reduced, and attenuations lengths are about five times longer than those quoted above for lower energy neutrons. Shield thicknesses of twenty feet and more are therefore commonly used near the median planes of the large synchrocyclotrons. Far from the median plane, the problem is similar to that in lower energy machines, so that thicknesses of five feet or less usually suffice.

Access through shield walls is generally by removable doors or mazes. The latter have not proved completely successful, since the passageway sizes required for easy access are large enough to allow a considerable neutron leakage. On the other hand, some sort of maze is necessary to bring in the various utilities, so that it is frequently convenient to combine this with a personnel maze. The latter may be supplemented by a relatively light weight, manually operated door.

35. Hazards from radioactivity. Protection from the radiations emitted by the radioactive residual nuclei is often a very difficult problem. In the worst case, every nuclear reaction leaves a residual radioactive nucleus. Where the bombardment has been carried to saturation, the decay rate at the end of bombardment is just equal to the number of nuclear reactions per second during the bombardment. To convert to curies, the usual unit of radioactive source strength, these rates should be divided by 3.7×10^{10}. Thus, in the typical examples used in the precious section, the target radioactivities would be about 100, 2000, and 200 curies respectively. Since bombardments are not usually carried to saturation, and since many of the products are either not radioactive or have very short or very long (and thus non-saturated) half-lives, these estimates of target radio-acticity are somewhat high, but targets in the 100 curie range are not uncommon in cyclotrons. When such a target is handled at a distance of one meter, the dosage rate in roentgen per hour is approximately equal to the source strength in curies, so that target handling is frequently a very difficult problem. In some cyclotrons, facilities are available for transferring targets from the cyclotron into a shield from a remote and shielded location.

Another health hazard which is even more difficult to overcome is the radio-activity in the cyclotron itself, induced by the accelerated particles striking the dees and deflection apparatus, and by neutrons emitted from the target. The radioactivity from the charged particles can frequently be avoided by covering the affected parts of the dees with a material giving only short lived activities. Carbon is suitable for this purpose, especially since it can withstand the heat load. Another simple solution is to use aluminum dees. In any case, some measures must often be taken, since protons or deuterons striking copper induce the long lived (250 day) activity of Zn^{65}. The deflector septum is frequently a very serious problem since it is usually made of tungsten or tantalum, both of which can be activated by long lived activities. At lower energies however, P_r, for these heavy elements is quite small.

The radiation arising from neutron-induced activities is best reduced by a careful choice of materials. Fortunately, neither iron nor copper, the two materials which are most difficult to avoid, have large cross sections for the production of long lived activities. One serious *potential* hazard of this type is from the use of large amounts of silver solder, as for example in soldering cooling tubes to the dees. This is especially serious since it is frequently necessary to do mechanical repairs and alterations on the dees.

The situation as regards relatively short lived activities is not nearly so favorable. The five minute activity from $Cu^{65}(n, \gamma) Cu^{66}$ is very frequently an important source of exposure for perhaps fifteen minutes after a high intensity bombardment. After that, the 2.6 hour activity from $Fe^{56}(n, p) Mn^{56}$ and $Mn^{55}(n, \gamma) Mn^{56}$ (the latter reaction is from the manganese alloyed in most magnet steels to the extent of a few tenths percent) is the most serious radiation; it is sometimes necessary to delay operations for several hours while waiting for it to decay. The 12.8 hour activity from $Cu^{63}(n, \gamma) Cu^{64}$ may also cause some difficulty.

h) Controls and interlocks.

36. Controls. Cyclotrons are operated from a control point outside the shield wall and often some distance from the machine. This control point has indicating instruments and adjusting mechanisms for each of the components described in Sects. 18 through 35. While there is considerable variation among the various installations, almost all of the following items are generally present:

Magnetic field. A total current meter and coarse and fine current adjustments. The latter typically changes the field in about 0.1 and 0.01 % steps respectively. Usually large adjustments must be made at the motor generator set.

Radiofrequency system. Meters to indicate dee voltage and oscillator plate voltage and current; plate voltage on-off and raise-lower controls; dee trimmer adjustment and position indicator; a frequency measuring device; and in synchrocyclotrons, a timing device.

Ion source. Gas feed adjustment, filament and arc voltage and current meters, and on-off and raise-lower controls; and ion source positioning adjustments.

Vacuum system. One or more of the vacuum gages; and a diffusion pump valve closing device.

Deflection system. Deflector voltage meter and deflector positioning controls; controls for beam focusing apparatus.

Radiation monitors. Usually a recorder showing the current in an ionization chamber near the cyclotron.

Beam monitors. Beam current meter and recorder showing temperature rise in target cooling water.

In addition, there are usually many other meters and controls to suit the special needs of particular installations. For example, if the dees carry a D.C. bias, voltage and current meters, and on-off and raise-lower adjustments for this bias are necessary; if means are available for adjusting the positions of the dees, drive controls and position indicators for these are provided; etc.

37. Interlocks. In order to safeguard both personnel and equipment, a large number of interlocks are provided. Most of these prevent the oscillator high voltage from being turned on, although sometimes radiation protection devices are made to disable the ion source arc. In most cyclotrons, interlocks prevent application of oscillator voltage unless all cooling circuits have adequate flow rates, the oscillator housing is closed, and appropriate measures are taken to assure that personnel are outside the shield wall (e.g., the door is closed, a warning signal is given, etc.). The oscillator is also automatically turned off when there is excessive drain from its plate voltage supply; this occurs quite frequently when sparking is present.

Since there are a large number of individual water cooling circuits, a separate indicator is generally provided for each to show when the flow is deficient. It usually turns on a light, rings an alarm, or both.

The complete control and interlock system of most cyclotrons includes many miles of wiring, hundreds of relays, and dozens of meters, drive motors, flow indicators, etc. However, the simplicity of operation and maintenance thereby achieved is usually well worth the effort.

C. Miscellaneous topics.

38. Operation. Cyclotron controls are usually sufficiently simple that the machine can be operated by a single technician. If each of the components is functioning properly, a beam will be achieved very rapidly. A typical start-up procedure would be to install the target, check the cooling water flow and the vacuum, set the magnetic field approximately, turn on the oscillator and slowly raise the dee voltage to the operating point, and turn on the ion source. To achieve a beam, it is necessary to vary the magnetic field until the ionization chamber indicates that radiation is present; further adjustment then gives a beam current reading. The beam can then be maximized by adjusting first the magnetic field or dee trimmer condenser, and then the various ion source and deflector controls. Usually the dee voltage is gradually worked upward to its maximum practical value.

During operation, it is necessary for the operator to make constant adjustments of either the magnetic field or the dee trimmer condenser to compensate for changes in the resonant frequency caused by thermal expansions and contractions of the dee system. In a few cases, systems for making these adjustments automatically have been developed, but these are not in general use. In a standard cyclotron, the oscillator is turned off by sparking at irregular intervals and must be turned on again by the operator. When the vacuum is marginal, or when the cyclotron has recently been open to air, this may occur as often as every few seconds; under very favorable conditions, it may not occur for several hours. In synchrocyclotrons, constant attention from the operator is not necessary. The dee voltage is lower so that sparking is not a problem, and small changes in the physical dimensions of the dees merely shift the operation to a different part of the frequency modulation cycle.

The principal routine maintenance operations are filling the liquid nitrogen traps every few hours, and changing the ion source filament at 50 to 100 hour intervals. Other frequent maintenance jobs include finding and repairing vacuum leaks, and maintaining the various electronic equipment. The target preparation and changing activities are, of course, dependent on the experimental program.

All in all, a cyclotron is a relatively stable and smoothly operating device, once the sources of trouble have been found and corrected. This typically requires a year or two of operation. After a few years, beam innages of close to 90% are usually maintained. Major repairs requiring shutdowns of a few weeks occur perhaps once a year on an average.

39. Characteristics of various standard cyclotrons. In 1955, Dr. F. T. HOWARD of Oak Ridge National Laboratory made a survey of the world's fixed frequency cyclotrons [13]. Extensive literature searches were carried out, and questionaires were sent to all laboratories. He has kindly made his work available and parts of it are reproduced below.

The basic information is shown in Table 1. Since it is somewhat conventional to name a cyclotron by its pole-tip diameter, that dimension is listed first, and cases where the commonly used name does not coincide (except where it is named for the institution) are explained in footnotes. Next are listed the name of the institution to which the cyclotron is attached, the location, and the date of completion (usually the date when the first circulating beam was achieved). Where extensive remodeling has been carried out, the date of the most recent remodeling is listed in brackets. Column 6 and 7 shows the principal magnet data. The magnet weight includes both iron and conductor. The conductor is copper in all but numbers 8, 9, 12, 18, 26, and 40 where it is aluminum. In the latter cases, it accounts for about 10% of the weight, and where copper is used, it generally accounts for 10 to 16% of the weight except in number 3 where it is only 7%. The magnetic field listed is the maximum used in operating with protons or deuterons. Where considerably higher fields are obtainable, they are reported in footnotes. Data on radiofrequency systems is very difficult to obtain since the names of the various types of oscillators are not well standardized, and reliable measurements of dee voltage are not often available. The most reliable index of performance is therefore the oscillator output power; this is listed in column 8, and the frequency is given in column 9. Since almost every cyclotron uses an ion source of its own design or with modifications from other designs, and since the naming is not standardized, data on ion sources is very difficult to collect, and thus none is given in Table 1. Column 10 gives the shielding material and thickness. Where less shielding is used in some directions, this is given in brackets.

Columns 11 and 12 give the principal types of particle accelerated and their maximum energy. Only ions accelerated from the ion source are included. In addition, several cyclotrons (especially numbers 3, 4, 5, and 7) make use of electron stripping collisions in the course of the acceleration to obtain very high energy heavy ions; however, the beams are very small, and very inhomogeneous in energy. They are useful in producing small quantities of isotopes which cannot be made in other ways, but practically no physical measurements have been made with these heavy ions.

Since the resonant frequency (2.4) is the same, any cyclotron which accelerates deuterons to energy E can also accelerate alpha particles to energy $2E$ and molecular hydrogen ions to energy E (and thus, each of the two protons of which it is composed to energy $E/2$) by substituting helium or hydrogen respectively for the deuterium fed into the ion source. A slight change in tuning is necessary, and in some cases this requires slight changes in shimming. For cyclotrons where this is done, column 11 lists "deuterons" and an asterisk is attached. This procedure is followed even in cases where alphas or protons are accelerated more frequently than deuterons. In a few of these cyclotrons, protons rather than hydrogen molecule ions are accelerated by reducing the magnetic field to half its usual value. This would ordinarily require extensive changes in the magnetic shimming.

Several recently built or recently remodeled cyclotrons have the feature of variable energy. This requires that there be a simple method of changing the frequency of the R.F. voltage applied to the dees, continuously and over a wide range, and also that the magnetic field be variable without changes in shimming. Cyclotrons having this feature are numbers 1, 23, 26, 43, and 47. It adds considerably to their value in doing quantitative research in nuclear reactions and scattering. Most of these cyclotrons have a single dee as this eases the problem of shifting the frequency.

The principal common feature missing from some cyclotrons is a deflected beam. Numbers 3, 13, 28, 29, 33, 44, and 46 did not have deflected beams by late 1955, although some of these were actively planning to install deflection equipment.

A few other special features might be noted. The Los Alamos cyclotron uses a small amount of azimuthally varying field focusing. The Livermore and the three Oak Ridge cyclotrons are vertically mounted, i.e., their magnetic median planes and dees are vertical rather than horizontal. The Livermore cyclotron has a cam-shaped magnetic field which allows the deflected beam to emerge more nearly normal to the magnetic field. This avoids the rather strong divergence of beam by the fringing field found in most cyclotrons, and thus improves deflection efficiency.

40. Characteristics of various synchrocyclotrons. In April 1955, Dr. Howard instituted a survey of synchrocyclotrons similar to the standard cyclotron survey described in Sect. 39. The information received by the deadline for submission of this article is summarized in abbreviated form in Table 2. Some of the differences in column headings between Table 2 and 1 should perhaps be explained. The magnet power is listed in connection with synchrocyclotrons because it is a major cost item in many of these. The R.F. voltage is low enough to be measured meaningfully by standard techniques; it is therefore included in Table 2. The repetition rate is an important quantity in synchrocyclotrons which, of course, has no meaning on standard cyclotrons. The shielding is not listed in Table 2 partly for lack of information, but principally because it is too complex to describe briefly. All of the large synchrocyclotrons have thick concrete shields. A dash in column 5 of Table 2 indicates that construction is not yet complete.

Table 1.

(1) No.	(2) Pole face diameter (inches)	(3) Institution	(4) Location	(5) Date of first operation	(6) Magnet weight (tons)	(7) Operating magnetic field (kG)	(8) RF power (kW)	(9) Frequency (mc/sec)	(10) Shielding	(11) Principal particles accelerated	(12) Energy (Mev)
1	90	Univ. of California Radiation Lab.	Livermore, California	1955	(350)	9	380	4—9.5	5 [2] ft concrete	protons deuterons*	14
2	86	Oak Ridge National Laboratory	Oak Ridge, Tennessee	1950 [1955]	400	8.8	250	13.4	5 ft concrete	protons	22
3	83^1	Nobel Institute	Stockholm, Sweden	1951	400	10.7^2	180	8.1	underground	deuterons*	22
4	72^3	Univ. of California	Berkeley, California	1939 [1953]	250	16.2^4	250	12.3	5 ft water	deuterons*	24
5	70	Centre d'Etudes Nucléaires	Saclay, France	1953	270	18	40	10.6	6 ft concrete	deuterons	(25)
6	63	Oak Ridge National Laboratory	Oak Ridge, Tennessee	1952	—^{5}	15	25	5.1	none	nitrogen ions	29
7	61.5^6	Univ. of Birmingham	Birmingham, England	(1948)	300	13.5^7	140	10.2	4 ft concrete [2.5 ft water]	deuterons*	20
8	60	Argonne National Laboratory	Lemont, Ill.	1951	280	14.9^8	(130)	11.2	7 [4] ft concrete	deuterons*	21.6
9	60	Brookhaven National Laboratory	Upton, N.Y.	1951	240	15	(65)	11.5	5 ft limonite concrete	deuterons*	20
10	60	National Advisory Committee for Aeronautics	Cleveland, Ohio	1953	250	14.6	400	11	1 ft concrete +4 ft earth	deuterons*	21
11	60	Univ. of Washington	Seattle, Wash.	1951	218	15	100	11.6	4 ft concrete [1 ft concrete +3 ft water]	deuterons*	21
12	60	Carnegie Institute	Washington, D.C	1943	—	12.7	55	9.7	underground	deuterons*	15
13	50	Hammersmith Hospital	London, England	1955	122	15	75	11.3	6 ft concrete	deuterons*	20

No.		Institution	Location	Year					Shielding	Particle	
14	47	Ohio State Univ.	Columbus, Ohio	1940 [1954]	80	14.5	40	11.0	none	deuterons*	12
15	47	Univ. of Pittsburgh	Pittsburgh, Pennsylvania	1946	100	15.5	30	11.8	water	deuterons*	16.2
16	45	Univ. of Indiana	Bloomington, Indiana	1941	87	14.6	40	10.4	5 [3] ft water	deuterons*	11
17	45	Washington Univ.	St. Louis, Missouri	1941	82	14.3	70	10.8	underground	deuterons*	10.2
18	44.5	Council for Scientific and Industrial Research	Pretoria, South Africa	1955	82	18	40	13.7	12 ft earth +concrete	deuterons*	20
19	44⁵	Oak Ridge National Laboratory	Oak Ridge, Tennessee	1956	—⁵	6.3	(300)		none	protons	5
20	44	Osaka Univ.	Osaka, Japan	1954	45	14	45	11.6	concrete	deuterons*	11.5
21	42.5¹¹	Mass. Inst. of Technology	Cambridge, Mass.	1940 [1947]	96	17.4	70	13.1	4 [3] ft concrete	deuterons*	15.8
22	42.5	Univ. of Kyoto	Kyoto, Japan	1955	82	17.5	(100)	13.1	1.7 meters concrete	deuterons*	15.3
23	42	Los Alamos Scientific Lab.	Los Alamos, New Mexico	1955⁹	100	18.7	85	8.6—14	5 ft concrete	deuterons*	17.5
24	42	Univ. of Michigan	Ann Arbor, Michigan	1936 [1950]	?	12.5	35	10.2	3 ft concrete [2½ ft water]	deuterons*	7.8
25	41¼	Univ. of Illinois	Urbana, Illinois	1942	70	15.8	25	12.2	4 ft water	deuterons*	12
26	40	Univ. of Melbourne	Melbourne, Australia	(1955)	45	14	50	(4.3—10.7)	2.5 ft brick and sand	deuterons* protons	6.3 12.5
27	40	Max Planck Institut	Heidelberg, Germany	1941	80	17	(65)	13.0	5 ft water +3 ft concrete	deuterons*	13
28	38.5	Univ. of Louvain	Louvain, Belgium	1951	62	18	(18)	13.7	4 ft concrete	deuterons	13.4
29	38	Biochemical Research Foundation	Newark, Delaware	1939 [1955]	50	15	15	12.5	3 ft water	deuterons	10
30	37	Oregon State College	Corvallis, Oregon	1955	50	14	(20)	(10.7)	concrete	deuterons*	7.5
31	37	Cavendish Lab.	Cambridge, England	1938	51	16	100	13.0	3 ft water	deuterons*	8

11*

Table 1. (Continued.)

(1) No.	(2) Pole face diameter (inches)	(3) Institution	(4) Location	(5) Date of first operation	(6) Magnet weight (tons)	(7) Operating magnetic field (kG)	(8) RF power (kW)	(9) Frequency (mc/sec)	(10) Shielding	(11) Principal particles accelerated	(12) Energy (Mev)
32	36.5	Purdue Univ.	Lafayette, Indiana	1937 [1952]	43	16.2	(25)	12.4	3 ft water	deuterons*	10
33	36	Columbia Univ.	New York, N.Y.	1936 [1955]	?	17.5	30	13 — ?	4 ft water	deuterons / proton	8 / 16
34	36^{10}	Univ. of Liverpool	Liverpool, Engl.	1939	54	16	20	12.5	4 ft concrete	deuterons*	9
35	35.5	Inst. of Theoretical Physics	Copenhagen, Den.	1939	—	—	—	—	concrete	deuterons*	9
36	35.5	Collège de France	Paris, France	1939	27	17	(13)	11.5	1 M. water	deuterons	6.7
37	33	Physikalisches Institut der E. T. H.	Zürich, Switzerland	1944	60	17	50	13	2 M. concrete [1 M. water]	protons / alphas	7.2 / 17.2
38	32.5	Collège de France	Paris, France	1938	26	15	20	11.5	1 meter water	deuterons*	7
39	32^1	Nobel Institute	Stockholm, Sweden	1939	17	15	20	12	3 ft concrete	deuterons*	7
40	31	Australian National Univ.	Canbarra, Australia	1955	31	12.6	50	19.6	20 in. concrete	protons	8
41	28	Yale Univ.	New Haven, Conn.	1938	20	14.5	20	10.6	16 in. concrete	deuterons*	4.1
42	27	Stanford Univ.	Stanford, California	1941	11.5	12.3	6	9.4	4 ft water	deuterons	3
43	26.4	Univ. of Rochester	Rochester, New York	1954	17	14.5	25	9 — 22	3 ft limonite concrete	deuterons* / protons	4.3 / 8
44	25	Scientific Research Inst.	Tokyo, Japan	1952	22	14.8	12	11.3	2.6 ft concrete	deuterons	4
45	18	Brookhaven National Laboratory	Upton, N.Y.	1952	6	12	20	18.5	none	protons	2.5
46	17	Cornell Univ.	Ithaca, N.Y.	1935	3	12	2.5	(9.2)	none	deuterons / protons	1.4 / 1.2
47	16	Univ. of Tokyo	Tokyo, Japan	1955	7	18.5	6	11 — 14	2 ft concrete [2 ft water]	deuterons* / protons	2 / 4

Footnotes of Table 1.

( ) Indicates data was not obtained directly but was calculated from other information.
* Also accelerate alphas to twice the listed energy, and protons to half the listed energy. See text.

[1] The Nobel Institute Cyclotrons are known as the "80 Cm" and "225 Cm" cyclotrons.

[2] Fields up to 20000 gauss are available and are frequently used for acceleration of heavy ions.

[3] Known as "Crocker Laboratory 60-Inch Cyclotron". Originally, the poles were tapered to give a 60-inch pole tip.

[4] Maximum available field is 19700 gauss. Fields up to 19100 are used for accelerating heavy ions (especially oxygen-16).

[5] The Oak Ridge 63-Inch and 44-Inch Cyclotrons are in a large calutron track. The actual pole faces of the 44-inch are large and non-circular. The name is derived from the size of a magnet that would ordinarily be used for such a machine.

[6] Known as the "Nuffield Cyclotron".

[7] Maximum available field is 18000 gauss; fields up to 17700 are used for heavy ion acceleration.

[8] Maximum available field is 20000 gauss.

[9] Magnet was originally constructed and used for the former Harvard University Cyclotron, and was later used in a previous Los Alamos cyclotron.

[10] This cyclotron is known as the "37-inch cyclotron".

[11] Known as the "Markle Cyclotron".

41. Status and prospects.

41. Status and prospects. Standard cyclotrons are used principally for production of radioactive isotopes, for studies of nuclear reactions and scattering, and of nuclear energy level structure and character, for a neutron source, and for radiation damage studies.

The best of the standard cyclotrons can produce many radioactive isotopes at a cost which is considerably less than the chemical processing costs. Since the latter must also be applied to reactor produced isotopes, the total cost of reactor and cyclotron produced isotopes are quite similar. Since most nuclear reactions in cyclotrons lead to elements different from the one bombarded [e.g., $(p, 2n)$, (d, n), (d, α), etc.], the radioactive element can be chemically separated from the target material, thus giving very much higher specific activities than can be obtained in isotopes produced by (n, γ) reactions in reactors. Also, special applications of the annihilation radiation from positron activities have been developed. Thus, it would seem that the standard cyclotron should have a continuing and ever-increasing application for production of radioactive isotopes.

The standard cyclotron is, in some ways, not so ideally suited to nuclear physics studies. In general, these use only a fraction of a microamp of beam, but require high energy resolution, and, for many applications, variable energy. By careful magnetic analysis, relatively good energy resolutions have been achieved, and recently, several cyclotrons have been built or modified to have variable energy. However, in neither case is the performance comparable to that of Van der Graaf generators. With the development of higher energy Van der Graaf's, it is possible that this application of cyclotrons will eventually becomes less useful. Over the next several years, however, cyclotrons certainly have much to contribute.

The use of a standard cyclotron as a neutron source has been in the past and continues to be quite important. As a high energy neutron source, it is unsurpassed except perhaps in nuclear bomb explosions. As a source of low energy neutrons for cross section work, it seems to be capable of higher resolution than choppers operating on pile neutron beams.

Radiation damage studies with standard cyclotrons are still in a primitive state. However, there would seem to be many applications of the large currents of high energy particles available only from standard cyclotrons to both solid state and radiation chemistry problems.

Table 2.

(1) No.	(2) Poleface diameter (inches)	(3) Institution	(4) Location	(5) Date of first operation	(6) Magnet weight (tons)	(7) Magnet power (kW)	(8) Operating magnetic field (kG)	(9) R.F. voltage (kv)	(10) R.F. power (kW)	(11) Freq. range (mc/sec)	(12) Repetition rate (pulses/sec)	(13) Particle	(14) Energy (Mev)
1	237	Big. Volga Lab.	U.S.S.R.	?	7000	—	—	—	—	—	50—80	protons	680
2	197	CERN	Geneva, Switzerland	—	2500	750	18.8	12	25	16.6—28.7	55	protons	600
3	184	Univ. of California	Berkeley, California	1946	4300	750	15.0	17	(17)	9.8—22.9	60	protons deuterons alphas	350 195 390
4	170	Univ. of Chicago	Chicago, Illinois	(1950)	2073	650	18.6	—	12	11.8—28.6	60	protons deuterons alphas	450 256 512
5	164	Columbia Univ.	New York, N.Y.	(1950)	2150	414	17.4	10	—	11.3—28	60—120	protons	385
6	158	Univ. of Liverpool	Liverpool, England		1650	860	18.9	6.5	10	29.2—18.9	110	protons	410
7	141	Carnegie Institute of Technology	Pittsburgh, Pennsylvania	1952	1590	430	20.5	15	35	19.8—31.5	180—200	protons	450
8	130	Univ. of Rochester	Rochester, New York	1948	1080	160	17.0	8	(15)	19.5—26.3	150—300	protons	240
9	110	AERE	Harwell, England	1949	(720)	300	16.2	10	(12)	18.9—26.3	100	protons	175
10	95	Harvard Univ.	Cambridge, Mass.	1949	800		16.0				170	protons	100
11	90.5	Univ. of Uppsala	Uppsala, Sweden	1951	650	440	21.5	10	(15)	25.3—33.3	240	protons	192
12	82	McGill Univ.	Montreal, Canada	1947	265	100	16.4				500	protons	100
13	75	Univ. of Bonn	Bonn, Germany	—	200	100	14.5	14	8	10.6—11.2	2500	deuterons	33

No.	Institution	Location	Year										
14	Inst. for Nuclear Research	Amsterdam, Netherlands	1949	170	90	13.6	25	8	10.05—10.34	2000	deuterons alphas	28 56	71
15	Com. Nacional de la Energia Atomica	BuenosAires, Argentina	1954	196	240	14.6	17	(15)	10.46—10.88	1950	deuterons alphas	30 56	71
16	Istanbul Univ.	Istanbul, Turkey	—	200	90	14.5	15	3.5	10.3—10.7	2000	deuterons	27	71
17	Univ. of California	Los Angeles, California	1946	80	60	16.6	9	6	22—25.6	1000	protons	20.5	41
18	Princeton Univ.	Princeton, N.J.	1951	48	20	19	7	(4)	25—28	2000	protons	19.6	35

It thus seems that standard cyclotron development in the future will be pointed toward higher currents, although there is, at present, considerable interest in the variable energy feature.

Synchrocyclotrons have been used for all the above applications, and also as a meson source. They have opened up the fields of high energy nuclear physics and meson physics, and have been probably the most important contributors to the advance of nuclear physics since their development. Their future, however, seems to be somewhat in doubt because of the development of A.V.F. cyclotrons. The latter would seem to be cheaper to construct because they use far less iron to attain a given energy. In addition, they should be capable of much higher currents since they can operate continuously rather than pulsed. The principal problems are the difficulties in constructing the magnetic field, and in accelerating through resonances. These problems are being attacked vigorously by several groups, and it seems reasonable to expect relatively satisfactory solutions. In any case, it seems quite certain that cyclotrons of some type will continue to be extremely useful in the energy region between 25 and 1000 Mev. If methods for constructing azimuthally varying fields become highly developed, the A.V.F. cyclotron may usurp much of the field of standard cyclotrons. On the other hand, it may not be unreasonable to expect other developments in standard cyclotrons to provide an easier method of achieving higher currents at energies below 25 Mev.

42. History and bibliography. The standard cyclotron was developed by E. O. LAWRENCE and coworkers [16], [17] at Berkeley beginning in 1930. The synchrocyclotron was proposed independently by VEKSLER [38] and McMILLAN [28] in 1945. The first working model was completed at Berkeley in 1946 [30]. The azimuthally varying field cyclotron was first proposed by THOMAS [36] and rediscovered by McMILLAN in 1949 and by the Mura group in 1954.

Some of the more important publications about cyclotrons are listed below. References [6], [10], [11], [19], [21] to [24], [26], [29] and [39] are review articles on cyclotrons as a whole. Articles devoted to general cyclotron theory are references [4], [7], [27], [31], [36], and [40], whereas theoretical discussion of specialized portions of the theory may be found in references [8], [9], [12], [15], [18], [35], and [37]. Various cyclotron components are discussed in references [14], [20], [25], and [34]. Some interesting aspects of operation are presented in references [1], [32], and [33] and references [3] and [5] describe some highly developed systems for using cyclotron beams.

Appendix. — Derivation of Eq. (6.13)

Eqs. (6.12) are of the form

$$\frac{d^2x}{d\vartheta^2} + (A + B \cos p\,\vartheta)x = 0. \tag{A.1}$$

In accordance with Floquet's theorem[1], the solution of (A.1) is of the form

$$x = e^{i\nu\vartheta}u(\vartheta) \tag{A.2}$$

where ν is small compared to p and $u(\vartheta)$ is periodic with period $p\vartheta$ and can thus be expanded as

$$u = \sum_\varkappa u_\varkappa e^{i\varkappa p\vartheta}. \tag{A.3}$$

Inserting (A.2) and (A.3) in (A.1) and using the identity

$$\cos p\,\vartheta = \frac{e^{ip\vartheta} + e^{-ip\vartheta}}{2}$$

gives a series of equations typified by

$$-(\varkappa p + \nu)^2 u_\varkappa + A u_\varkappa + \frac{B}{2}u_{\varkappa-1} + \frac{B}{2}u_{\varkappa+1} = 0. \tag{A.4}$$

Assuming that the expansion (A.3) converges very rapidly so that only u_{-1}, u_0, u_{+1} are different from zero, (A.4) reduces to

$$\left.\begin{aligned}
[\nu^2 - A]\,u_0 &= \frac{B}{2}u_1 + \frac{B}{2}u_{-1}, \\[4pt]
[(\nu + p)^2 - A]\,u_1 &= \frac{B}{2}u_0, \\[4pt]
[(\nu - p)^2 - A]\,u_{-1} &= \frac{B}{2}u_0.
\end{aligned}\right\} \tag{A.5}$$

Neglecting ν and A with respect to p in the second and third, and substituting values for u_1 and u_{-1} from these into the first, gives

$$\nu^2 = A + \frac{B^2}{2p^2}. \tag{A.6}$$

Inserting the values of A and B from (6.12) into (A.6) gives (6.13).

Acknowledgment.

The author would like to acknowledge the very extensive assistance of T. A. Welton, M. R. Donaldson, R. J. Jones, F. T. Howard, E. D. Hudson, R. S. Lord, R. S. Livinston, A. Zucker and J. A. Martin in the preparation of various sections of this article.

Bibliography.

[1] Alvarez, L. W.: High Energy Carbon Nuclei. Phys. Rev. 58, 192 (1940).
[2] Anderson, H. L., J. Marshall, L. Kornblith jr., L. Schwarcz and R. H. Miller: Synchrocyclotron for 450-Mev Protons. Rev. Sci. Instrum. 23, 707 (1952).
[3] Bender, R. S., E. M. Reilley, A. J. Allen, R. L. Ely, J. S. Arthur and H. J. Hausman: The University of Pittsburgh Scattering Project. Rev. Sci. Instrum. 23, 542 (1952).
[4] Bohm, D., and L. L. Foldy: Theory of the Synchrotron. Phys. Rev. 70, 249 (1946). Theory of the Synchrocyclotron. Phys. Rev. 72, 649 (1947).
[5] Boyer, K., H. E. Gove, J. A. Harvey, M. Deutsch and M. S. Livingston: Instrumentation of the MIT Cyclotron for the Study of Nuclear Reactions. Rev. Sci. Instrum. 22, 310 (1951).
[6] Chu, E. L., and L. I. Schiff: Recent Progress in Accelerators; Cyclotron. Ann. Rev. Nucl. Sci. 2, 82 (1953). Recent Progress in Accelerators; Synchrocyclotron. Ann. Rev. Nucl. Sci. 2, 83 (1953).
[7] Cohen, B. L.: The Theory of the Fixed-Frequency Cyclotron. Rev. Sci. Instrum. 24, 589 (1953).

---

[1] E. T. Whittaker and G. N. Watson: Modern Analysis, p. 412. Cambridge: Univ Press 1946.

[8] CREWE, A. V., and K. J. LeCouteur: Extracted Proton Beam of the Liverpool 156-Inch Cyclotron. Rev. Sci. Instrum. 26, 625 (1955).
[9] Foss, M. H.: Design of Cyclotron Magnets. Report NP-426.
[10] FREMLIN, J. H., and J. S. GOODEN: Cyclic Accelerators. Rep. Progr. Phys. 13, 295 (1950).
[11] GALLOP, J. W., London, Hammersmith Hospital: Notes on a Tour of American Fixed-Frequency Cyclotrons in the Autumn of 1950, Report NP-3537.
[12] HOUGH, P. V. C.: Radial Oscillations in the Cyclotron. Rev. Sci. Instrum. 24, 42 (1953).
[13] HOWARD, F. T.: ORNL-2119, A Survey of Cyclotrons.
[14] JONES, R. J., and A. ZUCKER: Two Ion Sources for Production of Multiply Charged Nitrogen Ions. Rev. Sci. Instrum. 25, 562 (1954).
[15] KERST, D. W., and R. SERBER: Theory of the Betatron. Phys. Rev. 60, 53 (1941).
[16] LAWRENCE, E. O., and N. E. EDELFSEN: On the Production of High Speed Protons. Science 72, 376 (1930).
[17] LAWRENCE, E. O., and M. S. LIVINGSTON: A Method for Producing High Speed Hydrogen Ions without the Use of High Voltage. Phys. Rev. 37, 1707 (1931).
The Production of High Speed Protons without the Use of High Voltage. Phys. Rev. 38, 834 (1931).
The Production of High Speed Light Ions without the Use of High Voltage. Phys. Rev. 40, 19 (1932).
The Multiple Acceleration of Ions to Very High Speeds. Phys. Rev. 45, 608 (1934).
The Production of 4 800 000 Volt Hydrogen Ions. Phys. Rev. 42, 212 (1933).
[18] LeCouteur, K. J.: The Regenerative Deflector for Synchrocyclotrons. Proc. Phys. Soc. (Lond.) B 64, 1073 (1951).
Perturbations in the Magnetic Deflector for Synchrocyclotrons. Proc. Phys. Soc. (Lond.) B 66, 25 (1953).
[19] LIVINGSTON, M. S.: The Cyclotron. J. Appl. Phys. 15, 2, 128 (1944).
[20] LIVINGSTON, M. S.: Ion Sources for Cyclotrons. Rev. Mod. Phys. 18, 293 (1946).
[21] LIVINGSTON, M. S.: Particle Accelerators. Adv. Electronics 1, 269 (1948).
[22] LIVINGSTON, M. S.: Standard Cyclotron. Ann. Rev. Nucl. Sci. 1, 157 (1952).
[23] LIVINGSTON, M. S.: Synchrocyclotron. Ann. Rev. Nucl. Sci. 1, 163 (1952).
[24] LIVINGSTON, M. S.: High-Energy Accelerators. Interscience Tracts on Physics and Astronomy, No. 2. New York: Interscience Publishers 1954.
[25] LIVINGSTON, R. S., and R. J. JONES: High-Intensity Ion Source for Cyclotron. Rev. Sci. Instrum. 25, 522 (1954).
[26] MANN, W. B.: The Cyclotron. London: Methuen & Co. 1940.
[27] McMILLAN, E. M.: Nuclear Physics for Engineers. MDDC-1014
[28] McMILLAN, E. M.: The Synchrotron.—A Proposed High Energy Particle Accelerator. Phys. Rev. 68, 143 (1945).
[29] PICKAVANCE, T. G.: Cyclotrons. Progr. in Nucl. Phys. 1 (1950).
[30] RICHARDSON, J. B., B. T. WRIGHT, E. J. LOFGREN and B. PETERS: Development of the Frequency-Modulated Cyclotron. Phys. Rev. 73, 424 (1948).
[31] ROSE, M. E.: Focusing and Maximum Energy of Ions in the Cyclotron. Phys. Rev. 53, 392 (1938).
Magnetic Field Corrections in the Cyclotron. Phys. Rev. 53, 716 (1938).
[32] ROSSI, G. B., W. B. JONES, J. M. HOLLANDER and J. G. HAMILTON: The Acceleration of Nitrogen-14 (+6) Ions in a 60-Inch Cyclotron. Phys. Rev. 93, 256 (1954).
[33] SEWELL, D. C., L. HENRICH and J. VALE: Some Operating Phenomena Associated with 184-Inch Cyclotron. Phys. Rev. 72, 739 (1947).
[34] SHOEMAKER, F. C., R. J. BRITTEN and B. C. CARLSON: A New Method for Focusing Ion Beams. Phys. Rev. 86, 582 (1952).
[35] TEICHMANN, T.: Beam Oscillation in F-M Cyclotron. J. Appl. Phys. 21, 1251 (1950).
[36] THOMAS, L. H.: The Paths of Ions in the Cyclotron. Phys. Rev. 54, 580 (1938).
[37] TUCK, J. L., and L. C. TENG: Regenerative Deflector for Synchrocyclotron. Phys. Rev. 81, 305 (1951); U. of Chicago 170-Inch Synchrocyclotron Progress Report III, NP-1767 (1950).
[38] VEKSLER, V.: A New Method for Acceleration of Relativistic Particles. C. R. Acad. Sci. USSR. 43, 329 (1944).
On a New Method of Acceleration of Relativistic Particles. C. R. Acad. Sci. USSR. 44, 365 (1944).
A New Method of Acceleration of Relativistic Particles. J. Phys. USSR. 9, 1953 (1945).
Concerning Some New Methods of Acceleration of Relativistic Particles. Phys. Rev. 69, 244 (1946).
[39] WALKER, D., and J. H. FREMLIN: Acceleration of Heavy Ions to High Energies. Nature Lond. 171, 189 (1953).
[40] WILSON, R. R.: Theory of the Cyclotron. J. Appl. Phys. 11, 781 (1940).

Electron Synchrotrons.

By

ROBERT R. WILSON.

With 5 Figures.

1. Introduction. The electron synchrotron was invented in 1945 almost simultaneously by VEKSLER[1] in Russia and by McMILLAN[2] in the USA. It is a particular application of their more general principle of phase stability; in the case of the synchrotron, electrons are kept on an orbit of constant radius. This is done by the use of a magnetic guide field in which the field strength is periodically raised from a low to a high value. At the low field, low energy electrons are injected into the orbit and are trapped by the synchrotron mechanism. The electrons remain on the orbit as the magnetic field is increased so that their energy must necessarily increase also. Eventually, when the energy has become very large, the electrons are withdrawn or are caused to strike a target which produces an x-ray beam. The energy of the electrons is supplied by a radiofrequency electric field applied across a gap which the electrons must cross on successive revolutions of the orbit. The frequency of revolution of the electron in its orbit and the frequency of the radiofrequency field are made the same, and, according to the principle of phase stability, the energy of the electrons is automatically increased at just the right rate so that the electrons stay on the proper orbit. The electrons are usually injected into the synchrotron at an energy of several Mev and thus their velocity is very nearly that of light, hence the radiofrequency, which determines the size of the orbit, is held constant during the acceleration. It is this property that distinguishes the electron synchrotron from the proton synchrotron where the frequency must be varied because the speed of the protons changes during the acceleration process.

The essence of the synchrotron is that the magnetic guide field need extend only over a very narrow region along the electron orbit instead of over the whole area defined by the orbit. Thus a major economy in the use of iron can result, for the guide magnet can indeed be made very small—the limitation being the degree of collimation and the uniformity in energy of the initial source of electrons. The Van de Graaff is ideal in supplying such a narrow monoenergetic beam and hence makes an excellent injector as does also the electron linear accelerator.

2. Phase stability for synchrotrons. The principle of phase stability is easy to describe as it applies to the synchrotron. Consider an electron travelling in a circular orbit of constant radius and, for the moment, of constant magnetic field. Now let us apply an r.f. accelerating electric field at some point on the orbit and such that the radiofrequency corresponds exactly to the frequency of resolution of the electron in its orbit. Let us suppose first that the phase of the

[1] V. I. VEKSLER: J. Phys. USSR. **9**, 3 (1945).
[2] E. M. McMILLAN: Phys. Rev. **68**, 143 (1945).

electron is such that it comes through the accelerating gap just as the voltage is zero and falling. Then, of course, nothing happens to the electron and, assuming it loses no energy by radiation, it remains on its orbit. Next, let us assume that the phase of the electron with respect to the r.f. voltage is such that it receives a positive increment of energy; hence its radius of curvature becomes larger; hence it takes longer to traverse its orbit; and so its phase will have lagged at the second crossing of the r.f. gap. On successive turns, the energy, and thus the radius, continues to grow; the phase lags become larger and larger until we reach the point where the phase has changed so much that the acceleration voltage has changed sign. The radius now decreases on each successive turn until it has reached its original value. But during all this time the phase has been lagging more and more each turn and the phase now is such that the voltage is still negative, hence the electron continues to lose energy. This makes it go on to smaller radii which now cause the phase to advance. It is clear that the radius of the orbit will oscillate back and forth about the initial orbit which we call the synchronous orbit. These oscillations are called synchrotron oscillations and imply oscillations of the energy and the radius as well as of the phase of the electron. Because of the emission of electromagnetic radiation by the electrons, the oscillations will eventually damp to zero. Expressions for the period and damping of synchrotron oscillations will be given in Sect. 4.

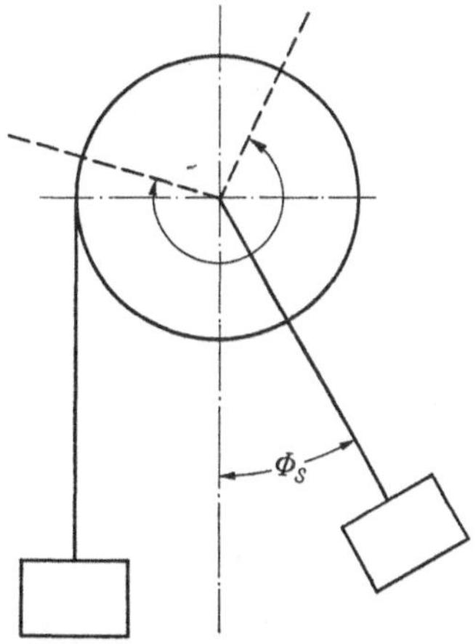

Fig. 1. Mechanical analogue for synchrotron oscillations. The equation of motion of the pendulum of moment of inertia I with constant torque T and restoring torque $G \sin \varphi$ is $\dfrac{d}{dt}\left(I \dfrac{d\varphi}{dt}\right) = T - G \sin \varphi$. The equation of phase for the synchrotron is
$$2\pi \frac{d}{d\vartheta}\left(E_s^{\frac{3}{2}} \frac{d\varphi}{d\vartheta}\right) = E_s\left[\frac{1}{f}\frac{dE_s}{dt} - eV\sin\varphi\right],$$
where E_s is the total synchronous energy of the electron. The analogous variables are $I = 2\pi E_s^{\frac{3}{2}}$, $t = \vartheta$ the azimuthal position of the electron, $T = (E_s/f)\,(dE_s/dt)$, $G = E_s\,eV$, and $\varphi = \varphi$ the synchronous phase. The t in dE_s/dt is not to be confused with the time variable, t, of the pendulum. For the example shown, the limits of stable oscillation are indicated by the dashed lines.

Next suppose that the magnetic field is increased at a uniform rate such that the particle must receive an energy eV at each revolution if it is to stay on the same orbit. Now if the electron crosses the accelerating gap at a phase φ_s such that $V_0 \cos \varphi_s$ is equal to V_1, then it will remain on its synchronous orbit. But if the voltage is higher or lower than V_1, then by exactly the same argument used above for the constant field case, the electron will make oscillations about the phase φ_s. If the rate of rise of the magnetic field changes, then the phase φ_s will automatically adjust to the proper value. Thus, if the peak r.f. voltage V_0 is always greater than V_1, then electrons will either stay on the synchronous orbit defined by the radiofrequency or they will make synchrotron oscillations about that orbit. In any case their energy will increase as the magnetic field increases.

Not all phases of the electron lead to stable synchrotron oscillations. The region of phase stability can best be visualized in terms of a simple mechanical analogy which consists of a pendulum with constant torque such as that shown in Fig. 1. The torque is analogous to the energy gain per turn which is necessary to keep the electron at constant radius. For example, we can visualize from the model, for the case of a steady field, that all phases from $-\pi$ to $+\pi$ are stable. However, as the rate of rise of the field is increased, i.e. as φ_s in the diagram is increased, the range of phase where stable oscillations can occur decreases, becoming zero when V_1 just equals V_0.

3. Synchrotron magnets. In principle, the magnet of an electron synchrotron need not differ from that of a proton synchrotron and, in fact, it should be possible to accelerate either particle in either machine, assuming that the r.f. problem

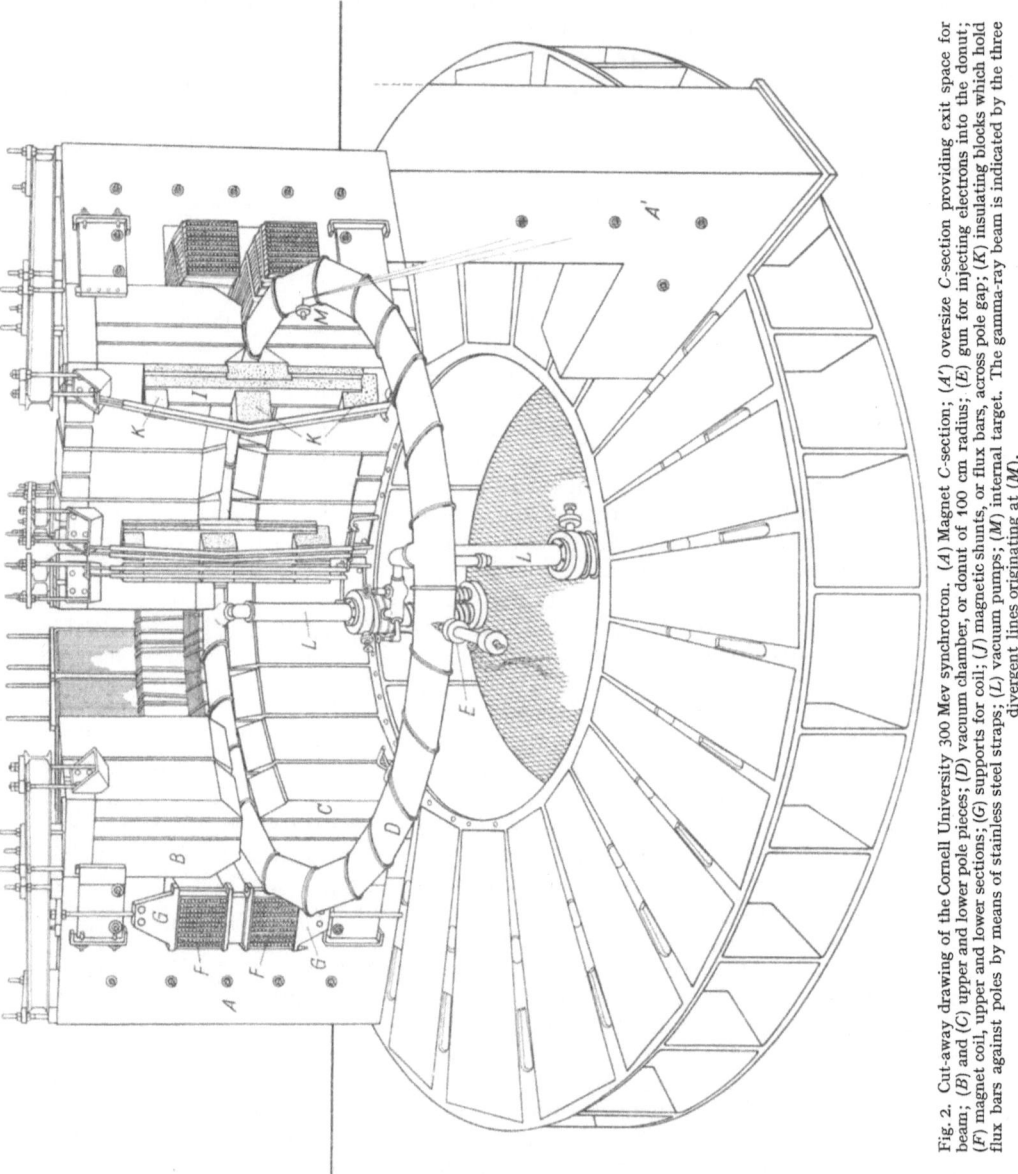

Fig. 2. Cut-away drawing of the Cornell University 300 Mev synchrotron. (A) Magnet C-section; (A′) oversize C-section providing exit space for beam; (B) and (C) upper and lower pole pieces; (D) vacuum chamber, or donut of 100 cm radius; (E) gun for injecting electrons into the donut; (F) magnet coil, upper and lower sections; (G) supports for coil; (J) magnetic shunts, or flux bars, across pole gap; (K) insulating blocks which hold flux bars against poles by means of stainless steel straps; (L) vacuum pumps; (M) internal target. The gamma-ray beam is indicated by the three divergent lines originating at (M).

is solved appropriately in each case. In practice, there are some differences, and electron magnets have developed quite differently from their proton cousins. The main difference comes about because it is convenient to inject electrons into the synchrotron at a lower momentum than in the case of the more massive protons. Thus the injection field for most electron synchrotrons corresponds to 10 or 20 gauss while for most proton machines it is more likely to be a few hundred gauss. We shall also see that the radiation of quanta which occurs for electrons but not for protons makes it desirable to increase the magnetic field rapidly in the electron case so as to minimize the number of turns. Thus most electron synchrotrons work at higher repetition rates than do proton synchrotrons.

Most of the early synchrotrons were designed for an energy of about 300 Mev and, of course, were of the weak focusing type. These machines had in common that injection was at about 60 kv and that the preliminary acceleration was

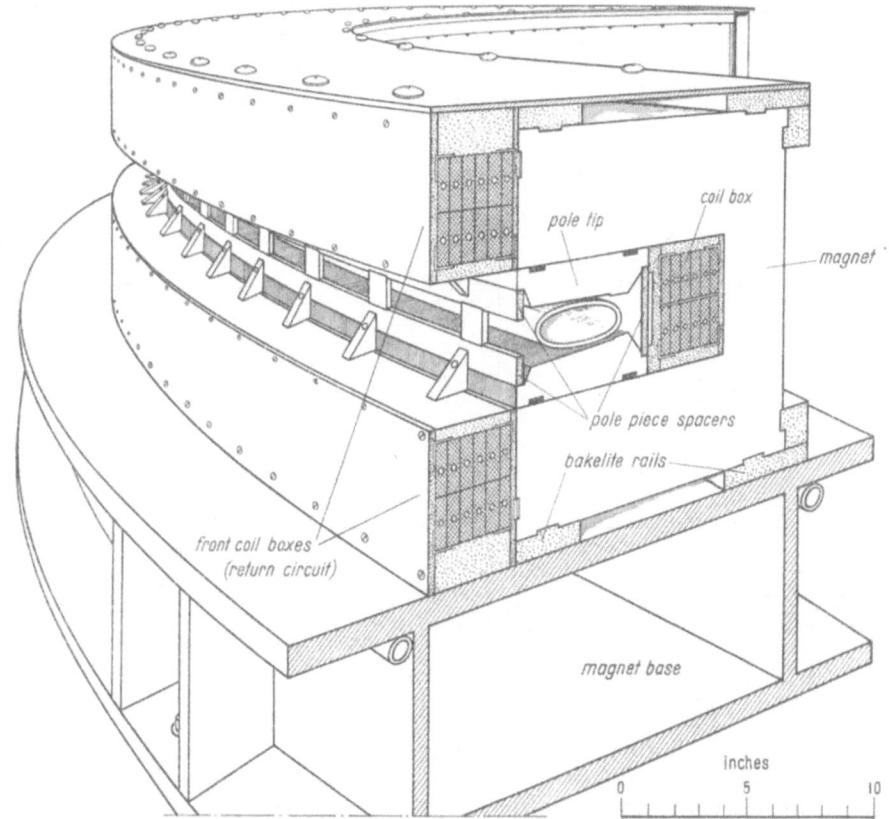

Fig. 3. One quadrant of the Cornell 1.5 Bev synchrotron. The orbit radius is 150 inches and the quadrants are separated by straight sections 36 inches in length.

by the betatron method which lasted until the electrons reached a completely relativistic energy such as 5 or 10 Mev. At this energy, the injection energy into the synchrotron operation, the flux bars, which produced the betatron acceleration and which were of rather small cross section, saturated. Just before this saturation set in, the radiofrequency voltage was turned on, and the electrons were accelerated to their final energy by the synchrotron process. Inasmuch as the excitation frequency of the magnetic field is 30 or 60 cycles per second, the magnets were constructed of laminated transformer steel (0.015 inch thick) in order to reduce eddy current effects to a minimum. The Cornell 300 Mev synchrotron is characteristic of these synchrotrons and is illustrated in the cutaway drawing of Fig. 2. The magnets are not fundamentally different from the betatron magnets described in the betatron section and the various magnetic effects due to saturation and to eddy currents which perturb the electron orbits are completely described in the accompanying betatron article.

One can distinguish another type of electron synchrotron in which the electrons are injected at much higher energy from a Van de Graaff or linear accelerator. Most of these synchrotrons were designed much later and since the beam from a Van de Graaff, for example, is so small and is so well collimated it is possible to use a much smaller magnet aperture. Hence this type of synchrotron is considerably more efficient in the uses of iron and power. The new Cornell 1.5 Bev strong focusing synchrotron is shown in Fig. 3 and is typical of these machines. Twenty tons of iron are used in the new magnet compared to eighty tons for the magnet of the old 300 Mev Cornell synchrotron.

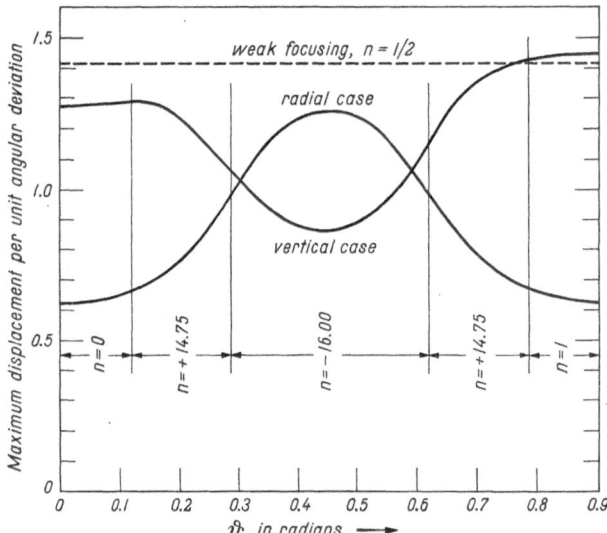

Fig. 4. The maximum displacement that results from the betatron oscillation induced by a unit vertical angular deviation of an electron on a central orbit is plotted as a function of the azimuthal position of the point where the angular deviation occurs. The example is for the Cornell strong focusing synchrotron. Only one "octant" of the magnet is shown inasmuch as the rest of the quadrant is a reflection of the "octant" shown and the other quadrants are identical.

Both strong focusing and weak focusing have been used in electron synchrotrons. Below about 1 Bev, it appears wisest to use weak focusing. Above this energy, strong focusing is indicated. The first consideration in choosing between weak and strong focusing has to do with the question of the aperture of the magnetic guide field. By aperture let us understand that we mean the maximum angular displacement that a particle on a central orbit can experience and still remain on a stable orbit that is contained in the region of good magnetic field and within the vacuum chamber or "donut". In a strong focusing machine the aperture as just defined will change as we go around the orbit and as we depart from the central orbit. We can then consider either an "average aperture" or a "minimum-value aperture". Let us not worry about this refinement, however. Strong focusing, in general, tends to improve the aperture for a given size of "donut" or region of good magnetic field. This is because the betatron oscillations have a shorter wavelength in strong focusing than in weak focusing. On the other hand, because of the strong focusing fields, the betatron oscillations are no longer sinusoidal in shape but rather have complicated shapes that appear somewhat like sine waves with higher harmonic peaks or waves superimposed. In any case, for a given angular displacement, a much larger lateral displacement results in the case of strong focusing than for weak focusing if we consider an oscillation of the same wavelength. In Fig. 4 the maximum displacement due to a unit angular displacement is plotted as a function of azimuthal position for the case of the Cornell strong focusing synchrotron. Also plotted is the same quantity for the equivalent case of weak focusing. We see that the aperture for radial oscillations is better than the equivalent aperture which obtains for the weak focusing case, but that the aperture for vertical oscillations is about the same or only slightly better.

Now as we change the radius of the machine, i.e. change the energy, the form of the strong focusing betatron oscillations remain essentially unchanged, but the number of oscillations per turn and hence the relative wavelength does change. For weak focusing, the number of betatron oscillations per turn is independent of the radius of the machine. Thus for smaller machines, the aperture for the strong focusing case becomes worse, but for larger machines it becomes much better.

Another consideration in choosing a magnet has to do with the momentum compaction, i.e. the change of the average radius of an equilibrium orbit, dr, produced by a given change in the momentum, dp, of a particle. This is defined as $\alpha = p\, dr/r\, dp$ and is a measure of the distance between successive turns of a coasting beam. For the Cornell example used above, α is 0.28 as compared to the value of α of 3 which would obtain for the weak focusing case ($n = \frac{2}{3}$). A large momentum compaction, i.e. small α, is desirable at injection for it tends to reduce the sensitivity to energy variations. It also admits of the possibility of multiple-turn injection. A large momentum compaction is also desirable in that it decreases the amplitude of synchrotron phase oscillations and hence increases the acceptance of electrons into stable synchrotron oscillations at injection for a given size of donut.

We shall see in the section on radiation effects that radial and phase oscillations due the emission of radiation are much reduced by strong focusing. One result of this should be remarked on here. The amplitude of the r.f. voltage used for the synchrotron acceleration is much less for the strong focusing case than for weak focusing. Thus we shall see that the C.I.T. weak focusing synchrotron needs 200 kv to attain 1.5 Bev whereas the Cornell strong focusing synchrotron requires only 140 kv for the same energy—the difference being caused by radiation-induced synchrotron oscillations.

It is clear that the ultimate development in synchrotrons has not yet been reached in that the aperture of magnet has not been reduced to its lowest practical value. In the Cornell machine, the beam of electrons is quite visible because the electrons emit light due to their centripetal acceleration by the magnetic field. At high energy the beam appears to be one or at most two millimeters in diameter. The beam can still be seen at lower energies by simply cutting off the r.f. accelerating voltage at earlier times in the cycle. The beam is seen to remain in exactly the same position in the donut at different energies and it remains the same size down to about 100 Mev where it is no longer visible. Thus in the Cornell machine, except at injection, the donut cross section which is roughly 3″ by 1″ is not used and a great amount of magnetic energy is uselessly stored in a region that is never explored by electrons.

On the other hand, at injection when the beam does fill the donut, the magnetic field due to the magnet is at its worst, consisting largely of out-of-phase components, and it must be corrected or essentially built up by the use of currents in various kinds of correction coils. Clearly the acceleration process should be further separated. For example, the injection magnetic field could be produced in a different manner than the ultimate high energy field. Various suggestions have been put forward: saturating pole tips such that the final field is much smaller than the injection field, CRANE's suggestion; the injection field to be produced without iron but by suitably arranged coils which extend a small region of iron-produced field, SALVINI's suggestion; tandem machines, i.e. the injection is to be made in one magnet of large aperture and then the beam is to be transferred to a second magnet of small aperture which is tangent to the first, the transfer to be made after the beam has become well bunched and has damped to its ultimate small size, the author's suggestion.

4. Injectors. The ideal injector should produce an intense pulsed beam of monoenergetic electrons. The beam must also be narrow and parallel and the energy should not be too low. Fortunately, there are a number of different kinds of injectors available that at least partially satisfy the above desiderata. The Van de Graaff can produce a short pulse of current of the order of 0.5 ampere. The beam is roughly 1 cm in diameter and the spreading angle is of the order of 1 milliradian. A 2 Mev Van de Graaff, for example, is used as the injector of the Cornell 1 Bev synchrotron. The principal criticism of such an injector is that the energy is too low. Thus the magnetic field at injection is small and is seriously affected by eddy currents and residual magnetism. Further, the velocity of the electrons is still 2% different from the velocity of light and this necessitates some modulation of the frequency of the accelerating voltage. More serious is the effect of the space charge of the beam and of the perturbing effect of the magnetic field due to the beam itself. Thus at Cornell, the high energy beam is not proportional to the injected beam for currents greater than tens of milliamperes.

An injector giving about 10 Mev would be much preferable: at that energy, all the previously mentioned effects would become much less important. A linear electron accelerator or a microtron both give such an energy. The linear accelerator gives a high current pulse, a narrow parallel beam, but it does not give very monoenergetic electrons. The half width of the energy spread may be as much as 20% while the energy acceptance of a synchrotron is of the order of a few tenths percent. Thus, of the current of several hundred milliamperes from the linac, only a few milliamperes are useful. The microtron is less developed and may eventually fulfill all the criteria of a high energy synchrotron injector. It is also possible with further development that the energy spread of the linac will decrease. A 20 Mev synchrotron is under construction at Cornell as an alternate injector. It has the advantage of giving a phase-bunched beam to the large synchrotron.

Of the low energy injectors, a pulse transformer giving 1 Mev is used for the C.I.T. synchrotron. A pressurized Cockcroft-Walton machine of 2 Mev is planned as an injector for the Italian 1 Bev electron synchrotron. Both the Swedish and Japanese 1 Bev synchrotrons are designed to use high energy injectors; a microtron for the former, a linac for the latter. The Cambridge 6 Bev synchrotron is designed to make use of a 40 Mev linac.

5. Pressure effect in electron synchrotrons. Pressure effects are somewhat different, and much simpler to treat, in a small-aperture electron synchrotron than in a proton machine. This is because single scattering only is important in the electron case, whereas multiple scattering is dominant when protons are used as has been shown by COURANT[1]. This difference comes about because of the shielding of the nuclear electric field by the atomic electrons which imposes an "effective minimum angle" of scattering, $\vartheta_{min} = \lambda/a$, where λ is the de Broglie wavelength of the scattered particle and a is the Fermi-Thomas radius of the atom, $\hbar^2/m e^2 Z^{\frac{1}{3}}$. For electrons, ϑ_{min} can easily be greater than the angle defined by the aperture of the donut; while for protons ϑ_{min} is usually very much smaller because of the greater mass and hence the smaller λ.

For simplicity, let us assume the condition that ϑ_{min} is much larger than the angle of the aperture. Then the total cross section for scattering out of the

---

[1] N. BLACHMAN and E. COURANT: Phys. Rev. **74**, 140 (1948).

donut per atom is

$$\sigma = \frac{8\pi Z^2 e^4}{p^2 v^2} \int_{\vartheta_{\min}}^{\vartheta_{\max}} \frac{d\vartheta}{\vartheta^3} \tag{5.1}$$

or

$$\sigma = \frac{4\pi Z^2 e^4}{p^2 v^2} \frac{1}{\vartheta_{\min}^2}. \tag{5.2}$$

Substituting

$$\vartheta_{\min} = \frac{m e^2 Z^{\frac{1}{3}}}{\hbar p} \tag{5.3}$$

we get

$$\sigma_0 = 4 Z^{\frac{1}{3}} (Z+1) \beta^{-2} \pi \left(\frac{\hbar}{m c}\right)^2 \tag{5.4}$$

where the $(Z+1)$ term has been written instead of Z in order to include the scattering by the atomic electrons. In the relativistic region, this cross section is independent of the energy up to the energy at which $\vartheta_{\min}$ becomes comparable to that defined by the donut aperture. Thus, for a first approximation, we can assume that the scattering cross section is constant as given by (5.4) up to a critical energy, E_s, and then, because the scattering is predominantly forward, is zero beyond that. E_s will be given approximately by the condition that $\vartheta_{\min}$ is just equal to the angle of the aperture of the donut, δ, i.e.

$$\delta = \frac{m e^2 Z^{\frac{1}{3}}}{\hbar p_s} \tag{5.5}$$

where the momentum, p_s, corresponds to E_s. For relativistic energies, we can write (5.5) as

$$E_s = Z^{\frac{1}{3}} m c^2 \frac{e^2}{\hbar c} \Big/ \delta$$

and if δ is expressed in milliradians this becomes for air

$$E_s = \frac{7.2}{\delta} \text{ Mev}. \tag{5.6}$$

The attenuation of the beam due to scattering will then be given by

$$i = i_0 e^{-N\sigma_0 c t} \tag{5.7}$$

where ct is the total distance travelled by the electrons between the energy of injection and the critical energy and hence t is the time the electron takes to reach E_s. N is the number of gas atoms per cm³. For air, we can write (5.7) as

$$i = i_0 e^{-0.67 p t} \tag{5.8}$$

where p, the pressure, is measured in μ-mm Hg and t is in milliseconds.

For the case of the Cornell synchrotron, the acceptance angle depends on the tuning conditions but is of the order of a few milliradians, hence E_s is about 5 Mev. It takes about 0.3 msec for the beam to get to 5 Mev after injection, and so for an average pressure of 6 μ-mm Hg the beam intensity will be decreased to about $1/e$ of its original value.

The above crude treatment can be made a little more precise. The Born approximation solution for a plane wave of electrons being scattered from a potential of the form $V = Z e^2 e^{-r/a}/r$ is

$$\frac{d\sigma}{d\omega} = \frac{4 Z^2 e^4}{p^2 v^2 (\vartheta^2 + \vartheta_0^2)^2} \tag{5.9}$$

where $\vartheta_0 = \vartheta_{\min} = \lambda/a$. Now, if an angle of scattering greater than δ, the donut aperture, loses the electron from the beam, then the total cross section for loss per atom is obtained by integrating (5.9) over angles greater than δ which gives

$$\sigma = \frac{4\pi Z (Z + 1) e^4}{p^2 v^2 (\vartheta_0^2 + \delta^2)} \tag{5.10}$$

which is to be compared to (5.2).

Remembering that ϑ_0 is a function of the energy as given by (5.3) and that the energy can be considered to rise linearly during the small time after injection that is important in this process, we get the total loss by integrating (5.10) over the path of the electrons as the energy increases over the injection value E_0 at the rate $(dE/dt)_0$. If we express the result as $i = i_0 e^{-ap}$, where p is in μ-mm Hg, then a has the form:

$$a = \frac{5.3}{\delta (dE/dt)_0} \left[\frac{\pi}{2} - \tan\left(\frac{\delta E_0}{7.2}\right) \right] \tag{5.11}$$

where δ is in milliradians, E_0 is in Mev and $(dE/dt)_0$ is in Mev/millisecond. It is only necessary to use (5.11) when E_s as given by (5.6) is close to or less than E_0.

The problem has been given much more careful treatment by MORAVCSIK and SELLEN[1] and by C. BERNARDINI[2], both giving similar results.

It appears from the above that the loss is negligible at a pressure of 1 μ-mm Hg and is appreciable above 10 μ-mm Hg. The loss can be seen for the Cornell synchrotron on the induction detector which displays the beam as a function of the time on an oscilloscope. A rapid, and vaguely exponential, decrease in the beam is observed during the first millisecond, after which the beam remains absolutely constant throughout the next 14 milliseconds of the acceleration process. Deliberately raising the pressure to above 10 μ-mm Hg emphasizes the loss and indeed causes the high energy beam to decrease by several orders of magnitude. The exact loss depends very much on the tuning of the machine, i.e. the actual magnetic aperture as determined by various correction currents, and on the distribution of the pressure in the donut which is not at all necessarily constant. Usually the pressure is a few μ-mm Hg and then the corresponding pressure loss is not more than about 50%.

In the glass donut of the Cornell synchrotron, such pressures are easy to obtain using normal rubber gasket techniques and small oil pumps. One pump at each straight section is adequate and since the pumping speed of a quadrant of the donut is only a few liters/sec there is not much point in using a pump larger in diameter than 10 cm.

The interior walls of the vacuum chamber must be made conducting, about 100 Ω per square, just as in the case of the betatron.

6. Emission of radiation by electrons. Electrons being accelerated in a synchrotron emit electromagnetic radiation because they are subject to a strong centripetal acceleration caused by the guiding field. The phenomenon needs study not only because it has intrinsic interest but because it gives rise to synchrotron and betatron oscillations as well as damping of these oscillations; it also sets a limit to energy that can be achieved by synchrotron acceleration.

[1] M. MORAVCSIK and J. SELLEN: Rev. Sci. Instrum. **26**, 1158 (1955).
[2] C. BERNARDINI: Proceedings of the CERN Symposium, 1956, p. 463.

The emission of radiation is dramatically visible to the eye. If one looks along a tangent to the electron orbit of an operating machine, a blinding blue light will be seen. On moving the eye a few degrees from the plane of the orbit, the light changes color somewhat and then disappears, thus demonstrating the forward distribution of the radiation.

The effect is essentially classical and one can readily derive from LARMOR's formula for the radiation from an accelerated charge that the total energy loss by an electron per revolution is given by[1]

$$\frac{\Delta E}{m c^2} = \frac{4\pi}{3} \frac{r_0}{R} \left(\frac{E}{m c^2}\right)^4 \tag{6.1}$$

where $m c^2$ and r_0 are the rest energy and classical radius of the electron and E and R are the energy and radius of curvature of the electron. A more practical formula is

$$\Delta E_{\mathrm{kev}} = \frac{88.5 E_{\mathrm{Bev}}^4}{R_{\mathrm{meters}}} . \tag{6.2}$$

If we consider the angular distribution of the radiation in the system moving with the electron, we find the familiar pattern of an oscillating charge, but in our system the 90° points of symmetry of the moving system are transformed forward to an angle of $m c^2/E$; hence about 50% of the radiation emitted will fall within this very narrow angle.

We can also get an idea of the frequencies of the emitted radiation. From the above argument it can be seen that the electrons will emit light in the direction of an observer while travelling a distance in their orbit of $R m c^2/E$ and hence for a pulse-time of $R m c^2/E c$. The Fourier spectrum of this pulse will contain frequencies up to about $E c/R m c^2$, and these will be Doppler shifted by an additional factor of $(E/m c^2)^2$ because the source is approaching the observer with a velocity close to that of light. Thus we can expect angular frequencies up to about $(E/m c^2)^3 (c/R)$ and down to about the repetition rate of c/R.

SCHWINGER[2,3] has derived formulae for the energy and angular distributions of the radiation. The instantaneous power emitted per unit frequency interval at the frequency ω is given by

$$p(\omega) = \frac{3^{\frac{3}{2}}}{4\pi} \frac{e^2}{R} \left(\frac{E}{m c^2}\right)^4 \frac{\omega_0 \omega}{\omega_c^2} f\left(\frac{\omega}{\omega_c}\right) \tag{6.3}$$

where ω_0 is the orbital angular frequency c/R and ω_c is a frequency not far from the peak of the spectral distribution and is given by

$$\omega_c = \frac{3}{2} \omega_0 \left(\frac{E}{m c^2}\right)^3 \tag{6.4}$$

which is very close to that obtained by the very rough argument in the preceding paragraph. In more practical units, (6.4) becomes

$$\lambda_c = \frac{5.6 R_{\mathrm{meters}}}{E_{\mathrm{Bev}}^3} \tag{6.5}$$

where λ_c is given in angstrom units. Expressions for $f(\omega/\omega_c)$ are given in SCHWINGER's paper[2], but TOMBOULIAN and HARTMAN[4] have used SCHWINGER's result

---

[1] A. LIÉNARD: L'Eclairage Elec. 16, 5 (1898); see also D. IWANENKO and I. POMERANCHUK, Phys. Rev. 65, 343 (1944), who first pointed out the relevance to electron accelerators.

[2] J. SCHWINGER: Phys. Rev. 75, 1912 (1949).

[3] L. ARZIMOVITCH and I. POMERANCHUK: J. Phys. USSR. 9, 267 (1945).

[4] D. TOMBOULIAN and P. HARTMAN: Phys. Rev. 102, 1423 (1956).

to compute a universal curve which is reproduced in Fig. 5 and in which the instantaneous power per unit wavelength at wavelength λ is given. The distribution drops off rapidly below the peak of the curve which occurs at a wavelength of 0.42 λ_c. Thus the electrons in the Cornell synchrotron provide a continuous source of radiation which extends below a wavelength of 10 Å, i.e. roughly 1 kev X-rays.

TOMBOULIAN and HARTMAN have also evaluated SCHWINGER's formula for the angular distribution of the radiation. They give curves for the electron energies of 240 and 320 Mev and for various wavelengths from 100 to 400 Å. Their distributions are complicated but are roughly consistent with gaussianlike curves, i.e.

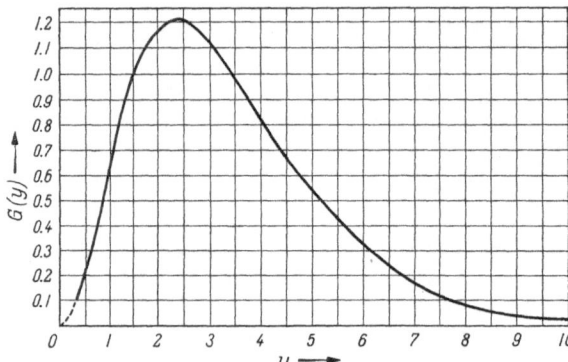

$$p(\psi, \lambda) = p_0\, e^{-(\psi/\psi_c)^2} \quad (6.6)$$

where $p(\psi, \lambda)$ is the power per unit angle, ψ, relative to the orbital plane and per unit wavelength, λ. The constant, ψ_c, arrived at empirically, is roughly given by

$$\psi_c \approx \frac{m\,c^2}{E}\left(\frac{\lambda}{\lambda_c}\right)^{\frac{1}{2}} \quad (6.7)$$

Fig. 5. Universal spectral distribution curve for the radiation from electrons. $P(\lambda)$, the instantaneous power radiated per angstrom is given by $P(\lambda) = (3^{\frac{5}{2}}/16\pi^3)\,(e^2 c/R^3)\,(E/m_0 c^2)^7 G(y)$, where $y = \lambda_c/\lambda$ and where $\lambda_c = 5.59\,R_{\mathrm{Mev}}/E^3_{\mathrm{Bev}}$. The curve was kindly prepared for me by Professor PAUL L. HARTMAN.

where λ_c is given by (6.5). p_0 is the intensity in the forward direction and is not evaluated here.

A number of experimental investigations have confirmed SCHWINGER's classical theory in detail. In the practical operation of high energy synchrotrons, it has been found necessary to impart an energy to the electrons at each turn of at least that given by Eq. (6.1) in order to keep the electrons on a constant radius orbit. CORSON[1] has measured the total energy loss per turn of electrons in the 300 Mev Cornell synchrotron by observing the rate of shrinkage of the electron orbit after the r.f. accelerating voltage has been turned off. ELDER et al.[2] investigated the spectral distribution of the radiation in the visible region using the 70 Mev General Electric synchrotron. TOMBOULIAN and HARTMAN[3] have made a particularly detailed study of the spectral distribution of the radiations from the Cornell 300 Mev synchrotron in which their measurements extended down to 60 Å. They have also confirmed the angular distribution predicted by the classical theory. KIMURA[4] has also measured angular distributions.

The above discussion refers only to the radiation by individual electrons. Coherent effects can increase the rate of energy loss if the dimensions of a bunch of electrons becomes small enough. In that case the wavelengths that are enhanced by coherent effects will be of the same order of length as that of the bunch of electrons doing the radiating. SCHIFF[5] has given an expression for the case in

[1] D. R. CORSON: Phys. Rev. 90, 748 (1953).
[2] ELDER, LANGMUIR and POLLACK: Phys. Rev. 74, 52 (1948).
[3] See footnote 4, p. 179.
[4] KIMURA et al.: Science Reports of the Tôhoku University Series I, Vol. XL, No. 4, p. 224, 1957.
[5] L. I. SCHIFF: Rev. Sci. Instrum. 17, 6 (1946).

which N electrons per bunch are spread in a Gaussian distribution having an angular width of φ between $1/e$ points. Then the additional loss per revolution for each electron due to coherent radiation is, in the absence of metallic shielding,

$$\Delta E_{coh} \approx \frac{r_0}{R} N \varphi^{-\frac{4}{3}} m c^2. \tag{6.8}$$

Now inasmuch as the enhanced radiation is near the microwave region, it can be greatly reduced by metallic shielding. SCHIFF also gives an expression for the coherent energy loss for the case that the orbit lies midway between two parallel metallic sheets of infinite extent and which are each a distance a from the orbit. He finds that (6.7) is then replaced by

$$\Delta E_{coh} = 5 \left(\frac{a}{R}\right)^2 \frac{r_0}{R} N \varphi^{-\frac{8}{3}} m c^2. \tag{6.9}$$

NODVICK and SAXON[1] have made detailed calculations of the suppression of the coherent radiation for the case of finite metallic shields. To summarize their results rather roughly, it can be said that the above formula (6.9) is valid when the width of the plates is a few times their separation. Of course, the expression (6.1) for the radiation by single electrons is unaffected by shielding.

We can ask if the emission of radiation imposes a limitation on the energy that can be achieved by a synchrotron. For the Cornell machine of 3.8 meters radius, the energy loss per turn, which of course must be compensated at each turn by an equivalent energy gain from the r.f. accelerator, is 24 kev at 1 Bev and is 120 kev per turn at 1.5 Bev. The Cambridge 6 Bev accelerator is designed for an orbit radius of about 26 meters and has a maximum radiation loss of 4.4 Mev per turn. At 7.5 Bev in the same machine, the radiation loss would amount to 10.7 Mev/turn which would require about 750 kva for the oscillator which is to be compared to about 650 kva for the magnet excitation. Thus radiation effects are becoming serious indeed.

To consider the limitation of electron energy imposed by radiation loss, let us compute the ratio of the total energy lost by radiation to the kinetic energy stored in the beam, for when this ratio becomes large the synchrotron can be considered impractical. The ratio is easily computed by integrating (6.2) over the cycle of acceleration and we get that

$$\frac{E_{tot.\,rad}}{E} \approx 1.6 \cdot 10^3 \frac{E^3}{\omega R^2} \tag{6.10}$$

for the case where the magnetic field varies sinusoidally with an angular frequency ω and where E_0 is the final energy in Bev, and R is the orbit radius in meters [the coefficient in (6.10) is $3.8 \cdot 10^3$ for the case of a completely biased sinusoidal variation of the magnetic field]. Since R usually varies linearly with E_0, the ratio increases only as the first power of E_0.

For the Cornell machine the above ratio is about 1.4 at 1 Bev and about 7 at 1.5 Bev, and for the Cambridge machine at 7.5 Bev it will be about 12. The ratio would be unity for a hypothetical machine with an energy of 2.2 Bev and with a 60 cycle magnetic field variation; this assumes that the magnetic field would reach 10 kg, i.e. a radius of 7.3 m.

There are two obvious ways of reducing the fractional amount of radiation at higher energies. A larger relative orbit radius can be used so that the maximum field would be less than the usual design value of 10^4 gauss. For example, if we

---

[1] J. NODVICK and D. SAXON: Phys. Rev. 96, 180 (1954).

let R vary as $E_0^{\frac{3}{2}}$, the ratio (6.10) will remain constant. Alternatively, we can let R vary linearly with E_0, as usual, but increase ω. Thus, if we let ω vary as E, the ratio (6.10) will then remain constant with increasing E.

The r.f. voltage required at the beginning of the cycle is given by

$$e\,V = \frac{2\pi\,\omega\,R\,E_0}{c}\,. \qquad (6.11)$$

It is interesting to note that if we let both ω and R vary as E_0, then the initial r.f. voltage varies as E_0^3 which is exactly the same variation of the final value of the r.f. voltage required by radiation losses at the end of the cycle. Thus the r.f. voltage remains roughly constant during the cycle which is good because it tends to utilize the r.f. oscillators more efficiently.

We see then that if one compares the total radiation loss to the energy stored in the beam, there need be imposed no limit to the energy attainable by the synchrotron method. The limitation, if any, is most likely to be the practical one of holding the very close tolerances necessary in making a large-orbit magnet or, alternatively, of being able to provide the requisite accelerating voltage in the space that is available in the straight sections.

To get an idea of the order of magnitude of the difficulties, let us consider a few examples. We have seen that the total radiation loss is just equal to the beam energy for a synchrotron which would give 2.2 Bev and which works at 60 cycles and which reaches a magnetic field of 10^4 gauss ($R = 7.35$ m). For higher energies, we can now let R increase as $E^{\frac{3}{2}}$ in order to keep the relative losses the same. Thus we can have 20 Bev if we make $R = 200$ m in which case $E_r = 70$ Mev/turn maximum. If we scale up the 26.2 m orbit Cambridge lattice (keeping the length of the unit lenses the same) we go from 48 straight sections to 366 straight sections. They plan to get an acceleration energy of 0.7 Mev/turn per cavity, hence we would need only $70/0.7 = 100$ of the sections for r.f. cavities. Hence the machine could be made considerably smaller, although the radiation losses would then be increased.

We can also consider a 50 Bev machine. By the same logic, the orbit radius would be 800 m, and there would be a maximum energy loss of 700 Mev/turn. The number of straight sections would be about 1500 of which only 1000 need be used for cavity accelerators. The maximum magnetic field would be about 2100 gauss. The electrons would make about 200 turns with a period of roughly 20 μsec per turn. A peak power of about 100 MW and an average power of about 20 MW might be required. The magnet might consume only 1 MW which points up the significance of our r.f. problems.

Even 100 Bev can be considered, but we begin to lack enough space in the straight sections for cavities of the present conventional design.

The above is meant only as a comment on the radiofrequency problem; it does not mean to say that other problems such as magnet tolerance would not become formidable.

7. Synchrotron oscillations. Synchronous phase oscillations have already been treated in Sect. 6 of COHEN's article, in this volume. The emission of radiation by electrons, however, introduces a number of modifications in that theory, and these modifications will be considered here. In the first place, a strong damping of the synchrotron oscillations sets in at high energy and we shall begin by a calculation of this effect.

Consider an electron whose radius of curvature r differs from that of the synchronous orbit, r_s. Because of the different radius, the phase of the electron ψ

measured with respect to the phase φ_s of the synchronous orbit can advance or retard by

$$\frac{d\psi}{dn} = -\frac{2\pi k}{\lambda} \frac{(r-r_s)}{r} \tag{7.1}$$

where n is the number of turns, $\lambda = 1 + l/2\pi r$ corrects for the straight sections of total length l, and k is the harmonic order of the r.f. For relativistic electrons, the definition of the momentum compaction α allows us to write

$$\frac{(r-r_0)}{r} = \alpha \frac{E'}{E} \tag{7.2}$$

where E' is the difference, $E - E_s$, between E the energy of the electron and E_s the energy of an electron on a synchronous orbit. Thus we can write (7.1)

$$\frac{d\psi}{dn} = -2\pi \frac{k}{\lambda} \alpha \frac{E'}{E} \tag{7.3}$$

and differentiating gives us

$$\frac{d^2\psi}{dn^2} = -2\pi \frac{k}{\lambda} \frac{\alpha}{E} \frac{dE'}{dn}. \tag{7.4}$$

Now

$$\left.\begin{aligned} \frac{dE'}{dn} &= \frac{dE}{dn} - \frac{dE_s}{dn} \\ &= (V\sin\varphi - E_r) - (V\sin\varphi_s - E_{rs}) \\ &= V\psi\cos\varphi_s - \frac{dE_r}{dE} E' \end{aligned}\right\} \tag{7.5}$$

where $V\sin\varphi$ is the energy gain per turn from the r.f. and E_r is the energy loss per turn by radiation at r and is given by (6.1) which allows us to evaluate dE_r/dE in (7.5). E_{rs} refers to the radiation energy loss at the synchronous orbit. Thus

$$\frac{dE'}{dn} = V\psi\cos\varphi_s - 4\left(1 - \frac{\alpha}{4}\right) E_r \frac{E'}{E}. \tag{7.6}$$

Substituting this into (7.4) and evaluating E'/E from (7.3) gives

$$\frac{d^2\psi}{dn^2} - 4\left(1 - \frac{\alpha}{4}\right) \frac{E_r}{E} \frac{d\psi}{dn} + \frac{2\pi k \alpha V\cos\varphi_s}{\lambda E}, \qquad \psi = 0 \tag{7.7}$$

which is the familiar equation for damped simple harmonic motion. The last coefficient of (7.7) gives us the period of the synchrotron oscillations, P_s, the square of which is

$$P_s^2 = \frac{2\pi \lambda E}{k \alpha V\cos\varphi_s} \tag{7.8}$$

where the period is measured in turns per synchrotron oscillation. The second coefficient gives the damping which in terms of the amplitude decrement is

$$\frac{d\psi}{\psi} = -2\left(1 - \frac{\alpha}{4}\right) \frac{E_r}{E} \tag{7.9}$$

per turn. The factor $(1 - \alpha/4)$ differs appreciably from unity only for the case of weak focusing in which case $\alpha = 1/(1-n)$, n now being the field index, so for weak focusing

$$\frac{d\psi}{\psi} = -\frac{E_r}{E} \frac{3-4n}{2(1-n)}. \tag{7.10}$$

This result was first given by BOHM and FOLDY[1].

---

[1] D. BOHM and L. FOLDY: Phys. Rev. **70**, 249 (1946).

In the Cornell synchrotron at the injection energy of 2 Mev, the synchrotron oscillation period is about 5 μsec and the radiation damping is negligible. At 1 Bev the period is about 1 millisecond and the 1/e period for damping is $2 \cdot 10^4$ turns or 2 milliseconds, all of which is to be compared to the total acceleration time of about 15 milliseconds. For the weak focusing synchrotron of the California Institute of Technology at the same energy the damping time is 5 milliseconds. The new Cambridge synchrotron will have a damping time of 350 turns or 260 μsec at 6 Bev, to be compared to 15 milliseconds total acceleration time.

We see, therefore, that the damping can become strong at high energy. Originally it was believed that because of such strong damping the bunches of electrons would become minuscule and in that case coherent effects as given by (6.7) could become important in increasing the energy loss per turn.

SANDS[1] first pointed out that a quantum effect would act to keep the bunches from becoming small. The radiation by the electrons occurs in discrete steps which leads to a departure from the average behavior of the beam. This is a random walk problem and fractional deviations ΔE from the normal energy E are given by

$$\overline{\Delta E^2} = n\,\varepsilon^2 \tag{7.11}$$

where n is the number of quanta of average energy ε emitted by the electron during the time that $\overline{\Delta E}$ accrues.

For the problem of phase oscillations, the characteristic time to consider is the natural damping time which, according to Eq. (7.9), is that corresponding to $E/2E_r$ turns [neglecting the $(1 - \alpha/\gamma)$ factor] and the energy lost in this time will be $E_r(E/2E_r) = E/2$. Hence, the average number of quanta emitted is $E/2\varepsilon$ so that (7.11) becomes

$$\frac{\overline{\Delta E^2}}{E^2} = \frac{\varepsilon}{2E} \tag{7.12}$$

and evaluating ε from Eq. (6.4), i.e.

$$\varepsilon = \frac{3}{2}\,(\hbar\,4R_0)\left(\frac{E}{m\,c^2}\right)^3 \tag{7.13}$$

we get

$$\frac{\overline{\Delta E^2}}{E^2} = \frac{1}{2}\,\frac{\hbar}{m\,c}\left(\frac{E}{m\,c^2}\right)^2. \tag{7.14}$$

These energy variations correspond to synchrotron oscillations of radial amplitude given by (7.2) to be

$$\left.\begin{aligned}
\frac{\overline{\Delta r^2}}{r^2} &= \frac{\alpha^2\,\varepsilon}{2E} \\
&= \frac{1}{2}\,\alpha^2\,\frac{\hbar}{m\,c}\left(\frac{E}{m\,c^2}\right)^2
\end{aligned}\right\} \tag{7.15}$$

which correspond to phase oscillations which are given by

$$\overline{\psi^2} = \frac{k\,\alpha}{\lambda(1 - \alpha/4)}\,\frac{\varepsilon}{V\cos\varphi_s} \tag{7.16}$$

where the symbols have the same meaning as in Eq. (7.3)[2]. Eq. (7.16) is actually the result of a more rigorous calculation of OSBORNE and RITSON for the strong

---

[1] M. SANDS: Phys. Rev. **97**, 470 (1955).
[2] Eq. (7.16) is readily derived from Eqs. (7.12), (7.7) and (7.3) but the result differs by a factor of 2π from that given by (7.16).

focusing case, and it reduces to the equation originally given by SANDS for weak focusing[1].

The phase spreading by radiation brings about two salutary effects. It prevents the bunches from contracting to a size which would result in serious coherent radiation losses. Also, because of the $V \cos \varphi_s$ term in the denominator of (7.15), it becomes much easier to extend the x-ray beam in time. Thus, as the r.f. voltage, V, is dropped, φ_s approaches 90°, the phase oscillations increase tremendously and the beam spills out of synchronization. Generally, it has been found much easier to "shape" the beam to a long duration at high energy than at low—just the opposite than one would expect if only radiation damping obtained.

The phase spreading causes the beam to spread radially but not by a very large amount. For a given machine, Eq. (7.15) tells us that the radial excursions due to the phase oscillations increase linearly as the energy. However, as machines are made larger, the momentum compaction varies roughly as $5/n$ where n is the field index which usually varies approximately linearly with R. Thus the final amplitude of radial oscillations does not increase as the radius is made larger, hence we need run into no limitation of the synchrotron energy from this cause.

The phase oscillations, as we can see from (7.16) decrease with increasing energy. Thus, if we make $\varphi_s = 60°$, since $V \sin \varphi_s$ must equal E_r, we can write $V \cos \varphi_s = 0.57 E_r$. But ε/E_r varies as $1/E$, hence the oscillations will decrease with increasing E. As machines are made larger, the product $k\alpha/\lambda$ remains roughly constant.

At low energies where the radiation-induced phase oscillations can be significant, the r.f. voltage must be made somewhat greater than would be necessary were the oscillations damped to zero in which case φ_s could be allowed to approach 90°, the most efficient angle for acceleration. For the case of the California Institute of Technology 1.5 Bev machine where the effect is rather large because of the use of weak focusing, SANDS computes $\psi_{rms} \approx 30°$ and $\Delta r_{rms} = 0.7$ cm ($r = 375$ cm, $k = 4$, $n = 0.6$, $\lambda = 1.25$). At full energy, according to SANDS, φ_s must be about 30° to avoid loss, hence the r.f. voltage must exceed 200 kv instead of 120 kv which is the value of E_r given by (6.2). In the case of the Cornell machine at 1.5 Bev, $\varepsilon \approx 2$ kev, $E_r \approx 120$ kev, and if $\varphi_s = 60°$, $V \cos \varphi_s = 68$ kv and $\psi_{rms} \approx 4°$. V would have to be 140 kv and there would be no loss of intensity. If we tried to run with $\varphi_s = 80°$ instead, then V could be reduced to 127 kv, but then ψ_{rms} would be 24° and hence some of the electrons would be lost from synchronization.

8. Betatron oscillations. The emission of radiation in discrete quantities also induces betatron oscillations as OSBORNE and RITSON[2], as well as KOLOMENSKY and LEBEDEV[3], have shown. Again let us make a highly simplified calculation of this effect. Radial variations are related, at relativistic energies, to energy variations by

$$\frac{\Delta r}{r} = \frac{\alpha E}{E}. \tag{8.1}$$

Once more we can sum the squares of the n individual energy fluctuations, ε, that occur per turn to get

$$\frac{\overline{\Delta r^2}}{r^2} = \frac{\alpha^2 n \varepsilon^2}{E^2}. \tag{8.2}$$

---

[1] See footnote 1, p. 184.

[2] L. OSBORNE and D. RITSON: Cambridge Accelerator Project Report No. 7, Aug. 11, 1955.

[3] A. KOLOMENSKY and A. LEBEDEV: Abstract of Reports of All-Union Conference on High-Energy Physics, Academy of Sciences of the USSR, Moscow Meeting, May 14, 1956.

Now $n\varepsilon$ is equal to E_r, hence

$$\frac{\overline{\Delta r^2}}{r^2} = \frac{\alpha^2\,\varepsilon\,E_r}{E^2}$$

or

$$\overline{\Delta r^2} = 2\pi\alpha^2\,\frac{\hbar c}{e^2}\,r_0^2\left(\frac{E}{m\,c^2}\right)^5. \tag{8.3}$$

This amount of fluctuation in the radial position of the electrons wil be contributed at each turn and must now be summed over all the turns, i.e.

$$\overline{\Delta r^2}(E_0) = \int\limits_0^t 2\pi\alpha^2\,\frac{\hbar c}{e^2}\,r_0^2\left(\frac{E}{m\,c^2}\right)^5\frac{E}{E_0}\,e^{Dn}\,dn \tag{8.4}$$

where the term (E/E_0) is the familiar damping factor due to the increase of the mass of the electron from the energy E at which the fluctuation takes place to the final energy E_0. The term e^{Dn} is an additional anti-damping factor due to radiation and has been treated by KOLOMENSKY and LEBEDEV[1] and by ROBINSON and RITSON[2]; it will be discussed presently. In the integral, E is a function of n, the number of turns and this function will depend on the nature of the time-variation of the magnetic field, for example, on the amount of D.C. biasing current.

It will turn out in our discussion of the anti-damping factor e^{Dn} that D has the approximate value $E_r/2E$. It will also be shown later on that the emission of radiation brings about a damping of vertical oscillations with the same coefficient, i.e. $-E_r/2E$. Now it is possible to couple vertical and radial oscillations either by the choice of the magnet lattice (choosing $q_r = q_v$, for example) or by using specially shaped lenses or fields in the straight sections[3]. ROBINSON suggests a quadrupole lens which has been rotated 45° with respect to the guiding field direction. In either case the vertical damping will just balance the radial anti-damping and we can ignore the e^{Dn} term in (8.4). The integral then can easily be solved with the result that

$$\overline{\Delta r}_{\mathrm{rms}} = 6.8\,\alpha\,(E^5_{\mathrm{Bev}}/\omega\,R_M)^{\frac{1}{2}} \tag{8.5}$$

in cm for the case of a sinusoidally varying magnetic field of angular frequency ω and where f is the rotational frequency of the electrons. For a fully biased sinusoidal field the coefficient is 8.2 instead of 6.8.

ROBINSON[3] also shows that it is possible to couple synchrotron and betatron oscillations, for example, by modifying the magnet structure such that the radiation loss decreases with increasing radius. In that event, some of the strong damping of the synchrotron oscillations, i.e. $2E_v/E$, can be used to counteract the radial anti-damping or, in fact, even to give positive damping. ROBINSON also shows that it is not feasible to transfer damping rate from synchronous to betatron oscillations by shaping the r.f. electrodes.

As an example of the magnitude of the oscillations as given by (8.5), we get $\Delta r_{\mathrm{rms}} = 2.5$ mm for the Cornell synchrotron at 1.5 Bev and 1.5 cm for the California Institute of Technology machine. OSBORNE and RITSON give $\Delta r_{\mathrm{rms}} = 7$ mm for an example in which $\omega/2\pi = 15$ cycles, $R = 20$ m, $E = 6$ Bev and α is 0.047. For the case of the 50 Bev machine considered in Sect. 6, Δr_{rms} would be about 0.5 cm, depending, of course, on the lattice chosen.

---

[1] See footnote 3, p. 185.

[2] K. ROBINSON and D. RITSON: Cambridge Accelerator Project Report No. 14, Nov. 30, 1955.

[3] L. W. ROBINSON: CEA-14, May 14, 1956.

It is clear that Δr_{rms} would become seriously large if the anti-damping were not neutralized by one of the above-mentioned devices, for the e-folding time would be of the same order of magnitude as the $1/e$ time for the synchrotron oscillations which we have seen to be small at high energy.

Let us now return to a discussion of the nature of the anti-damping of radial oscillations. ROBINSON and RITSON[1] give a simple classical explanation of the anti-damping factor D. They ask us first to consider an electron losing radiation at a constant rate in which case its equilibrium orbit would be shrinking uniformly except for the compensation of energy by the r.f. acceleration which keep the electron on a constant orbit radius[2]. Next, suppose that for some reason the electron loses more energy in one region A of its orbit than in another region B. Because it has lost more energy in the region A, it must necessarily have a *smaller* radius than for the normal case in region B due to the previous abnormal large loss at A. This corresponds to a betatron oscillation that has been introduced by the abnormal energy loss. But just such a loss will be caused by a betatron oscillation itself. The electron will radiate more energy during the betatron oscillation when it is outside the equilibrium orbit, i.e. has a smaller radius of curvature, than when it is inside the equilibrium orbit in which case the curvature of the betatron oscillation subtracts from that of the average orbit. By the above mechanism, if a betatron oscillation exists, it will be subsequently enhanced and the beam will expand in size.

Quantitatively, suppose that the average curvature of an orbit is σ_0, and that the extra average curvature due to one-half of a betatron oscillation is σ_1. Then when the electron is outside the equilibrium orbit, its curvature is on the average $\sigma_0 + \sigma_1$ and while it is inside the curvature is $\sigma_0 - \sigma_1$ except for the effect that we are calculating in which case instead of losing a normal amount of energy $E_s = k\sigma_0$ during the first half oscillation [Eq. (6.2) written in a different form], the electron will lose an amount $k(\sigma_0 + \sigma_1)$ and its energy will differ from E_0 by an amount $k\sigma_1$ from what it would be had we neglected the extra radiation. Hence in the second half of the oscillation the average curvature will be reduced by an additional factor $(1 - k\sigma_1/E)$, where E is the average energy of the electron in the orbit. Thus, the curvature of the betatron oscillation will be increased by an amount $k\sigma_1/E$, and a proportional increase of the amplitude will occur. Clearly the same argument applies to the next half of the oscillation and to the subsequent oscillations, so that if we add up the increase of the amplitude for all the oscillations in a turn we shall find it proportional to the total radiation per turn. Thus, the increase per turn in E_r/E where E_r is the radiation per turn as given by (6.2).

A damping effect due to the electric field of the r.f. accelerator must also be included. If there is a betatron oscillation, the electron will, on the average, not be parallel to the average electric field (which we will assume to be parallel to the equilibrium orbit) as it crosses the acceleration gap or gaps. Because of the acceleration, the angle ϑ of the orbit of the electron with respect to the electric field will be reduced at each crossing by an amount $eV\vartheta/E$ where V is the voltage across the gap. Assuming that the magnet lattice is not at a resonance, the phase of the betatron oscillation will be different at each crossing and ϑ will vary from zero to a maximum value, the average being $\vartheta_{\mathrm{max}}/2$. Then the average reduc-

---

[1] See footnote 2, p. 186.
[2] We must define our equilibrium orbit in terms of quasi-spiral orbits. The deviation from a circle is roughly $\dfrac{\pi}{2} \dfrac{R}{N^3} \dfrac{E_r}{E}$ where it is assumed that the r.f. voltage is applied at N equally spaced points. The deviation is negligible for $E < 100$ Bev.

tion of the angle per turn will be $(E_r \vartheta_{max}/2E)$ because eV per turn is essentially equal to E at high energy and, since the amplitude is proportional to ϑ_{max}, the decremental damping will be given by $(E_r/2E)$, just one-half of our result for the anti-damping. Thus the total anti-damping is $(E_r/2E)$ which is the value of D to be used in Eq. (8.4). This result agrees with the more detailed calculations of ROBINSON and RITSON[1] or of KOLOMENSKY and LEBEDEV[2] and applies to strong focusing. For the case of weak focusing we must multiply the above result by the ubiquitous factor $n/(1-n)$ which comes about because the wavelength of the betatron oscillations is large compared to $2\pi r$ and hence the momentum compaction factor must be included in the above calculation.

The damping of the vertical oscillations has also been calculated independently by KOLOMENSKY and LEBEDEV[2] and by ROBINSON and RITSON[1]. We can understand their result on the following rough argument. Because the electrons are oscillating vertically, there will be a corresponding curvature of the orbits perpendicular to the average orbit curvature. The electrons will radiate extra quanta because of the vertical motion and we can approximate the extra energy radiated by averaging the curvature of the roughly sinusoidal vertical motion and substituting this for the curvature factor $(1/R^2)$ which appears in Eq. (6.1)[3]. For a vertical betatron oscillation having an amplitude a and q_v vibrations per turn we get for the extra power radiated:

$$P_v = \frac{1}{2}\frac{a^2 q_v^4}{R^2}E_r.\qquad(8.6)$$

This is to be compared to the energy inherent in the betatron oscillations which is

$$U = \frac{1}{4}a^2 q_v^2 \frac{E}{R^2}.\qquad(8.7)$$

The damping per turn is $P_v/2U$ which is

$$D_v = -\frac{q_v^2 E_r}{E}\qquad(8.8)$$

which differs only slightly from the more rigorously computed result previously mentioned, namely $D_v = -E_r/2E$. The extra term, q_v^2, in Eq. (8.8) is probably a result of assuming simple sinusoidal motion for the betatron oscillations.

9. Radiofrequency. The older synchrotrons were constructed very much like betatrons and consequently the guide field had the same characteristic circular shape. A circular or donut-shaped vacuum chamber with walls made of glass or ceramic was then placed in the guide field. The accelerating voltage was developed across the open end of a quarter-wave resonator so constructed that it formed a section of the donut. This was accomplished by building up a resonant cavity by plating silver strips on the inside and outside of a piece of quartz which was identical in conformity to the rest of the donut, the strip construction serving to reduce eddy current heating due to the A.C. magnetic field. The length of the quartz piece was chosen to be that of a quarter-wave line loaded with quartz dielectric. For a synchrotron of one meter radius, the arc of the donut occupied by the resonator was about 45°. Such a resonator was then driven to a peak voltage of about 2000 volts by means of a 47 Mc oscillator.

The construction of such resonators turned out to be a very complicated art, to put it mildly, and it was decidedly a great step forward when CRANE sug-

---

[1] See footnote 2, p. 186.
[2] See footnote 3, p. 185.
[3] Eq. (6.1) refers to the energy radiated per turn and varies as $1/R$. To get the energy radiated per second, we divide by the period, $2\pi R/c$.

gested the racetrack form which most subsequent synchrotrons have taken. The racetrack form usually consists of magnetic quadrants separated by straight sections in which injectors, resonators, pumps, detectors, etc., can be placed. Now the cavity resonator can be made of solid copper sheet fabricated in a convenient form without having to solve at the same time complicated problems concerning the vacuum and magnet systems.

The first choice in designing the radiofrequency system concerns which harmonic of the fundamental rotational frequency of the electrons in their orbits will be used. One advantage of making the frequency high is that the resonator becomes smaller and hence easier to construct and install. Another advantage of using a high harmonic is that the fraction of electrons that are captured into synchronous orbits at injection for a given aperture increases as the square root of the harmonic number. In the case of the new Cornell synchrotron, the 8th harmonic was chosen with the above considerations in mind. The resulting frequency of about 80 Mc is one which lends itself to the development of radio tubes and techniques made in the television industry. The resonator is about 1 m in diameter and thus is not too large compared to the magnet. Other advantages of choosing a high harmonic come about because of radiation effects which have already been treated in the previous sections.

A disadvantage of going to a high harmonic number is that the sensitivity to the energy of injection is increased. This effect is not yet well understood. If linear accelerators are to be used as injectors, then the sensitivity to energy can be important in determining the fraction of the injected electrons that are captured into stable orbits.

In the old synchrotrons, the first stage of acceleration was accomplished by a betatron action. After the electrons were at an energy of about 10 Mev, this action stopped because of saturation of the flux bars and a fixed frequency r.f. voltage was applied to the acceleration gap. It was found that if about twice the minimum necessary voltage was required to optimize the acceptance from the betatron to the synchrotron phase of operation. Indeed, nearly all the beam could be transferred in this way.

Crane pioneered another system in which the betatron acceleration was eliminated. He injected at an energy of several hundred kilovolts and then turned on a radiofrequency voltage of variable frequency which tracked the changing rotational frequency of the electrons as their speed increased to that of light. When this condition obtained, i.e. the electrons had reached an energy of several Mev, he turned off the frequency modulated resonator and turned on a fixed frequency oscillator which carried out the subsequent acceleration of the electrons to high energy. He found that there was no difficulty in transferring the acceleration process from one cavity resonator to another and that no beam was lost in the transfer. The reason for separating the F.M. (frequency modulation) phase from the F.F. (fixed frequency) phase, was that the F.M. oscillator-cavity system was necessarily lossy or wide-band. Thus the inefficient oscillator system needed only to operate for a few hundred microseconds during which time the initial acceleration to 10 Mev occurred; then the efficient narrow-band, F.F. oscillator could be turned on for the rest of the acceleration which might take many milliseconds.

It is this system that has been adopted with the new Cornell synchrotron and with many others of the newer synchrotrons. Those synchrotrons that intend to use high energy injectors such as the Japanese, or the Swedish, or the Harvard-M.I.T. machines will conveniently circumvent the F.M. problem entirely.

For synchrotrons of energy up to a few Bev, the r.f. problem is a serious one, but the expense is small compared, for example, to that of the magnet. At higher energies, the r.f. problem becomes dominant. For example, in the Harvard-M.I.T. synchrotron[1], sixteen resonant cavities located in different straight sections are required to compensate for the 4.4 Mev radiation loss per turn which occurs at 6 Bev, the total voltage of the cavities amounting to 6.2 Mev. They foresee a power requirement in their r.f. equipment of about 800 kW which is to be compared to about 580 kW for the magnet system. They have chosen to work on the 360th harmonic (476 Mc) of the orbital frequency of the electrons (1.33 Mc).

As synchrotrons are pushed to even higher energies the r.f. part of the equipment will become even more dominant until ultimately the magnet will be dispensed with, thus arriving at the linear accelerator as a limiting case of the synchrotron.

10. Calibration of the synchrotron. The output of the synchrotron must be calibrated before it is a useful tool for research. The energy of the electrons as they strike the target, the time distribution of the resulting x-ray beam pulse, the x-ray energy spectrum, the total x-ray intensity and finally the spatial distribution of the x-rays are all quantities that must be measured.

The energy of the electrons striking the target can be determined by measuring the magnetic field at the orbit which intersects the target and the radius of curvature of the orbit. It is more practical to measure the magnetic field at the central orbit at the peak of the field and then to compute the energy. The Cornell synchrotron has an electronic meter which senses the A.C. voltage of the magnet and then integrates it between two signals, one given by a peaking strip as the field goes through zero value and the other by a delayed pulse the time of which can be set by means of an oscilloscope to correspond to the center of the x-ray pulse. The reading of the meter has been calibrated by various magnetic measurements and gives the energy of the electrons directly in Bev. For a given setting of the magnet conditions, any energy less than the peak value is readily obtained by turning off the r.f. accelerating voltage at an appropriate earlier time.

The time distribution of the x-ray pulse is observed on an oscilloscope which displays the response of a scintillation counter placed near the target. It has been found that it is easy to shape the x-ray pulse in time by shaping the envelope of the r.f. accelerating voltage at the time the beam pulse is to be formed in such a manner that the beam "leaks" out of the synchrotron and strikes the target. The rate of leakage is controlled by varying the magnitude of the acceleration voltage. The beam pulse can be made very short, a few microseconds, by dropping the voltage rapidly, or it can be made several milliseconds long by dropping the voltage very slowly toward the amount that is necessary to counteract the radiation loss of the electrons. It is the synchrotron oscillations induced by the radiation loss that permit the beam to be spread out so easily. Usually the width of the beam pulse is determined by the energy spread that can be tolerated in a given experiment. The energy width can be computed from the sinusoidal variation of the magnetic field and a knowledge of the timing and time duration of the beam pulse.

The x-ray energy distribution resulting from electrons striking a thin target is the so-called bremsstrahlung spectrum[2]. This spectrum has been verified by

[1] Proceedings of the CERN Symposium 1956. M. S. Livingston, p. 439.
[2] H. Bethe and W. Heitler: Proc. Roy Soc. Lond. **146**, 83 (1934). See also L. I. Schiff: Phys. Rev. **83**, 252 (1951) and A. Sirlin: Phys. Rev. **106**, 637 (1957).

De Wire and Beach[1] for 300 Mev electrons and by Lawson for 80 Mev electrons. There is no reason to expect deviations in the theoretical shape as the energy is increased. Usually, however, a thick target is used in which case there is some degradation of the energy of the electrons in the target and multiple scattering becomes important. The soft quanta become somewhat enhanced relative to the hard quanta, the beam spreads out, and the spectral shape depends slightly on the angle of emission with respect to the initial direction of the electron. Generally, the only safe procedure to be followed in the case of a thick target is to use a pair spectrometer to determine the energy distribution of the x-rays.

For experiments which do not require a high intensity, it is possible to select monoenergetic photons[2]. Thus after a photon of energy W has been radiated by an electron of initial energy E_0, the electron can still have an energy $E = E_0 - W$ and the angle of the electron after the collision will be of the order mc^2/E_0. It is possible to select the electrons of energy E, for example, with a scintillation detector placed in a magnetic field. One then looks for a coincidence between this electron and a particle which has resulted from a reaction produced by the accompanying photon of energy $W = E_0 - E$. The method is particularly useful for calibrating gamma ray counters because it is absolute, namely, there is a one-to-one correspondence between the electrons counted and the number and energy of the accompanying x-rays.

The total x-ray intensity in the bremsstrahlung beam of a synchrotron is normally specified in "equivalent quanta", Q, which are defined as the total energy of the photons, U, divided by the maximum single photon energy, W_0, i.e.

$$Q = \frac{U}{W_0} = \frac{1}{W_0} \int\limits_0^{W_0} W\, n(W)\, dW$$

where W is the photon energy and $n(W)$ is the number of photons between W and $W + dW$. A rough approximation to the bremsstrahlung spectrum is $n(W) = k/W$ in which case $Q = k$, that is to say, Q is roughly the number of photons per logarithmic interval. More generally, a specification of Q and the spectral shape allows an exact determination of $n(W)$. Two standard instruments in the form of ionization chambers have been used for measuring beam intensity. At low energy, i.e. below 500 Mev, the so-called Cornell thick chamber was used. The chamber walls were made of one-inch thick copper so that the charge collected on a central plate was largely due to ionization by electrons of a shower near the shower maximum. The number of electrons of a shower at the shower maximum is roughly proportional to the photon energy, hence the charge collected is roughly proportional to the total energy in the beam, U. The charge collected was then measured by a vacuum tube electrometer[3]. The thick chamber was calibrated at Cornell against a pair spectrometer which was used as an absolute instrument by counting the pairs produced by the x-ray beam in a radiator of known thickness and for which the cross section of pair production has been measured by an absorption method. It was also calibrated by the shower method developed at California[4]. A calorimetric method devised by Edwards and Kerst[5] has also been used in the calibration. All the methods have given agreement to within

---

[1] J. De Wire and L. Beach: Phys. Rev. **83**, 476 (1951). — J. Lawson: Phys. Rev. **75**, 433 (1949).

[2] J. Weil and B. McDaniel: Phys. Rev. **92**, 391 (1953).

[3] R. Littauer: Rev. Sci. Instrum. **25**, 148 (1954).

[4] Blocker, Kenney and Panofsky: Phys. Rev. **79**, 419 (1950).

[5] P. Edwards and D. Kerst: Rev. Sci. Instrum. **24**, 490 (1953).

about 5% at 300 Mev, but some discrepancy between calibration methods has been noted at 500 Mev [1].

Since the calibration of the thick chamber depended on the energy and since the design was such that recombination prevented it being used at high intensity, a new ionization chamber has been devised which should be absolute, for which the calibration is independent of the energy, and which can be used at high intensity. This ionization chamber, the Quantameter [2], consists of many copper plates, in which a shower is developed and automatically integrated so that the ionization charge is strictly proportional to U independently of the energy of the photons.

Where the beam intensity varies appreciably over the face of an irradiated target, it is sometimes important to know the angular distribution of the x-rays. There is a typical distribution given by bremsstrahlung theory [3] which applies for very thin radiators and which is characterized by an average width of the order of magnitude of mc^2/E_0. However, for targets of a practical thickness the angular distribution is determined also by multiple scattering of the electrons in the target. SCHIFF [4] has computed this effect and finds that the ratio of intensity at an angle ϑ to the intensity at $\vartheta = 0$ is per unit solid angle in terms of an exponential integral:

$$n(\vartheta)/n(0) = [-\operatorname{Ei}(-\vartheta^2/2\vartheta_0^2)]/[\ln(2\vartheta_0^2 E_0^2/m^2 c^4) - 0.5772]$$

for ϑ larger than mc^2/E_0, where ϑ_0 is given by $(9.2 Z e^2/E)(N t)^{\frac{1}{2}}$ for electrons of energy E_0, penetrating a thickness t of target containing N atoms of nuclear charge Z per unit volume.

[1] R. L. WALKER et al.: Phys. Rev. 99, 210 (1955).

[2] R. R. WILSON: Nuclear Instrum. 1, 101 (1957).

[3] See footnote 2, p. 190.

[4] L. SCHIFF: Phys. Rev. 70, 87 (1946). See also A. SIRLIN: Phys. Rev. 101, 1219 (1956).

The Betatron.

By

D. W. KERST.

With 13 Figures.

I. Introduction.

1. The principle of induction acceleration. The betatron is a charged particle accelerator which uses for its accelerating mechanism the electric field produced by a time-varying magnetic flux linking the orbit of the charged particle. In the case of the betatron, the particle makes many revolutions about the accelerator while the flux linkage is changing. The slowly varying flux distinguishes betatron acceleration from the acceleration employed in synchrotrons, which is sometimes due to radiofrequency variation of flux linkage with a period the same as or shorter than the period of particle revolution. Betatron acceleration is especially suitable for the acceleration of electrons, and it is for this reason that the name of the β particle was used in the name of the accelerator.

Early studies of accelerators using the induction acceleration principle were carried out by J. SLEPIAN in 1922[1], who suggested a transformer core linking an orbit focused by permanent magnets, which provided a weak magnetic field for a small diameter orbit and a strong magnetic field for a large diameter orbit of a high energy particle. Experimenters who subsequently worked with the method were G. BREIT and M. A. TUVE, R. WIDERÖE, E. T. S. WALTON, W. W. JASSINSKY, M. STEENBECK, and many others[2].

In principle, acceleration by magnetic induction is based on LENZ's law which would require that the induced current (the particle beam) would increase in such a manner that it opposes the changing current in the inducing electric circuit. The magnitude of the electric field providing the acceleration is $\frac{1}{c}\frac{\partial \boldsymbol{A}}{\partial t}$ where $\boldsymbol{A}$ is the vector potential due to the currents in the inducing circuit. Most of the suggested acceleration schemes have a central iron core. This may provide all or a portion of the electric field. In any event the voltage per turn around the orbit is $\dot{\Phi}/10^8$ or the electric field is $\dot{\Phi}/(2\pi r \cdot 10^8)$. Consequently, as long as the flux linkage can rise, the particles are accelerated. The acceleration period is followed by a period of returning the currents and flux to low values so new particles can be accelerated.

II. Theory of orbits.

a) Characteristics of the orbits[3].

2. If we want to use magnetic induction to accelerate a particle at a fixed radius in a rotationally symmetrical magnetic field, we must have the magnetic

---

[1] U.S. patent 1 645 304.

[2] A summary of these early developments is found in Nature, Lond. **157**, 90 (1946). Historical Development of the Betatron.

[3] This discussion is based on D. W. KERST and R. SERBER: Phys. Rev. **60**, 53—58 (1941).

field at the orbit rise in proportion to the momentum gained by the time increase of accelerating flux linkage. The electric field causing the momentum to rise is $E = \dot{\Phi}/(2\pi r_0 c)$, and the rate of gain of momentum is $eE = \dot{p}$; so

$$p = \frac{e}{c} (\Phi - \Phi_b)/(2\pi r_0)$$

is the momentum which must be held to a fixed radius by the magnetic field. It is equal to $\frac{e}{c} B r_0$. This gives the flux condition which must be met at all times

$$\Phi - \Phi_b = 2\pi r_0 B \tag{2.1}$$

Φ_b, the integration constant, is a bias flux through the orbit center, so the flux linkage change within the orbit from the time of zero momentum is twice the flux which would exist if the field $B(r_0)$, extended uniformly throughout the orbit.

For focusing, the radial decrease of B must be less rapid than $1/r$, the radial decrease of required centripetal force; so that particles with $r > r_0$ will experience more inwardly directed magnetic force than that necessary to keep them at r, and particles at $r < r_0$ will be acted upon by too weak a magnetic force to hold them within r_0. This field dependence will cause particles to hunt radially about r_0.

Axial restoring forces are assured by requiring that the field decrease with radius. This causes the field above and below the median plane to have radial components which force circulating particles toward the median plane.

The conditions for restoring forces are thus that the magnetic field decrease with radius but that it decrease less rapidly than $1/r$.

The restoring force directed toward the median plane is $\frac{e}{c} B_r v$, where the radial component of the magnetic field B_r is found from the radial gradient of the vertical component by

$$\vec{V} \times \vec{B} = 0 \quad \text{or} \quad \frac{\partial B_z}{\partial r} = \frac{\partial B_r}{\partial z}.$$

Thus $B_r = Z \frac{\partial B_z}{\partial r}$ since $B_r = 0$ at $z = 0$. If we suppose that $B_z \sim 1/r^n$ at $r = r_0$, then $B_r = -Z n B_0 / r_0$ and the restoring force is $-\frac{e}{c} n B_0 v Z / r_0$. With $v/r_0 = \omega$, the angular velocity of the particle about the machine, the differential equation for a non-relativistic electron becomes

$$\ddot{z} = - n \omega \frac{e}{mc} B_0 Z = - n \omega^2 Z. \tag{2.2}$$

Thus the angular frequency of axial betatron oscillations is

$$\omega_z = \omega \sqrt{n}. \tag{2.3}$$

The radial oscillations frequency can be found similarly since it is produced by the difference between the centripetal force required to hold a particle at its radius, r, mv^2/r and the actual magnetic force at r, $(e/c) B_z v$. If x is the radial departure from the equilibrium orbit,

$$m\ddot{x} = \frac{mv^2}{r} - \frac{e}{c} B_z v$$

since

$$B_z = B_0 - \frac{n B_0 x}{r},$$

we have

$$\ddot{x} = -\left(\frac{e}{mc}\right) B_0 v \left(\frac{r_0}{r^2} - \frac{n}{r}\right) x = -(1-n)\,\omega^2 x, \qquad (2.4)$$

and the angular frequency of radial oscillation is

$$\omega_r = \omega \sqrt{1-n}. \qquad (2.5)$$

Since B_0 or ω increase with time, the differential equations (2.2) and (2.4) are like the equation $\ddot{y} = -K(t)\,y$ for an oscillation with a slowly stiffening spring constant. The solution to such an equation,

$$y = y_0(t) \sin \omega_y t$$

has its amplitude, y_0, inversely proportional to the fourth root of the adiabatically changing spring constant:

$$y_0 \sim K(t)^{-\frac{1}{4}}.$$

This means that the axial and the radial betatron oscillations decrease in amplitude as the magnetic field rises. Thus

$$\left. \begin{array}{l} X_0 \sim B(t)^{-\frac{1}{4}}, \\ Z_0 \sim B(t)^{-\frac{1}{4}}. \end{array} \right\} \qquad (2.6)$$

Fig. 1 a—d. The figures represent the developed paths of particles. r_0 is the fixed equilibrium orbit position, and r_i is the position of the instantaneous equilibrium orbit. The injector is at O. (a) Path of a particle injected tangentially on its instantaneous equilibrium orbit. It approaches r_0 without oscillation. (b) Path of a particle with its instantaneous equilibrium orbit coincident with the fixed equilibrium orbit. Oscillation occurs about r_0. (c) Path of a particle whose instantaneous equilibrium orbit does not pass through O at the time of injection but is between O and r_0. The oscillation is about r_i while r_i approaches r_0. (d) A real beam from the injector showing image formation at 1, 2, 3, and 4. The instantaneous equilibrium orbit is coincident with the fixed equilibrium orbit.

This adiabatic damping of the oscillations makes possible the maintenance of a small beam of electrons in spite of the long path (300 kilometers) which they traverse through the residual gas of the vacuum tube. However, when gas pressures exceed $\sim 2 \times 10^{-5}$ mm, most betatrons begin to lose yield rapidly by scattering.

The focusing properties of the magnetic field result in the formation of a succession of images of the injector along the orbit. The case $n = \frac{1}{2}$ produces images by axial focusing and radial focusing at the same point. This type of field dependence has been used in Beta-ray spectrographs for the formation of source images[1].

In addition to the damping of the motion of particles oscillating about the equilibrium orbit, a particle injected tangentially where $\frac{e}{c} B(r_1)\,r_1 = p_1$ and $r_1 \neq r_0$ will find either too much or too little flux change to satisfy (2.1). If the particle is at $r_1 < r_0$, then $\Delta \Phi > 2\pi r^2 B(r)$ provided $0 < n < 1$ and the accelerating flux provides momentum more rapidly than that which the rise in the local field can maintain in a circular orbit. So the particle spirals asymptotically out toward the equilibrium orbit. If the particle is at $r_2 > r_0$ and $\frac{e}{c} B(r_2)\,r_2 = p_2$, then the orbit again asymptotically approaches r_0 with time.

---

[1] K. SIEGBAHN and N. SVARTHOLM: Nature, Lond. **157**, 872—873 (1946).

In general, particles are injected with an angular spread and at times when the injection radius is not at the radius of momentum balance. Then the particles oscillate about the locus of momentum balance, the "instantaneous circle", while this circle gradually approaches r_0 and while the amplitude of the oscillations about this instantaneous circle are adiabatically damped. The situation is shown in Fig. 1.

The adiabatic shift of the particles toward the equilibrium orbit and away from the injecting structure is in some cases sufficient to allow particles to avoid striking the injector on successive revolutions. In most betatrons, however, the influence of space charge, space current, and transient electric or magnetic fields provides the mechanism for avoiding the injection structure. These effects will be considered later.

b) The dynamical equations in linear approximation.

3. Instantaneous and fixed equilibrium orbits. The Hamiltonian for the general case of the field being independent of φ and for the radial component of the motion is:

$$\mathscr{H} = c\left(m^2 c^2 + p_r^2 + \left[\frac{p_\varphi}{r} + \frac{e}{c} A\right]^2\right)^{\frac{1}{2}} \tag{3.1}$$

where the circumferential component of mechanical momentum is

$$p = \frac{p_\varphi}{r} + \frac{e}{c} A = \frac{m r \dot{\varphi}}{\sqrt{1 - \dfrac{v^2}{c^2}}} \tag{3.2}$$

and $A = A(r, z, t)$ is the vector potential having only a φ component. p_φ is a constant of the motion, since φ does not occur in the Hamiltonian.

First, consider an orbit for which

$$\dot{p}_r = 0. \tag{3.3}$$

This orbit toward which the restoring forces are directed is called the "instantaneous circle" or instantaneous equilibrium orbit. This locus can be a closed orbit if a second condition

$$\dot{r} = 0 \tag{3.4}$$

is satisfied, or it can be a locus of slowly or rapidly changing radius in which cases a closed curve will not strictly result.

The consequences of (3.3) alone will now be found. Since

$$\dot{p}_r = -\frac{\partial \mathscr{H}}{\partial r},$$

$$\dot{p}_r = c^2 \frac{p}{\mathscr{H}} \frac{\partial p}{\partial r}.$$

Thus on the instantaneous equilibrium orbit

$$\frac{\partial p}{\partial r} = 0 = \frac{p_\varphi}{r^2} - \frac{e}{c} \frac{\partial A}{\partial r} \tag{3.5}$$

and thus

$$\frac{c}{e} p_\varphi = r^2 \frac{\partial A}{\partial r} = \text{const} \tag{3.6}$$

along the instantaneous equilibrium orbit for a particle with a given p_φ, and this orbit is defined by (3.6) as A changes with time. Particles will attempt to follow this path by oscillating about it. The closeness with which they follow will be discussed later.

To find the physical meaning of $r^2 \partial A / \partial r$ being constant, notice that the flux included within the instantaneous circle is

$$\Phi = \int_\Sigma \boldsymbol{B} \cdot d\boldsymbol{\sigma} = \int_\Sigma \nabla \times \boldsymbol{A} \cdot d\boldsymbol{\sigma} = \oint_\Theta \boldsymbol{A} \cdot d\boldsymbol{s} = 2\pi r A$$

and since

$$B_z = \frac{1}{r} \frac{\partial (r A)}{\partial r} = \frac{A}{r} + \frac{\partial A}{\partial r}$$

then

$$r^2 \frac{\partial A}{\partial r} = r^2 B_z - \frac{\Phi}{2\pi}$$

or

$$2\pi r^2 B(r t) - \Phi(r, t) = \text{const}. \tag{3.7}$$

This shows the requirement on flux linkage and field strength which holds at an instantaneous equilibrium orbit as it changes with time. Under some circumstances (3.7) may be satisfied at several values of r.

The flux, Φ, can be composed of a constant central part which does not contribute field at the orbit if A has a term having a $1/r$ dependence. This can be a central biasing flux.

In (3.5) if $A(r, t)$ goes through zero simultaneously at all points in space, then $p_\varphi = 0$ and $\partial A / \partial r = 0$ or

$$2\pi r_0^2 B = \Phi \tag{3.8}$$

for all time, or the instantaneous orbit becomes a fixed equilibrium orbit, and it is located where the vector potential is an extremum (minimum).

If there is a bias flux, Φ_b, in the center of the orbit,

$$A = A_1(r, t) + \frac{\Phi_b}{2\pi r},$$

then

$$r^2 \frac{\partial A_1}{\partial r} = \frac{\Phi_b}{2\pi} + \frac{c}{e} p_\varphi$$

and $\partial A_1 / \partial r \neq 0$; but if $A_1(t)$ is zero everywhere at some instant, then $\partial A_1 / \partial r = 0$ always at the instantaneous orbit, and

$$\frac{\partial^2 A}{\partial r \partial t} = \frac{\partial^2 A_1}{\partial r \partial t} = \frac{\partial E}{\partial r} = 0 \tag{3.9}$$

or the instantaneous orbit is located where the electric field is a minimum and it has becomed fixed.

In general, if we require a second condition $\dot{r} = 0$, giving a fixed equilibrium orbit, then by (3.5)

$$\frac{\partial A}{\partial r} = \text{const on } r_0$$

and

$$\frac{\partial E}{\partial r} = 0,$$

or the electric field E must be a minimum at r_0 without the restriction on the time dependent part of the vector potential being zero at all radii simultaneously as in (3.8) and (3.9).

This provides a very convenient way of locating the orbit in a betatron. Several wires of equal length are wound in the orbital plane to form coils at different radii, and thus each has a different number of turns. The coil with the lowest voltage induced in it is at the location of the equilibrium orbit.

The instantaneous equilibrium orbit radius may be increasing with time if there is an excessive amount of flux accumulating within the orbit. This is the usual method of moving the orbit out to the target after acceleration has occurred. Similarly, the instantaneous orbit can be moved inward by an insufficient increase of flux. This is the orbit contraction method commonly used to rapidly move particles away from an injecting electrode structure which they might hit after several revolution.

Two particular examples of the spacial dependence of the vector potential will be discussed. In the first example, let

$$B_z \sim 1/r^n$$

with

$$A \sim 1/r^{(n-1)}$$

without an $1/r$ part. Then

$$r^2\, \partial A/\partial t = \text{const}$$

means

$$r(t)\, p(t) = \text{const}. \tag{3.10}$$

Thus for a $1/r^n$ field the momentum of the particle is inversely proportional to the radius of the instantaneous orbit toward which it is focused, or the instantaneous orbit encloses a constant amount of flux. This is the type of acceleration tested by BREIT and TUVE in 1926[1].

In the second example, the time dependent part of the vector potential is all of $1/r$ dependence for the case where there is only a time varying central flux. The particle focusing field is constant in time. This is the case of a fixed field alternating gradient (FFAG) betatron[2] which has the ability to hold both small and large momenta in its focusing magnet simultaneously while the rising flux causes the instantaneous orbit to spiral inwardly or outwardly depending on the radial gradient of the average field at the orbit being negative or positive respectively. In this case it follows directly from (3.6) that

$$\Phi_c - 2\pi r^2(t)\, B(r)\, \frac{k+1}{k+2} = \text{const} \tag{3.11}$$

or

$$\Phi_2 - \Phi_1 = (k+1)\, \Delta\Phi_{\text{guide}}$$

where Φ_2 and Φ_1 are the final and initial central accelerating fluxes, Φ_c, and $\Delta\Phi_{\text{guide}}$ is the flux in the focussing gap contained between the initial and final orbits. k is the field index used in FFAG terminology,

$$k \equiv \frac{R}{\bar{B}}\, \frac{\partial\bar{B}}{\partial R},$$

with $\bar{B}$ the average field along the equilibrium orbits including the effects of the angles of the magnet edges. (It is like $-n$.) It is evident from (3.11) that in principle the smallest $\Phi_2 - \Phi_1$, or the smallest central core results when $k < 0$, that is, when the higher energy orbits have smaller radii, because $\Delta\Phi_{\text{guide}} = 2\pi \int rB\, dr$ will be negative and smaller in absolute magnitude if the high field is at the low radius. In practical designs, this gain may not always be achieved.

[1] Carnegie Year Book, 1926.
[2] K. R. SYMON, D. W. KERST, L. W. JONES, L. J. LASLETT and K. M. TERWILLIGER: Phys. Rev. **103**, 1855 (1956).

Formula (3.7) is used in designing the orbit expanders to move the particles to the target after acceleration. It is evident that if part of Φ is a separately controllable central flux, as in the example of the fixed field case (3.11) or of a betatron orbit expander, then no matter what the different time dependences of $B(r, t)$ and of the separately controlled central flux may be, the same change of central flux will finally be required to bring a particle to a predetermined radius and momentum. This is because only the initial and the final states of B, r, Φ are involved according to (3.7) and the route by which the orbit gets from r_1 to r_2 and from p_1 to p_2 does not matter when the flux is supplied to expand the orbit to the vicinity of the target, — a well known characteristic of orbit expanders.

A serious difficulty with betatrons is that if they are used for electrons a large radiation loss occurs due to the central acceleration[1]. This becomes important at energies of 100 Mev, and within limits it can be compensated for by the addition of expanding fluxes. If the energy of the particle is proportional to the $\sin(ft)$, then the fraction of energy lost by the time the particle has reached its peak energy is

$$\Delta E/E = E^3/(300 r^2 f).$$

4. The oscillatory and transient aspects of the motion.
We want to describe the deviations from r_i, the instantaneous equilibrium orbit. Let $r = r_i + x$ where x will be a small deviation from r_i. Keeping terms no higher than quadratic in x and p_r, the Hamiltonian (3.1) becomes

$$\mathscr{H} = \varepsilon + \frac{1}{2\varepsilon} \left[c^2 p_r^2 + 2 c^2 p_i \, d p_i \right]$$

where

$$p_i = \frac{p_\varphi}{r_i} + \frac{e}{c} A$$

and

$$\varepsilon = c \, (m^2 c^2 + p_i^2)^{\frac{1}{2}}.$$

With prime denoting differentiation with respect to r

$$d p_i = p_i' + \tfrac{1}{2} p_i'' \, x^2;$$

but

$$p_i' = 0 \quad \text{on} \quad r_i$$

by (3.5); so

$$\mathscr{H} = \varepsilon + \frac{c^2 p_r^2}{2\varepsilon} + \frac{c^2 p_i}{2\varepsilon} \left(\frac{2 p_\varphi}{r_i^3} + \frac{e}{c} A'' \right) x^2. \tag{4.1}$$

The resulting equations of motion are

$$\left. \begin{aligned} \dot{x} &= \frac{c^2 p_r}{\varepsilon} - \dot{r}_i, \\ \dot{p}_r &= - \frac{p_i c^2}{\varepsilon} \left(2 \frac{e}{c} \frac{A'}{r_i} + \frac{e}{c} A'' \right) x. \end{aligned} \right\} \tag{4.2}$$

Eliminating p_r we have

$$\ddot{x} + \frac{p_i c^3 e}{\varepsilon^2} \left(2 \frac{A'}{r_i} + A'' \right) x + \dot{x} \frac{\dot{\varepsilon}}{\varepsilon} = - \dot{r}_i \frac{\dot{\varepsilon}}{\varepsilon} - \ddot{r}_i \equiv F. \tag{4.3}$$

The forcing terms, F, on the right represent the non-adiabatic corrections to the motion.

---

[1] For an extensive treatment see J. Schwinger: Phys. Rev. **75**, 1912—1925 (1949).

If we substitute $x = (mc^2/\varepsilon)^{\frac{1}{4}} u$ and omit the forcing terms for the moment, (4.3) becomes

$$\ddot{u} + \omega_r^2 u = 0$$

where

$$\omega_r^2 = \frac{p_i c^3 e}{e^2} \left(2\frac{A'}{r_i} + A'' \right) + \frac{1}{4} \left(\frac{\dot{\varepsilon}}{\varepsilon} \right)^2 - \frac{1}{2} \frac{\ddot{\varepsilon}}{\varepsilon}.$$

The last two terms are smaller than the first by a factor of approximately $(c\,p_m/\varepsilon)^2 f^2/\omega^2$ where ω is the angular velocity of the particle, and f is the angular frequency of the magnetic field, and p_m is the maximum momentum reached by the particle. Omitting these terms, we obtain

$$\omega_r^2 = \left(\frac{v}{r} \right)^2 \frac{2A' + r A''}{B_z} = \left(\frac{v}{r} \right)^2 \left(1 + \frac{r}{B_z} \frac{\partial B_z}{\partial r} \right)$$

where v is the tangential speed of the particle. We define

$$\frac{r}{B_z} \frac{\partial B_z}{\partial r} \equiv -n$$

which may or may not be a constant. Then

$$\omega_r = \omega \sqrt{1 - n}. \tag{4.4}$$

For the forces to be restoring forces we see that in general

$$2\frac{A'}{r_i} + A'' > 0$$

must hold on an instantaneous equilibrium orbit in addition to (3.6). The previous assertion that the electric field at the equilibrium orbit is a minimum in betatrons with bias follows from this new condition.

The asymptotic form of the solution, u, for $\omega_r t \gg 1$ and $t = 0$ when $A = 0$ is

$$u = \frac{\alpha}{\sqrt{\omega_r}} \sin \left[\int_0^t \omega_r dt + \gamma \right]$$

where α and γ are arbitrary constants. Converting to displacement from the instantaneous orbit,

$$x = \alpha \sqrt{\frac{mc}{e B_z \sqrt{1 - n}}} \sin \left[\int_0^t \omega_r dt + \gamma \right] \tag{4.5}$$

which shows that the amplitude of the oscillation is damped as $B_z^{-\frac{1}{4}}$. In terms of the voltage gained by the particle any time in its acceleration period or particularly immediately after injection this means

$$\delta a/a = -\tfrac{1}{4} (\delta V/V) \tag{4.6}$$

where $\delta a/a$ is the fractional change in the amplitude while the particle of energy eV gains δV electron volts. If the particle is relativistic, the factor $\tfrac{1}{4}$ is to be replaced by $\tfrac{1}{2}$. This damping in some cases is sufficient to allow a particle to lose amplitude rapidly enough so that it does not strike the electrode structure from which it started. For example, for the 300 Mev betatron at the University of Illinois $\delta V = 3000$ volt/turn, $V \sim 100$ kev, $a \sim 8$ cm, so $\delta a \sim 0.8$ mm per turn. A large fraction of the particles revolve about five to ten times before the phase of the betatron oscillation brings them close to the injecting electrode structure. Thus the total $\delta a \sim 4$ to 8 mm which is sufficient to escape striking the gun.

In addition to the damping, the instantaneous equilibrium orbit moves, and this can carry the particles away from the injector so that they are trapped. To evaluate this effect we take the total derivative of (3.6) and find

$$\dot{r}_i = -\frac{\partial A'/\partial t}{2\dfrac{A'}{r_i} + A''} \tag{4.7}$$

for the motion of the instantaneous equilibrium orbit. To put this in terms of field strength and flux linkages we substitute

$$A' = B_z - \frac{\Phi}{2\pi r_i^2}$$

and

$$2\frac{A'}{r_i} + A'' = (1 - n)\frac{B_z}{r_i}$$

giving

$$\dot{r}_i = -\frac{r_i}{(1 - n)B_z}\frac{\partial}{\partial t}\left[B_z(r_i, t) - \frac{\Phi(r_i, t)}{2\pi r_i^2}\right] \tag{4.8}$$

being sure to take only partial derivatives with respect to time.

For the example of a fixed focusing field but time varying central flux as in FFAG betatron (4.8) becomes

$$\delta r_i = \frac{\delta \Phi_c}{2\pi(k + 1)B_z r_i} \tag{4.9}$$

where δr_i occurs while the central flux increases by $\delta \Phi_c$ as we saw in (3.11).

Another example is that of the usual betatron with a fixed equilibrium orbit where $r_0 =$ const. Remembering that $2\pi r_0^2 B_z = \Phi(r_0, t)$ we have by (4.8)

$$\dot{r}_i = -\frac{r}{(1 - n)B_z}\frac{\partial}{\partial t}\left[(r_i - r_0)\frac{\partial B_z}{\partial r} - (r_i - r_0)\left(\frac{1}{r_i}B_z - \frac{\Phi}{\pi r_i^3}\right)\right] \tag{4.10}$$

or $\delta r_i = -(r_i - r_0)\delta B_z/B_z$ with δB_z due to time alone.

It is this shift of the instantaneous equilibrium orbit away from the injecting structure in addition to the damping of the orbital oscillations about the instantaneous orbit which combine to trap particles in orbits which do not return to the injector.

The closeness with which a particle will follow the shift of the instantaneous orbit must be examined, since we have ignored the forcing terms in the differential equation. If r_i changes too rapidly, we might expect non-adiabatic effects. In (4.3) the forcing terms, F, on the right side can be shown equal to

$$-\frac{\dot{B}}{B}\dot{r}_i - (1 - \beta^2)(1 - n)\frac{\dot{r}_i^2}{r_i}$$

or to

$$-\left(\dot{B} + \frac{\Phi}{2\pi r_i^2}\right)\frac{\dot{r}_i}{B} - \beta^2\left(\dot{B} - \frac{\Phi}{2\pi r_i^2}\right)\frac{\dot{r}_i}{B}.$$

If we inject at t_1, and if $\omega_r t \gg 1$, then the asymptotic solution of (4.3) due to the forcing terms can be formed from solutions (4.5):

$$x = \int_{t_1}^{t}\left\{\frac{\sqrt{1 - n(s)}\, p_i(s)\, r_i(t)}{\sqrt{1 - n(t)}\, p_i(t)\, r_i(s)}\right\}^{\frac{1}{2}}\frac{F(s)}{\omega_r(s)}\sin\left[\int_{0}^{t}\omega_r\, dt'\right]ds$$

$$= -\left\{\frac{\sqrt{1 - n(t_1)}\, p_i(t_1)\, r_i(t)}{\sqrt{1 - n(t)}\, p_i(t)\, r_i(t_1)}\right\}^{\frac{1}{2}}\frac{F(t_1)}{\omega_r(t_1)^2}\cos\left[\int_{t_1}^{t}\omega_r\, dt'\right] + \frac{F(t)}{\omega(t)^2}$$

which gives a maximum displacement

$$x_m = -2F(t_1)/\omega_r(t_1)^2.$$

If δB, $\delta \Phi$ and δr_i are the increments of B and Φ in one revolution due only to their time dependence and not to the radius change, and, if δr_i is the radius change then

$$x_m = \frac{\delta r_i}{2\pi^2(1-n)B}\left\{\left(\delta B + \frac{\delta \Phi}{2\pi r_i^2}\right) + \beta^2\left(\delta B - \frac{\delta \Phi}{2\pi r_i^2}\right)\right\}.$$

At relativistic speeds this reduces to

$$x_m = \frac{\delta r_i \, \delta B}{\pi^2(1-n)B}.$$

At classical speeds in a betatron with a fixed equilibrium orbit with $\Phi_0 = 2\pi r_0^2 B$, the result is the same. But with a fixed field magnet used for focusing, at relativistic speeds $x_m \to 0$ since $\partial B/\partial t = 0$ and then the particles have no forced oscillation, but at non-relativistic speeds

$$x_m = \frac{(\delta r_i)^2}{2\pi^2 r},$$

in this approximation.

For all these cases the orbit follows so closely to the instantaneous equilibrium orbit that a fairly sudden contraction or expansion of this orbit will not generate serious orbital oscillations, and thus a sudden shift induced by coils can be used to augment the natural damping of the oscillations to assist in escaping impact with the injecting structure. For betatrons in which the accelerating voltage all appears at an accelerating gap in a conducting vacuum tube coating, more of an oscillation would be excited. With four accelerating gaps in the circumference as in the 300 Mev betatron at the University of Illinois, there are about six gaps per oscillation wave length which would excite very little oscillation.

As mentioned before, vertical focusing results from the radial component of the magnetic field reversing sign on opposite sides of the orbital plane. The vector potential becomes

$$A_\varphi = A(r, t) - \tfrac{1}{2}A_v(r, t)\,z^2$$

where

$$A_v = \left(\frac{\partial B_z}{\partial r}\right)_{z=0} = \frac{c}{e}\left[\frac{1}{r}\,(r\,p)'\right]_i'.$$

The Hamiltonian (4.1) has $\mathscr{H}_z$ added to it.

$$\mathscr{H}_z = \frac{c^2}{2\varepsilon}\,p_z^2 - \frac{c^2}{2}\,p_i\left[\frac{1}{r}\,(r\,p)'\right]_i'\frac{z^2}{\varepsilon}$$

and the equations of motion are

$$\dot{z} = \frac{c^2}{\varepsilon}\,p_z,$$

$$\dot{p}_z = \frac{ec}{\varepsilon}\,p_i\left[A'' + \frac{A'}{r_i} - \frac{A}{r_i^2}\right]z.$$

These equations differ from (4.2) in the absence of a forcing term and in the replacement of $A'' + \frac{2}{r}\,A'$ by

$$A'' + \frac{A'}{r_i} - \frac{A}{r_i^2}.$$

Thus the solutions are the same as (4.5) with $\omega_z = \omega \sqrt{n}$ or $\sqrt{1-n}$ replaced by $\sqrt{n}$. The adiabatic damping of vertical oscillations is the same as that of the radial oscillations.

5. Imperfect fields. The orbits which have been discussed until now have been in an ideally shaped magnetic field. If there are inhomogeneities in this field, the disturbance to the orbit can be estimated in the following way[1]. A particle following a circular path will receive a transverse impulse $\frac{e}{c} \Delta B_z v t$, where t is the time of passage. If the inhomogeneity has the length W, then the impulse is $\Delta p_r = \frac{e}{c} \Delta B_z w W$. The angle through which the particle has been bent is $\Delta p_r / p_r = \Delta B_z W / B_z r$, and the ampli-tude of oscillation which is generated by this impulse or angular deflection with each passage through the region of ΔB_z is $\lambda \Delta p_r / p$, where λ is the wavelength of the betatron oscillation. Consequently, the oscillation generated by each pas-sage of the bump can be represented by a rotating vector of length $Y = \Delta B_z W / (B_z \sqrt{1-n})$. The phases with which successive rotating vectors are added are determined by the ratio ω_r/ω. The method used in optics to sum sine waves by a Cornu spiral is useful in determin-ing the sum of the oscillations produc-ed. The vector diagram or spiral for such an excited motion is shown in Fig. 2. The effect of the bump is decomposable into a steady displacement, D, of the average position of the orbit, a force os-

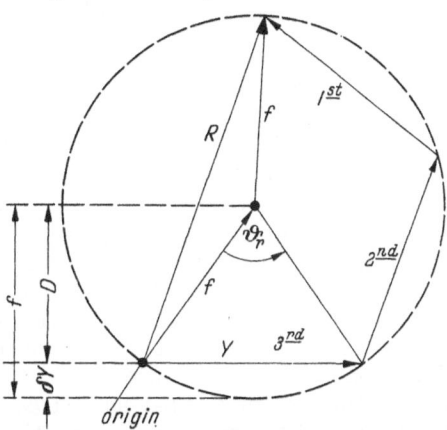

Fig. 2. Vector spiral for summing oscillations produced by passage of a particle through a field error ΔB_z. The vectors created by the first, second, and third passage through the bump are shown. R is their resultant which gets another Y vector added to it on each passage. The vector diagram all rotates through ϑ_r between passages through the bump.

cillation, δY, producing a kinked equilibrium orbit, and a free oscillation, f, about the new closed orbit. From the figure geometry the average displacement is

$$D = \frac{Y}{2} \cot \frac{\vartheta_r}{2} = \frac{\Delta B_z W \cot (\pi \sqrt{1-n})}{2 B_z \sqrt{1-n}}$$

and the forced oscillation is

$$\delta Y = \frac{Y}{2} \tan \frac{\vartheta_r}{A} = \frac{\Delta B_z W \tan \left(\frac{\pi}{2} \sqrt{1-n} \right)}{2 B_z \sqrt{1-n}}$$

while the free oscillation is

$$f = \delta Y + D.$$

6. Space charge effects. It is easy to make electron guns which emit large currents, and if these currents are sufficiently well collimated or if injection takes place for a sufficiently large number of revolutions enough particles from these guns can be captured to result in very noticeable and important space charge and space current effects. First of all, it should be noticed that at relati-vistic energies space charge effects are completely balanced by the magnetic focusing of a beam, provided there are no ions or electrons which accumulate

[1] D. W. KERST, G. A. ADAMS, H. W. KOCH and C. S. ROBINSON: Rev. Sci. Instrum. **21**, 479 (1950).

due to residual gas. This can be seen for the electrical force on a stray electron at a distance Δ from the center of the cylindrical beam is $e\varepsilon = 2\sigma e/\Delta$, where σ is the linear charged density in e.s.u./cm. The magnetic attraction on a straying particle due to the main current in the beam is $ev B/c = (r/c)^2 2\sigma e/\Delta$. This shows that when the velocity of light is approached the magnetic pull of the beam just equals the electrostatic repulsion for the stray particle. At injection energies this cancellation does not occur, and a crude estimate of the charge which can be held by the accelerator can by made ba calculating the number of particles which would be necessary to produce an electrostatic repulsion just equal to the magnetic focusing force at the edge of the beam. This calculation shows:

$$I = 5 f e r_0 \Delta^2 B_z^2 (1 - n)/m c^3)$$

for non-relativistic electrons. f is 2π times the magnet frequency, B_z is the magnetic field at injection time, I is the target current in amperes. The real space charge problem is less ideal. Electrostatic images of the beam on the conducting coat of the vacuum tube must be taken into account. The same is true for magnetic images of the beam in the pole faces.

These space charge and space current effects which occur during injection cause transient effects which can be several orders of magnitude larger than the transient effects already calculated for single particle trajectories at injection time. Some measurements have been made of the magnetic effects of the captured beam[1,2]. There are several effects, each of which is large, and the effects can sometimes hinder capture so the result is a difference between large effects. The proportions of the various effects are hard to calculate. One effect is simply due to the fact that if particles stream into the accelerator and circulate many times around the machine before their orbit returns them to the injector, they have been building up the electromagnetic energy in the self field of the beam. This electromagnetic energy increases as thesquare of the current which is building up during multi-turn injection, and the energy of this field must come from the kinetic energy of the electrons as they are being injected. Consequently, an extremely rapid change in the momentum of the injected particles results. One portion of this electromagnetic energy, namely, the reverse flux, which an injected beam creates, due to its own ampere turns, has been observed in the case of a 22 Mev betatron. In this example, about 130 Maxwells of flux were withdrawn from within the orbit between the time of the first and the last injected electrons. This flux change could contract the orbit away from the injector structure by more than a millimeter, which is about 100 times more than the single particle orbit transients would displace a particle in ten revolutions. External electrical circuits are commonly used to produce such a transient contracting pulse of flux so that the particles are rapidly shifted away from the injector. In the case of the 22 Mev betatron, this external contractor makes only about a 30% increase in the yield when a large beam of electrons is emitted from the injector; but, if only a small beam of electrons is injected, then the electrical contraction circuit produces an order of magnitude increase in yield, implying that the large circulating beam from the injector helps greatly in the capture process.

Another effect which can be quite large is capture by the sacrifice of circulating space charge. We have shown earlier [Eq. (4.5)] that $a \approx \omega_r^{-\frac{1}{2}}$. The effect of space charge is to change ω_r. If ω_r is very small because there is a large circulat-

---

[1] See footnote 1, p. 203.
[2] D. W. Kerst: Phys. Rev. **74**, 503 (1948).

ing current within the tube, and if subsequently many of the circulating particles are intercepted by the injector structure and the space charge decreases, then ω_r will increase, and the amplitude of oscillation of the remaining particles in the beam will be damped. We known from measurements that many particles circulate from ten to 100 times before they are intercepted by the injector, so that damping due to space charge can be adiabatic. Estimates of this damping effect in the case of a 22 Mev betatron show that the current which must be injected to produce the 130 Maxwells observed is several times larger than the current which finally reaches the target, and thus the injected current has approached the space charge limit. The subsequent sacrifice of some of this circulating current can produce a damping of approximately two millimeters which again is orders of magnitude greater than the single particle transient effect. These examples can merely indicate the magnitude of some of the effects, but since the importance of them and the effect for particles which are injected early and particles which are injected late is different, the final result of these processes would require much more study.

A process was pointed out by BARDEN[1] in which the beam of electrons, which is bent by the field across the orbits in the vacuum chamber after it has injected particles into the orbit, produces a lens-like space charge, which suddenly disappears when the injected beam is cut off. The resulting transient in the electrons' focusing would shift the orbits and help them escape interception by the injector. Some people have reported[2] that particles are only captured if the injected stream of electrons is turned off and that, therefore, BARDEN's process may be important. However, an important point which must be cleared up before these experiments can be evaluated is the effect of wall potentials resulting from a high resistance wall coating struck by the injected beam. Any potential in the orbital region of the order of 100 volt is known to affect the yield. Therefore, if a high resistance coating is struck, the resulting wall potential may be equivalent to a large magnetic bump which disappears when the beam is turned off, and consequently only particles injected when the beam is turned off can be captured. In the betatrons at the University of Illinois, low resistance coatings of the order of 10 to 75 ohm are used in the vacuum chamber, and an approximately rectangularly shaped voltage pulse on the injector has demonstrated no beneficial effect of injecting when the voltage is being turned off. On the contrary, it has been demonstrated that injection on the front of the voltage pulse, where the voltage rise fits the acceptable energy of the rising magnetic field, results in a several fold increase in yield. In summary, any transient effect, such as the space charge effect, or the previously mentioned contraction coils, or transient bumps created in the magnetic field, is helpful in trapping electrons. Recent experiments mentioned in a CERN Conference report[3] bear on this question.

III. Practical considerations.

7. Types of betatrons. If the magnet of a betatron is made of iron which has an upper limit on the flux density, there are three different types of structures[4,5] (Fig. 3). The first is the conventional type with an air gap making the central

[1] S. E. BARDEN: Proc. Phys. Soc. Lond. B **64**, 85, 579 (1951).
[2] R. WIDERÖE: J. Appl. Phys. **22**, 362 (1951).
[3] LOBANOV, LOGUNOV, OVCHINNIKOV, PETUKHOV, RABINOVICH and RUSANOV: CERN Symposium Reports 1956.
[4] See footnote 1, p. 203.
[5] D. W. KERST: Phys. Rev. **68**, 233 (1945).

flux proportional to the field at the orbit. The second type has some direct current in the field coils but only alternating current in the flux coils[1]. This biases the field and allows it to go from a small value to a large value while the flux goes from a negative value to a positive value. This allows double the energy of a conventional betatron for the same size magnet. The wave form for this "field biased" type is shown in Fig. 4. The third type is a flux biased type which must be unidirectionally pulsed. The central iron core is biased with direct current, Fig. 5. It is much easier to bias a closed iron core than it is to bias a field magnet.

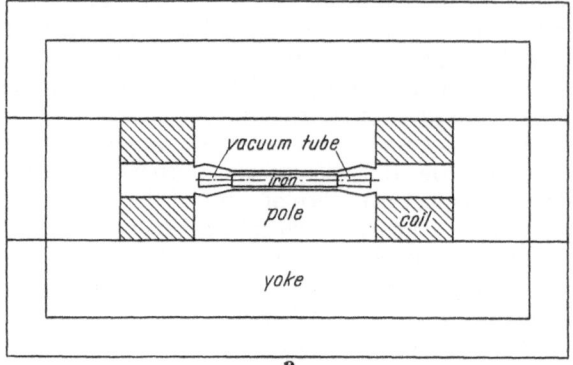

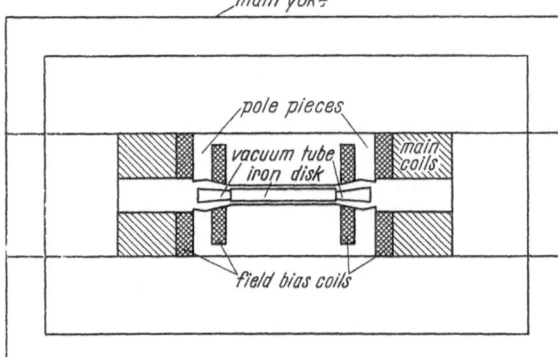

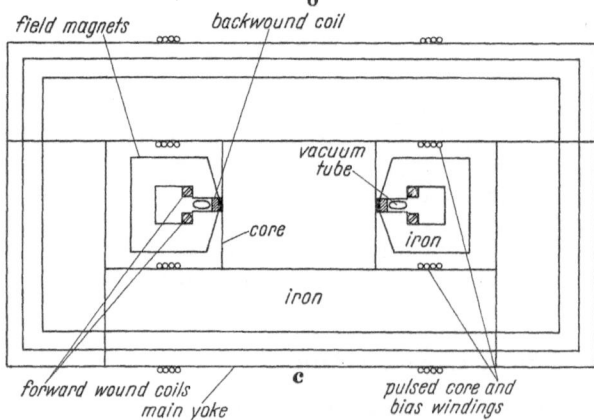

Fig. 3 a—c. Types of Betatrons. (a) Conventional betatron with air gap. (b) Field biased betatron with biasing coils and air gap. (c) Flux biased betatron.

8. Beam extraction processes. Several methods of extracting beams from betatrons will be mentioned. A simple magnetic channel, placed at a radius greater than the radius of the beam, into which the orbit can be expanded has been used quite successfully[2]. The channel is laminated, and it reduces the field in its interior so that particles thrown into the slot follow a nearly straight path and emerge from the fringing flux (Fig. 6). These so-called peelers appear to be between 20 to 50% efficient in extracting the beam, but their presence in the vacuum chamber is a hindrance to operation, for it cuts down the number of particles accepted in injection by more than a factor of ten. Accelerators giving a time average electron current of 10^{-7} amperes without this magneto-static peeler can give 2×10^{-9} amperes in a spot focused to a millimeter four meters away with a magneto-static peeler for extracting the

[1] W. F. Westendorp: J. Appl. Phys. **16**, 657 (1945).
[2] L. S. Skaggs, G. M. Almy, D. W. Kerst and L. H. Lanzl: Phys. Rev. **70**, 95 (1946).

beam. Although the current is low from this peeler, the system is capable of providing the pulse about 300 microseconds long by slowly expanding the orbit into it. This is sufficiently long for convenient counter measurements.

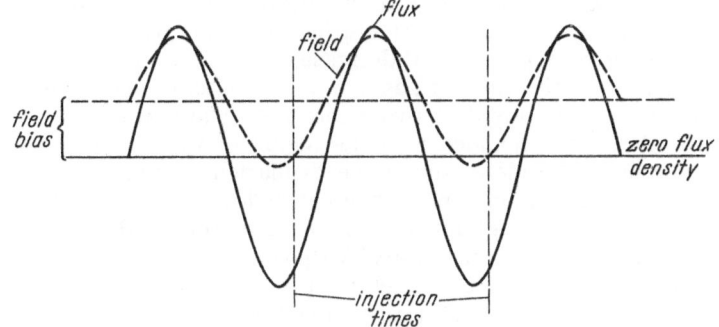

Fig. 4. Field and flux wave form in a field biased betatron.

Another extraction system developed by L. SKAGGS, LANZL, ROBINSON, and FOOTE, and applied by FOOTE and KOCH, was a pulsed system of coils producing

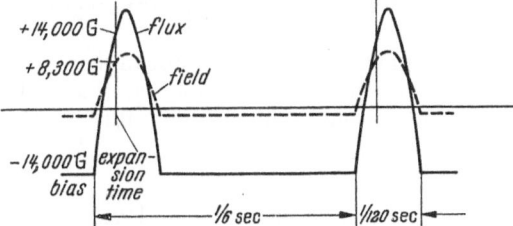

Fig. 5. Field and flux wave form in a flux biased betatron.

a magnetic field similar to that of an iron peeler. The virtue of the pulse system is that high electron currents are possible, since the peeler does not influence

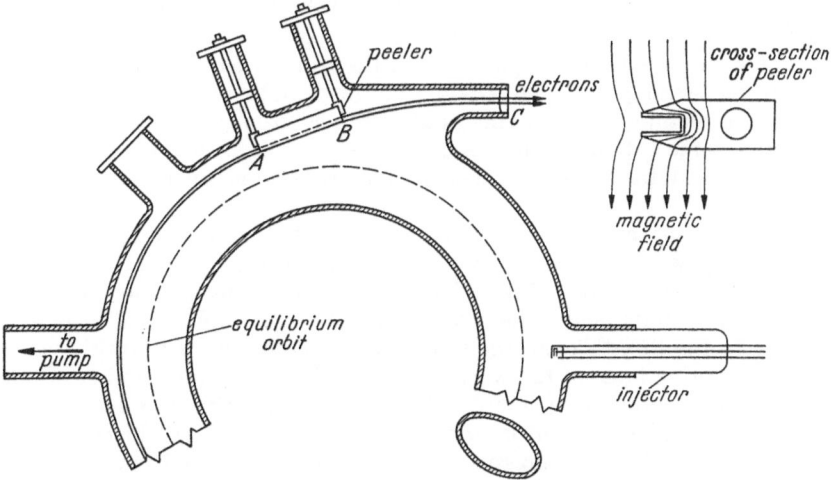

Fig. 6. Arrangement for extracting a beam by a magnetostatic peeler.

orbits until it is turned on after acceleration. Well shaped beams are produced, but one difficulty is that the time duration of the beam is necessarily short.

An unplanned aid to extraction by the old peeler method comes from an instability studied and further exploited by HAMMER and BUREAU[1]. Since the peelers had been used on accelerators with n near to $^3/_4$, a half betatron wave length is equal to the circumference of the orbit. This means that a variation in n at one azmuth in the machine can either cause betatron oscillations to exponentially grow or to exponentially decrease, depending upon the phase of the betatron oscillation at the time of passage of the inhomogeneity in n. This instability provides a very rapid stepwise increase in amplitude in the betatron oscillations as they are expanded into the region of the magneto-static peeler. Thus rapid rise in amplitude of successive oscillations is important in throwing particles well into the channel where they can follow a well shielded path without having had to pass through much of the fringing flux of the peeler. HAMMER and BUREAU generated this instability with coils and demonstrated that the beam could be thrown out of the accelerator in about three oscillations.

Probably the best way to extract a beam will be to use the type of coils which HAMMER and BUREAU employ to throw a particle into a magnetic channel such as a magneto-static peeler, placed at a radius great enough so that the magnetic channel will not inhibit capture of particles at the injection time. In addition, the arrangement should besuch that the coils would produce the variation of n only at a large radius so a slow expansion into this region should allow one to produce an extracted beam of long time duration.

Another successful beam extraction mechanism has been used by GUND[2] in which a thin foil scatters the electrons and an electrostatic deflector pulls the scattered electrons out through the fringing flux of the magnet. This is very succesful in extracting a large proportion of the beam which is expanded into contact with the foil.

9. Magnetic transients. When a magnetic field varies rapidly with time, as it does in an iron magnet driven with a 180 cycle per second or 60 cycle per second wave form, there are important magnetic effects which disturb the orbit other than the ordinary saturation effects which occur at high magnetic field strength in the case of direct current magnets. These disturbing magnetic fields are caused by the fact that the magnetic flux through the iron pole lags slightly in phase behind the alternating current in the coils. The result of this is that when the magnetic field in the gap, due to the iron, is passing through zero, the current in the driving coils already has a finite magnitude and therefore an accompanying magnetic field. This magnetic field from the coil penetrates into the air gap to some extent and is present as a distortion in the field gradient when the magnetic field from the iron has risen to the point where injection should occur.

The coil current can be considered as composed of two components: 1. a component in phase with the magnetic field in the iron and therefore 90° out of phase with the induced voltage. This is a wattless current and merely stores energy in the magnetic field. The second component is in phase with the induced voltage due to the flux through the iron and, consequently, is a working current. The magnetomotive force of the working current has the same number of ampere turns as the magnetomotive force of a disturbed effective loss current in the iron. This effective loss current is composed of two parts: first, the eddy currents which are largest when the magnetic field is passing through zero, and which run mainly around the edges of laminations, and the ampere turns equivalent

[1] C. E. HAMMER and A. S. BUREAU: Rev. Sci. Instrum. **26**, 594, 598 (1955).
[2] K. GUND and H. BERGER: Strahlentherapie **92**, 489 (1953).

to the hysteresis phase shift in the iron, that is, ampere turns equivalent to the coercive force distributed on the surface of the iron. The flux plot of the magnet can be considered as composed of two flux plots 90° out of phase, corresponding to the components of current. The wattless flux plot is the one which would exist with the magnetomotive force surfaces originating on currents in the coil at one side of the pole and extending through the air gap to terminate at the oppositely directed positions of the same current at the other side of the pole. The loss current— working current flux plot is composed of magnetomotive force surfaces which originate on the working current within the coil around the pole's edge and terminate on the equal and oppositely directed loss current distributed on the outside magnet surface. The flux plot, due to this combination of currents, is completely different from the wattless flux plot upon which it is imposed.

Fig. 7 exhibits this simple point of view, and in the diagram shown it is evident that at injection time a position of zero field exists near the pole rim which produces a high negative gradient. In some cases, this disturbance does not penetrate sufficiently to seriously influence the orbits, but if the frequency of alternation of the magnetic field is raised, or if injection at much lower energies is attempted, the field distortion at injection could make the beam unstable.

In case the magnetic field is pulsed, and the current suddenly starts to rise from zero, the the hysteresis portion of the loss current appears in the form of remnant magnetic field. The magnitude of remnant magnetic field is given roughly by

$$B_R = \frac{L}{G} H_c$$

where L is the effective length of the flux path in the iron, G, the magnitude of the air gap, and H_c is the coercive force of the iron for the magnetic cycle

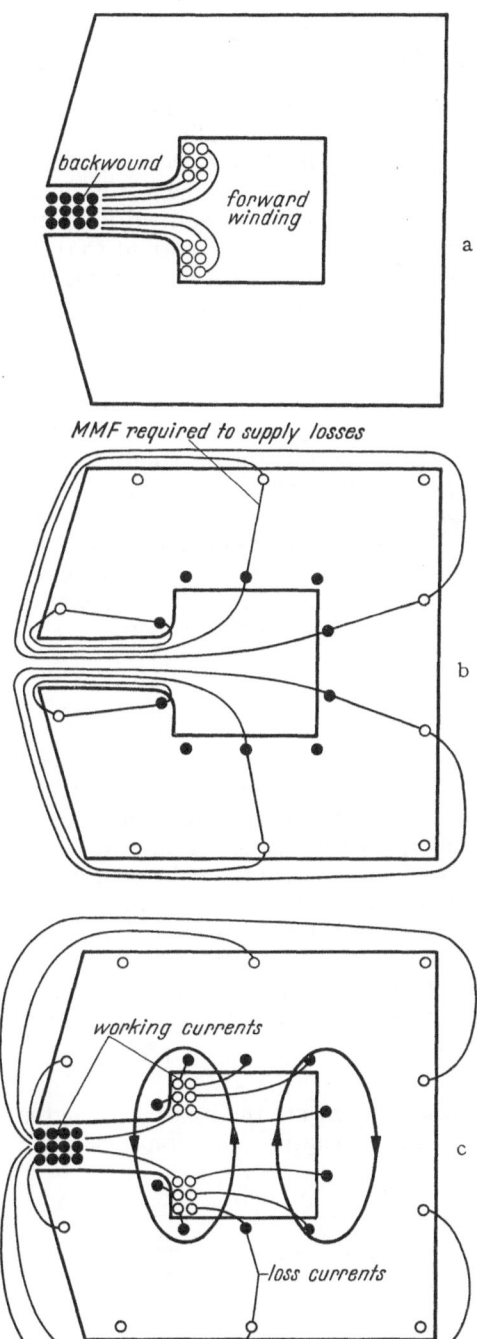

Fig. 7 a—c. Flux plot due to working currents and loss currents. (a) Working current plot. (b) Loss current plot. (c) Combination plot near injection time showing reversal of field just outside pole rim.

employed. Coercive forces are of the order of 0.2 Oersted, giving of the order of two gausses of residual field in typical magnets. However, for pulsed magnetic fields, cases have been found in which the residual magnetic field in the gap is less when the magnet is pulsed to high amplitude than it is when the magnet is pulsed to a lesser amplitude.

In the pulsed case, the eddy currents must be established in the iron before it appears to be ferro-magnetic, during the period of the establishment of eddy currents equal and opposite driving ampere turns in the coil must rapidly rise too. The time for this eddy current to rush into the coil or for the penetration and the establishment of the eddy current has been calculated for purposes of design of pulse transformers[1].

$$t = \frac{(\pi \mu d)^2}{3 \times 10^9 \varrho}$$

where t is the time for the eddy currents to reach 95% of their value, ϱ, the resistivity of the iron in ohm centimeters, d, the lamination thickness in centimeters, and μ is the differential permeability in iron. Heavily biased iron starts more rapidly when the pulse of current is applied because of its low differential permeability. In the Illinois 300 Mev betatron, the center biased accelerating core requires only about ten microseconds to establish a good equilibrium orbit in the gap of the magnet. In the case of focusing magnets with an air gap, there is a time of establishment of eddy currents in the iron which can be short because of the presence of the air gap.

This is

$$T = \mathscr{L} \frac{d^2 L}{4 \varrho A}$$

where $\mathscr{L}$ is the power line inductance plus leakage inductance, d is the lamination width, L is the iron flux path length, A, the air gap area, and ϱ is the resistivity.

If the magnetic circuit feeding flux at various azmuths around the orbit not perfectly symmetrical, and if some of the iron flux paths are longer than others, then when the pulse comes on, the long iron paths will be slower in getting a start than the short iron paths. This will produce a magnetic field bump which will disturb the orbit, and the bump has its largest effect at injection time because it amounts to a large fractional error only at that time. The field bump, $\varDelta B$, which is produced, is the difference between the field in the two paths, and it is approximately

$$\varDelta B = \dot{B} \tau \left(1 - \frac{L_1 A_2}{L_2 A_1}\right) (1 - e^{-t/\tau})$$

where L_1, L_2 are the iron path lengths, and A_1 and A_2 are the iron areas for the two path lengths. The time constant τ is the time for the field error to build up,

$$\tau = \frac{\pi d^2 L}{\varrho G \cdot 10^9}$$

where G is the air gap. The field error reached for $d = 0.014$ inches is

$$\varDelta B = \dot{B} \tau F = 5 \times 10^{-8} \dot{B} F L/G$$

where F is the fractional difference in iron reluctance.

If a section of magnet must have a long iron return path with a resultant field error at the orbit due to its lagging flux, or if some section of the magnet

---

[1] See, for example, H. L. REHKOPF: The Equivalent Circuit of a Pulse Transformer Core Radiation Laboratory, Massachusetts Institute of Technology, Report 666.

has a larger phase shift than another because of higher losses, the various portions of the magnet can be brought together in time by the use of correcting coils, either distributed on the pole faces and supplying a small bias flux or distributed over the iron path, supplying the needed ampere turns to advance the driving magnetic field on that portion of the iron circuit. Such corrector coils have been very important in the operation of iron core accelerators.

In order to establish the relation (3.8) required for a fixed equilibrium orbit, when magnetic structures are used for supplying different portions of the field or flux, two simple ways have been used. The first method was to let the reluctance of the gap in all magnetic circuits cause the magnetic field and the flux to retain the required proportionality as the current rises. If some portions of the magnetic flux carrying circuit have no air gap (Fig. 3 c), then a method of flux forcing coils can be used to eliminate the effect of the varying reluctance of the closed iron path. Flux forcing coils are merely coils connected in parallel and of the right number of turns so that the flux which they are supposed to inclose produces equal voltage. When the coils can be connected in parallel, there eas to be the proper flux in the various parts of the magnetic circuits affected (Fig. 8). The only error in this flux forcing arrangementis due to the fact that the resistive voltage loss in the coils which are connected in parallel may not be the same fore ach coil. Consequently, the fluxes are not quite forced to equality.

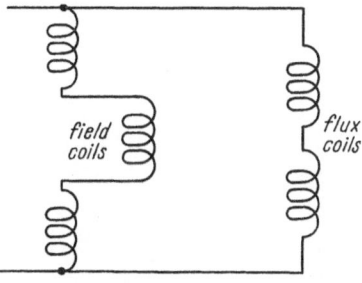

Fig. 8. Flux forcing coils. The parallel connection insures proportionality of flux and field except for resistive voltage drops.

Not only does the iron of the magnet have to be laminated for the reduction of eddy currents but also the copper used in the coil must be subdivided to a size which does not produce troublesome heating or troublesome eddy current magnetic fields. The ratio of eddy current loss to circuit loss for a conductor in a fringe magnetic field can be estimated by the formula

$$B_0^2 f^2 W^2/(24 \times 10^{16} \varrho^2 \xi^2)$$

where B_0 is the magnetic field in which the wire is situated, f is the circular frequency of the magnetic field, W is the width of the copper bar in centimeters, ϱ is the resistivity in ohms centimeters, and ξ is the current density in amperes per cm². Copper bars, close to the orbit and 0.23 cm thick, in a direction perpendicular to the 10000 gauss, 60 cycle per second wave form magnetic field in the 300 Mev betatron have given no trouble. Bigger copper bars have been used on coils in field biased magnets in which the magnetic field starts off from the through of a biased sine wave with a very low rate of rise and, consequently, with very small disturbing eddy currents. This is the method employed by R. R. WILSON for the Cornell University electron synchrotron.

10. The vacuum chamber. α) *Vacuum vessels.* Pyrex glass, plate glass, and porcelains have been used for the vacuum envelopes of betatrons, since they are all insulators and produce no eddy current effects. However, to avoid surface charges on the interior of the chamber, a conducting coating sufficient to carry off wall charges but insufficient to conduct seriously large eddy currents is needed. An accelerating tube will be described of the type developed at the University of Illinois, which has been capable of manufacture in quantities sufficient for commercial betatrons. The vacuum envelope was developed by

Professor R. K. HURSH[1], of the Ceramics Engineering Department at the University of Illinois. The general development of the accelerating tube was carried out by Professor G. M. ALMY, of the Physics Department of the University of Illinois. After considerable experimentation, the ceramic body chosen for slip casting of the porcelain envelope was composed of 15% English ball clay, 40% English

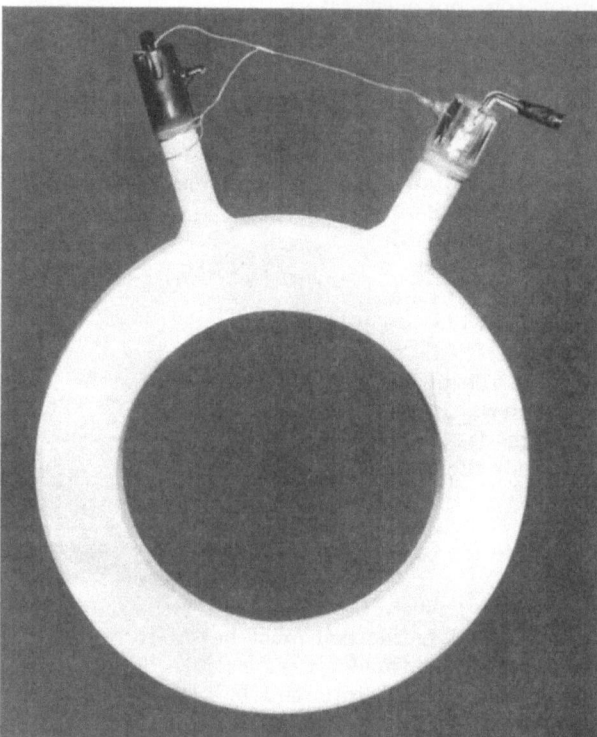

china clay, 25% feldspar, and 20% flint. This choice of composition gave thermal expansion properties close to that of Corning glass No. 705, which is sealed directly onto the finished porcelain envelope. In addition, this body was capable of being heated several times to temperatures of 400° to 700° C for purposes of glazing, glass sealing, interior coating, and vacuum bake-out. In addition, the body must not sag and distort seriously during its initial firing, and it must be vacuum tight. The body is fired to cone 9 or 10. The glaze is sprayed onto the outside of the tube and fired at cone 10. The molar composition chosen for the glaze was 3.50 SiO_2, 0.20 K_2O, 0.40 CaO, 0.30 ZnO, 0.10 MgO and 0.35 Al_2O_3, 0.20 B_2O_3. The 705 glass was sealed on over the glaze, and injectors or pumping tubes can connected. A barium getter is usually used at the time of seal-off for

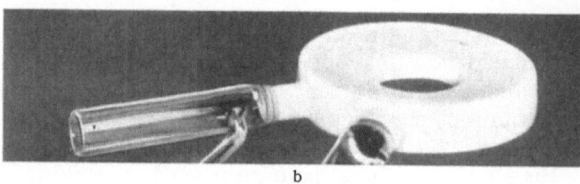

Fig. 9 a and b. 4.5 and 24 Mev betatron porcelain vacuum tubes.

a permanently evacuated tube. These tubes appear to be as good as glass tubes for permanently sealed off applications (Fig. 9).

β) *Wall coating.* The conducting coating on the wall has a resistance which is kept below 75 ohm on a square. The reason for this is that injected beams of the order of a tenth of an ampere to an ampere striking the wall will produce a potential where the wall is struck, depending on the magnitude of resistance of the wall. We know from experience that potentials of the order of 200 volts on electrodes inside the vacuum chamber cause important disturbances in the orbit and yield. Consequently, this source of resistance was thought sufficiently

[1] R. K. HURSH: J. Amer. Ceram. Soc. **432**, 5 (1949).

low to be safe. If the resistance is less than 25 ohm, heating of the vacuum tube due to currents circulating around the whole circumference is objectionable, and the circumferential path must be cut to form an accelerating gap. In the 300 Mev betatron, there are four accelerating gaps in the circumference, and the wall coating has a resistance of 10 ohm per square.

Once when deterioration of the conducting coating on the 300 Mev betatron occured, the out-put of the accelerator dropped to two percent of its normal value, due to charges accumulating, and causing an electrostatic field bump. This was completely compensated empirically by a magnetic field bump, and the location of the required magnetic field indicated exactly where the coating had deteriorated. Repair of the conducting surface which had deteriorated immediately restored the X-ray out-put to its full value without the necessity for further magnetic correction. There are, however, indications that some synchrotrons or betatrons with coating resistances of the order of 1000 ohm are not as sensitive to disturbances of uncoated areas. It seems there may be resistive potential drops which would assist in keeping the beam from striking the bare spot sufficiently to prevent the bare spot from acquiring an exceedingly high voltage. Some of these accelerators have somewhat different injection characteristics. For example, the beam cannot be captured until the injected beam is being turned off. As has been previously mentioned, an injection beam crossing the tube and striking a high resistance coat could produce a disturbance which would disappear when the injected beam is turned off, allowing particles to be captured while the injection voltage is going to zero. This behavior has not been observed in accelerators with low resistance coatings.

The method of coating the interior of unglazed porcelain vessels is to use Hanovia Liquid Bright Palladium No. 62, which is diluted with acetone to a specific gravity of 0.97 to 1.0. This solution is shaken inside of the vacuum tube, drained off, and rotated slowly on a machine, with a slight draft of air passing through the tube. This rotation prevents the wet coating from draining to one side, and in two or three hours the rotation can be stopped, and the coating material is sufficiently dry to stay in place. After two or three days of further drying on the shelf, it is brought up to 500° C in a furnace at about 6° per minute, with a small stream of air circulating through the doughnut. It is held at the peak temperature for about 20 min, and then the furnace is shut off. This produces a palladium coat of the proper resistance, which is quite satisfactory and which can have burnished silver paste baked onto it for connections. The only aspect of this coat which has caused trouble is that any high voltage spark or discharge which may strike this coating destroys a small patch of it. Under proper conditions, this deterioration does not occur.

γ) Injectors. Fig. 10 shows typical injector structures which emit a rather large spray of electrons and consequently do not need critical alignment. More than 95% of the emitted electrons pass out into the vacuum vessel, and the injectors for the 20 Mev betatron are usually run at 100 to 200 milliamperes emission and have a life-time of 100 to 500 hours. The commercial version of this injector has a thoriated tungsten filament, installed by the Macklett Company and it has a much longer life-time. In about 1951, the same company sealed into their vacuum doughnuts the regenerative peelers so that X-ray beams or electron beams could be obtained with the sealed off commercial tubes used in the betatrons manufactured by the Allis-Chalmers Company. The 20 Mev injectors can be run up to 60 or 75 kilovolt for approximately two microsecond pulses of 100 to 200 ma. The large injectors for the 300 Mev betatron can be pulsed to

150 kilovolt and are used with emission currents generally of the order of 250 milliamperes, although these injectors have been shown capable of injecting 10 amperes into a vacuum tube when a slightly larger filament is used.

δ) *The fixed field alternating gradient betatron.* The motion of the instantaneous equilibrium orbit for a betatron with direct current magnets has been described,

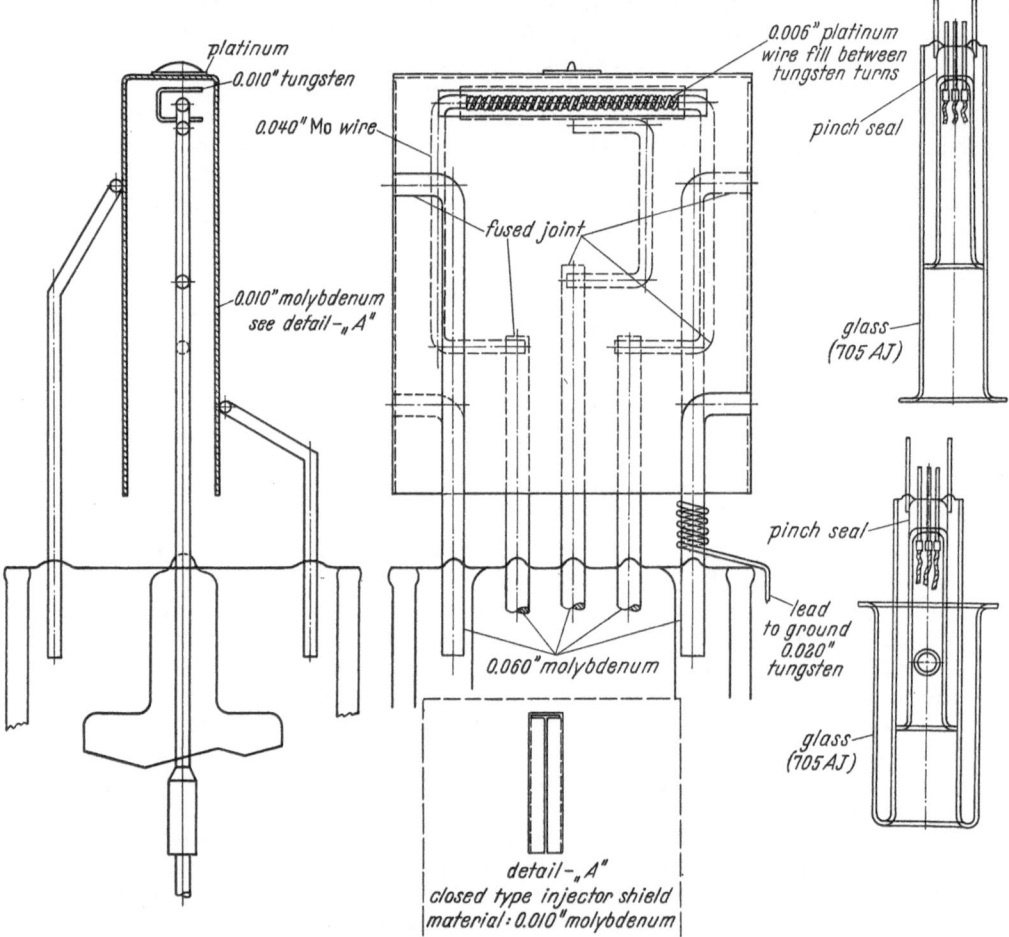

Fig. 10. A typical injector for a 24 Mev betatron.

and the Eqs. (3.11) and (4.9) relate to it. The only time varying flux is in a central core. If this core operates on a sine wave voltage, we have Fig. 11, showing the long injection time possible and the long period of emission of X-rays or electrons after acceleration. If there is enough iron in the core, the portion of the time when yield is emitted can be approximately 25%. Consequently, the out-put of this type of betatron occurs for a greater portion of time than does the out-put from a self-rectifying X-ray tube. Since the injection time can be milliseconds per cycle instead of 10^{-7} sec per cycle, we see that this type of betatron should give yields many times greater than the yield from A.C. field betatrons. 24 Mev betatrons have 100 to 200 r/mm of X-rays at one meter or about two watts of electron beam. The FFAG type should be able to produce kilowatts. A small model has been constructed which verifies the production of X-rays for about 25% of the time[1].

To produce focusing in this accelerator we use the alternating gradient focusing principle[2, 3]. The reason for this is that we want the average magnetic field, $\bar{B}_z$, to increase like r^k with $|k| \gg 1$ so that both low energy and high energy orbits can be held within a small annular ring magnet excited by a constant current. But such a large value of k would not conform to our limits, $-1 < k < 0$ for an azimuthally uniform field. The alternating gradient principle allows large k by requiring the gradient of magnetic field to reverse periodically around the orbit. This magnetic focusing system is analogous to an optical system with a sequence of focusing and defocusing strong lenses which can be focusing in its final effect.

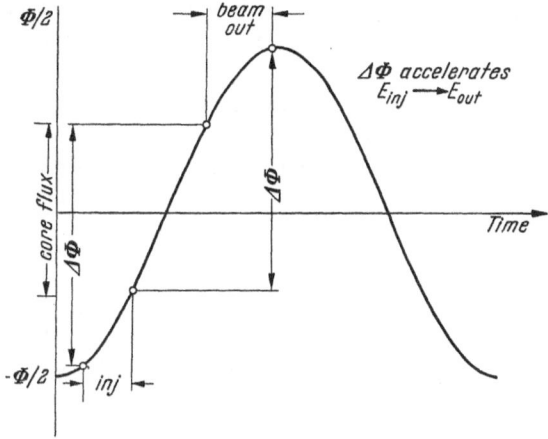

Fig. 11. Time dependence of FFAG betatron flux, injection, and yield. The duty factor is $D = (1/2\pi) \arccos [\Delta\Phi/\Phi_0 - 1]$.

The general theory and description of FFAG accelerators appears elsewhere[3], but an approximate method of treating the problem will be given here. There are two characteristic types of focusing magnets: the radial sector type (Fig. 12), and the spiral sector type (Fig. 13). These have a field fixed in time, and they surround the accelerating flux core.

Consider focusing by radial sector magnets in which the orbit passes through fields which reverse in direction. (We omit the straight section of zero field for simplicity.) We define:

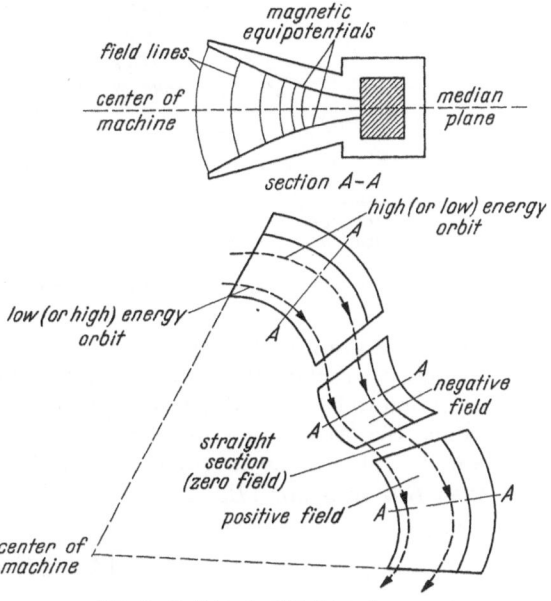

Fig. 12. Radial sector FFAG focusing magnets.

ν_r = the number of radial betatron oscillations around the orbit,

ν_z = the number of axial betatron oscillations around the orbit,

N = the number of sectors in the accelerator. A sector is composed of a positive and a negative field magnet.

[1] Jones, Terwillinger and Haxby: Rev. Sci. Instrum. 27, 651 (1956).
[2] E. D. Courant, M. S. Livingston and H. S. Snyder: Phys. Rev. 88, 1190 (1952).
[3] K. R. Symon, D. W. Kerst, L. W. Jones, L. J. Laslett and K. M. Terwillinger: Phys. Rev. 103 1837 (1956).

$k = \dfrac{r}{\bar{B}_z}\dfrac{d\bar{B}_z}{dr}$ where $\bar{B}_z$ is the average field along the orbit,

$C = B_{z\,\mathrm{maximum}}/\bar{B}_z$, the circumference factor,

q = fraction of path in a sector with one polarity of field; thus $(1-q)$ is the fractional path length in the other polarity,

f = the flutter factor in $B_z(N\vartheta) = B_0[1 + fg(N\vartheta)]$ following the equilibrium orbit,

$g(N\vartheta)$ = a function having zero mean value and normalized so that its mean square is $\frac{1}{2}$.

The description of the field variation along the orbit by the factors g and f is necessary for the following formulas. Then the radial and axial focusing wave numbers are:

$$\nu_r^2 = k + 1 + \pi^2(k+1)^2 f^2 q(1-q)/(6N^2),$$

$$\nu_z^2 = -k + f^2/2 + \pi^2(k+1)^2 f^2 q(1-q)/(6N^2)$$

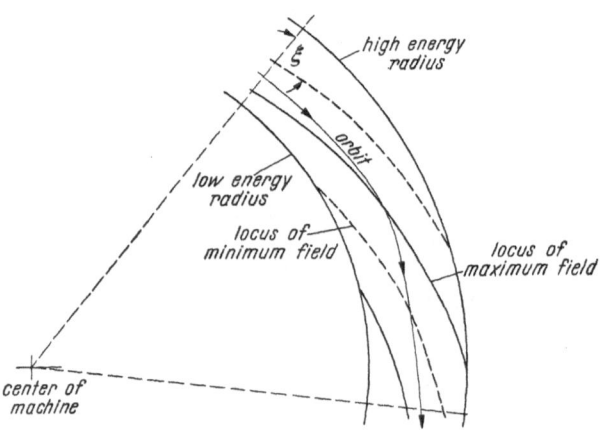

Fig. 13. FFAG spiral sector focusing field distribution.

according to Symon's smooth approximation. For stability of motion it is fundamentally necessary that both $2\nu_r$ and $2\nu_z$ be less than N, and, for the smooth approximation to be good, it is best that they be less than $N/2$. We must adjust f and q along the orbit to get the ν_r and ν_z we want. Usually k is chosen to give the desired width of the ring-magnet. If the flutter f is very large compared to 1, then $\bar{B}_z \ll$ maximum B_z. This leads to a very large circumference factor.

We have

$$C = \sqrt{1 + \tfrac{1}{2}f^2}.$$

It is possible to have $C \sim 5$ or larger. This means that the machine is five times the radius which it would have if the field were not reversed along part of the path. In machines with a small number of sectors $C < 5$ is possible because the orbit leaves the magnet edges at an angle sufficient to provide extra axial focusing.

It is possible for radial sector accelerators to have $k < 0$ and $|k| \gg 1$.

Now consider the spiral sector case. We can have field variations indicated in Fig. 13, or we can have separated spiral sectors with spaces between. In either case, if

$$B_z(N, \vartheta, r) = B_0 (r/r_0)^k (1 + f \sin [N \vartheta - (r - r_a)/W])$$

on the orbital plane, then

$$\nu_r^2 = 1 + k,$$

$$\nu_z^2 = - k + f^2/2 + f^2/(N^2 W^2).$$

Also $1/(N W) = \tan \xi$, ξ is the angle between the spiral edge and a radius.

Again $2\nu < N$ for stability, and $4\nu \lesssim N$ for accuracy of this approximation.

When ν is an integer, it is possible to have a resonant build up of oscillation due to a field imperfection. When ν is a number of half integers, field gradient imperfections excite resonances. When ν is a number of $\frac{1}{3}$ integers, quadratic non-linearities in the magnetic field can cause instability. Sometimes $\frac{1}{4}$ integers are also bad if there are cubic non-linearities. There are additional coupling resonances between axial and radial oscillations which must be avoided.

The difficulties of making the focusing field for an FFAG betatron are great but the accelerator has great simplicity of operation and has the very attractive high duty factor.

The Proton Synchrotron.

By

G. K. GREEN and ERNEST D. COURANT[1].

With 98 Figures.

A. Introduction.

1. Proton synchrotron has become the generic name for magnetic particle accelerators which produce proton beams in the Bev energy range. Originally the proton synchrotron was distinguishable from other particle accelerators by its pulsed ring magnet and its swept accelerating radio-frequency. However, electron synchrotrons have been developed in forms essentially identical to those of the proton synchrotron. The only major difference is the problem of radiative loss which is still unimportant in circular proton accelerators, but it is now limiting the size of circular electron types. The ring magnet remains but it is not necessarily pulsed, since the advent of the fixed-field alternating-gradient concept. There is no reason, in principle, that prevents the "proton" synchrotron from accelerating deuterons, alpha-particles, or heavier ions. A straightforward re-arrangement of electrical equipment would permit acceleration of such heavier particle, but the research requirements of particle physics want the highest energy per bombarding nucleon and the simplest collision situation. Consequently, there has been little demand for acceleration of ions other than protons to energies above 1 Bev.

Three general forms are described in this chapter. The original one is the constant-gradient, or low-n, synchrotron of which four are operating and at least five are under construction. Extension of energy made the constant-gradient magnets appear too large and expensive, so the alternating-gradient principle was developed. Three alternating-gradient proton synchrotrons are being constructed. The fixed-field alternating gradient (FFAG) synchrotron has been explored theoretically in extensive detail. Working models have been operated but no full scale machine has been authorized (1957).

2. Operating principles. An annular region of magnetic field, whose shape is focussing over the required aperture, is provided. (The term "focussing" is used with the connotation of stability, rather than that of image-formation.) In pulsed machines the magnetic field strength begins near zero and is raised rapidly. At the proper instant a pulse of protons at moderate energy is emitted by the pre-accelerator (Fig. 1) and deflected into stable orbits by the inflector fields. The protons then circulate inside the main vacuum chamber and are accelerated each revolution by radio-frequency fields which must be synchronized with the particle rotation frequency. At a desired energy the protons impinge on a target, or are perhaps ejected from the ring. The general form of field, frequency and proton kinetic energy is plotted in Fig. 1. Although conventional usage defines the "pulse length" as the length of time for the field to rise, nearly an equal time is required to discharge the magnet. After this the equipment is

---

[1] Parts A, B and D by G. K. GREEN, Part C by E. D. COURANT.

automatically prepared for the next accelerating cycle. Acceleration is done at approximately constant orbit radius, and only particles within a narrow range of energy are stable at any given time.

A fixed-field synchrotron would have a ring of field with somewhat greater radial extent, and the field would be so shaped that particles over a wide range of momentum would have stable orbits within the aperture. It is possible to arrange field shapes such that the high-energy orbits lie either at the maximum or at the minimum radius. Particles injected on the low-energy orbits are accelerated by varying radio-frequency. Since the magnetic guide field is not

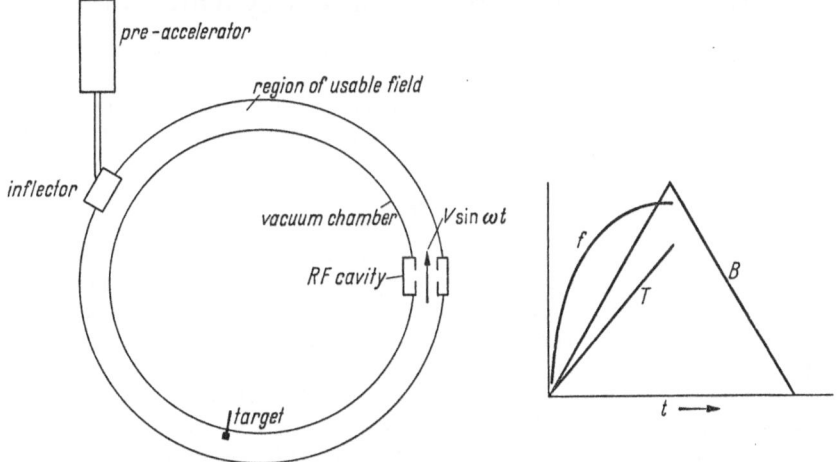

Fig. 1. Schematic plan view of proton synchrotron. General nature of the increase of frequency, magnetic field and energy (f, B, T) is indicated as a function of time.

time-varying the acceleration may be stopped and re-started at will, and numerous different bunches of particles, at different energies, can be contained and accelerated simultaneously.

The principle of phase stability is fundamental in all synchrotron operation since it requires that, within limits, particles must remain synchronous with the accelerating radio-frequency.

3. Installations. A proton synchrotron was the first accelerator to exceed an energy of 1 Bev (10^9 ev). The Brookhaven National Laboratory (USA) *Cosmotron* operated at 2.3 Bev in 1952 (and later at 3 Bev); the University of Birmingham (Great Britain) Proton Synchrotron at 1 Bev in 1953; the University of California (USA) *Bevatron* at 6.2 Bev in 1954; and the Institute of Nuclear Research (USSR) *Synchrophasotron* at 10 Bev in 1957. A 10 Bev Air-Core Proton Synchrotron is scheduled for completion in 1958 at the Australian National University, and a 3 Bev Synchrotron à Protons is expected to operate at Centre d'Études nucléaires, Sacley (France) in 1958. Construction is expected to begin in 1957-8 on a 2 to 3 Bev rapid-cycling type at Princeton University (USA); a 12.5 Bev wedge-focusing type at Argonne National Laboratory (USA); and a 7 Bev machine at AERE, Harwell (Great Britain).

Three alternating-gradient proton synchrotrons are under construction. The 100 meter radius installation at CERN, Geneva (Switzerland) should exceed 25 Bev, and the 421 foot (138 m) radius AGS at Brookhaven National Laboratory (USA) should exceed 30 Bev. A 40 meter radius version for 6 to 7 Bev being built by the Academy of Sciences, USSR, will use inverse field magnet elements to eliminate the phase transition. The Academy of Sciences, USSR, also contemplates building a larger size, perhaps 50 to 60 Bev.

If one considers the complexity and cost of proton synchrotrons, the foregoing is a rather impressive list. Numerous additional machines will no doubt be built, but it is questionable whether much extension of size and energy is practicable with the ferrous-core design. Several interesting possibilities for obtaining greater energy have been suggested. For examples, see Kerst ([5], p. 36), Budker [5], p. 68) and Veksler ([5], p. 80).

B. The constant gradient proton synchrotron.

I. Particle orbits.

4. Equations of motion. The designation "constant-gradient" can be rigorously applied only to magnetic fields which are independent of azimuth. In cylindrical coordinates r, ϑ, z the field takes the form

$$\boldsymbol{B} = \boldsymbol{B}(r, z)$$

and the field gradient is not a function of angle ϑ. However, it is common practice to insert field-free spaces or "straight-sections" into the field such that the gradient dB_z/dr is either $-nB_z/r$ or zero [cf. Eq. (4.6)]. The circular machine will be examined first, and then corrections for straight sections noted in Sect. 7, since the particle motions are not essentially different. The term "weak-focussing" is often used, and is probably a better description than "constant-gradient". Both apply to accelerators in which the *momentum compaction factor* α is less than one. α is defined by $dp/p = \alpha\, dr/r$. Inversely, the "alternating-gradient" or "strong-focussing" accelerators have the momentum compaction greater than one, and usually operate with several free, or betatron, oscillations per revolution of the particles.

The equilibrium orbit of a particle, momentum p, in the field is defined as that orbit which closes on itself after one revolution. In an axially symmetrical field with a plane of symmetry perpendicular to the z-axis (the median plane) the equilibrium orbit is by definition a circle $z = 0$, $r = p/eB$. It can be proved [28] that an equilibrium orbit exists in the more general magnetic field. If the synchrotron is to operate it is necessary that the equilibrium orbit lie within the space bounded by the vacuum chamber, or by the extent of properly shaped magnetic field. A particle whose path deviates from the equilibrium orbit by a small displacement and angle must remain near the equilibrium orbit, and within the bounded space, for all subsequent time of the accelerating process.

If the field B is increasing with time the momentum of a proton must increase if its equilibrium orbit is to remain at constant, or approximately constant, radius. Apply a radiofrequency field across a narrow gap with peak potential V and angular velocity ω_r. The equations of motion of a proton of charge e and total mass m are[1] [9], [12]

$$\left. \begin{aligned} \frac{d}{dt}(m\dot{r}) &= mr\dot{\vartheta}^2 - er\dot{\vartheta}B_z, \\ \frac{d}{dt}\left(mr^2\dot{\vartheta} - \frac{e\Phi}{2\pi}\right) &= \frac{eV}{2\pi}\sin\left(\vartheta - \int\omega_r\,dt\right), \\ \frac{d}{dt}(m\dot{z}) &= er\dot{\vartheta}B_r. \end{aligned} \right\} \tag{4.1}$$

---

[1] An alternate formulation is given in the Appendix of [23].

For protons note that B_z must be oriented opposite to the angular velocity ω. In the second equation of (4.1) Φ is the total flux linking the orbit. The term in Φ is the "betatron acceleration" and is generally small compared to the energy gain per revolution so it will be dropped. The right side of the second equation represents the radiofrequency acceleration and only the synchronous case will be considered, namely, $\int \omega_r \, dt$ changes almost exactly 2π per revolution of the proton. If a proton continues to move on an equilibrium orbit, radius r_0, in a cylindrically symmetrical field the quasi-steady state solution of (4.1) is the familiar set of relations

$$B_0 = \frac{\sqrt{T^2 + 2\,T E_0}}{e\,r_0\,c} = \frac{\sqrt{E^2 - E_0^2}}{e\,r_0\,c}, \tag{4.2}$$

$$\omega_0 = \frac{e\,B_0}{m_0 \sqrt{\left(\frac{e\,r_0}{m_0 c}\right)^2 B_0^2 + 1}} = \frac{c}{r_0}\sqrt{1 - \frac{E_0^2}{E^2}}, \tag{4.3}$$

in which ω_0 and r_0 refer to the proton; E, T, and E_0 are the total, kinetic, and rest energy of the proton. The subscript 0 (excepting on E_0) will refer to the equilibrium orbit values. If r_0 is constant a combination of (4.2), (4.3) gives the required energy gain per turn

$$\Delta E = 2\pi e\,r_0^2\,\frac{dB_0}{dt} \tag{4.4}$$

and the eV of (4.1) will henceforth be assumed to exceed this value of ΔE.

For convenience[1] let the magnetic field at $z = 0$ vary with radius in the neighborhood of r_0 as

$$B_z = B_0 \left(\frac{r_0}{r}\right)^n \tag{4.5}$$

from which, differentiating, one obtains the field index n,

$$n = -\frac{r}{B_z}\,\frac{dB_z}{dr}. \tag{4.6}$$

Following the notation of [12], define the small amplitude motions about the steady-state solution by:

$$\left.\begin{aligned}
x &= r - r_0, \\
\dot\varphi &= \dot\vartheta - \omega_0, \\
\dot\psi &= \dot\vartheta - \omega_r, \\
z &= z, \\
\psi &= \vartheta - \textstyle\int \omega_r \, dt,
\end{aligned}\right\} \tag{4.7}$$

and (4.1) can be reduced to the first order equations

$$\frac{d}{dt}\left(\frac{B_0\,\dot x}{\omega_0}\right) = r_0\,\omega_0\,B_0\left[\gamma^2 - (1-n)\right]\left(\frac{x}{r_0} + \eta\,\frac{\dot\varphi}{\omega_0}\right), \tag{4.8}$$

$$\frac{d}{dt}\left[\frac{r_0}{\omega_0}\,\frac{d}{dt}\left(\frac{B_0\,\dot x}{\omega_0}\right) + (1-n)\,r_0\,B_0\,x\right] = \frac{V}{2\pi}\,(\sin\psi - \sin\psi_0) \tag{4.9}$$

with

$$\eta = \frac{\gamma^2}{\gamma^2 - (1-n)}; \qquad \gamma = \frac{e\,B_0}{m_0\,\omega_0} = \frac{m}{m_0}; \qquad \Delta E = eV\sin\psi_0.$$

---

[1] The power law field shape was introduced by Kerst and Serber: Phys. Rev. **60**, 53 (1941).

η is a slowly varying function of time, and the time variation of the right side of (4.9) is very slow with the usual parameters, so (4.9) becomes:

$$\frac{d}{dt}\left(\frac{B_0 \dot{x}}{\omega_0}\right) + (1-n)\,\omega_0\,B_0\,x = 0. \qquad (4.10)$$

The third equation of (4.1) becomes

$$\frac{d}{dt}\left(\frac{B_0}{\omega_0}\,\dot{z}\right) + n\,\omega_0\,B_0\,z = 0. \qquad (4.11)$$

These are the equations of betatron oscillation about the equilibrium orbit, with frequencies of the order of the frequency of revolution ω_0. In the approximation for the slow time variation of ψ, too slow to be influenced by the rapid betatron variation of $\dot{x}$, (4.8) and (4.9) become

$$\left.\begin{aligned}
\frac{x}{r_0} + \eta\,\frac{\dot{\varphi}}{\omega_0} &= 0, \\
(1-n)\,r_0^2\,\frac{d}{dt}\left(\frac{\eta\,B_0}{\omega_0}\,\dot{\varphi}\right) &= \frac{V}{2\pi}\,(\sin\psi_0 - \sin\psi),
\end{aligned}\right\} \qquad (4.12)$$

the equations for the phase oscillations. In the proton synchrotrons to be described, the betatron and phase oscillations differ in frequency by at least two orders of magnitude and, except for certain special circumstances, the small coupling between them will be disregarded. Using the solution of TWISS and FRANK [12], if a proton is initially displaced in r and $\dot{r}$ from the equilibrium orbit by amount x_i and $\dot{x}_i$ at B_i and ω_i its subsequent deviation from r_0 will be

$$x(t) = \left(\frac{B_i}{B_0}\right)^{\frac{1}{2}}\left[x_i \cos\left(\sqrt{1-n}\int\omega_0\,dt\right) + \frac{\dot{x}_i}{\sqrt{1-n}\,\omega_i}\sin\left(\sqrt{1-n}\int\omega_0\,dt\right)\right] \qquad (4.13)$$

and a like equation applies to a deviation $z(t)$ with $\sqrt{1-n}$ replaced by $\sqrt{n}$. If the motion is to be stable the arguments of cos and sin must be real, so one obtains the familiar KERST-SERBER stability requirement: $0 < n < 1$. The motion of the betatron oscillations is thus to a first order sinusoidal with frequencies conveniently expressed as

$$\omega_r = \omega_0\sqrt{1-n}, \qquad \omega_z = \omega_0\sqrt{n} \qquad (4.14)$$

and amplitude damped as $B_0^{-\frac{1}{2}}$. Although this is not true damping, in the usual sense of dissipation of oscillatory energy, the decrease of amplitude of oscillation with increasing B_0 is one of the most important factors in successful operation of a proton synchrotron.

5. Phase oscillations. If a proton does not cross the accelerating gap at the stable phase angle ψ_0 when the radiofrequency $eV\sin\psi$ equals the ΔE of (4.4), then it will execute phase oscillations about ψ_0. The equation of phase motion (4.12) involving ψ can be reduced to the form of a harmonic oscillator in the adiabatic approximation since $\eta\,B_0/\omega_0$ may be considered constant over a phase oscillation cycle. Setting

$$\eta = [1 - (1-n)\,E_0^2/E^2]^{-1},$$

$$\sin\psi_0 - \sin\psi \approx (\psi_0 - \psi)\cos\psi_0,$$

and using (4.2), (4.3), (4.12) reduces to

$$\ddot{\psi} + \frac{c^2 eV\cos\psi_0}{2\pi r_0^2 E}\left(\frac{1}{1-n} - \frac{E_0^2}{E^2}\right)(\psi - \psi_0) = 0 \qquad (5.1)$$

from which the small amplitude angular frequency of the phase oscillation about ψ_0 is (see also [10], [13])

$$\Omega = \frac{c}{r_0} \left[\frac{eV \cos \psi_0}{2\pi E} \left(\frac{1}{1-n} - \frac{E_0^2}{E^2} \right) \right]^{\frac{1}{2}}. \tag{5.2}$$

If ψ exceeds $\pi - \psi_0$ the "restoring force" of the "oscillator" is reversed and, since B, E and ω are increasing with time, the proton will lose phase stability beyond recovery. By setting $\eta B_0/\omega_0$ constant (4.12) can be integrated directly

$$\dot\psi^2 - \dot\psi_i^2 = \frac{2\Omega^2}{\cos \psi_0} (\cos \psi + \psi \sin \psi_0 - \cos \psi_i - \psi_i \sin \psi_0) \tag{5.3}$$

where subscript i indicates initial values. Substitution of the limiting value $\psi = \pi - \psi_0$ for ψ_i and $\dot\psi_i = 0$ gives the other angular limit of stability, ψ_1, as a root of

$$\cos \psi_1 + \psi_1 \sin \psi_0 = (\pi - \psi_0) \sin \psi_0 - \cos \psi_0. \tag{5.4}$$

A WKB approximation for (4.12), when considered in detail, shows that the phase oscillations are slightly undamped if $n < \frac{2}{3}$ in the non-relativistic starting period of the accelerating cycle [12], [13]. The relativistic damping of ψ is proportional to $B_0^{-\frac{1}{2}}$. Although some non-relativistic region damping is possible, in principle, by increasing the radiofrequency voltage with time, the utility of the principle is severely limited by radiofrequency power considerations. The very small total phase oscillation damping is a serious source of proton loss because small disturbances of the phase during acceleration can easily drive the protons beyond the limits of phase stability.

The limit of phase stability in ψ has corresponding limits of energy and radial deviation ΔE_m and Δr_m. Since a proton executing phase oscillations must deviate in frequency from the radiofrequency by $\Delta \omega_0$,

$$\frac{d\psi}{dt} = \Delta \omega_0 = \frac{\partial \omega_0}{\partial E} \Delta E$$

and $\partial E/\partial \omega_0$ obtained from (4.2), (4.3) and (4.5) gives

$$\Delta E = - \frac{r_0 \sqrt{E^2 - E_0^2}}{c} \left(\frac{1}{1-n} - \frac{E_0^2}{E^2} \right)^{-1} \dot\psi. \tag{5.5}$$

The limit $\dot\psi_m$ can be obtained from (5.3), by setting $\psi = \psi_0$, $\psi_i = \pi - \psi_0$ and $\dot\psi_i = 0$, and combined with (5.2) to give the maximum stable energy excursion

$$\Delta E_m = \left\{ \frac{eV}{\pi} \frac{E^2 - E_0^2}{E \left(\frac{1}{1-n} - \frac{E_0^2}{E^2} \right)} [2 \cos \psi_0 + (2\psi_0 - \pi) \sin \psi_0] \right\}^{\frac{1}{2}}. \tag{5.6}$$

Since

$$\frac{dr}{dE} = \frac{rE}{(1-n)(E^2 - E_0^2)} \tag{5.7}$$

the limiting radial phase oscillation becomes

$$\Delta r_m = r_0 \left[\frac{eV}{\pi} \frac{2 \cos \psi_0 + (2\psi_0 - \pi) \sin \psi_0}{(1-n) E \left(1 - \frac{E_0^2}{E^2} \right) \left[1 - (1-n) \frac{E_0^2}{E^2} \right]} \right]^{\frac{1}{2}}. \tag{5.8}$$

The radial phase oscillations [13] are damped as $T^{-\frac{1}{2}}$ during the non-relativistic part of the acceleration cycle, and as $E^{-\frac{3}{4}}$ during the relativistic part. Considerable radial aperture must be allowed for the Δr_m of injected particles, but the radial phase oscillation damps to a very small amplitude at high energy. This latter characteristic is useful when it is required to strike a target or to eject the proton beam.

6. Disturbances. A physical synchrotron cannot be constructed with ideal symmetry. The long wavelength of the betatron oscillations (greater than the circumference) indicates that the restoring forces are very weak and that relatively small disturbances in the magnetic field can well produce lateral motions larger than a reasonable magnet and vacuum chamber size. Eqs. (4.10) and (4.11) can be written (disregarding the slow phase oscillations)

$$\begin{cases} \ddot{x} + \omega_0^2 (1 - n)\, x = 0, \\ \ddot{z} + \omega_0^2\, n z = 0. \end{cases} \tag{6.1}$$

If $B_z = B_0 \left(\dfrac{r_0}{r}\right)^n$ has small azimuthal variations it can be approximated by $B_z = \left(1 - \dfrac{n x}{r_0} + a_k \cos k\vartheta\right) B_0$ and the first equation (4.1) becomes

$$\ddot{x} + \omega_0^2 (1 - n)\, x = a_k r_0 \omega_0^2 \cos k \omega_0 t \tag{6.2}$$

which is the familiar equation of forced oscillations [10]. The particular integral of (6.2) has the forced oscillation amplitude

$$\frac{1}{(1 - n) - k^2}\, a_k r_0 \tag{6.3}$$

and a similar amplitude is given for z by replacing $1 - n$ by n. If $n = 0$ or 1 the radial or vertical oscillations respectively will; equal the revolution frequency ω_0, be in integral resonance, and grow to unlimited amplitude if there is any first harmonic, or $k = 1$, variation in B_z or B_r. Since $n = 0, 1$ are also the stability limits, the integral resonance is here a matter of limited interest. However if the first harmonic content of the magnetic field is a few parts per thousand of B_0 the forced oscillation amplitude at $n = 0.6$ is many centimeters, and can scarcely be tolerated.

If there are twists in the median plane there will be coupling between the r and z modes. The simplest case is given by a uniform tilt angle α in the surface defined by $B_r = 0$ (a conical median surface). The field can be approximated by

$$B_z = B_0 \left(1 - \frac{n x}{r_0} - \frac{\alpha n z}{r_0}\right),$$

$$B_r = B_0 \left(-\frac{n z}{r_0} + \frac{\alpha n x}{r_0}\right),$$

and (6.1) takes the form, for $\alpha \ll 1$,

$$\begin{cases} \ddot{x} + \omega^2 [(1 - n)\, x - \alpha n z] = 0, \\ \ddot{z} + \omega^2 (n z - \alpha n x) = 0. \end{cases} \tag{6.4}$$

Representing the motion by

$$x = X\, e^{i q \omega t}; \qquad z = Z\, e^{i q \omega t} \tag{6.5}$$

q must satisfy (6.4) by substitution

$$q^2 = \tfrac{1}{2} \pm \sqrt{(n - \tfrac{1}{2})^2 + \alpha^2 n^2} \tag{6.6}$$

and

$$\frac{Z}{X} = \frac{\alpha n}{n - q^2}. \tag{6.7}$$

If $\alpha = 0$ (6.6) reduces to the results of (4.14), $q^2 = n$, $1 - n$. However, if $n = \frac{1}{2}$ then

$$q = \frac{1 \pm \alpha}{2}; \qquad \frac{Z}{X} = \pm 1$$

and an initial amplitude in one coordinate will couple into the same amplitude in the other. Just after injection, a large fraction of the protons can have radial amplitude greater than the vertical aperture, so operation at $n = 0.5$ can cause large beam loss. When, later in the accelerating cycle, the betatron amplitudes have damped to a fraction of the aperture, operation on the coupling resonance is possible. If n differs from 0.5 the coupling diminishes rapidly. At $n = 0.6$ the amplitude beating factor (6.7) is 3α and the back-and-forth amplitude transfer will be small since α will not exceed a few milliradians in a properly constructed magnet. If a magnet is rotated about the ϑ axis by angle α the median surface will translate in z an amount $\alpha r_0/n$. This large multiplier on α imposes more stringent tolerances than coupling considerations do. However, a localized "bump" in the median surface will give powerful coupling between the r and z modes (see Sect. 13).

If n is a function of ϑ then (6.1) becomes a Mathieu equation [2] and it will be found (see methods of Sect. 34) that large amplitudes will be excited by the first ϑ harmonic in dB_z/dr if $n = \frac{3}{4}, \frac{1}{4}$. This condition corresponds to betatron oscillation frequencies of half the revolution frequency, or half-integral resonances. The field can be approximated by $B_z = B_0 \left(1 - \dfrac{n - b \cos \vartheta}{r_0} x\right)$ from which the first equation of (4.1) to first order is

$$\ddot{x} + \omega^2 (1 - n) x = - \omega^2 b x \cos \vartheta \tag{6.8}$$

and substituting $x = X \cos \sqrt{1 - n}\, \omega t$ and $\vartheta = \omega t$ gives forcing terms on the right hand side including

$$\frac{- b X \omega^2}{2} \cos \left(1 - \sqrt{1 - n}\right) \omega t.$$

If $\sqrt{1 - n} = 1 - \sqrt{1 - n}$ a resonance occurs with $n = \frac{3}{4}$ and the corresponding z resonance is found at $n = \frac{1}{4}$. The build-up in amplitude with slow crossing of this resonance has been treated by Courant [15]. Betatron oscillation amplitude attained in a synchrotron on a half-integral resonance is very sensitive to non-linearities in the field, that is, departures of B_z from $B_0 \left(1 - \dfrac{n x}{r_0}\right)$. With the proper amount of $d^2 B_z/dr^2$ the orbit can lock-in on the half-integral resonance and be stable[1].

The resonances listed thus far are

$$\omega_r, \omega_z = \omega_0, \qquad n = 0, 1,$$

$$\omega_r = \omega_z, \qquad n = 0.5,$$

$$\omega_r, \omega_z = \omega_0/2, \qquad n = \tfrac{3}{4}, \tfrac{1}{4}.$$

Higher orders are possible at $\omega_r = \frac{1}{3}\omega_0$, $\omega_r = 2\omega_z$, $\omega_r + 2\omega_z = 2\omega_0$, $\omega_z + 2\omega_r = 2\omega_0$, etc. The relative excitation of such higher order resonances[2] depends on the detailed field shape and, if precise magnetic field data are available, may require numerical integration of the equations of motion.

---

[1] J. D. Lawson: Nature, Lond. 165, 109 (1950).
[2] Losses on such resonances are described by G. C. Baldwin, F. R. Elder and W. F. Westendorp: JAP 25, 1553 (1954).

The phase oscillations have a very low frequency and consequently a weak restoring force, in addition to their weak damping. Disturbing effects in the phase oscillation frequency band (of the order of kilocycles) are produced by ripple and noise in the radiofrequency amplitude, V, the magnetic field, B_0, and in the radiofrequency, ω_r. Let these take the form $\varepsilon_1 V \sin \omega_1 t$, $\varepsilon_2 B_0 \sin \omega_2 t$ and $\varepsilon_3 \omega_0 \sin \omega_3 t$. If ω_0 is considered slowly-varying the equations of phase motion can be approximated by equating the rate of change of ΔE, the deviation of proton energy from its equilibrium orbit energy, to the rate of energy gain at the radiofrequency gap plus the variation in radiofrequency amplitude

$$\left. \begin{aligned} \frac{d}{dt}(\Delta E) &= \frac{\omega_0}{2\pi}\left[e\,V\,\varphi\cos\psi_0 + e\,\varepsilon_1 V\sin\omega_1 t\sin\psi_0\right] \\ \varphi &= \psi - \psi_0. \end{aligned} \right\} \tag{6.9}$$

with

The angular velocity rate is equated to the normal (5.5) and forcing terms

$$\frac{d\varphi}{dt} = \frac{\partial\omega_0}{\partial E}\Delta E + \frac{\partial\omega_0}{\partial B_0}\varepsilon_2 B\sin\omega_2 t + \varepsilon_3\omega_0\sin\omega_3 t. \tag{6.10}$$

From (4.3), (5.2) and (5.5)

$$\Omega^2 = -\frac{eV\omega}{2\pi}\frac{\partial\omega}{\partial E}\cos\psi_0 \tag{6.11}$$

and the two equations of motion (6.9) and (6.10) reduce to

$$\ddot{\varphi} + \Omega^2\varphi = -\Omega^2\tan\psi_0\,\varepsilon_1\cos\omega_1 t + \frac{\omega_0}{1-n}\varepsilon_2\omega_2\cos\omega_2 t + \varepsilon_3\omega_0\omega_3\cos\omega_3 t. \tag{6.12}$$

The amplitudes of forced oscillation due to disturbances in V, B_0, and ω_r respectively are

$$\frac{\Omega^2\varepsilon_1\tan\psi_0}{\Omega^2-\omega_1^2}, \qquad \frac{\dfrac{1}{1-n}\varepsilon_2\omega\omega_2}{\Omega^2-\omega_2^2}, \qquad \frac{\varepsilon_3\omega\omega_3}{\Omega^2-\omega_3^2}.$$

If ω_1, ω_2 or ω_3 are near Ω the first is far less serious because it involves Ω^2 while the other two are of the order of $\omega\Omega$ and $\omega/\Omega \sim 10^3$. These approximations do not hold at resonance because $\sin\psi - \sin\psi_0$ was set equal to $(\psi-\psi_0)\cos\psi_0$.

The phase oscillations diminish in frequency rapidly with large amplitude. Numerically integrated frequency-amplitude functions can be found in Fig. 6 of [12]. The problem of magnet power-supply ripple passing through resonance has been treated analytically by BLACHMAN [17] who has also calculated the resonance build-up on a digital computer and found, as expected, that the nonlinear treatment gives much less amplitude build-up. Phase noise in the radio frequency can force phase oscillations of individual protons (the beam is always a collection of protons spread over a range of ψ) beyond the limits of stability. This phenomenon is one of the most serious sources of beam loss during the accelerating cycle [17].

7. The effects of straight sections. Field free regions are most useful in a synchrotron for placing apparatus to inject, accelerate and eject the beam[1]. The straight sections also permit expanded sections of vacuum chamber without increase in the magnet gap. Although the insertion of straight sections violates the specification of constant gradient, the proton orbits are similar in form to those of the circular machine. Betatron oscillation frequencies, relative to rotation frequency, are increased and the amplitudes increase.

---

[1] Proposed by H. R. CRANE: Phys. Rev. **69**, 542 (1946).

Introduce N field-free sections of length L between N circular arcs. The radius of the equilibrium orbit in the circular arcs is r_0, and the field is shaped according to (4.5). The average angular frequency of revolution will be smaller than (4.3) in the ratio $\left(1 + \dfrac{NL}{2\pi r_0}\right)^{-1}$. There is no change in the momentum compaction because the straight sections of the constant-gradient machine do not alter the relationship between B and r, (4.2), and from (4.5)

$$\frac{dp}{p} = \alpha \frac{dr}{r} = (1 - n) \frac{dr}{r}, \tag{7.1}$$

so the momentum compaction factor $\alpha = 1 - n$. The energy gain per turn is increased over that of the circular machine, (4.4), by the ratio of ω_0 and becomes

$$\Delta E = \left(1 + \frac{NL}{2\pi r_0}\right) 2\pi\, e\, r_0^2\, \frac{dB_0}{dt}. \tag{7.2}$$

Straight-section modification of the betatron oscillations has been obtained by BLACHMAN and COURANT, using recursion relations [13], [16]. The damping of the betatron oscillations remains proportional to $B^{-\frac{1}{2}}$, but the frequencies increase to

$$\left.\begin{aligned}
\omega_r &= \left(1 + \frac{NL}{4\pi r_0}\right)\sqrt{1 - n}\,\omega_0, \\
\omega_z &= \left(1 + \frac{NL}{4\pi r_0}\right)\sqrt{n}\,\omega_0,
\end{aligned}\right\} \tag{7.3}$$

if NL is small compared to $4\pi r_0$. It is convenient to define $\nu = \omega_\beta/\omega_0$, the number of betatron wavelengths per revolution. (European literature uses the symbol Q for this basic parameter.) If the phase shift of the betatron oscillation per period, i.e., one arc plus one straight section, is μ, it is shown in [13] that

$$\cos\mu = \cos\left(\frac{2\pi\sqrt{1-n}}{N}\right) - \frac{L\sqrt{1-n}}{2r}\sin\left(\frac{2\pi\sqrt{1-n}}{N}\right) \tag{7.4}$$

for the radial mode. Replacing $\sqrt{1-n}$ by $\sqrt{n}$ in (7.4) gives the relationship for the vertical mode. Since

$$\nu = \frac{\omega_\beta}{\omega_0} = \frac{N\mu}{2\pi}, \tag{7.5}$$

this, with (7.4), leads to the approximation (7.3). It was noted by SERBER[1] that μ must be real if the orbits are to be stable. He also observed that the amplitude of betatron oscillations, for a given displacement, is increased by the introduction of straight sections[2]. (The particle orbits are readily obtained by using matrix solutions of differential equations with periodic coefficients, more recently developed by COURANT, SNYDER and others. A derivation of the straight-section relations will be found in Sect. 32.) Amplitudes grow so large for $N = 2$ that the use of two straight sections is generally inadvisable. However, four are convenient and increase amplitudes by only a few percent if $L/r < \frac{1}{2}$. Although L/r ratios much greater than one are possible with real μ (Table 1 of [13]), such ratios require large and uneconomical apertures.

The pattern of resonances must be considered carefully when setting the operating point of a synchrotron. If k, l, m are integers then $\pm k\nu_r \pm l\nu_z = m$ is the condition for resonance. The lines representing the resonances to order 2

---

[1] R. SERBER: Phys. Rev. **70**, 434 (1946).

[2] See also D. M. DENNISON and T. H. BERLIN: Phys. Rev. **70**, 764 (1946).

are shown in Fig. 2 on a v_z vs. v_r plot, with higher orders omitted because they excite little amplitude growth unless large azimuthal harmonics are present. A constant-gradient synchrotron cannot operate at any point in the v_z, v_r plane, but must lie on a locus which is determined by N, L and r. Fig. 2 has the loci for the circular machine, $L/r = 0$, and for $N = 4$, $L/r = \frac{1}{3}$. The shaded area is the preferred working region because it tends to vertical amplitudes smaller than radial ones ($v_z > v_r$), thus conserving magnet gap length, and in addition, the anti-damping of the phase oscillations [12], [13] at low energy is small if $n > 0.6$. The lower resonance for both circular and straight-section synchrotrons is then

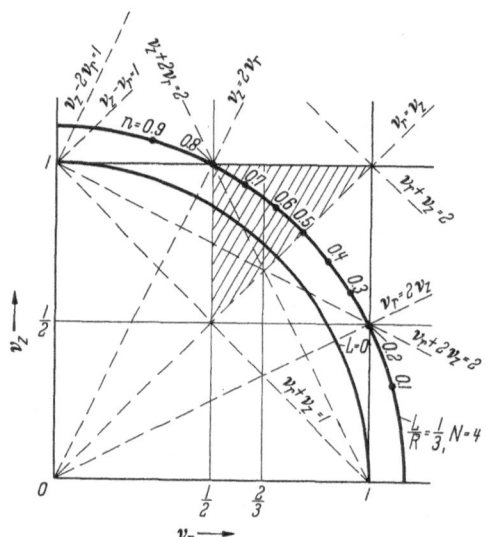

$v_r = v_z$ at $n = 0.5$, the coupling resonance. Large beam losses can occur here just after injection, when many of the protons have large radial amplitude which couples completely into vertical amplitude at $n = 0.5$ (Sect. 6). Losses of two orders of magnitude near or at $n = 0.5$ have been observed in the Bevatron. Since n normally increases at saturation, this resonance is not important at high energies, although it can probably be crossed with impunity when the oscillations have damped.

The higher order resonances, $v_z = \frac{2}{3}$ and $v_z + 2v_r = 2$ have not been observed in the four operating proton machines, although they have been found in betatrons with sizable second harmonics in the field and its derivatives. The circular machine will encounter the half-integral resonance $v_r = 0.5$ at $n = 0.75$ and, un-

Fig. 2. Resonance pattern of the constant gradient machine. Operating point locus shown for circular magnet, and magnet with four straight sections. Lines of constant n would radiate from origin.

less it is crossed rapidly, a small first ϑ harmonic in n will blow up the beam. With $L/r = \frac{1}{3}$ (the Cosmotron) three resonances are near $n = 0.8$ (Fig. 2). These invariably lose all the beam in the Cosmotron, if the magnet is not corrected to keep $n < 0.79$. Of the three, the integral $v_z = 1$ is most serious because amplitudes are excited by first harmonics of the field and of all its z derivatives. A working point beyond $n = 0.75$ or 0.8 is not desirable because radial amplitudes due to small angular disturbances (4.13) become unduly large. As straight-sections are lengthened the operating locus in Fig. 2 moves away from the origin and the working region diminishes. The Synchrophasotron, $L/r = 0.35$, is nearly the same as the Cosmotron, while the Bevatron, $L/r = 0.4$, encounters $v_z = 1$ at $n = 0.78$. An $L/r \gtrsim 0.5$ would be very convenient in experimental use of a proton synchrotron, but may well never be used. Present models were designed conservatively, before extended experience, and designs in process tend to minimize aperture as much as feasible in order to obtain more rapid pulse repetition rates.

The straight sections modify the phase oscillations by altering the value of $\partial \omega_0 / \partial E$ from (5.5) to

$$\frac{\partial \omega_0}{\partial E} = - \frac{c}{\left(r_0 + \frac{NL}{2\pi}\right)\sqrt{E^2 - E_0^2}} \left[\frac{1}{(1-n)\left(1 + \frac{NL}{2\pi r_0}\right)} - \frac{E_0^2}{E^2} \right], \qquad (7.6)$$

and (6.11) becomes

$$\Omega = \frac{c}{r_0 + \dfrac{NL}{2\pi}} \left[\frac{eV\cos\psi_0}{2\pi E} \left(\frac{1}{(1-n)\left(1+\dfrac{NL}{2\pi r_0}\right)} - \frac{E_0^2}{E^2} \right) \right]^{\frac{1}{2}}. \tag{7.7}$$

The limiting excursions of energy and radius for phase stability follow from the same considerations which led to (5.6) and (5.8):

$$\Delta E_{max} = \left\{ \frac{eV(E^2 - E_0^2)[2\cos\psi_0 + (2\psi_0 - \pi)\sin\psi_0]}{\pi E\left[\dfrac{1}{(1-n)\left(1+\dfrac{NL}{2\pi r_0}\right)} - \dfrac{E_0^2}{E^2} \right]} \right\}^{\frac{1}{2}}, \tag{7.8}$$

$$\Delta r_{max} = r_0 \left\{ \frac{eVE[2\cos\psi_0 + (2\psi_0 - \pi)\sin\psi_0]}{\pi(E^2 - E_0^2)(1-n)\left[\dfrac{1}{1+\dfrac{NL}{2\pi r_0}} - (1-n)\dfrac{E_0^2}{E^2} \right]} \right\}^{\frac{1}{2}}. \tag{7.9}$$

The phase damping can be derived from a WKB approximation [13] which shows the phase angle amplitude to be proportional to:

$$\left[\frac{\tan\psi_0}{E\dot{B}_0} \left(\frac{1}{(1-n)\left(1+\dfrac{NL}{2\pi r_0}\right)} - \frac{E_0^2}{E^2} \right) \right]^{\frac{1}{4}}. \tag{7.10}$$

The weak phase angle damping is slightly diminished by the straight section Set $C = \left[(1-n)\left(1+\dfrac{NL}{2\pi r_0}\right)\right]^{-1}$ and $\dfrac{d}{dE}$ of the function in the brackets of (7.10) is proportional to

$$-\frac{C}{E^2} + 3\frac{E_0^2}{E^4}, \tag{7.11}$$

if $\tan\psi_0/\dot{B}_0$ is assumed constant. At injection, where $E \approx E_0$, this is negative only if $C > 3$. In general, angular damping is obtained if $C > 3\frac{E_0^2}{E^2}$, indicating that the operating point should be set at the highest practicable n value. There is strong damping of the proportional energy spread, $\Delta E/E$, and of the radial amplitude through the relationships shown in (7.8) and (7.9).

The radiofrequency of the accelerating system can be set at an integral multiple, h, of ω_0. Since an angular variation of the proton from synchronism is then multiplied h times in its angular variation along the radiofrequency wave, the $eV\cos\psi_0(\psi - \psi_0)$ term of (5.1) is multiplied by the harmonic order, h. The phase oscillation frequency (7.7) will be increased by the ratio $h^{\frac{1}{2}}$. Angular limits of stability, with reference to the radiofrequency wave, are the same as those for fundamental operation ($h = 1$) but the energy and radial excursions (7.8) and (7.9) are reduced by the ratio $h^{-\frac{1}{2}}$. Thus the radial amplitude of the phase oscillations can be reduced by increasing the harmonic order although a radiofrequency system at higher frequency and with greater band-width is required. There will be h discrete "bunches" in the synchrotron.

8. Gas scattering. Collision of the protons with gas molecules will cause an angular deviation of the proton path, seldom large enough to deflect the proton directly into the aperture wall[1]. However, the amplitude of the betatron oscillation will be altered, and the mean amplitude will grow with multiple collisions.

---

[1] J. M. GREENBERG and T. H. BERLIN: Rev. Sci. Instrum. **22**, 293 (1951).

Simultaneously, the oscillation amplitude is damped as $B^{-\frac{1}{4}}$ and the scattering cross section decreases with energy gain. The betatron oscillations increase to a maximum amplitude at approximately four times injection energy and then diminish to a small value at high energy. Derivation of a gas scattering theory which gives reasonable fit to observed proton loss has required the use of refined forms of the particle scattering relationships and careful evaluation of the cut-off angles [14]. For losses less than approximately one-half, reference [34], p. 836, gives the probability of loss as

$$L(\eta) = \sqrt{\frac{2\pi}{\eta}}\, e^{-\frac{1}{2\eta}} \tag{8.1}$$

$$\eta = \frac{7.1\,\pi^2\,p\,r_0^3\,S^2 Z^2 e^4 \cdot 10^{16}}{4\,n\,A^2\,T_i\,eV} \left[2\ln\left(\frac{5.3\,A\,\sqrt{n}\cdot 10^{-9}}{1.2\sqrt{S\,r_0}\,\lambda Z^{\frac{1}{3}}\sqrt{1 + 3.333\,(Z/137\,\beta)^2}} \right) - 1 \right]. \tag{8.2}$$

p is the pressure in mm of Hg of atomic number Z and S is the straight section factor $1 + NL/2\pi r_0$. A is the semi-aperture, T_i the injection energy and eV the energy gain per turn. λ is $1/2\pi$ of the de Broglie wavelength of the proton and β is v/c, both evaluated at $2\,T_i$. Eq. (8.1) takes amplitude damping into account but assumes that the initial vertical amplitude is small compared to the aperture[1]. η is defined as

$$\eta = \frac{\langle z^2 \rangle}{4\,A^2}$$

with the numerator representing the mean-square amplitude caused by multiple gas-scattering.

II. General description.

9. Four synchrotrons. The first group of proton synchrotrons, operating in 1957, consists of the Brookhaven Cosmotron, the Birmingham synchrotron, the Berkeley Bevatron, and the Dubno Synchrophasotron. These four machines are identical in basic design—all have pulsed magnet with $n \approx 0.6$, electrostatic injector, and controlled radiofrequency acceleration in the fundamental mode. However, they differ materially in size and in the execution of the design. Observed particle orbits are in accord with the orbit theory. The size, cost, and engineering difficulties do not permit allowance of much factor of safety. The process of acceleration lasts one to three seconds, during which the protons traverse a path of several hundred thousand kilometers. An orbit disturbance of the order of a tenth meter will lose the protons, and it is difficult to prove *a priori* that small systematic and random perturbations will not destroy the beam of particles. Experience has shown that the inherent stability of the orbits is adequate to overcome the disturbances—and has also shown that constant, meticulous adjustment and maintenance are necessary to preserve the intensity of accelerated beam. In order to minimize the risks involved in building the full size synchrotrons, a one-quarter scale model of the Bevatron was constructed and was successfully tested in 1949 [44].

The essential things required of the apparatus of the proton synchrotron are:

a) Shaping of the magnetic field at all working values so that stable orbits exist within the aperture,

b) attainment of pressures low enough to keep most of the protons from being scattered out of the aperture,

c) injection of protons into stable orbits, and subsequent missing of the injector and other obstructions,

---

[1] Corrections for initial amplitude are given by L. B. Mullett: AERE Report GP/R 2072, 1956.

Fig. 3. View of the Cosmotron, overlooking the injection system.

Table 1. *Design parameters of* C-G *proton synchrotrons.*

	Birmingham	Cosmotron	Bevatron	Synchro-phasotron	Units
Peak energy	1	3	6.2	10	Bev
Radius, r_0	4.50	9.14	15.24	28	m
No. straight sections, N . .	0	4	4	4	
Length straight sections, L	—	3.05	6.10	8	m
Magnet cross section . . .	2.44 × 2.44	2.38 × 2.38	6.25 × 2.9	7.5 × 5.3	m
Weight of steel	800	1650	9700	36000	tons
No. of exciting turns . . .	24	48	88	44	
Weight of copper	10	70	350	460	tons
Peak current	12500	7000	8300	12800	amps
Initial magnet E	1100	5400	14000	11000	volts
Peak stored energy	7	12	80	148	10^6 joules
Rise time	1	1	2	3.3	sec
Repetition rate	6	12	10	5	per min
Injection field	210	295	300	150	gauss
Peak field	12.6	13.8	15.4	13	kgauss
Magnet gap	0.21 × 0.5	0.24 × 0.92	0.33 × 1.68	0.4 × 2.0	m
Approx. aperture (i) . . .	0.1 × 0.35	0.16 × 0.65	0.30 × 1.17	0.36 × 1.5	m
Design n	0.67	0.6	0.6	0.65	
Injection energy	0.46	3.6	9.9	9.0	Mev
Approx. dB_i/dt	15	16	9	5	kgauss/sec
Approx. Δr_i/turn	3.1	4.1	4.2	15	mm
Approx. pressure/10% loss .	1	4	10	7	10^{-6} mm Hg
Initial frequency	300	360	355	182	kc
Final frequency	9.7	4.18	2.47	1.45	mc
Approx. energy gain/turn .	0.2	1	1.5	2.5	kev
Harmonic order, h	1	1	1	1	
No. of accelerating stations	1	1	1	2	

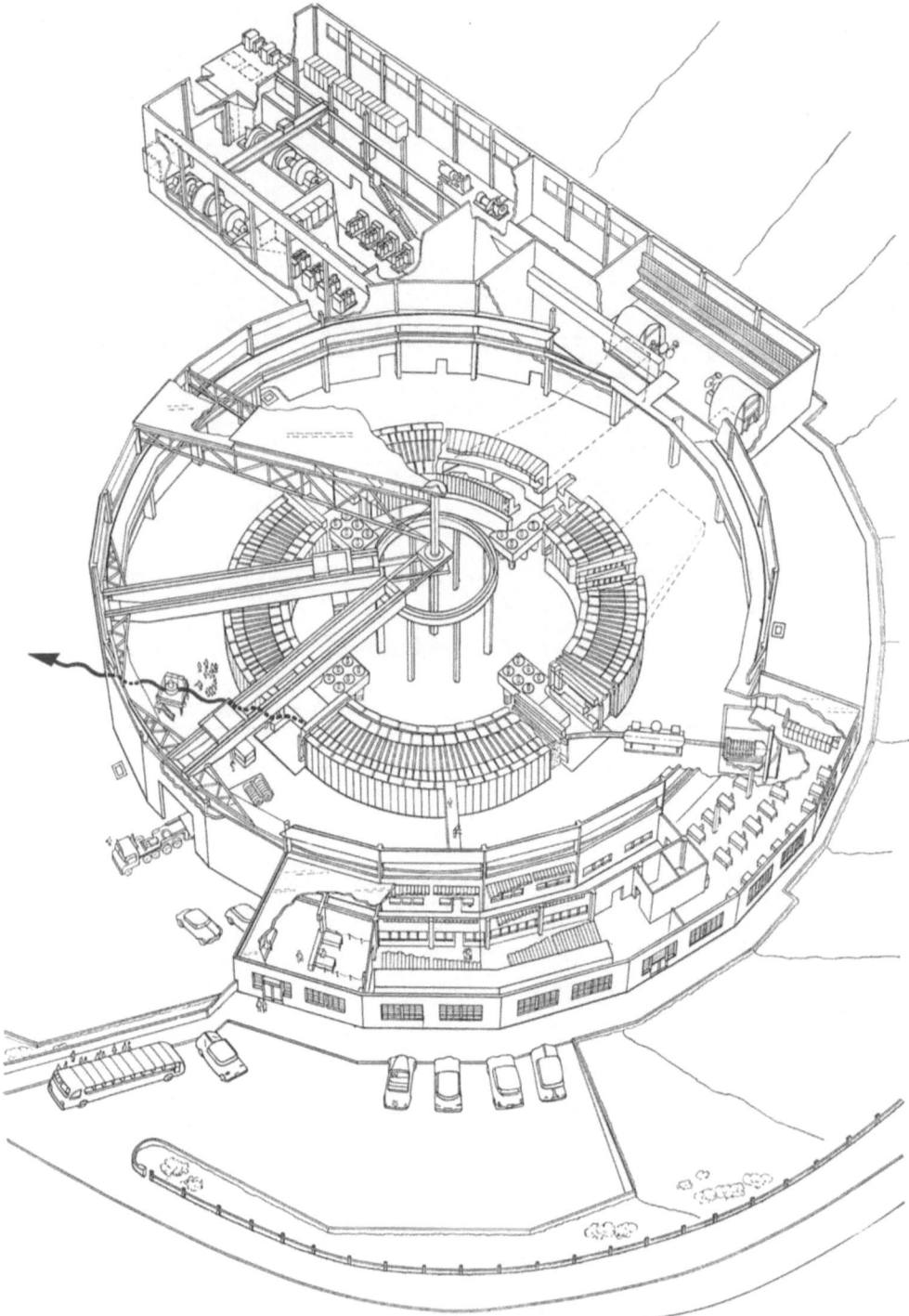

Fig. 4. Sectional drawing of the Bevatron, with power room at top.

d) application of a synchronized accelerating radiofrequency potential, with noise level so low that some protons do not exceed the stability limits of the phase scoillations.

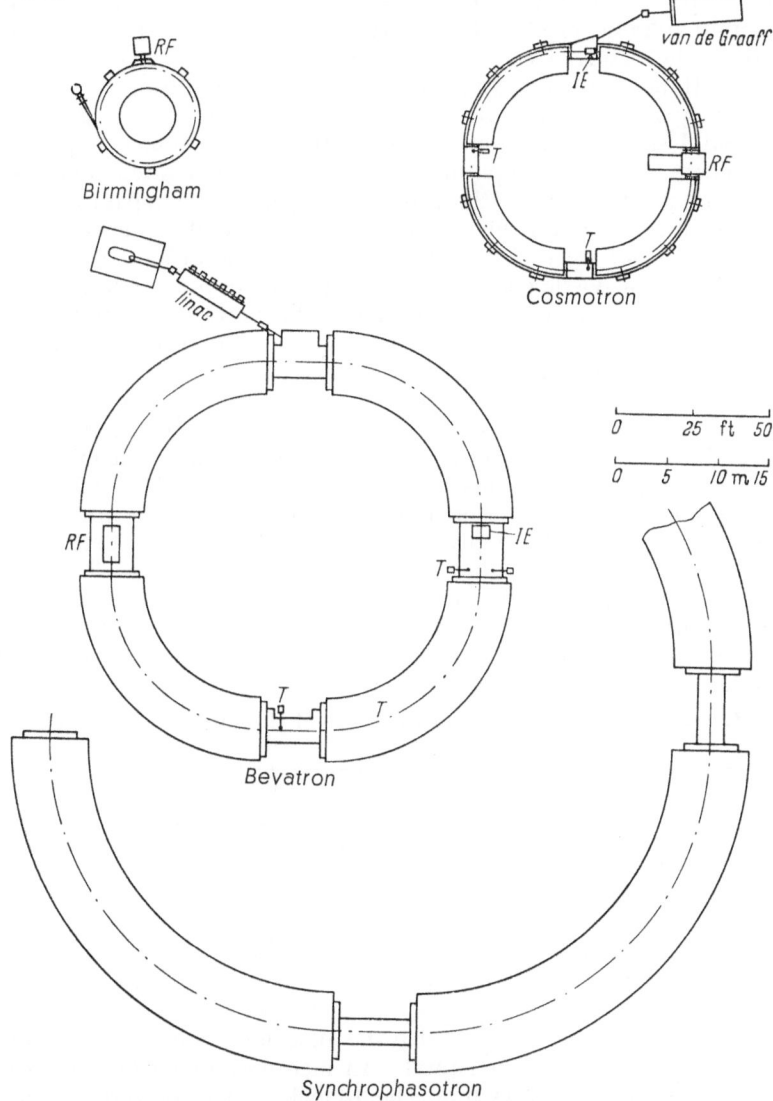

Fig. 5. Plan views, to same relative scale, of four operating proton synchrotrons. *T*-target, *IE*-induction electrode, *RF*-accelerating electrode.

Successful operation of a proton synchrotron is made possible by the strong damping of the betatron oscillations (as $B^{-\frac{1}{4}}$) and the phase stability with respect to the radiofrequency accelerating waveform.

The general configuration of proton synchrotrons can be seen in the photograph of Fig. 3 and the drawing of Fig. 4. The plan views, Fig. 5, compare sizes of the four machines and indicate the layout of components. Table 1 lists the design parameters. All four machines operate in the manner outlined in the introduction. The details are best described by considering the components singly.

III. Magnets.

10. Magnet parameters. The magnet of the synchrotron must maintain, from injection to peak strength, a field configuration over the required aperture which contains stable orbits. The vertical semi-aperture must accommodate at injection; the betatron oscillation amplitude corresponding to a reasonable vertical admittance (see Sect. 23 for definition), the amplitude of forced oscillations caused by ϑ variation of B_r, and the amplitude increase by gas scattering. The horizontal semi-aperture must contain the corresponding amplitudes plus the radial phase oscillation amplitude. The motions enumerated are anharmonic, so the maximum excursion is the arithmetic sum of the separate amplitudes. The large excursions of the phase oscillation considerably increase the radial aperture requirement over that of the vertical. In addition, phase oscillation damping at non-relativistic energies requires a value of n as high as is consistent with stability requirements, and this makes the vertical betatron oscillations somewhat smaller than the horizontal. Fringing distortions of the magnetic field are most easily minimized by choosing a gap width several times the gap length, and ampere-turn excitation is proportional to gap length. All these considerations lead to minimization of the vertical semi-aperture. To obtain the gap length there must be added to the vertical aperture the vacuum chamber wall and pole-face winding thicknesses plus allowances for mechanical assembly. The gap width will exceed the required horizontal aperture by an allowance for fringing field which is a function of the detailed design of the magnet. Side walls of the vacuum chamber may, or may not, be placed within the gap. The estimation of aperture required is seriously affected by the injection process, since protons are injected with displacement from their equilibrium orbit, and aperture must be allowed in the vertical or radial direction to contain the betatron oscillations corresponding to the displacement.

For an example, the aperture requirements of the Cosmotron at injection are:

	Vertical semiaperture	Radial semiaperture
Admittance $= 10^{-3}$ cm rad	0.6 in	0.75 in
0.1% 1st harmonic, B_z		0.5 in
0.05% 1st harmonic, B_r	0.6 in	
Radial phase oscillation		8
Arithmetic sum	1.2 in	9.2 in
$\times 2 =$ total	2.4 in	18.4 in

The gas scattering loss was estimated to be 10% for a 5 inch vertical aperture, which allows twice the estimated total. An inch was added for possible use in injection, and general safety factor. The 18 inch radial aperture was increased by 8 inches for injection displacements. The net available aperture in the Cosmotron, $6\frac{1}{4} \times 26$ inches, is obtained in a magnet gap 9.35×36 inches. This gap provides the allowances for vacuum chamber, pole-face windings, assembly clearances, and fringing field. (The choice of aperture for a given set of parameters is not based solely on logic. The intuition of the designers plays an important part and, as more experience is accumulated, apertures can be diminished with less risk.)

During the pulse of increasing field the usable magnetic aperture diminishes, at first slowly, then rapidly as the magnet iron saturates. (In a non-ferrous, or "air-core", machine the large mechanical deflections at high fields may have a similar effect.) The oscillations damp rapidly and the relatively small required space at high fields is largely a function of the magnitude of azimuthal harmonics (which is probably uncorrelated with that at low fields) and target insertion needs. During the field increase the field index must not cross any of the low-order resonance values or the beam will surely be lost.

Eddy currents in the magnet and vacuum chamber distort the field shape rather seriously at low fields but, since $\dot{B}$ is roughly constant, the eddy current distortions become quite negligible at high fields. The thickness of the magnet laminations can usually be chosen for approximate eddy current compensation of remanent field effects. Eddy currents, radiofrequency accelerating voltage, and power supply rating vary directly as $\dot{B}$ while gas scattering and I^2R losses vary inversely (the variation is not necessarily linear). If $\dot{B}$ and the repetition rate are quite high then hysteresis losses in the iron also become important. With respect to injection, a low value of $\dot{B}$ permits long injection time, but also makes injector-missing more difficult. Selection of a value of $\dot{B}$ (or rise time) requires detailed study of the cost as well as the physical factors. It is generally desirable to adopt a value of $\dot{B}$ as high as is consistant with feasible sizes of magnet power supply and radiofrequency system.

Specific design proceeds from the fundamental selection of air-core or ferrous-core type magnet; and if ferrous core, selection of H or C shape, and pole or pole-less construction. The air-core magnet is attractive because it eliminates the bothersome non-linearities of iron, and permits extremely high field strengths. (Excepting specialized circumstances, the air-core magnet is not indicated below 15 to 20 kilogauss.) However, at high fields mechanical forces, exciting currents, and stored energy are enormous, although the latter can be reduced somewhat by an iron enclosure or "return path".

The choice of H or C shape encounters two opposing factors. The H is magnetically more efficient than the C. However, the C provides an accessible gap from which particle beams can be extracted with little restriction, a factor which is important in the use of a machine. The choice of "poleless" or "pole-type" magnet cross section involves compromise of mechanical and magnetic design requirements. "Poleless" construction is magnetically efficient but can result in difficult mechanical restrictions.

11. Air-core magnet. An air-core proton synchrotron is under construction at the Australian National University[1]. The magnet is designed to produce a peak field of 80000 gauss on an orbit radius of 480 cm—corresponding to a proton energy of 10.6 Bev. Four 250 cm straight sections are provided ([5], p. 344). Although numerous configurations of conductors can be devised to give the required field shape, it is necessary that the arrangement be compact so that the enormous forces can be restrained. Two overlapping ellipses, with the common area removed, and with opposite currents in the two conductors, give a field in the hole with a constant n. Bending the conductors into a circle, in plan, alters the n value. The Canberra magnet cross section, Fig. 6, shows the elliptical conductors broken up into 1.15 inch square bars. The n value is adjusted to about 0.55. Separate bars are transposed and crossconnected in the entire magnet so that there are four turns, and 1.6×10^6 amperes per turn is required to produce the 80000 gauss. The net usable aperture is circular and 22 cm diameter. Currents of such magnitude (50000 amp per bar) produce a force of repulsion between the conductors at each side of the aperture of 40 tons per inch,

---

[1] M. L. OLIPHANT: Proc. Roy. Soc. Lond., Ser. A **234**, 441 (1956).

and a radial bursting force of 4 tons per inch of circumference. The duralumin $\frac{1}{2}$ inch enclosing plates are packed almost solidly together, and with various clamps and enclosures, restrain the forces. Deflections are estimated to be quite large but well within elastic limits. The eddy currents in the copper bars perturb the field rather badly at injection (850 gauss) and are compensated by correcting windings, in addition to the considerable compensation by the eddy currents in the duralumin plates.

Common types of electrical machinery are not suitable for exciting such a magnet. The Canberra group has constructed a homopolar generator in which four 19 ton steel discs, 139 inches diameter, contra-rotate at 900 rpm in the gap of a magnet originally intended for a synchrocyclotron. When the generator is connected to the magnet a damped current oscillation ensues, reaching 1.6×10^6 amps peak at 0.8 sec and returning to zero at about 1.8 sec at which time it is proposed to open the circuit and leave the rotors spinning oppositely from the start. The rotors are then accelerated as motors by connecting them to a dc source. Contact is made to the rotors with liquid NaK jets, protected by an inert atmosphere and supplied by a recirculating system. The generator no load emf is 800 volts. Its stored energy is 5×10^8 joules, of which about one half is stored in the peak field of the magnet—of this half 27% is stored in the volume inside the conductors, 28% inside the conductors, and 45% outside the magnet proper. It is evident that a magnet of this type must be designed in combination with its generator. The greatest disadvantage of this ingenious design is the long repetition period—approximately 10 min between pulses. A more complete description will be found in p. 344 — 358 of [5].

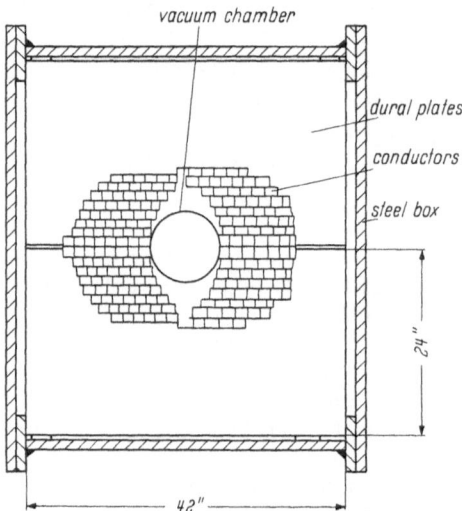

Fig. 6. Cross section of air-core magnet of Australian National University Synchrotron. From Blamey [5], p. 34

12. Steel. The introduction of iron in the synchrotron magnetic circuit adds the complexities due to eddy currents in the iron, and dependence of μ on B. The choice of steel type is strongly influenced by two requirements; low coercive force and good high-induction permeability. The first minimizes the remanent magnetic field, which is always troublesome, and the second assures the highest energy for a given structure size. Although the electrical grades of silicon-steels show very low coercive force and high initial permeability, they saturate at comparatively low induction. Standard low-carbon steel plate seems the best magnetic compromise for magnets with a low $\dot{B}$, and is also one of the cheapest of the steels. If obtained with less than 0.1% carbon, and the lowest practicable limits of other impurities, it will have a coercive force less than 2 oersteds, a permeability greater than 500 at B 100 gauss and greater than 300 at 18000 gauss. The low resistivity will make eddy currents greater than those in silicon steel, but compensation is relatively simple and core heating is negligible. If a rapidly pulsed machine, perhaps excited with ac, is designed it is necessary to use thin silicon-steel laminations to keep core heating within bounds, but high-field performance will deteriorate badly. It has been found empirically that the high induction μ of commercial low-carbon steel diminishes as the steel is rolled thinner, so slowly-

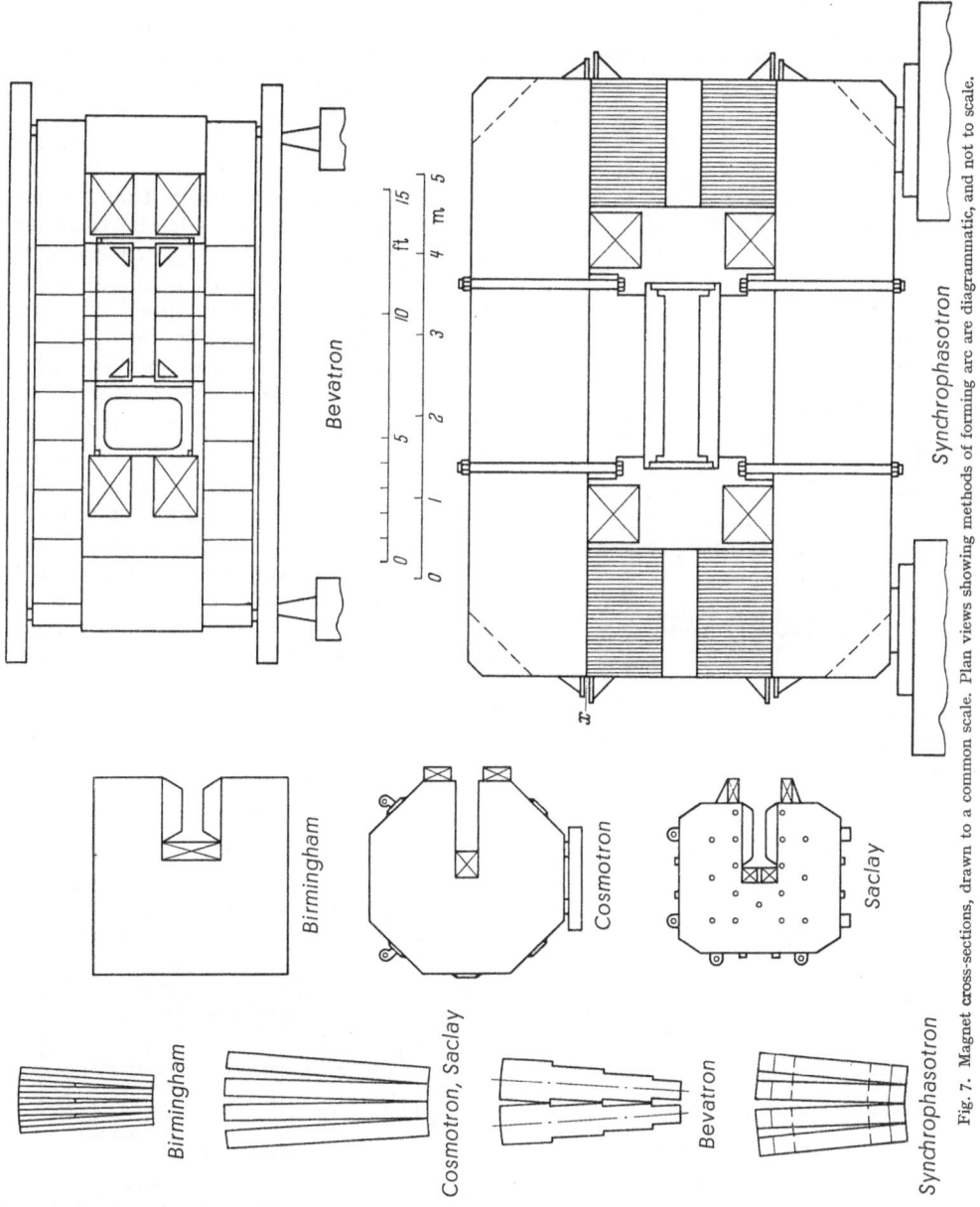

Fig. 7. Magnet cross-sections, drawn to a common scale. Plan views showing methods of forming arc are diagrammatic, and not to scale.

pulsed machines are preferably constructed of plate as thick as eddy-current considerations will allow.

The magnetic properties of steels are a complex function of processing as well as of chemical composition[1]. Details of steel properties and selection of steel are outside the scope of this paper.

---

[1] R. M. Bozorth: Ferromagnetism. New York: Van Nostrand 1951.

13. Magnetic performance. The characteristics of existing magnets will illustrate the performance of typical iron-core designs. Five existing cross-sections are drawn to relative scale in Fig. 7.

It is convenient, and conventional, to evaluate the performance of a synchrotron magnet by examining the gradient dB_z/dr along the median plane. This gradient is related to field strength by

$$n = -\frac{r}{B_z}\frac{dB_z}{dr}. \qquad (13.1)$$

In all the magnets of this section n must be greater than 0.5, the value for the $r-z$ coupling resonance. The upper limit on n is set by the half integral or integral resonance; at $n = 0.75$ for Birmingham, 0.79 for the Cosmotron, 0.77 for the Bevatron and Saclay, and 0.8 for the Synchrophasotron. On the geometrical median plane the radial component of field, B_r, must not exceed a few parts in 10^4 of B_z.

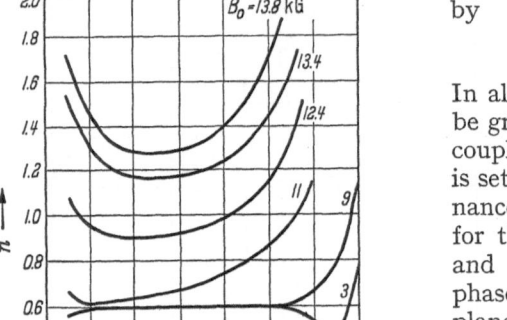

Fig. 8. Field index n as a function of radius in the Cosmotron for B_0 field strengths at 360 inches as indicated.

The Cosmotron magnet is a "poleless C" made of blocks about $6\frac{1}{2}$ inches thick arranged outside an arc of $23\frac{1}{2}$ feet radius. The resulting wedge shape gaps are unfilled and may be considered as diluting the iron. At intermediate field strengths (Fig. 8) the usable region extends radially from the face of the coil to $\frac{2}{3}$ gap from the open side, some 30 inches out of a 36 inch "pole". However, the coil clamps, pole-face return windings and vacuum chamber use 5 inches, leaving 25 inches of usable radial aperture. At increased fields the radial function

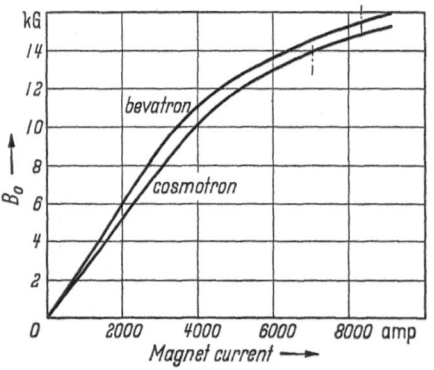

Fig. 9. Central field *vs.* exciting current in Cosmotron and Bevatron.

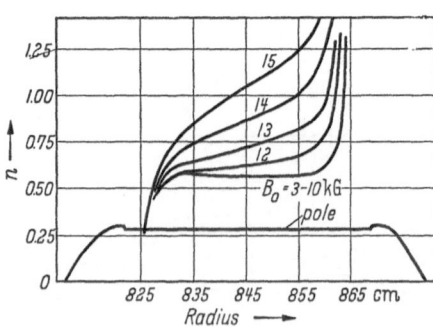

Fig. 10. Field index n *vs.* radius at indicated central field strengths for Saclay magnet. From Bruck [5], p. 331.

of n rises and becomes bowl-shaped (Fig. 8) until at some 11 500 gauss a value of $n = 0.82$, corresponding to an integral resonance, is reached. At full design field of 13 800 gauss $n = 1.3$, although there is a 12 inch width with $\Delta n = 0.1$. Current vs. B_0, (Fig. 9) indicates the diminishing returns encountered if this structure is pushed beyond 14000 gauss.

The CEA (Saclay) magnet has short poles so that a "window" is left for the inner coil. Since the reluctance path in the iron varies less between the edges of the poles than it does in the Cosmotron magnet, effects of saturation are quite different (Fig. 10). There is a usable region at 13 000 gauss, although this region has a considerable second derivative of field. The radial width of good field is diminished by the amount lost to fringing at the inside edge of the pole, but it becomes feasible to excite to higher fields than those of the poleless C. The usable width of 37 cm (at a few thousand gauss) is 0.5 times the pole base, compared to the poleless Cosmotron (net) 25 inches which is 0.7 times the pole width. (The basis of comparison is somewhat arbitrary.)

Saturation characteristics of an H magnet differ from those of a C because the symmetry of the H preserves the field derivative at the

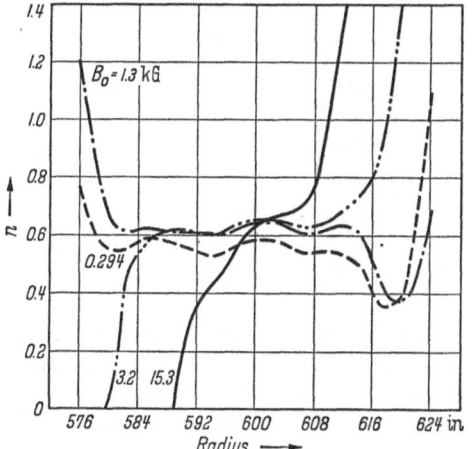

Fig. 11. Field index n vs. radius in Bevatron magnet for B_0 values shown. From G. R. LAMBERTSON (unpublished).

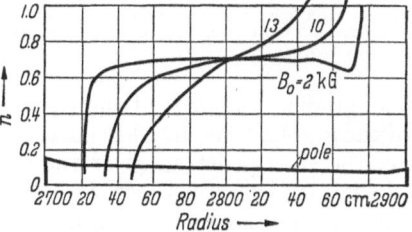

Fig. 12. Field index n vs. radius in Synchrophasotron. From ZHURAVLEV [5], p. 340.

center of the pole. As saturation sets in the n value falls at one edge and rises at the other, until finally no working region is left. *The B_0 vs. I characteristic of the Bevatron magnet* (Fig. 9) is not markedly different in character from that of the Cosmotron. However, the $n(r)$ characteristics are quite different (Fig. 11) and show usable working region at 15300 gauss. At low fields the working region of some 45 inches is about 0.68 times the pole width. The effect of the cut-outs in the corners of the pole tips is evident at 3200 gauss, where saturation produces the character of tapered poles, while at low fields the poles appear, magnetically, full width.

Saturation characteristics of another H magnet, that of the Synchrophasotron (U.S.S.R.), Fig. 12, are similar. The long poles and lack of a uniform pole taper (due to provisions for mounting the vacuum chamber parts) limit the peak usable field to 13 000 gauss.

At low field strengths, where particles must be injected, the largest aperture of the accelerating cycle is required, and the effects of remanent field and eddy currents are worst. Eddy currents depress the field strength. Although the decrease in field is of no great consequence as such, it is important that eddy current paths be uniform around the periphery of the magnet to prevent intolerably large azimuthal harmonics. Alteration of the field shape by eddy currents in the thick iron plates is so large that little or no stable region would exist for injection without compensation. This is illustrated by Fig. 13 which shows the $n(r)$ at 270 gauss B_0 for the Cosmotron magnet when pulsed from a demagnetized initial state. Although no usable injection region would remain in a C magnet under such conditions, an H magnet would have a small, non-linear, region such as that of the Synchrophasotron (Fig. 14) when pulsed from the demagnetized state.

The unidirectional current pulse (Sect. 19, Fig. 23) leaves a strong remanent field. After a full pulse to peak field the remanent B_0 in the various magnets is:

Birmingham	43 gauss,
Saclay	14 gauss,
Cosmotron	19 gauss,
Bevatron	33 gauss,
Synchrophasotron	40 gauss.

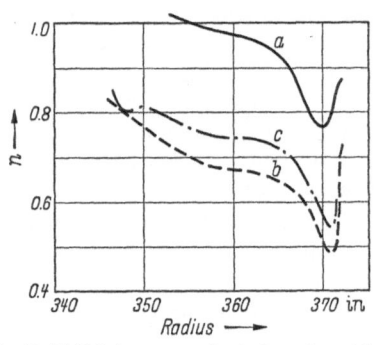

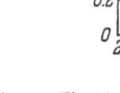

Fig. 13. Field index n vs. radius in Cosmotron at injection ield, $B_0 = 270$ gauss. (a) Pulsed from demagnetized state; (b) after short magnet pulse; (c) after pulse to full field.

Fig. 14. n vs. radius in the Synchrophasotron at $B_0 = 150$ gauss. (a) Uncorrected during pulse from demagnetized state; (b) corrected by pole face windings.

The gradient, dB_z/dr, of the remanent field has a variation vs. r inverse to that of the gradient due to eddy currents. The long time constant of the eddy currents, i.e. the slow penetration of the field into the thick iron plates[1], leaves regions of the magnet return legs with reverse magnetization. Selection of plate thickness (1 to 3 cm in the magnets being considered) can give almost exact compensation of eddy current gradients by remanent gradients at injection field.

Remanent field strength and gradient are functions of the waveshape of the

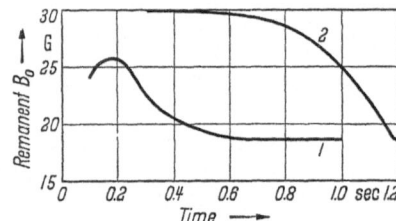

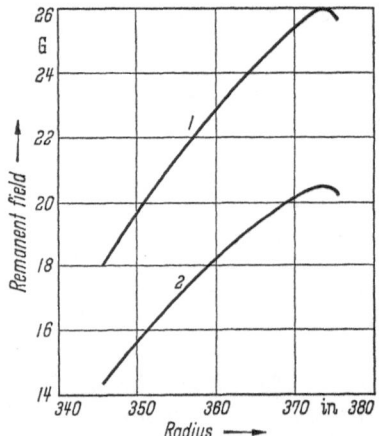

Fig. 15a and b. (a) Remanent field in Cosmotron vs., 1, length of rise time of pulse; 2, time of shorting magnet after peak of pulse. (b) Remanent field of Cosmotron as a function of radius after pulses of rise time 1, 0.1 sec; 2, 1.0 sec. After L. W. SMITH [34], p. 753.

exciting current pulse. Variations in the poleless C of the Cosmotron are exaggerated, and will serve to illustrate the behaviour of the remanence. If the Cosmotron magnet is slowly raised to peak field and then slowly reduced to zero current the remanent B_0 is 29 gauss. If it is excited by triangular pulses the remanent B_0 as a function of the rise time of the pulse (dI/dt held constant) is shown in Fig. 15a. The remanence is apparently inversely proportional to peak induction. If the magnet is short-circuited on the negative slope of the current wave, at

---

[1] See p. 857 of [34].

various times after peak current, the remanence varies vs. delay after peak to shorting as the second curve of Fig. 15a. Thus a very slow decrease of field (time constant about 6 sec) does not produce eddy currents large enough to reverse the flux in the iron, although it will be noted that the effect extends from shorting at nearly half peak field. Remanent B vs. r is plotted in Fig. 15b for two pulse lengths—the n value is about -4.5.

A remanent phenomenon in the Cosmotron, whose origin has never been satisfactorily explained, is shown in Fig. 16. The localized "bump" in the height of the median surface (defined as that surface on which $B_r = 0$), near 372 inches radius, is due entirely to a radial component of the remanent field and is remarkably uniform at all azimuths. Since no B_r disturbance is found in the dynamic field, the "bump" disappears rapidly as the field increases.

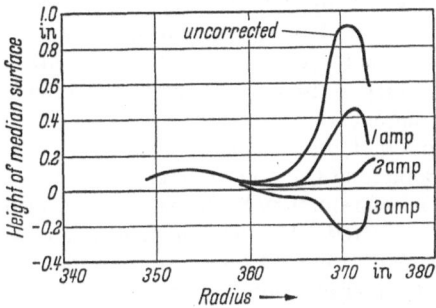

Fig. 16. Height of Cosmotron median surface at injection, uncorrected and for three values of correcting current in pole face windings. From L. W. Smith [*34*], p. 775.

14. Correction of field. Successful operation of a proton synchrotron requires that defects in the magnetic field be compensated so that an adequately large stable working region is available during the entire accelerating cycle. Field defects can be considered in two categories—incorrect gradient, and azimuthal harmonics of the field strength on the central orbit.

The constant-gradient synchrotron will tolerate larger azimuthal harmonics in gradient than in field strength (especially if not set close to $\nu = 0.5$). Consequently, if good mechanical tolerances are held on the magnet gap, corrections of gradient, or field shape, are made uniformly around the entire periphery. Shape corrections are readily made with azimuthal windings on the face of the poles. Equal currents in the same direction in pole face windings on both top and bottom poles will not shift the median surface. If the total pole face current in the ϑ direction is σ amps per unit r, evaluation of a line integral around a rectangle with sides spaced $\varDelta r$, gives $\varDelta B_z/\varDelta r = 0.4\pi\sigma/l$ with l the gap length. The change in n will be approximately $\varDelta n = 0.4\pi\sigma r/B_z l$. (See p. 773 of [*34*] for a more accurate calculation which takes the images into account.) A non-uniform distribution of currents will of course give a radial variation of the n correction. The pole face winding returns are laid inside the gap, outside the gap, or part each, depending on the voltage required to be induced by the primary magnetic flux. If the windings are self-powered, perhaps controlled by an adjustable resistor, the correction $\varDelta B_z/\varDelta r$ will be proportional to $\dot B$, and approximately constant, thus applicable to compensation of remanent and eddy current effects. If variable and readily controllable corrections in n are required, the pole face windings are energized by dc generators. Current sheets flowing in opposite ϑ directions on the top and bottom poles, σ' amps per cm, will produce a radial field component approximately $B_r = 0.4\pi\sigma'$. Since $B_r = nzB_z/r$, the median surface will shift by $\varDelta z = 0.4\pi\sigma' r/nB_z$. It is evident that many arrangements of pole face windings can be devised to fit the various circumstances. Regulation of total current density, σ, will control the shape, while differential current density control in the same windings will shift the median surface.

Azimuthal harmonics of B_z can be corrected by energizing pole face windings at the edges of the poles. However, it is usually more convenient to separate control functions and use windings around the pole bases, or around the return legs. Azimuthal harmonics of B_r are most easily inserted with differential currents in top and bottom pole face windings. Since the first and second ϑ harmonics are all that should need correction (Sect. 6) division of azimuthal trimming coils into four quadrants is adequate for adjustment of phase and amplitude.

The use of dipoles and quadrupoles in four straight sections will provide corrections for any magnet errors likely to be encountered. Space is at such a premium for auxiliary and research equipment in the present synchrotron straight sections that pole face windings are preferable, even though area in the gap is extremely expensive. Future designs may give consideration to division of the magnet into octants and allowance of short straight sections for correcting lenses.

A well designed, carefully fabricated and erected constant-gradient magnet will not require correction at intermediate field strengths—where the permeability is high, and remanent effects are relatively negligible. All the present magnets require correction at either, or both, low and high fields. The Birmingham magnet, with short poles, has a remanent $n = 0.45$. The pole laminations are $\frac{1}{4}$ inch so the eddy current increase of n is well compensated by the low n-value of the remanence. No shape corrections are necessary, and no pole face windings are fitted. During construction it was necessary to assemble the steel on the ring as it came from the mill, and no mixing, or shuffling, was done. The remanent B_z (average value 43 gauss) had a first harmonic somewhat greater than 1 gauss amplitude, and a second harmonic about 5 gauss amplitude. Self-excited windings around the poles are used to correct these harmonics—since $\dot{B}$ is roughly constant, the correcting currents are nearly constant as required.

The Cosmotron magnet requires shape correction at both low and high field intensities. At injection fields, about 300 gauss, the eddy current gradients are almost completely compensated by the remanence left from a previous short pulse (dashed curve, Fig. 13). The remanence compensation is barely adequate after a long pulse (broken curve, Fig. 13) and the n value reaches 0.8 just inside the vacuum chamber. Under normal operation protons with small oscillation amplitudes cannot reach the $n = 0.8$ region so resonance effects are not encountered.

The pole face windings of the Cosmotron are one-piece moulded polyester-glass sheets, each covering one quadrant, and containing 56 conductors ([34], p. 777). Return windings are provided at the front and rear of the gap (Fig. 28). The pole face windings are not needed for alteration of n at injection, but six conductors at the outer radius are reserved for correction of the bump in the median surface (Fig. 16). Four conductors on top and four on the bottom are connected in series forming a long coil whose axis is then in the r direction. Since the disturbance is due to remanence, a constant correction is sufficient, and the coil can be powered by induced voltage, which is roughly constant. This is done by offsetting the top conductors with respect to the bottom ones ($\frac{1}{2}$ inch) so a small amount of flux is included. All quadrants are connected in series and with a series resistor to set the current. At some 2 amp the correction is quite good (Fig. 16). Adjustment of this median plane disturbance is very convenient for setting the coupling of betatron oscillations just after injection. At some 11 500 gauss the n-value rises to 0.8 and the beam is lost through resonances. Pole face currents rising to about 200 amp/inch on each pole at peak field of 13 800 gauss are required to maintain stable orbits ($n < 0.79$). Forty-eight turns on each pole are connected in series with return windings part inside, and part outside, of the gap so that little voltage is induced by the main field. The four quadrant pole face windings are connected in series with four 40 kw d.c. generators (Fig. 17). Unless great care is taken to adjust the current to zero at low fields the beam will be destroyed at or shortly after injection. The series contactor is left open until the small induced voltage passes through zero, approximately 0.7 sec after the cycle begins. The contactor is then closed and after a short delay a step voltage is applied to the generator fields. Inductance of the circuit shapes the current pulse which rises from 0 at 0.8 sec to 100 amp at 1.0 sec (in 2 conductors per inch on each pole). The generator potential is reversed rapidly and the contactor opened as the current passes through zero so that heating of the windings is minimized. The adjustments are not critical because it is only necessary to hold n between 0.5 and 0.79. However, when beam is being ejected (Sect. 30) more accurate control is required.

The Cosmotron did not need any azimuthal trimming for the first five years of operation. In 1957 short circuits in the magnet cores, inserted during the course

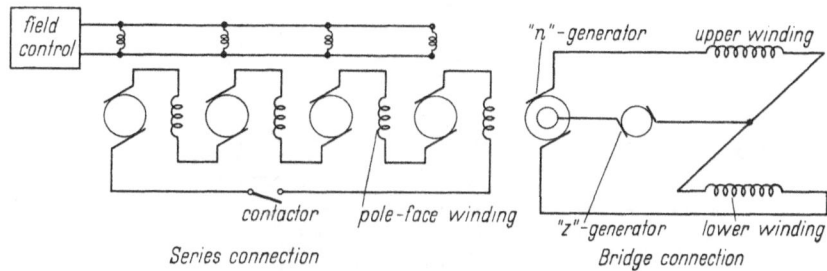

Fig. 17. Cosmotron pole-face winding generator connections.

of repairs, produced non-uniform eddy currents which shifted the equilibrium orbit a few inches. To avoid dismantling the assembly, coils around the back legs of two quadrants were self-powered and the harmonic error was cancelled by adjusting series resistors. Final adjustments were derived by plotting the orbits of the protons. Azimuthal adjustments could be made by connecting each quadrant of pole face winding separately to its generator in the bridge circuit of Fig. 17. The small generator connected to the center-tap slip rings of the "n" generator will permit the median surface to be adjusted independently.

Eddy currents in the Bevatron magnet give almost complete compensation for the remanent field at the relatively high injection field (294 gauss curve of Fig. 11). The Bevatron has 42 pole face windings clamped to the face of the pole tips (inside the vacuum and water cooled), uniformly distributed in radius, half on top, half on the bottom. The n value at injection is quite near 0.5, the value for the coupling resonance (Fig. 11). Large beam losses observed with z-direction escape indicated that the $r - z$ coupling was strongly excited. Alternate pole face turns on all pole tips were connected in series with return connections outside the gap, and a series resistor, so that a self-excited current of 3 to 4 amps would increase n to about 0.65. Since the correction is compensating effects of remanence and eddy currents, and is itself relatively constant, the corrected n-value will remain nearly constant for a considerable time after injection. The importance of such correction of magnetic field shape is shown by the subsequent increase of two orders of magnitude in the Bevatron beam intensity. There are also windings around the bases of the poles. These are not used in normal operation because the azimuthal disturbances in the Bevatron have very small low order harmonics. On occasion, the stray field of an analyzing magnet placed very close to the yoke of the Bevatron has inserted too much first harmonic.

16*

Energization of a few adjacent pole base windings will correct the disturbance and eliminate beam losses.

The Synchrophasotron has an H magnet and is usable to 13 000 gauss without compensation (Fig. 12) but pole face windings can be used to obtain more radial

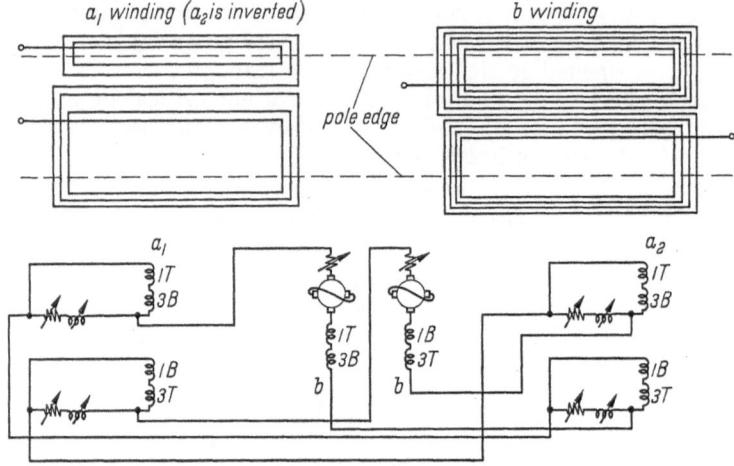

Fig. 18. Synchrophasotron pole-face windings and connections. Windings are shown separated for clarity. In connection scheme shown 1 and 3 refer to quadrant numbers, T and B to top and bottom. A similar set is provided for quadrants 2 and 4. From Zhuravlev, [5], p. 343.

aperture at 13 000 gauss. The remanent-eddy current compensation is not adequate, and large azimuthal harmonics are present in both r and z components of the field. The magnet is demagnetized after each acceleration pulse by a quasi damped-wave produced by seven successively smaller pulses of alternating sign.

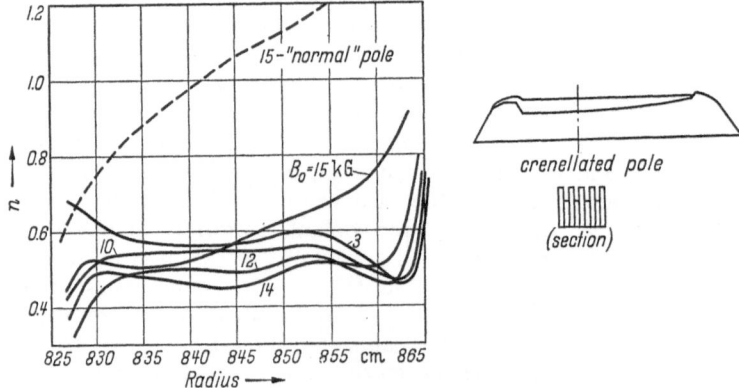

Fig. 19. Crenellated pole shape and n vs. r at values of B_0 shown, in Saclay model. Dashed curve is for normal pole at 15 kgauss in same model yoke. From Bruck, [5], p. 335.

The remanence is reduced to about 3 gauss, at which value remanent gradient and azimuthal variations are relatively unimportant. Two sets of pole face windings are installed in the vacuum chamber, along the face of the pole tips. One set has non-uniformly spaced conductors at the sides of the pole tips, and the other set is uniformly spaced along the center of the pole (Fig. 18). The first set is connected in two independent groups with non-symmetrical returns so that the induced voltage will give a current with time-constant regulated by the series resistor and inductor. Proper setting gives the well-corrected field shape compared in Fig. 14 to the uncorrected shape. The uniform windings are returned symmetrically so that little emf is induced, and are connected to amplidynes (special type of dc generator). The value of n can be shifted, and differential currents in top and bottom windings can be adjusted to correct the median plane. The windings of opposite quadrants were interconnected in the manner shown in Fig. 18. The two quadrature groups can be adjusted independently so that the amplitude and phase of first harmonic corrections can be made, if necessary, a function of time by programming the amplidynes. It is evident that local reconnections can be made to alter the correction in detail ([5], p. 324).

A series of calculations and model experiments at CEA (Saclay) has produced a design which minimizes the worst fault of the C magnet, the extreme field distortion with saturation ([5], p. 330). First experiments were done with a wedge-shaped air gap at the base of the pole. Although this diminished the increase of n at 14000 gauss by perhaps 0.15, the increase in ampere-turns was some 10%. A very great improvement was made by the design of crenellated poles. These consist of alternate laminations of different profile (Fig. 19) such that, at low fields where the permeability is high, the field shape is governed by the "high" laminations; but at high fields and low permeability the field is modified by the changed magnetic shape of the pole tip. Fig. 19 shows model tests of a crenellated pole tip usable to 15000 gauss with the striking improvement of $\Delta n = -0.4$ over uniform pole tips. The cost in ampere-turns is only some 2.3% at 15000 gauss.

15. Construction. The Birmingham magnet was designed earliest, and simplicity of construction was emphasized. Considerable general mechanical simplification was obtained by omitting return windings at the outer periphery of the gap. This omission increases the stored energy by a large amount and would scarcely be practical in a larger magnet because the power supply would be too large for a reasonable repetition rate and rise rate. I^2R losses in the return windings are readily supplied. The large stray field is rather disadvantageous for many experimental problems.

The Birmingham magnet is made with one piece block-C laminations, 8 ft. square and $\frac{1}{2}$ inch thick (Fig. 7). Pairs of these plates are set on an arc, touching at the inner radius. The sector is then filled, above and below, by 1 inch plates extending inward to the rear of the cutout, and planed to a trapezoidal shape. The empty wedge remaining between every second plate is used as an air duct for cooling the coils. The assembly is fixed by continuous weldments front, rear, top and bottom, the front ones not being too near the gap. Pole tips are short sectors of $\frac{1}{4}$ inch plate held with bolted clamps to the main structure. Plates of yoke and pole tips are insulated with thick coats of paint and, in the pole tips, interleaved paper. The magnet rests on two circular steel rails embedded in a concrete foundation, and level was adjusted by shimming. This magnet, unlike the other proton synchrotrons, has no straight sections. The coil is wound with 144 turns of stranded cable, 380/0.036 inch strands, supported on insulating slats and cooled by an air blast through the wedge spaces, with the necessary baffles and shrouds. Some difficulty was originally had with buckling of the cables, but additional supports were added and the coil has operated 4 to 5 years without failure. After first assembly, a number of pole tips tore off upon pulsing. The air gap between pole tips and yoke was carefully adjusted and stronger clamps were fitted. It must be noted that forces of many tons are applied and relieved on these synchrotron magnets during each pulse. Since millions of pulses are expected (the life should extend to a few times 10^7 pulses) a certain amount of difficulty will be encountered due to loosening of metal parts, plastic flow in insulation, and metal fatigue.

The Cosmotron magnet is a one-piece C made of 288 blocks 72, per quadrant. Each block is somewhat over 6 inches thick and made of 12 one-half inch plates welded together at the places indicated in Fig. 7. The front welded tie straps cut $\frac{1}{4}\%$ of the total flux at saturation, and do not appreciably disturb the gap field. Insulation between the plates is "fish paper", a tough, impregnated cardboard. The blocks are set on steel base plates in groups of 12 and leveled with shims. Each group of 12 is fastened together with steel straps welded to the steel ties of the blocks. The sectors between blocks are left open because the most serious saturation is in the region of the yoke adjacent to the coil. Increasing the amount of iron within the poles only would give little benefit in the poleless-C, as well as making serious mechanical problems. The introduction of straight sections adds the problem of restraining the end blocks of the quadrant, which tend to be pushed into the straight section. Massive spreader bars above and below the straight section tank brace the quadrant ends, and the pressure within the quadrants is taken by jack screws in all the air wedges, adjacent to the gap, which prevent the plates from bowing. End blocks must have insulated bolts near the gap to keep the laminations from fanning.

The coil of the Cosmotron has 48 turns of copper bar, spread at the ends and front of each quadrant to give access to the gap. The inner coil bars are $\frac{3}{4} \times 1\frac{7}{8}$ inch, and the outer bars $\frac{3}{4} \times 2\frac{13}{16}$ inch. The copper is insulated by wrapping individual turns, is embedded in plastic ground insulation, and is clamped to the yoke with stainless steel frames. Each bar has a $\frac{3}{8}$ inch hole and water is circulated through the 384 individual bars in parallel. The quadrant ends are quite complicated, with the bends, bolted turn connections, and water leads. Three major breakdowns have occurred. A water leak from a cracked water tube caused a turn-to-turn short, requiring dismantling of the quadrant coil and replacement of the two bars involved. After 7×10^6 pulses, a bend cracked from fatigue and required extensive rebuilding, which included additional clamping of the coil in a manner to reduce localized stresses rather than gross motion. Another fatigue break has led to replacement of the entire coil, with design improvements.

The Bevatron magnet has 36 "sectors" per quadrant. Each yoke sector of the H is composed of eight pieces, two of each; cross pieces, end uprights, pole bases and pole tips (Fig. 7). The pole tips are $\frac{1}{4}$ inch plate insulated with vitreous enamel, the other pieces $\frac{1}{2}$ inch plate insulated with fish paper. The cross pieces and poles are roughly wedge-shaped, done by stepping the sides with various lengths of plate. The waves in the $n(r)$ of Fig. 11 are due to the step structure. The radial thicknesses of the two end uprights are inverse to their azimuthal thickness in order to make the areas equal. All the steel sector parts are fastened with insulated through — bolts or rivets and the entire structure is held together by an extensive system of tie rods and bolted clamps which essentially draw the top and bottom cross-pieces together. The final restraint (which establishes the required assembly precision) is a set of stanchions which supports the edges of the pole tips. These stanchions bear the weight of the upper parts, the clamping stress and the vacuum and magnetic forces. Azimuthal ties are bolted from sector to sector, and the massive straight section tanks include compression members which restrain the quadrant ends during pulses. The coil is wound with 88 turns, each turn consisting of two insulated 1.31 sq. inch copper cables, transposed at the quadrant ends, and paralleled quadrant by quadrant. The cables are supported by moulded plastic strips and are air cooled through a system of shrouds which conduct air from a large duct under the entire magnet. Upper and lower coils are separated by strong springs, and are held in the radial direction by braces and tie straps. The vacuum chamber encloses the pole tips and is an integral part of the structure. The proportioning of the structure allowed for installation of wide, short pole tips to give a 2×8 foot aperture for initial operation. The first-stage assembly was never made and the final pole tips were installed with the large stanchions providing the "crawl space" at the inner radius. The structure is supported on jacks which permit independent levelling of each sector. There has been one major breakdown of the Bevatron magnet after a few million pulses. Steel angles which fastened the air shrouds applied cyclic forces to their copper rivets, many of which fatigued and broke. Some of the angles were torn off and lodged among the cables so it was feared that shorts might develop. The magnet was partially dismantled and the angles replaced by non-magnetic materials.

The Synchrophasotron magnet has eight pieces in cross section, but is assembled differently from the Bevatron. Each quadrant has 12 "blocks" and each block has four top and four bottom cross pieces. These cross pieces are rectangular and are spread at the outer radius, leaving air wedges between. The end uprights of each block are made in two pieces with cutouts such that a hole is available in the radial direction for pump lines and general access. To do this, the uprights are laminated in the azimuthal direction. Flux patters at the corner joint where the cross laminations butt against the vertical ones at right angles are complex and difficult to reproduce. In addition, the upper yoke gap (X in Fig. 7) was varied to equalize the mechanical stresses. These variations are probably the reason for the azimuthal harmonics in this magnet, which are larger than those of the other proton synchrotrons. Pole bases are bolted to the yoke, and the pole tips (inside the vacuum chamber) are held in compression on the side walls of the vauum chamber by the clamp bolts of the structure. The entire magnet is made of 1 cm thick low—silicon steel and insulated with varnished cardboard. The pole tips each contain six 4 cm plates and have the proper width and taper to provide a continuous support for the vacuum chamber. The magnet pieces are held together with insulated bolts, and additional fastening is applied to the outer yoke pieces by welded straps. The coil has 44 turns of copper strap, 25×125 mm, to one edge of which is soldered a 24 mm diameter tube which carries the cooling water. The bars are insulated, enclosed in a dural jacket and have clamps bolted through the cross bars of the yoke. The magnet is supported on a massive concrete ring foundation, 9.2×3.5 m in cross section, and is leveled with shims.

16. Tolerances. The erection of a constant-gradient magnet requires tolerances which are readily met by standard engineering techniques. Of greatest importance in setting tolerances is the requirement that the equilibrium orbit lie in the center of the aperture. If large displacements of the equilibrium orbit are to be avoided then B_z at r_0 must have small first, second, and perhaps third, Fourier components in ϑ; and B_r must be very small (comparted to B_z) on the geometric median plane. The parts of the magnet must be set in place within the necessary tolerances; and the parts must mechanically and magnetically meet close tolerances. From (6.3) the amplitude of forced oscillation x_k produced by a harmonic of the field a_k is

$$x_k = \frac{a_k r}{k^2 + n - 1}; \qquad B(\vartheta) = B_0 (1 + \Sigma\, a_k \cos(k\vartheta + \delta)), \qquad (16.1)$$

but a displacement in radius, Δr, is equivalent to a field error

$$a = \frac{\Delta B}{B} = \frac{n \Delta r}{r},$$

and

$$x_k = \frac{n \Delta r_k}{k^2 + n - 1}, \tag{16.2}$$

also

$$z_k = \frac{n \Delta z_k}{k^2 - n}. \tag{16.3}$$

For $k = 1$ and $n = 0.6$ a forced oscillation amplitude, and semi-aperture loss, of 1 cm requires a first harmonic amplitude $\Delta r = 1$ cm and $\Delta z = 0.7$ cm. Erection tolerances are generally set much less than such values to allow for settling of foundations and similar disturbances during operation. If a magnet is tilted about the ϑ-axis through angle α the median plane will shift by $\Delta z = r\alpha/n$. The multiplier of α ranges from 1 to 5×10^3 cm, so the leveling of the magnets along the radius vector must be controlled to a fraction of a milliradian.

If the disturbance of the orbit is to be less than $10^{-3}r$, which is about $1/_{10}$ the vertical semi-aperture of the existing machines, then the a_1 of (16.1) must be less than 10^{-3}. At intermediate fields the permeability of the iron is of the order of 3000 and, if the magnetic length in the iron is 10 times the air-gap length, the reluctance of the iron is $1/_{300}$ of the total. Thus the permeability of the iron can vary some 30% and affect the field only 1 part in 10^3. However, the length of the gap must be machined and assembled to a tolerance of the order of 1 in 10^3 to give uniform field strength. It is observed that the permeability of a given steel, without special selection, will vary some 15% at low and intermediate induction; and about half as much at high induction. The coercive force will vary some 20%. While mechanical tolerances primarily determine $B(\vartheta)$ at intermediate fields, the variations of the magnetic properties of the steel give the largest harmonics at injection field, and somewhat less at saturation. The variations can be smoothed by thorough mixing or "shuffling" of the steel of the numerous discrete batches in which it is made. Mixing is most effective if it is possible to accumulate all the steel prior to fabrication and then to mix the steel in a manner which minimizes the important harmonics, which are the first and second in a constant-gradient synchrotron. This mixing procedure was done on the Bevatron magnet. Individual magnetic testing of the several pieces making up the cross section of the H is so difficult and time-consuming as to be impracticable. However, the one-piece C magnets, composed of 200 to 300 blocks, can be measured and smoothed directly. A complete block-testing program was carried out for the Cosmotron ([*34*], p. 760) and for the CEA magnet [*50*]. When the field characteristics are known for each block, a stacking arrangement can be made by pairing adjacent blocks with corresponding high and low values of the most troublesome parameter, then pairing pairs, etc. The worst variation in the Cosmotron blocks was in the remanent field, leading to some 2% variation in B_i. Very little disturbance of the median plane was observed, at most the order of 0.1 inch, because the properties of each single steel plate are relatively uniform within the plate. Matching of the components of an H magnet to obtain symmetry above and below the median plane requires careful mixing of the steel. The CEA block measurements were made by difference methods and gave quite accurate results [*50*]. The greatest variations of the Saclay magnet blocks were in the remanent field, as expected. When pulsed, the blocks varied ± 1 gauss at $B_0 = 300$ gauss, rms variation 0.41 gauss. It is estimated that the erection order determined from the measurements gives probability of 99% that the maximum equilibrium orbit deviation will be less than 0.76 cm.

Measurements of the assembled Cosmotron showed very small low order harmonics, and observations of orbits confirmed the measurement results.

Since there is fringing field at the quadrant ends, it is necessary to foreshorten the length of steel in order to obtain a "magnetic length" of 90° in ϑ. The Cosmotron "quadrant" length was set at 88.4° by integrating the stray field along the orbit at low field strengths, and setting $\int B\, ds/B_0 = \pi r_0/2$ for one quadrant. At peak field the stray field is quite different, and the error in quadrant length is about 0.2°, which error causes a loss of radial aperture of some 0.2 inches. The Bevatron magnetic quadrant length is greater than the angle subtended by the pole tips by 0.5° at B_i, 1.5° at 3000 gauss, and 0.2° at saturation. The pole tips were fore-shortened 1° per quadrant.

17. Design and measurements. Design of a synchrotron magnet is best approached by a simultaneous program of calculations and model magnet measurements. Pole shapes can be calculated very accurately by the relaxation method ([34], p. 741) and eddy currents can be estimated by one-dimensional analogies. The use of electronic computers for calculation of complex shapes is beginning to be well developed. (Some very irregular shapes have been computed successfully by the Midwestern Universities Research Association using an IBM-704, but are yet unpublished.) The calculations become tedious when three-dimensional structural details and eddy currents are inserted, and when the approximation of infinite μ is replaced by an empirical $\mu(B)$. It is also difficult to be sure that all necessary parameters have been inserted in the calculation. For these reasons models of the magnet are constructed and measured in detail. It is usually convenient to model a short sector, only long enough to give a usable region of azimuthal uniformity, at fractional scale. However, a short sector at full scale is useful in many measurements since the size of the measuring heads of the magnetic instruments is less restricted than at small fractional scale.

Although absolute values of field strength and gradient must be obtained, it is most important that the shape of the field be precisely measured, and the shape is a slowly varing function of the field strength. Search coils are well adapted to this class of measurements for, although it is difficult and time-consuming to measure the absolute effective area (or area-turns) of a multi-layer coil to better than $\pm \frac{1}{4}\%$, the ratio of areas of two coils can readily be determined to a part in 10^4, and two aged coils can be made and held equal in area to within a few parts in 10^5 ([34], p. 743). Search coils can be flipped in a steady field and their output read on a fluxmeter, or they can be connected to an electronic integrator and the variation of field with time can be read on an oscilloscope, or photographed[1] [34]. The remanent field is added to the integrator output as the constant of integration. Thus, two matched, spaced coils can be used to determine gradient, and their output can be ratioed to that of a single coil by null readings to obtain gradient/field. It is preferable to construct the search coils as long rectangles which subtend two or more "blocks" so the coil ends can be placed at the center of a block where $dB_z/d\vartheta$ is small. Often very long coils are made to facilitate azimuthal integrations, or to smooth local variations with ϑ. The median surface can be located by using a large coil with its axis in the r direction. The high angular precision necessary can be obtained by adjusting the coil so its output is the same after rotation through exactly π around any of the three orthogonal axes.

---

[1] R. E. RICHARDSON and D. C. SEWELL: Rev. Sci. Instrum. **22**, 697 (1951). — E. W. FULLER and L. U. HIBBARD: J. Sci. Instrum. **31**, 36 (1954). — Onde Electrique, p. 1076, Nov. 1955. Search coil design is discussed in M. W. GARRETT: J.A.P. **22**, 109 (1951). — A general survey is given by J. L. SYMONDS: Rep. Progr. Phys. **18**, 83 (1955).

A permalloy strip magnetometer is useful for exploration of the remanent field. The nuclear resonance, or proton-moment, flux-meter is superior for measurement of absolute values of fields ([5], p. 404), but is hard-put to compete with search coils for gradient measurement. In the constant-gradient synchrotron magnet, B_z varies about 1 in 10^3 per inch so to determine the exponent n to 1% with measurements spaced an inch requires a precision of order 10^{-5}. Search coil groups accurately matched and checked by interchanging their positions can quickly and accurately measure such small gradients, but the use of instruments requiring subtraction of point measurement values is very exacting.

IV. Magnet power supply.

18. Design. The magnet requires very large power machinery for its excitation, and practical considerations involved in the power supply have considerable bearing on selection of synchrotron parameters. Most important are magnet

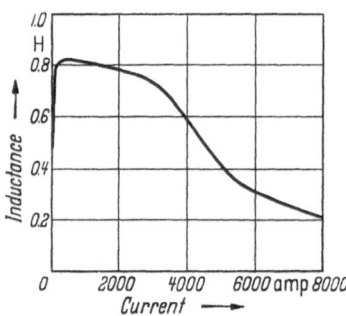

Fig. 20. Inductance of Cosmotron magnet as a function of current at $\dot{B} \approx 15\,000$ gauss/sec.

stored energy, pulse rise time, repetition period, and magnet losses. The stored energy is, to a first order, proportional to $B_0^2 r_0 A$ (A is the gap area). If the relative aperture is selected then $A = K^2 r_0^2$, and since $p \propto B_0 r_0$, the stored energy will be proportional to $p^2 r_0 K^2$. The designer will reduce the relative aperture as much as is consistent with orbital estimates in order to reduce the size of both power supply and magnet. Stored energy is further reduced by working at as high a peak field, and small a radius, as can be attained. However, in the iron circuit the stored energy increases more rapidly than B_0^2 as the magnet saturates, the simple relationship is no longer valid, and a limit of increase in B_0 is soon reached. (If there is any excess the designer normally uses it up by increasing p.) (An air-core magnet would be designed with the largest possible B_0, and smallest possible r, limited primarily by the strength of the materials employed.) The peak power rating of the supply is inversely proportional to the rise time, or charging time, of the pulse. If the rise time is made long, to lower the peak power, then the magnet $I^2 R$ losses will be large, and little stored energy will be recovered from the magnet. The repetition period cannot be shorter than about twice rise time, although it is usually limited by the thermal ratings of the machinery and cooling of the magnet. A short rise time is desirable to diminish gas scattering, but requires high radiofrequency accelerating potentials. Conversely, a long rise time is useful for increasing intensity with multi-turn injection (unless space charge limits are approached). Long injection times can also be obtained by slowing the initial rate of rise of the current pulse. Saturable reactors have been proposed for such purpose, and some experiments have been done with diminished starting voltage.

When the magnet geometry and peak field are determined, the stored energy and flux linkage are specified. The area of copper allowed then determines the ratio of inductance to resistance, L/R. The copper may then be subdivided into the number of turns which gives the most favorable volt/amp ratio for the power supply. The current variation is described by the familiar equation

$$L(I) \frac{dI}{dt} + IR = E(I, t) \tag{18.1}$$

in which L is a highly variable function of current. Fig. 20 for the Cosmotron is illustrative. The power supply terminal voltage, $E(I, t)$, must usually be allowed to decrease with increasing current for reasons of economy. Linear approximations of power, energy and losses from a simplified cycle such as shown in Fig. 1 are too inaccurate, so it is necessary to integrate (18.1) numerically using empirical or estimated functions for L and E (see Fig. 23). Extensive compromise of characteristics is obviously necessary. Characteristics selected for the four operating machines are given in Table 2.

Table 2. *Magnet power supply parameters.*

Parameters	Birmingham PS	Cosmotron	Bevatron	Synchro-phasotron	Units
B_{max}	12600	13800	15400	13000	gauss
I_{max}	12500	7000	8333	12800	amp
Stored energy	7	12	80	148	10^6 joules
Rise time	1	1	1.9	3.3	sec
Initial E	1.1	5.4	14	11	kv
Peak power	10	26	100	141	Mva
Repetition period	10	5	6	12	sec
Ave Magnet loss	230	650	3500	4000	kw
Peak proton T	1	3	6.2	10	Bev

The energy and power quantities of Table 2 show the engineering and economic difficulties involved in pushing the energy of the constant-gradient synchrotron much over 10 Bev. The fluctuations in power during a pulse are far too large to allow direct connection to a public utility power network, so some form of energy storage is necessary. A rotating mass is the most practicable storage device, since the amount of condensers required for these machines would be enormous.

19. Description. The Birmingham proton-synchrotron power supply uses dc generators [*36*]. Two are mounted on a common shaft with a 36 ton flywheel and 1500 hp induction motor. To produce a pulse the series ignitron is fired (Fig. 21) as the contactor starts moving. The ignitron gives a noise-free accurately-timed contact and is shorted by the contactor before its current-rating is exceeded. The pilot exciter, controlled by synchronized cam switches, is reversed to produce the wave forms of Fig. 21, and the contactor is opened as the current returns through zero. The control voltages at the end of the pulse are adjusted by observing a magnetometer and regulating the pilot exciter to give a standard remanent field value.

The dc generator system is relatively simple, and has been trouble-free in operation, but it becomes cumbersome when extended to higher power ratings. The three larger synchrotrons use flywheel-alternators and ignitron rectifier-inverters. Alternators can be built in very large ratings, and such a system can be voltage reversed more rapidly and controlled more flexibly.

The Cosmotron magnet power supply [*34*] uses a single 21000 kva rated alternator coupled to a 43 ton flywheel and 1750 hp induction motor, running at 900 rpm. This rotating system stores about 120×10^6 joules. The alternator is wound with 12 phases, connected as 4Y-s which supply two six-phase, half-wave rectifiers. The two banks of twelve ignitrons (two in parallel per phase) are displaced 30° to produce a 12-phase ripple fundamental. The basic circuit (Fig. 22) will rectify if the ignitrons are fired when their phase voltage is positive. If the firing angle is retarded about 150° the ignitrons will strike when their phase voltage is negative, the striking of each phase will extinguish the preceding-phase tubes, and the dc terminal voltage will reverse, reversing the power flow.

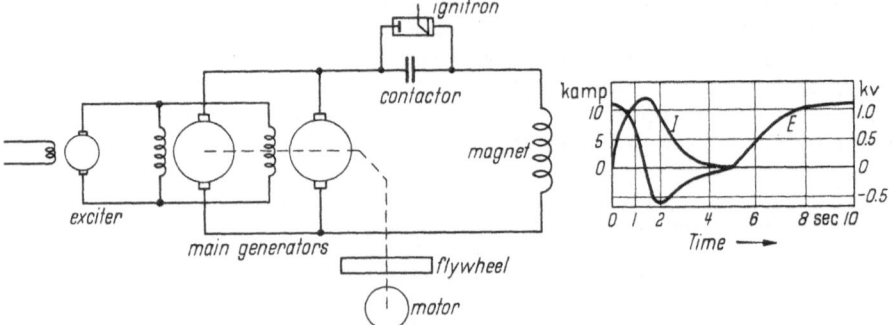

Fig. 21. Schematic diagram of Birmingham synchrotron magnet power supply, and cycle of voltage and current *vs.* time. After [*36*].

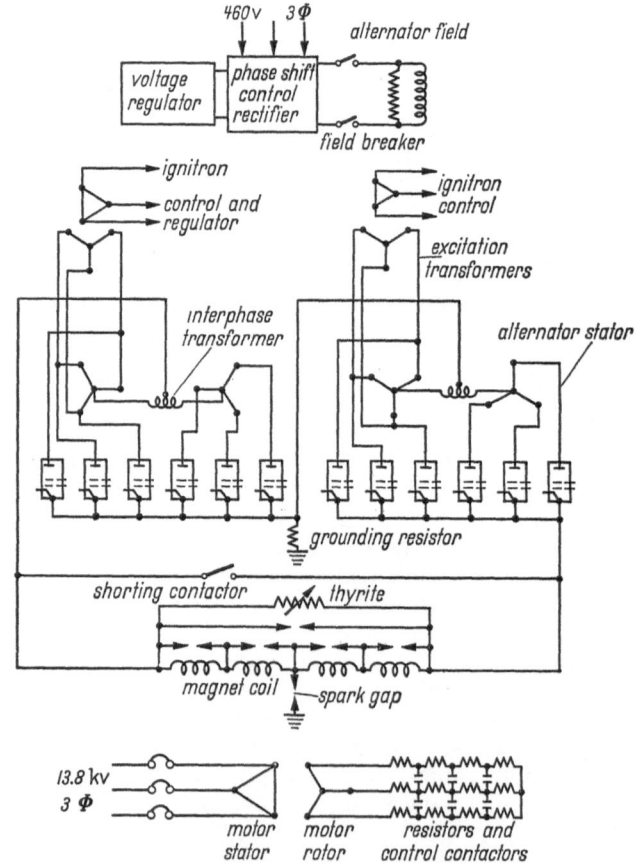

Fig. 22. Schematic diagram of the Cosmotron magnet power supply. From [*34*].

A typical cycle is shown in Fig. 23. The set is changed over from rectification to inversion in two steps to halve the torque shock on the shaft and foundation, the step interval being set to one-half period of the shaft torsional oscillation.

This split arrangement also permits one six-phase bank to be held on invert while the other is rectifying so as to produce a "flat top" on the magnet current during which protons can steadily be ejected for experiments wanting a long output pulse of particles. Rectification is always started with the same phase to control the initial wave shape, and thus reduce field variations at injection. Control of the ignitrons is done by voltages derived from excitation transformers connected to two alternator Y-s. An ignitron exciter for the alternator field is electronically controlled to reproduce the magnet cycle from pulse to pulse. The use of taps permits the voltage, and consequently $\dot{B}$, to be adjusted from $\frac{1}{2}$ to full value, and the control circuits permit any desired pulse length to be selected. With short pulses the

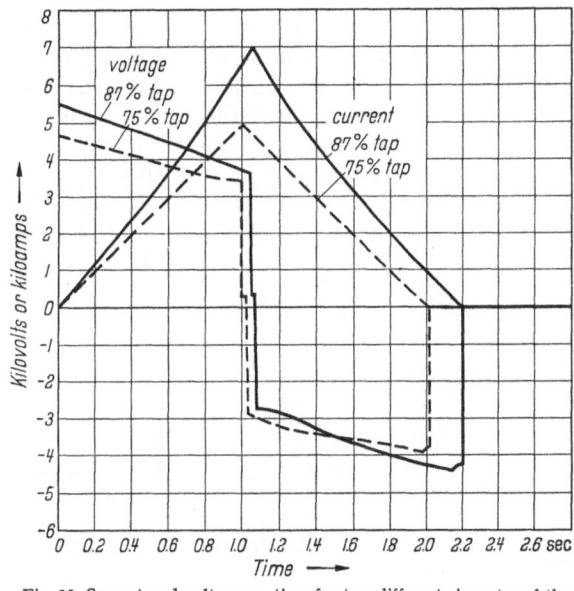

Fig. 23. Current and voltage *vs.* time for two different rise rates of the Cosmotron. From [*34*].

repetition rate can be increased. The large stored energies demand meticulous design of protection schemes. The simple system fitted to the Cosmotron has also been adopted by the Bevatron and Synchrophasotron. Since circuit breakers

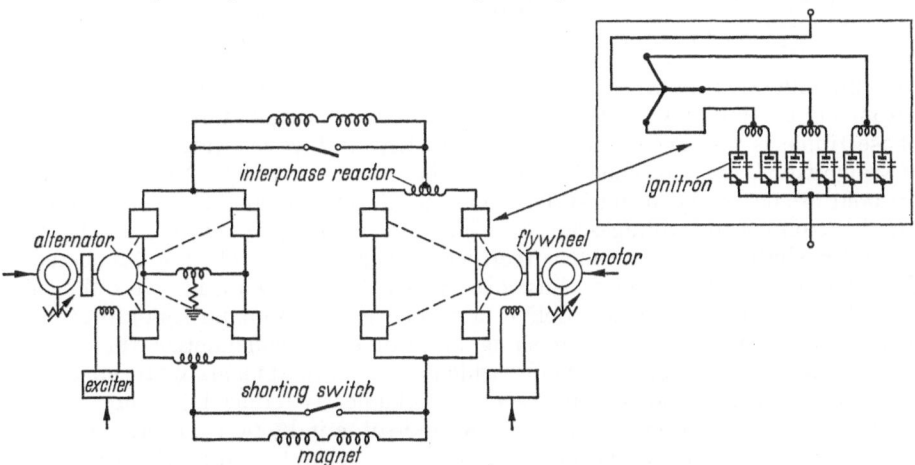

Fig. 24. Schematic diagram of Bevatron magnet power supply. The four Y's on each alternator are successively displaced 30°.

cannot be opened on an inductance as great as a synchrotron magnet, when serious trouble is encountered a shorting contactor is closed across the magnet (Fig. 22). The magnet has adequate thermal capacity for its selfdischarge; the short-circuit on the power supply is cleared by shorting the igniters and stopping commutation. Added protection is given by simultaneous tripping of a high-speed field breaker. The magnet coil is also protected by spark gaps and a thyrite (non-linear) resistor.

The Bevatron power supply[1,2] is similar but, due to the greater power required, is equipped with two 46000 kva generators, each on its own shaft with a 67 ton flywheel and a 3600 hp induction motor (Fig. 24). The two power units and two halves of the magnet are series connected as shown to keep the voltage to ground to a minimum. (The connection of footnote 1 did not operate properly because the sets would not parallel reliably during inversion, and the power supply was reconnected as in Fig. 24.) The rectifiers use 48 ignitrons.

Power requirements of the Synchrophasotron led the designers to use four 37500 kva alternators on separate shafts, each with its 50-ton flywheel and

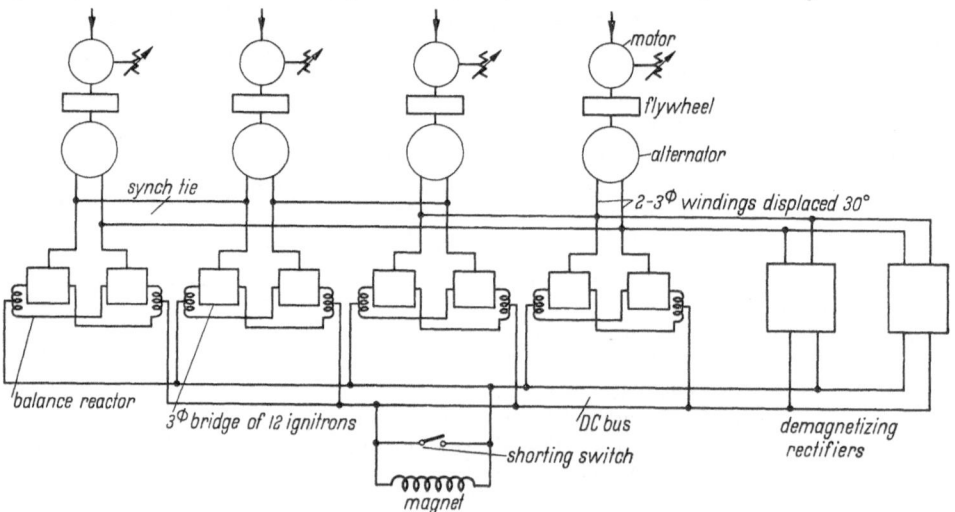

Fig. 25. Diagram of Synchrophasotron magnet power supply.

2500 kw induction motor (p. 378, [5]). The alternators are paralleled by a synchronizing tie (Fig. 25). Each alternator has two windings at 30° so that twelve-phase ripple is produced. Four main rectifiers, full-wave bridges using a total of 96 ignitrons, are paralleled on the dc bus to the magnet. Numerous circuit breakers are necessary in this system in addition to the magnet shorting contactor. The ignitron control voltages are derived from auxiliary generators on the main generator shafts. Two auxiliary rectifiers, 12 ignitrons each, are connected to the alternators and to the dc bus, one with polarity reversed. After the main pulse these auxiliary rectifiers deliver a sequence of current pulses of alternating polarity and successively diminishing amplitude to demagnetize the magnet. The process requires some 5 seconds, which makes rapid pulsing with short pulses difficult. However, the remanence of the magnet is reduced to some 3 gauss.

The preceding power supplies use wound-rotor induction motors with rotor resistance controlled by switched resistors or variable liquid rheostats. The programming of variation in rotor resistance while the speed is changing during a pulse limits peaks and slow transients in the power drawn from the source.

20. Ripple. The final phase oscillation frequency, $\Omega_\infty/2\pi$, can be crudely approximated from (5.2) by setting the energy gained per turn, $eV \sin \psi_0$, equal

---

[1] J. V. Kresser: The Bevatron power plant. Electr. Engng. **71**, 338 (1952).
[2] J. L. Boyer and C. R. Marcum: Ignitron Converters for High-Energy Particle Accelerators. Trans. Amer. Inst. Electr. Eng. **69**, 1110 (1950).

to $E/f_\infty \tau$ (E is the final energy, τ the rise time, and the final revolution frequency $f_\infty = c/2\pi r_0$). Then if $\sin \psi_0 \approx \cos \psi_0$, and $n = 0.6$, (5.2) reduces to

$$\frac{\Omega_\infty}{2\pi} \sim \frac{1}{2\pi} \sqrt{\frac{2.5\,c}{r_0 \tau}}\,. \tag{20.1}$$

This frequency is a slowly varying function of design; the pulsed synchrotrons just described have a final phase oscillation frequency between $\frac{1}{2}$ and 3 kc, and an initial value 2 to 3 times the final value. During acceleration the phase oscillation frequency will sweep through the ripple harmonic frequencies of the power supply. If there is ripple in the magnetic field the phase oscillations will be excited by it, mostly when the frequencies coincide (see Sect. 6 and [17]).

If the rectifiers are six-phase the ripple has components of frequency $350\,k$ cps, where $k = 1, 2, 3, \ldots$ (for a 60 cps machine running somewhat under synchronous speed) and if twelve-phase of frequency $700\,k$ cps. Thus a twelve-phase rectifier has only two or three potentially resonating harmonics at 1400, 2100, and perhaps 700 or 2800 cps, while a six-phase system would produce five or six at least. Since numerous ignitrons are required to carry the heavy current it is good engineering practice to use a twelve-phase circuit. One of the most convenient methods for regulating the voltage of an ignitron rectifier is firing delay, or phase shift control. This method is probably not usable during the accelerating part of the cycle due to the large increase it produces in the ripple voltage. During inversion the ripple voltages are very large. The circuits of Figs. 22 and 24 have been used to give approximately constant current at high fields, or slow rise (low $\dot{B}$) during injection, by putting half the ignitrons in the rectifying condition while leaving the other half inverting. The dc component of voltage can then be adjusted by phase-shifting the rectifier and thus controlling $\dot{B}$. If the phase oscillation frequency is not equal to a ripple harmonic during such period of operation, beam losses are not excessive.

When frequencies of the order of kilocycles are considered the magnet is, electrically, a lossy delay line and the ripple component of the magnetic field will vary azimuthally. Such variation can couple into either betatron or phase oscillations. Considerable and erratic beam loss in the Bevatron was traced to azimuthal ripple variation which shifts as the relative phase of the two independent alternators shifts. A rather elaborate synchronizer was constructed to keep the two machines at such relative phase angle that the 700 cycle ripple component cancelled around the series circuit.

The ripple frequencies are much higher than the frequency components of the slow basic pulse so filters are quite practical. Conventional filters between power supply and magnet are straightforward in principle but are large because the series elements must carry the heavy currents. The transients produced in a filter (such as the simple low-pass type of Fig. 26a) by the initial step wave of applied voltage propagate through the system and may largely destroy the beam at injection. Transients can be reduced by precharging the condensers and using a high-speed switch, or diodes, to connect them as the pulse begins. (This has been done on the Bevatron.) Since the total ripple power in the magnet is of the order of kilowatts it is not difficult to use an electronic feed-back filter. The type of Fig. 26b was fitted to the Cosmotron early in 1952 when magnet ripple was suspected of destroying the beam, but when difficulties were found elsewhere the filter was disconnected. The coupling winding of Fig. 26b consisted of four turns of pole face returns around the magnet gap itself. Ripple was reduced more than a factor of 5, even though the equipment was hurriedly constructed. The scheme of Fig. 26c is proposed for the CEA magnet ([5], p. 376). Fig. 26d is similar but uses the voltage induced in a coil in the magnet gap to provide the signal for an audio amplifier which then has its output connected in inverse polarity to a coil wound around the gap. This system has been used in the Synchrophasotron and reduced the ripple by a factor of 5 or 6 ([5], p. 381). A simple internal feedback circuit is used successfully on the Bevatron (Fig. 26e). Two turns around the yoke are capacity coupled to one turn around the gap composed of the outermost pole face windings. The ripple is thus forced into the stray field and is reduced in the gap by a factor of 4 to 8 (H. G. Heard: unpublished). The circuit is heavily damped so the initial transient is attenuated by injection

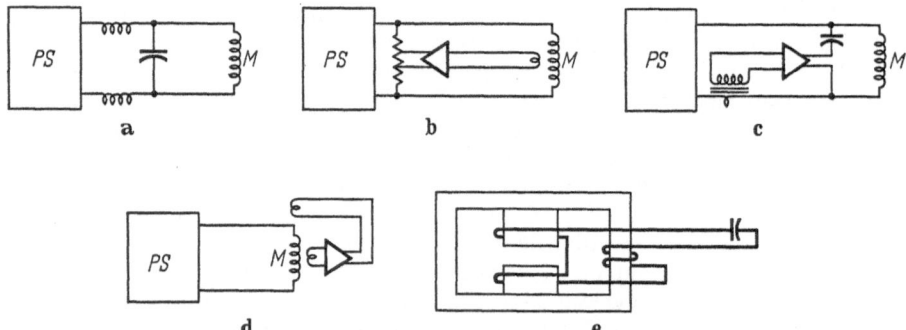

Fig. 26 a—e. Magnet ripple filter types. *PS*-magnet power supply, *M*-magnet exciting coils.

time. Filters are useful for improving the injection and acceleration characteristics, but equally important, they reduce the magnet ripple structure in particle beams from targets (see Sect. 30).

V. Vacuum system.

21. Vacuum chamber. The vacuum chamber of a proton synchrotron would ideally; withstand the external pressure, occupy none of the area of the magnet gap, neither transmit nor evolve any vapor, present a conducting surface to the beam, and conduct no eddy currents to disturb the shape of the magnetic field. It is probably impossible to satisfy all these requirements, and it is interesting to note that the vacuum chambers of the synchrotrons are quite individual compromises.

The problem of insulating surfaces presented to the beam has no easy numerical solution. If charged surfaces produce a known electric field in the orbit space, the forced oscillations are easily calculable. As a simple example, a vertical electric field of about 50 v/cm in two successive quadrants of the Cosmotron would drive the beam completely out of the aperture. This field requires roughly 5×10^{12} charges, a quantity lost during acceleration in only one or two minutes of operation. However, the estimation of charge collection on an insulating surface, multiplication of secondaries, etc., depends largely on the assumptions made. It is, therefore, wise to keep the area of non-conducting surfaces facing the aperture at a minimum. No experimental data is at hand, but there are numerous instances known of complete beam loss in electron machines, where the problem is much more serious due to the low velocity and field at injection. Vacuum chambers are best illustrated by using the existing designs as examples.

The Birmingham proton synchrotron is fitted with a ceramic vacuum chamber which, considering vacuum properties and eddy currents, is an excellent solution of the problem. Unfortunately, as one goes to larger machines, the aperture area taken by the thick ceramic walls becomes prohibitively expensive, and it is doubtful whether sections of adequate strength can be cast. (Test sections fabricated for the Cosmotron in 1949 collapsed at about one atmosphere.) The Birmingham chamber is cast in 60 sectors of electrical porcelain, with average wall thickness 2.8 cm. Vertical inside height is about 10 cm, and inside width is 35 cm for 42 sections, 47 cm for the remainder. These wider sections are arranged in three groups of six so that the 12 cm step in the outer periphery can be pierced with a hole for injection and for removal of particle beams. In addition, five sets of 6 sections have side T's for connection to the pump manifolds. Fig. 27 shows a typical arrangement, at the injector azimuth [37], [39]. A platinum

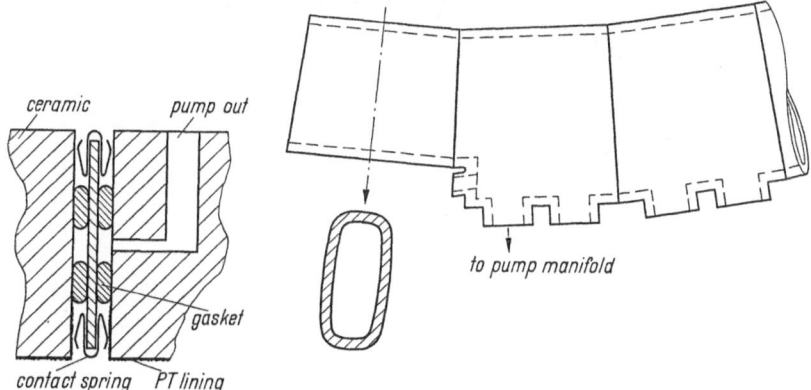

Fig. 27. Ceramic vacuum chamber sections of Birmingham proton synchrotron. Inset shows section of gasket seal between adjacent pieces. From RIDDIFORD [37].

conducting coating (order of an ohm per square) is applied to the entire inner surface of the chamber, over a slightly conducting glaze, and broken where desired by scribing. The sections are sealed together with rubber gaskets bonded to both sides of a metal washer having the same section as the chamber. Connections to the platinum coating are made with springs (see Section of Fig. 27). The chamber also serves as the accelerating electrode. Approximately 100° are sprayed on the outside with copper stripes 0.002 in. thick and an inch wide to form the "Cee". The outer and inner coatings are connected so that little radiofrequency field will exist in the body of the ceramic to cause heating. At the ends of the Cee the outer conducting stripes are carried across the end of the section to the inner coating, which is broken to form the two accelerating gaps. The spacing of the vacuum chamber from the poles averages some 2 cm in order to keep the Cee capacity reasonably small. The chamber is assembled in place and then mechanically pre-compressed radially by arms connected to the magnet, after which it can be evacuated without further movement. The side tubes are connected to the manifolds with flexible stainless-steel bellows. Five 16-inch diffusion pumps having a speed of 2000 l/sec and liquid-air traps maintain a pressure below 10^{-6} mm of Hg in this relatively "clean" system.

A limited attempt to design a ceramic chamber for the Cosmotron was stopped when it appeared that risk of collapse was too great. The final design in use consists of one inch stainless-steel side walls joined with tapered stainless "grid-bars" about two inches wide, spaced $\frac{1}{16}$ inch ([34], p. 821). These grid-bars (see Fig. 28) are insulated from the side walls at both ends with Kel-F boots. To seal this assembly it is necessary that a gas-tight blanket be stretched across the side walls on top and bottom. Thin Inconel sheets could be used, but reliable vacuum-tight seams were not available. The distortion of $n(r)$ caused by 0.008 inch

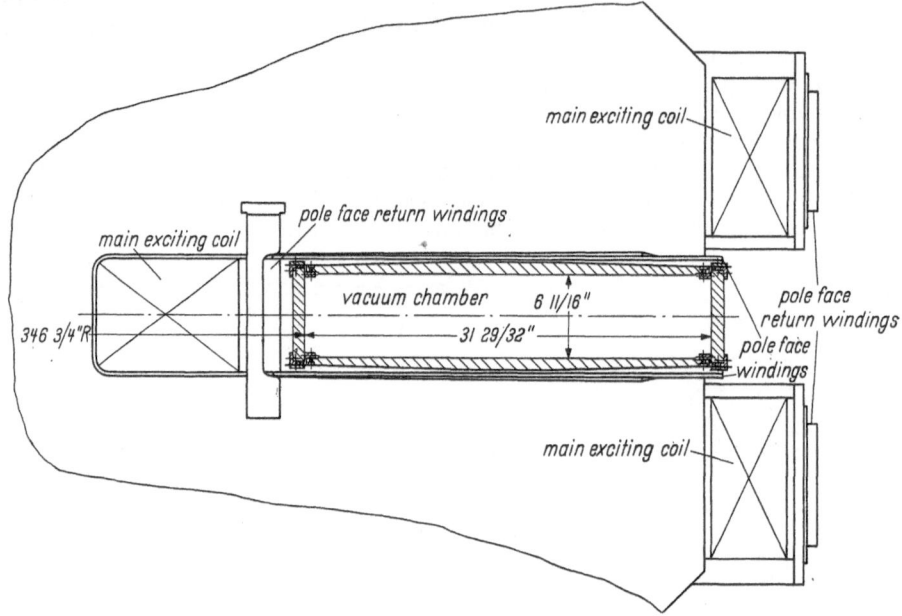

Fig. 28. Cross section of Cosmotron vacuum chamber, showing location of pole face windings. From Jacobus and Polk, [34], p. 822.

sheets of Inconel (7 milliohms/square) one quadrant long is shown in Fig. 29. Such gradient alterations could, if necessary, be corrected by pole-face windings. One thin plastic cover material tested, Kel-F (monochlortrifluorethylene) has vacuum properties nearly as good as those of metal sheets ([34], p. 827). Unfortunately, the material is so inert that

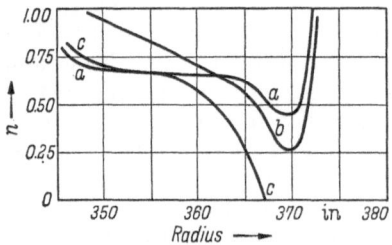

Fig. 29. n vs. r in Cosmotron. (a) Without vacuum chamber membrane; (b) 0.008 in inconel membrane insulated from walls; (c) connected to walls. From L W. Smith (unpublished.)

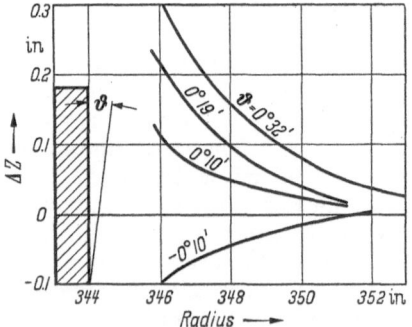

Fig. 30. Shift in median surface, Δz, for rotations of rear wall of vacuum chamber. From W. H. Moore, [34], p. 751.

vacuum-tight seams cannot be reliably produced. The Cosmotron sealing blankets are Myvaseal, a low vapor-pressure rubber. The grid bar surfaces are ground and polished so the rubber blanket will seal to them and only the cracks will present a gassing surface. Gas evolution is then small enough for a high-speed pumping system to achieve the necesary pressure. The Myvaseal is sufficiently conducting to drain charges from the bars. The large quantities of stainless-steel in the gap are type-305, whose normal magnetic permeability is 1.01, and does not exceed 1.1 after severe cold-working. Eddy currents in the rear wall of the chamber will seriously distort the median plane if the wall is tilted (Fig. 30), and careful control of alignment is necessary. The ends of each quadrant of chamber are connected to an "end box". The closure involves a three-way joint, which is rather common in accelerators, and is made with a moulded T whose arms are vulcanized to the necessary lengths of extruded gasket to produce the full gasket "harness". This is drawn schematically in Fig. 31.

The vacuum chamber deflects $\frac{3}{8}$ inch under load and leaves a net vertical aperture of $6\frac{5}{16}$ inch. The pole face windings, outside the chamber, are inserted in the magnet gap first. An entire quadrant chamber is assembled in one piece on a fixture, and then is inserted into a magnet quadrant gap by the use of an elaborate set of handling jigs. The straight sections are filled by stainless-steel boxes of simple construction with an aperture (2×4 feet) considerably larger than that of the chamber. Twelve 20 inch oil-diffusion pumps, each with speed of about 1500 l/sec at the chamber wall, were installed. One pump was removed to make way for ejection equipment and the remaining eleven will hold the average pressure at about 3×10^{-6} mm of Hg. If more than two or three of the remaining pumps are turned off, the gas scattering makes operation unsatisfactory. No liquid nitrogen is required, but the pump baffles are refrigerated to $-20°$ C.

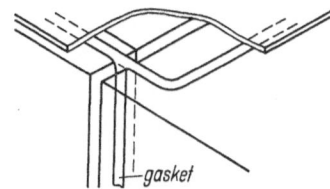

Fig. 31. Schematic diagram of three-way vacuum joint made with moulded "T"

The large H-magnets require so much steel and copper, and such large power supplies, that additional gap area for the vacuum chamber can hardly be afforded. The Bevatron vacuum chamber is composed of large sheets of stainless-steel, about $\frac{1}{16}$ inch thick, located well away from the gap (Fig. 32). The sides are supported against the stanchions, which also act as pole tip separators, and the top and bottom are supported by the split pole pieces and the large inner stanchions. The corners are sealed by double gaskets to $1\frac{1}{2}$ inch square corner bars. The large top and bottom sheets must be broken into short azimuthal lengths. This is done by cutting the sheets radially every second magnet sector and sealing them together with insulated radial bars and rubber gaskets (see plan view, Fig. 32). These bars run in grooves left in the pole tips and bases. Similar vertical bars subdivide the side walls. The vacuum joint where azimuthal, vertical and radial bars join is made with a set of four moulded-T gaskets using a rather complicated version of the technique shown in Fig. 31. Since most of the vacuum joints are buried in the magnet structure, all gaskets are spaced double-type and a complete set of pump-out tubes is brought to accessible locations. The eddy currents in this chamber do not affect the shape of the field in the gap to any great extent because the metal sheets are so deep in the poles. The field strength is reduced by the eddy currents; it is found that the field in the two center sectors of a quadrant, where the top and bottom sheets span one sector rather than the two elsewhere, differs about 5 gauss. Pole base windings have been self-energized for correction. At the straight sections the quadrants seal to massive steel "tangent tanks" some 7×9 feet in cross section. Six 32 inch oil diffusion pumps are attached to each tangent tank and will hold the pressure within the tank to about 10^{-6} mm of Hg. Shorts in the vacuum chamber joints, caused by chips or accidental touching of adjacent parts, can set up eddy currents of destructive magnitude. A complete electrical monitoring system is used to detect such short circuits.

17*

The Synchrophasotron has a double vacuum chamber with rough vacuum in the outer chamber, which can be made of materials having a gassing rate too large for high-vacuum use, and a thin inner chamber which need not withstand

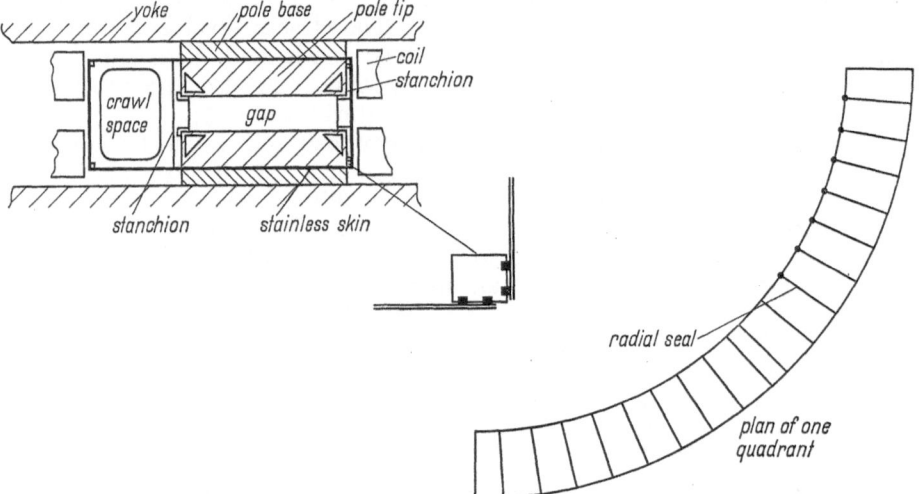

Fig. 32. Cross section of Bevatron vacuum chamber and plan showing insulated radial joints.

large forces. The complexity and poor accessibility of a double chamber are balanced by its advantages of small space requirements, good vacuum properties and small eddy currents. The rough vacuum chamber is supported between the pole tips and bases (Fig. 7), and on the 75 mm thick stanchions which space the poles. The top and bottom are textolite sheets and the sides are rubber blankets bent over the textolite sheets and sealed to them by wedges under the pole bases. Radial gaskets seal the joints of the textolite sheets. Reduction of large eddy currents in the stanchions requires holes to be milled, which are then filled with textolite to support the side rubber blankets. Each pole tip is divided into three azimuthal portions by 4 cm thick plates interleaved with the 1 cm

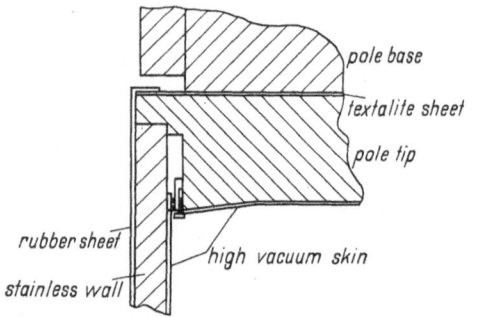

Fig. 33. Corner seals of Synchrophasotron vacuum chamber showing rubber sheet sealing rough vauum and stainless skin sealing high vacuum.

laminations, and a stainless strip is gasketed to the edge of each pole tip. Three stainless-steel sheets 35 × 200 cm, and 0.15 mm thick, are sealed to each pole tip to form the top and bottom of the high-vacuum chamber. These stainless sheets are carefully insulated and are clamped to their gaskets by strips bolted to the edge strips and the thick interleaved plates. Adjacent pole tips are then sealed by radial gaskets. The vertical walls of the high vacuum inner chamber are 4 mm thick; pressed against the side strips of the pole on sealing gaskets. (A highly simplified sketch of one corner is shown in Fig. 33.) The T joint formed at the junction of pole tips and side walls is quite complicated (see [5], p. 385). The side wall pieces span two pole tips, and half have oval pipes welded to them for vacuum pump connections, targets, etc. The pipes pass through, and are sealed to, the rubber side blanket. The rough vacuum chamber is pumped to about 1 mm of Hg with mechanical pumps, then the high-vacuum chamber is evacuated by 56 38-cm oil-diffusion pumps with an unbaffled speed of 5000 l/sec. With liquid nitrogen traps, a pressure of about 5×10^{-6} mm of Hg is maintained.

22. Gas scattering observations. The predictions of loss due to gas scattering (Sect. 8) can be compared with synchrotron operation by two relatively simple experiments: the proportion of accelerated beam lost is measured as a function of pressure in the vacuum chamber; or the proportion lost is measured as a function of aperture at constant pressure. There are several serious uncertainties in the experiments. Pressures are measured with air-calibrated ion gauges, but the mixture of gas in the

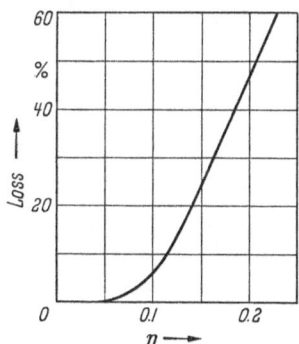

Fig. 34. Gas scattering loss *vs.* η.

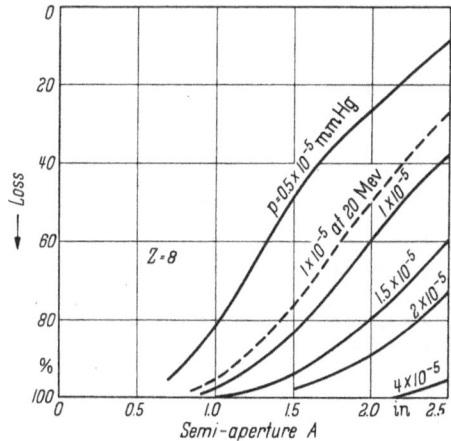

Fig. 35. Gas scattering loss in the Cosmotron as a function of semi-aperture, and at full energy. Dotted curve at 20 Mev.

synchrotron contains large fractions of heavy molecules, which scatter more effectively than air molecules. FOUND and DUSHMAN[1] made measurements which show that the ion current of the gauge equals kpN (k is an electrical constant, p the pressure, and N the molar number) and note that "the number of ions produced per unit electron current is (in most gases) approximately proportional to the total number of electrons per unit volume". (They also found that for hydrogen the ion current is too small, compared to air, by about a factor of two.) Thus the ion gauge compensates rather well, and the loss percentage is of the right order of magnitude[2]. Gauges on a machine are necessarily located at discrete points, and their average reading may be quite different from $\frac{1}{2\pi} \int_0^{2\pi} p\, d\vartheta$ which is the proper value. Eq. (8.1) is derived on the basis of no initial betatron oscillation amplitude, so the effective semiaperture, A, is reduced from the "mechanical" value by initial amplitude and also by errors in the median surface height. Coupling between horizontal and vertical oscillations can make the phenomena more complex, especially if there are also horizontal aperture limitations. Variation of aperture is done by inserting a rod or vane, usually at one azimuth. An analysis by MULLETT[3] of various refinements of the theory introduces corrections which are probably less than the experimental errors of present measurements.

[1] C. G. FOUND and S. DUSHMAN: Studies with Ionization Gauge. Phys. Rev. **23**, 734 (1924).

[2] See also L. RIDDIFORD: Notes on the Ionization Gauge. J. Sci. Instrum. **28**, 375 (1951) and [37].

[3] L. B. MULLETT: Gas Scattering in Proton Synchrotrons. AERE GP/R 2072, 1956.

The loss as a function of η is plotted in Fig. 34. Comparative characteristics of the four operating synchrotrons are:

Table 3. *Characteristics relating to gas scattering. Symbols from Sect. 8.*

Parameter	Birmingham	Cosmotron	Bevatron	Synchro-phasotron	Units
A	5	7.5	15	20	cm
R	450	915	1520	2800	cm
$1 + \dfrac{NL}{2\pi R}$	1	1.212	1.254	1.181	
T_i	0.46	3.6	9.8	9	Mev
eV	0.2	1	1.75	2.5	kev
η	7	1.5	0.5	0.9	$10^4\,p_{mm}$
η for $A/\sqrt{2}$*	12	2.8	0.9	1.6	$10^4\,p_{mm}$
η for $A/2$*	19	4.9	1.7	2.9	$10^4\,p_{mm}$

$\eta = 0.115$ *for* 10% *loss, the corresponding pressure:*

	Birmingham	Cosmotron	Bevatron	Synchro-phasotron	Units
at A, $p =$	1.7	7.5	23	13	10^{-6} mm of Hg
at $A/\sqrt{2}$, $p^* =$	1	4.1	13	7	10^{-6} mm of Hg
at $A/2$, $p^* =$	0.6	2.3	7	4	10^{-6} mm of Hg

* i.e. for the semi-aperture reduced by a factor of $\sqrt{2}$ or 2.

The general nature of the gas scattering losses shown in Fig. 35 is calculated for the Cosmotron. Observed losses can be expected to be greater than those

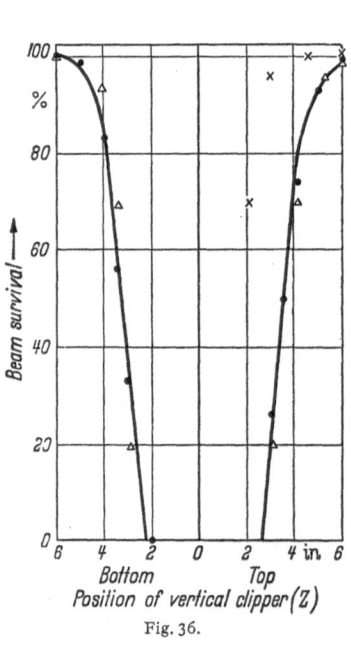

Fig. 36.

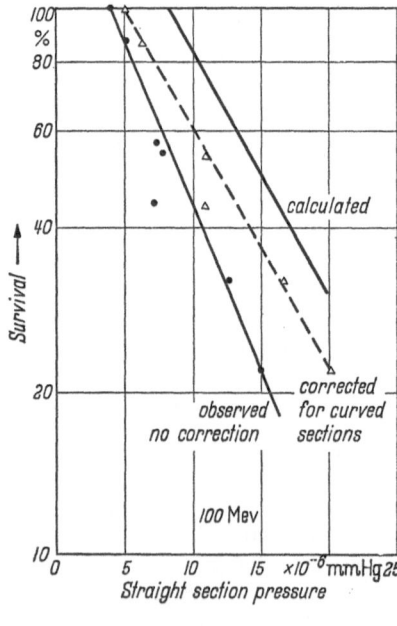

Fig. 37.

Fig. 36. Effect of obstruction at height indicated in Bevatron. Dots are experimental. × calculated for 5×10^{-8} mm of Hg. $\varDelta$ calculated for 1×10^{-8} mm of Hg and A reduced 1 inch. Experimental data from E. J. Lofgren and H. G. Heard (unpublished).

Fig. 37. Beam survival at 100 Mev vs. pressure in Bevatron. E. J. Lofgren and H. G. Heard (unpublished).

calculated, for the reasons stated before (all of which increase the observed scattering) and the loss of beam vs. aperture will be steeper for the same reasons.

E. J. Lofgren and H. G. Heard (unpublished) have measured the loss of beam vs. aperture in the Bevatron using shutters which extend across the aperture radially and are variable in z position (Fig. 36). The pressure was quoted as $5-7\times10^{-6}$ mm of Hg, measured at the straight sections. If one applies the data of Table 3 directly in (8.1) and (8.2) the points x of Fig. 36 are obtained, in disagreement with the measured data. However, if the loss is computed with the semi-aperture reduced 1 inch (for initial oscillations) and the pressure corrected (for gradient in the unpumped curved sections) to 10^{-5} mm of Hg the Δ points of Fig. 36 are obtained. The agreement is fortuitously good, but the corrections are reasonable. Another set of Bevatron data (Fig. 37) relating loss to pressure was obtained by Lofgren, Cork and Chupp (unpublished) who corrected for the unpumped

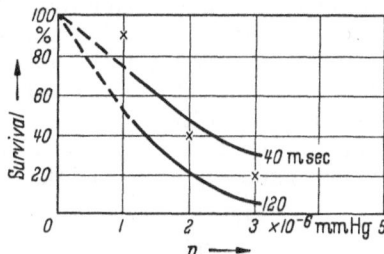

Fig. 38. Beam survival *vs.* pressure in the Birmingham synchrotron. × points calculated for A reduced 1 cm, and at final energy. From Riddiford, van der Raay and Coe [40].

curved sections by measurements with adjacent straight sections valved off from their pumps. The calculated loss, corrected to $A=5$ inches, disagrees with the observed data by almost a factor of two in pressure at small losses.

Measurements on the Birmingham synchrotron [40] of loss vs. pressure at two successive energies are compared in Fig. 38 with points calculated for vertical A reduced 1 cm and final energy. The Birmingham diaphragming experiments indicate scattering losses in the radial direction also (Fig. 39). The radial phase oscillations fill

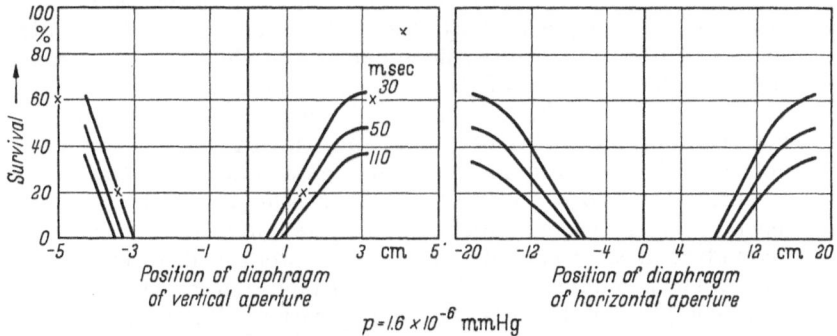

$p = 1.6 \times 10^{-6}$ mmHg

Fig. 39. Beam survival as a function of aperture in the Birmingham synchrotron. × points calculated for loss at full energy, and corrected for median plane displacement. From Riddiford, van der Raay and Coe [40].

the aperture, and in this machine there is coupling between radial phase and betatron oscillations, as well as probable coupling between vertical and radial betatron motions [40]. Consequently, gas scattering is critical and an increasing pressure will quickly cause large beam intensity losses. Eqs. (8.1) and (8.2) give the calculated points in Fig. 39 for loss at full energy of 1 Bev. The observed loss is greater but, due to the complications introduced by coupling, the agreement can be considered good.

In general, it can be concluded that designs should make an allowance of at least a factor of two in pressure in order to keep gas scattering losses reasonably small.

VI. Injection.

23. Deflection and focussing systems. Injection into a proton synchrotron involves three successive processes; pre-acceleration of the protons to energy T_i, conduction through an electromagnetic optical system which launches the particles on an orbit stable within the existing magnetic field, and control of the entire synchrotron apparatus so that the protons do not collide with the injector or chamber walls on any succeeding revolution.

A particle beam emerging from a pre-accelerator is evaluated by the amount of current, and the magnitude of the emittance. To make a convenient definition of emittance, consider an orthogonal coordinate system x, z, s in which s is the central ray and $x' = dx/ds$, $z' = dz/ds$. The emittance in the xs plane is defined as the area in the phase plane xx' covered by the emerging particles, and similarly in the zs plane. The *admittance* of the synchrotron ring is defined similarly as the area in the phase plane xx' or zz', at the injection azimuth, which contains the protons that can be accepted into the aperture. In a circular machine (4.13) can be rewritten for z, neglecting damping,

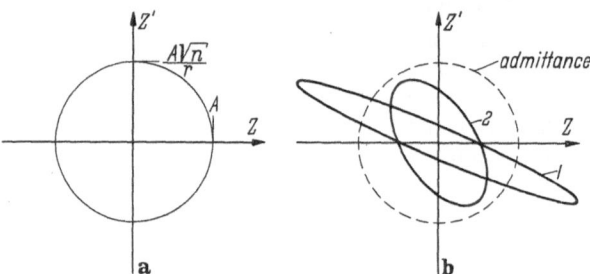

$$z = z_i \cos \sqrt{n}\,\vartheta \\ + \frac{z'_i\, r}{\sqrt{n}} \sin \sqrt{n}\,\vartheta. \quad \Bigg\} \quad (23.1)$$

If z is not to exceed the semi-aperture A then z_i, z'_i must lie within the boundary of Fig. 40a, and the admittance is $\pi A^2 \sqrt{n}/r$. A similar admittance could be written for x. However, gas scattering, obstruction by the inflector, and orbit distortions reduce the admittance well below the value given simply by the semiaperture (Sect. 22, 24). Straight sections reduce the admittance by the ratio in which they increase ν and by an amplitude factor given in Eq. (7) of [13]. Although the emittance of the pre-accelerator, evaluated at any image plane, cannot be reduced without acceleration, its shape can be changed by suitable lenses. Thus the emittance of curve 1, Fig. 40b, can be modified by a lens to the shape of curve 2, which would fall within the assumed admittance shown. Lenses are usually required to "match" the pre-accelerator to the synchrotron, and are generally astigmatic to satisfy differing requirements on the x and z axes. It is evident that serious abberations can produce distorted emittance shapes, parts of which could lie outside the admittance curve. Detailed calculations are also complicated by nonuniform population of the emittance area.

A good pre-accelerator is obviously one with high beam current and small emittance. Its energy must be high enough to permit injection at sizeable magnetic fields, so that field distortion can be minimized, and high enough to limit gas scattering at the attainable working pressure. Cyclotrons have often been proposed; a small one was used in the Bevatron model [44] and an 8 Mev cyclotron is being constructed for the Canberra machine. Although the emittance of a cyclotron can, in principle, be small, the necessary techniques have not been fully developed. Cockcroft-Walton sets, Van de Graaff generators, and linear accelerators have been chosen thus far. Pre-acceleration is done with a 460 kev Cockcroft-Walton set at the Birmingham PS; a 3.6 Mev Van de Graaff generator at the Cosmotron (a similar one at CEA); a 9.9 Mev linear accelerator at the Bevatron; and a 9 Mev linear accelerator at the Synchrophasotron. Details of these types of accelerator are given elsewhere in this volume and need not be repeated.

Plan views of the injection systems at Birmingham, Bookhaven Cosmotron, and Berkeley Bevatron, Figs. 41 to 43, show the elements of the ion optics.

These systems must be accurately aligned by mechanical adjustments, and if necessary realigned, before the electromagnetic trimming devices can be used. The ion source in these injectors is run at low dc current output, and pulsed to high current during the injection period. The Birmingham machine is the simplest (Fig. 41) [*41*]. Adequate aperture is available in the accelerating tube for small changes in beam position and the limiting stop of the system is the entrance to the inflector, an opening some 0.3×0.38 inches. Final precise steering of the pre-accelerated beam is done by crossed electrostatic deflector plates placed just after the ion source extractor. The protons must pass through the intense stray field of the circular magnet, and are magnetically shielded, by μ-metal, from the accelerating tube to the inflector entrance. The electrostatic inflector lies in a field, at injection, of about 210 gauss at the exit and 140 at the entrance. The plates are divided into three sections so the potential on each can be adjusted to compensate for the field. The inflector electrodes are in a grounded stainless steel shield. Focussing is done by four convergent electrostatic lenses [*41*]. The first lens is formed between the ion source canal (extractor) and a cylinder whose potential can be varied. This is normally set to form a real image of the source at the deflecting

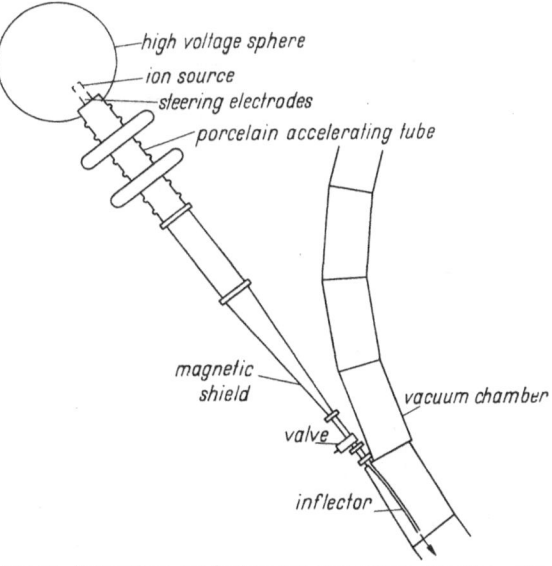

Fig. 41. Birmingham synchrotron injection system, plan view. After RIDDIFORD [*37*].

plates. The three gaps in the accelerating tube then form a real image at the inflector. The inflector is a positive lens in the r direction, and nearly neutral on the z axis, so this makes the system astigmatic, which is probably desirable. The energy of the pre-accelerator is held to a variation of less than 1 % by a regulator in the primary of the rectifier transformer. During the injection pulse the energy can be raised by applying a rising voltage to a pulse transformer with its secondary in series with the output filter condenser [*41*].

The Cosmotron injection system [*34*] is complicated somewhat by the inclusion in the optics of a system for regulating the energy of the Van de Graaff pre-accelerator (Fig. 42). The protons are first deflected through 25° by an analyzing magnet with wedge-shaped double-focussing poles, which bends the mass-2 beam about 17.7° (molecular ions comprise about half the small idling beam of the ion source). The mass-2 beam falls on differential slits connected to the Van de Graaff corona energy control which, being a degenerative feedback system, keeps the beam centered on the slits. The analyzing magnet is regulated to high precision and its field strength sets the energy, since the deflection angle is fixed. A horizontal fine motion of the differential slit system initially adjusts the angle of the beam so it strikes the center of the inflector entrance. The electrostatic inflector has its negative plate enclosed in the grounded positive plate so that stray fields will not disturb the circulating protons. The inflector spacing is $\frac{1}{2}$ inch, and the structure is 3 inches high, forming an almost complete obstruction in the z-direction. Vertical position of the beam is approximately set by moving the Van de Graaff bodily, then fine control is obtained by adjusting the shoes of the analyzer magnet. These shoes control the wedge angle of the four pole

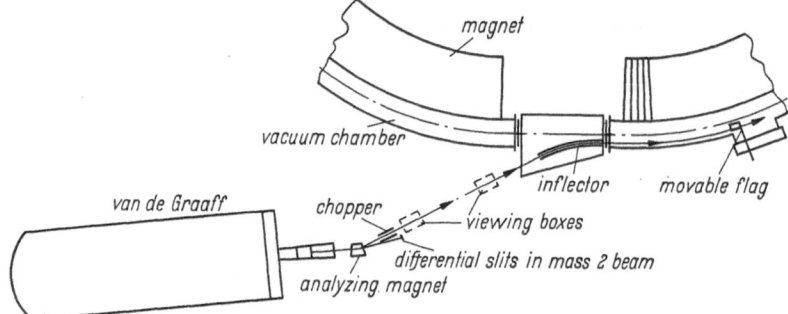

Fig. 42. Plan view diagram of Cosmotron injection system. See also [*34*], p. 817.

edges and so affect the focussing properties of the magnet. The proton beam is focussed by electrostatic lenses just following the ion source, and by the accelerating tube, to a real image some 30 to 50 feet beyond the analyzing magnet. The analyzing magnet is set converging and forms an image of the source between the viewing boxes, about 8 feet ahead of the inflector. The beam is then horizontally focussed by the inflector and leaves it roughly parallel; in the vertical direc-

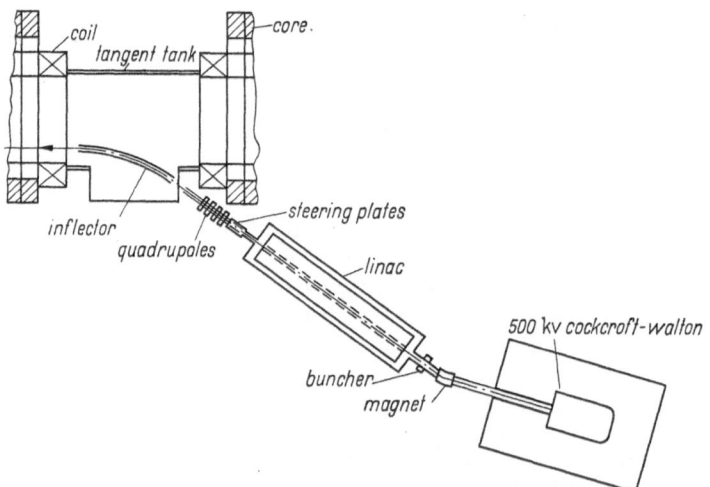

Fig. 43. Plan view of Bevatron injection system. After Cork [*45*].

tion the beam diverges some $1\frac{1}{2}$ milliradian, corresponding to about 0.6 inches of vertical oscillation amplitude. The "viewing boxes" contain fluorescing quartz plates and micrometer operated cross wires for determining the position and shape of the beam, necessary items for proper adjustment of the system. If a potential of about 2.5 kv is applied to the "chopper" plates the beam is deflected out of the inflector aperture. The potential can then be removed and reapplied by hydrogen thyratrons to pass a short, accurately timed pulse of protons. Pulses of about $\frac{1}{3}$ and 1 revolution in length (roughly 1 and 3 microsec duration) are very useful for investigations of particle dynamics. One serious disadvantage of the standard Van de Graaff is the fall in energy during the injection pulse. With a terminal capacity of 500 μμf the Cosmotron injector drops 10 kev at 100 μsec when delivering 40 ma (of which 10 ma are protons). A system is being constructed by C. M. TURNER in which an insulated liner in the pressure tank will be pulsed positively with a triangular waveform to hold the output energy constant, or to give it a rising characteristic.

The Bevatron injection system, Fig. 43 [45] begins acceleration with a 460 kv voltage multiplier and conventional accelerating column. The beam is deflected 20° by a magnet, passes through a radiofrequency buncher and enters the 9.9 Mev linear accelerator. The linear accelerator output is focussed by a set of four quadrupoles into the 35° electrostatic inflector, which has a $\frac{7}{8}$ inch plate spacing on an 18 foot radius. The electrostatic lenses of the pre-accelerator and the deflecting magnet focus the beam on the linac entrance. Shoes on the wedge-magnet can be rotated to alter the wedge angle and thus the focal length. Beam position is adjusted by moving the magnet and its shoes. The linear accelerator is a 43 drift tube, 202.5 mc single-cavity type and is grid focussed. Its energy distribution is estimated to be $\pm 0.3\%$. The quadrupoles image the beam somewhere in the inflector plates, and can be differentially adjusted to set different focal lengths on the r and z axes. A small amount of steering is obtained with electrostatic plates placed just after the linac.

The Synchrophasotron 9 Mev linear accelerator is set nearly at right angles to the straight section axis, and the beam is then deflected some 75° by a magnet, after which the five-section electrostatic inflector brings the protons tangent to the ring orbits. There is not yet enough operational experience available to permit an assessment of the performance.

24. Missing the inflector. Successful injection will be achieved only if the protons conveyed into the ring by the optical system miss all obstructions on succeeding revolutions. This missing cannot be done in a static system. A theorem by Poincaré shows that if trajectories have no branches to infinity then an orbit issuing from a small volume will traverse this volume infinitely often[1]. Consequently, a proton in a static system will collide with the launching machinery, and it is necessary to vary the electromagnetic fields with time. Numerous schemes of injection can be devised. A number of possibilities are described in Sect. V of [12], and the peeler-regenerator system is noted in [5], p. 147. Injection and stripping of molecular ions is a promising method which has not yet been tried (P. B. MOON: [5], p. 231). The pulsed synchrotron can take advantage of the positive $\dot{B}$ and leave the momentum of the protons fixed during an injection period of many revolutions (neglecting the small inverse betatron effect). If not operating on a resonance (an advisable restriction) a few revolutions will occur before the betatron oscillations are in phase with the inflector, and the orbit radius will have diminished. Thus protons can miss the inflector. All four of the operating machines have the inflector on the outside of the aperture, inject on the median surface, and employ "radial missing".

At the azimuth of the inflector there will exist an equilibrium orbit corresponding to total injection energy E_i (or kinetic energy T_i) with radius of curvature given by

$$B r = \frac{\sqrt{E_i^2 - E_0^2}}{e c} \sim \frac{\sqrt{2 T_i E_0}}{e c} . \tag{24.1}$$

---

[1] WHITTAKER: Analytical Dynamics, Sect. 185.

If the field on the central equilibrium orbit of radius r_0 is B then the field at the inflector is

$$B_d = B_0 \left(\frac{r_0}{r_d}\right)^n \approx B_0 \left(1 - \frac{nA}{r_0}\right),$$ (24.2)

with r defined to adequate approximation for a machine with straight sections by the inset of Fig. 44. At time t the radius of a proton is given by

$$B(t)\, r = B_0(t)\, r_0^n\, r^{1-n} = \text{const} \approx \frac{\sqrt{2TE_0}}{ec},$$ (24.3)

$$\frac{dr}{r} = -\frac{dB_0}{(1-n)\,B_0} = \frac{1}{1-n}\frac{dp}{p},$$ (24.4)

and the shrinkage of r per turn is

$$\Delta r = -\frac{r\dot{B}_0}{(1-n)\,B_0\,f} = -\frac{2\pi r^3 e\dot{B}}{1-n}\frac{E_i}{E_i^2 - E_0^2}\left(1 + \frac{NL}{2\pi r_0}\right)$$
$$\approx -\frac{\pi r^3 e\dot{B}}{(1-n)\,T_i}\left(1 + \frac{NL}{2\pi r_0}\right),$$ (24.5)

while

$$\dot{r} = -\frac{r\dot{B}_0}{(1-n)\,B_0}.$$ (24.6)

Suppose a proton to be injected from the center of the inflector and parallel to the equilibrium orbit, which orbit must be presumed to have only moderate distortion due to inhomogeneities in the magnet. The proton will oscillate about the equilibrium orbit with amplitude $D = r_d - r$ (nomenclature is defined by Fig. 44) and frequency;

$$\omega_\beta = \alpha \sqrt{1-n}\,\omega = \nu\omega; \qquad \alpha = 1 + \frac{NL}{4\pi r}.$$ (24.7)

The radius r will diminish Δr per turn (24.5). Phase shift of the oscillation per revolution is $2\pi\nu$ so if the proton is to miss the inflector on the k-th passage it is required that

$$b < D(1 - \cos 2\pi k\nu) + k\,|\Delta r|.$$ (24.8)

The order of magnitude of the quantities involved in this oversimplified description is listed in Table 4. The semi-aperture of the inflector is $\bar{b}$ and D_1, D_2 are the minimum amplitude values necessary to miss the inflector after the first and second revolutions with $n = 0.7$. The value of D_1 is controlling.

Table 4. *Injection parameters (see text for definitions).*

	Δr	$\bar{b}$	D_1	D_2	ν $(n=0.7)$	Δt
	mm	mm	mm	mm		μsec
Birmingham PS	3.1	4	0.5	0	0.55	140
Cosmotron	4.1	6	1.1	0	0.61	200
Bevatron	4.2	11	4	2.7	0.62	330
Synchrophasotron . .	15				0.60	240

The radius is of the order of 10^4 mm in these synchrotrons so it is evident that missing the inflector is marginal and that further details must be considered.

Filling the aperture at constant injection energy requires the pulse of protons to begin when $r = r_d$ and to continue until $r = r_0$. If $r_0 = \frac{1}{2}(r_d + r_i)$ protons injected at the latter instant will oscillate directly into the inside wall. The length of the usable injection pulse, Δt in Table 4, is approximated from (24.6).

$$\Delta t = \frac{(1-n)\, B_0}{r \,\dot{B}_0}\,(r_d - r_0).\tag{24.9}$$

When the equilibrium orbit corresponding to T_i is at radial distance D (Fig. 44) from the point of injection the protons will oscillate about this orbit (Sect. 33) with radial motion

$$x = C_1 \sqrt{\beta}\,\cos\left[\int \frac{ds}{\beta} + C_2\right].$$

It is convenient to evaluate the matrix elements at the center of the straight section where β has the value given by Eq. (33.14), $\alpha = 0$, $\beta' = 0$, and $\gamma = \dfrac{1+\alpha^2}{\beta} = \dfrac{1}{\beta}$. The invariant of the motion given by Eq. (33.18) then becomes

$$\left. \begin{aligned} U &= \frac{1}{\beta}\,(x^2 + \beta^2\, x'^2) \\[4pt] &= \frac{1}{\beta}\,(x_0^2 + \beta^2\, x_0'^{\,2}) \end{aligned} \right\} \tag{24.10}$$

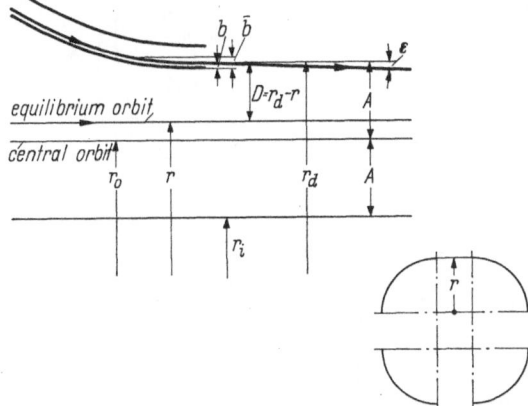

Fig. 44. Definition of injection coordinates. Inset shows definition of "r" at straight section.

with $x' = dx/ds$ and $x_0' = \varepsilon$. The maximum amplitude at the azimuth of the inflector becomes $\sqrt{D^2 + \beta^2 \varepsilon^2}$. At a given azimuth the phase of the oscillation changes per revolution

$$\mu = 2\pi\nu \approx 2\pi\left(1 + \frac{NL}{4\pi r}\right)\sqrt{1-n},\tag{24.11}$$

giving

$$x = \sqrt{B^2 + \beta^2\,\varepsilon^2}\,\cos(k\mu + C_2);\quad \tan C_2 = \frac{\beta\varepsilon}{D},\tag{24.12}$$

where k is the revolution number. In a circular machine the motion is sinusoidal and

$$\beta = \frac{R}{\sqrt{1-n}} = \frac{\lambda}{2\pi};\quad \mu = 2\pi\sqrt{1-n},\tag{24.13}$$

where λ is the wavelength of the betatron oscillation. In the Cosmotron $\beta = 700$ inches, a similar circular machine would have $\beta = 650$ inches.

The successive transits of an injected proton, $D = 1$ inch, are indicated in Fig. 45 for $\varepsilon = 0$, ± 1, ± 2 milliradians. This figure is a discontinuous phase plot. The inner edge of the inflector is shown at positions $D - b + k\Delta r$ so it is necessary for the survival of a proton that its phase point on any revolution k be to the left of the corresponding inflector line. Divergence of the injected beam not only increases the amplitude but advances or retards the phase and affects the coincidence with the inflector. For the small-oscillation frequency of the Cosmotron at the inflector radius, $\mu \approx 260°$ ($n \approx 0.59$) and it is seen in Fig. 45 that protons with $\varepsilon = 0$, ± 1 milliradians miss, but those with $\varepsilon = 2$ are lost at the first revolution, and with $\varepsilon = -2$ on the third. Orbits traced by inserting fluorescent flags (see [34], p. 861) into the Cosmotron, Fig. 46, illustrate the effects. The full angular aperture, for small oscillations, is then about 2.5×10^{-3} radians, and inspection of Fig. 46 shows that the full width of the inflector,

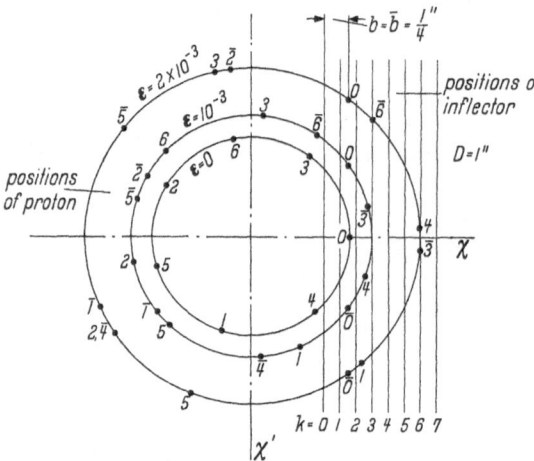

Fig. 45. Phase plot of proton on successive transits compared to position of inflector on successive transits. A bar over a proton transit number indicates it is associated with negative ε.

1.27 cm, can be used if $\varepsilon = 0$. The admittance for horizontal oscillations is then 0.8×10^{-3} cm rad expressed as radius times divergence angle or 2.5×10^{-3} cm rad expressed as area in phase space. As D increases, the permissible ε will increase somewhat less than in proportion. This is useful in the Cosmotron because the energy of the injector diminishes with time (due to the discharge of the capacity of the Van de Graaff generator) and ε increases during the injection pulse. However, the radial admittance is limited by multiturn injection to a small fraction of the radial admittance of the aperture, and it is evident that the angle of injection must be precisely controlled. As D increases the equilibrium orbit lies at different radii and may be disturbed variably as a function of radius. This effect must be kept small for good orbit characteristics, and is allowable to a small extent due to the increase in permissible ε with B. How-

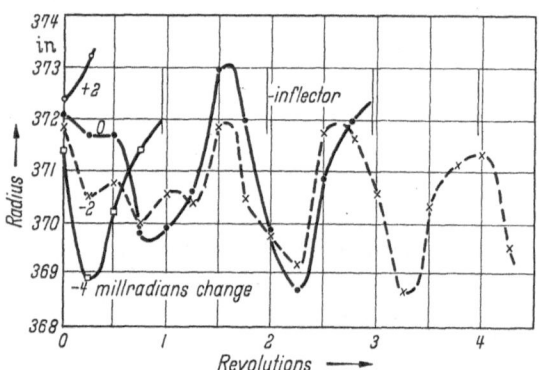

Fig. 46. Proton orbits in Cosmotron for different injection angles, ε. (No acceleration.) After BLEWETT, GREEN, KASSNER, MOORE, SMITH and SNYDER, [34], p. 861.

ever, the value of n may vary markedly with radius(see Fig. 13 for the Cosmotron) causing μ to vary with radius and introducing non-linear terms in the equations of motion so that μ is dependent on amplitude.

Typical orbits for five values of D, Fig. 47, show the complexities introduced by small magnet imperfections, and by radial-vertical mode coupling. Such coupling can be used to assist in missing the inflector by transferring energy from the radial to the vertical mode such that the radial amplitude is diminished at the critical turns, but by the time the inverse transfer occurs the orbit radius has shrunk enough for definite missing. The maximum to minimum amplitudes will be small if n is well removed from 0.5, the value for the coupling resonance. The "bump" in the Cosmotron median plane (Fig. 16) makes a convenient coupling mechanism, acting directly to cause "forced oscillation" type of coupling from radial to vertical mode. The strength of coupling is easily adjustable by varying the correcting current; and the injected beam moves inward from the "bump", diminishing the coupling to a negligible value after several turns. An orbit of several turns, with coupling adjusted to a value approximately optimum, was traced in the Cosmotron, Fig. 48. The effect of the coupling in helping to miss on turns 3, 4, and 7 is evident.

In operation, it is necessary to adjust the time of injection, the injector energy, and the inflector potential to give the least particle loss. The proper timing matches the energy to the field, and is critical only if energy is varying during the pulse (provided the ion source pulse is longer than the acceptance time). The inflector potential vs. the energy sets the angle of entry, ε. It is also necessary to align the system and to adjust such auxiliary things as coupling and field corrections. Orbitplotting is tedious and lengthy so extended experimental surveys have been made with "survival curve families". A pulse of protons about one-third circumference (1 μsec) long is injected and observed on the induction electrode (Sect. 26).

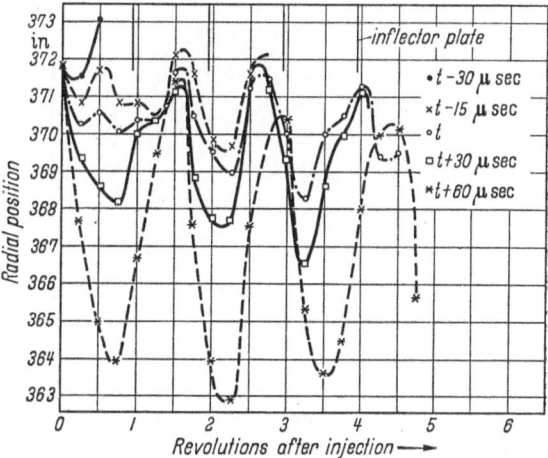

Fig. 47. Proton orbits in Cosmotron for injection at various field strengths. i.e. various values of D. $\dot{B} = 16\,000$ gauss/sec.

Successive revolutions through the electrode appear on the oscilloscope as discrete rectangles; any loss shows as a decrease in amplitude, and the revolution number of the loss can be noted. The survival time expiring before the bunch is collected on the inner wall is measured. The inflector voltage is fixed, simply because it can be monitored with a precise potentiometer, and the timing of injection is varied to obtain a function of survival of the circulating pulse vs. position of the equilibrium orbit. A family of such curves is then obtained for a series of injection energies, Fig. 49. The operating point can be selected as indicated for maximum injected charge. The curves of Fig. 49a were run by keeping constant time separation between pulsing of the ion source and pulsing of the injection chopper, so the

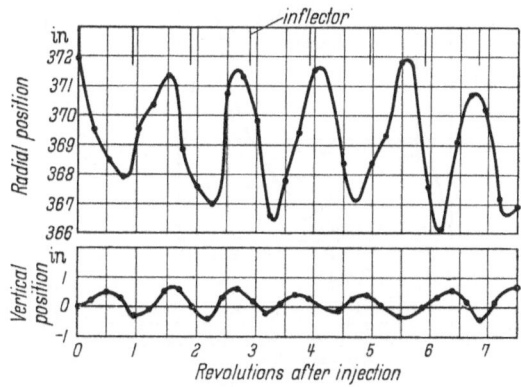

Fig. 48. Orbit of injected protons for several revolutions in Cosmotron. Coupling of r to z modes is approximately optimum for missing inflector. After BLEWETT, GREEN, KASSNER, MOORE, SMITH and SNYDER, [34], p. 861.

energy is constant on any one curve. During normal injection the Van de Graaff energy will fall; injection then follows an operating line of a type indicated by the dashed curve of Fig. 49a. If adjustments are not properly made losses will enter large areas of the "survival family" as shown in Fig. 49b which represents the condition of little coupling between radial and vertical modes.

The Birmingham proton synchrotron has $n \approx 0.75$ at injection, obtained by measuring $\dot{r}$ [41], so $\nu \approx \frac{1}{2}$. The phase of the radial oscillations then approximately coincides with the inflector on the second turn resulting in considerable loss to

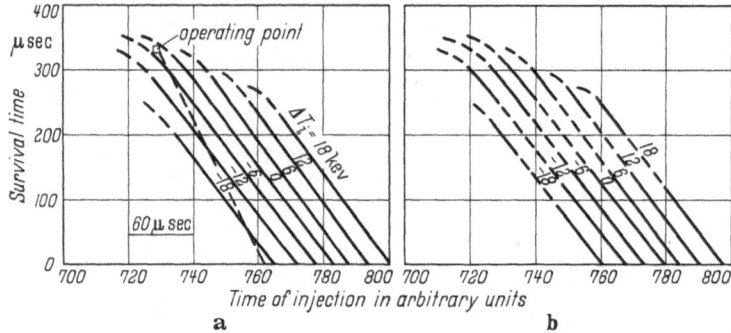

Fig. 49a and b. Family of curves showing survival time of injected beam in the Cosmotron *vs.* time of injection for several energies. Dotted curve indicates less than 100% survival to back wall. (a) Good adjustment, (b) poor adjustment of pole face corrections.

the inflector. The low injection energy gives a relatively large Δr per turn of 3.1 mm (Table 4), compared to the 4 mm half-aperture of the inflector, and this permits missing. To explore the injected beam a probe was inserted at an azimuth approximately $\frac{1}{2}$ radial betatron wavelength from the inflector, and the pulse was made long enough to bracket the entire injection period. The quantity of charge collected was observed on an oscilloscope. When the probe is at radius greater than r_0, protons are collected directly from the inflector during the first oscillation—the "secondary spike" of Fig. 50. At smaller

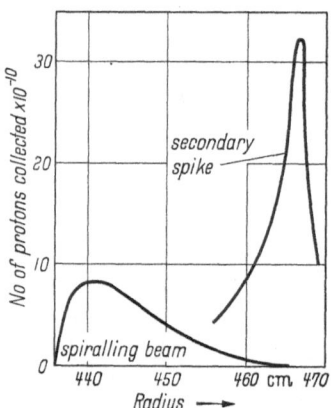

Fig. 50. Protons collected on a probe as a function of radius. Collection after injection in Birmingham synchrotron, no acceleration. After RAMM, COE and VAUGHAN [41].

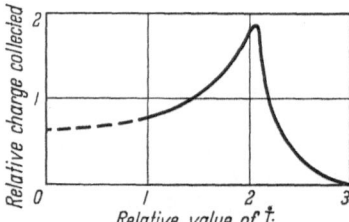

Fig. 51. Relative charge collected on inner probe vs. $\dot{T}_i$. No acceleration, Birmingham synchrotron. After RAMM, COE and VAUGHAN [41].

radii the protons must spiral to reach the probe, and just outside the inner chamber wall the entire spiralling beam will be collected, Fig. 50. The measurements shown in Fig. 50, and other supporting data [41], show that 75% of the beam is lost to the inflector. If the "rising voltage" equipment increases the energy during injection the equilibrium orbit will shrink more slowly, and may even be held stationary near the inflector. Thus particles will not be lost until their orbit has crossed the aperture, and Δt can be doubled relative to the case with T_i constant. The amplitude of radial betatron oscillations is smaller and more charge is collected if $\dot{T}_i$ has the proper value, Fig. 51. Since the Birmingham inflector is in a strong stray field (averaging some 100 gauss) the increasing T_i with increasing B will tend to make injection geometry stationary. Timing of the rising voltage unit, and of radiofrequency turn-on, is done from the signal obtained when protons strike a probe at radius slightly greater than r_d. Such timing gives first order correction for variations of T_i but does not correct for variations of ε with T_i. The Birmingham inflector bends the beam only a little more than 10° so a relative change in inflector voltage vs. T_i of 1% changes ε only some 2 milliradians.

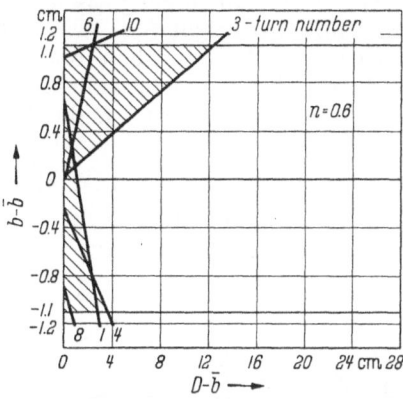

Fig. 52. Survival of injected beam in Bevatron as a function of position in inflector. Protons in shaded area are lost. L. SMITH and E. H. SCHWARTZ (unpublished).

The Bevatron acceptance has been calculated by L. SMITH and E. H. SCHWARTZ (unpublished). They determined which protons would be lost after each turn as a function of D and b using a method analogous to that of Fig. 45. The geometry of the long inflector permits ε to be related to b by $\varepsilon = -1.35 \times 10^{-3} (b - \bar{b})$. Fig. 52 shows the losses as shaded areas for each turn. The losses can be normalized, if the incoming beam is assumed to have uniform density, to the area extending to $D = A$. Normalized losses for various n, Table 5, show high values if $\nu_r \approx \frac{3}{4}$ or $\frac{2}{3}$; most of the beam is absorbed on the fourth or third revolution respectively when the radial betatron phase comes into coincidence with the inflector. Loss to the inflector with operation at $n \approx 0.6$ is observed to be about 10%, in good agreement with the calculations. No detailed injection orbit tracing has been done.

A 180 Mev model of the Synchrophasotron was used for several injection experiments. Injected beam into this model, from a Van de Graaff, had a radial width of 4 mm and divergence of 1 milliradian. The parameters were: $N = 4$,

Table 5. *Bevatron losses on the inflector according to L. Smith and E. H. Schwartz (unpublished).*

n	ν_r	Normalized loss
0.55	$0.76 \sim \frac{3}{4}$	49.3%
0.60	0.72	9.9%
0.65	$0.67 \sim \frac{2}{3}$	44.9%
0.70	0.62	8.3%
0.74	0.58	7.7%

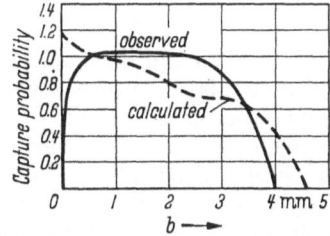

Fig. 53. Capture probability *vs.* distance from inner inflector plate in model of Synchrophasotron. From DANILKIN, ZINOVIEV, PETUKHOV and RABINOVICH, [5], p. 515.

$r_0 = 2$ m, $L = 0.67$ m, $n = 0.69$, $\dot{B} = 2 \times 10^4$ gauss/sec, $T_i = 0.75$ Mev, $\Delta E/$turn $= 66$ v. The Δr per turn was very small, about $\frac{1}{4}$ mm, so the injected beam was guided quite close to the inner inflector plate. The value of $n(r)$ was determined by measuring $\dot{r}$ and a numerical integration of injection efficiency was done. Calculation of losses on the inflector ([5], p. 513) was similar to methods described previously, and gave an efficiency of 59%, compared to 60% measured. A further calculation of efficiency vs. b is shown in Fig. 53, and compared to experimental results.

Considerations of phase oscillation damping and Δr (24.5) combined with the necessity of avoiding resonances, indicate that $n > 0.6$ is advisable in the proton synchrotron. Thus ν_r is restricted to a range of order 0.5 to 0.75 (see Fig. 2) and possibilities for staying out of phase with the inflector are severely limited in this range. It is possible to consider "vertical missing" since it is quite practical to set $\nu_z = 0.9$ and to obtain ten turns before coincidence with the inflector. However, "vertical missing" is expensive in gap length since the gap must equal the inflector height plus twice the admittance allowance plus allowance for gas scattering. It would be economically practical only with high injection energy. In general, for radial missing of the inflector, vertical amplitudes are kept small to reduce gas scattering, and to keep radial-vertical coupling effects under control.

VII. Radiofrequency system.

25. Conditions for acceleration. After injection most of the radial aperture will be filled with protons. In the synchrotrons being described, the vertical

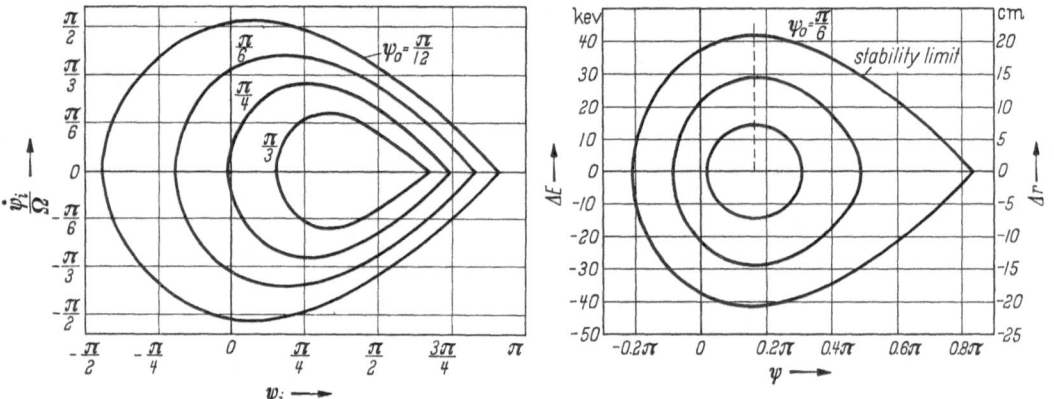

Fig. 54. Stable region of the phase oscillations for various values of the stable phase angle, ψ_0. From Frank and Twiss [12].

Fig. 55. Phase oscillations in the Cosmotron at $\sin \psi_0 = 0.5$, $T_i = 3.6$ Mev.

extent of the beam will be small and later will grow with gas scattering. A fraction of the circulating protons will be captured into stable phase motion and accelerated—the remainder will spiral into the inner wall and be lost. When the radio frequency is turned on, the ribbon of protons is uniformly spread over 2π of ϑ and 2π of phase angle, ψ. To find which particles can be accepted, the limit of stability $\psi = \pi - \psi_0$, $\dot\psi = 0$ is substituted in (5.3) and ψ_i vs. $\dot\psi_i$ plotted to obtain the boundary of stable initial conditions for a given stable phase angle, ψ_0 (Fig. 54). The ordinates of Fig. 54 in units of $\dot\psi_i/\Omega$ can be expressed as energy deviation, ΔE, by combining (5.2) and (5.5) or, for straight section machines, (7.6) and (7.7). Alternatively, application of (5.7) will give the ordinates in units of radial excursion Δr. A value of $\psi_0 = \pi/6$ is commonly used because a large acceptance area is obtained, and the factor of two in radiofrequency peak voltage over the volts per turn required ($\sin \psi_0 = 0.5$) gives adequate margin. It is doubtful whether the improvement obtained by decreasing ψ_0 to $\pi/12$ would be worth the large increase which would be necessary in radiofrequency power. Phase vs. energy and radial deviation in Fig. 55 are plotted for the Cosmotron with $\psi_0 = \pi/6$ and $T_i = 3.6$ Mev.

The radius of the defined central orbit, r_0, will be related to the instantaneous magnetic field by (4.2),

$$r_0 = \frac{\sqrt{T_0^2 + 2 T_0 E_0}}{e c B_0} . \tag{25.1}$$

Each proton will be oscillating about an orbit which is displaced in radius from r_0 by

$$\Delta r = \frac{r_0 \Delta T}{2(1-n)T_0},\tag{25.2}$$

obtained from (5.7) by setting $T_0 \ll E_0$, where ΔT is the proton's kinetic energy difference from the defined energy T_0. The radio frequency determines a synchro-

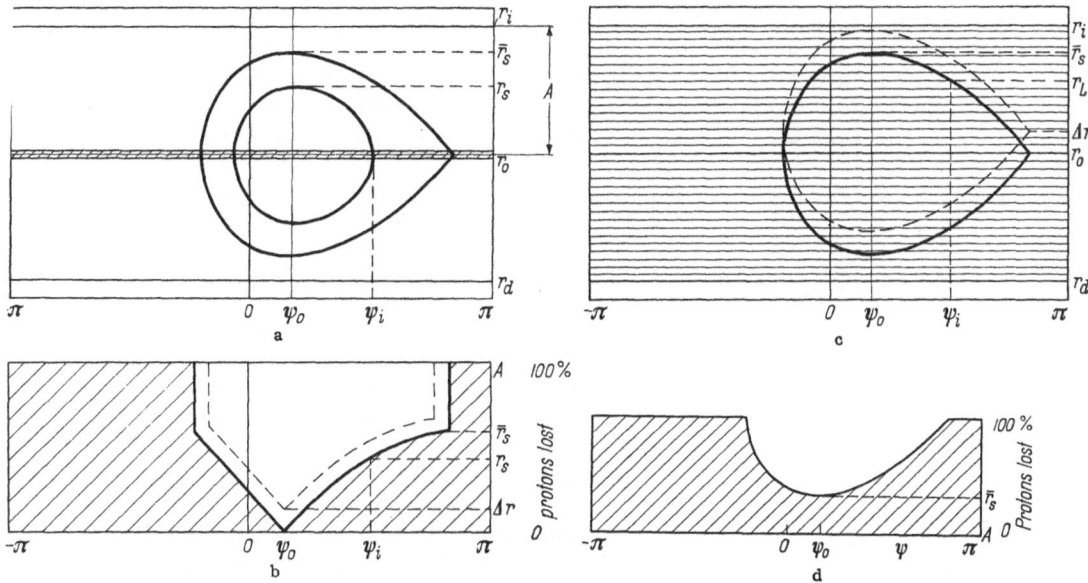

Fig. 56a—d. Acceptance patterns in $r-\psi$ space, (a) and (b) for constant T_i, (c) and (d) for modulated T_i. Shaded areas in (b) and (d) represent particles lost.

nous orbit radius in the magnetic field which is given by (4.3) and the straight section correction

$$\omega = \frac{ecB}{\sqrt{e^2 B^2 r^2 + m_0^2 c^2}} \left(\frac{1}{1 + \frac{NL}{2\pi r}} \right).\tag{25.3}$$

If the value of B from (4.5) is substituted in (25.3) then differentiation gives, around r_0,

$$\frac{\Delta \omega}{\omega_0} = \frac{\Delta r}{r_0} \left\{ \frac{(1-n)m_0^2 c^2}{e^2 B_0^2 r_0^2 + m_0^2 c^2} - \frac{1}{1 + \frac{NL}{2\pi r_0}} \right\}.\tag{25.4}$$

When protons are injected at constant energy, T_i, they will have a common equilibrium orbit, and injection will begin when this orbit radius reaches the inflector radius, r_d. After interval Δt given by (24.9), the equilibrium orbit radius will be at r_0, and the protons lie in the central band of the phase plot, Fig. 56a. Those injected at the beginning of the interval will have very small betatron oscillations; those injected at the end of Δt will be performing the maximum allowable radial betatron oscillations of amplitude A. The injection pulse consisted of a discrete number of turns, however, to a good approximation it may be considered that the protons in an interval $\Delta \psi_i$, at ψ_i, are continuously distributed over betatron amplitudes from 0 to A. Let the radio-frequency acceleration be turned on with $\Delta \omega = 0$ (25.4) at $\Delta T = 0$ (25.2). The protons will

18*

begin phase oscillations which will follow paths in the phase plane for each ψ_i (Fig. 56a). These paths are given by (5.3) with $\dot{\psi}_i = 0$. After a quarter phase oscillation the particles at ψ_i will have reached orbit radius r_s, and those with betatron amplitude greater than $A - r_s$ will have been lost to the edge of the aperture. The fraction lost, r_s/A, is plotted vs. ψ_i as the solid line in Fig. 56b, with arbitrary $\bar{r}_s/A$ indicated.

The energy T_i can be increased during the injection pulse so that the equilibrium orbit remains stationary, just inside r_d. After $2\Delta t$ the protons will be spread across the aperture on a closely spaced distribution of orbits (Fig. 56c) with small betatron amplitudes. If the radiofrequency acceleration is turned on, all those protons inside the envelope of limiting stability can be captured. At

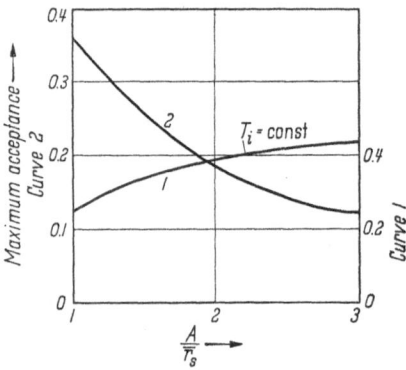

any ψ_i fraction r_L/A can be captured. However, twice as much charge as the previous case will have been injected, at the same input current. Loss vs. ψ_i is shown in Fig. 56d. In either case phase acceptance extends over a maximum spread in ψ_i of $1.05\,\pi$, o-189°, for $\psi_0 = \pi/6$. By numerical integration of Fig. 56 it can be determined that if T_i is constant and $A = \bar{r}_s$, the maximum fraction of injected protons which can be accepted by suddenly applied radiofrequency acceleration is 0.24. The corresponding fraction for T_i modulated upward at the greatest permissable rate is 0.36. Since the injection interval for the latter is twice as long, three times as many protons will be accepted. If $A > \bar{r}_s$, the constant T_i efficiency improves, while the modulated T_i efficiency diminishes. Maxi-

Fig. 57. Maximum attainable acceptance at radiofrequency turn-on as a function of the ratio of semiaperture to maximum radial phase oscillation. Curve 1 for T_i constant, curve 2 for T_i increased at the maximum allowable rate. Ordinates are doubled for curve 2 in order to compare charge efficiency.

mum capture fractions vs. $A/\bar{r}_s$ in Fig. 57 have the ordinates on the modulated T_i curve doubled to indicate relative charge efficiency. The two methods are equivalent at $A \sim 2\bar{r}_s$. Intermediate rates of energy modulation will evidently lie between the two cases plotted in Fig. 57. If the frequency at the moment of turn-on is in error an equivalent radial error, Δr of (25.4), will increase the loss of Fig. 56b as indicated by the dashed line. The frequency error shown by the dashed envelope of Fig. 56c will not increase the loss of protons unless $\Delta r > A - \bar{r}_s$. This tolerance of initial frequency error is an advantage of the modulated energy method.

Capture efficiency in synchrotron operation is complicated by factors not considered in the preceding paragraph. Loss to the inflector will not, in general, be the same for different rates of variation of T_i. Coupling between the betatron and phase oscillations is not completely negligible and, since there is essentially no damping of either during the first few phase oscillation cycles, such coupling will give additional loss. If there is a variation in the radiofrequency accelerating field with radius considerable coupling can be induced. It has been shown[1] that a radial variation of the accelerating field will damp either the betatron or synchrotron oscillation amplitude, but only at the expense of the other. Gas scattering will also cause loss by increasing those radial amplitudes which already nearly equal A. The phase oscillations are non-linear [Eq. (5.1) is valid only

[1] Unpublished reports by L. Smith, A. A. Garren, R. L. Gluckstern and L. R. Henrich; also E. A. Crosbie and M. Hamermesh.

for small amplitudes] and their period increases with amplitude[1]. After a "quarter-cycle" the phase plot of Fig. 56a will have changed to the form shown in Fig. 58. Losses during the capture process will appear at odd quarter-cycles, but will not be sharply defined. If the shape of the proton bunch (charge density *vs.* phase angle) is observed by induction electrodes, the amplitude and angle subtended will fluctuate at twice the phase oscillation frequency. However, the skewing of the phase plot, to which will be added disturbances from noise in the radiofrequency and magnetic field, will smooth the fluctuations in the bunch within ten to twenty cycles (see Fig. 61).

An improvement in capture efficiency can be obtained by slowly applying the radiofrequency accelerating voltage, meanwhile tracking the mean orbit with the frequency. This consideration is somewhat academic in the conventional synchrotron because the radiofrequency must be turned on within a phase oscil-

lation period or the beam will be lost to the inner wall. Qualitative adjustment of the Cosmotron, with the frequency tracking r_0, indicates that best capture is affected when turn-on is started about 20 μsec before the particle orbits reach r_0 and is completed (to $\sin \psi_0 = 0.5$) in about 40 μsec. The adjustment of slope is not critical, but the relative timing is. If the particle equilibrium orbit could be held fixed by betatron acceleration while the radiofrequency was turned on adiabatically a large gain could be made in capture efficiency (Sect. IV of [12]) and phase damping would be increased.

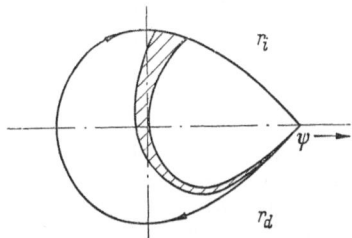

Fig. 58. Typical distortion of shape of bunch after a quarter cycle of phase oscillation.

It is observed in electron synchrotrons (where betatron start is used to avoid frequency variation) that the transfer from betatron to synchrotron operation involves little particle loss. Such a betatron core for the Cosmotron would have a cross sectional area of about 1 m².

The frequency set by the radiofrequency control system will establish, with the corresponding value of B, an instantaneous equilibrium orbit radius about which the radial phase oscillations take place. During the early part of the acceleration cycle best use of the radial aperture is obtained by fitting ω of (25.3) to B_0 and r_0. Radial error, from (25.4), is approximately

$$\Delta r \approx -\frac{r_0}{n}\frac{\Delta\omega}{\omega_0},\tag{25.5}$$

and will produce particle losses in the manner indicated by Fig. 56. The starting frequency (especially with constant T_i) must generally be accurate within 0.1%. To avoid serious transients the frequency should be started sweeping before the accelerating voltage is turned on. If the radius being tracked by the frequency changes quickly (with respect to a phase oscillation period) the stability region will be displaced, similar to Fig. 56c, and protons outside the new stability limits will be lost. As acceleration proceeds and oscillation amplitudes damp there will be working room within the aperture and the radius can slowly be changed, without particle loss, by altering the frequency tracking. A slow rate of change of tracking changes ψ_0 and the stability limits, but if rate of change of radius is small compared to the $\dot{r}_s$ of the phase oscillations no particles will be lost after some damping has been obtained. At relativistic energies (25.4) approximates

---

[1] See Fig. 6 of Twiss and Frank [12].

to $\Delta r \approx -r_0 \Delta\omega/\omega_0$, oscillation amplitudes are small compared to A, and there is considerable freedom for alteration of the tracking radius. Radial changes are useful for following the best shaped region of magnetic field, and for moving the beam toward targets.

If the frequency tracked the central radius smoothly, and any changes made in radius were strictly adiabatic, the phase stable particles $\left(\dfrac{\pi}{6}, 0 \text{ in the phase}\right.$ plot, Fig. 55$\Big)$ would not gain any phase amplitude, and other particles would rotate about the origin of the phase plot, shrinking slowly in ψ amplitude, and rapidly in ΔE and Δr amplitude. However, there is noise in the radiofrequency phase and amplitude, and ripple in the magnetic field. These disturbances will excite increased phase amplitude in all the particles and, early in the accelerating cycle, drive many of them beyond the limit of phase stability. Effects of disturbances in the linear approximation were treated in Sect. 6. Fortunately, the "detuning" effects of the non-linear phase oscillation diminish the amplitude increase produced by a sinusoidal driving term. This is favorable, since the phase oscillation frequency falls through the magnet power supply ripple frequencies during acceleration, as well as through harmonics of the power frequency always present to some extent in radiofrequency phase and amplitude. The problem of passage of the phase oscillation through resonance with magnet ripple, including numerical integration of the equations with sinusoidal restoring force, has been published by Blachman[17]. The amplitude reached by the phase synchronous particle when excited by sinusoidal phase modulation of the radiofrequency

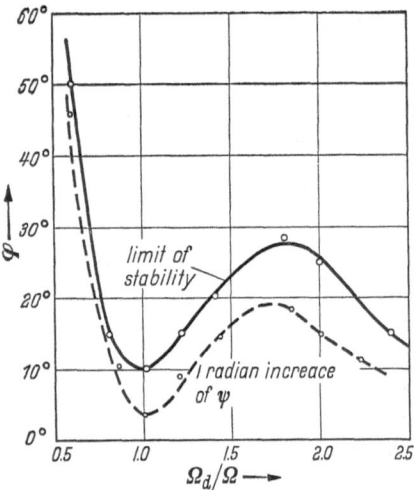

Fig. 59. Amount of phase modulation required to exceed limit of stability as a function of the ratio of disturbing frequency to phase oscillation frequency. From Lloyd Smith (unpublished).

has been calculated on a differential analyzer by Lloyd Smith (unpublished). Fig. 59 shows the limit at which the phase synchronous particle becomes unstable plotted as the depth of phase modulation, φ, vs. the ratio of disturbing frequency to phase oscillation frequency Ω_d/Ω. There is little resemblance to the resonance curve of a linear oscillator. The dashed curve of Fig. 59 is taken from a table by Smith, and shows the φ required to impart one radian of phase amplitude to the synchronous particle. Losses would be large with the latter disturbance, and total with the former. The noise in the frequency (or phase) will have frequency components over the entire range of Ω, and will drive the phase oscillations at random. Phase amplitudes produced by noise tend to be large because there is no "detuning" effect, and the noise acts during the entire accelerating cycle. If the rms fluctuation of $\omega_{\text{radiofrequency}}$ is W over radiofrequency control system band-width B, then the spectral density in the phase oscillation region is $W/\sqrt{B}$. During acceleration period τ, Smith shows that rms phase amplitude induced will be

$$\psi_{\text{rms}} = \frac{W}{2}\sqrt{\frac{\tau}{B}}. \tag{25.6}$$

This corresponds (with different nomenclature) to Eq. (25) of [17].

Radiofrequency acceleration is applied at one or more gaps around the ring. The total energy gained by a phase synchronous particle in one revolution must equal the required energy gain per turn set by $\dot{B}$, Eq. (7.2). If one gap is used, with sinusoidal voltage having peak amplitude V, then (7.2) becomes

$$eV \sin \psi_0 = \left(1 + \frac{NL}{2\pi r_0}\right) 2\pi e r_0^2 \frac{dB}{dt} = C_0 r_0 e\dot{B}, \qquad (25.7)$$

where C_0 is the orbit circumference. If N_1 gaps are provided with their electrical phase angles differing by their angular displacement along the orbit, each gap need supply only V/N_1 peak amplitude. Since power, at constant electrical Q, is proportional to V^2, the total output power of the radiofrequency system can be reduced by using multiple accelerating stations.

26. Observations of acceleration. Quantitative observation of the processes of capture and acceleration is difficult, and data is somewhat scanty. Capture is

Table 6. *Radiofrequency and phase oscillation parameters calculated for $n = 0.7$, $\psi_0 = \pi/6$, and average $\Delta E/turn$ listed.*

Parameter	Birmingham	Cosmotron	Bevatron	Synchro-phasotron	
$1 + \dfrac{NL}{2\pi r_0}$	1	1.212	1.254	1.181	
Initial frequency . . .	300	360	355	182	kc
Final frequency . . .	9.7	4.18	2.47	1.45	mc
Mean $\Delta E/turn$	0.2	1.0	1.5	2.5	kev
Imitial $\Omega/2\pi$	4000	3100	2100	1700	cps
Final $\Omega/2\pi$	3100	1800	970	600	cps
Initial $\Delta r_{max} = \bar{r}_s - r_0$	10	18	23	54	cm
Radial A	13	32	58	75	cm
Final Δr_{max}	0.21	0.64	0.95	1.8	cm
Initial ΔE_{max}	5.9	42	88	103	kev
Final ΔE_{max}	0.2	0.8	1.3	2.0	Mev

complicated by uncertainty as to the distribution of the protons after injection, and the degree of coupling between radial betatron and phase oscillations. Acceleration is disturbed by the noise level of the synchrotron system; quantitative

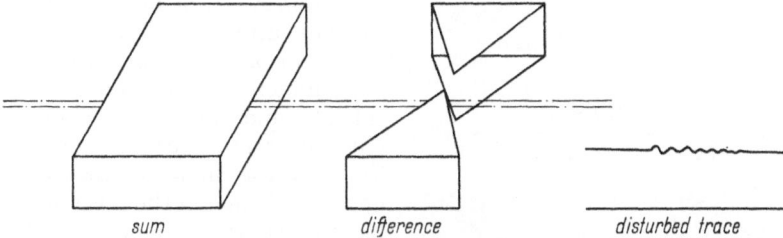

sum   difference   disturbed trace

Fig. 60. Isometric sketch of sum and difference type pickup electrodes, with sketch of disturbed scope trace showing coherent phase oscillations on envelope.

determination of the effective noise power (in radiofrequency system plus magnetic field) has been virtually impossible.

The approximate radio frequency characteristics of the four synchrotrons, calculated by the relations of Sect. 7, are given in Table 6.

The most useful devices for observing the characteristics of acceleration of the beam are the induction electrodes, or "pick-up" electrodes [33] and ([34], p. 85). With them the beam can be observed without destroying or disturbing it. These electrodes are placed inside the vacuum chamber, a straight section making the best location, and take the general form of Fig. 60. The proton bunch

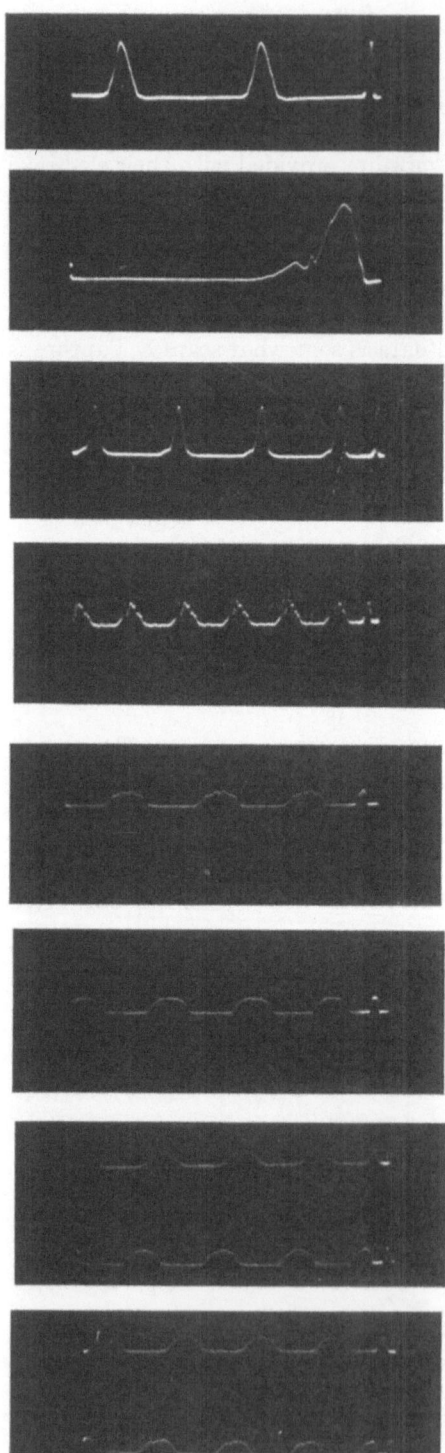

induces charge on the insulated electrodes, and the resulting voltage pick-up can be amplified and presented on an oscilloscope. The "sum electrode" signal on a slow sweep gives a trace with the amplitude roughly proportional to the beam intensity, and indicates sudden losses of beam clearly. Examples of such traces can be seen on p. 258 and 502 of [5]. A transient in the radiofrequency, or magnetic field, will excite coherent phase oscillations in the proton bunch; this can be seen as a perturbation on the trace at twice phase oscillation frequency, as sketched in Fig. 60. Wide band amplifiers will display the shape of the proton bunch and permit study of damping, beam loss, and disturbances. A series of single sweep oscillograms of the Cosmotron bunch, Fig. 61, shows the erratic nature of the phase oscillation "damping" (C. E. Swartz, unpublished). A violent disturbance has occurred just prior to the 200 msec trace. Swartz has also fitted an integrating circuit at the Cosmotron so that the area of the bunches is presented on indicators calibrated in number of protons circulating. The signals from the "difference electrodes" are subtraced electronically and normalized by the "sum signal" to indicate radial (or vertical) position of the beam.

Proton synchrotrons lose quantities of particles after initial capture through the loss of phase stability by protons initially undergoing large phase oscillations and subsequently pushed beyond the stability limits by disturbances in the system. Study of the shape of the bunch as a function of time in the Cosmotron shows that the captured bunch, subtending π soon after radiofrequency turn-on, is reduced to about $\pi/2$ after a few tens of phase oscillations (Fig. 61). The reduction in subtended angle is due to pinching off and loss of the particles with large excursions in ψ. Although the phase angle of the bunch should damp down by

Fig. 61. Oscillograms of pick-up electrode trace showing shape of proton bunch in Cosmotron at 20, 25, 100, 200, 300, 400, 600 and 900 milliseconds. C. E. Swartz (unpublished).

at least a factor two during the remaining acceleration, it is observed that the angle subtended remains approximately constant, any damping being counteracted by the noise of the system. In average operation the Cosmotron loses the order of 80% of the captured beam during acceleration, most or all of the loss occurring in the first fifth of the period.

Injection and capture in the Birmingham synchrotron are complicated by coupling. The n-value is 0.75 (determined by measuring $\dot{r}$) so $\nu_r = \frac{1}{2}$, and large radial betatron oscillations will be discriminated against. There is coupling of betatron to phase oscillations and there also may be a small tilt in the median plane coupling vertical to radial betatron oscillations. If the accelerated beam intensity is observed at 8 msec as a function of inflector voltage [41] cut-off is found at about $\pm 2\frac{1}{2}\%$ (Fig. 62a). Since betatron amplitudes are increased by alteration of inflector voltage, the minimum radial amplitude after injection

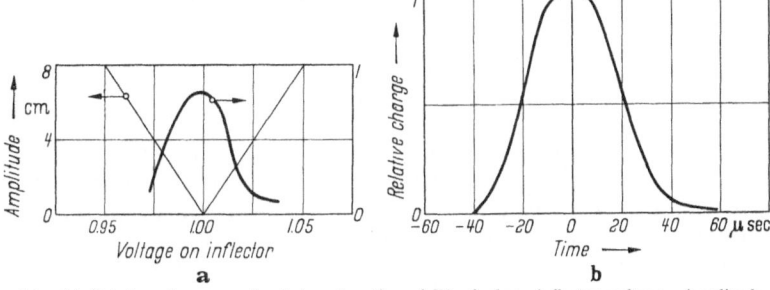

Fig. 62a and b. (a) Relative charge accelerated as function of Birmingham inflector voltage. Amplitude of smallest betatron oscillations superimposed. (b) Relative charge accelerated as function of radiofrequency turn-on time. After RAMM, COE and VAUGHAN [41].

can be plotted in Fig. 62a. The cut-off would indicate that protons with radial betatron amplitude greater than 4 cm are unsuitable for acceleration. The measurement is somewhat unfavorable because large deviations in inflector voltage will cause increased loss to the tip of the electrodes. If the accelerated beam at 8 msec is measured $vs.$ radiofrequency turn-on time the variation of Fig. 62b is obtained with injection energy constant. With the turn-on mistimed 40 μsec, corresponding to $\Delta r \approx 4.5$ cm, essentially all the particles are lost. The frequency tracks the central orbit so the maximum phase oscillation amplitude captured is 4.5 cm, a maximum ψ_i of about 40°. In reference [40] it is estimated that about 145° of beam are captured, of which some 50% is lost during the first $\frac{1}{4}$ phase oscillation, leaving those protons with $\psi_i < 35°$. If the spiraling charge collected on the inside probe is measured as a function of pre-accelerator energy modulation, $\dot{T}_i$, the injected charge will be about twice as large when the initial equilibrium orbit is held just inside the inflector as when $T_i = \text{const}$, Fig. 51 [41]. Then if $\dot{T}_i$ is set at the peak of Fig. 63a the aperture should remain full of protons with a phase plot similar to Fig. 56c and, if the frequency tracks r_0, the time of radiofrequency acceleration turn-on should be immaterial. Fig. 63a shows the insensitivity of accelerated beam intensity to variation of radiofrequency turn-on time under such conditions. Intensity of captured and accelerated beam as a function of $\dot{T}_i$, Fig. 63b, shows a maximum nearly $2\frac{1}{2}$ times the intensity accelerated with $T_i = \text{const}$. (The curve of intensity $vs.$ $\dot{T}_i$ rises from zero because the radiofrequency turn-on is set too late for operation at $T_i = \text{const}$.) Since the Birmingham semi-aperture is little larger than the maximum phase oscillation amplitude, this ratio is in good agreement with the conclusions if Fig. 57. It will be noted from the values of A and $\Delta r_{\max}$ in Table 6 that (according to Fig. 57) T_i modulation is definitely indicated in the Birmingham synchrotron, constant T_i is better for the Bevatron, and either is usable on the Cosmotron. The slight advantage of modulated energy at the Synchrophasotron is not important, since the linear accelerator injector is inherently a constant energy device.

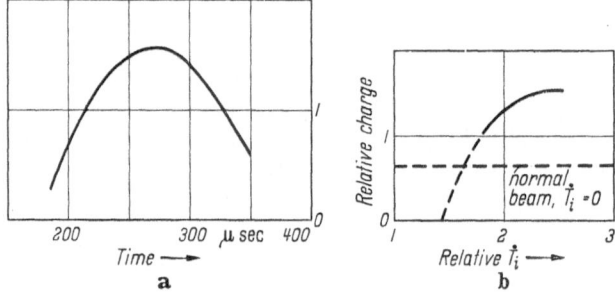

Fig. 63a and b. Relative charge accelerated in Birmingham synchrotron (a) as function of radiofrequency turn-on time with optimum $\dot{T}_i$, (b) as function of $\dot{T}_i$ with turn-on set at peak of (a).

Experiments in beam loss were made on the Bevatron by frequency modulating the master oscillator. (E. J. Lofgren and H. G. Heard, unpublished.) The phase oscillation frequency can be determined by noting the perturbing frequency at which maximum beam loss occurs with small modulation. Just after injection $\Omega/2\pi$ was observed to be 1.2 kc for a peak accelerating voltage just greater than the required volts per turn, and 2.3 kc at twice the volts per turn. A perturbing amplitude of $\Delta\omega/\omega = 0.1/\%$, corresponding to $\Delta\psi \sim 10°$ would lose the beam (see

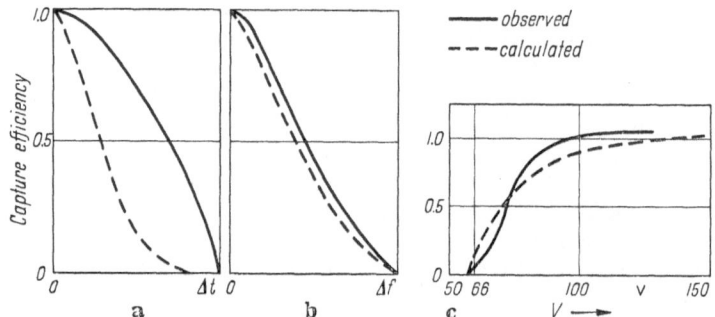

Fig. 64a—c. Capture efficiency in model of the Synchrophasotron. (a) vs. radiofrequency starting time; Δt is time for beam to cross aperture. (b) vs. frequency error; Δf is deviation equivalent to limit of phase stability. (c) vs. radiofrequency amplitude; 66 v per turn are required. From Danilken, Zinoviev, Petukhov and Rabinovich, [5], p. 516.

Fig. 59). However the loss would occur within 2 to 3 phase oscillation cycles at the lower accelerating voltage compared to 5 to 7 cycles at the higher voltage. The range of frequencies around $\Omega/2\pi$ that would seriously perturb the phase oscillations was $\pm 30\%$ for the lower voltage, and $\pm 12\%$ for the higher, showing the expected advantage of smaller ψ_0. With $\sin\psi_0 \approx \pi/6$ modulating frequencies at 60, 360 and 720 cps, roughly subharmonics of $\Omega/2\pi$, did not give observable excitation of the phase oscillations at deviations as large as 0.2%. Larger deviations than this at low frequencies cause beam loss due to variation of the equilibrium orbit. At twice $\Omega/2\pi$ coupling into the phase oscillations could be observed for 0.2% deviations, but beam losses were small. When random noise was used to modulate the master oscillator it was found that an rms frequency deviation of 0.01 % for one second would cause some beam loss. The amount of loss was a function both of the deviation and of the application time. At $\omega \approx 400$ kc and a bandwidth of 20 kc, Eq. (25.6) would give a ψ_{rms} of 8°, enough to lose some beam. The Bevatron system noise level was estimated at about 0.01 % rms, since little additional beam survival occurred when the disturbing noise was reduced below this level. Modulation of the amplitude of the Bevatron master oscillator would produce beam loss at lower than expected levels, due to the conversion of amplitude disturbances to phase disturbances by the electronic circuits. Proportional amplitude disturbances need to be at least two orders of magnitude greater than proportional frequency disturbances for equivalent effects on the beam.

A series of measurements on a 180 Mev model of the Synchrophasotron were made to compare observed and calculated capture efficiency ([5], p. 513). Fig. 64 shows the captured intensity as a function of error in turn-on time of the radio-frequency, error in matching equilibrium orbits corresponding to frequency and T_i, and radiofrequency accelerating amplitude. The calculated curves, normalized, were obtained by integrating phase space areas in a manner analogous to that illustrated in Fig. 56.

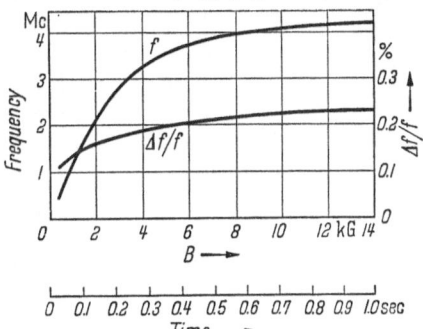

Fig. 65. Cosmotron radiofrequency *vs.* magnetic field and time. Frequency deviation is the $\Delta f/f$ for one inch change in orbit radius. After J. P. Blewett, [*34*], p. 779.

27. Radiofrequency apparatus.

The radiofrequency system divides naturally into two parts—the low-level or frequency controlling equipment, and the highlevel power amplifiers and accelerating electrodes. The frequency control equipment is located in the control room and connected by cables to the high power parts which are adjacent to the synchrotron ring. The frequency determines the radius of the equilibrium orbit and so must track accurately with the rising magnetic field. For a period of the order of a tenth second after injection it is desirable that the frequency track the central orbit with an accuracy of one or two cm $(d f/f < 10^{-3})$ in order to capture and hold a large number of protons. After this the frequency tolerance could be relaxed if only intensity were concerned. However, it is usually necessary to control the radius accurately when striking targets or ejecting the beam, for research purposes, and tolerance of the order of 10^{-3} should be maintained to provide good experimental conditions. The precise control of a rapidly varying frequency is difficult. Straightforward programming on a time axis is not feasible because the magnet pulse is not sufficiently reproducible, and it is therefore necessary either to tie the frequency to the magnetic field strength by a computer, or to compare frequency and field at closely spaced points and make corrections controlled by the error signals. The frequency changes very rapidly just after injection, the rate of change diminishes quickly, and finally the frequency becomes asymptotic to the limiting relativistic value. Frequency as a function of time is shown for the Cosmotron in Fig. 65, along with the relative error for a one inch shift in radius.

The Birmingham synchrotron has a unique mechanically-controlled radiofrequency system [*42*]. The frequency must be swept over the wide range of 300 kc to 9.7 mc, a range most readily covered with a beat-frequency oscillator. The variable frequency oscillator, Fig. 66, is tuned by a rotating condenser with

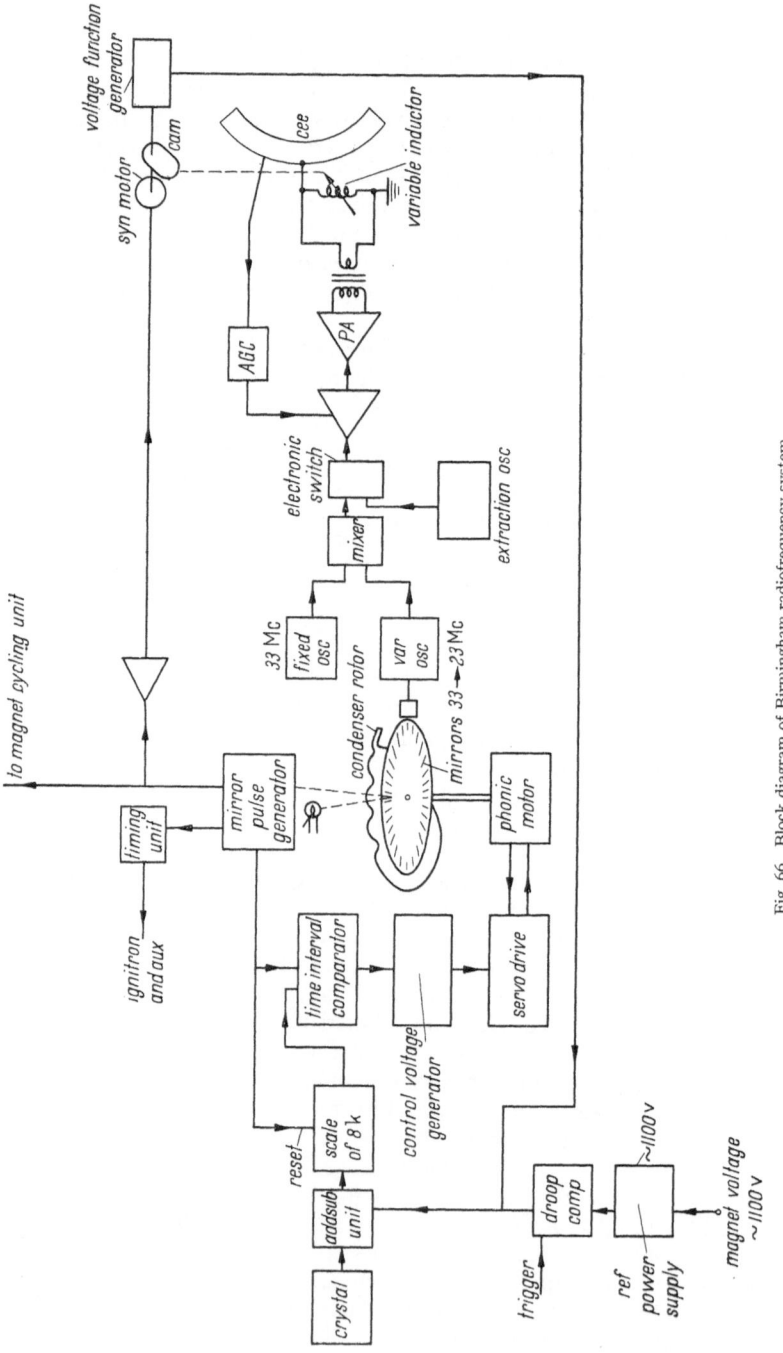

Fig. 66. Block diagram of Birmingham radiofrequency system.

a highly specialized shape, driven by a servo controlled phonic motor so that the rotational speed is controllable over a 10% range by a d c signal input. To monitor the rotation 120 mirrors are set on the condenser rotor to an accuracy of 4 sec of arc, and actuate a photoelectric mirror-pulse generator. The mirrors are set at 3° intervals and produce pulses every 10 msec. Each mirror pulse simultaneously triggers a time interval comparator and resets a scale of 8000 which counts pulses from an 800 kc crystal generator. When the scaler fills, its output pulse is compared to the occurrence of the next mirror pulse and any error in coincidence is converted by the control voltage generator to a d c error signal which corrects the servo drive on the condenser motor. If the comparison time interval from the scaler is shortened or lengthened by adding or substracting input pulses from the crystal (which is done by electronic gates) the condenser velocity will increase or decrease to match. The condenser rotation must track the magnetic field, and this is accomplished by observing the voltage applied to the magnet. One of the mirror signals triggers the ignitron which initiates the magnet pulse (Sect. 19). The magnet voltage is then opposed by a standard reference voltage and any difference signal actuates the add-subtract unit which then alters the rotational speed of the condenser. Since the magnet voltage falls during the pulse (Fig. 21), a generator droop compensator adds a matching voltage-time function for the first half second of the pulse, after which the system is clamped. The beam radius can be programmed, and small changes can be introduced in the field-frequency relationship, by inserting waveforms from the voltage function generator into the add-subtract unit. The voltage function generator consists of suitable potentiometers rotated by a synchronized shaft. In order to move the beam outward onto a target the frequency is clamped. The fixed-frequency extraction oscillator is switched in by an electronic switch, in place of the variable oscillator system. Frequency and phase synchronism are obtained by injecting a little radiofrequency from the mixer into the extraction oscillator in order to lock it[1]. The low-level radiofrequency signal is then transmitted to the power amplifier and the output of this amplifier is transformer coupled to the resonant accelerating-electrode circuit. It is necessary to resonate the accelerating electrode to raise its impedance, which is only $-2.5j$ ohms at 9 mc. The tuning inductor originally installed was a helix of variable pitch which was plunged into a mercury pool by a cam driven linkage. A synchronous motor which drives the cam is powered at amplified mirror pulse frequency so it will track with the rotating condenser. The cam shaft also carries the voltage function generator and switches for auxiliary functions. Another synchronous motor, in the generator room, operates cam driven switches which program the generator exciters to give the magnet cycle[2].

The condenser disc servo has a time constant of about 20 msec, giving response to above 10 cps. The system is well damped and has adequate response to follow the variations of normal magnet cycle operation. The speed control is limited to a range of about 2% by diodes in the circuits, although the inherent range is 10%. Sparking in the mercury tuning inductor caused noise which would excite phase oscillations, and lose beam [40]. The plunging helix was replaced by a saturable ferrite reactor, tuned by saturating current controlled from a phase detector ([5], p. 512). The beam improved in both intensity and reproducibility. The mechanically tuned frequency system has an advantage in its inherently low phase-noise level; and has the disadvantage that the programming of the acceleration cycle is quite inflexible.

Three types of accelerating electrode have been used, and many variations are applicable. The Birmingham synchrotron has a 102° "Cee", formed by a silver coating fired on the outside of the ceramic vacuum chamber [42]. This

---

[1] D. E. Caro and L. U. Hibbard: Phil. Mag. **44**, 964 (1953).

[2] Detailed descriptions of components of this system are given by L. U. Hibbard, D. E. Caro, L. Riddiford, J. Y. Freeman and W. Raudorf in J. Sci. Instrum. **29**, 366, 403 (1952); **30**, 245, 378 (1953); **31**, 139, 170 (1954) and Brit. J. Appl. Phys. **4**, 147 (1953).

coating is broken into 1 inch wide strips to reduce eddy currents due to $\dot{B}$. The coating on adjacent chamber sections is joined by springs on the gasket plate (Fig. 27), and at the end of the Cee the silver coating is connected to the internal platinum coating. The accelerating gaps are then made in the internal coating, and are inclined at about 45° to the radius (Fig. 67). Radial gradient of the ac-

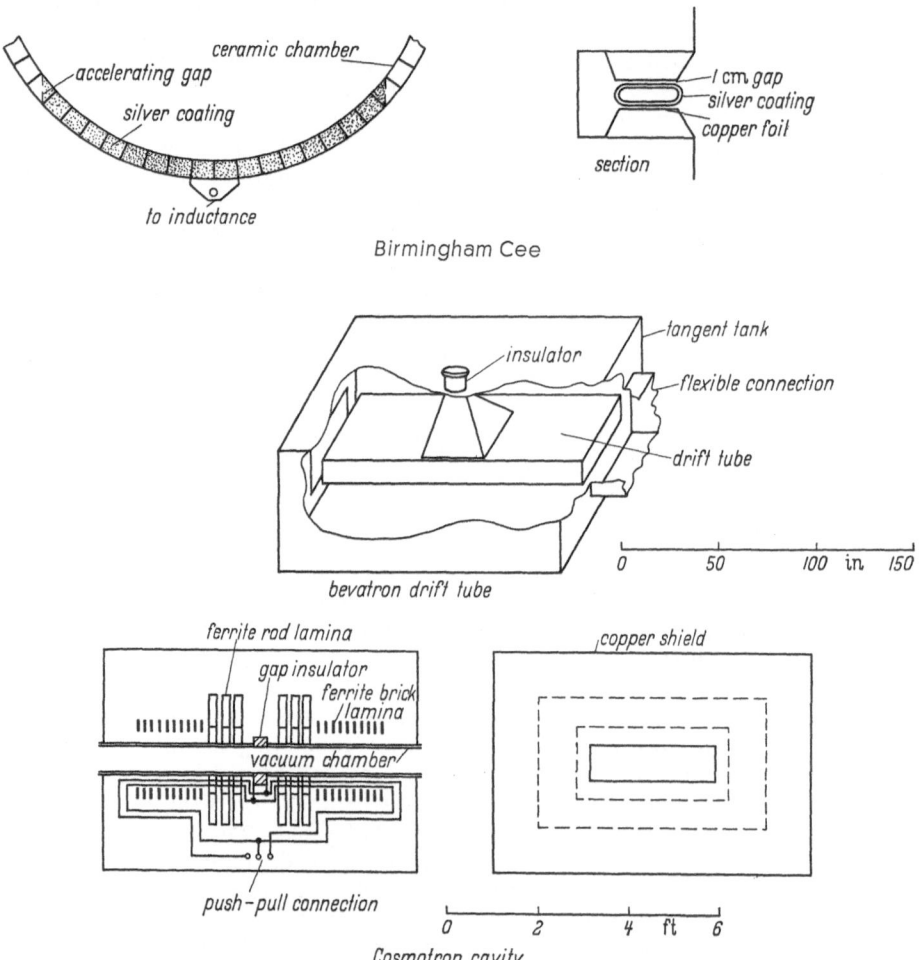

Fig. 67 a—c. Accelerating electrodes. (a) Long drift tube. (b) Short drift tube. (c) Ferrite loaded cavity.

celerating potential diminishes the phase oscillations, but at the expense of larger betatron oscillations. This may be necessary in this machine, in which the semi-aperture is about the same as the limiting radial phase excursion. If the Cee were 180° long it would resemble a cyclotron Dee, but at peak frequency it would be near a half wave long and its impedance would be very critical with respect to both frequency and mechanical variations. If the peak radiofrequency voltage on the Cee is V, the Cee subtends angle ϑ, and a proton passes the center at phase ψ, the proton will pass through potential difference.

$$V\left[\sin\left(\psi + \frac{\vartheta}{2}\right) - \sin\left(\psi - \frac{\vartheta}{2}\right)\right] = 2V\cos\psi\sin\frac{\vartheta}{2}. \tag{27.1}$$

As the Cee is shortened, the radiofrequency voltage V must be increased, but the capacitance diminishes. A minimum in reactive power is obtained with $\vartheta = 135°$, however the Birmingham Cee was reduced further to 102° in order to keep it appreciably shorter than one-half space wavelength at the maximum frequency. Its capacity is about 6500 μμf. In order to minimize losses, the magnet poles were covered with insulated copper foil which makes the return connection to the inductor. Power required is less than 5 kw.

The Cee can be foreshortened to a "drift tube" in a straight section machine. If ϑ is very short (27.1) is approximately equal to $V \vartheta \cos \psi$ and the peak voltage on the drift tube must be $1/\vartheta$ times the peak potential gain. However the drift tube can be located in an expanded straight section box so its capacity, and reactive power, will be relatively small. The Bevatron drift tube is a $13 \times 55 \times 131$ inch open-ended box (Fig. 67) with a capactiy of about 400 μμf. It subtends a ϑ about $\frac{1}{6}$ radian, and so requires a peak voltage of order 20 kv to give a $\psi_0 = \pi/6$. The capacitance is resonated with a saturable ferrite core inductor; a peak potential of 35 kv can be obtained with less than 30 kw of radiofrequency power [46]. The Synchrophasotron uses two drift tubes, in opposite straight sections, driven out of phase.

The Cosmotron power amplifier is untuned (aperiodic) and energizes an untuned cavity. A cavity with $\mu = \varepsilon = 1$, built in the space available, would have an impedance less than one ohm over portions of the $\frac{11}{1}$ frequency range. In order to raise the impedance to some 50 ohms, which can be driven to 2 kv by practical amplifiers, the cavity is loaded with 2800 lbs of magnetic ferrite (Fig. 67), in the form of frames which surround the vacuum chamber on either side of the gap insulator. The ferrite core is excited by a distributed primary winding, and can be considered a transformer core with the proton beam as the secondary "winding". The ferrite has a permeability at the radio frequencies involved of about 500 and a Q, at operating levels of flux density, ranging from 20 at 370 kc to 1.2 at 4 Mc. The loaded cavity reactance varies from $46j$ ohms at 370 kc to $-150j$ ohms at 4 Mc ([34], pp. 795 and 800). It is driven by an aperiodic amplifier with power output exceeding 100 kva.

All the accelerating cavities have impedance variations over the frequency range. The amplifier systems are fitted with automatic gain control loops to keep the accelerating potential reasonably constant. The gain control circuits can also be used to shape the radiofrequency turn-on, and to control radiofrequency turn-off for various modes of beam ejection.

The Cosmotron frequency is controlled from the magnetic field by an electronic analogue computer ([34], p. 782). A voltage proportional to $\dot{B}$ is obtained from the "$\dot{B}$-coil", incorporated in the pole face windings, and surrounding the central half of the magnet pole area in each quadrant. This voltage is integrated by an electronic integrator (Fig. 68). To set in the constant of integration, the electronic switch which connects the input to the integrator is triggered by a pulse from a peaking strip ([34], p. 848) when the magnetic field corresponds (at radius r_0) to the oscillator starting frequency, f_0. The integrator output potential is then proportional to B. A small ferrite ring core, with a few turns of figure-8 winding, is the inductance of the L-C master oscillator. The bias winding on the core, decoupled from the radiofrequency winding, will reduce the core permeability by saturation and will vary the oscillator frequency over the needed range of 340 kc to 4.2 Mc[1]. A control amplifier with heavy negative feed-back drives the bias winding. Since the relationship between frequency and magnetic field,

---

[1] A. I. PRESSMAN and J. P. BLEWETT: Proc. Inst. Radio Engrs. **39**, 74 (1951).

and between control amplifier input and frequency, are known, a function can be derived which relates integrator output to control amplifier input. The system constants are proportioned such that the latter function has everywhere a negative second derivative. The function is then generated electrically by the diode network of Fig. 68. When the resistors of the diode network are properly adjusted (at present 40 junction diodes are used) the frequency will track the magnetic field at r_0. Since the respective diodes conduct successively as the acceleration cycle progresses, selected diode resistors can be adjusted to change the radius of the beam at controlled times. Between cycles automatic servos reset the integrator zero and adjust the oscillator to f_0. The output of the master oscillator is amplified by an aperiodic chain of several stages. The final stage is a 100 kw,

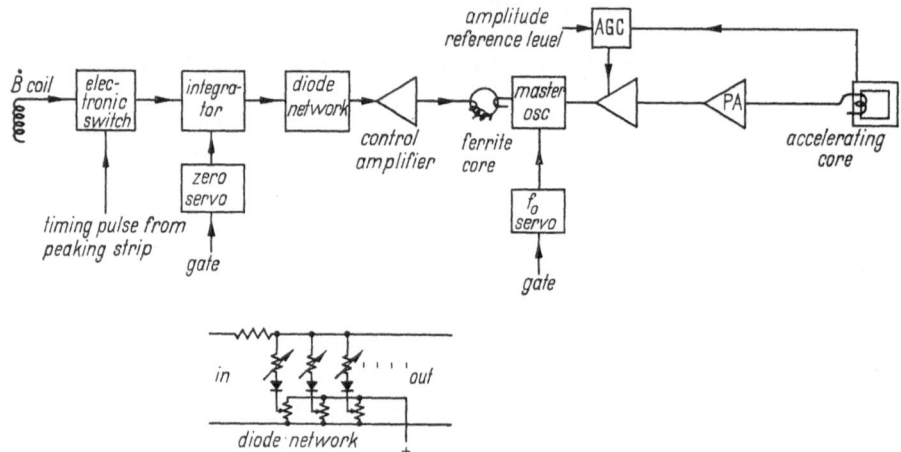

Fig. 68. Block diagram of Cosmotron radiofrequency system.

push-pull, 4 Mc band-width amplifier which excites the ferrite accelerating core ([34], p. 789). The automatic gain control loop compares the gap voltage to an amplitude reference signal, permitting shaping of radiofrequency turn-on and turn-off. The electronic circuits of the Cosmotron radiofrequency system have a wide effective noise band-width. Meticulous care is necessary in reducing the noise level, and hum and noise pick-up, to the minimum attainable limits. The original frequency control system has been replaced by one carefully refined on the basis of operating experience. Reduced noise level resulted in large improvement of beam intensity and reproducibility. This electronic frequency control system will follow the magnetic field without readjustment over a wide range of magnet pulse lengths and repetition periods. A small readjustment is necessary if large changes are made in $\dot{B}$.

The Bevatron frequency control derives its primary signal from the magnet current [46]. One half of the magnet current flows through a shunt and through a series of peaking transformers which generate 29 current marker pulses, Fig. 69. The primary core of the L-C master oscillator is so designed that, with proper back bias, a fraction of the magnet current will saturate the core and generate the required frequency as a function of B to within a few percent. A small ferrite "shaper core" in series with the primary core supplies the inductance correction for accurate frequency tracking. The shaper core inductance is varied by a bias winding and bias power supply whose current output can be set independently in each of 30 intervals by the "curve corrector", the intervals being determined by the time of occurrence of a set of definite magnet current values. It is important that the corrections be introduced very smoothly, for rapid changes in frequency slope will excite phase oscillations and spill protons from the beam. The 30-point curve corrector has, for most purposes, been replaced by a simple waveform generator using gated R-C networks. An alternate frequency generator, activated by $\dot{B}$ is also available. The low-level signal is amplified by a multi-stage wide-band chain which drives the final power amplifier. This final stage is single ended and can supply over 50 kw to the resonant drift tube circuit. The drift tube capacitance is resonated by a multiple ferrite core inductor, adjusted by bias current from a 1000 amp saturating supply. Reson-

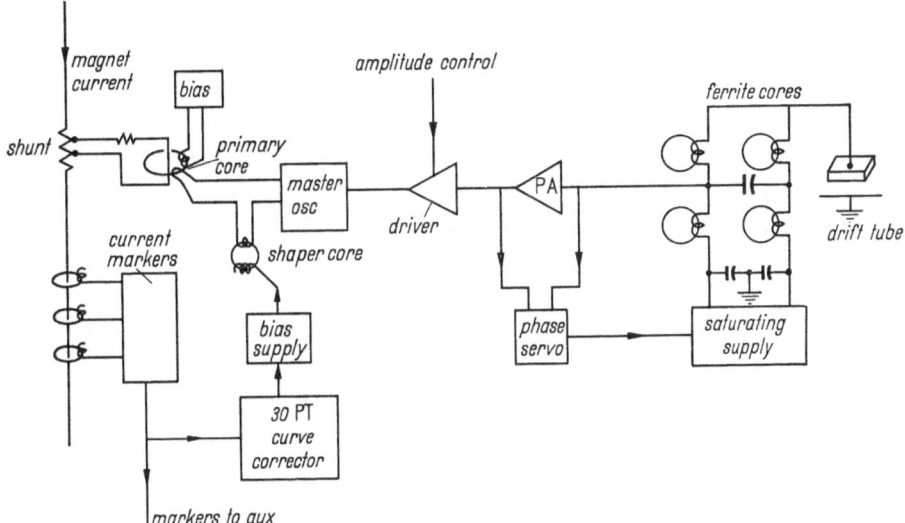

Fig. 69. Block diagram of Bevatron radiofrequency system.

ance is maintained as the frequency sweeps by servo-controlling the saturating supply. The inductor is connected so as to decouple the bias and radiofrequency currents [46].

The Synchrophasotron was fitted with two frequency control systems ([5], p. 429). One is basically the same as the Cosmotron system. The other is similar to the Bevatron scheme, but differs in that the primary ferrite core is saturated by an auxiliary magnet in series with the synchrotron magnet. Corrections are supplied by an integrator and diode shaping-network operating on a heterodyning oscillator. The final amplifiers feed two drift tubes.

28. Beam controlled frequency systems. The circulating particle bunch in a synchrotron generates on the pick-up electrodes a voltage waveform whose fundamental frequency is necessarily synchronous with the particle rotation frequency[1]. It was independently proposed by CH. SCHMELZER and the author that the accelerating waveform be derived from the bunched beam, rather than an independent electronic oscillator. A straightforward method for doing this is shown in Fig. 70. The fundamental is extracted from the sum electrode signal by a filter and is then amplified and applied to the accelerating electrode. A phase shift must be introduced to account for: the angular displacement of the

---

[1] See also p. 395 of Ref. [5].

pick-up electrode ahead of the accelerating electrode, $h\vartheta$; for the phase difference between the radiofrequency peak and center of the bunch, $\pi/2 - \psi_0$; and for the phase delay of the electrical system including its cables, $2\pi l/\lambda$ (where l is the effective electrical length between the electrodes, λ is the free-space wavelength). If k is an integer, the phase retardation to be inserted is

$$\varphi_r = (\pi/2 - \psi_0) + h\vartheta - \frac{2\pi l}{\lambda} + k\,2\pi. \tag{28.1}$$

From (25.7) it is required that

$$\sin \psi_0 = C_0 r_0 \frac{\dot{B}_0}{V} \tag{28.2}$$

and approximately that, if the center of the bunch is not at r_0,

$$\sin \psi_0 \propto r^2 \frac{\dot{B}}{V}. \tag{28.3}$$

If the phase shift of the electronic system is not correct, the beam radius will change, and the particles will drift out of the aperture. The control can be given

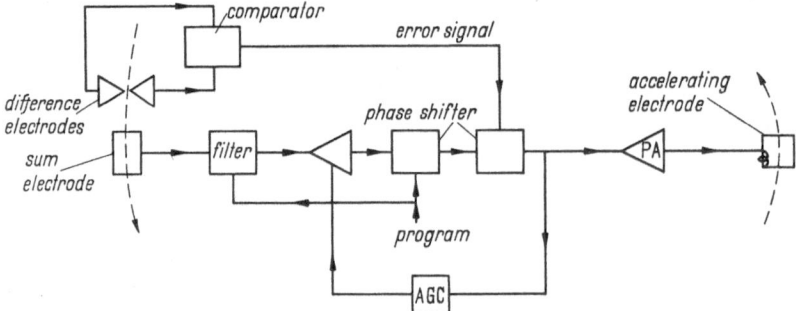

Fig. 70. Block diagram of feedback-type beam-controlled accelerating system.

long-time stability by obtaining an error signal propertional to the Δr of the center of the bunch, and causing this error signal to control a phase shifter in the amplifier chain (Fig. 70). The pass band of this error servo should extend well beyond the phase oscillation frequency, and should cut off below the radio-frequency; a reasonable requirement because ω/Ω is two or three orders of magnitude. Such an error signal is readily obtained from difference electrodes by comparing the signals, rectifying, and filtering with proper time constants.

There is an important difference between the phase-controlled, or phase-lock, method and the programmed method described in Sect. 27. When the frequency is controlled by an external generator, the relative phase of the radiofrequency and of the beam is correlated only by the phase stability through the mechanism of radial variation. $\dot\psi$ must be proportional to Δr, and particles are lost if $\dot\psi$ exceeds the stability limits, Fig. 54. The phase plot can be drawn more completely in the form of Fig. 71a, with the zero of ordinates representing the equilibrium orbit value $r_{\rm rf}$ defined by the radiofrequency, $\omega_{\rm rf}/2\pi$, and B. The relationship of (28.3) will readjust itself if $r_{\rm rf}$ changes slowly, but there is no direct relationship of control between ψ and ψ_0. If there is a step in the phase or frequency of the master oscillator, the particles in the shaded areas of Fig. 71a will be displaced in phase space in phase or in radius. Particles thrust outside the separatrix will be lost along the indicated paths. (It should be noted that a return step could re-enclose particles in the separatrix.) A phase-lock system controls ψ from ψ_0, and thus from $\dot{B}/V$. To first order, $\dot{\psi}$ is not related to Δr. Let the system of Fig. 70 be accelerating particles in the initial portion of the $r - \psi$ phase space of Fig. 71 b, and apply a step function in ψ_0 corresponding to radial change Δr. Although the final position of the bunch lies outside what would be the stability limit in Fig. 71 a, the initial-final difference in $\dot{\psi} - \psi$ space is very small, and within stability limits. Bunch and radiofrequency phases are locked together and the equilibrium orbit radius will spiral in. The change in orbit radius per turn $d \Delta r \propto \Delta r$, so the particles will asymptotically change from the initial to the final positions along paths like the one indicated in Fig. 71 b, somewhat in the fashion of an overdamped oscillator. There are second order effects in the process because the shape of the bunch changes slightly. In the preceding discussion ψ_0 has been taken as the phase of the maximum of the pick-up electrode signal, i.e., the phase of maximum particle density, which may not exactly coincide with the phase of the equilibrium particle.

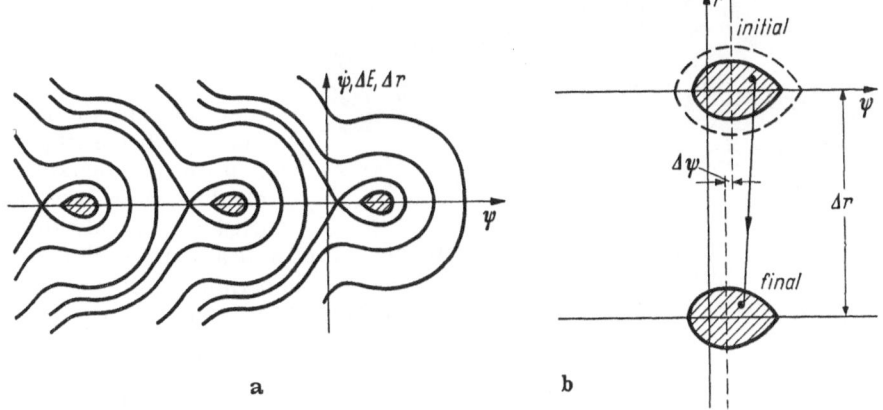

Fig. 71 a and b. Phase diagrams of (a) externally controlled system, (b) phase-lock system. The dotted stability lim *would* apply if r were proportional to $\dot{\psi}$.

For a beam-controlled system to operate it is evident that the beam must be bunched. This can be accomplished by starting with a programmed oscillator, or by "chopping" the injected beam at the starting radio frequency rate. The filter and one phase shifter of Fig. 70 must be programmed to follow the frequency, but with only moderate precision. The term $2\pi l/\lambda$ will vary over the frequency band by a large amount (measured as $10\frac{1}{2}\pi$ at the Cosmotron) so the programmed phase shifter is technically difficult. The system of Fig. 70 was constructed with a bandwidth of some half mc and tried on the Cosmotron by switching from programmed oscillator to beam-control at about 0.5 sec, when little frequency change remains. (This avoided the necessity of programming the filter and phase shifter for the experiment.) Operation was very stable[1]. If a step in radius control of nearly 6 in. was inserted the beam moved without loss, and without any indication of induced coherent phase oscillations. An oscillogram of the signals from the outside and inside radial pick-up electrodes (Fig. 72) shows the movement of the beam. Simultaneous observation of the sum electrode found no beam loss as the radius changed. The control phase shifter was limited to $\pm 70°$. A similar beam-controlled radiofrequency system was used on the Brookhaven Electron Analog, an electron synchrotron operating at 7 Mc with $\Delta f/f$ about

---

[1] G. K. GREEN and E. C. RAKA: 1953, unpublished.

6% [1]. No auxiliary oscillator was needed since the bunch was formed by injecting one-half revolution of particles, the revolution frequency then being immediately available for the radiofrequency accelerating system ($h = 1$).

A phase comparison method, suggested independently by Ch. Schmelzer and H. S. Snyder, will provide phase-lock control of acceleration (Fig. 73). A

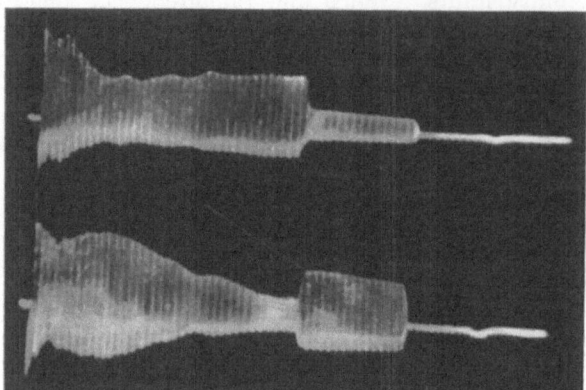

programmed master oscillator, designed for low noise level and moderate tracking accuracy, will bunch the beam at radiofrequency turn-on. Equal length cables carry signal from the sum electrode and accelerating gap to a phase discriminator which compares the inputs and produces an error signal proportional to phase difference. This error signal corrects the master oscillator; being superimposed on the program control. The phase servo loop must have a bandwidth ex-

Fig. 72. Oscillogram of inner and outer difference electrodes in the Cosmotron. Sweep lasts 1 sec, and beam is being accelerated with phase-lock system of Fig. 70. Radial step was caused by applying a step function to the comparator.

tending well beyond the phase oscillation frequency, and as near to the radiofrequency as moderate filtering of the discriminator output will permit. In order to correct slow radial variations another error signal proportional to Δr modifies the relative

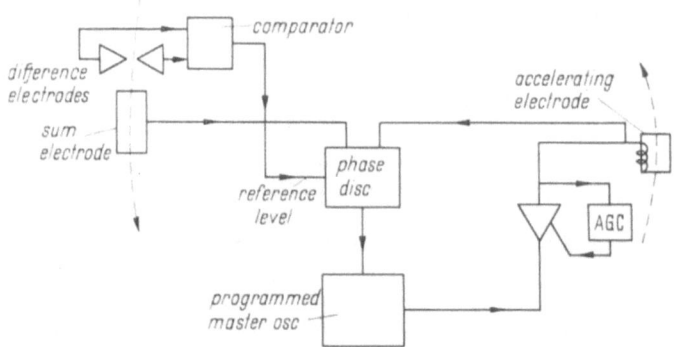

Fig. 73. Block diagram of phase comparison beam-control system.

reference level of the phase discriminator; alternatively it could control a narrow range phase shifter in one of the discriminator inputs, or the automatic gain control servo and consequently the voltage level on the gap. The radial servo loop will have a restricted bandwidth. The choice of system (type of Fig. 70 or Fig. 73) depends on details of higher order effects in change of bunch shape, and on the signal to noise ratios of specific apparatus.

Error signal from radial pick-up electrodes can be used to control the master oscillator directly, and to correct variations in radius due to errors in the oscillator program. The loop response includes the beam and is subject to the delay

---

[1] G. K. Green, M. Plotkin and E. C. Raka: Unpublished. See also p. 104 of [5].

of the phase oscillations so carefully designed anti-hunt networks are required. The advantages of phase-lock, as illustrated in Fig. 71, are not obtained. This type of self-tracking system has been used on the Bevatron and operates satisfactorily, but does not improve beam intensity.

A variant of the phase comparison method was operated in the Cosmotron radiofrequency system by E. J. ROGERS (in press). The phase comparison loop was closed to an auxiliary control winding on the master oscillator core in the manner shown in Fig. 73. The loop response extended from a few hundred cycles to many times the phase oscillation frequency. A slow radial loop was not fitted, consequently the circuit would not control the radial position of the beam. The phase lock, at a rate higher than $\Omega/2\pi$, damped coherent phase oscillations which were arising from unknown disturbances. Beam losses diminished and pulse-to-pulse reproducibility was improved.

VIII. Proton synchrotron operation.

29. General characteristics. The large number of components of a pulsed proton synchrotron, and precision of timing, require automatic operation. Magnet power is so great that servo control of the magnetic field cycle to a predetermined waveform is difficult and expensive. The magnetic field is therefore usually taken as the independent variable during the pulse. However, the B vs. t cycle reproduces quite well under controlled conditions and direct use of time for actuating many devises is practicable. At the Cosmotron a gated quartz-crystal clock with settable pre-determined counters triggers the components whose timing is not critical ([34], p. 846). Peaking transformer cores, around the magnet bus, biased for control of their pulse will give markers which indicate the occurrence of definite values of I, and consequently of B. A set of 29 such markers has been fitted to the Bevatron magnet circuit. Interpolation can be done with simple time delay units. (Variation of bias current is another possible method.) Very accurate field markers for injection and start of frequency sweep are provided by peaking strips ([34], p. 848). Peaking strips made with thin wire cores will reproduce within long time limits of $1/10^4$ and short time limits of $1/10^5$. Other precise triggering methods are proton or electron resonance and fm oscillator-discriminators ([5], p. 432). The various triggers are required for initiating injection and frequency sweep, turning radiofrequency amplitude on and off, timing magnet correction circuits, actuating target motions and gating research equipment.

The most flexible control arrangement treats each accelerating pulse as a free-standing event, initiated by the application of voltage to the magnet. This arrangement makes the synchrotron independent of repetition rate and permits alternate pulse operation of multiple experiments. Different experiments may require different pulse length, intensity, radiofrequency timing, target actuation, mode of striking target and auxiliary magnet excitation. Switching systems at the Cosmotron and Bevatron automatically change the necessary parameters to supply perhaps three different experiments.

Monitoring of acceleration is done by observing various oscilloscope displays of the pick-up electrode signals. Jitter (pulse-to-pulse intensity variation) is a serious problem; continuous search for causes of jitter, and continuous refinement of apparatus are necessary. Under good operating conditions the jitter can be kept below 10%. Intensities reproducible to some 1 to 2% have been obtained at the Cosmotron by injecting noise in the radiofrequency system early in the cycle until the beam is reduced to a standard intensity, somewhat less than the expected minimum. Monitoring of waveform and timing of numerous components, and careful adjustment of them, are necessary to maintain good intensity and low jitter. As many as one or two dozen simultaneous oscilloscope traces may be employed. The long time, or day-to-day, variation of beam intensity is erratic and will sometimes be an order of magnitude. ("Disasters", such as forgotten objects in the aperture, are not unknown.)

The maximum energy and repetition period of typical proton synchrotrons are given in Table 1. When operating at energies less than maximum the repetition period of the Cosmotron and Bevatron can be shortened, a factor of two being readily obtainable at half energy. The magnet systems of the Birmingham synchrotron and the Synchrophasotron are somewhat less flexible. The attainable intensity is best expressed in terms of protons per pulse, and is approximately (1957):

	Injected current	Protons accelerated
Birmingham . . .	700 μa	10^{10} per pulse
Cosmotron	4 ma	10^{11}
Bevatron	350 μa	10^{11}
Synchrophasotron.	(in starting-up period)	

Improvements in intensity will be made in these machines, largely by refinements of the injection and radiofrequency systems. Intensity will eventually be limited by space charge. Space charge modifies the electromagnetic fields in the aperture and will, as the quantity of charge increases, eventually force particles outside stability limits of betatron or phase oscillations, or move the operating point into destructive resonance bands[1]. Space-charge limits imposed[1] by stability limits on betatron and phase oscillations are of the same order of magnitude. In the machines under consideration the space-charge limit just after injection for a bunched beam is of order 10^{12} to 10^{13} protons.

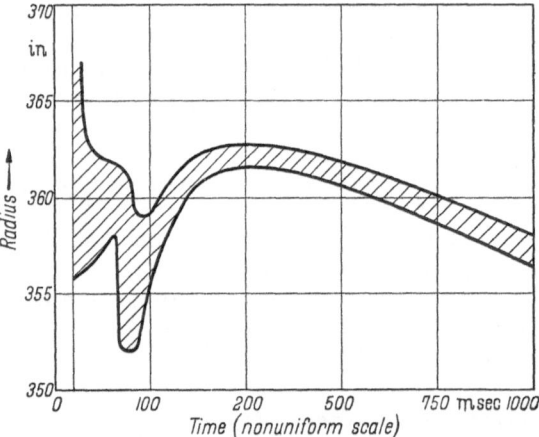

Fig. 74. Radial edges of the beam, as a function of time, in the Cosmotron. Erratic changes of mean radius have since been largely eliminated.

30. Methods for striking targets. When the beam of a proton synchrotron has been accelerated to peak energy the lateral extent of the particle bunch is small. Betatron oscillations have been reduced as $B^{-\frac{1}{2}}$, radial phase oscillations are limited to about a cm and the energy spread $\Delta E/E < 10^{-3}$ (see Δr_{max} and ΔE_{max} in Table 6). The size, which is twice oscillation amplitude, has been observed to be about 2×2 cm at Birmingham, 1 in. (z) $\times2$ in. (r) in the Cosmotron, and 2×4 inches in the Bevatron. At one fourth maximum energy the beam size is about two times larger. Measurement of the radial extent of the protons has been done in the Cosmotron and Bevatron by flipping a vane into the aperture. By varying the timing and radius of the vane, and noting the time of loss on

---

[1] See Ref. [22] and P. B. Moon: Proc. Phys. Soc. Lond. **69**, 153 (1956).

the sum electrode trace, the contour of the beam as a function of time can be plotted, Fig. 74. Beam can be put on target over a wide range of energies, one of the best characteristics of the pulsed proton synchrotron. In addition, $B_0 \propto p$ so particle paths are relatively independent of extraction energy. For analogous situations at different energies only a small correction is necessary for change of shape of the magnetic field with saturation.

If the radiofrequency is turned off while B is rising the equilibrium orbit radius will decrease at a Δr per turn and radial velocity given by (24.5) and (24.6). The beam will spiral inward and can be made to strike an "inside" target. If acceleration is carried past the peak of the magnet cycle and then stopped when $\dot{B}$ is negative the beam will spiral outward and can strike an "outside" target. The orbit radius can also be manipulated by the radio frequency control; beam-controlled systems should be especially useful for this purpose. The rate at which the orbit can be moved by the radiofrequency depends on ψ_0 (i.e. V) and if $\sin \psi_0 = 0.5$ the equilibrium particle can be moved outward at the rate of (24.6) or inward at three times this rate. (Protons at other phases would be lost.) If the frequency is clamped the radius will expand. Differentiating (4.3) with the straight section correction, and using (4.5) for expansion around r_0,

$$\left[\frac{1}{\left(1 + \dfrac{NL}{2\pi r_0}\right)(1-\beta^2)} - (1-n) \right] \frac{\dot{r}}{r_0} = \frac{\dot{B}}{B_0} \qquad (30.1)$$

for $\omega = \text{const}$. Representative values of beam motion at peak energy using the averaged parameters of Tables 1 and 6 are:

Table 7. *Radial motion at peak energy.*

	Δr/turn $(V=0)$ 10^{-4} cm	$\dot{r}$ $(V=0)$ 10^3 cm/sec	$\dot{r}$ $(\omega=\text{const})$ cm/sec
Birmingham	− 1.5	− 1.5	110
Cosmotron	− 7.3	− 3.0	70
Bevatron	− 7.7	− 1.9	17
Synchrophasotron . . .	− 16.6	− 2.4	8

The preceding values can be changed by manipulating $\dot{B}$.

Inside targets are the most straightforward and, to date, the most used. A target could be thrust through the beam, since adequate time is available, but precise timing would be difficult. It is equivalent to position the target just inside the beam and to diminish the orbit radius by altering the frequency tracking. The bunch cross section is traced out by protons performing betatron and phase oscillations over a range of amplitudes and phases. If not at a resonance (which is assumed) the oscillations and rotation frequency are anharmonic, and the betatron frequency is much higher then the phase oscillation frequency. Particles with the largest total amplitude will touch the target first, followed by those with progressively smaller amplitudes until, when the orbit has shrunk half the width of the beam, all have struck the target. The emerging particles, secondaries and scattered primaries, will have radiofrequency structure, that is, will emerge in bursts of order 90° at the radiofrequency rate. If no recent disturbances have excited coherent phase oscillations there will be little variation at phase oscillation frequency. (The converse is often seen.) Magnetic field ripple causes the equilibrium orbit to expand and contract about its mean value, and consequently there will be fluctuations at the ripple frequencies.

Geometrically, it is evident that full penetration of a thick target is not probable with Δr per turn less than $1/_{100}$ mm. Radii of curvature are very large, in the straight sections infinite. Protons grazing the target will be scattered, and some will penetrate on a succeeding revolution. In order to assure target penetration a lip, or thin projecting piece, preferably a light element to minimize scattering, is attached. The operation of the lip is diagramed in Fig. 75. When a proton penetrates the lip it is at the maximum of a radial betatron oscillation, the energy loss displaces the equilibrium orbit toward the lip, and the proton proceeds with reduced betatron amplitude. When the proton is again in phase with the lip the process will repeat, until the equilibrium orbit is drawn into the lip. After that, the orbit will continue inward but the oscillation amplitude will increase until the proton penetrates the thick target. Since a stepwise Δr of the order of mm is used, full thick target penetration can be obtained[1]. The proton may make multiple traversals of the target, unless it is thick enough

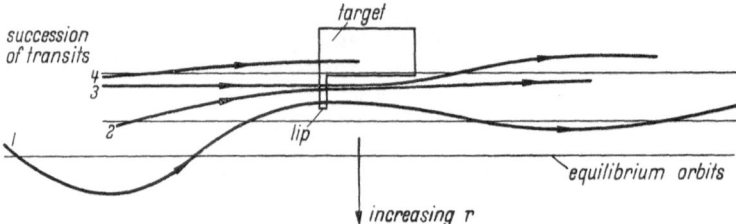

Fig. 75. Diagrammatic sketch of action of target lip on proton paths. Several revolutions may occur between the transits shown, during which the betatron oscillations are out of phase with the target.

to assure rapid loss. If multiple traversals are troublesome a "clipper", or thick loss target, can be inserted at a different azimuth.

The target should not block appreciable aperture at injection, unless a serious decrease in intensity can be tolerated for a particular experiment. On the other hand, as the magnet saturates there are n-values on either aide of the beam at which resonances occur. If the beam lies on the other side of a resonance from the target, it will be lost before reaching the target. (Although beam could be thrown into the target by a radial resonance, provided the azimuthal phases were adjusted properly.) The Birmingham orbits encounter a resonance above 800 Mev at $r = 440$ cm, 4 cm greater than the inner aperture stop. The Cosmotron field (Fig. 8) can be corrected by the pole face windings so that the $\nu_z = 1$ resonance is less than 3 or 4 inches from the inner vacuum chamber wall (at $r = 347$ in.). Fixed targets are usable to $r = 354$ in. because the aperture is not fully occupied by injection, however, this will not obtain when the decaying injection energy is corrected. Most targets are rammed just before beam contraction by compressed-air cylinders operating through sliding vacuum seals on the inside covers of the straight sections. Many experiments are troubled by the background from spillout, which is less concentrated at the location of the experimental port if the target is retracted, and particle paths may be more favorable if the target is at a larger radius than is wanted at injection. The saturation characteristic of the Bevatron H-magnet (Fig. 11) leaves the usable region in the middle of the aperture. Targets must, in most instances, be moved. In addition, to miss the outer leg of the magnet it is necessary to locate targets inside the quadrants for many experiments. There is limited use of compressed air rams on the straight sections. Other air-cylinders operate linkages through the "crawl-space" (Fig. 32)

[1] E. M. McMillan: Rev. Sci. Instrum. 22, 117 (1951).

to ram quadrant targets into the aperture. A third mechanism, the "flip-up" target, carries the target proper on a pantograph linkage so that the target lies normally on the chamber bottom, and is propelled upward by passing current through an actuating coil in the magnetic field ([5], p. 499).

The beam is best delivered to the target, after the latter has been positioned, by turning the radiofrequency amplitude off. After drifting for a phase oscillation period, or longer, the protons will be uniformly spread in azimuth and will be collected on the target with little or no structure. The collection pulse is of the order of a millisecond long and is too short for many experimental purposes. The pulse on target can be lengthened by slowly reducing the radiofrequency amplitude and gradually spilling protons out of phase stability. Radiofrequency, phase oscillation, and magnet ripple structure will be present, and although the pulse can be 0.1 sec long a large energy variation will occur during the pulse. Instead of slow turn-off, perturbing modulation can be applied to the frequency to shake protons out of phase stability. This method is useful for multiple, or "parasitic" operation wherein two experiments are operated during one acceleration cycle. The target for the first experiment is inserted and struck with a fraction of the beam by modulating the frequency (or tracking the beam briefly into the target) and then is retracted in favor of the second target which receives the remainder of the protons. The energy difference between the two target pulses is limited only by the speed of operation. The Bevatron flip targets can be interchanged in less than 0.1 sec. Typical oscillograms of secondary beam structure follow p. 502 of [5].

Thick outside targets are not used because a method has not yet been devised to obtain full traversal. All outside targets must be rammed, for even thin foils would interfere with injection. It would be possible to warp the equilibrium orbit after injection in such a way as to strike a fixed outside target, but the movement of resonance limits into the gap makes such a process very difficult. An outside thin foil target, rammed after injection, is struck by tracking the radius outward (or running over the top of the magnet cycle). The protons will penetrate the foil on an outward oscillation maximum, the energy loss will shift the equilibrium orbit inward, with a like increase of betatron amplitude. The process will continue until the oscillations increase enough to strike an inner obstruction, or if the foil is very thin, until phase stability is lost and the protons drift to smaller radius. Secondary particles can be obtained from the thin outside target, especilly scattered-out protons. If an inside target is positioned it will intercept the protons after they penetrate the outside foil, and provide a second particle source. A thin foil, destroying phase stability will deliver a beam on an inside target which will have no radiofrequency or phase oscillation structure and can have a long duration.

Very long secondary beam pulses can be produced by flat-topping the magnet pulse, i.e., holding the current approximately constant. Large magnet ripple structure is present. The difficulties produced by ripple in many experiments using different target methods make it increasingly important to provide filters for the magnet current.

The various target methods described have all been employed in one or more of the proton synchrotrons[1]. Numerous variations have also been used, and new methods are continually invented to meet research requirements.

---

[1] See p. 493, 496 and 511 of [5], [39]. — N. E. BOOTH and G. W. HUTCHINSON: Nucl. Instr. 1, 80 (1957). — S. J. GOLDSACK: Nucl. Instr. 1, 90 (1957). Much material of this section has been obtained from unpublished reports of E. J. LOFGREN and H. G. HEARD.

It is advantageous for many experiments to use a beam external to the synchrotron. An ejected beam has been obtained at the Cosmotron by use of a lipped target[1], or "jump target". A magnet with shielded gap is placed in a straight section (Fig. 76) with its gap centerline at a few inches greater radius than the inner wall. After the beam is damped down the jump target is rammed into position at 4 inches greater radius than the magnet gap (which is 1×3 inches cross section). The protons delivered to the lip are then made to penetrate the 1 inch Be target as described for Fig. 75. The energy loss displaces the orbit 2 in.; after one additional revolution the protons enter the magnet gap, are

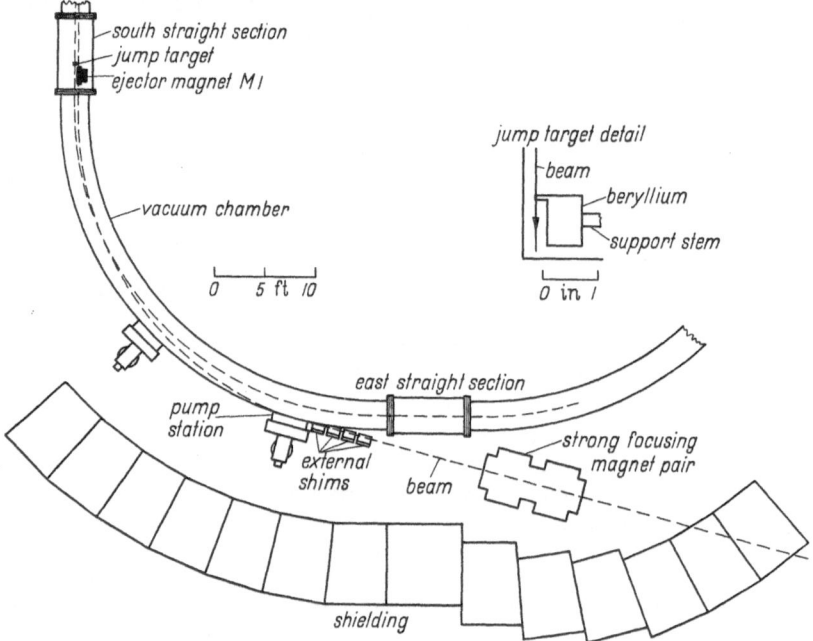

Fig. 76. Cosmotron beam ejection system. From Collins, [5], p. 129.

deflected outward, and emerge from the chamber. The pole face windings are adjusted so that ν_r just exceeds $\frac{1}{2}$, ν_z is then just under 1, and since the target and magnet are spaced 2π in ϑ the beam from the target is double-focussed on the ejection magnet gap. The focus minimizes loss due to scattering in the target. The beam emerges through a thin window in a vacuum chamber pump line and is again focussed by magnetic shims which alter the stray field. After leaving the field of the ring the beam is nearly parallel in the horizontal plane, and slightly converging vertically. A crossed-quadrupole lens produces a "spot" some 2 to 3 inches in diameter at 60 feet. Ejection of 50% of the internal beam has been done at energies from 1 to 3 Bev. The ejection magnet is pulsed in order to avoid disturbing the beam with large first harmonic at low energy, however, the internal beam intensity is reduced to about one-half and it will be necessary to compensate the harmonic. It should also be feasible to eject protons by using the regenerative, or "peeler-regenerator" system in which a radial resonance is excited by carefully shaped field perturbations ([5,] p. 140).

---

[1] O. Piccioni, D. Clark, R. Cool, G. Friedlander and D. Kassner: Rev. Sci. Instrum. **26**, 232 (1955). — B. T. Wright: Rev. Sci. Instrum. **25**, 429 (1954).

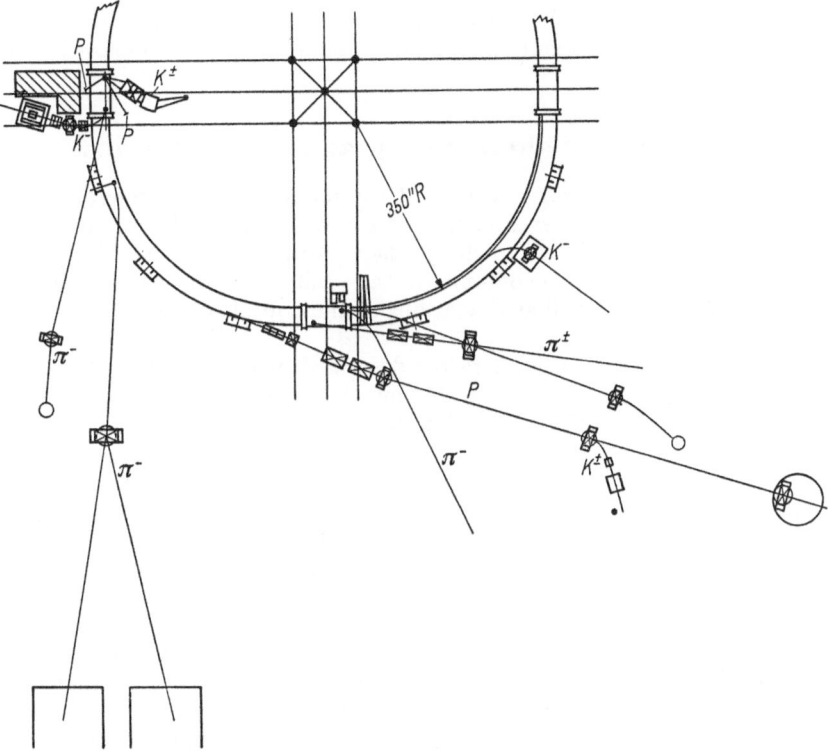

Fig. 77. Various target locations and particle beams at the Cosmotron. From L. W. SMITH, [5], p. 495.

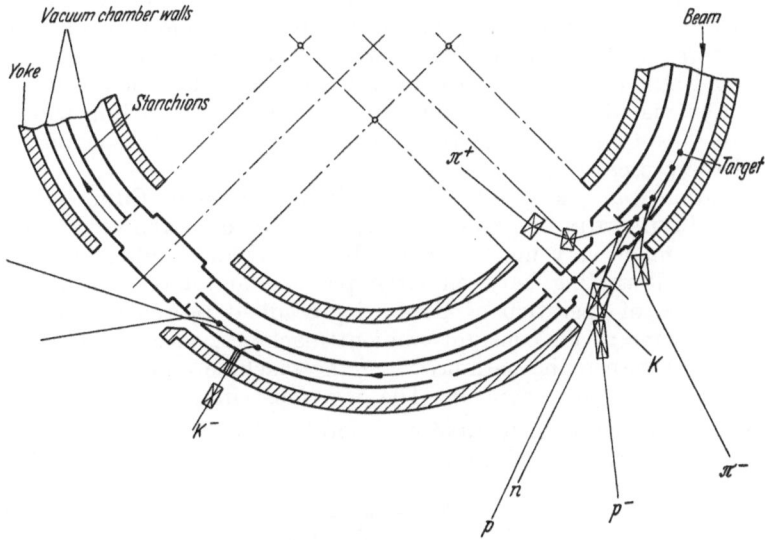

Fig. 78. Various target locations and particle beams at the Bevatron. Thin windows in the vacuum chamber walls transmit the particles. See also LOFGREN, [5], p. 499.

Fig. 77 shows some of the particle beams that are used at the Cosmotron. The open C-magnet permits access to the beams at numerous azimuths. Typical arrangements of secondary particle paths at the Bevatron are drawn in Fig. 78.

Many targets are used in the quadrants of the H-magnet, and an opening has been made in one of the yoke legs. Multiple operation of experiments is common on both machines.

C. Theory of alternating-gradient accelerators.

31. Introduction. The cross section of the magnet of an accelerator is determined by the size of the aperture, that is the region near the equilibrium orbit which must be made available to the particles during the course of acceleration. This region is necessary to accommodate the oscillations of the particles around the equilibrium orbit, as well as deviations of the equilibrium orbit from its ideal location.

In the constant-gradient accelerators described in Part B, stability of vertical and radial oscillation is obtained by shaping the magnetic guide field so that the field decreases radially. The gradient index

$$n = -\frac{r}{B}\frac{\partial B}{\partial r},$$

must be positive to provide vertical focussing, but less than 1 to provide horizontal focussing. The resultant focussing forces are rather weak; the strength of the vertical focussing forces is limited by the requirement of horizontal stability, and vice versa. The wavelength of both modes of oscillation is therefore greater than the circumference of the machines. This results in the requirement that the aperture be at least several percent of the radius.

While the Cosmotron and the Bevatron were being constructed, it was realized that the magnets for machines for substantially higher energies than the 6 Bev of the Bevatron would be prohibitively expensive if the aperture were scaled up in proportion to the radius. Detailed calculations show that, even with constant-gradient focusing, the linear dimensions of the aperture could be made to increase somewhat more slowly than the first power of the radius, but the mass of the magnet would still have to increase with about the square of the momentum. Thus a 30 Bev machine would require about 100000 tons of iron as compared to less than 2000 tons for the 3-Bev Cosmotron, the most economically constructed of the present-day proton synchrotrons.

This situation was greatly altered when it was found by Christofilos[1] and by Courant, Livingston and Snyder [19] that the radial and vertical focussing forces could be made very much stronger by letting the gradient index n vary with azimuth, alternating between large positive and large negative values.

The principles of operation of alternating-gradient proton synchrotrons are essentially the same as for constant-gradient machines, except that the quantitative relations between the parameters of the machine and the characteristics of the particle orbits are changed. The two most important differences are:

1. The ratios of the frequencies of horizontal and vertical "betatron" oscillations to the frequency of revolution

$$\nu_r = \frac{\omega_r}{\omega_0}, \qquad \nu_z = \frac{\omega_z}{\omega_0}, \tag{31.1}$$

can be made large compared to unity, while in the constant-gradient accelerator $\nu_r = \sqrt{1-n}$, $\nu_z = \sqrt{n}$ [see Eq. (4.14)], so that both are necessarily less than unity.

---

[1] Unpublished manuscript 1950. See also Courant, Livingston, Snyder and Blewett: Phys. Rev. **91**, 202 (1953).

2. The dependence of mean orbit radius on particle momentum in a given field is changed in such a way that the "momentum compaction factor"

$$\alpha = \frac{p}{R} \frac{\partial R}{\partial p} \tag{31.2}$$

is small compared to unity. As a result the phase oscillation frequencies and amplitudes are changed; the effect is to replace the factor $1/(1-n)$ occurring in the expressions in Sect. 5 by α. This reduces the radial excursion associated with the phase oscillations. Another result of the small value of α is that the stable phase angle shifts from ψ_0 to $\pi - \psi_0$ at a "transition energy"

$$E_t = \frac{E_0}{\sqrt{\alpha}}. \tag{31.3}$$

The complications arising from this fact will be dealt with in Sect. 36.

These features of the "alternating-gradient" or "strong-focussing" scheme make it possible to increase the "betatron" oscillation frequencies, thus reducing the amplitudes of these oscillations, and also reduce the radial extent of the synchrotron oscillations. As a result, the aperture, and therefore the magnet cross section, can be made smaller than with constant-gradient fields; and it becomes economically feasible to design accelerators for higher energies than could be considered otherwise. Two large proton synchrotrons in the range of 25 to 30 Bev are now under construction; both of these have apertures of only a few centimeters, magnet cross sections a fifth as large as that of the Cosmotron, and magnet weights in the neighborhood of 4000 tons.

The strong momentum compaction (small α) associated with the alternating-gradient geometry led SYMON et al. [53], [54] to propose "fixed-field alternating-gradient" (FFAG) accelerators, in which the magnetic field is constant in time, and arranged so that stable orbits exist simultaneously for all energies from injection energy to the output energy. The advantage of this scheme is that the limitations imposed on the performance of a synchrotron by the magnet pulse cycle are removed. This makes possible a higher pulse repetition rate—limited only by radiofrequency considerations—and consequently a high average beam intensity, comparable to that obtained in synchrocyclotrons.

32. Stability of betatron oscillations. In an alternating-gradient accelerator, the magnet is composed of a fairly large number N of sections, designed to be identical, in each of which the field gradient index n varies azimuthally between positive and negative values.

Just as in constant-gradient machines, it is legitimate to consider separately the betatron oscillations, whose behavior is governed by the properties of the guide field and which are independent of the accelerating field, and the phase or synchrotron oscillations, which arise from the accelerating process.

To obtain the betatron oscillations, we consider a field in which particles of appropriate momentum can move on a certain closed curve, which we call the equilibrium orbit. We assume that this orbit lies in a horizontal plane. Then the derivation of Eqs. (4.10) and (4.11) for the betatron oscillations remains valid even if the field and the gradient vary along the equilibrium orbit, provided x and z are interpreted as the horizontal and vertical components of the distance from the orbit; ω_0 as the *local* angular velocity on the orbit:

$$\omega_0 = v/\varrho, \tag{32.1}$$

(ϱ being the radius of curvature of the orbit). B_0 is the field on the orbit, which satisfies

$$B_0 = \frac{pc}{e\varrho} \tag{32.2}$$

and n is

$$n = -\frac{\varrho}{B}\frac{\partial B}{\partial x}, \tag{32.3}$$

evaluated on the equilibrium orbit. If we transform to the independent variable

$$s = \int v\, dt \tag{32.4}$$

(the distance traveled by the particle along the equilibrium orbit) we find that the equations of betatron oscillation become

$$\frac{1}{p}\frac{d}{ds}\left(p\frac{dx}{ds}\right) + \frac{(1-n)}{\varrho^2}\, x = 0, \tag{32.5}$$

$$\frac{1}{p}\frac{d}{ds}\left(p\frac{dz}{ds}\right) + \frac{n}{\varrho^2}\, z = 0, \tag{32.6}$$

where p is the particle momentum.

Since the particle momentum changes only slowly, we may, for the moment, neglect the variation of p with s, so that the equations reduce to

$$x'' + \frac{(1-n)}{\varrho^2}\, x = 0, \tag{32.7}$$

$$z'' + \frac{n}{\varrho^2}\, z = 0, \tag{32.8}$$

where the prime denotes differentiation by s.

The coefficients of x and z in these equations are periodic functions of s, with period equal to $1/N$-th of the circumference C of the equilibrium orbit. Both equations are thus examples of Hill's equation, i.e., linear equations with periodic coefficients and without first derivative terms. We write both equations in the form

$$\frac{d^2 y}{ds^2} + K(s)\, y = 0, \tag{32.9}$$

with the periodicity condition

$$K(s+L) = K(s), \tag{32.10}$$

where $L = C/N$ is the length of one period. We now ask: How can $K(s)$ be chosen so that the solutions of (32.9) represent stable oscillations, with "small" amplitudes? The solution of (32.9) at any particular s is uniquely determined by the initial conditions at $s = s_0$:

$$\left.\begin{aligned} y(s) &= a\, y(s_0) + b\, y'(s_0), \\ y'(s) &= c\, y(s_0) + d y'(s_0), \end{aligned}\right\} \tag{32.11}$$

or, in matrix notation

$$\begin{bmatrix} y(s) \\ y'(s) \end{bmatrix} = \mathsf{M}(s\,|\,s_0) \begin{bmatrix} y(s_0) \\ y'(s_0) \end{bmatrix}. \tag{32.12}$$

The matrix $\mathsf{M}(s\,|\,s_0)$ effects a linear transformation of (y, y') phase space at s_0 into phase space at s. The matrix has unit determinant because of the absence of first derivative terms in (32.9). For a sector with constant K, the matrix is

$$M(s \mid s_0) = \begin{bmatrix} \cos \varphi & \dfrac{l}{\varphi} \sin \varphi \\ -\dfrac{\varphi}{l} \sin \varphi & \cos \varphi \end{bmatrix},$$ (32.13)

where $l = s - s_0$ and $\varphi = K^{\frac{1}{2}} l$. [Where K is negative or zero, (32.13) is more conveniently written in the form

$$\begin{bmatrix} \mathrm{Cos} \mid \varphi \mid & \dfrac{l}{\mid \varphi \mid} \mathrm{Sin} \mid \varphi \mid \\ \dfrac{\mid \varphi \mid}{l} \mathrm{Sin} \mid \varphi \mid & \mathrm{Cos} \mid \varphi \mid \end{bmatrix}$$ (32.14)

or

$$\begin{bmatrix} 1 & l \\ 0 & 1 \end{bmatrix},$$ (32.15)

respectively.]

The matrices of particular interest are those which characterize motion through one or more whole unit cells. We denote the matrix for a whole unit cell, starting at s and ending at $s + L$, by

$$M(s + L \mid s) = M(s).$$ (32.16)

We now define the parameters $\mu, \alpha, \beta, \gamma$ by writing $M(s)$ in the form

$$M(s) = I \cos \mu + \begin{pmatrix} \alpha(s) & \beta(s) \\ -\gamma(s) & -\alpha(s) \end{pmatrix} \sin \mu$$ (32.17)

where I is the unit matrix, and where $\beta \gamma - \alpha^2 = 1$, since $\det M = 1$. The ambiguity of sign of $\sin \mu$ is resolved by requiring that β be positive if $\mid \cos \mu \mid < 1$ and positive imaginary if $\mid \cos \mu \mid > 1$.

The parameters α, β, γ are periodic functions of s, while μ is independent of s. The latter fact is most easily seen by noting that $M(s_1)$ and $M(s_2)$ are related by a similarity transformation. The trace of M is $2 \cos \mu$, and its eigenvalues are $e^{i\mu}$ and $e^{-i\mu}$.

The matrix for passage through any number k of periods is

$$[M(s)]^k = I \cos k\mu + \begin{bmatrix} \alpha & \beta \\ -\gamma & -\alpha \end{bmatrix} \sin k\mu.$$ (32.18)

In order for the motion to be stable, it is necessary and sufficient that the elements of M^k remain bounded as k increases. This is the case when μ is real, i.e., when

$$-1 < \cos \mu = \tfrac{1}{2} \mathrm{Tr} \, M < 1,$$ (32.19)

both for the matrices characterizing horizontal and vertical oscillations.

The quantity μ may be interpreted as the phase shift of the betatron oscillation in one period. The number of wavelengths per revolution is then

$$\nu = N \frac{\mu}{2\pi}$$ (32.20)

where N is the number of periods in a revolution.

If the accelerator is composed of pieces in each of which $K(s)$ is constant, the matrices are easily calculated by taking the appropriate products of matrices of the forms (32.13) to (32.15).

In the simplest alternating-gradient configuration [19] there are sectors of equal length $L/2 = \pi R/N$ with n alternately equal to $+n_0$ and $-n_0$. If $n_0 \gg 1$, the two Eqs. (32.7) and (32.8) are nearly identical if the origin of s is shifted by $L/2$. The phase shift per period, for each mode, is given by

$$\cos \mu = \cos \left(n^{\frac{1}{2}} \frac{\pi}{N} \right) \operatorname{Cos} \left(n^{\frac{1}{2}} \frac{\pi}{N} \right). \tag{32.21}$$

This determines μ as a function of $\varphi = n^{\frac{1}{2}} \pi/N$, as shown in Fig. 79. The number of wavelengths per revolution, according to (32.20), increases proportional to N if n is made to increase proporational to N^2 so as to keep φ constant. For example, if $n = N^2/4$, $\cos \mu = 0$, and

$$\nu = \tfrac{1}{4} N = \tfrac{1}{2} \sqrt{n}. \tag{32.22}$$

By making N and n large enough, one can obtain large values of ν, and correspondingly short wavelengths of betatron oscillations.

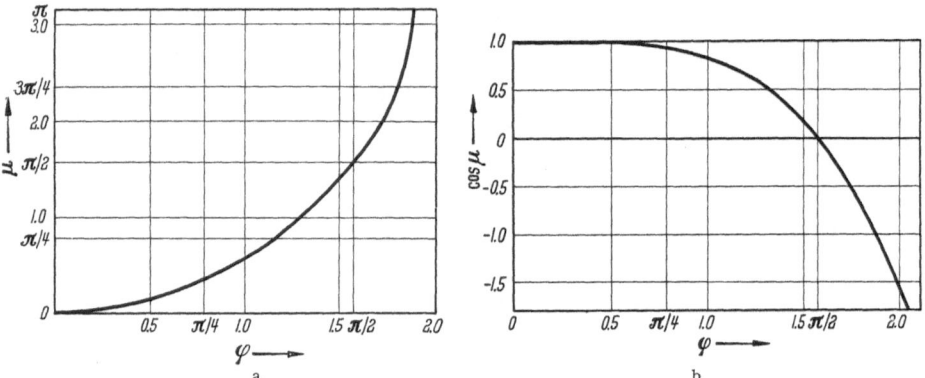

Fig. 79a and b. Phase shift per period, μ, and $\cos \mu$ vs. $\varphi = n^{\frac{1}{2}}\pi/N$ for configuration with equal and opposite gradients without straight sections.

In constructing actual alternating-gradient accelerators, it is advantageous to insert field-free straight sections between the magnet sectors. These will modify the quantitative relation (32.21) generally in the direction of larger values of μ for given n and N. However, the scaling law $\nu \sim N$ continues to apply.

33. Amplitude of betatron oscillations. The advantage of the increase in the oscillation frequencies ν that can be obtained with alternating gradients is that the amplitudes of oscillations, and therefore the aperture requirements, become smaller as ν is increased.

To give a quantitative meaning to this statement, we introduce the concept of "admittance", which is defined as follows: If the effective semi-aperture of the vacuum chamber is a, the admittance is defined as the area of that region of (y, y') phase space for which any particle injected with initial values of the displacement y and derivative y' within the region will execute oscillations whose maximum excursion is less than a. Here y may represent either the horizontal or vertical mode; correspondingly there is both a horizontal and a vertical admittance (assuming the two modes remain separated).

If the oscillations are sinusoidal with wavelength

$$2\pi \lambda = 2\pi R/\nu \tag{33.1}$$

($2\pi R =$ circumference of the orbit), the maximum excursion of a particle with initial values y_0, y_0' is given by

$$y_{\max}^2 = y_0^2 + \frac{R^2}{\nu^2} y_0'^2. \tag{33.2}$$

Therefore, the admittance is the area of the ellipse

$$y^2_{\max} = a^2,$$

which is

$$A = \frac{\pi a^2 \nu}{R}. \tag{33.3}$$

If the oscillation in an alternating gradient configuration were sinusoidal in form, this relation would continue to hold, thus leading to admittance increasing with ν for a given aperture, or required aperture decreasing with $\nu^{-\frac{1}{2}}$ for a given admittance.

Actually, the form of the oscillation is not sinusoidal. Therefore, a form factor must be inserted in (33.3):

$$A = \frac{\pi a^2 \nu}{F R} \tag{33.4}$$

with $F > 1$.

To evaluate the form factor F, we recall FLOQUET's theory of HILL's equation[1]. The general solution of (32.9) is a linear combination of

$$y_1(s) = P(s)\, e^{i \mu s/L} \tag{33.5}$$

and its complex conjugate, where $P(s)$ is a (generally complex) function of s which is periodic with period L.

It is convenient to write

$$P(s) = w(s)\, e^{i \chi(s)} \tag{33.6}$$

where w and χ are *real* periodic functions of s, and to introduce

$$\psi(s) = \frac{\mu s}{L} + \chi(s), \tag{33.7}$$

so that the normal solution (33.5) becomes

$$y_1(s) = w(s)\, e^{i \psi(s)}. \tag{33.8}$$

If the motion is stable, μ is real, and therefore w and ψ are both real. Substituting (33.8) into the differential equation, and separating real and imaginary parts, we find that (33.8) is a solution of (32.9) if w and ψ satisfy

$$\psi' = \frac{k}{w^2}, \tag{33.9}$$

and

$$w'' + K w - \frac{k}{w^3} = 0, \tag{33.10}$$

where k is an arbitrary constant, which we may set equal to 1.

We can now express the matrix $\mathsf{M}(s)$ for the motion through one period in terms of the functions w and ψ. We obtain

$$\left.\begin{aligned}
M_{11} &= \cos\mu - w\, w' \sin\mu, \\
M_{12} &= w^2 \sin\mu, \\
M_{21} &= -\frac{1 + (w\, w')^2}{w^2}\, \sin\mu, \\
M_{22} &= \cos\mu + w\, w' \sin\mu,
\end{aligned}\right\} \tag{33.11}$$

where

$$\mu = \psi(s+L) - \psi(s) = \int_{s}^{s+L} \frac{d s}{w^2}. \tag{33.12}$$

---

[1] See, for example, WHITTAKER and WATSON: Modern Analysis, 4th edit. Chap. XIX. Cambridge: Cambridge Univerity Press 1927.

Comparing (33.11) with (32.17) we see that

$$\alpha = -w\,w', \quad \beta = w^2, \quad \gamma = \frac{1 + (w\,w')^2}{w^2} = \frac{1 + \alpha^2}{\beta}. \tag{33.13}$$

It may be verified that α, β, γ satisfy the system of differential equations

$$\left. \begin{aligned} \alpha' &= K\beta - \gamma, \\ \beta' &= -2\alpha, \\ \gamma' &= 2K\alpha. \end{aligned} \right\} \tag{33.14}$$

We may regard (33.12) as the definition of the phase shift μ; it has the advantage of being unique, while (32.17) only defines μ modulo 2π.

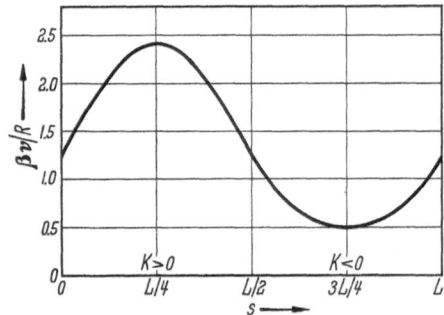

Fig. 80. $\beta(s)$ in units of R/ν for one period of configuration of Fig. 79 with $\varphi = \pi/2$.   Fig. 81. Form factor F vs. φ for configuration of Fig. 79.

If we consider again an accelerator with N identical sectors of total length $C = NL$, we have

$$\nu = \frac{N\mu}{2\pi} = \frac{1}{2\pi} \int\limits_{s}^{s+C} \frac{ds}{\beta}. \tag{33.15}$$

The general solution of (32.9) may now be written in the form

$$y(s) = a\,\beta^{\frac{1}{2}} \cos\left[\nu\,\varphi(s) + \delta\right] \tag{33.16}$$

where

$$\varphi(s) = \frac{1}{\nu}\,\psi(s) = \int \frac{ds}{\nu\,\beta} \tag{33.17}$$

is a function which increases by 2π every revolution, and whose derivative is periodic.

The amplitude function β can be obtained from the transformation matrix, and the "frequency" ν and the phase function φ are derived from β by means of (33.15) and (33.17). Fig. 80 shows β as a function of s for one unit cell of the configuration considered in (32.21), with $n = N^2/4$.

We now return to the admittance problem. From the form (33.16) of the solution it follows that the quantity

$$W = \frac{1}{\beta}\left[y^2 + (\alpha\,y + \beta\,y')^2\right] = \gamma\,y^2 + 2\alpha\,y\,y' + \beta\,y'^2 \tag{33.18}$$

is constant, independent of s. Therefore the largest displacement is attained where $\beta(s)$ has its maximum value, and equals

$$y_{\max} = (\beta_{\max}\,W)^{\frac{1}{2}}. \tag{33.19}$$

Identifying this with the semi-aperture a, we see that the admittance equals the area in phase space of the ellipse

$$\beta_{max}\,W = \frac{\beta_{max}}{\beta}\,[y^2 + (\alpha\,y + \beta\,y')^2] = a^2,$$

which equals

$$A = \pi\,a^2/\beta_{max}. \tag{33.20}$$

From (33.15) we see that the mean value of $1/\beta$ is just ν/R. If, as is generally the case, β is not constant, its maximum value is therefore greater than R/ν. Thus the form factor F which governs the admittance, Eq. (33.4) is just

$$F = \beta_{max}\,\nu/R > 1. \tag{33.21}$$

A useful alternating-gradient accelerator must be designed so that F does not exceed 1 by too much. In the simple case considered previously in Eq. (32.21) we have

$$F = \frac{\text{arc cos } (\cos\varphi\,\text{Cos }\varphi)}{[1 - (\cos\varphi\,\text{Cos }\varphi)^2]^{\frac{1}{2}}} \times \frac{\text{Sin }\varphi + \sin\varphi\,\text{Cos }\varphi}{2\varphi} \tag{33.22}$$

with $\varphi = n^{\frac{1}{2}}\pi/N$. This function is graphed in Fig. 81. Observe that $F \to 1$ for $\varphi \to 0$, i.e., when the alternations of the gradients are so rapid as to make μ small, and that $F \to \infty$ as the stability limit $\mu = \pi\,(\cos\varphi\,\text{Cos }\varphi = -1)$ is approached. These two features remain for more complicated configurations.

As the particle energy increases during the acceleration, the actual equations of motion are (32.5) and (32.6) rather than (32.7) and (32.8). The effect is, just as in constant-gradient accelerators, that the amplitudes damp proportional to $p^{-\frac{1}{2}}$, assuming that $n(s)$ and $\varrho(s)$ remain constant in time.

If, on the other hand, there are slow changes in the field configurations, so that $K(s)$ changes with time, the quantity W defined in (32.18) is an adiabatic invariant [28]. Therefore, the maximum amplitude varies as $\beta_{max}^{\frac{1}{2}}$. Thus, if both energy and focusing field change slowly, the maximum amplitude varies as

$$(\beta_{max}/p)^{\frac{1}{2}}.$$

This is, of course, only true as long as β_{max} does not vary too fast; therefore, it will cease to be true when a limit of stability is approached, since there $\beta_{max} \to \infty$, as we have seen.

34. Magnet imperfections. The fields in actual magnets will differ somewhat from the ideal design. The magnets can be manufactured to specifications only within some finite tolerances, and further small errors arise in placement of the magnets. The magnet is still usable provided that

(a) there always exists a closed orbit, the "displaced equilibrium orbit", along which the particles can move, and

(b) oscillations about this orbit are stable.

α) *Displaced equilibrium orbit.* If the field on the ideal equilibrium orbit differs from the design value, the right-hand sides of the equations of motion (32.7) and (32.8) must be changed from zero to

$$\frac{\varDelta B}{B\,\varrho}$$

where $B\varrho$ is the magnetic rigidity of the particle, and $\varDelta B$ the field error; $\varDelta B = B_r$ for vertical oscillations and $\varDelta B = B - B_0$ for horizontal.

The equations of motion are then of the form

$$\frac{d^2 y}{d s^2} + K(s)\, y = F(s) = \frac{\Delta B}{B\, \varrho},\tag{34.1}$$

if non-linear terms are neglected. Here $K(s)$ has the period $L = C/N$, while $F(s)$ has in general only the period C, i.e., the circumference of the machine.

Eq. (34.1) can be solved in terms of the solutions of the homogeneous equation, which is (see Sect. 33)

$$a\, \beta^{\frac{1}{2}} \cos\left(\int \frac{d s}{\beta^{\frac{1}{2}}} + \delta\right) = a\, \beta^{\frac{1}{2}} \cos\left(\nu\, \varphi + \delta\right).\tag{34.2}$$

We assume $\beta(s)$ to be known and introduce the new variables

$$\eta = \beta^{-\frac{1}{2}} y$$

$$\varphi = \int \frac{d s}{\nu\, \beta},$$

instead of y and s. The differential equation (34.1) transforms to

$$\frac{d^2 \eta}{d\varphi^2} + \nu^2 \eta = \nu^2 \beta^{\frac{3}{2}} F(s).\tag{34.3}$$

Here β is periodic in φ with period $2\pi/N$, while $F(s)$, regarded as a function of φ, is periodic with period 2π. The displacement of the equilibrium orbit is that solution of (34.3) which is periodic.

Since (34.3) is just the equation of the forced harmonic oscillator, it possesses a periodic solution *unless* ν is an integer and the right-hand side contains a Fourier component of order ν. If $F(s)$ arises from random construction errors, it will contain all Fourier components, and therefore the displacement of the equilibrium orbit becomes infinite for all integral values of ν. Thus, the design of the accelerator must avoid integral ν for both modes.

If ν is not integral, the periodic solution of (34.3) is

$$\eta(\varphi) = \frac{\nu}{2 \sin \pi \nu} \int\limits_{\varphi}^{\varphi + 2\pi} f(\psi) \cos \nu\, (\pi + \varphi - \psi)\, d\psi\tag{34.4}$$

where $f(\varphi) = \beta^{\frac{3}{2}} F(s)$.

This may be analyzed in terms of Fourier components. Let

$$f(\varphi) = \sum_k f_k\, e^{i k \varphi}$$

with

$$f_k = \frac{1}{2\pi} \int\limits_{0}^{2\pi} f(\varphi)\, e^{-i k \varphi}\, d\varphi = \frac{1}{2\pi \nu} \int\limits_{0}^{C} \beta^{\frac{1}{2}} F(s)\, e^{-i k \varphi}\, d s.$$

Then the periodic solution of (34.3) is

$$\eta = \sum \frac{\nu^2 f_k}{\nu^2 - k^2}\, e^{i k \varphi}.\tag{34.5}$$

This formula exhibits the resonance properties of the orbit. The orbit is most sensitive to those components of the perturbation whose order is close to the free oscillation frequency; the Fourier components are to be taken with respect to the phase variable φ, rather than the geometric variable s.

If the perturbing function $F(s)$ arises from random errors, we do not know it in detail, but know only its statistical properties. Consider, for example, a machine with M magnet sectors, with a field error ΔB_r at the ideal orbit position of the r-th magnet (assumed constant throughout that magnet). Let us define

$$\Delta y_r = \frac{F_r}{K_r} = \frac{1}{B\varrho}\frac{\Delta B_r}{K_r}, \tag{34.6}$$

where we also assume $K(s) = K_r$ is constant in the magnet. The significance of Δy_r is that it is that displacement from the ideal position of the magnet which will cause a field error ΔB_r, regardless of whether that error is actually caused by a magnet displacement or by an error in the intrinsic characteristics of the magnet. If we assume further that the errors in successive magnets are uncorrelated, with the mean-square error

$$\langle \Delta y^2 \rangle_{\mathrm{Av}} = \delta^2, \tag{34.7}$$

one finds, by using (34.4), that the mean-square displacement of the equilibrium orbit is approximately

$$\langle y^2 \rangle_{\mathrm{Av}} = \frac{\pi^2}{2\nu^2 \sin^2 \pi \nu}\frac{R^2}{\varrho^2}\frac{n^2}{M}\delta^2. \tag{34.8}$$

Here $R = C/2\pi$ is the mean radius of the machine, ϱ is the radius of curvature in the magnets ($\varrho < R$ because of straight sections); n is assumed to have the same absolute value in all magnets.

Practically, the important number is not the mean-square displacement, but some maximum displacement. The maximum square displacement exceeds the mean (34.8) by the product of the following factors:

— a factor 2 to convert the mean-square value of a sine wave to the square of its amplitude.

— a factor $F(=\beta_{\max}\nu/R)$ to account for the non-sinusoidal shape of oscillations.

In addition, another factor is necessary because (34.8) is an ensemble average rather than the average in any one particular machine. We ask: By what factor G must (34.8) be multiplied in order that the probability that $\langle y^2 \rangle_{\mathrm{Av}}$ in a particular machine be less than G times (34.8) is less than some reasonably small number? It has been shown by Lüders [23] that the probability distribution $\langle y^2 \rangle_{\mathrm{Av}}$ is exponential. Thus, the probability that $\langle y^2 \rangle_{\mathrm{Av}}$ exceeds four times (34.8) is $e^{-4} \approx 0.02$. Thus, the probability is 98% that the maximum displacement will not exceed

$$P_{0.98} = (4F)^{\frac{1}{2}} \cdot \frac{\pi}{\nu |\sin \pi \nu|}\frac{R}{\varrho}\frac{|n|}{M^{\frac{1}{2}}}, \tag{34.9}$$

times the root-mean-square magnet displacement δ. For the Brookhaven alternating-gradient synchrotron (see Part D of this article), $F \approx 1.5$, $\nu = 8.75$, $R/\varrho = 1.5$, $|n| = 357$, $M = 240$, so that $P_{0.98} \approx 44$. Thus, if the r-m-s magnet error is, say, 0.02 inches, the probability is 98% that the equilibrium orbit nowhere deviates by more than 0.88 inches from the ideal position.

β) *Errors in field gradients.* The field *gradients* along the equilibrium orbit may also deviate from their ideal values. As a result the periodicity relation

$$K(s + L) = K(s)$$

is no longer exactly satisfied; however, the weaker periodicity condition

$$K(s + C) = K(s)$$

remains valid. As a result, the formal considerations of Sects. 32 and 33 remain valid, but the unit cell that must be considered is the whole revolution, rather than its N-th part, as in the ideal machine.

The matrix for a whole turn in the ideal machine is

$$M_C = (M_L)^N \tag{34.10}$$

where M_L is the matrix for a unit cell. Its trace is

$$\mathrm{Tr}\, M_C = 2 \cos (2\pi\nu) = 2 \cos N\mu, \tag{34.11}$$

where μ is the phase shift for one unit cell. If the N unit cells differ slightly from one another, (34.10) is replaced by the product of N slightly unequal matrices. The elements of M_C will differ slightly from those of $(M_L)^N$. In particular, the trace will be different.

If $\mathrm{Tr}\, M_C$ differs substantially from 2 or -2, the change caused by small errors in the function $K(s)$ will be small, and cannot effect stability. But if $|\mathrm{Tr}\, M_C| \approx 2$, it only takes a small perturbation to push the trace outside of the range between -2 and $+2$, leading to instability. This happens when $2\pi\nu$ is an integral multiple of π, i.e., when ν is either integral or half-integral.

Detailed analysis [23], [28] shows that errors in $K(s)$ generate narrow *bands* of instability, called *stopbands*, lying near these integral and half-integral values of ν. The width of the stopband near $\nu = k/2$ (k integral) equals [28]

$$\left.
\begin{aligned}
\delta\nu &= \frac{1}{2\pi} \left| \int_0^C \beta(s)\, \Delta K(s)\, e^{ik\varphi(s)}\, ds \right| \\
&= \frac{1}{2\pi} \left| \int_0^{2\pi} \nu\, \beta^2\, \Delta K\, e^{ik\varphi}\, d\varphi \right|,
\end{aligned}
\right\} \tag{34.12}$$

where ΔK is the error in the focusing function K, and φ is the phase variable introduced in Sect. 33 and previously employed in the problem of orbit deviations.

Even outside the stopbands, gradient errors have deleterious effects. It will be recalled from Sect. 33 that the amplitude form factor $F[\nu/R$ times the maximum of the amplitude function $\beta(s)]$ goes to infinity at the limit of stability $\mu = \pi$. The edges of the stopbands are limits of stability, and F will generally become infinite at the edges of each stopband. Outside the stopbands, F is finite but is increased by the irregularities; it is shown in references [23] and [28] that F is increased approximately by a factor

$$G = 1 + \frac{\delta}{|2\nu - k|}, \tag{34.13}$$

in the vicinity of the stopband corresponding to the resonance $\nu = k/2$.

Uncorrelated random errors of the field gradient n in an accelerator composed of M sectors are shown in reference [28] to lead to a r-m-s stopband width of

$$\langle \delta\nu \rangle_{\mathrm{rms}} = \frac{|n|}{\nu\, M^{\frac{1}{2}}}\, \frac{R}{\varrho}\, \left\langle \frac{\Delta n}{n} \right\rangle_{\mathrm{rms}}. \tag{34.14}$$

$\gamma)$ *Coupling between horizontal and vertical oscillations.* Imperfections may also destroy the assumed medium plane symmetry, causing coupling between the horizontal and vertical modes of oscillation. Then the equations of motion of the two modes are no longer separate, but are coupled.

Since the motion of particles in an electromagnetic field is derivable from a Hamiltonian, it is convenient to use the canonically conjugate pairs of variables x, p_x; z, p_z $\left(\text{where } p_x = p\, \dfrac{dx}{ds} + \dfrac{e}{c}\, A_x, \text{ etc.}; \boldsymbol{A} \text{ is the vector potential}\right)$.

The two matrix equations of form (32.12) applying to the horizontal and vertical modes are now replaced by the single equation

$$
\begin{pmatrix} x(s) \\ p_x(s) \\ z(s) \\ p_z(s) \end{pmatrix} = \mathsf{M}(s \mid s_0) \begin{pmatrix} x(s_0) \\ p_x(s_0) \\ z(s_0) \\ p_z(s_0) \end{pmatrix} \tag{34.15}
$$

where $\mathsf{M}(s \mid s_0)$ is a 4×4 matrix. Because of the Hamiltonian character of the problem, this matrix must be symplectic; this means that it must satisfy

$$
\widetilde{\mathsf{M}} \mathsf{S} \mathsf{M} = \mathsf{S}. \tag{34.16}
$$

where $\widetilde{\mathsf{M}}$ is the transpose of M, and

$$
\mathsf{S} = \begin{pmatrix} 0 & -1 & 0 & 0 \\ 1 & 0 & 0 & 0 \\ 0 & 0 & 0 & -1 \\ 0 & 0 & 1 & 0 \end{pmatrix}. \tag{34.17}
$$

This arises from the Poisson bracket relations

$$
[q_i, q_k] = [p_i, p_k] = 0; \quad [q_i, p_k] = \delta_{ik}
$$

where

$$
[q_i, q_k] = \sum_r \left[\frac{\partial q_i(s)}{\partial q_r(s_0)} \frac{\partial q_k(s)}{\partial p_r(s_0)} - \frac{\partial q_i(s)}{\partial p_r(s_0)} \frac{\partial q_k(s)}{\partial q_r(s_0)} \right]
$$

etc.; these relations are valid for any Hamiltonian system with pairs of conjugate coordinates and momenta q_i, p_i.

The motion of the system will be stable, just as in the one-dimensional case, as long as none of the eigenvalues of the matrix M exceed 1 in absolute value. From the symplectic condition (34.16) it follows [28] that the eigenvalues are pairwise reciprocal. Therefore, we have stability only if all four eigenvalues lie on the unit circle.

Since the matrix is real, the eigenvalues must either be real or pairwise complex conjugate. Therefore, there are four possibilities:

(a) All four eigenvalues are on the unit circle, forming two complex conjugate and reciprocal pairs.

(b) One reciprocal pair is real, the other complex and on the unit circle.

(c) Two real reciprocal pairs.

(d) One eigenvalue, say λ_1, complex and not on the unit circle; the other three are $\lambda_2 = 1/\lambda_1$, $\lambda_3 = \lambda_1^*$, $\lambda_4 = 1/\lambda_1^*$.

Only situation (a) is stable. Cases (b) and (c) correspond to one or both modes being unstable in the uncoupled case. Case (d) can only arise with coupling, and represents the way in which coupling can produce instability.

Since we assume that the coupling is slight, the matrix, and its eigenvalues, must be close to the matrix and the eigenvalues of the unperturbed problem. Situation (a) can be changed into (d) by a *small* shift only if the eigenvalues of the two separate unperturbed modes are (approximately) equal, i.e., if

$$
\cos 2\pi \nu_x = \cos 2\pi \nu_z. \tag{34.18}
$$

This means that either

$$
\nu_x + \nu_z = \text{integer}, \tag{34.19}
$$

or

$$
\nu_x - \nu_z = \text{integer}. \tag{34.20}
$$

It turns out that instability cannot occur in case (34.20), but will in general occur in case (34.19). This can be seen by considering invariant quantities analogous to (33.18). The expression (33.18) can be written in matrix notation as

$$W = \frac{1}{2 \sin (k \pi \nu)} \left[Y, (\mathsf{S}_2 \mathsf{A} - \tilde{\mathsf{A}} \mathsf{S}_2) Y \right],$$

where A is the transformation matrix for one complete revolution for the y-mode; Y is the vector

$$\begin{pmatrix} y \\ y' \end{pmatrix}$$

and S_2 is the 2×2 matrix

$$\mathsf{S}_2 = \begin{pmatrix} 0 & -1 \\ 1 & 0 \end{pmatrix}.$$

Analogously, it follows from the symplectic condition (34.16) that

$$V = [X, (\mathsf{S} \mathsf{M} - \tilde{\mathsf{M}} \mathsf{S}) X] \tag{34.21}$$

is invariant, where X is now the vector (34.15). The most general invariant can be shown to be

$$V_k = [X, (\mathsf{S} \mathsf{M}^k - \tilde{\mathsf{M}}^k \mathsf{S}) X], \tag{34.22}$$

where k is any integer. Now for the uncoupled case, this reduces to

$$V_k = 2W_x \sin (2\pi k \nu_x) + 2W_z \sin (2\pi k \nu_z), \tag{34.23}$$

where W_x and W_z are the forms (33.18) for the x and z modes. In case (34.20), $\sin (2\pi k\nu_x) = \sin (2\pi k\nu_z)$ for all k, and V_k is a definite quadratic form (since W_x and W_z are positive definite for stable systems).

Therefore, none of the coordinates can increase indefinitely, and the system is stable. On the other hand, if $\nu_x + \nu_z$ is integral, $\sin (2\pi k\nu_x) = -\sin (2\pi k\nu_z)$ for all k, and the form (34.23) is indefinite; in this case, a small coupling perturbation can cause instability.

This instability will, just as in the case of gradient errors, manifest itself in stopbands of small but finite width near the lines $\nu_x + \nu_z =$ integral. In the case of an accelerator with M magnets, each of which is tilted around the orbit axis by a random error angle ϑ, with no correlation between the errors, the root-mean-square stopband width is

$$\langle \delta \nu \rangle_{\mathrm{rms}} = \frac{4 |n|}{\nu M^{\frac{1}{2}}} \frac{R}{\varrho} \langle \vartheta \rangle_{\mathrm{rms}}.$$

For the Brookhaven parameters this comes to

$$\langle \delta \nu \rangle_{\mathrm{rms}} = 16 \langle \vartheta \rangle_{\mathrm{rms}}.$$

35. Non-linear effects. The actual equations of motion will be non-linear, and therefore the linear theory presented in the previous sections is only an approximation. Qualitatively, one may expect non-linearities to modify the behavior of the oscillations in a number of ways.

(a) The frequencies of oscillation become dependent on amplitude. Therefore the resonance conditions which, in the linear theory, cause infinite orbit displacements or build-ups of amplitude will cease to be satisfied when the amplitude has increased to a finite value; thus resonances may lead to finite instead of infinite amplitudes. Conversely, if the zero-amplitude frequencies are away from resonance, there may exist finite amplitudes for which the resonance conditions are fulfilled; this may lead to a limitation on allowable amplitude.

(b) New resonances, not predicted by the linear theory, may occur. In principle, these may occur whenever there exists a relation of the form

$$a\nu_x + b\nu_z = q, \tag{35.1}$$

with integral a, b, q, as first pointed out by DENNISON and BERLIN[1]. The resonant build-up is expected to be very slow for large a and b, so that only low order resonances of this type are expected to be important. [Incidentally, the linear resonances considered in the previous section correspond to (35.1); for $a=1$, $b=0$ or $a=0$, $b=1$, one obtains the orbit displacement resonance; for $a=2$ or $b=2$ the half-integral stopbands, and for $a=b=1$ the coupling resonance.]

A theory of both the "detuning" and the non-linear resonances, as applied to particle motion in synchrotrons, has been developed in recent years by a number of authors [27], [29]. We shall give a brief account of this theory, based on a method developed by MOSER[2] and extended by HAGEDORN [27]. STURROCK [29] uses a slightly different but equivalent approach.

The treatment is based on the fact that the motion in the vicinity of an equilibrium orbit in a static magnetic field may be treated by a Hamiltonian formulation in which the independent variable is not the time but rather the distance s traveled along the equilibrium orbit; the Hamiltonian will be a periodic function of s and will, in general, depend on the coordinates representing the deviations from the equilibrium orbit and their canonically conjugate momenta[3]. These coordinates may be taken to be

$x =$ component in the osculating plane of the equilibrium orbit of the vector from the nearest point on the equilibrium orbit to the given point;

$z =$ component of the same vector perpendicular to the osculating plane.

We choose the scale of the independent variable so that it is

$$\vartheta = \frac{2\pi s}{C}, \tag{35.2}$$

where C is the circumference of the equilibrium orbit. Then the Hamiltonian is of the form

$$H(x, p_x, z, p_z \vartheta) = H_2 + H_3 + H_4 + \cdots, \tag{35.3}$$

where H_n is a polynomial of the n-th degree in the variables x, p_x, z, p_z with coefficients which are periodic functions of ϑ with period 2π. The equations of motion are, as in all Hamiltonian systems,

$$\left.\begin{array}{ll} \dfrac{dx}{d\vartheta} = \dfrac{\partial H}{\partial p_x}; & \dfrac{dp_x}{d\vartheta} = -\dfrac{\partial H}{\partial x} \\[2ex] \dfrac{dz}{d\vartheta} = \dfrac{\partial H}{\partial p_z}; & \dfrac{dp_z}{d\vartheta} = -\dfrac{\partial H}{\partial z}. \end{array}\right\} \tag{35.4}$$

MOSER'S procedure is to construct a sequence of canonical transformations which eliminate the ϑ-dependent parts of $H_2, H_3, H_4 \ldots$ in turn. The resultant Hamiltonian G is independent of ϑ and is therefore a constant of the motion[4].

---

[1] D. M. DENNISON and T. H. BERLIN: Phys. Rev. **70**, 764 (1946).

[2] J. MOSER: Nachr. Akad. Wiss. Göttingen, IIa **1955**, No. 6, 87—120.

[3] See appendix to [28] or G. D. BIRKHOFF: Dynamical Systems, Chap. IV. New York: American Mathematical Society 1927.

[4] MOSER has shown (loc. cit.) that the power series defining his transformation is in general not convergent, and therefore the constant of the motion G does not strictly exist for infinite time. We shall ignore this difficulty here.

If the surfaces $G = $ const in phase space are closed, the motion must be stable; otherwise instability is possible.

We assume that the solution of the linear problem is known, and is stable. Then a linear transformation can be constructed from the form (33.8) of the solution such that H_2 becomes

$$H_2 = \tfrac{1}{2}(v_x^2 x^2 + p_x^2) + \tfrac{1}{2}(v_z^2 z^2 + p_z^2), \tag{35.5}$$

while $H_3, H_4 \ldots$ are still polynomials in (x, p_x, z, p_z) with periodic coefficients. We assume, then, that H is the form (35.3) with H_2 given by (35.5).

We may attempt to reduce the rest of H to ϑ-independence by a transformation which leaves (35.5) unchanged and which is periodic in ϑ. It turns out that this can be done, up to the terms H_k, provided no resonance relation of the form (35.1) holds with $q \leq k$. In particular, if v_x and v_z are irrational and incommensurable, such transformations can be found to all orders. The transformed Hamiltonian is of the form

$$G = v_x J_x + v_z J_z + G_4(J_x, J_z) + G_6(J_x, J_z) + \cdots \tag{35.6}$$

where, in terms of the new variables X, P_x, Z, P_z,

$$J_x = \frac{1}{2v_x}(v_x^2 X^2 + P_x^2); \qquad J_z = \frac{1}{2v_z}(v_z^2 Z^2 + P_z^2), \tag{35.7}$$

and G_{2n} is a polynomial of order n in J_x, J_z, with constant coefficients.

The variables canonically conjugate to J_x, J_z are

$$\varphi_x = \arctan\frac{P_x}{vX}; \qquad \varphi_z = \arctan\frac{P_z}{vZ}. \tag{35.8}$$

Since these do not appear in (35.6), the "amplitude variables" J_x, J_z are separately constant, and the motion is stable.

The ϑ-dependence of H_k, with $k = |a| + |b|$, can no longer transformed away by a transformation periodic in ϑ with period 2π. Suppose H has been reduced to the form (35.6) up to order $k-1$. Then H_k can be written as a polynomial of order k in $J_x^{\frac{1}{2}}$, $J_z^{\frac{1}{2}}$, with coefficients that are periodic functions (period 2π) in φ_x, φ_z and ϑ; the term in $J_x^{r/2} J_z^{s/2}$ contains only Fourier harmonics of orders $r, r-2, r-4, \ldots, -r$ in φ_x and $s, s-2, s-4, \ldots -s$ in φ_z. The nall ϑ-dependent terms of H_k can still be transformed away by a periodic transformation, except the terms

$$h\, J_x^{|a|/2} J_z^{|b|/2}\, e^{i(a\varphi_x + b\varphi_z - q\vartheta)} + \text{complex conjugate.} \tag{35.9}$$

These terms can be made formally independent of ϑ by the transformation

$$\varphi_x = \psi_x + \alpha\,\vartheta, \qquad \varphi_z = \psi_z + \beta\,\vartheta, \tag{35.10}$$

with any α and β for which

$$\alpha a + \beta b = q. \tag{35.11}$$

This transformation adds

$$-\alpha J_x - \beta J_z$$

to the Hamiltonian, so that the transformed Hamiltonian is of the form

$$\left.\begin{array}{l} G = (v_x - \alpha) J_x + (v_z - \beta) J_z + G_4 + G_6 + \cdots \\[4pt] \quad + 2|h|\, J_x^{|a|/2} J_z^{|b|/2} \cos(a\psi_x + b\psi_z + \delta) + \cdots. \end{array}\right\} \tag{35.12}$$

This quantity will be an invariant of the motion up to order k. Since this is true for any α and β satisfying (35.11), it follows that

$$F = b \, J_x - a \, J_z, \tag{35.13}$$

is constant. If a and b are of opposite sign, this implies limits on the amplitudes in both dimensions, and therefore stability.

If either a or b is zero, say $b = 0$, (35.13) implies $J_z = \mathrm{const}$ but permits J_x to be large. Stability is then determined by whether a small value of G (35.12) is compatible with a large value of J_x. We must now take $\alpha = q/a = \nu_x$. Now if $a = 3$, the term containing the cosine factor is the lowest order non-zero term in (35.12). This term is compatible with large J_x (argument of the cosine $= \pi/2$); hence we have instability.

If $a = 4$, the cosine term and G_4 are of the same order. Therefore we have stability if the coefficient of J_x^2 in G_4 is greater than $|2h|$ and instability if it is less. For $a \geqq 5$, G_4 is of lower order than the cosine term; therefore we have stability at small amplitudes, unless the coefficient of J_x^2 in G_4 happens to be exactly zero.

If a and b are both different from zero and of the same sign, similar arguments lead to the conclusion that for $a + b = 3$ there is instability, for $a + b \geqq 5$ the term G_4 dominates the cosine term and in general leads to stability for small amplitudes, and for $a + b = 4$ stability depends on whether $|h|$ is less or greater than an appropriate combination of the coefficients of G_4.

We conclude that non-linearities will cause instability when ν_x or ν_z is one-third of integer, or $\nu_x + 2\nu_z$ or $2\nu_x + \nu_z$ is integral, and may do so if $4\nu_x$, $3\nu_x + \nu_z$, $2\nu_x + 2\nu_z$, $\nu_x + 3\nu_z$ or $4\nu_z$ is an integer. In each of these cases the instability is excited by the appropriate Fourier component of the second or third derivatives of the field on the orbit. Thus if the magnet is designed with N identical periods, the field will contain mostly components of order $0, N, 2N, \ldots$; consequently third-integral values of ν that must be avoided are $N/3$, $2N/3$, $\ldots$ (assuming N is not divisible by 3, in which case these would coincide with integral resonances that must be avoided anyway). Similarly, $\nu_x + 2\nu_z = N$ or $2N$, and $2\nu_x + \nu_z = N$ or $2N$ must be avoided.

Other third order subresonances are excited only by imperfections, i.e., by deviations from identicality of the magnets. The effects of these will be limited by the constant term G_4 of (35.12), arising from the third derivative of the field, to rather small amplitudes[1]. The fourth-order subresonances arising from imperfections are likely to be less than G_4 terms at all amplitudes, and therefore will not occur at all, at least in proton synchrotrons of the Brookhaven type.

The above theory of non-linear effects applies only to static conditions. Actually, the field shapes will change with time. Thus, the subresonances may be crossed during the acceleration cycle. On the other hand, amplitudes are damped by the adiabatic damping factor $p^{-\frac{1}{2}}$. It appears possible that if synchrotron oscillations sweep the beam back and forth through a third-order imperfection resonance the resulting rate of growth of amplitude exceeds the damping rate [29]. It is therefore advisable to avoid third-order subresonances.

36. Phase oscillations. The mechanism of phase stability in alternating-gradient synchrotrons is essentially the same as in the constant-gradient case treated in Sects. 4 and 5. The difference arises from the fact that the field index n is not constant, and therefore n must be replaced by a suitable average wherever it appears in the equation of synchrotron oscillations. It turns out that the

---

[1] E. D. Courant, [5], p. 254.

quantity η occurring in Eq. (4.12) changes sign at a certain energy, known as the "transition energy", and this leads to a shift in the equilibrium phase angle.

Phase stability in any accelerator with radiofrequency acceleration arises from the combination of two effects: The energy gain of a particle in one traversal of the accelerating gap depends on the phase of the radio-frequency at the instant when the particle crosses the gap, and the rate at which this phase changes from one traversal to the next depends, among other things, on the energy of the particle. It is the latter relation, between energy and rate of change of phase, that is affected by the magnet structure and is different in alternating-gradient and constant-gradient machines.

Consider a particle whose momentum differs by Δp from the momentum p at which it could travel on the standard equilibrium orbit at a given value of the magnetic field. It will travel on a displaced equilibrium orbit with circumference $C + \Delta C$, and its speed will differ by Δv from the speed of a particle with momentum p. For small Δp, both ΔC and Δv are proportional to Δp; we may define the "momentum compaction factor" α by

$$\frac{\Delta C}{C} = \alpha \frac{\Delta p}{p}, \tag{36.1}$$

while, from relativistic kinematics,

$$\frac{\Delta v}{v} = \frac{1}{\gamma^2} \frac{\Delta p}{p},$$

where $\gamma = E/Mc^2$ is the ratio of total energy to rest energy. Thus the time for one revolution is changed by ΔT, with

$$\frac{\Delta T}{T} = \left(\alpha - \frac{1}{\gamma^2}\right) \frac{\Delta p}{p} = \zeta \frac{\Delta p}{p}. \tag{36.2}$$

Thus $\zeta = \alpha/\eta$ in the notation of Sects. 4 and 5. In the constant gradient case, $\alpha = 1/(1-n)$. In the alternating-gradient case we shall see that $\alpha \ll 1$, so that $\zeta = 0$ at the energy

$$E_t = M c^2/\sqrt{\alpha}. \tag{36.3}$$

Replacing $(1 - n)$ by $1/\alpha$ and η by α/ζ in (4.12) and the equations of Sect. 5, we see that (4.12) becomes

$$R^2 \frac{d}{dt} \frac{B_0}{\zeta \omega_0} \dot{\psi} = \frac{V}{2\pi} (\sin \psi_0 - \sin \psi), \tag{36.4}$$

and the circular frequency of small phase oscillations is

$$\Omega = \frac{c}{R} \left(\frac{he V \zeta \cos \psi_0}{2\pi E}\right)^{\frac{1}{2}}, \tag{36.5}$$

where $R = C/2\pi$, C being the circumference of the equilibrium orbit.

We also have taken the applied frequency to be h times the angular velocity of the particles; h is known as the "harmonic order".

The limits of phase stability (5.4) are unchanged, but the corresponding limits of energy and radial excursions (5.6) and (5.8) become

$$\Delta E_m = \left\{ \frac{eV}{\pi} \frac{E^2 - E_0^2}{h \zeta E} \left[2 \cos \psi_0 + (2\psi_0 - \pi) \sin \psi_0\right] \right\}^{\frac{1}{2}} \tag{36.6}$$

and

$$\Delta r_m = \alpha R \left[\frac{e V E}{E^2 - E_0^2} \frac{2 \cos \psi_0 + (2\psi_0 - \pi) \sin \psi_0}{h \pi \zeta} \right]^{\frac{1}{2}}. \tag{36.7}$$

We see that the frequency Ω is real, and therefore the oscillations are stable, only if ζ and $\cos \psi_0$ are of the same sign. In the constant-gradient case, ζ is always positive, and therefore ψ_0 must be chosen as that root of $eV \sin \psi = \Delta E \, (= \text{energy}$ supplied to the synchronous particle per turn by the radiofrequency) for which $\cos \psi_0$ is positive[1]. In the alternating-gradient case, ζ is negative at energies below E_t, so that ψ_0 must be between $\pi/2$ and π, while above E_t ψ_0 must be between 0 and $\pi/2$. Thus the phase of the applied radiofrequency must be changed by $\pi - 2\psi_0$ at transition energy in order that phase stability be maintained.

Even if this phase shift is accomplished, it appears from (36.6) and (36.7) that the energy and radial excursions may become infinite at transition energy, when $\zeta = 0$. But actually these quantities remain finite, and in fact manageably small. To see this, we write (36.4), for small phase deviations, in the form:

$$\frac{d}{dt} \frac{1}{\Omega^2} \frac{d\psi}{dt} + (\psi - \psi_0) = 0. \tag{36.8}$$

The amplitude of the oscillations in ψ is damped, according to the adiabatic (WKB) approximation, proportional to

$$\Omega^{\frac{1}{2}} \sim (\zeta/E)^{\frac{1}{4}}. \tag{36.9}$$

The radial excursion is, according to (4.12), equal to

$$\Delta r = - R \frac{\alpha}{\zeta} \frac{\dot\psi}{\omega_0} = - \frac{R^2 \alpha}{c \beta \zeta} \dot\psi,$$

and therefore varies with energy proportional to

$$\frac{\Omega^{\frac{3}{2}}}{\beta \zeta} \sim \frac{1}{\beta (\zeta E^3)^{\frac{1}{4}}}. \tag{36.10}$$

But the adiabatic approximation is not valid when ζ approaches zero. A better approximation is obtained by replacing ζ/E by a linear function of time. This leads to solutions of (36.8) in terms of Bessel functions of order $\frac{2}{3}$, approaching finite limits as ζ approaches zero at a finite rate. The result is [28] that the ratio of the radial displacement at transition energy to the displacement at injection is

$$\frac{\Delta r_t}{\Delta r_i} = 1.259 \beta_i \left[\frac{E_i^9 |\zeta_i|^3}{E_0^4 E_t^4 e V} \frac{h \pi \cos \psi_0}{2 \sin^2 \psi_0} \right]^{\frac{1}{12}}, \tag{36.11}$$

while the ratio of phase displacement at transition to the amplitude of phase oscillations at injection is

$$\frac{(\psi - \psi_0)_t}{(\psi - \psi_0)_i} = \frac{2}{\sqrt{3}} \frac{\beta_i E_i}{E_t} \frac{\Delta r_i}{\Delta r_t}. \tag{36.12}$$

Numerically, for the Brookhaven or CERN machines, the ratio (36.11) for the radial amplitude is about 0.5, and the ratio (36.12) for the phase amplitude is about 0.1.

Thus the maximum radial excursion will occur near injection. Because of the factor α in (36.7), this maximum radial excursion is much smaller than in constant gradient machines, so that the radial aperture allowance for synchrotron oscillations can be small.

---

[1] Note that the definition (4.7) of the phase is the negative of the definition used in many references, such as [10], [13], [28]. In terms of the definition used there, ψ_0 lies between $\pi/2$ and 0 below the transition energy, and between $\pi/2$ and π above it.

If the phase is switched just at the transition energy, the amplitudes continue to follow the adiabatic law; the ratio of amplitudes at higher energies to the amplitudes at phase transition are given by the reciprocals of (36.11) and (36.12) with E_t, ζ_i, β_i replaced by E, ζ, β at the higher energies.

If the phase is switched at the wrong time, or by the wrong amount, additional synchrotron oscillations will be excited. An error $\delta\varphi$ in the *magnitude* of the phase shift excites an oscillation of amplitude $\delta\varphi$ at transition. An error δt in the time of transition causes the amplitude to grow by approximately

$$\exp\left[\int \Omega \, dt\right].$$

The exponent equals unity when the energy differs from transition energy by

$$\Delta E = \left(\frac{9\sin^2\psi_0}{16\pi\,h\cos\psi_0}\right)^{\frac{1}{3}}\left(\frac{E_t\,eV}{E_0^2}\right)^{\frac{1}{3}}E_t. \tag{36.13}$$

Apart from unimportant numerical factors, this agrees with an expression found by Goldin and Koskarev [31].

For the Brookhaven machine, this is about 100 Mev. Thus errors in the timing of phase transition should be smaller than the time during which particles are accelerated by 100 Mev, which is about three milliseconds at Brookhaven. This should not be a difficult requirement to fulfill.

This conclusion is borne out by experimental investigations. Bodenstedt [30] simulated phase oscillations with a mechanical analogue computer, and found that the timing of phase transition was not critical. At Brookhaven, electrons were accelerated through the phase transition in the "Electron Analog", a small electrostatic alternating-gradient synchrotron; it was found that some electrons could be accelerated without any phase shift at the transition energy, and that there was no detectable loss when the transition phase shift was executed within ± 100 microseconds, in agreement with (36.13)[1].

The quantity α which determines the transition energy is found by using the methods of Sect. 34. A particle with momentum p and Δp satisfies the equation of horizontal betatron oscillation

$$\frac{d^2x}{ds^2} + \frac{(1-n)}{\varrho^2}\,x = \frac{1}{\varrho}\,\frac{\Delta p}{p}. \tag{36.14}$$

We obtain the displaced equilibrium orbit, i.e., the periodic solution of (36.14), by the Fourier harmonic method of Sect. 34. The solution is

$$x = \frac{\Delta p}{p}\,\beta^{\frac{1}{2}}\nu^2\sum_k \frac{a_k e^{ik\varphi}}{\nu^2 - k^2} \tag{36.15}$$

with

$$a_k = \frac{1}{2\pi\nu}\int_0^C \frac{\beta^{\frac{1}{2}}}{\varrho}\,e^{-ik\varphi}\,ds. \tag{36.16}$$

Therefore the change in the circumference of the equilibrium orbit is

$$\Delta C = \int_0^C \frac{x}{\varrho}\,ds = 2\pi\nu^3\frac{\Delta p}{p}\sum_k \frac{|a_k|^2}{\nu^2 - k^2}, \tag{36.17}$$

and

$$\alpha = \frac{1}{2\pi R}\,\frac{\Delta C}{\Delta p/p} = \frac{\nu^3}{R}\sum_k \frac{|a_k|^2}{\nu^2 - k^2}. \tag{36.18}$$

---

[1] G. K. Green, [5], p. 103. — Raka, Kassner, Plotkin and Spiro: Bull. Amer. Phys. Soc. II **1**, 177 (1956).

In most accelerator designs, the dominant term will be the one with $k=0$. If the magnet is composed of sectors in all of which the radius of curvature ϱ is the same, separated by straight sections, and if β can be replaced by its average value R/ν (see Sect. 33), we have approximately

$$a_0 = \frac{R^{\frac{1}{2}}}{\nu^{\frac{3}{2}}},$$

so that

$$\alpha = \frac{1}{\nu^2}, \tag{36.19}$$

and the transition energy is

$$E_t = \nu E_0. \tag{36.20}$$

VLADIMIRSKII and TARASOV [32] have devised a magnet structure in which (36.19) is not valid, but rather α is zero or negative. This is done by inserting a number K of "compensating magnets" with reversed fields but with the same gradients as would be called for without compensating magnets. If ν is slightly less than K, the terms $k=\pm K$ in (36.18) are large and negative, and can be made to cancel the leading term $k=0$. It is expected that this configuration will be used in the 50 Bev proton synchrotron being planned in the USSR [52]. The advantage is that the problems associated with the transition energy simply do not arise; the disadvantage is that a larger circumference is needed for a given peak particle energy.

D. Alternating-gradient proton synchrotrons.

37. General considerations. The pulsed alternating-gradient (AG) synchrotron operates in the same basic way as the constant-gradient type. However, the particle orbits in the alternating-gradient field alter design requirements, and permit greater design flexibility. The magnet ring will be composed of magnet units with n positive, units with n negative, and straight sections. These may have several different lengths, and some magnet units may even have reversed fields for special purposes (Sect. 36). The number of choices of "lattice structure" is unlimited, but the region of choice is considerably reduced by criteria imposed for physical and engineering reasons.

The simplest magnet structure would consist of N identical unit cells, each containing arc s of field with index n and arc s with index $-n$. For an example, such a structure is indicated schematically in Fig. 82. The inset shows representative shapes of magnetic fields, assumed to have constant gradient over the working aperture and, for simplicity, sharp transition at the interfaces. Typical orbits are plotted with modes $\mu = \pi/2$ and $\pi/5$, from which $\sqrt{n}/N$ is determined by Fig. 79 and (32.21) to be 0.5 and 0.33. The orbital oscillations are seen to consist of a "smooth approximation" sinusoid on which the "wiggly motion", or alternating-gradient structure is superimposed. It will be noted that if μ (the phase shift of the betatron oscillation per period) is small, the fine structure is small. From Fig. 80, the form factor for the examples shown is 2.4 and 1.3. The particle path is readily described by the phase-amplitude form of solution, rewriting (33.16)

$$y(s) = a\beta^{\frac{1}{2}} \cos\left[\int \frac{ds}{\beta} + \delta\right]. \tag{37.1}$$

The function $\beta(s)$ can then be obtained using the relationships of Sect. 32. Some simplification of the calculations can often be made by use of (33.14). If an observer notes successive transits at any s, and if the system is not on a resonance, amplitudes will be noted up to a maximum value proportional to $\beta^{\frac{1}{2}}$. If $\sqrt{\beta}$ is then plotted vs. s (Fig. 82) the product of $\sqrt{\beta}$ and an arbitrary constant gives the aperture required for the betatron oscillations as a function of azimuth. Maximum excursions occur in the center of "focussing" sections, and the $\beta(s)$ may be applied to the z direction by slipping the function one-half unit in azimuth. The arbitrary constant [a in (37.1)] is determined so as to accommodate admittance requirements of the injector, gas scattering and magnet imperfections. A synchrotron also requires the addition of radial aperture for the phase oscillations (36.6).

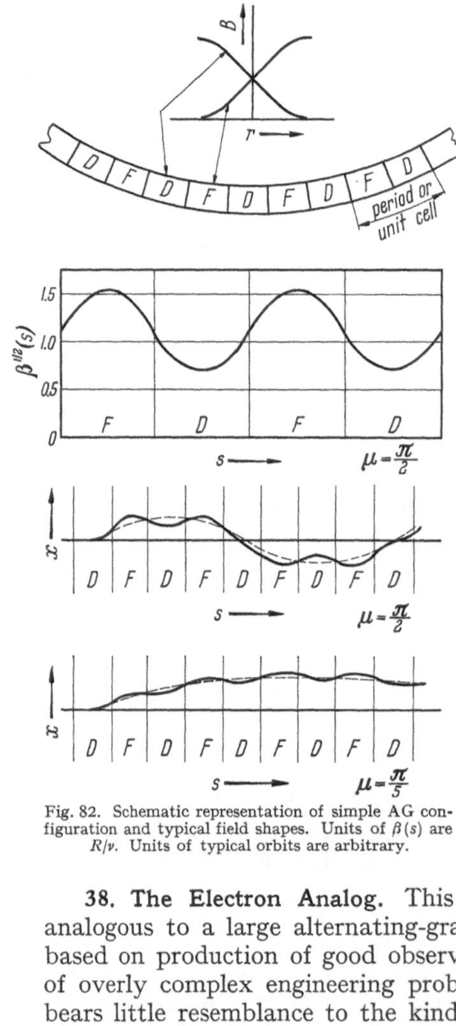

Although the simple alternating unit cell gives unity circumference factor (ratio of orbit circumference to $2\pi\varrho$) field free spaces, or straight sections are usually required for insertion of various auxiliary devices. Designate half of a focussing unit by F, half of a defocussing unit by D, and a straight section by O. Unit cells with useful properties are $FFODDO$, $FOFODODO$, and $FOFDOD$. The last conserves circumference factor, has a reasonable form factor, and permits insertion of gradient-modifying elements in the center of lenses, where such elements have the greatest effect. The $FOFDOD$ unit cell was used in the "Electron Analog", a working synchrotron built at the Brookhaven National Laboratory in 1954 to explore physically the properties of alternating gradient machines. Since the Electron Analog is the only pulsed AG synchrotron which has been operated with passage through the phase transition, and with wide variations of operating point, a description of it will illustrate general properties of such machines.

Fig. 82. Schematic representation of simple AG configuration and typical field shapes. Units of $\beta(s)$ are R/ν. Units of typical orbits are arbitrary.

38. The Electron Analog. This electron synchrotron was designed to be analogous to a large alternating-gradient proton synchrotron. The design was based on production of good observational characteristics without introduction of overly complex engineering problems. Such aims resulted in a design which bears little resemblance to the kind of 10 Mev electron synchrotron one would build as a routine source of particles. The ν was selected at about $6\frac{1}{2}$, a value high enough to give good momentum compaction ($\alpha = 0.024$), and phase transition at about $2\frac{3}{4}$ Mev. Fig. 83 shows the arrangement of the Analog, with unit cell $FOFDOD$, and elements F, O, and D equal in length (7 inches). Parameters are:

ϱ, radius of curvature	15 ft.	μ, mode	$\sim\pi/3$
R, physical radius	22.5 ft.	T_i, injection energy	1 Mev
a, aperture	0.8 in.	E_{max}, maximum energy	10 Mev
N, number of unit cells	40	Rise time	5 millisec
n,	225	f_∞, revolution frequency	7 Mc.

The multiplier (34.9) relating orbit displacement to field imperfections is of the order of 40. To hold the probable orbit displacement less than 0.2 inches then requires r and z positioning of the fields to be done with an accuracy better

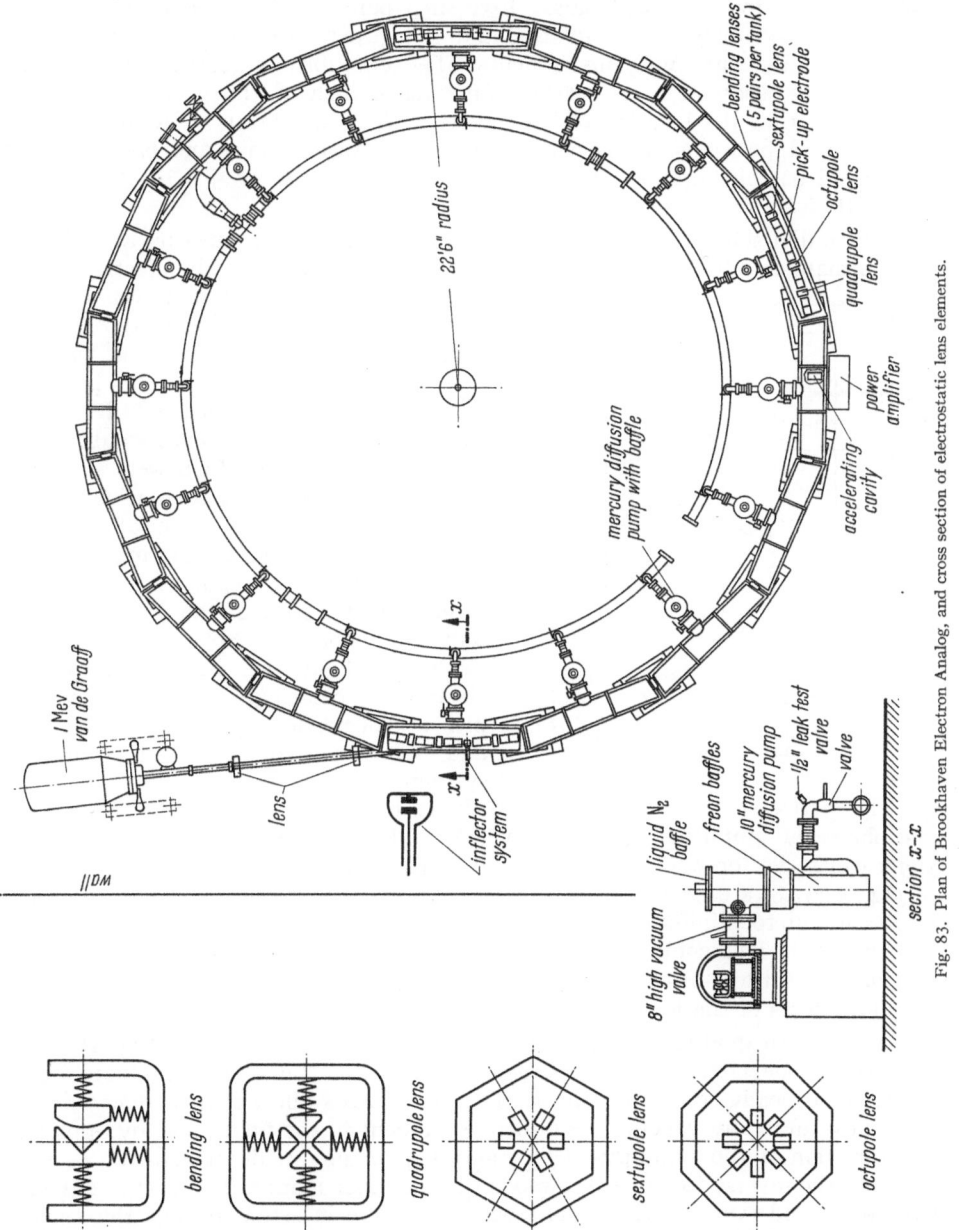

Fig. 83. Plan of Brookhaven Electron Analog, and cross section of electrostatic lens elements.

than 0.005 inch rms. Azimuthal errors, $r \Delta \vartheta$, can be allowed to be about an order of magnitude greater. Many difficulties are involved in producing very precise, weak, pulsed magnetic fields, but the low energy and large radius of curvature permit use of readily obtainable electric fields. The electric fields were produced between accurately machined stainless-steel electrodes, mounted on ground ceramic insulators in assemblies which were carefully gauged. The structure is shown in Fig. 83. Main bending and focussing lenses are hyperbolic in cross section (it was convenient to make one electrode the degenerate hyperbola) and are joined in FD pairs. Five such pairs were erected on each girder using jig-bored reference templates, and the girders were then aligned with a radius bar and surveying instruments. The equilibrium particle orbit of the Analog was observed to be within some tenth inch of the geometrical orbit, confirming that the location of the electric fields was within a few mils rms of the idealized positions. Lens potentials were balanced above and below ground and could be constant for observation of coasting beam, or ± 0 to 30 kv linear sawtooth form for accelerating beam.

The radial setover was only some two mils per turn, so missing an inflector in the manner described in Sect. 24 was quite impractical. The electron beam from a 1 Mev Van de Graaff was bent by the electrostatic inflector field onto the orbit (Fig. 83). A second transit through the inflector would force the particles beyond the aperture, but a full circumferential ribbon of particles would exist. Then if the inflector field were switched off in a fraction of a revolution period the complementary fraction of the particles would be left circulating. It was normal practice to regulate the relative timing of the start of the Van de Graaff pulse and the inflector turn-off so as to leave about one half revolution of electrons circulating. This gave a strong pick-up electrode signal for observation, and excitation of the radiofrequency system.

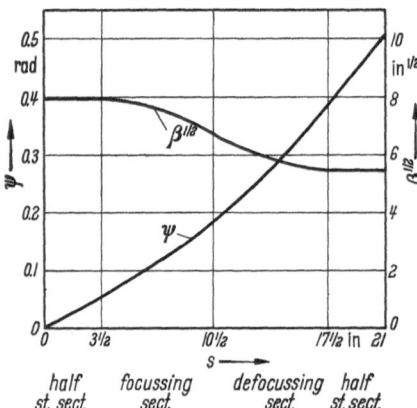

Fig. 84. Phase shift Ψ and amplitude function $\beta^{\frac{1}{2}}$ for Electron Analog.

The lattice structure properties are shown in Fig. 84 by $\psi(s) = \int ds/\beta$ and $\beta^{\frac{1}{2}}$[1], the phase and amplitude functions for (37.1). For most purposes the same functions will serve for both r and z motion since, with $n = 225$, the n and $|1 - n|$ of (32.7), (32.8) are nearly equal. The form factor (33.21) $F = 1.51$ and the ratio of maximum to minimum aperture requirements is $\beta_{max}/\beta_{min} = 1.47$. Thus the aperture of 0.8 inch could be constricted to 0.54 inch in the center of a defocussing section without loss of particles. The ν value was calculated to be 6.47, neglecting end effects of the lenses, not far in error as seen in Fig. 85.

In order to modify the operating point, sixteen quadrupoles (Fig. 83) were inserted in the lattice, eight between horizontally focussing lenses, and eight between vertically focussing lenses. The two classes allow independent adjustment of ν_r and ν_z by varying the effective gradient in the fields, and by applying the variation at the azimuths of β_{max} and β_{min} for one mode of betatron oscillation and the inverse for the other mode. In a similar manner two classes of eight sextupoles and two classes of eight octupoles were inserted to vary the second and third derivatives of the field (quadratic and cubic non-linearities). The multipole lenses, similar to the main or dipole lenses, could be excited by constant potential, or by linear sawtooth voltage waves of adjustable slope. Individual elements of the quadrupoles were separately connected so that small difference

---

[1] L. J. Laslett: Unpublished report.

potentials could be applied to the vertical or horizontal electrode pairs to produce dipole "kickers", The "kickers" were supplied from a resistor matrix which would apply 6th, 7th, or 8th azimuthal harmonics, adjustable for compensation of small misalignments of the main lenses [see (34.5)]. When any of the multipole lenses were energized the periodicity of the machine was reduced from 40 to 8, so that strong resonances, or "super stop-bands" would be introduced for $\nu = 4, 8, 12 \ldots$.

Fig. 85 shows a resonance diagram for the Electron Analog. This diagram was obtained by observing patterns on the pick-up electrodes produced by a

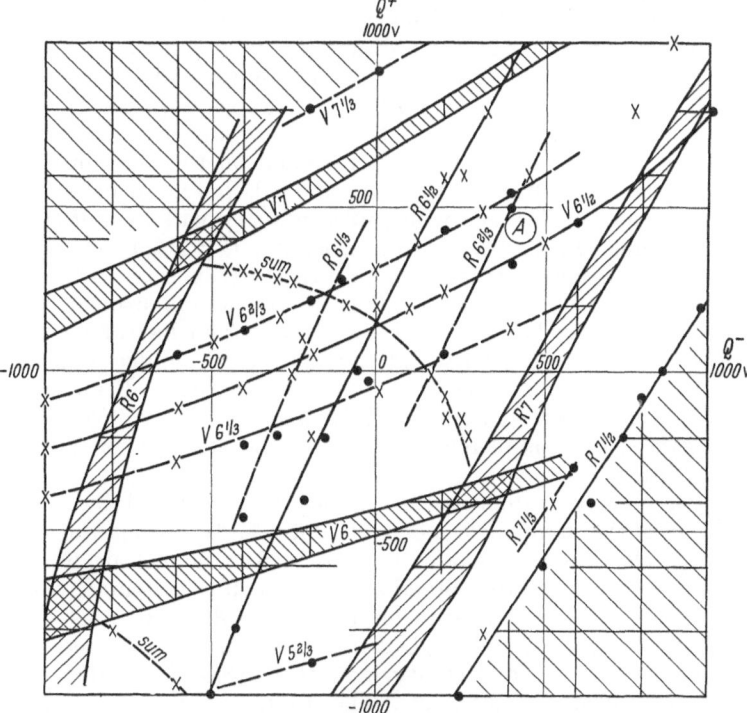

Fig. 85. Experimental resonance diagram of Electron Analog. Heavy bars and $\times$ indicate beam loss. Integral stop-bands are shaded. V indicates vertical, and R radial, ν values of resonances. Approximate dynamic operating point was usually at A.

half-revolution of injected beam, which would coast for a few milliseconds with fixed potential on the lenses. Settings of kickers and sextupoles were optimised and the region was searched by varying the potential on the quadrupoles, in many regions by sweeping the quadrupole potential rapidly. The axes in Fig. 85 represent the applied potential to eight Q^+ located between vertically focusing lenses and to eight Q^- between horizontally focusing lenses. It will be noted that if ν is an integer the beam is lost, and is lost over a stop-band of considerable width. The loss occurs within a few turns so that an attempt to cross the integral resonance was fruitless even at rates of 100 v/μsec. The very wide stop band at $\nu = 8$ can be seen extending almost to the adjacent half-integral $\nu = 7\frac{1}{2}$. The half-integral resonances are seen to be narrow, and did not invariably lose the beam. If the proper quadratic non-linearity was applied the beam would "lock on" the half-integral resonance, taking up an equilibrium orbit which closed on itself in two revolutions, and circulate more than 10^4 times without loss (see [5], p. 258). The sum resonances, $\nu_r + \nu_z =$ integer, were very narrow but produced complete beam loss when the operating point crossed them. Difference resonances, of which the "principal diagonal" $\nu_r = \nu_z$ is most important, were not found, in agreement with the theoretical result (Sect. 34) that these should not lead to instability but should couple the r and z oscillation modes. Sum and difference resonances are difficult to pinpoint because they do not show a characteristic pattern on difference induction electrodes. The various resonances with rotation frequency show a plain characteristic pattern (see [5], p. 258) and are readily identified when observing a coasting beam.

The equilibrium orbit for a fixed injection energy could be shifted by varying the main lens potential (with coasting beam) and the corresponding shift in ν observed from the beat pattern on difference induction electrodes. The uncorrected shift of ν with momentum, due to non-linear end effects of the lenses, was $d\nu_r/\nu_r = -3\,dp/p$ and $d\nu_z/\nu_z = +1.8\,dp/p$. Proper adjustment of sextupole voltages would make $d\nu/dp \approx 0$, and such sextupole voltages were used advantageously when accelerating. Some losses were found on the one-third integral resonances. The cause was traced to non-linear fields produced by the "kickers", which were not good dipoles, and which were adjusted with azimuthal harmonics for orbit smoothing. Since this class of resonances is excited by azimuthal harmonics of the second field derivative, a confirmatory experiment was performed by exciting two opposite sextupoles either in like or unlike polarity. The like polarity, producing even harmonics, would excite beam loss at $\nu_r = \frac{20}{3}$ but not at $\frac{19}{3}$, while the unlike polarity did the inverse. A similar experiment was done with two opposite octupoles with opposite polarities. Beam losses, up to perhaps 30%, were observed at $\nu_r = \frac{23}{4}, \frac{25}{4}$ and $\frac{27}{4}$. Even effects of the one-quarter resonances are completely masked by the stronger integral and half-integral resonances.

The radiofrequency system of the Analog was operated by beam control (Sect. 28). Since the frequency increased only 5.5% during the accelerating cycle, a simple electronic system with band width about 1 Mc was sufficient. The injected beam, adjusted to subtend about one-half the circumference, induced a radiofrequency signal on the sum electrode, Fig. 86, which signal was then amplified, phase controlled, and applied to the power amplifier and accelerating cavity. A simple ferrite-loaded cavity with Q reduced to about 6 was used so tuning over the band was unnecessary. The rise rates used were normally set to require 200 to 250 volts/turn so the cavity gap voltage was set at 300 to 350 volts peak. The first phase shifter was programmed by an exponential, set to compensate approximately for the variation of phase shift with frequency of the system from sum electrode to accelerating gap. A manually adjusted phase shifter set the proper starting condition. To give the system long-time stability (more than several revolutions) a radial error signal corrected the phase by actuating the second phase shifter. Acceleration begins with phase stability on the positive slope of the accelerating wave, but at the transition energy where $\alpha = \gamma^{-2}$ in (36.2) there is no phase stability, and the stable phase angle at higher energy transfers to the negative slope of the radiofrequency wave. The kinetic energy at the phase transition can be approximated from (36.20),

$$T_t \approx (\nu - 1)\,E_0.$$

The phase oscillation frequency of the analog was about 15 kc initially, fell to zero at about 1.1 msec after injection, and then rose to about 3 kc (Sect. 36). At phase transition the third phase shifter was triggered and advanced the phase $\pi - 2\psi_0$ (about 100°) in a time of 30 μsec. The radiofrequency wave was thus moved to the succeeding position of stability. Since the sense of the radial control reverses with the reverse in slope on the accelerating wave, an electronic reversing switch in the radial control loop was actuated simultaneously with the phase jump (Fig. 86). Most of the accelerated beam could be carried through phase transition, and mistiming of the phase jump could be as large as ± 100 µsec before serious beam loss was encountered (see also Sect. 36).

The Analog lens system was housed in magnetic tanks which were remagnetized by applying and removing a.c. fields so as to nullify the earth's field. A careful program of "degaussing" reduced magnetic fields at the orbit to 5 to 10 milligauss. The system was evacuated with mercury pumps and liquid nitrogen traps to operating pressure of 3 to 5×10^{-7} mm Hg, which gave reasonably small gas scattering losses. Great care was taken to prevent entry of oil or organic vapors which might coat the lenses with microscopic insulating films and seriously distort the electric fields. No effects were noted which could be traced to charged

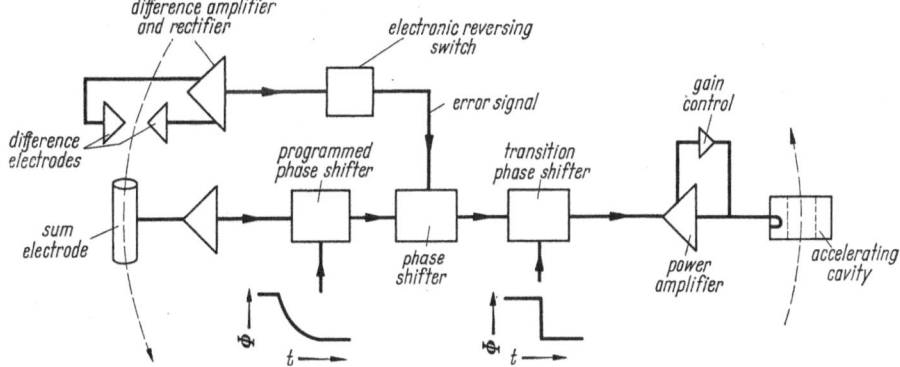

Fig. 86. Block diagram of Electron Analog radiofrequency system.

insulating films, although an ungrounded pick-up electrode would quickly charge up enough to destroy the beam by its azimuthal perturbation.

39. Pulsed proton synchrotrons. Alternating-gradient focussing permits the use of very small apertures. Consequently AG synchrotrons can be built with magnets of small cross section and, from a cost viewpoint, with radii of perhaps hundreds of meters. Two conflicting requirements enter into the design of large synchrotron rings. Firstly, as ν is increased the aperture required for betatron oscillations due to injector emittance, etc. (33.3), (33.18), and for phase oscillations (36.7) diminishes; but secondly, as ν is increased the aperture required for magnet misalignment increases. [In (34.9), if μ is held approximately constant to stay in a favorable region of the stability diagram[1], then as ν increases, N and M will increase proportionately, and n as ν^2, so $n/\nu M^{\frac{1}{2}}$ will increase as $\nu^{\frac{1}{2}}$.] Examination of the relations of Sects. 34 and 35 indicates that the difficulties arising from magnet errors and misalignments increase alarmingly at very large values of n and ν. Furthermore, if ν is large the proportional variation of field shape between resonances is small, presenting a difficult magnet design problem when field strength must vary by a factor of the order of a hundred. If n is very large the design of ferrous-core magnets becomes difficult, since the field on the median plane, if strictly linear, would fall to zero at r/n from the equilibrium orbit on one side, and double at the same radial increment on the other side. (This distance is about 25 cm in machines under construction.) Very high n,

---

[1] See Fig. 1 of [28].

and large gradient, requires a lowered maximum orbit field because the magnet will saturate badly on the high field side, and this in turn means increased radius for a given energy.

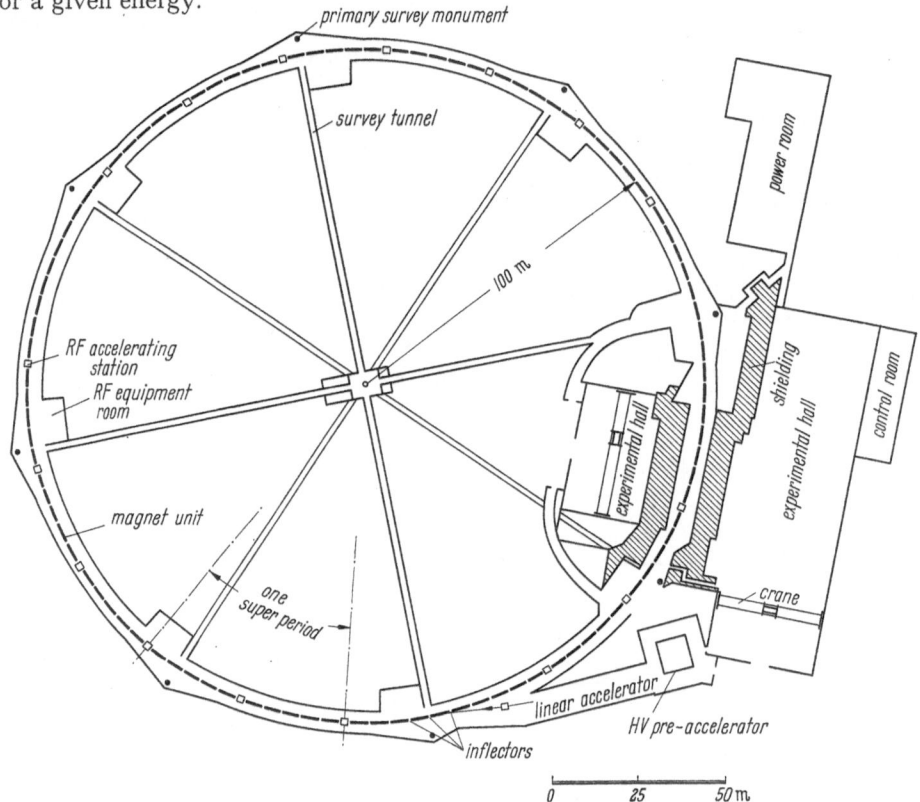

Fig. 87. Plan of CERN PS.

Two pulsed synchrotrons designed to cross the transition energy are being constructed. The Brookhaven AGS has a radius of 421.45 ft. (128.5 m) and the CERN PS a radius of 100 m. Both will be in the energy range 25 to 30 Bev. Constant gradient types have a relative aperture (aperture divided by radius of curvature) of 0.01 to 0.02 vertical and 0.04 to 0.08 radial. By comparison the AG types are designed with relative aperture 0.0008 vertical and 0.0014 to 0.0018 radial. The general advantages and limitations of alternating-gradient machine design will be illustrated by a brief description.

A plan of the CERN PS, Fig. 87, shows the general proportioning of what might be called a "typical" alternating gradient proton synchrotron. The initial basic parameters which must be chosen for a design are energy, radius of curvature (or peak field strength), μ, ν, and lattice structure.

	CERN PS	Brookhaven AGS
Max. energy (Bev) . . .	24.3	24.6
at field (kgauss) . . .	12	10
Possible max, energy (Bev)	28	32
at field (kgauss) . . .	14	13
ϱ (m)	70.08	85.37 (280 ft)
R (m)	100	128.5 (421.5 ft)
μ (rad)	$0.25\,\pi$	$0.29\,\pi$
ν	$6\frac{1}{4}$	$8\frac{3}{4}$
n	288	357

The PS lattice structure, Fig. 88, has $N = 50$ periods, each period composed of two $\frac{1}{2}F\frac{1}{2}D$ magnet units. Ten "super-periods" are formed by the insertion of "long" straight-sections, so each super-period has ten magnet units, two

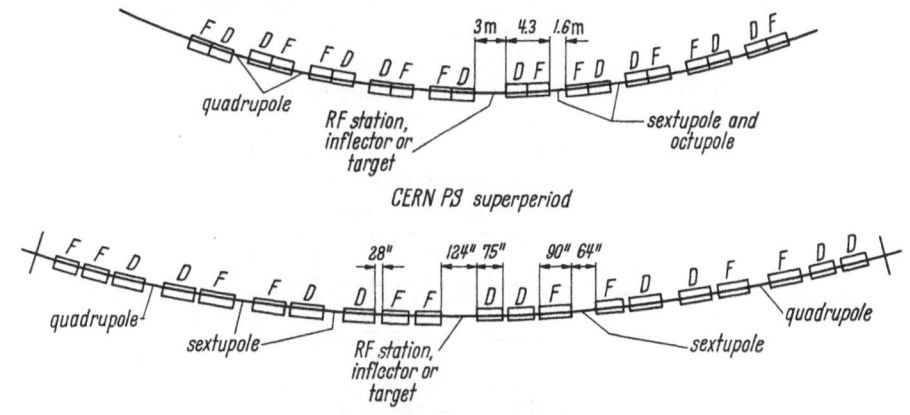

Fig. 88. Lattice structure of CERN and Brookhaven National Laboratory synchrotrons. F and D refer to radial oscillations—the inverse apply to vertical oscillations.

3.0 m and eight 1.6 m straight sections. Back-legs of the **C**-magnets are alternated in order to give small angle access to the aperture, required for injection of protons and extraction of high energy particle beams.

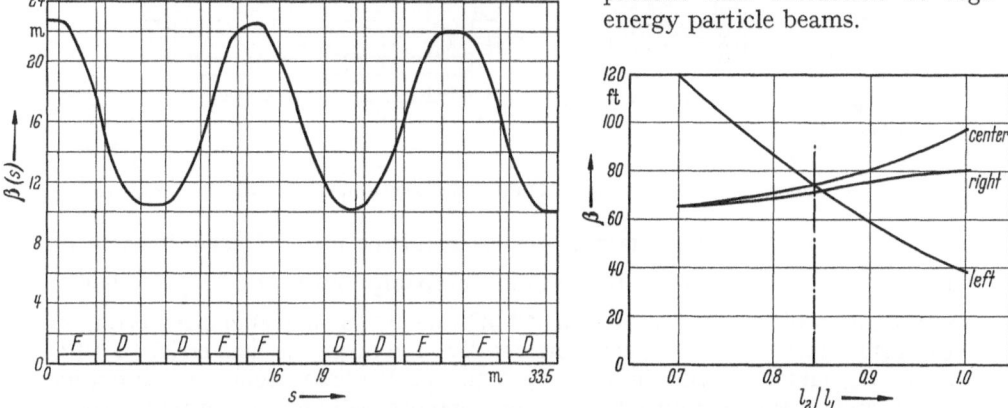

Fig. 89. Amplitude function of the AGS for one half super-period. The function for the adjacent half superperiods is symmetrical about the end ordinate axes.

Fig. 90. Amplitude function in terms of ratio of length of foreshortened magnets to length of other magnets. "Left", "center" and "right" are curves of maximum β of corresponding peaks in Fig. 89.

The Brookhaven AGS lattice structure, Fig. 88 has $N = 60$ periods, each period composed of four magnet units, two $\frac{1}{2}F$ and two $\frac{1}{2}D$. There are twelve super-periods, each containing 20 magnet units, two 10-foot, six 5-foot, and twelve 2-foot straight sections. (The iron-to-iron length of the straight sections is 4 inches greater than their nominal length, just given.) The amplitude function, $\beta(s)$, is plotted in Fig. 89 for the AGS. The form factor (30.21) is 1.53. β_{max} was minimized by foreshortening the two magnets on either side of the 10-foot straight section. These "short" magnets are 75 inches long, compared to 90 inches for the others. The influence of such factors on aperture, or conversely on admittance with constant aperture, can be seen in Fig. 90 where β is shown as a function of l_2/l_1, the magnet length ratio. The three curves are the maximum values of β corresponding to the three peaks of Fig. 89. A pronounced minimum in β_{max} occurs at $l_2/l_1 = 0.84$ with considerable improvement over the equal length value. The square root of the maximum to minimum β of Fig. 89 is 1.5. Noting (37.1) this means that, if the available radial magnet aperture is 6 inches, the aperture at the azimuths of β_{min} can be diaphragmed to 4 inches without interference with the beam. Such permissible aperture restrictions are especially useful for design of induction electrodes. Aperture requirements for the AGS at injection energy (where the maximum is necessary, since gas scattering will be small) can be estimated from the relationships of Sects. 33, 34, 36. The semi-

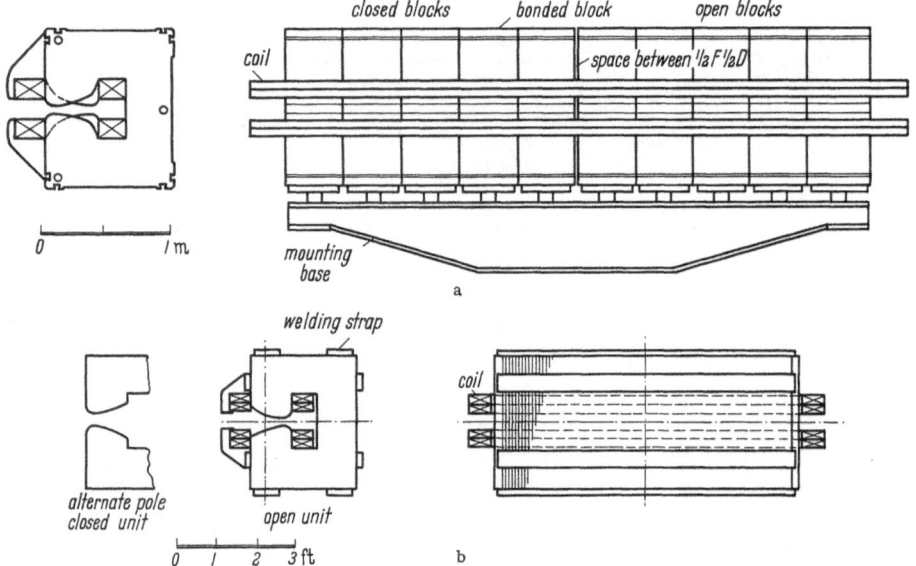

Fig. 91 a and b. AG synchrotron magnet units. (a) CERN PS magnet unit; (b) Brookhaven AGS magnet unit.

aperture, A_1, occupied by betatron oscillations of the injected beam is based on an injector emittance of $\frac{\pi}{2} \cdot 10^{-3}$ cm rad, probably optimistic. A_2 is allowed equilibrium orbit distortion [P_{98} of (34.9)] due to magnet alignment errors of 0.01 in rms vertically, 0.02 in rms horizontally; and magnetic field errors of 0.1% rms in B_r, and 0.2% rms in B_z. A_3 is the peak phase oscillation amplitude ($h = 12$) and A_4 the allowance for vacuum chamber wall and low current correct-iug coils. Estimated values are then:

	Vertical (z)	Horizontal (r)
A_1	0.5	0.5 inches
A_2	0.6	1.3
A_3	0	0.8
A_4	0.1	0.1
Total	1.2	2.7 inches

The total required aperture is about $2\frac{1}{2} \times 6$ inches. Values for the CERN PS are similar.

Insertion of the "long" straight sections reduces the periodicity of the structure from 60 periods to 12 super-periods. The super-periods must be quite identical to one another. The ν value should be well removed from the resonance values $N_1/2$, N_1, $3N_1/2$ ... where N_1 is the super-period number. In the PS the $N_1 = 10$ produces dangerous resonances at $5, 10, 15 ...$, and the ν is set at $6\frac{1}{4}$. The corresponding resonances of the AGS are $6, 12, 18 ...$ and ν is $8\frac{3}{4}$.

Magnet design for an alternating-gradient synchrotron with $\nu \gg 1$ is complicated by the large gradient which must be inserted in the field. Remanent field effects are complex because the various parts of the magnet core traverse very different hysteresis cycles, and eddy currents continue beyond the end of the magnet pulse. Remanent difficulties lead to a demand for injection energies as high as practicable, in the PS and AGS at 50 Mev, corresponding to 147 and 121 gauss respectively. The configuration of the PS and AGS magnets is shown in Fig. 91, where the major difference will be seen to be the combination of $\frac{1}{2}F\frac{1}{2}D$ in units for the PS and separation of the units for the AGS. Although the simplest (and most symmetrical) magnet structure would use one pole form and reverse the return leg from outside to inside the orbit, the necessity of extracting very high energy particles at small angles to the orbit re-quires a more complex lattice.

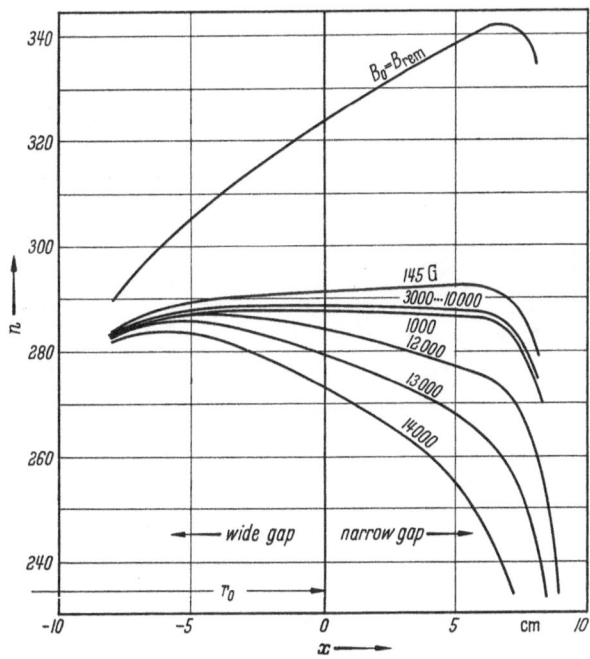

Fig. 92. Gradient, expressed as index n, as a function of radius and central field strength in the CERN PS magnet unit.

In the PS three successive unit return legs are aligned on one side of the orbit each superperiod, and in the AGS the backleg is reversed every ten units.

The necessity of small angle access and large change in flux density across the poles makes the H magnet form unattractive. A linear gradient field of the form

$$B = B_0\left(1 + \frac{n\,x}{r_0}\right)$$

is seen to result from hyperbolic poles. The physical magnets have poles which are hyperbolic in the central region, around r_0, but flare more widely than a hyperbola at the wide gap side and more narrowly at the closed gap. A neutral pole, or slab of iron, at the wide gap asymptote would improve the low field side; however, the high field side is severely limited by saturation. Gradient shapes vs. radius for various field strengths, measured at the azimuthal center of the PS magnet[1] are shown in Fig. 92. The distorting effects of remanence and eddy currents can be seen at low fields, and of saturation at high B_0.

There is complex fringing at the ends of the magnets. Since the units are short compared to a betatron oscillation wavelength a good approximation can

---

[1] Unpublished reports by the CERN Magnet Group.

be made by assuming the field to have a rectangular distribution along the s direction at radius r, with the central amplitude $B_0(r)$ and semilength

$$\frac{\int_0^\infty B(r)\, ds}{B_0(r)}.$$

This is equivalent to adding to each end of the magnet a length

$$L_B = \frac{\int_0^\infty B(r)\, ds}{B_0(r)} - l$$

where l is the semilength of the iron core. A similar length L_G can be derived for the gradient $\partial B_z/\partial r = G(r)$. By differentiation, the two lengths are related by

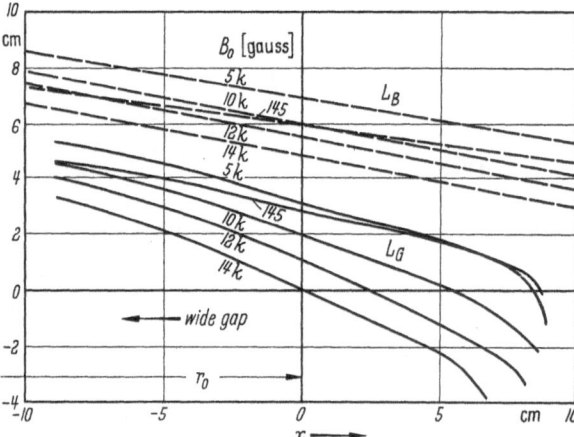

$$L_G - L_B = \frac{B}{G}\frac{\partial L_B}{\partial r}.$$

Thus if L_B is not constant with radius, and it is not with a rapidly flaring gap, the "focussing length" L_G is less than the "bending length" L_B.

It is necessary to increase the n value associated with the pole shape about $1\frac{1}{2}\%$ in both designs to obtain the desired ν. Second derivative effects are also inserted by the end fringing, shown by the variation of L_B and L_G with r in the PS

Fig. 93. Bending length, L_B, and focussing length, L_G, to be added to each end of CERN PS magnet unit. From reports of CERN Magnet Group.

magnet, Fig. 93. Another formulation of field shape can be expressed by expanding

$$\bar{B}(x) = \int_{-\infty}^\infty B(x)\, ds$$

as a power series

$$\bar{B}(x) = \bar{B}(0)\left[1 + a_1 x + a_2 x^2 + a_3 x^3 + \cdots\right].$$

The integrals were intercompared in the AGS magnets by using long (10 ft.) search coils, and have the coefficients (for units of inches)[1].

B_0 (gauss)	a_1	a_2	a_3
120	0.1045	-0.00051	
250	0.1053	-0.00034	
5000	0.1060	-0.00017	
10000	0.1056	-0.00026	
12000	0.1039	-0.00080	-0.00009
13000	0.1024	-0.00118	-0.00012

There are also small differences in the field shapes of the open and closed sections. Variation of the a_1 coefficient causes variation of ν; the "non-linear"

---

[1] Unpublished reports, Brookhaven Magnet Group.

coefficients a_2 and a_3 affect the variation of ν with momentum (or radius) and with oscillation amplitude. Corrections can be made with pole-face windings and with quadrupoles, sextupoles and octupoles. The multipoles are best placed in the center of main lens elements, where amplitudes and consequently effects are greatest, and at azimuthal positions which provide the least interference effects between the horizontal and vertical sets (i.e., sets in horizontally or vertically focussing magnets). Both the PS and AGS lattices are arranged for insertion of the multipole correcting lenses between $\frac{1}{2}F - \frac{1}{2}F$ elements. The quadrupoles are used as shown in Fig. 85. Even if the field were "linear" the sextupoles, or equivalent pole-face windings, would be needed. In a linear machine ν varies approximately as

$$\frac{d\nu}{\nu} \approx -\frac{dp}{p}$$

so that variation of momentum within the aperture can drive the operating point onto a resonance. End effects of the magnets, and distortion of the field by remanence, saturation etc., will in general increase the variation of ν with p. The sextupoles (or pole face windings) can be energized so as to reduce the ν variation across the working aperture to a small quantity. Third derivative in the field can be compensated by octupoles. The operating point variations due to end effects can be diminished by connecting $\frac{1}{2}F \frac{1}{2}D$ units and selecting the proper spacing. This procedure is being done in the CERN PS with a spacing determined to be 2 cm (see [5], p. 307).

The high sensitivity of the orbits to small random misalignments, Sect. 34, imposes severe tolerances on the location of the magnet unit reference marks with respect to the field, and on the setting of the magnets with respect to the orbit (which, with straight sections, is not a circle). The relationship of the field to the mechanical reference marks must be the same for all corresponding magnets units at all field strengths. Differences between classes of magnet units, such as open and closed, affect $\nu(p)$ but if the units in a class are identical the equilibrium orbit will not be disturbed, although the periodicity of the machine will be reduced. The PS and AGS lattices are so designed that the reduced periodicity is higher than that of the super-periods, and is not near 2ν.

In both the PS and AGS the magnet laminations (1.5 mm and 0.031 inch thick respectively) were cut with precision dies in order to maintain uniformity of shape. Steel cannot be produced with the required uniformity of magnetic characteristics. The entire quantity of steel blanks for a magnet class is accumulated and then shuffled, or mixed, such that the various heats, rollings and annealings are distributed into each magnet unit. The laminations are stacked in registry on precise fixtures and fastened, by bonding for the PS, and by welding to steel straps for the AGS. It is found that if magnet units are compared by finding the line in each at which $\int B \, ds$ is the same for intermediate fields, say 5 kilogauss, then the integrals will track within a part in 10^3 from injection to peak field. The variation of remanence is somewhat larger, and is reduced in effect by arranging the units on the ring in an order giving a high periodicity of the variation. The magnets are supported on deep columns or piles carried to the underlying strata, independent from the building structure and overburden. Design of the supporting ring considers details of the support distortions and minimizes the resulting orbit distortions by avoiding harmonics near ν. The figure of the magnets is determined by advanced geodetic survey methods; a precision of 1 in 10^6 in angle and length is desirable.

The vacuum chamber design selected for both PS and AGS uses a nearly elliptical tube of high resistance, non-magnetic alloy (resistivity somewhat over 100 μohm-cm) with axes about 7×15 cm and 2 mm wall. Eddy currents in the vacuum chamber distort the shape of the field, requiring some compensation by pole face windings or multipoles, and reduce the field strength appreciably. The latter can cause serious orbit disturbances at low fields unless the resistance of the chamber sections is made quite uniform. A pressure below 10^{-5} mm Hg will hold gas scattering to unimportant amplitude. Such pressures are readily attained with oil-diffusion pumps (PS), or titanium evaporation pumps (AGS), but the low conductance of the vacuum chamber requires that pumps be close spaced, and some 50 pump stations are required.

The vacuum chamber effects can be reduced by slowing the initial rate of rise of the magnetic field. This is to be done in the AGS by inserting a large saturating reactor in series with the magnet such that the initial $\dot{B}/B$ is reduced about a factor of four. The reactor also helps to reduce the initial transient azimuthal distribution. The magnet is a delay line with a long propagation time, and resistors across the units are necessary to reduce the transient propagation time. Some filtering of power supply ripple is also advisable for reduction of the large azimuthal variation of ripple current caused by the sizeable shunt capacitances.

The power supplies of the PS and AGS resemble that of the Cosmotron (Sect. 19) and are only slightly larger in rating, about 30 Mw peak. Repetition period of 3 sec, and relatively large ohmic losses in the magnets require drive motors of 4 to 5×10^3 hp. A one second rise time of the pulse appears to be a good compromise.

The injector for a large AG synchrotron should have an output at high energy and small emittance. Injection at fields much less than 100 gauss in the magnet ring would encounter formidable difficulties with remanent field variation and shape, eddy currents and magnet transient response. Both the PS and AGS will use 50 Mev linear accelerator injectors ($B_0 = 147$ and 121 gauss respectively). The linear accelerator can have an emittance of about $\pi\,10^{-3}$ cm rad, which occupies a reasonable portion of the ring aperture. Momentum spread in the linear accelerator output becomes radial spread of equilibrium orbits in the ring, producing greater oscillation amplitude and lesser phase acceptance. Betatron oscillations can be minimized by inserting a series of electromagnetic lenses designed to offset momenta horizontally at the same rate as the dr/dp of the ring. (See [5], pp. 179 and 200.) A momentum spread $\pm\frac{1}{2}\%$ corresponds to a radial spread of about ± 1 cm, half the limiting phase oscillation amplitude. Although the linear accelerator output can be expected to be somewhat better, perhaps $\frac{1}{2}$ to $\frac{2}{3}$ of the foregoing, a momentum spread approaching $\pm\,0.1\%$ is desirable. The proton bunches from the linear accelerator can be permitted to drift some 10 to 15 m, after which the momentum will be correlated with time of transit and passage through a rapidly varying longitudinal field will reduce the momentum spread. A debuncher cavity, with radiofrequency supplied at the proper phase from the linear accelerator, can reduce momentum spread by a factor of three or four. The transverse emittance (Sect. 23, especially Fig. 40) of a well-adjusted linear accelerator will be nearly symmetrical; however, the synchrotron admittance will generally be quite different in the r and z planes at the inflector, even if nearly equal in area. The necessary astigmatism is readily inserted by spaced quadrupoles (see [5], p. 184 seq.) and these focussing quadrupoles can also match the injector to the ring. The transverse wavelength of a proton linear accelerator is considerably shorter than that of the synchrotron, so that it is necessary to alter the shape of the injector emittance, even though the area is invariant in the connecting drift spaces. Proper beam focussing for optimum injection thus bears little relation to the more usual problem of obtaining minimum spot size. The injected beam passes through magnet stray fields with large higher derivatives, producing undesirable smearing of the injector emittance shape. This can be compensated by inserting sextupole, and even octupole, non-linear lenses. At the entry to the ring the injected beam must be deflected about one degree. The problem is similar to that of the Electron Analog, since the radial change per turn at injection is of the order of a mm. Straightforward inflector missing (Sect. 24) seems impractical at a Δr per turn much smaller than the input beam diameter. The PS and AGS will use "one-turn" injection, in which electrostatic deflection is applied to bend the beam onto the equilibrium orbit and, after an injection time at least equal to one revolution period, the electrostatic field is switched off. The switching process must occupy a small fraction of a period, and must switch the inflector fields accurately to zero. A typical injection electromagnetic optical system is shown in Fig. 94, which outlines the plans for the AGS. Alternate methods for accomplishing multi-turn injection have been proposed. One would use molecular hydrogen ions, at twice the proton energy, stripped at the injection point, so the sudden change in radius of curvature would provide the necessary deflection angle (Sect. 24). Another method is similar to the peeler-regenerator extraction systems

	PS	AGS
Stable phase angle . . .	$\pi/6$	$\pi/6$
Total peak voltage (kv) .	110	180
Harmonic order	20	12
No. of double cavities .	16	12
Starting frequency (Mc)	2.9	1.4
Final frequency (Mc) . .	9.5	4.5

of synchro-cyclotrons. One or more bumps could be introduced in the magnet field such that the distorted equilibrium orbit so produced passes initially through the inflector. The bumps could then be diminished, drawing the beam away from the inflector (see notes p. 147, [5]).

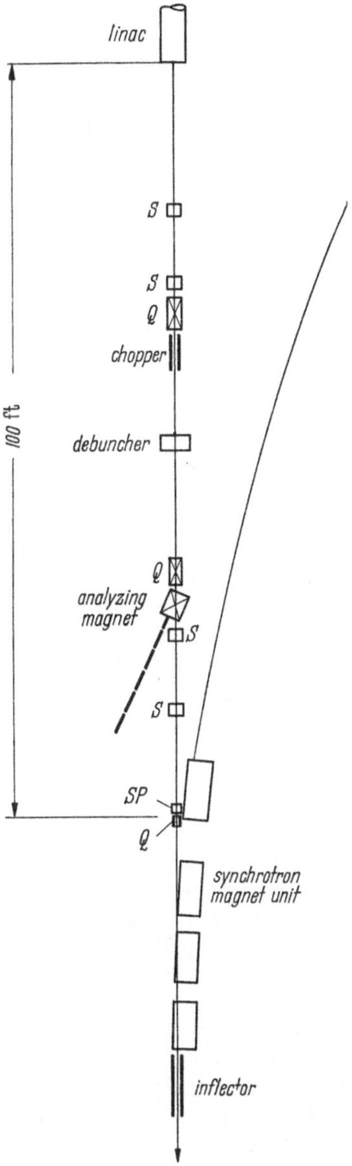

Acceleration is required at the rate of some 50 to 60 kv/turn in the PS and 80 to 90 in the AGS, Eq. (25.7). The accelerating systems are designed for the characteristics in the Table.

Fig. 94. AGS injection optics. S steering coils, Q focussing quadrupoles, SP sextupole.

Accelerating cavities are coaxial type, loaded with ferrite to reduce the resonant size for insertion in the long straight sections. The ferrite is saturated by phase-servo controlled magnetic fields to reduce the inductance and maintain the cavities in resonance with the varying frequency. Setting the harmonic order to an integral multiple of the super-period number permits the power amplifiers which excite the cavities to be driven in or out of phase from a common low-level source.

The low-level radiofrequency system must supply a relatively noise-free excitation to the power amplifiers, with frequency matched to the rising magnetic field (Sect. 25). If (33.1) and (33.3) are combined with (33.2),

$$\frac{\Delta f}{f} = \left[\left(\frac{E_t}{E}\right)^2 - 1\right]\frac{\Delta r}{R},$$

and the relative frequency error in the AGS for $\Delta r = 1$ cm is:

$T = E - E_0$	$\Delta f/f$	$T = E - E_0$	$\Delta f/f$
50 Mev	5.3×10^{-3}	$7\frac{3}{4}$ Bev	0
1 Bev	1.4×10^{-3}	10 Bev	-2.9×10^{-5}
3 Bev	2.9×10^{-4}	20 Bev	-6.5×10^{-5}
5 Bev	8.8×10^{-5}	30 Bev	-7.2×10^{-5}

Since the frequency is proportional to v/c, or $\sqrt{1 - (E_0/E)^2}$, most of the frequency sweep takes place in the first tenth second of a one-second rise. During the initial acceleration the radiofrequency can be supplied by a programmed oscillator, tracked with the magnetic field. Such an oscillator can readily be made with $\Delta f/f$ precision of 1×10^{-3}, which is adequate to bunch the beam during the first few phase oscillations (at most a few milliseconds). However, as phase transition is approached the frequency tolerance becomes unmanageably small and, at phase transition, frequency error will simply be integrated as phase error. A programmed frequency system would probably leave the beam at an unstable phase shortly after the phase transition. The phase-lock, or beam-controlled, system described in Sect. 28 is suitable for acceleration during the latter part of the cycle. Such a phase-lock system will be switched in instead of the programmed oscillator at an appropriate early time. It will resemble the block diagram of Fig. 86, including the phase jump at transition and reversal of the sense of the radial position correction. The simple programmed phase shifter of Fig. 86 would involve many radians of phase shift due to the long cables required, so it will be replaced by a heterodyne arrangement which can be made to compensate for cable delays.

The phase oscillation frequency falls rapidly after injection to zero at transition, then rises to a low value (36.5). This is shown for the PS and AGS in Fig. 95. It is evident that care must be taken to minimize harmonics of power frequency in the magnetic field and radiofrequency system. The shape of the variation of radial and phase amplitude, explained in Sect. 36, is shown in Fig. 96. The sharp increase in phase oscillation radial amplitude at transition diminishes rapidly soon afterward, but may make experimental use of the AG ring at energies just above E_t subject to large phase oscillation modulation of the beam on a target (Sect. 30).

The phase transition energy can be increased to a very large value, or even to infinity, by making the momentum compaction, α as defined by (36.1) zero or negative (Sect. 36). Phase stability then depends only on the change of velocity with momentum. One method for reducing α to zero is incorporated in a

6 to 7 Bev pulsed proton synchrotron being built in the USSR ([*32*] and [*5*], p. 118). The magnet ring has $N = 56$ periods and $N/4$ magnet units with "normal" gradient but reversed field. Fig. 97 shows the superperiod arrangement in which every fourth horizontally focussing unit has radius of curvature

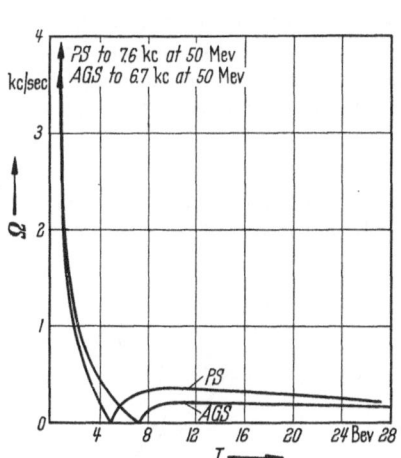

Fig. 95. Phase oscillation frequency vs. energy.

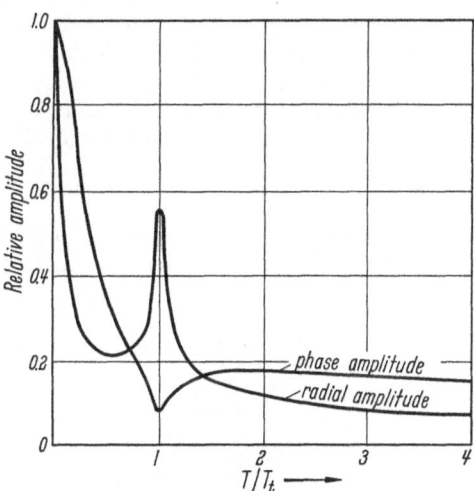

Fig.96. Relative amplitude of phase and radial excursions of the phase oscillations as a function of energy in units of transition energy.

— 53.5 m, compared to 27.8 m for the other units. The dashed orbit of Fig. 97 indicates the effect of the reversed units in maintaining constant orbit length with respect to momentum. The circumference factor (ratio to circumference of ring without reversed sections) is 1.25 and the radial momentum compaction is about one fourth that of a corresponding unreversed AG ring. A ν of $12.75 = 0.226\ N$ [*32*] was chosen. Expected tolerances and admittance are to be contained in an 8×11 cm vacuum chamber. The high gradient led to use of neutral poles on the 98 C-magnets, and asymmetric quadrupoles for the 14 reversed field

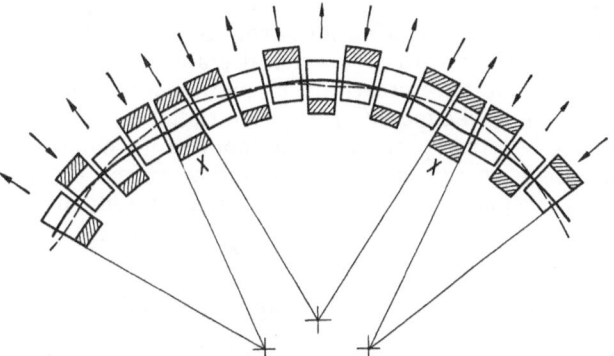

Fig. 97. Lattice structure of AG synchrotron with no phase transition. Small arrows indicate direction of gradient. $\bar{X}$ magnets have reversed field direction. [*5*], pp. 118, 122.

magnets, in which the gradient-to-field ratio is nearly twice that of the C's. The phase oscillation frequency will be relatively low, about 3.3 kc at injection and 136 cps at full energy; while the frequency tolerances are not sensibly different from those of the uniform AG ring. The increased size of the compensated synchrotron is accompanied by an advantage—the radial variation with momentum in the compensating sectors is considerably larger than the dr/dp of an uncompensated machine. This could make multi-turn injection feasible, and should assist in ejection of an external beam of protons.

40. Fixed field alternating gradient synchrotron (FFAG). The radial momentum compaction possible with alternating gradient focussing makes it interesting to consider a sychrotron ring with static field in which orbits are stable at all energies from injection to maximum. If acceleration is to span a large range of energy (several rest masses) it is not advisable to sweep over a series of resonances and so a field is required in which ν is independent of momentum. A number of field configurations with the desired properties have been devised[1]. Two field types, radial sector and spiral ridge, which have been extensively studied have orbits which scale, i.e., differ only in magnification but not in shape with momentum variation. The radial sector FFAG has a field which varies as

$$B = B_0 \left(\frac{r}{r_0}\right)^k,$$

but its N periods are each composed of a pair of magnets, one with reversed field (Fig. 98). Thus the gradient is reversed each sector to obtain AG focussing

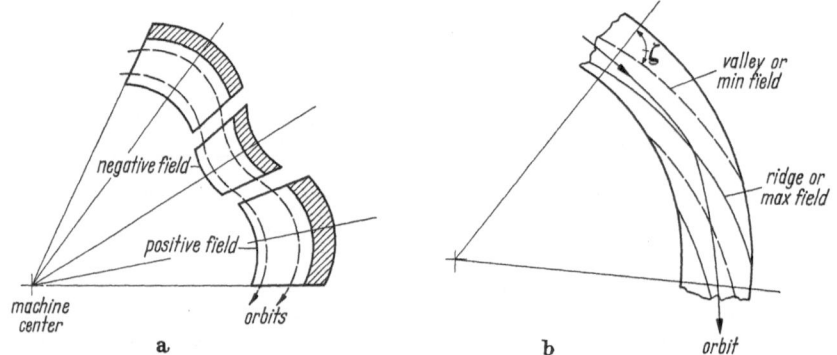

Fig. 98 a and b. Schematic representation of two FFAG arrangements. (a) radial sector type. (b) spiral ridge type [53]

[53], [54]. The equilibrium orbit is scalloped and the circumference factor will be at least five to give vertical stability, but possible use of high mean fields could make the "practical" circumference factor smaller than five in comparison to the pulsed type AG ring. The radial sector type has a large admittance and the interesting property that momentum compaction can have either sign, so the high energy orbits can lie either at the outer or inner radius. If an orbit length equals $2\pi R$ (R is the equivalent radius) then for scaling orbits [53],

$$p = p_0 \left(\frac{R}{R_0}\right)^{k+1}.$$

The betatron oscillations are approximated by

$$\nu_r^2 = k + 1 + \{AG\},$$
$$\nu_z^2 = -k + \{AG\}$$

in which $\{AG\}$ is an alternating gradient focussing term (see Eq. (6.7) and (6.8) of [53]) subject to manipulation by variation of the sector lengths and relative reversed field strength. A radial sector machine has moderate inherent non-linearity and must avoid resonances of the same nature and general severity as the pulsed AG ring. The M.U.R.A. group has built an electron model of the radial sector type with $N = 8$ and $k = 3.36$. An experimental resonance diagram is shown in [55].

---

[1] The most extensive studies have been conducted by the Midwest Universities Research Association (M.U.R.A.) and are described in privately circulated reports.

Another scaling type of FFAG, the spiral ridge, has a median plane field of the form

$$B = B_0 \left(\frac{r}{r_0}\right)^k \left[1 + f \sin\left((N\vartheta - \frac{1}{w} \ln \frac{r}{r_0}\right)\right]$$

in which f is the flutter factor, $1/w = N \tan \zeta$; and ζ is the spiral angle. The orbits crossing the alternate ridges and valleys of the field (Fig. 98) are then passing through alternating gradients. Betatron oscillation frequencies are approximated by $\bigl($see Eqs. (6.24), (6.25) of $[53]\bigr)$

$$\nu_r^2 = k + 1,$$

$$\nu_z^2 = -k + \frac{f^2}{2} + \{AG\},$$

so the vertical oscillations are directly affected by the alternating gradient term, while the radial oscillations, unlike the radial sector type, are almost unaffected by the AG terms. The spiral ridge synchrotron has a circumference factor of the order of $1 + f$, so it can be made with about the same radius as an AG pulsed ring, and much smaller than a radial sector synchrotron. However, the non-linearity of the spiral ridge machine is inherently large and sector resonances, at which the betatron wavelength spans a small integral number of sectors, must be avoided. The non-linearities impose a limiting amplitude for stability on the betatron oscillations, especially unfavorable if the betatron wavelength spans approximately three sectors. An electron model of the spiral ridge type has been operated successfully by M.U.R.A., and shows the expected resonances.

A spiral ridge synchrotron would have a relative radial aperture of at least 0.03, a radial sector synchrotron at least 0.02. This compares to some 0.002 for the pulsed AG ring, indicating the relatively massive aspect of the FFAG. The potential advantage of an FFAG lies in its ability to contain particles at any, or several, momenta within its range. The cycling rate is dependent on the capabilities of the radio frequency system. Multiple radiofrequency systems can be used, in which particles are accelerated to an intermediate energy and left circulating, later to be captured by another system and accelerated onward ($[5]$, p. 44)[1]. Such a process could leave high density beams circulating at full energy, with very refined vacuum systems, for hours. The FFAG has a phase transition at

$$E_t = \sqrt{k+1}\, E_0$$

and, since $d\omega/dE$ changes sign at the transition, the radiofrequency first increases, then above transition decreases.

The possibility of storing intense high-energy beams of small cross section leads to the further possibility of constructing two FFAG rings tangent so that the rotating beams would collide and give $p - p$ interactions with very high center-of-mass energy ($[5]$, p. 36). An alternate intersecting beam machine has been proposed[2] in which both beams would circulate in a radial sector synchrotron with equal positive and negative field sectors. Several intersection regions could be arranged, with the two beams displaced vertically between. Problems of space charge, beam interactions and vacuum technique are formidable. It has also been proposed that storage rings be constructed to take the output of pulsed synchrotrons ($[5]$, p. 64) or linear accelerators.

---

[1] Terwilliger, Jones and Pruett: Beam stacking experiments in an electron model FFAG. Rev. Sci. Instrum. **28**, 987 (1957).

[2] T. Ohkawa: Two-beam FFAG accelerator. Rev. Sci. Instrum. **29**, 108 (1958).

41. Modified synchrotrons. It is evident that particle acceleration can be accomplished in a great variety of electromagnetic fields. The desire for greater intensity, greater energy, and enhanced experimental utility has produced many proposals for modification of the more conventional types, and for construction of hybrid types. A modification for increased intensity is incorporated in the constant-gradient synchrotron being assembled at Princeton University (U.S.A.). The ring is designed with an aperture only $2\frac{1}{2} \times 8$ inches at some 3 Bev, the minimum indicated by experience and orbital calculations. Stored energy is reduced to such a degree that the magnet can be resonated with condensers and driven at 20 cps to give a pulse rate 60 times that of the Cosmotron ([5], p. 525). A constant-gradient field was selected to permit multi-turn injection, and a high harmonic order will minimize radial phase oscillations. Another constant-gradient modification under construction at the Argonne National Laboratory (U.S.A.) will use a pulsed H-magnet with uniform field, and obtain focussing by the small wedge angles of the ends of the eight magnet sectors. By careful minimization of aperture, and use of high maximum field, it is expected that 12.5 Bev can be attained with only a few thousand tons of magnet weight. Multi-turn injection will be used in order to attempt to load the ring to its space-charge limit at the initial energy of 50 Mev ([5], p. 42).

The constant-gradient synchrotron can further be modified by designing the magnet with a double aperture ([5], pp. 9 and 40), a large low field aperture for injection and initial damping, a small aperture for final acceleration, and a smooth transition between. The one would permit injection of large quantities of particles and favorable space-charge conditions, while the other would require minimum magnetic energy storage. Such a combination could give high beam intensity at a rapid cycling rate. Another modification which can reduce the magnetic stored energy consists of a combination of spiral ridged poles in an otherwise conventional constant-gradient ring ([5], p. 9). As noted for the FFAG, the spirals have little effect on the radial focussing and thus permit multi-turn injection, but give strong vertical focussing, so a short gap is possible. The reduction in stored energy would then permit rapid cycling.

Hybrids, or combinations of machines, have been proposed for the purpose of injecting into a ring at quite high energy, so that aperture requirements will be reduced. An early proposal of this kind by OLIPHANT[1] visualized a synchro-cyclotron surrounded by a constant-gradient synchrotron. Detailed studies have been made by L. C. TENG and T. A. WELTON (unpublished) on the problems of initial acceleration in a spiral-ridge cyclotron to perhaps 1 Bev, and transfer to a compact AG pulsed ring for further acceleration. The transfer would involve peeler-regenerators, and requires additional study and experimental work. When hybrids involving the numerous FFAG configurations are considered, an unlimited set of combinations can be imagined.

Acknowledgment.

A large portion of the detailed literature of proton synchrotrons is found in the privately circulated reports. These reports are available in only a few laboratories, and are not referenced because they cannot be consulted by many readers. We are deeply indebted to the colleagues in many accelerator centers who have prepared these reports, as indeed are most of the authors cited in the bibliography, and we regret that it is not practicable to cite reports and credit their authors.

---

[1] Nature, Lond. **165**, 466 (1950).

Bibliography.

I. General references.

[1] STRUTT, M. J. O.: Lamésche-Mathieusche und Verwandte Funktionen in Physik und Technik. Berlin: Springer 1932 (also Ann Arbor: Edwards 1944).

[2] STOKER, J. J.: Nonlinear Vibrations. New York: Interscience Publishers 1950. — This and the preceding book introduce the mathematics of differential equations with periodic coefficients.

[3] LIVINGSTON, M. S.: High Energy Accelerators. New York: Interscience Publishers 1954. — An elementary introduction to particle accelerators.

[4] STURROCK, P. A.: Static and Dynamic Electron Optics. Cambridge 1955. — The theory of particle orbits using the formalism of generalized mechanics.

[5] Proceedings—CERN Symposium on High Energy Accelerators—Vol. I. CERN, Geneva, 1956. — A valuable collection of papers on high energy accelerators by an international group of authors. Individual papers are referenced throughout the text.

[6] FREMLIN, J. H., and J. S. GOODEN: Cyclic Accelerators. Rep. Progr. Phys. **13**, 295 (1950).

[7] BLEWETT, J. P.: Recent Developments in Proton Synchrotrons. Ann. Rev. Nucl. Sci. **4**, 1 (1954).

[8] BLEWETT, J. P.: The Proton Synchrotron. Rep. Progr. Phys. **19**, 37 (1956).

II. Theory of particle orbits in the synchrotron.

Numbers 9 to 18 present the theory of orbits in constant-gradient synchrotrons. 9, 10, 12 and 13 contain the basic relationships. The orbit theory of alternating-gradient machines is presented in numbers 19 to 32. General treatment of the fundamentals will be found in 24, 27, 28 and 29. A previous edition of 28 has had wide private circulation.

[9] FRANK, N. H.: Stability of Electron Orbits in the Synchrotron. Phys. Rev. **70**, 177 (1946).

[10] BOHM, D., and L. FOLDY: Theory of the Synchrotron. Phys. Rev. **70**, 249 (1946).

[11] GOODEN, J. S., H. H. JENSEN and J. L. SYMONDS: Theory of the Proton Synchrotron. Proc. Roy. Soc. Lond. **59**, 677 (1947). — The first paper published on proton synchrotron orbits.

[12] TWISS, R. Q., and N. H. FRANK: Orbital Stability in a Proton Synchrotron. Rev. Sci. Instrum. **20**, 1 (1949).

[13] BLACHMAN, N. M., and E. D. COURANT: Dynamics of a Synchrotron with Straight Sections. Rev. Sci. Instrum. **20**, 596 (1949).

[14] BLACHMAN, N. M., and E. D. COURANT: Scattering of Particles by Gas in a Synchrotron. Phys. Rev. **74**, 140 (1948); correction Phys. Rev. **75**, 315 (1949) and extension Rev. Sci. Instrum. **24**, 836 (1953).

[15] COURANT, E. D.: Resonance Effect in the Synchrotron. J. Appl. Phys. **20**, 611 (1949).

[16] BLACHMAN, N. M.: Forced Betatron Oscillation in Synchrotron with Straight Sections. Rev. Sci. Instrum. **22**, 569 (1951).

[17] BLACHMAN, N. M.: Synchrotron Oscillation Resonance. Rev. Sci. Instrum. **21**, 908 (1950); **23**, 250 (1952).

[18] BALDIN, A. M., V. V. MIKHAILOV and M. S. RABINOVICH: Method of Envelopes for Free Oscillations in Accelerators. Journ. Exp. Theor. Physics **4**, 857 (1957). [Russian edition **31**, 993 (1956)].

[19] COURANT, E. D., M. S. LIVINGSTON and H. S. SNYDER: The Strong-Focussing Synchrotron. Phys. Rev. **88**, 1190 (1952); **91**, 202 (1953).

[20] LUNDQUIST, S.: Stability of Orbits in a Strong Focussing Synchrotron. Phys. Rev. **91**, 981 (1953).

[21] BELL, J. S.: Stability of Perturbed Orbits in the Synchrotron. AERE Report T/R 1114, 1954.

[22] BARDEN, S. E.: Space-charge Forces in Strong Focussing Synchrotron. Phys. Rev. **93**, 1378 (1954).

[23] LÜDERS, G.: Über den Einfluß von Fehlen des magnetischen Feldes auf die Betatronschwingungen im Synchrotron mit Starker-Stabilisierung. Nuovo Cim. (10) **2**, Suppl. 4, 1075 (1955). In English CERN 56-8, Geneva 1956.

[24] SEIDEN, J.: La Stabilité des Orbites dans le Synchrotron a Forte Convergence. Ann. Phys. **10**, 259 (1955).

[25] ORLOV, I. Y.: Non-linear Theory of Betatron Oscillations in the Strong Focussing Synchrotron. Nuovo Cim. (10) **3**, 252 (1956).

[26] KOLOMENSKI, A. A.: Theory of Betatron Oscillations of Particles in Magnetic Systems. Ž. tekh. Fiz. **26**, 1969, 1978 (1956) [in Russian].

[27] Hagedorn, R.: Stability and Amplitude Ranges of Two Dimensional Non-linear Oscillations with Periodical Hamiltonian. Parts I and II. CERN 57—1, Geneva, 1957. — Hagedorn, R., and A. Schoch: Part III. CERN 57—14, Geneva, 1957.
[28] Courant, E. D., and H. S. Snyder: Theory of the Alternating-Gradient Synchrotron. Ann. Physics 3, 1 (1958).
[29] Sturrock, P. A.: Non-linear Effects in Alternating-Gradient Synchrotrons. Ann. Physics 3, 113 (1958).
[30] Bodenstedt, E.: Über die Phasenschwingungen beim Synchrotron mit starker Fokussierung. Ann. Physik (6) 15, 35 (1955). — Z. Naturforsch. 8a, 502 (1953).
[31] Goldin, L. L., and D. G. Koskarew: Synchrotron Oscillations in Strong Focussing Accelerator. Nuovo Cim. (10) 2, 1251 (1955).
[32] Vladimirskii, V. V., and E. K. Terasov: Possibility of Eliminating Critical Energy in a Strong Focussing Accelerator. Problems of the Theory of Cyclic Accelerators, USSR. Acad. of Sci. 1955, also Ž. tekh. Fiz. 26, 704 (1956) [in Russian].

III. Design and description of proton synchrotrons.

Numbers 33, 35 and 43 are the earliest descriptions of projected machines, and are now largely superceded.

[33] Livingston, M. S., J. P. Blewett, G. K. Green and L. J. Haworth: Design Study for a 3 Bev Proton Accelerator. Rev. Sci. Instrum. 21, 7 (1950).
[34] Cosmotron Staff: The Cosmotron. Rev. Sci. Instrum. 24, 723—870 (1953). — The most extensive description of an accelerator in the published literature. This issue has 27 technical papers on the Cosmotron.
[35] Oliphant, M. L., J. S. Gooden and G. S. Hide: Acceleration of Charged Particles to Very High Energies. Proc. Phys. Soc. Lond. 59, 666 (1947).
[36] Hibbard, L. U.: The Birmingham Proton Synchrotron. Nucleonics 7, No. 4, 30 (1950).
[37] Riddiford, L.: Vacuum System of the Birmingham Proton Synchrotron. J. Sci. Instrum. 28, Suppl. 1, 47 (1951).
[38] Proton Synchrotron of the University of Birmingham. Nature, Lond. 172, 704 (1953).
[39] Moon, P. B., L. Riddiford and J. L. Symonds: Experimental Characteristics of the Proton Synchrotron. Proc. Roy. Soc. Lond., Ser. A 230, 204 (1955).
[40] Riddiford, L., H. V. van der Raay and R. F. Coe: Some Beam Studies with the Induction Electrode. Proc. Phys. Soc. Lond. A 68, 489 (1955).
[41] Ramm, C. A.: Principles and Apparatus of the Injection System of the Birmingham Proton Synchrotron. J. Sci. Instrum. 33, 52 (1956). — Ramm, C. A., R. F. Coe and T. B. Vaughan: Analysis of Injection Phenomena, J. Sci. Instrum. 33, 102 (1956).
[42] Hibbard, L. U.: The radio-frequency system of the Birmingham Proton Synchrotron. J. Sci. Instrum. 31, 363 (1954).
[43] Brobeck, W. M.: Design Study for a 10 Bev Magnetic Accelerator. Rev. Sci. Instrum. 19, 545 (1948).
[44] Lofgren, E. J.: Proton Synchrotron. Science, Lancaster, Pa. 111, 295 (1950).
[45] Cork, B.: Proton Linac Injector for the Bevatron. Rev. Sci. Instrum. 26, 210 (1955).
[46] Winningstad, C. N.: Generating RF Energy for the 6 Bev Bevatron. Electronics 28, 164 (Feb. 1955).
[47] Rabinovich, M. S.: Certain Problems in the Theory of the Synchrophasotron of the Academy of Sciences (USSR). Soviet J. At. Eng. 5, 431 (1957) [in Russian].
[48] Veksler, V. I. and others: The 10 Gev Proton Synchrotron of the USSR Academy of Sciences. J. Nucl. Energy 4, 333 (1957). — Contains numerous references to papers in Russian on the 10 Gev machine.
[49] Bruck, H., et Lévi-Mandel: Sur le projet du Synchrotron á protons de Saclay. Nuovo Cim. (10) 2, Suppl. 1, 423 (1955).
[50] Bronca, G., H. Bruck, J. Hamelin, G. Neyret et T. Balzinger: Le Test des Blocs de L'Electro-Aimant du Synchrotron de Saclay. Nucl. Instr. 1, 123 (1957).
[51] Adams, J. B.: The Alternating Gradient Proton Synchrotron. Nuovo Cim. (10) 2, Suppl. 1, 355 (1955).
[52] Vladimirskii, V. V. and others: Basic Features of a Projected 50 to 60 Bev Strong Focusing Proton Accelerator. Soviet J. At. Eng. 4, 479 (1956).

IV. Fixed field alternating-gradient synchrotrons.

[53] Symon, K. R., D. W. Kerst, L. W. Jones, L. J. Laslett and K. M. Terwilliger: FFAG Particle Accelerators. Phys. Rev. 103, 1837 (1956).
[54] Laslett, L. J.: FFAG Accelerators. Science, Lancaster, Pa. 124, 781 (1956).
[55] Cole, F. T., R. O. Haxby, L. W. Jones, C. H. Pruett and K. M. Terwilliger: Electron Model FFAG Accelerator. Rev. Sci. Instrum. 28, 403 (1957). — Description of the first operating FFAG accelerator.

Linear Accelerators.

By

LLOYD SMITH.

With 26 Figures.

I. Introduction.

1. General considerations. The idea of producing energetic particles by passing them through a succession of low voltage gaps arranged in a straight line, is one of the oldest in the field of accelerators, dating back at least to 1924. This historical precedence is probably due more to the simplicity of the concept than to any intrinsic merit, but the method has unique features which establish for it an important role in medium and high energy physics, medicine, and certain industrial applications.

The chief advantage of the linear accelerator is the ease with which particles may be introduced at one end and extracted at the other. This feature makes it possible to utilize a source and d.c. pre-accelerator which can deliver to the low-energy end an intense beam of small diameter and energy spread. At the high energy end there emerges a beam far superior in energy definition and particle density to the beams extracted from magnetic accelerators. Unfortunately, the duty cycles of linear accelerators are usually quite low, for reasons to be discussed later; hence the time average current is not always impressive, but in situations requiring intense and well defined beams for short times, this type of machine is outstanding. For example, a 10-Mev electron accelerator delivering 100 milliamperes for one microsecond can produce an abundance of neutrons for time-of-flight work. Or again, proton accelerators of 10 Mev and greater are used almost exclusively as injectors for proton synchrotrons in the multi-Bev range, for these large machines require precise injection for times up to a millisecond.

Another application for which the linear accelerator appears most attractive is the production of electrons having energies substantially greater than 5 Bev, for in this method of acceleration, radiation loss is negligible, whereas in a synchrotron it becomes a dominant factor at high energies. The technology of electron linear accelerators has advanced to the point of permitting design studies up into the range of 50 Bev[1]; the cost would appear to be comparable to that of circular proton accelerators in the same energy range, that is, of the order of one million dollars per billion electron volts.

Finally, it has been demonstrated[2] that it is technically feasible to construct a proton or deuteron linear accelerator capable of producing time average beams of about one quarter of an ampere at energies as high as several hundred Mev. At the present time this is the only device known which could increase beam intensity by so many orders of magnitude in that energy range; unfortunately, cost and complexity would also increase considerably.

---

[1] R. B. NEAL and W. K. H. PANOFSKY: CERN Symposium, 1956.
[2] E. O. LAWRENCE: Science, Lancaster, Pa. **122**, 1127 (1955).

Because the particles in a linear accelerator traverse the machine only once, it is essential that strong accelerating fields be used to achieve the desired energy in a reasonable distance. For this reason it happens that, although the bunching and accelerating action is quite similar in principle to that of the synchrotron, the technical problems are very different. The design of a linear accelerator is dominated by the need for an efficient r.f. structure to favor the traveling electromagnetic wave on which the particles ride, and by the need for compact and reliable generators of r.f. power, problems which are of minor importance in most synchrotrons. As for confining the beam in the plane perpendicular to the direction of acceleration, one is concerned in the synchrotron with establishing forces which simply keep the particles on course for distances measured in thousands of miles; in the linear accelerator, distances are at most hundreds of feet, but the transverse electromagnetic forces associated with the accelerating field itself are so strong that they must be counteracted by some means, at least for non-relativistic energies. The importance of the r.f. system in the linear accelerator is further reflected in the fact that the difficult tolerances are generally associated with the maintenance of proper electric field distributions in the structure.

2. Historical review. The chronology of the significant steps in the development of linear accelerators as research machines to the end of 1954 is well presented in Ref. [1] to [3], which include extensive bibliographies of the original papers. In this article we shall content ourselves with a resume and an attempt to extend the description to the end of 1956.

In its earliest forms the linear accelerator consisted of a series of electrodes[1], alternate ones being connected electrically, with holes bored through them along the axis to permit the passage of an ion beam. The structure was driven at the resonant frequency of the mode in which successive electrodes were charged with opposite polarity, creating electric fields in the gaps between electrodes in the direction of the ion beam. By spacing successive gaps in proportion to increasing ion velocity, it was possible to have a certain fraction of the ions arrive at each gap when the electric field was accelerating. Somewhat later another scheme was devised[2] in which the electrode length was not so restricted, the proper timing of the gap fields being regulated by connecting the electrodes to suitably loaded transmission lines.

During the same years the Van de Graaff generator and the cyclotron were undergoing rapid development, with such successful results for the demands of the time that the technique of linear acceleration at radio frequencies received little attention. As in many other fields, World War II effected radical changes in the fortunes of the linear accelerator. The effort expended on mastering micro-wave techniques and developing high frequency power sources during the war made possible the construction of linear devices to aid in the drive for higher particle energies. Indeed, for a short time this method seemed the most attractive for very high energies[3]; while the arguments may still be true for electron acceleration, the newer types of synchrotron seem at this time more suitable for proton acceleration.

In the fall of 1945 a group at Berkeley set about applying the new high frequency technology to the acceleration of protons. In a few years they produced

---

[1] R. WIDERÖE: Arch. Elektrotechn. **21**, 387 (1928). — D. H. SLOAN and E. O. LAWRENCE: Phys. Rev. **38**, 2021 (1931).

[2] J. W. BEAMS and H. TROTTER Jr.: Phys. Rev. **45**, 849 (1934).

[3] L. I. SCHIFF: Rev. Sci. Instrum. **17**, 6 (1946).

a machine [*3*] which has become the prototype of all subsequent ion accelerators. Several groups began development of electron accelerators, all with essentially similar basic geometry (see Ref. [*2*] for a bibliography of electron accelerators). At present, the greatest activity in this field is centered at Stanford University (USA), where high energy work is emphasized, and at the AERE laboratory at Harwell (United Kingdom), where the application is to the medium energy range (5 to 30 Mev).

Ion accelerators are not as numerous, nor have they been as well documented in the literature. Therefore we list in Table 1 the ion accelerators operating or under construction as of the beginning of 1957. Those under construction are listed in the most probable order of completion.

Table 1.

Location	Type of ion	Energy (Mev)	Status	Purpose
Rad. Lab., Univ. of California	Proton	32	Operating	Research
Rad. Lab., Univ. of California	Proton	10	Operating	Injector for 6 Bev synchrotron
Rad. Lab., Livermore	Deuteron	7.5	Operating	Development for continuous operation at high current
Univ. of Minnesota	Proton	68	Operating	Research
USSR (Moscow)	Proton	40	Operating	Research
USSR (Big Volga)	Proton	9	Operating	Injector for 10 Bev synchrotron
USSR (Kharkov)	Proton	21	Operating	Development
Rad. Lab., Univ. of California	C, N, O, Ne	10/nucleon	Construction	Research
Yale University	C, N, O, Ne	10/nucleon	Construction	Research
AERE (Harwell)	Proton	50	Construction	Research and/or injector for synchrotron
CERN (Geneva)	Proton	50	Construction	Injector for 25 Bev synchrotron
Brookhaven Nat'l. Lab.	Proton	50	Construction	Injector for 30 Bev synchrotron

II. General principles.

3. Elementary description. In this section and the next, we shall present the features which are common to both electron and ion accelerators before proceeding to detailed discussion of either. The method of acceleration is, in principle, the same in both cases, but the difference in specific charge and the difference in behavior of relativistic and non-relativistic particles make for a wide divergence in technical approach.

The primary objective of the device is to maintain an electromagnetic wave traveling at a speed which matches the desired speed of the particles at each point along the length of the machine. If the wave has an electric field component in the direction of motion, particles will gain or lose energy, depending on their instantaneous position with respect to the crest of the wave. In particular, if the amplitude of the field exceeds a certain minimum value depending on the rate of increase of wave velocity, there will be two phases (ψ_s and $-\psi_s$ in Fig. 1) at which particles which have initially the same speed as the wave will remain as the wave moves along, gaining speed at the same rate as the wave. ψ_s, called the synchronous phase, is a stable point in the sense that particles ahead of it will gain speed less rapidly and thus drift back toward ψ_s, while particles following behind the synchronous phase will gain speed more rapidly and tend to catch up. There arises then a range of phase stability, indicated by the shaded area in Fig. 1, inside of which particles oscillate about synchronous phase relatively slowly compared to the frequency of the traveling wave, but gaining energy, on the average, at the desired rate. The quantitative relations will appear in the next section.

As the particles attain extreme relativistic energies oscillations cease, for particle speed becomes practically independent of energy. In this case, the wave speed is the speed of light and there is no synchronous phase or threshold amplitude; the final energy and energy spread depend only on the wave amplitude and the initial phase spread of the particles.

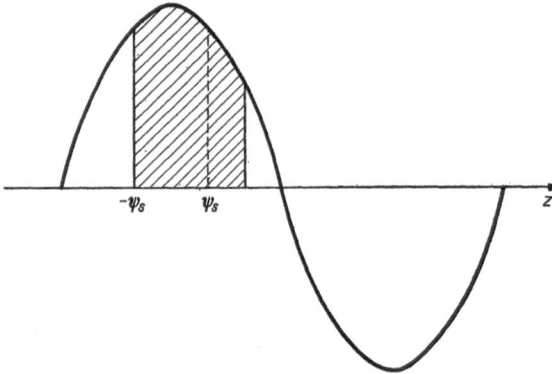

Fig. 1. One cycle of accelerating wave.

A traveling wave has, then, the intrinsic property of insuring phase bunching or, at worst, phase neutrality. Unfortunately, as will appear in the following section, the transverse forces acting on particles riding on the leading edge of the wave are defocusing; that is, an off-axis particle is deflected away from the axis. This phenomenon is apparent from the following consideration: in the reference frame in which the wave is at rest, the particles find themselves under the influence of an electrostatic potential. According to EARNSHAW's theorem, this potential cannot have an absolute minimum, so that if the synchronous phase is a stable point for the longitudinal motion, it must be an unstable point for motion in the transverse plane[1].

In summary, there are two problems which must be solved in devising a linear accelerator:

1. To provide an r.f. system which will propagate transverse magnetic waves of large amplitude at phase velocities less than or at most equal to the speed of light.

2. To provide means for confining a beam in the transverse plane.

4. Equations of motion. It is convenient to proceed from the following form of the equations of motion in which axial distance, rather than time[2], assumes the role of independent variable.

$$\frac{dt}{dz} = \frac{1}{v} = \frac{1}{c} \frac{\gamma}{\sqrt{\gamma^2 - 1}}, \tag{4.1}$$

$$\frac{dW}{dz} = e E_z(\varrho, z, t), \tag{4.2}$$

$$v \frac{d}{dz}\left[\frac{Wv}{c^2} \frac{d\varrho}{dz}\right] = e\left[E_\varrho(\varrho, z, t) - \frac{v}{c} H_\varphi(\varrho, z, t)\right] \tag{4.3}$$

---

[1] For a general proof see E. M. MCMILLAN: Phys. Rev. **80**, 493 (1950).
[2] Gaussian units are used throughout.

where ϱ, φ, z are the usual cylindrical coordinates,

$W =$ total particle energy,

$m_0 =$ particle rest mass,

$\gamma = W/m_0 c^2$.

We neglect the contribution of the radial motion to the total energy, and the small axial force proportional to $H_\varphi \frac{d\varrho}{dt}$. It is assumed that the electromagnetic field components are independent of azimuth, φ, the most usual case in practice. The field has the following form:

$$E_z = \varepsilon_0(z)\, I_0 \left[\frac{2\pi\varrho}{\beta_w\lambda}\, (1-\beta_w^2)^{\frac{1}{2}} \right] \cos\left[\omega t - \frac{2\pi}{\lambda} \int^z \frac{dz}{\beta_w} \right], \tag{4.4}$$

$$E_\varrho = -\frac{\varepsilon_0(z)}{\sqrt{1-\beta_w^2}}\, I_1\left[\frac{2\pi\varrho}{\beta_w\lambda}\, (1-\beta_w^2)^{\frac{1}{2}} \right] \sin\left[\omega t - \frac{2\pi}{\lambda} \int^z \frac{dz}{\beta_w} \right], \tag{4.5}$$

$$H_\varphi = -\varepsilon_0(z)\, \frac{\beta_w}{\sqrt{1-\beta_w^2}}\, I_1\left[\frac{2\pi\varrho}{\beta_w\lambda}\, (1-\beta_w^2)^{\frac{1}{2}} \right] \sin\left[\omega t - \frac{2\pi}{\lambda} \int^z \frac{dz}{\beta_w} \right] \tag{4.6}$$

where $\beta_w =$ wave velocity/light velocity

$\omega =$ angular radio frequency

$\lambda = \dfrac{2\pi c}{\omega} =$ free space wavelength

and I_0 and I_1 are Bessel functions with imaginary arguments.

These expressions constitute an approximate solution of MAXWELL's equations if ε_0 and β_w vary slowly with z[1]. We also neglect the effect of other waves which are necessarily introduced when a wave guide is loaded to depress the phase velocity of the desired mode. These other waves travel at a much different velocity, so that their average effect is negligible in most cases. In the occasional cases in which the intended fractional gain in energy per r.f. cycle is large, it is necessary to determine the exact field distributions and integrate Eqs. (4.1) to (4.3) numerically. However, for most purposes, the following treatment, including only the dominant component waves (4.4) to (4.6), is quite adequate.

We define first a synchronous particle as one for which

$$\beta_s(z) = \frac{v_s(z)}{c} = \beta_w(z). \tag{4.7}$$

Then the synchronous phase, ψ_s, of Fig. 1, is given by the relation:

$$\frac{d}{dz}\, W_s(z) = e\, \varepsilon_0(z)\, \cos\psi_s; \tag{4.8}$$

that is, the phase, relative to peak field, at which a particle with $\varrho = 0$ must travel to gain speed at the same rate as the wave[2]. To ψ_s corresponds a synchronous time defined by:

$$\psi_s = \omega\, t_s - \frac{2\pi}{\lambda} \int^z \frac{dz}{\beta_w}. \tag{4.9}$$

t_s is the time at which the synchronous particle arrives at the axial position z.

---

[1] For an investigation of the validity of this approximation, see Ref. [4].

[2] Note that our convention for zero phase differs from that of synchrotron theory by 90°. In linear accelerator literature, one finds both in use.

Then for any other particle, from Eq. (4.1)

$$\frac{d}{dz}(\psi - \psi_s) \equiv \omega\left(\frac{dt}{dz} - \frac{dt_s}{dz}\right) = -\frac{2\pi}{\lambda}\frac{\gamma - \gamma_s}{\gamma_s^3\beta_s^3} \tag{4.10}$$

to first order in the energy difference.

Eq. (4.2) may be written:

$$\frac{d}{dz}(\gamma - \gamma_s) = \frac{e\,\varepsilon_0}{m_0 c^2}[I_0 \cos\psi - \cos\psi_s] \tag{4.11}$$

where the argument of the Bessel function is omitted for the sake of brevity.

Combining (4.10) and (4.11), and changing the independent variable to number, n, of r.f. cycles spent by the synchronous particle in traversing distance, z,

$$dz = v_s\,dt_s = \beta_s\,\lambda\,dn, \tag{4.12}$$

we arrive at the equation governing phase motion:

$$\frac{d}{dn}\left[\gamma_s^3\beta_s^3\frac{d}{dn}(\psi - \psi_s)\right] = -\frac{2\pi e\,\varepsilon_0\lambda}{m_0 c^2}\beta_s[I_0\cos\psi - \cos\psi_s]. \tag{4.13}$$

Inserting (4.5) and (4.6) in (4.3) and using relation (4.12) yields the equation governing radial motion:

$$\frac{\beta}{\beta_s}\frac{d}{dn}\left[\frac{\gamma\beta}{\beta_s}\frac{d\varrho}{dn}\right] = \frac{-e\,\varepsilon_0\lambda^2}{m_0 c^2}\frac{I_1}{\sqrt{1-\beta_w^2}}(1-\beta\beta_w)\sin\psi \tag{4.14}$$

where β, γ refer to the instantaneous values for the particular particle, and again the argument of the Bessel function is omitted.

Usually the argument of the Bessel function is quite small, in which case,

$$I_0(x) \sim 1, \qquad I_1(x) \sim \frac{x}{2}$$

and (4.13) and (4.14) become

$$\frac{d}{dn}\left[\gamma_s^3\beta_s^2\frac{d}{dn}(\psi - \psi_s)\right] = -\frac{2\pi e\,\varepsilon_0\lambda}{m_0 c^2}\beta_s[\cos\psi - \cos\psi_s], \tag{4.13'}$$

$$\frac{d}{dn}\left[\frac{\gamma\beta}{\beta_s}\frac{d\varrho}{dn}\right] = -\frac{\pi e\,\varepsilon_0\lambda}{m_0 c^2\beta}(1-\beta\beta_s)\varrho\sin\psi. \tag{4.14'}$$

The phase motion is seen to be largely independent of the radial motion, but the radial motion depends intimately on the phase.

Consider the phase equation (4.13') for small deviations from synchronous phase. Then

$$\frac{d}{dn}\left[\gamma_s^3\beta_s^2\frac{d}{dn}(\psi - \psi_s)\right] = \frac{2\pi e\,\varepsilon_0\lambda}{m_0 c^2}\beta_s(\psi - \psi_s)\sin\psi_s \tag{4.15}$$

under the condition $|\psi - \psi_s| \ll 1$. If ε_0 and ψ_s are independent of n, (4.15) may be solved explicitly in the non-relativistic and extreme relativistic limits.

Non-relativistically, $\gamma \sim 1$ and, from (4.8) and (4.12), $\beta_s^{(n)} = \frac{e\,\varepsilon_0\lambda}{m_0 c^2}\cos\psi_s\cdot n$. The solution of (4.15) is:

$$\psi - \psi_s = n^{-\frac{1}{2}}\left\{A\,J_1[2\sqrt{-2\pi n\tan\psi_s}] + B\,N_1[2\sqrt{-2\pi n\tan\psi_s}]\right\} \tag{4.16}$$

where J_1 and N_1 are the usual Bessel functions and A and B are arbitrary constants. For large n, (4.16) represents an oscillation with amplitude decreasing as $n^{-\frac{3}{4}}$ (or $\beta_s^{-\frac{3}{4}}$) and instantaneous angular frequency $\sqrt{\frac{-2\pi\tan\psi_s}{n}}$. Note that ψ_s

must be negative for an oscillatory solution; this sign is consistent with the argument of Sect. 3 together with the phase convention adopted in expressions (4.4) to (4.6).

In the extreme relativistic case, $\beta_s \sim 1$ and, from (4.8) and (4.12), $\gamma_s(n) = \frac{e\,\varepsilon_0\,\lambda}{m_0\,c^2} \cos \psi_s \cdot n$. The solution of (4.15) is then:

$$\psi - \psi_s = n^{-1} \left\{ C\, J_2 \left[\frac{2}{\alpha} \sqrt{\frac{-2\pi \tan \psi_s}{n}} \right] + D\, N_2 \left[\frac{2}{\alpha} \sqrt{\frac{-2\pi \tan \psi_s}{n}} \right] \right\} \qquad (4.17)$$

where C and D are arbitrary constants and $\alpha = \frac{e\,\varepsilon_0\,\lambda}{m_0\,c^2} \cos \psi_s$. For large n, $\psi - \psi_s$ approaches a constant value:

$$\psi - \psi_s \to \frac{\alpha^2}{2\pi^2 \tan \psi_s} D$$

in accordance with the argument of Sect. 3.

With regard to the radial motion [Eq. (4.14')] we only note two features here; first, that for ψ negative (stable phase motion) the force is in the direction of increasing ϱ, as predicted in Sect. 3, and second, that its magnitude decreases rapidly as β and β_s approach unity, the focusing effect of the magnetic component of the wave compensating for the defocusing effect of the radial electric component.

III. Electron accelerators.

5. Basic cavity design. In this section we shall treat the problem of designing a simple and efficient system for producing the required traveling wave. Since electrons quickly approach the speed of light, the phase velocity in the main body of the accelerator can be taken equal to light velocity. We shall restrict discussion to this case for the present, postponing the problem of injection and early acceleration until a later section.

Perhaps the most straightforward scheme that comes to mind is to construct a series of short, electrically independent, cavities excited at resonance by admitting energy from a loaded transmission line, or by driving them with separate oscillators properly phased with respect to each other. The latter method has, indeed, been used[1], but the phasing and timing problems are difficult, and would become truly formidable for a long machine. Instead, one tries to combine accelerator and transmission line in a single structure, so reducing the problem of maintaining proper field amplitude and phase velocity to one of mechanical tolerances within the accelerator itself.

It would be convenient to be able to propagate a wave of the form described by expressions (4.4) to (4.6) in a uniform cylindrical wave guide which would also serve as accelerator. However, the boundary condition that E_z be zero at the cylinder wall, applied to expression (4.4), demands that β_w be greater than unity; a fact well-known in micro-wave theory. A more complicated structure is required to propagate waves at the speed of light or less.

A conceptually simple means for decreasing the phase velocity, one which has also been made to work, is to fill the cylinder with a dielectric such as quartz, with a small axial hole for the electrons to pass through. With appropriate choice of dielectric constant and cylinder radius a wave of given frequency can be made to travel at the required velocity[2]. However, at least three major

---

[1] For example, the Yale accelerator, H. L. SCHULTZ and W. G. WADEY: Rev. Sci. Instrum. **22**, 383 (1951).

[2] See Ref. [1], p. 13, for more detail; also G. T. FLESHER and G. I. COHN: Trans. Amer. Inst. Electr. Engrs. **70**, 887 (1951).

problems are raised by this approach, making it seem an unlikely solution, particularly for high energy accelerators, in which high fields must be used to keep the length within bounds:

1. Dielectric breakdown at high fields,

2. Dielectric heating, combined with generally poor heat-conducting properties,

3. Deposition of electrons from the beam on the surface of the hole, resulting in electric fields capable of deflecting the beam.

We are thus led to consider the method which has been adopted for most existing accelerators, and which promises to serve as a standard for future ones. In this scheme conducting disks with circular holes concentric with the axis are inserted at regular intervals along the cylinder (see Fig. 2). The effect of loading the guide in this fashion can be stated in various ways. For one, the progress of the wave down the guide proceeds by reflections and scatterings from the obstructions, resulting in a reduced average velocity. At a given frequency there exist simultaneously many component waves, traveling in both directions with various phase velocities. Only the component with $\beta_w = 1$ influences the electrons, but the energy carried by the other components is in this sense wasted. Alternatively, the guide now consists of a succession of simple cavities, coupled through the holes in the disks; by choice of cavity parameters and coupling strength, each cavity can be made to drive the next at the desired amplitude and phase. More formally, the fact that the metallic boundary is discontinuous at $\varrho = a$ permits one to retain a field component of type (4.4) to (4.6) for $\varrho < a$, adding other components in order to match at $\varrho = a$ the fields trapped in the region $\varrho > a$.

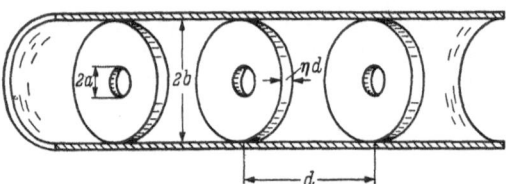

Fig. 2. Structure of disk-loaded accelerator showing important design dimensions [2].

The electrical characteristics of the coupled cavities differ in two important respects from those of the unloaded wave guide. Each TM_{010} mode of the single cavities is split by the coupling into many modes, ranging in order of increasing natural frequency from one in which all cavities are in phase (called the 0-mode in the literature) to one in which adjacent cavities are 180° out of phase (π-mode). Considered as a wave guide, the structure has a cut-off frequency close to that of the unloaded guide, but at a frequency corresponding to the π-mode, reflection from the disks is constructive, standing waves are formed, and no energy flows down the guide. The disk loading thus creates a stop band above the frequency of the π-mode. There exists a choice of possible ways in which the structure may be used: to propagate traveling waves between the 0-mode and π-mode cut-offs or to excite standing waves at one of the resonant frequencies. In the second case, of course, the wave traveling in the wrong direction dissipates electrical power but does not interfere with the acceleration process.

We now present an approximate method, due to Walkinshaw [7], for computing the field distribution in the loaded guide. More accurate methods exist [4], [8][1], but this one is sufficiently realistic to display the inter-relation of the guide parameters. The ultimate precision required, in any case, can only be achieved empirically.

---

[1] See also E. L. Chu and W. W. Hansen: J. Appl. Phys. **20**, 280 (1949).

The general solution of MAXWELL's equations for $\varrho < a$ (Fig. 2) in a periodic structure is:

$$E_z = e^{i(\omega t - k_w z)} \sum_{m=-\infty}^{\infty} A_m J_0(K_m \varrho) e^{-i \frac{2\pi m z}{d}}, \tag{5.1}$$

$$E_\varrho = i e^{i(\omega t - k_w z)} \sum_{m=-\infty}^{\infty} A_m \frac{k_w + \frac{2\pi m}{d}}{K_m} J_1(K_m \varrho) e^{-i \frac{2\pi m z}{d}}, \tag{5.2}$$

$$H_\varphi = i e^{i(\omega t - k_w z)} \sum_{m=-\infty}^{\infty} A_m \frac{k}{K_m} J_1(K_m \varrho) e^{-i \frac{2\pi m z}{d}} \tag{5.3}$$

where $k_w = \dfrac{2\pi}{\beta_w \lambda}$, $k = \dfrac{2\pi}{\lambda}$

$$K_m^2 = k^2 - \left(k_w + \frac{2\pi m}{d}\right)^2$$

and the A_m's are arbitrary constants. The desired accelerating wave is represented by the $m = 0$ term.

To determine A_m we assume a functional form for E_z at $\varrho = a$. The choice is not extremely critical, but should be as realistic as possible. For example, consider, for the n-th cavity,

$$E_z(\varrho = a) = \left\{1 - \frac{2 z_n}{(1 - \eta) d}\right\}^{-\frac{1}{4}} e^{i(\omega t - n k_w d)} \quad \text{for} \quad -\frac{(1 - \eta) d}{2} < z_n < \frac{(1 - \eta) d}{2} \tag{5.4}$$

and zero along the metal edges of the disks. z_n is the distance measured from the center of the n-th cavity. The exponential factor in (5.4) provides the proper phase shift of the structure per period, and the bracketed factor describes a field which is rather flat and becomes infinite as $z^{-\frac{1}{4}}$ at the corner of the disk, as it should.

The Fourier analysis of (5.4) yields:

$$A_m J_0(K_m a) = \frac{\pi}{2} (1 - \eta) J_0(X_m) \tag{5.5}$$

where $X_m = \dfrac{(1 - \eta)}{2} [k_w d + 2\pi m]$ †.

For $b > \varrho > a$, the fields are predominantly:

$$E_z = \frac{\pi}{2} \frac{F_0(k \varrho)}{F_0(k a)} e^{i(\omega t - n k_w d)}, \qquad E_\varrho = 0, \tag{5.6}$$

$$H_\varphi = i \frac{\pi}{2} \frac{F_1(k \varrho)}{F_0(k a)} e^{i(\omega t - n k_w d)}, \tag{5.7}$$

uniform in z, but varying in phase from one cavity to the next. Here

$$F_0(X) = J_0(X) N_0(k b) - N_0(X) J_0(k b),$$
$$F_1(X) = J_1(X) N_0(k b) - N_1(X) J_0(k b),$$

and the numerical coefficient was obtained by equating E_z to the average value of (5.4) at $\varrho = a$.

Finally, H_φ must be matched at $\varrho = a$. Since the z-dependence is different for the approximate solutions in the two regions, we choose to match the average values. Averaging (5.3) in z, inserting A_m from (5.5), and equating the result to (5.7) at $\varrho = a$, we find the relation:

$$\frac{F_1(k a)}{F_0(k a)} = (1 - \eta) \sum_{m=-\infty}^{\infty} \frac{k}{K_m} \frac{J_1(K_m a)}{J_0(K_m a)} J_0(X_m) \frac{\sin X_m}{X_m} \tag{5.8}$$

---

† Note that for $K_m d = \pi$ (π-mode), $X_m = -X_{-m-1}$, so that $A_m = +A_{-m-1}$, whence (5.1) to (5.3) represent a pure standing wave at that frequency, indicating the stop-band previously mentioned.

which provides a functional relationship between the parameters of the system, β_w, λ, a, b, d, and η. Examples of pertinent information derivable from (5.8)

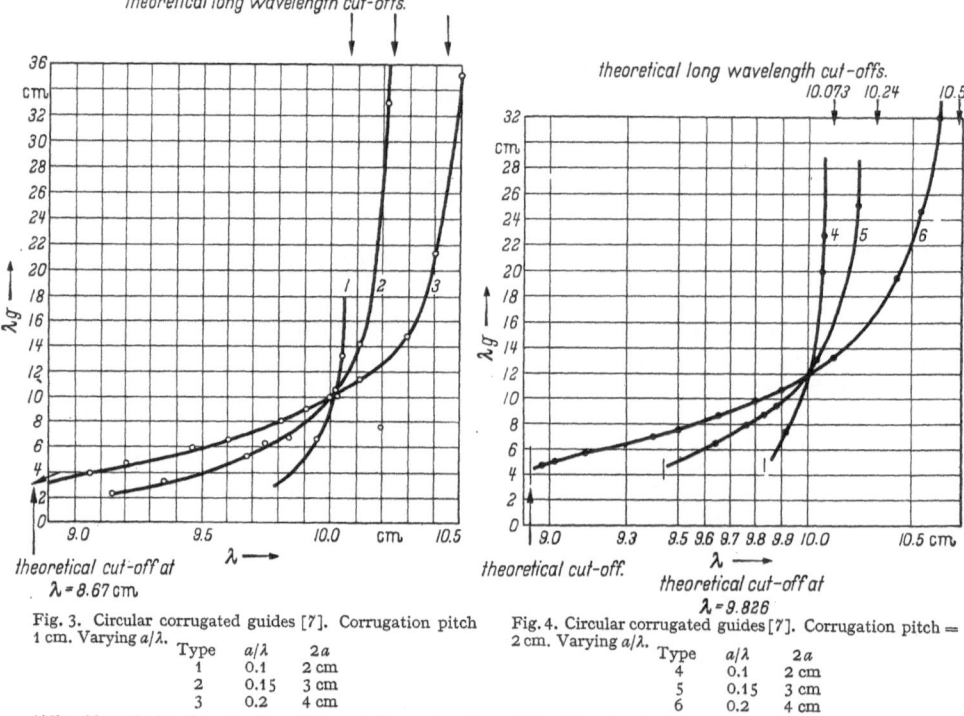

Fig. 3. Circular corrugated guides [7]. Corrugation pitch 1 cm. Varying a/λ.

Type	a/λ	$2a$
1	0.1	2 cm
2	0.15	3 cm
3	0.2	4 cm

(All guides calculated to give $\lambda_g = 10.0$ cm at $\lambda = 10.0$ cm.)

Fig. 4. Circular corrugated guides [7]. Corrugation pitch = 2 cm. Varying a/λ.

Type	a/λ	$2a$
4	0.1	2 cm
5	0.15	3 cm
6	0.2	4 cm

(All guides calculated to give $\lambda_g = 10.0$ cm at $\lambda = 10.0$ cm.)

are shown in Figs. 3 and 4—variation of guide wavelength, $\beta_w \lambda$, with λ for different a and d, with b having been chosen to make $\beta_w = 1$ at $\lambda = 10$ cm in each case. The curves were obtained from model measurements, agreeing well with values computed from (5.8). Fig. 5 shows another important relation, kb as a function of ka for $\beta_w = 1$, $k_w d = \pi/2$, and $\eta_w = 0$. The indicated points were computed from expressions somewhat more precise than (5.8).

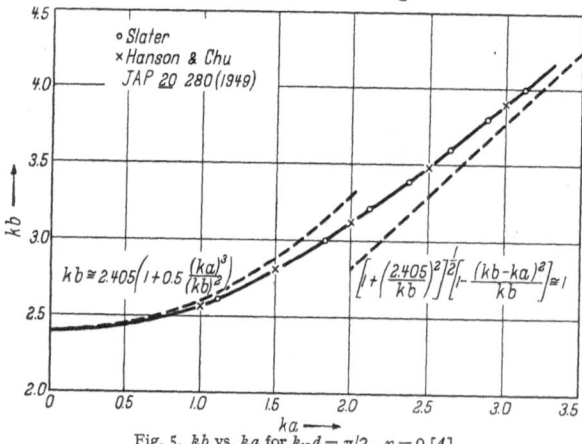

Fig. 5. kb vs. ka for $k_w d = \pi/2$, $\eta = 0$ [4].

6. Choice of cavity parameters. This section will be devoted to a discussion of the factors which determine the selection of λ, a, b, d, η (in the notation of Fig. 2) and mode of operation for the main body of the accelerator, in which $\beta_w = 1$. For this purpose, it is useful to introduce the following derived parameters:

$$Q = \frac{\text{stored energy}}{2\pi \times \text{energy loss per r.f. cycle}}, \tag{6.1}$$

a measure of the stored energy corresponding to a given power dissipation. Effective shunt impedance:

$$R = \frac{\varepsilon_0^2}{\text{power dissipated per unit length}} \dagger,\tag{6.2}$$

a measure of the accelerating field obtainable for given power dissipation. Group velocity:

$$v_g = c\,\frac{dk}{dk_w}\tag{6.3}$$

and attenuation length:

$$l_0 = \frac{v_g\,Q}{\omega},\tag{6.4}$$

the distance in which the energy in a traveling wave is reduced to e^{-1} due to wall losses.

We may first dispose of η, the ratio of disk thickness to disk spacing. As might be anticipated, η is not a critical parameter as long as it is small compared to unity. Its value is determined, therefore, on other grounds: it should be as small as possible, consistent with adequate mechanical rigidity and a radius of curvature at the inner corner large enough to discourage sparking. In existing machines it varies from about 0.10 to 0.25, reflecting variation in intended field strengths and engineering conservatism. It might be noted here that while disk thickness is not critical, it is possible in principle to reduce the amplitudes of the unwanted wave components and thus improve the effective shunt impedance (6.2) by changing the shape of the disks and cavity walls [1]. However, in good designs, only a small fraction of the electromagnetic power is carried in the unwanted waves, so that the gain would not justify the increased complexity of fabrication.

The free space wavelength, λ, which determines the scale of the structure, is chosen primarily on considerations of power and power sources. For similar structures made of the same material, the effective shunt impedance is proportional to $\lambda^{-\frac{1}{2}}$, and the stored energy required to maintain a given field on the axis is proportional to the square of the transverse dimensions, that is, to λ^{-2}. Both of these points favor a choice of as high a frequency as possible. However, availability of r.f. power sources must inevitably dominate the choice. Accordingly, most electron accelerators operate at about $\lambda = 10$ cm, corresponding to a frequency of 3000 Mc/sec, a range in which magnetrons and the recently developed Stanford klystrons [2] can produce the required power. To generate enough power at higher frequencies would be difficult in the present r.f. technology; moreover, as the dimensions of the waveguide are reduced, the space available for the electron beam is decreased and tolerance problems are magnified. Consequently, there appears to be little to gain in attempting to develop accelerators and power sources at higher frequencies.

Next, we consider d, the disk spacing. Two rather different arguments may be invoked to determine d, both of which lead to the same result. It was stated earlier that the succession of coupled cavities, considered as a resonant system, possesses a set of normal modes ranging in frequency from the 0-mode (infinitely many disks per wavelength) to the π-mode (two disks per wavelength). As is usual for such a system, the frequency spacing of the modes is much less at the edges of the band than at the center. If one wants to excite the guide with standing waves, it is essential to work in a frequency range in which the mode spacing is as large as possible, particularly if the number of cavities, and thus

---

† If the other wave components are negligible, then R is the usual shunt impedance.

the number of modes, is large. The favored choice for d is then $\lambda/4$, corresponding to the $\pi/2$ mode[1].

Whether standing or traveling waves are used, there is an optimum value of d from the point of view of power dissipation. The argument is the following:

To the extent that the coupling between cavities is small, that is, $a \ll b$, the fields in the n-th cavity are approximately those of the lowest TM mode of the cavity. In particular, the electric field is

$$E_{zn} = E_0\, J_0(k\,\varrho) \cos\left(\omega\,t - \frac{2\pi\,n\,d}{\lambda} + \varphi_0\right) \tag{6.5}$$

where E_0 is the on-axis field strength, independent of z. The energy gain per wavelength for an on-axis electron is:

$$\Delta W(\varphi_0) = e\,E_0\,\frac{\lambda}{d} \int\limits_0^d \cos\left(2\,\pi\,\frac{z}{\lambda} + \varphi_0\right) d\,z. \tag{6.6}$$

The maximum energy gain with respect to the entering phase φ_0 is:

$$\Delta W = e\,E_0\,\frac{\lambda^2}{\pi\,d} \sin\frac{\pi\,d}{\lambda}. \tag{6.7}$$

Since the fields in each cavity are known, the power dissipation can be easily computed. The power per wavelength has the form:

$$P = K\,E_0^2\left(1 + \frac{j_1}{4}\,\frac{\lambda}{\pi\,d}\right) \tag{6.8}$$

where K is a constant independent of d, and j_1 is the first zero of J_0. The second term in the brackets arises from losses on the disk surfaces.

Then

$$\frac{(\Delta W)^2}{P} = \frac{e^2\,\lambda^2}{K}\left(\frac{\lambda}{\pi\,d}\right)^2 \sin^2\left(\frac{\pi\,d}{\lambda}\right)\left[1 + \frac{j_1}{4}\,\frac{\lambda}{\pi\,d}\right]^{-1} \tag{6.9}$$

which has a maximum at $d \approx \dfrac{\lambda}{3.5}$.

The process of testing and adjusting a large number of cavities is considerably simplified if λ is an integral multiple of d, for then the resonant frequency of short isolated sections is an excellent measure of their condition (see Sect.10). Thus we are again led to the choice $d = \lambda/4$, the loss in energy for given power amounting to only a few percent as compared to (6.9).

The remaining parameters a and b are not independent, for it is still necessary to establish the correct phase velocity by requiring that a and b lie on a curve such as the one shown in Fig. 5. This requirement leaves one free parameter, but before deciding on how best to make the remaining choice we must first settle the question of traveling vs. standing waves.

In the early days of the art, the only power source available at 3000 megacycles was the magnetron. This type of tube functions best as a self-excited oscillator, so that in this application it is tempting to use the accelerator as a stabilizing resonant cavity and work with standing waves in order to maintain the necessary frequency stability. The standing wave method, however, is unattractive in two ways: first, the length of accelerator is severely limited by the decreasing frequency spacing of the resonant modes, and second, the extra

---

[1] The MIT accelerator was designed for the π-mode, but for the express purpose of studying the problem of mode separation, in preparation for a higher energy machine. See DEMOS, KIP and SLATER: J. Appl. Phys. **23**, 53 (1952).

power required by the wave component traveling in the wrong direction is substantially greater than the power which must be thrown away at the end of a well designed traveling wave tube. Since magnetrons with separate stabilizing cavities have been used successfully in the British traveling wave machines for many years, and the Stanford group has succeeded in developing high power klystron amplifiers, there seems to be little argument left for the use of standing waves at 3000 Mc/sec.

Having decided for a traveling wave machine, the concepts of group velocity (6.3) and attenuation length (6.4) become important. Q is rather insensitive to changes in a and b, so that by (6.4), l_0 and v_g are almost proportional. These two quantities, on the other hand, are quite sensitive to changes in a and b, as may be seen from the typical curves of Figs. 3 and 4, in which the group velocity appears as the reciprocal of the slope at $\beta_w = 1$. If a section of accelerator of length l is fed by a source capable of supplying an instantaneous power P_i there arises the choice of making the attenuation length large, so that the accelerating field does not decrease much in the distance l or of making v_g small, so that ε_0 will be large at the entering end[1]. There is a definite optimum for l_0, which may be seen as follows:

The energy gained by an electron riding the crest of the wave for distance l is:

$$\Delta W = \int_0^l \varepsilon_{0i} e^{-\frac{z}{2 l_0}} dz = 2 l_0 \varepsilon_{0i} \left[1 - e^{-\frac{l}{2 l_0}} \right] \tag{6.10}$$

where ε_{0i} is the amplitude of the accelerating field at the entering end. Also, by (6.2)

$$\varepsilon_{0i}^2 = R \left| \frac{dP_i}{dz} \right| = R \frac{P_i}{l_0} \tag{6.11}$$

where P_i is the power entering the accelerator. Then

$$\frac{(\Delta W)^2}{P_i} = 4 l_0 R \left(1 - e^{-\frac{l}{2 l_0}} \right)^2 \tag{6.12}$$

which has a maximum at

$$l_0 \approx 0.4 l \tag{6.13}$$

since R, like Q, is much less sensitive to changes in a and b than l_0.

If the relation (6.13) is imposed on a length, l, of accelerator fed by one power source, then a parameter which is almost independent of a and b is the filling time, t_F; that is, the time required for the incoming energy to fill the length of accelerator:

$$t = \frac{l}{v_g} = 2.5 \, Q/\omega . \tag{6.14}$$

This time is of the order of a few micro-seconds.

Now l is a parameter which is generally determined by factors extraneous to the present discussion, such as space requirements, physical size and disposition of power amplifiers, building costs, etc. Thus (6.13) largely serves to determine l_0, and thereby a and b.[2]

Finally, it is necessary to decide what to do with the wave after it has reached the end of its section of accelerator. In a low energy machine, with only one

---

[1] Since $P_i = v_g \times$ energy density $\sim v_g \varepsilon^2$.

[2] It is worth noting that if the optimum l_0 is inserted in (6.12), that expression is seen to increase indefinitely with l. Thus the energy obtainable for given power can be increased to the point where length-dependent costs are comparable to power-dependent costs.

feed point, it is usually absorbed in a lossy termination[1]. In a high energy machine, with many electrically separate sections and a need for simplicity of construction, it may be allowed to reflect at the end of its section, returning to the point of origin as the acceleration period ends.

7. Pre-accelerating and bunching. We have seen in Sects. 3 and 4 that the electrons first undergo an oscillatory motion which, as they near the speed of light, degenerates into an asymptotic approach to a certain phase, determined by the initial conditions. Thereafter, the electron's energy is given simply by

$$W(z) = W_a + \cos\psi_a \int^z e\, \varepsilon_0(z)\, dz \qquad (7.1)$$

where ψ_a is the asymptotic phase, and W_a the energy when the electron effectively reaches ψ_a.

The expression (7.1) is valid after the electrons have reached a few Mev, so that W_a does not represent an important contribution to the final energy in a long machine. On the other hand the asymptotic phase is of prime importance, for it determines the final energy of the individual electron. Since there is no need for phase stability, and a great need for high final energy with small energy spread, it is almost essential to devise a method of injection whereby as many electrons as possible have asymptotic phases within a few degrees of peak field.

The electrons are supplied continuously during the acceleration period by a conventional gun at an energy usually of the order of tens of kilovolts. The problem is then to convert this almost monoenergetic beam, uniformly distributed in time, into a beam moving almost at the speed of light and strongly bunched in time.

Before introducing a number of complications to bring this about, it is of interest to examine the possibility of injecting directly into the accelerator proper, in which β_w is unity, but the field strength is not necessarily equal to that in the rest of the machine. The equation of motion derived in Sect. 4 is not applicable in this case because the electron velocities differ greatly from the wave velocity and because the effects of the other component waves are not necessarily small if the electron energy varies substantially during an r.f. period. However, the general character of the motion can be ascertained by dealing only with the main wave component, neglecting the attenuation in amplitude. Then the equations of motion,

$$\psi = \omega t - k z = \frac{2\pi}{\lambda}(c t - z) \qquad (7.2)$$

$$\frac{dP}{dt} = e\, \varepsilon_0 \cos\psi \qquad (7.3)$$

$$\frac{dz}{dt} = c - \frac{\lambda}{2\pi}\frac{d\psi}{dt} = c P \left[P^2 + m_0^2 c^2\right]^{-\frac{1}{2}} \qquad (7.4)$$

are derivable from the time-independent Hamiltonian,

$$\mathscr{H} = \frac{2\pi}{\lambda} m_0 c^2 \left[\frac{P}{m_0 c} - \left(1 + \frac{P^2}{m_0^2 c^2}\right)^{\frac{1}{2}}\right] - e\, \varepsilon_0 \sin\psi \qquad (7.5)$$

in which P and ψ appear as canonical variables.

---

[1] Or in special cases, recirculated. See HARVIE and MULLETT: Proc. Phys. Soc. Lond. B **62**, 270 (1949).

All the pertinent information can be obtained from (7.5)[1]. Since $\mathscr{H}$ is a constant of the motion, the asymptotic phase may be found by substituting $P \to \infty$, $\psi = \psi_a$ and equating the result to $\mathscr{H}$ expressed in terms of the initial values P_i and ψ_i

$$\sin \psi_a - \sin \psi_i = \frac{2\pi m_0 c^2}{e \varepsilon_0 \lambda}\left[\left(1 + \frac{P_i^2}{m_0^2 c^2}\right)^{\frac{1}{2}} - \frac{P_i}{m_0 c}\right] = \frac{2\pi m_0 c^2}{e \varepsilon_0 \lambda}\left[\frac{1 - \beta_i}{1 + \beta_i}\right]^{\frac{1}{2}} \qquad (7.6)$$

where $-\frac{\pi}{2} < \psi_a < \frac{\pi}{2}$ in our notation, and β_i is injection velocity divided by light velocity.

The condition that electrons are bound to the wave is therefore:

$$\frac{e \varepsilon_0 \lambda}{\pi m_0 c^2}\sqrt{\frac{1 + \beta_i}{1 - \beta_i}} > 1 \qquad (7.7)$$

so that even for $\beta_i = 0$, the traveling wave w ll pull electrons into step if the amplitude is great enough. It is also encouraging to observe from (7.6) that if ε_0 or β_i is sufficiently large, all initial phases ψ_i lead to bound motion. However, the most disagreeable feature of this type of injection is also apparent from the same Eq. (7.6); to the extent that ε_0 or β_i can be made large enough to trap a large fraction of the electrons, the approximate relation appears:

$$\sin \psi_a \approx \sin \psi_i. \qquad (7.8)$$

Then

$$\psi_a = \psi_i \qquad \text{for} \quad -\frac{\pi}{2} < \psi_i < \frac{\pi}{2} \qquad \text{(initially accelerating phase)},$$

$$\psi_a = \psi_i - \pi \quad \text{for} \quad -\frac{3\pi}{2} < \psi_i < -\frac{\pi}{2} \quad \text{(initially decelerating phase)},$$

so that an incoming beam uniformly distributed in ψ_i becomes uniformly distributed in accelerating phases, ψ_a, resulting in a final beam composed of electrons of all energies, according to (7.1). A more precise evaluation of (7.6), as shown in Fig. 6, leads to a somewhat more optimistic result, in that about half of the trapped electrons would be within a few percent of peak energy for the optimum value of the parameter, α.

The disadvantage of a large energy spread can not be overstressed, for even though the injected current may be great enough to permit energy selection at the output end, the undesired electrons produce a background which is very difficult to eliminate. A straightforward method to increase the fraction of beam reaching peak energy is to interpose, between the electron gun and the accelerator, a bunching gap and drift space of more or less conventional klystron type. In this way the electrons can be prebunched into phases in the neighborhood of $-\frac{\pi}{2}$ at the beginning of the accelerator (see Fig. 6), with only a small spread in α corresponding to the spread in velocities introduced by the bunching gap.

However, a single gap with field sinusoidal in time will not bunch much than half of the electrons into a narrow phase packet, so that the gain over the indicated optimum in Fig. 6 would not be very great. One is then led to consider a system of many cavities, coupled to maintain a succession of amplitudes and phases which will give a better bunching performance. That this is possible is suggested by the equation of motion derived earlier:

$$\frac{d}{dn}\left[\gamma_s^3 \beta_s^2 \frac{d}{dn}(\psi - \psi_s)\right] = \frac{2\pi e \varepsilon_0 \lambda}{m_0 c^2}\beta_s(\psi - \psi_s)\sin \psi_s \qquad (4.15)$$

---

[1] See TERRALL and SLATER: J. Appl. Phys. **23**, 66 (1952), and also Ref. [5].

which, if γ_s, β_s, ε_0, and ψ_s vary slowly with distance, has the approximate solution

$$\psi - \psi_s = A \left[- \gamma_s^3 \beta_s^3 \varepsilon_0 \sin \psi_s \right]^{-\frac{1}{4}} \sin \left[\int \omega_\varphi \, dn + \alpha \right] \tag{7.9}$$

where

$$\omega_\varphi^2 = \frac{- 2\pi e \, \varepsilon_0 \lambda \sin \psi_s}{\gamma_s^3 \beta_s m_0 c^2} \tag{7.10}$$

and A and α are constants of integration.

The implication of (7.9) is that if the electrons are injected into a guide in which a wave is traveling, initially at the speed of the injected electrons, but gradually increasing in speed and amplitude, then the oscillations of the electrons

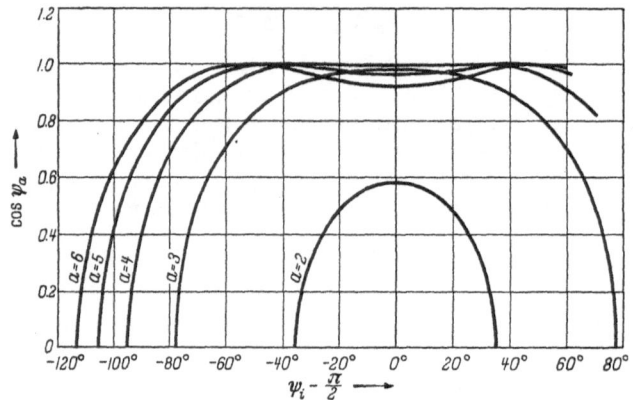

Fig. 6. cos ψ_a vs. ψ_i for $\beta_i = 0.5$, $\beta_w = 1$ and various values of $\alpha = \dfrac{e \, \varepsilon_0 \lambda}{m_0 c^2}$ [2].

in phase are damped in proportion to the change in the bracketed factor. While (7.9) is only valid for small deviations from synchronous phase, numerical solutions of the non-linear equation (4.13) indicate that large initial oscillation amplitudes are damped about as rapidly as small ones.

As an example of what one might expect from a bunching section of wave guide, suppose that the field amplitude, ε_{0i}, at the beginning of the buncher is one tenth of the amplitude, ε_{0f}, in the accelerator proper, and that the initial wave velocity, β_{wi}, and the initial electron velocity, β_i are both $\frac{1}{2}$ (80 kev electrons). Then, if the wave velocity is held constant for some distance, the initial value of ψ_s is $-\dfrac{\pi}{2}$ [Eq. (4.8)] and all initial phases lead to stable oscillations about ψ_s (see Fig. 1). Now let β_w and ε_0 begin to increase with distance, but in such a related way that $\psi_s(z)$ becomes, say, $\pi/3$, and γ_s becomes, say, 7 (3 Mev electrons) when ε_0 has increased to ε_{0f} and β_w almost to unity. At that point, the entire 360° initial phase spread will be compressed, according to (7.9), into a phase width

$$\Delta \psi_f = 360° \, (7)^{-\frac{3}{4}} \left(\frac{1}{2} \right)^{+\frac{3}{4}} (10)^{-\frac{1}{4}} \left(\sin \frac{\pi}{3} \right)^{-\frac{1}{4}} \sim 30° \tag{7.11}$$

centered about $\psi_s = -\dfrac{\pi}{3}$. The speed of the electrons is now essentially constant, so that the whole bunch may be shifted 60° in phase to the position $-15° < \psi_a < 15°$ by letting β_w exceed unity for a short distance, until the wave crest has caught up to the electron bunch. This state of affairs would then be most satisfactory, for all injected electrons would be caught and accelerated, with a final fractional energy spread, from (7.1), of $1 - \cos 15° = 0.034$ (provided that the final energy is great enough that the energy spread introduced by the buncher may be neglected).

If the buncher is confined to a moderate length, the parameters must change so rapidly with distance that (4.15) and (7.9) are of doubtful validity. The procedure which has been used in designing a buncher is to select a pattern of variation of ε_0 and β_w, guided by (4.15) as in the example of the preceding paragraph, but then to integrate numerically the exact equations of motion (4.1) and (4.2) to determine the properties of the device more precisely. Figs. 7 to 9 describe the buncher designed by the Stanford group for the 1 Bev accelerator[1]. The variation of ε_0 and β_w with distance have the general character of our example; Fig. 9 shows the result of numerical computations of the phase motion, however neglecting

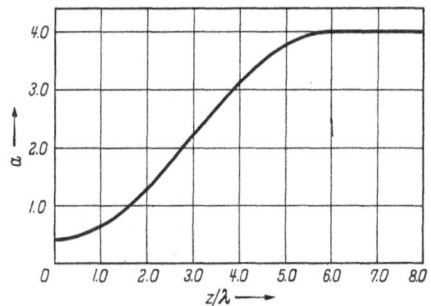

Fig. 7. $\alpha = \dfrac{e\,\varepsilon_0\,\lambda}{m_0 c^2}$ vs. z for the Stanford buncher [2].

all component waves but the desired one. Eqs. (4.15) and (7.9) appear to be in error in that not quite the full 360° of initial phase range is trapped and accelerated; however, the fraction trapped is bunched to about 20°, in reasonable agreement with (7.11).

The problem of determining the proper waveguide parameters for the buncher is somewhat more difficult than for the main accelerator, for the parameters

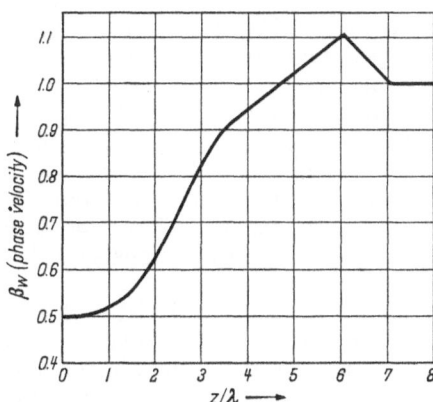

Fig. 8. β_w vs. z for the Stanford buncher [2].

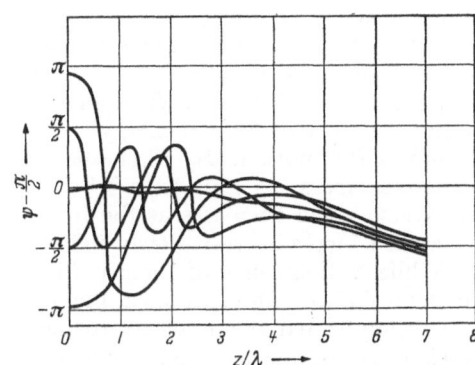

Fig. 9. Electron trajectories in the Stanford buncher [2].

must vary with distance to produce the desired variation of ε_0 and β_w. Details of the procedure are given in Refs. [5] and [7]; here we shall only mention that bunching sections so designed appear to behave according to expectations.

8. Transverse motion. Thus far there has been no mention of the second aspect of accelerator design mentioned at the end of Sect. 3, namely, the need for confinement of the beam in the transverse plane. The reason is that, while the question is important in principle, it is easily solved in practice for electron accelerators, and has no major influence on design.

---

[1] See also Ref. [7] for other designs for the non-relativistic range.

The differential equation for radial motion, derived in Sect. 4, is:

$$\frac{d}{dn}\left[\frac{\gamma\,\beta}{\beta_s}\frac{d\varrho}{dn}\right] = -\frac{\pi\,e\,\varepsilon_0\,\lambda}{m_0\,c^2\,\beta}\,(1 - \beta\,\beta_s)\,\varrho\,\sin\psi. \tag{4.14'}$$

In the main portion of an accelerator, where $\beta = \beta_s = \beta_w \sim 1$, and $\psi \sim 0$, the radial force is negligible and the equation of motion, in the absence of external focusing forces, is:

$$\frac{d}{dn}\left(\gamma\,\frac{d\varrho}{dn}\right) = 0 \tag{8.1}$$

which has the solution

$$\varrho = \varrho_i + W_i\frac{d\varrho_i}{dz}\int_{z_i}^{z}\frac{dz}{W(z)} \tag{8.2}$$

where ϱ_i is the displacement, $d\varrho_i/dz$ the angular deviation, and W_i the total energy at the initial distance z_i. If the acceleration is uniform,

$$W = \frac{W_f}{L}\,z$$

and

$$\varrho_f - \varrho_i = \frac{W_i}{W_f}\,L\,\frac{d\varrho_i}{dz}\,\log\frac{W_f}{W_i} \tag{8.3}$$

where W_f is the final energy and L the length of the accelerator. Thus, for example, if $\varrho_f - \varrho_i$ is to be less than 0.5 cm in a 1 Bev accelerator 200 feet long, then at the 5 Mev point, (8.3) demands:

$$\frac{d\varrho_i}{dz} < 3.2\times10^{-3}\ \text{radians}. \tag{8.4}$$

An angular divergence of the order of milliradians at 5 Mev is not an excessive requirement for an injection system, so that the possibility exists of doing without transverse focusing altogether for most of the distance.

It is important to note that (8.3), for a given rate of energy gain W_f/L, depends only logarithmically on the final energy, so that the tolerable angular divergence is quite insensitive to final energy. This fact results from the extreme relativistic nature of the motion. By (8.1), the transverse momentum is constant, but the transverse mass, as represented by γ, increases with distance. Therefore the transverse speed decreases as the total energy increases, and the transverse displacement increases very slowly indeed.

While a divergence of some milliradians might be easily achieved with d.c. acceleration to 5 Mev, as in a Van de Graaff generator, the situation is much more complicated if one elects to terminate d.c. acceleration somewhere below 100 kev and obtain longitudal bunching by one of the methods described in the preceding section. In the non-relativistic range the transverse force (4.14') is far from negligible; furthermore the force depends strongly on electron energy and phase so that a direct integration of (4.14') is impossible[1]. It is easy to see that some focusing force must be superimposed by a crude estimate based on (4.14'). For example, at the entrance end of the buncher described in Figs. 7 and 8, with $\beta_i = \beta_s = \beta_w = \frac{1}{2}$, $\alpha = 0.4$, and $\psi = \pi/2$, (4.14') becomes:

$$\frac{d^2\varrho}{dn^2} \sim 2\varrho \tag{8.5}$$

so that an initially divergenceless beam would double its size in a fraction of an r.f. cycle or a distance of a few centimeters.

---

[1] See Ref. [4] for extensive numerical calculations of some special cases.

The simplest method of providing focusing, and the method which is invariably used, is to introduce a time-independent axial magnetic field by means of coils wrapped around the accelerating tube. In this case, the radial equation of motion is:

$$\frac{d}{dt}(\gamma \dot{\varrho}) = \frac{e}{m_0}\left[E_\varrho - \frac{\dot{z}}{c}H_\varphi\right] + \frac{e}{m_0}\frac{\varrho \dot{\varphi}}{c}H_z + \gamma \varrho \dot{\varphi}^2 \tag{8.6}$$

in which the first two terms arise from the r.f. fields, as in (4.3), and the third from the presence of the coils. The centrifugal force term $\gamma \varrho \dot{\varphi}^2$ is now essential because of the rotatory effect of the axial magnetic field.

(8.6) is supplemented by the condition that the canonical angular momentum, P_φ is a constant of the motion, since the forces are independent of azimuth:

$$P_\varphi = m_0 \gamma \varrho^2 \dot{\varphi} + \frac{e}{c}\varrho A_\varphi = \frac{e}{c}\varrho_K A_{\varphi K} \tag{8.7}$$

where A_φ is the vector potential corresponding to H_z and the subscript K refers to values at the cathode surface of the electron gun.

If H_z varies slowly with axial distance, then

$$A_\varphi \approx \frac{\varrho}{2}H_z \tag{8.8}$$

and (8.6) becomes

$$\frac{d}{dt}(\gamma \dot{\varrho}) = \frac{e}{m_0}\left(E_\varrho - \frac{\dot{z}}{c}H_\varphi\right) - \frac{\varrho}{\gamma}\left(\frac{e H_z}{2 m_0 c}\right)^2 + \frac{1}{\gamma \varrho^3}\left(\frac{e \varrho_K A_{\varphi K}}{m_0 c}\right)^2 \tag{8.9}$$

or in the notation of (4.14') neglecting all but the principal traveling wave component,

$$\frac{d}{dn}\left[\gamma \frac{\beta}{\beta_s}\frac{d\varrho}{dn}\right] = -\frac{\pi e \varepsilon_0 \lambda}{m_0 c^2 \beta}\varrho(1 - \beta \beta_s)\sin \psi - \frac{\varrho}{\gamma}\frac{\beta_s}{\beta}\left(\frac{e H_z \lambda}{2 m_0 c^2}\right)^2 + \frac{1}{\gamma \varrho^3}\left(\frac{e \varrho_K A_{\varphi K}\lambda}{m_0 c^2}\right)^2 \tag{8.10}$$

The term in H_z^2 is the usual solenoidal focusing term[1], but it is also important to recognize the existence of the term in $A_{\varphi K}^2$, which arises if the electron gun is in a magnetic field. Since the effect of the latter is to decrease the focusing strength, one arrives at the non-obvious conclusion that some effort should be made to shield the gun magnetically from the focusing coils. Such shielding is also necessary if the coils are to terminate at the output end of the buncher; otherwise, according to (8.7), the electrons will suffer a deflection when A_φ drops to zero comparable in magnitude to that which the magnetic field had suppressed up to that point.

The magnitude of H_z required to confine the beam can be estimated by equating the right-hand side of (8.10) to zero for some representative values of ψ and β, say $\beta = \beta_s$ and $\psi = -\frac{\pi}{2}$. Such a procedure by no means guarantees a beam which will satisfy a criterion such as (8.4), because of the extreme variations in ψ and β for each electron during the bunching process. In practice, the focusing coils consist of several separate coils which are adjusted empirically to optimize the beam at the target. In low energy machines, the coils usually extend all the way to the target, for the required fields are quite moderate.

The fact that the transverse motion is force-free for most of the distance means that the beam is sensitive to effects which otherwise would be relatively unimportant. For example, if the accelerating tube is in a uniform transverse

---

[1] See, for example, ZWORYKIN et al.: Electron Optics and the Electron Microscope, Chap. 15. New York and London: John Wiley & Sons, Inc. 1945.

magnetic field, H_0, (8.1) becomes for $\gamma \gg 1$:

$$\frac{d}{dn}\left[\gamma \frac{d\varrho}{dn}\right] = \frac{e\,H_0}{m_0\,c^2} \tag{8.11}$$

which has the solution:

$$\varrho - \varrho_i = \frac{e\,H_0\,L^2}{W_f}\left[1 - \frac{W_i}{W_f}\left(1 + \log\frac{W_f}{W_i}\right)\right] \tag{8.12}$$

in the notation of (8.3), with $\frac{d\varrho_i}{dz} = 0$. For the same parameters as in (8.4), we find the requirement,

$$H_0 < 0.05 \text{ gauss} \tag{8.13}$$

so that in a long accelerator, even the earth's magnetic field must be suppressed by an order of magnitude with the aid of magnetic shielding and compensating coils.

Also, the alignment of the tube is critical. (8.3) may be used to illustrate this point, for a parallel beam passing a sudden bend through angle ϑ in the tube will acquire an angle

$$\frac{d\varrho_i}{dz} = -\tan\vartheta \sim -\vartheta \tag{8.14}$$

relative to the subsequent direction of the accelerating field. (8.4) then may be interpreted as the tolerance on directional alignment of the tube at the 5 Mev point. It is interesting to note that (8.3) has a maximum at $\frac{W_i}{W_f} = \frac{1}{e}$, where the tolerance is some 17 times more stringent.

An effective means of combating such errors is to introduce magnetic steering coils and/or focusing magnets at intervals along the tube. Using proper care in original alignment and de-gaussing permits the use of relatively few extra coils and magnets.

9. The operating cycle. To summarize the discussion of the preceding sections, the idealized electron accelerator consists of an electron gun, a bunching section, and a succession of identical sections of loaded wave guide, together with appropriate coils and magnets for transverse focusing. At the beginning of each section of accelerator a wave guide is coupled into the tube to transfer power from the r.f. generator for that section. A cycle of operation proceeds as follows:

1. R.f. power is turned on. If power amplifiers are used, they are all driven from a master oscillator, which determines the frequency, through a distributing wave guide system which includes elements to determine the relative phases and drive power of each. If magnetrons were to be used in an accelerator consisting of several sections, a more complex frequency and phase control system would be necessary.

2. After a time of the order of a microsecond, the wave front has reached the end of its section, and the proper field distribution exists throughout the machine. Then the electron gun is pulsed on and the acceleration period begins.

3. The acceleration period lasts for about a microsecond, at which time the gun is shut off, before the accelerating field has to be turned off. This time is determined by the limitations of the tubes and energy storage in their power supplies. In the case of machines in which the wave is allowed to reflect at the end of its section, it is also expedient to end the cycle before the returning wave can affect the impedance seen by the fed wave guide.

The focusing coils, requiring low power, may run continuously.

10. Tolerance requirements. If there are imperfections in the wave guide and in the r.f. driving system, the amplitude and phase velocity of the wave will deviate from the design values within each section, and sudden changes in phase and amplitude may occur in passing from one section to the next.

Since the ultimate purpose is to produce a beam of high energy and small energy spread, the tolerances placed on these deviations will refer to the maximum permissible deviation from peak energy and the maximum permissible energy spread.

The energy of an electron is now given by:

$$W(z) = e \int_{z_i}^{z} \varepsilon_0(z) \cos \psi(z) \, dz + W_i \tag{10.1}$$

rather than by the simpler expression (7.1). Within each section the phase of the electron, with respect to the crest of the wave, varies according to:

$$\frac{d\psi}{dz} = \frac{2\pi}{\lambda} \left[\frac{1}{\beta(z)} - \frac{1}{\beta_w(z)} \right]. \tag{10.2}$$

It might be well at this point to estimate the phase slip in the perfect machine arising from the fact that the electrons travel slightly slower than the nominal phase velocity, $\beta_w = 1$. Using $\frac{1}{\beta} \sim 1 + \frac{1}{2} \left(\frac{m_0 c^2}{W} \right)^2$, and assuming a uniform rate of increase of energy, the integral of (10.2) is:

$$\psi_f - \psi_i = \pi \frac{L}{\lambda} \frac{(m_0 c^2)^2}{W_i W_f} \left(1 - \frac{W_i}{W_f} \right) \tag{10.3}$$

where L is the accelerator length and W_i and W_f the initial and final energies. For a 1 Bev accelerator 200 feet long, the phase slip from the 5-Mev point on is about $5°$, resulting in a negligible loss in final energy for electrons riding near the crest of the wave. Therefore, in considering effects of errors in a long accelerator, it is legitimate to set β equal to unity.

We write

$$\varepsilon_0(z) = \varepsilon_{00}(z) \left[1 - h(z) \right] \tag{10.4}$$

where $\varepsilon_{00}(z)$ is the design value and $h(z)$ the fractional deviation of $\varepsilon_0(z)$. With

$$\cos \psi(z) = 1 - \tfrac{1}{2} \psi^2(z) \tag{10.5}$$

for small deviations in phase from the peak, the loss δW in energy gain is approximately,

$$\delta W = e \int_{z_i}^{z_f} \varepsilon_{00}(z) \left[h(z) + \tfrac{1}{2} \psi^2(z) \right] dz. \tag{10.6}$$

Similarly, (10.2) is

$$\frac{d\psi}{dz} = \frac{2\pi}{\lambda} \delta \beta_w \tag{10.7}$$

for small errors, $\delta \beta_w$, in phase velocity at $\beta_w = 1$.

Let us consider first an error in the applied frequency. Since, from (5.1) and following,

$$\beta_w = \frac{k}{k_w} \sim 1 \tag{10.8}$$

then

$$\delta \beta_w = \frac{\delta k}{k} \left[1 - \frac{dk_w}{dk} \right] = \frac{\delta \nu}{\nu} \left[1 - \frac{c}{v_g} \right] \tag{10.9}$$

where ν is the frequency and v_g the group velocity. From (10.7),

$$\psi(z) = \frac{2\pi}{\lambda} \left(1 - \frac{c}{v_g} \right) \frac{\delta \nu}{\nu} z. \tag{10.10}$$

The integration of (10.6) with $h=0$, is straightforward, even including the attenuation of $\varepsilon_{00}(z)$. For small attenuation, the fractional loss of energy gain in a section of length L is

$$\frac{\delta W}{W} = \frac{(2\pi)^2}{6} \left(\frac{c}{v_g} - 1\right)^2 \left(\frac{L}{\lambda}\right)^2 \left(\frac{\delta v}{v}\right)^2 . \tag{10.11}$$

Now v_g was determined in Sect. 6 [Eq. (6.13) and following] on the basis of optimizing energy gain for given power. Unfortunately, v_g must be quite low for that purpose—of the order of $c/100$—with the result that β_w is very sensitive to errors such as the one under immediate consideration (10.9). Even for short stretches of accelerator, (10.11) demands frequency stability to a few parts in 10^5.

(10.11) may also be interpreted in terms of temperature variations if the entire structure expands uniformly. A change in temperature ΔT gives rise to a frequency shift:

$$-\frac{\delta v}{v} = +\frac{\delta\lambda}{\lambda} = g\,\Delta T \tag{10.12}$$

where g is the coefficient of expansion of the waveguide material (1.6×10^{-5} per °C for copper).

Probably the most important contribution to poor performance comes from random and systematic errors in mechanical construction of the accelerating tube. If the traveling wave encounters a cavity with improper dimensions, part of the energy will be reflected, which in itself does not disturb the accelerator process, but the transmitted wave has its amplitude reduced and its phase shifted. For any single cavity the effect is small, but the total number of cavities is large and the errors are cumulative.

The effect on the electrons can be described in terms of the phase shifts, ϑ_n, in the various cavities and the associated losses, $\delta\varepsilon_{0n}$, in transmitted amplitude. In analogy with the behavior of voltage waves encountering a small shunt susceptance on a transmission line, or electromagnetic waves traversing a thin slab of slightly different dielectric constant, the loss in amplitude is related to the phase shift approximately as:

$$\delta\varepsilon_{0n} = \frac{\vartheta_n^2}{2}\,\varepsilon_{0n} \tag{10.13}$$

where ε_{0n} is the amplitude of the wave incident at the n-th cavity.

The point of emphasizing ϑ_n is that it is a quantity which can be measured and possibly corrected, for

$$\vartheta_n = d\,\delta_n(k_w), \tag{10.14}$$

since $k_w d$ is the phase advance in a single cavity. The error in k_w can be determined by measuring the change in frequency required to restore the proper guide wavelength; that is, by blocking off two adjacent cavities from the rest[1] and measuring their resonant frequency. At resonance,

$$\delta_n(k_w) + \frac{dk_w}{dk}\,\delta_n(k) = \delta_n(k_w) + \frac{2\pi}{v_g}\,\delta_n(v) = 0$$

so that

$$\vartheta_n = -\frac{2\pi d}{v_g}\,\delta_n(v) = -\frac{\pi}{2}\,\frac{c}{v_g}\,\frac{\delta_n(v)}{v}. \tag{10.15}$$

Here $\delta_n(v)$ is the deviation of the resonant frequency from the nominal value. Therefore the deviation in resonant frequency of a properly terminated section gives directly the net phase error, ϑ_n, caused by the mechanical errors in that

---

[1] Two adjacent cavities terminated by conducting walls at points of symmetry is the smallest unit which can resonate in the $\pi/2$-mode.

section. Moreover, the sensitivity of phase shift to various types of mechanical errors can be established by the same method of measuring the resonant frequencies of test cavities in which the guide parameters are intentionally varied.

In the approximation that each imperfect cavity causes a sudden change, the amplitude and phase vary in steps with z. After m cavities, the amplitude change is

$$h(m\,d) = \tfrac{1}{2}\sum_{n=1}^{m}\vartheta_n^2 \tag{10.16}$$

and the phase change

$$\psi(m\,d) = \sum_{n=1}^{m}\vartheta_n. \tag{10.17}$$

After N cavities, the error in energy gain (10.6) is, neglecting the normal attenuation:

$$\frac{\delta W}{W} = \frac{1}{2N}\sum_{m=1}^{N}\left[\sum_{n=1}^{m}\vartheta_n^2 + \left(\sum_{n=1}^{m}\vartheta_n\right)^2\right]. \tag{10.18}$$

If the phase errors are random, with mean square value $\overline{\vartheta^2}$,

$$\sum_{n=1}^{m}\vartheta_n^2 = \left(\sum_{n=1}^{m}\vartheta_n\right)^2 = m\,\overline{\vartheta^2} \tag{10.19}$$

so that the phase and amplitude errors contribute equally. Then[1]

$$\frac{\delta W}{W} \approx \frac{1}{2}N\,\overline{\vartheta^2}. \tag{10.20}$$

Eq. (10.20) applies to an accelerator section fed by one power source. If there are many such sections, the fractional energy error will be the same as for a single section, for the relative phases of successive power sources may be adjusted so that the electrons enter each section at the correct phase. In fact the phase shift contribution is probably reduced in practice by about a factor of 2 by having the electrons enter each section at a phase ψ_0

$$\psi_0 \equiv -\tfrac{1}{2}\sum_{n=1}^{N}\vartheta_n$$

so that they deviate from peak by the least possible amount in each section.

As an indication of the tolerances required by the foregoing arguments, let us take a machine built in 3 m long sections at $\lambda = 10$ cm. Then, at 4 cavities per wavelength, ($\pi/2$-mode), $N = 120$. If the energy loss is to be less than 1%, (10.20) gives:

$$\vartheta_{\text{rms}} < 0.013 \text{ radians} \tag{10.21}$$

and thus, for $v_g = 0.01\,c$, from (10.15)

$$\left(\frac{\delta v}{v}\right)_{\text{rms}} < 0.9 \times 10^{-4} \tag{10.22}$$

where the subscript denotes root mean square values. Some of the guide dimensions are more critical than others, but in order of magnitude the resonant frequency of a guide is inversely proportional to its size. Thus,

$$\frac{\delta q}{q} \sim \frac{\delta v}{v}$$

where q is any of the guide parameters. Since the cavity dimensions are of the order of inches, the mechanical precision required by (10.22) is of the order of a few ten-thousandths of an inch.

---

[1] Including the normal attenuation in (10.18) decreases (10.20) to some extent. See Ref. [5].

11. Details of construction. The electrical and mechanical techniques which various groups have developed to produce operating machines are described rather fully in the literature (see principally Refs. [1], [2], [6], and the specific

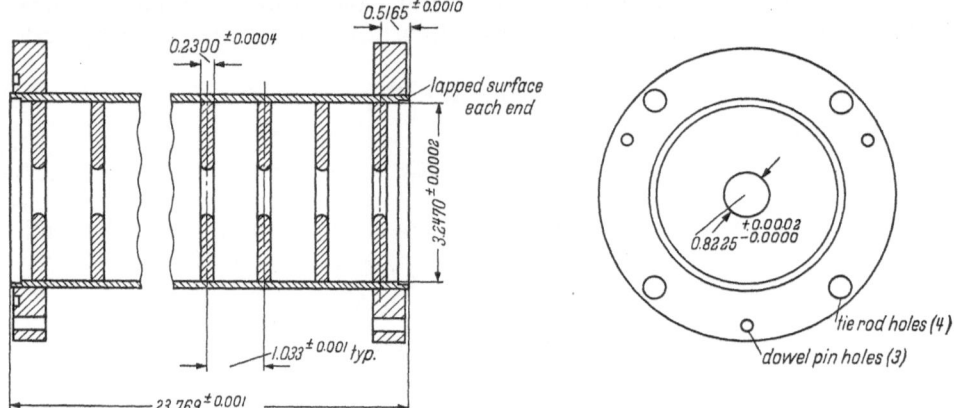

Fig. 10. Schematic diagram of Stanford accelerator subsection [2].

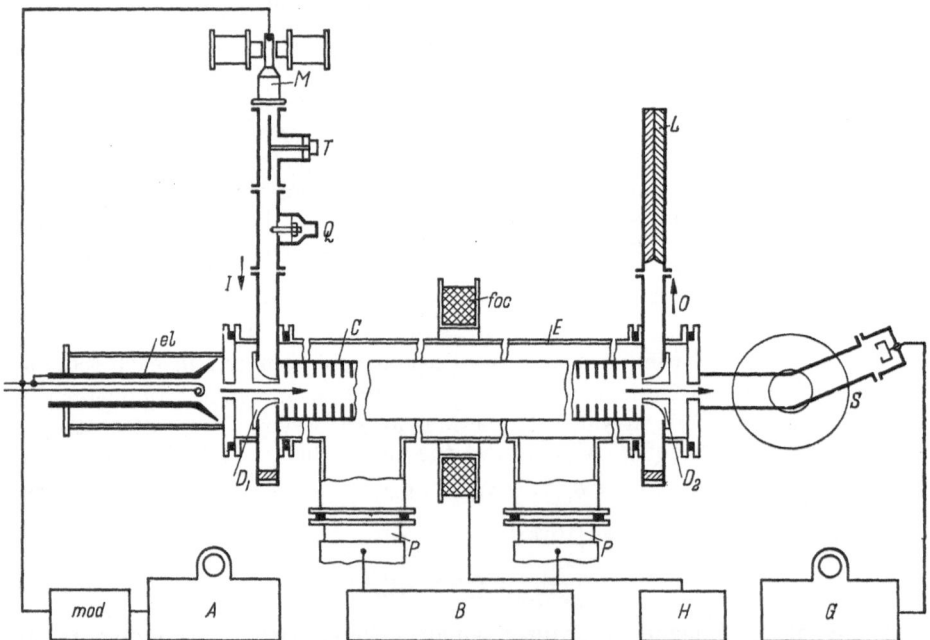

Fig. 11. Block schematic diagram of the 3.5 Mev linear electron accelerator built at A.E.R.E. C tapered corrugated waveguide in vacuum envelope E, el electron gun, I radio frequency input, D_1 door-knob feed, D_2 door-knob exit, O radio-frequency output, L steel load taking up remainder of wave power; M magnetron with modulator mod; T and Q line lengthener and probe for adjusting the magnetron operating frequency, A synchronising, pulse monitoring, safety interlock, remote control; Foc one of 15 focusing coils, H focusing current controls; P oil diffusion pumps, B mechanical rotary backing pump, vacuum pressure monitoring equipment and automatic control; S electron spectrometer with deflection coils and Faraday chamber, G spectrum analyser control and display unit [6].

papers referred to in them). We shall confine ourselves here to displaying a few sketches of general interest. Fig. 10 shows a cross-section of a part of the Stanford accelerating tube, indicating dimensions and tolerances. Their accelerator is made up of such basic units two feet in length, with power fed into every fifth

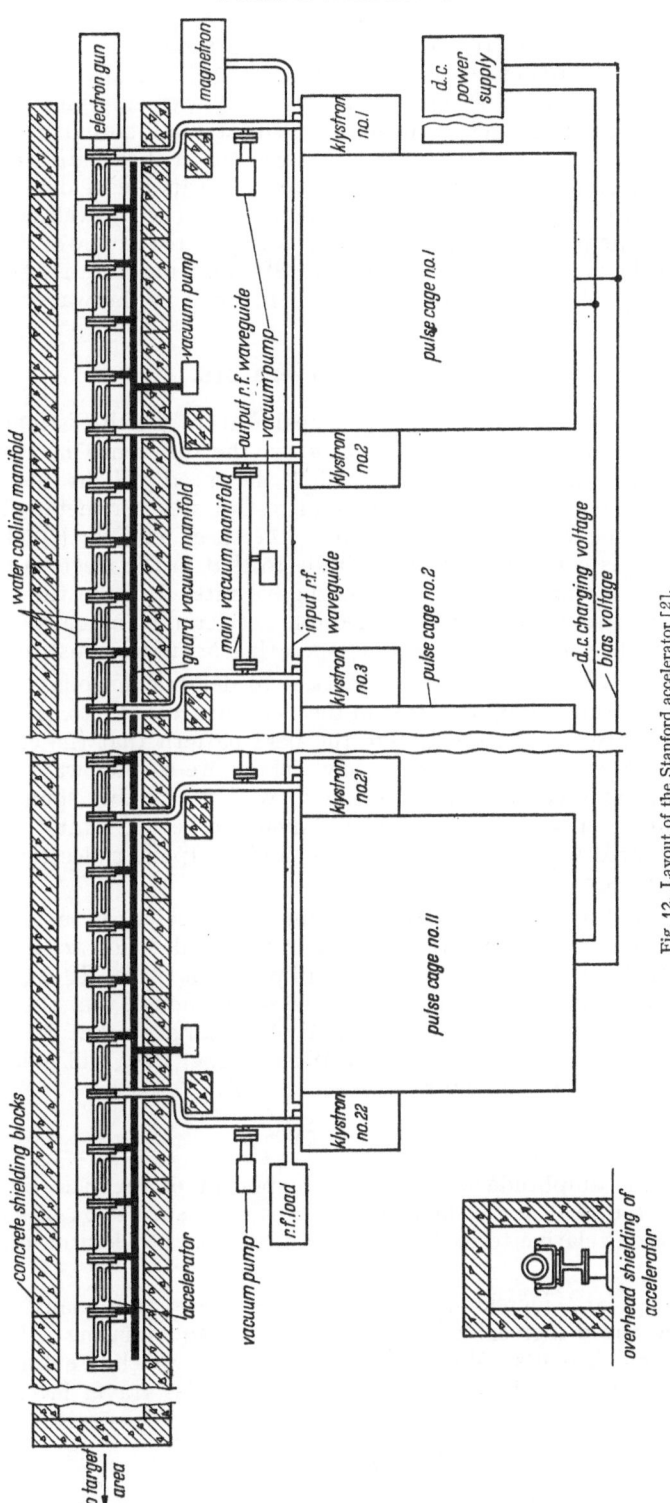

Fig. 12. Layout of the Stanford accelerator [2].

unit from individual klystron amplifiers. Fig. 12 shows an overall plan of that machine, which consists of 110 units representing 22 electrically separate accelerating sections, except that the individual klystron power amplifiers are driven by a common low power source, a magnetron with a small stabilizing cavity. The machine is intended to produce 1 Bev electrons, but at this writing the klystrons, which are still undergoing development, are only capable of delivering enough power to reach 630 Mev. Fig. 11 is a similar plan of a Harwell accelerator. It is fed at only one point, with a magnetron as the source of power. The principal superficial difference from a section of the Stanford machine is the addition of a terminating guide to absorb the energy remaining in the wave at the end of the section.

IV. Ion accelerators.

12. General considerations. The principal feature which distinguishes linear ion accelerators from electron accelerators is the difference in particle speed. Whereas electrons are injected at about half the speed of light into fields which increase their energy at the rate of several rest masses per meter distance, ions are most conveniently injected at about $\frac{1}{30}$ the speed of light (500 kev protons) or less into fields comparable in magnitude to that in the electron case, so that the total mass is thus increased at a negligible rate. In fact, the speeds of the low energy electrons injected into electron accelerators actually exceed those generated by the ion accelerators now operating or under construction.

Under these conditions, it is impractical to depress the phase velocity in a wave-guide by loading it in imitation of the electron accelerators. The undesirable wave components would so dominate the field pattern that the effective shunt impedance (6.2) would be extremely low. There are several ways in which the desired traveling wave can be established, with reasonable shunt impedance; here we shall discuss only the so-called Alvarez-type accelerator, which is the structure adopted in all present machines, and postpone a description of other possibilities to Sect. 20.

Consider first a doubly re-entrant symmetrical cavity, as in Fig. 13, excited in its lowest TM mode; that is the mode in which all fields are independent of azimuth with E_z in the same direction at all points in space. If charged particles are projected along the axis (through a hole small enough that the field pattern is not appreciably effected) at a speed sufficiently large so that the axial electric field does not change much during the transit time, they will gain an amount of energy given approximately by:

$$\Delta W = E_g g \cos \psi \tag{12.1}$$

where E_g is the amplitude of the axial component of the field on axis and ψ the phase of the field at the moment of transit. A succession of such cavities, properly phased relative to each other, constitutes an accelerator, either for ions or electrons[1].

To achieve a large increase in kinetic energy would require many such cavities, again raising the disagreeable problem of timing and phasing a large number of independent elements. We note instead that if we choose the length, L_n, of the n-th cavity equal to the distance the particles travel in one r.f. cycle:

$$L_n = \frac{v_n}{\nu} = \beta_n \lambda, \tag{12.2}$$

---

[1] See Sect. 5 and footnote 1, p. 347.

and choose, furthermore, to excite all the cavities at the same phase, then adjacent cavities will be contiguous. Since adjacent cavities are almost identical if the fractional change in velocity is small, the intervening end walls may be removed (except for a small structural member to support the adjoining re-entrant sections) and the entire string of cavities, now strongly coupled, will resonate at almost the same frequency and with almost the same field pattern. Thus we arrive at the configuration of Fig. 14, in which the symbols correspond to the ones in Fig. 13. Two important gains have been made—the power loss associated with currents flowing in the end walls has been eliminated, except for the first and last cell, and the entire structure is now a single electrical unit, with the desired phase relations between the various cells assured by the strong coupling.

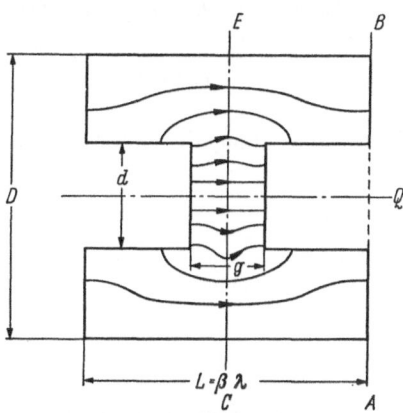

Fig. 13. Fields in "unit cell" of accelerator. Note that a conductor across EC would not change distribution [3].

Alternatively, the structure of Fig. 14 may be described as a single long cavity to which coaxial hollow cylinders, called drift tubes, have been added; these are of relatively small diameter, and their lengths increase with distance in such a way that the particles are electrically shielded while the axial electric field is in the wrong direction. In most machines these drift tubes have a small effect on the resonant frequency of the cavity and on the field distributions, except near the axis.

In waveguide terminology, the structure is one to be operated at the lower frequency limit of its pass-band, where the guide wavelength approaches infinity and the group velocity vanishes. Thus it is essential to operate with standing waves. It would appear, then, that this type of machine is much less efficient than those discussed in the previous chapter; this is indeed true for high velocities, but in the low velocity range no practical configuration has been devised which can claim a substantial advantage.

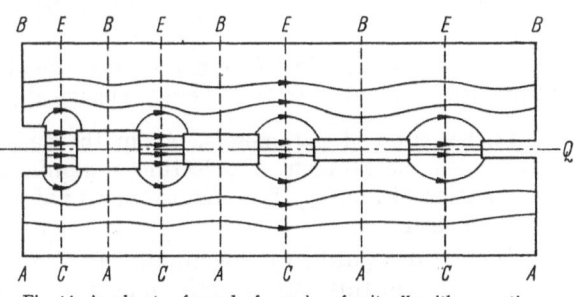

Fig. 14. Accelerator formed of a series of unit cells with separating partitions removed [3].

The presence of a traveling wave component of the type demanded in Sect. 4 may not have been apparent from the point of view adopted up to now in this section. It can be demonstrated in the following way:

Neglecting the slow increase in length of the drift tubes, the axial electric field has the form:

$$E_z(a, z) = \sum_{m=0}^{\infty} A_m I_0(K_m \varrho) \cos \frac{2\pi m z}{L} \cos \omega t \qquad (12.3)$$

where z is measured from the center of a gap, the A_m's are constants depending on drift tube geometry, and

$$K_m^2 = (2\pi)^2 \left[\left(\frac{m}{L}\right)^2 - \frac{1}{\lambda^2}\right] = \left(\frac{2\pi}{\lambda}\right)^2 \left[\left(\frac{m}{\beta_s}\right)^2 - 1\right] \qquad (12.4)$$

since the local period L is the distance traveled by the synchronous ion in one r.f. cycle. Eq. (12.3) describes the field in the neighborhood of the axis, out to the bore radius, a; that is, to the radius of the axial hole through which the particles pass.

Specifying E_z at $\varrho = a$ then determines the coefficients A_m. For example, if we assume

$$E_z(\varrho, z) = E_0 \frac{L}{g} \cos \omega t, \qquad nL - \frac{g}{2} < z < nL + \frac{g}{2} \tag{12.5}$$

and zero along the inner metallic wall of the drift tube bore, then

$$A_m I_0(K_m a) = 2 E_0 \frac{\sin (\pi m g/L)}{\pi m g/L}.$$

The term $m = 1$ is the one which contains the wave component we want, as well as its counterpart moving in the opposite direction. Thus the amplitude of the accelerating wave, ε_0 of Sect. 4, is

$$\varepsilon_0 = \frac{E_0}{I_0(K_1 a)} \frac{\sin (\pi g/L)}{\pi g/L}. \tag{12.6}$$

The constant E_0, called the average electrical gradient, is very closely

$$E_0 = \frac{2\pi}{\lambda L} \int_{\text{cell}} H_\varphi \, dr \, dz \tag{12.7}$$

that is, the peak gap "voltage" divided by the cell length. The quantity denoted by T,

$$T(\varrho) = \frac{I_0(K_1 \varrho)}{I_0(K_1 a)} \frac{\sin (\pi g/L)}{\pi g/L} \tag{12.8}$$

or more generally,

$$T(\varrho) = \frac{A_1}{2 E_0} I_0(K_1 \varrho) \tag{12.9}$$

is called the transit time factor, and measures the amplitude of the useful wave component compared to the average electrical gradient.

13. Selection of parameters. As in the case of electron accelerators, we are faced with the problem of selecting a number of parameters: λ, D, d, g/L (in the notation of Fig. 13), bore radius, a, rate of increase of cell length, L, and total cavity length.

First, with regard to λ, the arguments of Sect. 6 concerning power and stored energy apply in this case as well, indicating once more the choice of a high frequency. However, there are considerations unique to ion acceleration which establish an optimum frequency much lower than that for electrons. For one thing, injection from a d.c. accelerator dictates a certain minimum diameter of the order of 1 cm, for the drift tube bore. Now the quantity T, Eq. (12.8), for on-axis particles is insensitive to wavelength for $K_1 a \sim \frac{2\pi a}{\beta_s \lambda} \ll 1$, but decreases exponentially with frequency for $K_1 a \gg 1$; that is, the effective shunt impedance (6.2) drops rapidly. Thus to accelerate 500 kev protons with reasonable efficiency, λ should not be less than about a meter. Secondly, radial defocusing effects require auxiliary focusing elements which must fit inside the drift tube shells, and these demand considerable space, as will be seen later. Finally, since the structure is resonant, it is necessary that the frequency spacing of the normal modes be large compared to the width of the lowest mode. Neglecting the effect of the drift tubes, the frequency separation from the closest mode in a cavity of length L is

$$\frac{\Delta \nu}{\nu} = \frac{1}{2} \left(\frac{\lambda}{2L} \right)^2 \qquad (13.1)$$

which must be considerably larger than $1/Q$ for the modes to be well defined.

The first machine of this type was designed for a frequency of 200 megacycles ($\lambda = 1.5$m), primarily for the very practical reason that surplus radar equipment in that frequency range was available. The choice was, however, not far from optimum. Some later designs, in which a larger bore and more space within drift tube shells for focusing devices were required, have been based on frequencies as low as 50 megacycles ($\lambda = 6$m) but this figure probably represents a practical lower limit.

D, d, and g/L are not independent, for together they determine the resonant frequency of the unit cells (Fig. 13), the lengths of which are specified by the desired value of β_s. It is possible to obtain an approximate analytical representation of the fields in the unit cells by a technique similar to that described in Sect. 5[1], but the frequency precision required is so great that one must rely on models of unit cells to establish the relation of the para-

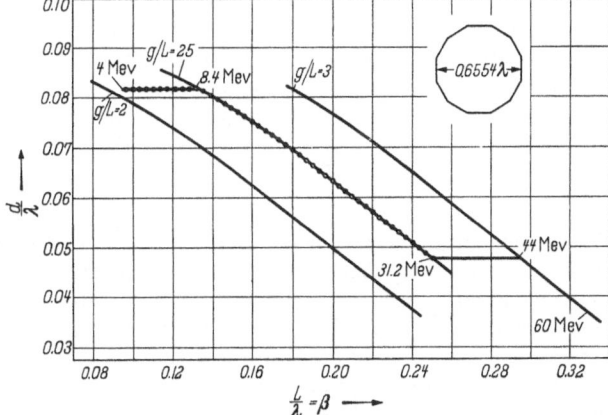

Fig. 15. Results of model tests on resonant frequencies of re-entrant cavities. Dotted curve indicates design points for 32-Mev accelerator, each dot corresponding to a unit cell. [OPPENHEIMER, JOHNSON, and RICHMAN, Phys. Rev. 70, 447 (A) (1946).]

meters for resonance. A few typical curves relating the parameters are shown in Fig. 15. These curves have the universal property that g/L must increase with L for constant cavity and drift tube diameter, or that drift tube diameter must decrease with L for constant cavity diameter and g/L. This type of accelerator has the unfortunate property of becoming increasingly inefficient with increasing particle velocity; as β_s approaches $\frac{1}{2}$, other structures become competitive[2].

The choice of the parameter, g/L, is restricted to a rather small range. For a given average gradient, E_0, the field in the gaps (12.5) becomes excessive if g/L is too small, while the transit time factor (12.8), and hence the effective shunt impedance, suffers if g/L is too large. Values ranging from about 0.20 to 0.40 represent the usual compromise between these factors. The problem, then, is to select a cavity diameter independent of β_s (not essential, but very convenient) and juggle drift tube diameter, d, and g/L to obtain a good shunt impedance, compatible with keeping d large enough to allow space for focusing elements in the drift tube shells. In the original Alvarez machine [3] g/L is held at a constant 0.25 for most of the length, with d decreasing with increasing β_s. In more recent designs, d is held constant as long as possible to simplify construction and alignment, changing discontinuously to a new value when the end of the

---

[1] See WALKINSHAW, SABEL and OUTRAM: AERE T/M 104 and 105.
[2] See preceding footnote and Sect. 20.

permissible range in g/L is reached. Of course, in an accelerator consisting of more than one cavity, one is free to make the cavities of different diameter for further optimization.

Recently, it has been pointed out[1] that, by first specifying the fields in a unit cell to be of a rather simple analytic form, and then fabricating the drift tubes in shapes which conform to the field pattern, it is possible to keep the cavity diameter constant and the drift tube diameter almost constant over a much larger range of β_s than with drift tubes of simple rectangular cross-section, while still maintaining a good shunt impedance.

The bore radius, a, has negligible influence on the resonant properties of the unit cell and on the field distribution, except on the axis. It is chosen as a compromise between the desire to allow the ions as much transverse room as possible, and the need for a transit time factor (12.8) close to unity. In addition, if the drift tubes contain quadrupoles or solenoids as focusing elements, the difficulty of providing the required fields increases rapidly with increasing bore area.

The rate of increase of cell length, or β_s, with axial distance is related to the choice of average electrical gradient, E_0, and synchronous phase, ψ_s, according to (4.8)[2],

$$\frac{dW_s}{dz} = e\, \varepsilon_0 \cos\psi_s = e\, E_0\, T(0) \cos\psi_s. \tag{13.2}$$

The choice of E_0 is one of the more difficult and elusive points in the design of an ion accelerator. With regard to r.f. exciting power, E_0 should be small and the tank length, L, large, for energy gain is proportional to $E_0 L$, while expended power varies with $E_0^2 L$. However, cavity, drift tubes, and building are sufficiently expensive that an economic optimization requires E_0 to be close to a maximum set by sparking in the gaps, excessive loading due to electrons emitted from drift tube surfaces, and related phenomena. It is still not clear how this limit depends on frequency, gap length, pulse length, and surface condition of the drift tubes[3], so that there is even yet an element of risk in designing an accelerator which differs in any significant way from one which already exists. This situation is in considerable part responsible for the fact that the accelerators used as injectors for larger machines tend to be copies of the original Alvarez machine, with a substantial reduction in design gradient, for in such application, reliability is far more important than cost or maximum energy.

Having decided on a nominal value for the electrical gradient, and for ψ_s, Eq. (13.2) determines β_s and L as functions of distance. The choice of ψ_s is not very critical; it should be small so that the synchronous energy will increase rapidly, but not so small as to restrict the range of stable phase (Fig. 1) unduly. Usual values lie between -20 and $-30°$. It should be noted that only the product $E_0 \cos\psi_s$ is prescribed by the rate of increase in the drift tube lengths (13.2), so that in a final machine, ψ_s may be varied at will from zero (threshold) by increasing E_0 from its threshold value to the sparking limit, wherever that may be.

The length of a single cavity for a given wavelength is subject to the considerations at the beginning of this section on separation of modes and to the closely related question of maintaining a uniform electric gradient (see Sect. 17). If the energy is to be low enough so that neither of these points is important, a

---

[1] N. C. Christofilos: Bull. Amer. Phys. Soc., Ser. II, 1, No. 6 (1956).

[2] Choosing to define synchronous conditions in terms of the transit time factor for $\varrho = 0$ is somewhat arbitrary; extensive numerical calculations of orbits for a variety of cases, however, seem to indicate that such a definition is at least satisfactory.

[3] See, however, W. D. Kilpatrick: UCRL-2321 "A Criterion for Vacuum Sparking".

single cavity is the simplest solution. For higher energies (of the order of 50 Mev at 200 Mc) one has to choose between careful construction of a single cavity and the complication of timing and phasing a number of separate cavities. Both approaches seem feasible; for example, the Minnesota 68 Mev machine is made up of three separate cavities, while the Brookhaven 50 Mev machine is designed as a single one, 100 feet in length.

14. Longitudinal motion. We have to deal here with the non-relativistic limit of the equations of motion derived in Sect. 4. Setting $\gamma_s = 1$, and adopting the notation of Sect. 13, (4.13) becomes:

$$\frac{d}{dn}\left[\beta_s^2 \frac{d}{dn}(\psi - \psi_s)\right] + \frac{2\pi e E_0 T \lambda}{m_0 c^2} \beta_s [\cos \psi - \cos \psi_s] = 0 \qquad (14.1)$$

where T means $T(0)$. The transverse motion has negligible effect on the longitudinal motion as long as $K_1 a$ is small, as it must be for T to be close to unity.

The symbols n and ψ lend themselves to a different interpretation in the geometry of the ion accelerator. n, the number of r.f. cycles spent by the synchronous particle, is also the index number of the unit cell through which the bunch is instantaneously passing; ψ, the phase of a particular ion with respect to the crest of the traveling wave, is also the phase of the accelerating field at the moment the particular ion passes the center of the n-th gap, by virtue of the conventions in z and t adopted in (12.3).

Since the machine parameters vary slowly with n, an approximate first integral of (14.1) can be constructed by treating the parameters as constants:

$$\frac{1}{2}\left[\frac{d\psi}{dn}\right]^2 + \frac{2\pi e E_0 T \lambda}{m_0 c^2 \beta_s}[\sin \psi - \psi \cos \psi_s] = \text{const.} \qquad (14.2)$$

Now, the non-relativistic limit of (4.10) is:

$$\frac{d\psi}{dn} = -2\pi \frac{\beta - \beta_s}{\beta_s} = -\pi \frac{w - w_s}{w_s} \qquad (14.3)$$

where w is the kinetic energy, $W - m_0 c^2$. Therefore (14.2) is equivalent to:

$$\left(\frac{w - w_s}{w_s}\right)^2 + \frac{4}{\pi} \frac{e E_0 T \lambda}{m_0 c^2 \beta_s}[\sin \psi - \psi \cos \psi_s - K] = 0 \qquad (14.4)$$

where K is an arbitrary constant.

(14.4) describes the way in which the energy and phase deviations from synchronism are related as the various ions oscillate about synchronous phase. The expression is useful in two ways. First the range of phases leading to stable phase motion, and thus steadily increasing energy, is found by setting $w = w_s$ at the other stationary point in the wave, $\psi = -\psi_s$ (see Fig. 1). For this special case,

$$K = \psi_s \cos \psi_s - \sin \psi_s, \qquad (14.5)$$

and the limit of the stable phase range ahead of the synchronous particle (again Fig. 1) is the other value of ψ for which

$$\sin \psi - \psi \cos \psi_s = -\sin \psi_s + \psi_s \cos \psi_s. \qquad (14.6)$$

The second solution of (14.6) is almost $2\psi_s$, even for ψ_s as large as $-60°$; thus the total phase stable range in ψ is very nearly $3|\psi_s|$. As ψ_s approaches $-90°$ (large E_0 or small rate of increase of cell lengths) oscillations can be sustained over the full $360°$ range.

24*

The second piece of information furnished by (14.4) is the maximum energy deviation corresponding to the limiting phase oscillation, a figure which establishes the permissible energy error or spread in the injected beam and the maximum possible energy spread at final energy. We note in (14.4) that $|w - w_s|$ has its maximum at $\psi = \psi_s$; using (14.5) we obtain:

$$\left| \frac{w - w_s}{w_s} \right|_{\max} = \left[\frac{8}{\pi} \frac{e E_0 T \lambda}{m_0 c^2 \beta_s} \right]^{\frac{1}{2}} [\psi_s \cos \psi_s - \sin \psi_s]^{\frac{1}{2}}. \tag{14.7}$$

Numerically, the permissible energy spread at injection is several percent for typical parameters inserted in (14.7), increasing as $|\psi_s|^{\frac{3}{2}}$ for small ψ_s and constant E_0. Fig. 16 shows the limiting curves defined by (14.5) for a few values of ψ_s.

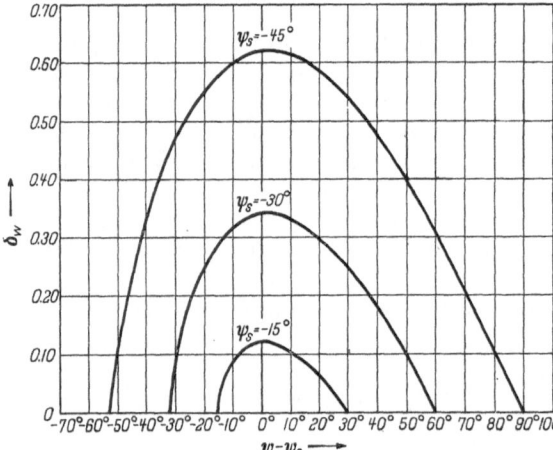

Fig. 16. Limiting curves of stable motion for various values of ψ_s. The ordinate is $\delta_w = \left[\frac{m_0 c^2 \beta_s}{e E_0 T \lambda} \right]^{\frac{1}{2}} \left| \frac{w - w_s}{w} \right|$.

For each ψ_s the family of bounded trajectories consists of curves inside the bounding curve and of similar shape, degenerating to ellipses centered at $\psi = \psi_s$ for small amplitude oscillations.

In Sect. 4, it was shown that an equation of the form (14.1) represents, for small deviations from synchronous phase, an oscillation of angular frequency,

$$\omega_\psi^2 = \frac{-2 \pi e E_0 T \lambda}{m_0 c^2 \beta_s} \sin \psi_s \tag{14.8}$$

with amplitude decreasing as $\beta_s^{-\frac{3}{4}}$ (or $w_s^{-\frac{3}{8}}$)[1]. The amplitude of the energy oscillations therefore varies as $\omega_\psi \beta_s^{-\frac{3}{4}}$ (or $w_s^{-\frac{5}{8}}$) because of relation (14.3). This damping is enough to be of considerable advantage in delivering a homogeneous final beam. For example, a 500 kev proton beam which develops an energy spread of a few percent in the course of the first phase oscillation (14.7) will have a relative energy spread which is smaller by a factor $\left(\frac{0.5}{50} \right)^{-\frac{5}{8}} = 18$ at 50 Mev. Such a beam is sufficiently mono-energetic for most research applications, though perhaps not for injection into a large synchrotron[2]. The damping is also of some help in the successful operation of machines consisting of several separate cavities, for in the drift space between cavities an energy spread gives rise to debunching of the ions.

Fig. 17 shows a typical set of phase oscillations for various initial phase deviations and zero energy deviations computed numerically not from (14.1) but from the corresponding cell-by-cell difference equations, which take some account of the complete axial field distribution, (12.3). Specifically, the parameters represent the 50 Mev Brookhaven machine (Table 1) with $E_0 = 1.9$ Mv/m, $\psi_s = -26°$, $\lambda = 1.5$ m (200 Mc), $T = 0.9$, and an injection energy of 750 kev. It will be noted that the large amplitude oscillations quickly pull away from the trailing

---

[1] The absolute length of the ion bunch, $\Delta z = \beta_s \lambda \frac{\Delta \psi}{2 \pi}$, then increases as $\beta^{\frac{1}{4}}$; that is, the bunch length remains almost constant as the unit cells increase in length.

[2] See, however, K. JOHNSEN, CERN internal report PS/KJ–29, for a method of decreasing the energy spread further.

edge of the accelerating wave (Fig. 1) and become symmetrical about ψ_s, so that the small amplitude approximation is good for most of the ions for most of the length.

15. Transverse motions. In contrast to the case of relativistic electrons, the transverse motion in an ion accelerator is an important aspect of design. The defocusing due to the accelerating field must be counteracted, and the geometry of the cavity requires that the focusing elements be integral with the drift tubes —the only available space near the ion beam free from r.f. fields. As a result, the fabrication, assembly and support of the drift tubes together with the enclosed

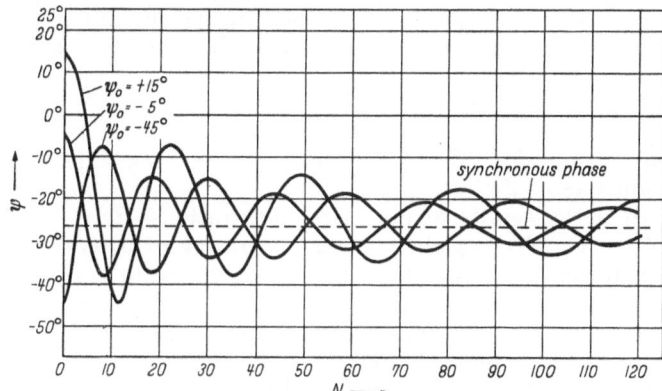

Fig. 17. Phase oscillations vs. drift tube number.

paraphernalia present some of the most difficult engineering problems in accelerator design.

We begin with Eq. (4.14'), with $\gamma_s = 1$, $\beta \sim \beta_s$, and using the notation of this chapter:

$$\frac{d^2}{dn^2}\varrho = -\frac{\pi e E_0 T \lambda}{m_0 c^2 \beta_s}\varrho \sin\psi. \tag{15.1}$$

Without additional focusing, the radial displacement of the synchronous ion would increase by a factor e every

$$\left[-\frac{m_0 c^2 \beta_s}{\pi e E_0 T \lambda \sin\psi_s}\right]^{\frac{1}{2}} \tag{15.2}$$

number of unit cells, neglecting the slow rate of increase of β_s. For the parameters of Fig. 17, the number (15.2) is about 4 at injection energy; it is evident that if one wants an intense, well-defined beam at the high energy end, 100 unit cells away, something must be done to prevent the beam from expanding into the drift tube surfaces. Several methods are available for focusing the ions: we shall present the principal ones in what is perhaps the order of increasing importance[1].

As in the electron accelerator it is possible to use an axial magnetic field. In this case, the field is produced by a succession of coils mounted inside the drift tube shells, the current being supplied by conductors running through the hollow stems which support the drift tubes. In analogy to Eq. (8.10), (15.1) now becomes:

$$\frac{d^2}{dn^2}\varrho = -\frac{\pi e E_0 T \lambda \varrho \sin\psi}{m_0 c^2 \beta_s} - \frac{1}{4}\left[\frac{e H(n) \lambda}{m_0 c^2}\right]^2 \varrho \tag{15.3}$$

---

[1] See also Ref. [1] for other proposals.

where $H(n)$ is the axial component of the magnetic field along the axis. Since the coils can only cover about half of the cell length, $H(n)$ varies considerably over a cell length, and the field is not uniform over the cross-sectional area of the bore. However, such local fluctuations are adequately accounted for by replacing $H^2(n)$ by its average value $\overline{H^2(n)}$ over a cell length; the remaining n-dependence refers to the strengths of successive solenoids, which we are free to choose.

From the form of (15.3) it is apparent that by making the magnetic field sufficiently strong, all ions will feel a net restoring force, regardless of phase. Unfortunately the field required to do so is prohibitively high in a typical accelerator, and it is more pertinent to look at the minimum requirement on field strength. This is obtained by asking that the synchronous ion feel no net radial force:

$$\overline{H^2_{\min}}(n) = -\frac{4\pi m_0 c^2}{e} \frac{E_0 T \sin \psi_s}{\beta_s \lambda} \tag{15.4}$$

which amounts to a peak field of 15 kilogauss at the low energy end of a machine with the parameters of Fig. 17.

Judging from the results of numerical integrations of (15.3), including the variations in particle phase, and from some operating experience, a beam can be adequately confined by fields as little as 10% in excess of (15.4). The net force constant in (15.3) can be changed at will by varying $\overline{H^2(n)}$, and thus the beam can be compressed or expanded during operation by a control which is independent of the high frequency fields. The disagreeable feature of solenoidal focusing is the large power dissipation necessary to produce the required fields. This method is therefore unattractive except in very special applications, either because of the expense of supplying power or the difficulty of removing the heat generated.

The only existing machine in which solenoids are used is the Livermore high current accelerator (see Table 1). In that device it is important to fill the entire entrance aperture with beam, and only the absolute focusing action of solenoids can prevent the ions from striking the bore. Moreover, the operating frequency is only 50 Mc, so that there is adequate space in the drift tubes and supporting stems to supply cooling water.

At the other extreme in engineering simplicity is the so-called "grid-focusing" method. In Sect. 3, EARNSHAW's theorem was invoked to demonstrate that radial defocusing is a necessary consequence of establishing axial (that is, phase) focusing. That consequence may be avoided by introducing an electric charge in the region traversed by the beam in an appropriate way, for under such conditions the theorem no longer applies. To see how this possibility may be exploited, consider again the field distribution specified by (12.5), but neglecting the radial dependence of E_z, as we have been doing in the intervening paragraphs:

$$E_z(\varrho, z) = E_0 \frac{L}{g} \cos \omega t \qquad -\frac{g}{2} < z_n < \frac{g}{2} \left.\begin{array}{c}\\\\\end{array}\right\}$$
$$= 0 \qquad\qquad \text{elsewhere} \tag{15.5}$$

where z_n is the distance from the center of the n-th gap.

Since div $E = 0$, the radial component of the field is:

$$E_\varrho = \frac{E_0}{2} \frac{L}{g} \varrho \left[\delta\left(z_n - \frac{g}{2}\right) - \delta\left(z_n + \frac{g}{2}\right)\right] \cos \omega t \tag{15.6}$$

where δ is the Dirac δ-function. Fourier analysis of (15.6) yields the force used in (15.1) (for $K_1 a \ll 1$); this step was unnecessary there because the general relation of E_ϱ to E_z had been already established in Sect. 4. However, in the form

(15.6) we note that if the entrance end of the farther drift tube were covered with a metallic foil, E_ϱ would be reduced to zero at that point by the charges induced on the foil and the first δ-function in (15.6) would be absent. Fourier analysis of the new E_ϱ then gives, instead of (15.1):

$$\frac{d^2\varrho}{dn^2} = -\frac{e\,E_0\,\lambda}{m_0\,c^2\,\beta_s}\,\frac{L}{2g}\,\varrho\cos\left(\psi - \frac{\pi g}{L}\right) \tag{15.7}$$

which is focusing for all phases

$$\psi > \frac{\pi g}{L} - \frac{\pi}{2}. \tag{15.8}$$

What has happened is illustrated in Fig. 18. Without the foil the ion sees first an inward and then an outward radial field, the outward force being stronger if the field is increasing in time as the ion traverses the gap ($\psi < 0$). With the foil in place, the force is always inward, provided that the ion enters the gap when the field is in the accelerating direction $\left(\psi - \frac{\pi g}{L} > -\frac{\pi}{2}\right)$. Thus, if it were possible to cover the entrance of each drift tube with a foil thin enough that effects of scattering and energy loss would not be serious, the focusing problem would be solved in a very simple way.

Unfortunately, it appears impractical to use sufficiently thin foils. Such

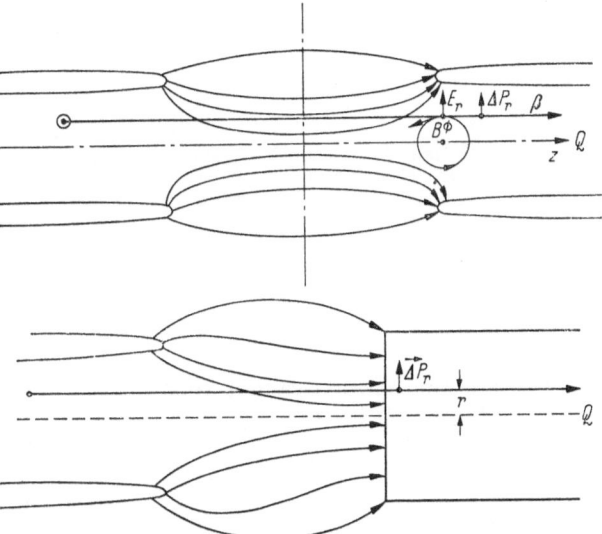

Fig. 18. Configuration of field lines at a gap, (a) without, and (b) with, a foil covering the following bore [3].

foils can be made [3], but are too delicate to withstand the forces to which they are subjected under typical operating conditions. In particular, they can be ruptured by even small amounts of sparking. One is forced to accept a not very satisfactory compromise in the form of thin ribbons set edgewise to the beam (Fig. 19). Since the ribbons must be so thick that they stop an ion which strikes them, they can not be allowed to obstruct more than a few percent of the area at each gap. As a result there is only a partial suppression of the radial field in the vicinity of the grid.

The condition (15.8) would indicate that for $g/L \sim 0.25$, the usual choice, the radial motion is absolutely stable if $\psi > -45°$. Numerical computations of orbits in fields produced by a grid such as that in Fig. 19 indicate that the radial motion is only satisfactory for $\psi > -20$ to $-30°$[1]. The observed behavior of grid-focused machines is consistent with this picture, for the optimum transmitted beam is of the order of 5% of the injected beam.

Grid-focusing is the most commonly used technique at present. However, the recently invented method of quadrupole focusing[2] promises to displace it

---

[1] L. SMITH and R. L. GLUCKSTERN: Rev. Sci. Instrum. **26**, 220 (1955).

[2] Its application to the linear accelerator was first discussed by J. P. BLEWETT: Phys. Rev. **88**, 1197 (1952).

for it should be possible to increase the transmission to about 50% by this method. We shall not attempt to explain the basic principles of strong focusing in this article, but proceed with their application to the problem at hand[1].

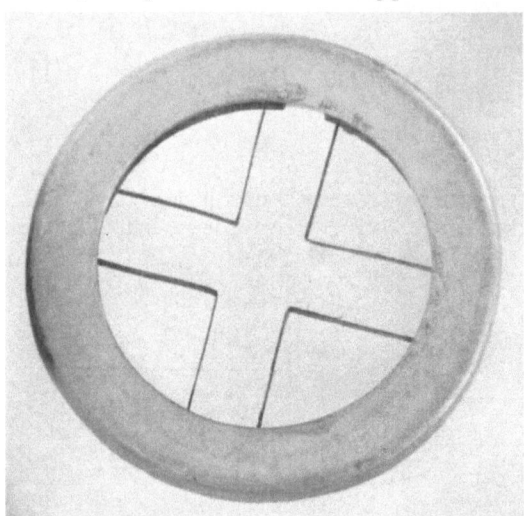

Fig. 19. A typical grid structure from the 10 Mev Berkeley accelerator. The tungsten ribbons are soldered into the circular piece, which is screwed into the drift-tube bore [3].

In each drift tube is mounted a symmetric quadrupole, which may be either electrostatic or magnetic in character. As in the case of solenoids, voltage or current, and possibly cooling water, are supplied through the stem supporting the drift tube. The quadrupoles are all oriented in the same way, thus defining two perpendicular planes in each of which the force is proportional to displacement, but alternating periodically between focusing and defocusing. The ions are thus subjected to a succession of transverse impulses, from the r.f. fields while traversing the gaps, and from the static fields while traversing the drift tubes.

The resulting motion cannot very well be analyzed analytically because the effect of the r.f. fields depends so strongly on the phase motion; in addition,

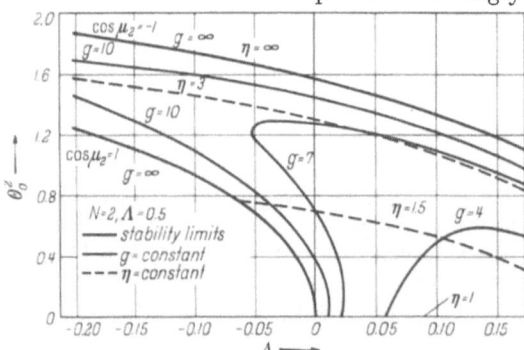

Fig. 20. Stability region and amplitude parameters for quadrupole focusing in a linear accelerator. After SMITH and GLUCKSTERN, cf. footnote 1 on p. 375.

$$\Delta = \frac{\pi e E_0 \lambda T \sin \psi}{m_0 c^2 \beta}, \quad \text{and} \quad \vartheta_0^2 = \frac{e H' \beta \lambda^2}{m_0 c^2} \quad \text{or} \quad \frac{e E' \lambda^2}{m_0 c^2}.$$

the parameters of the systems are not strictly periodic, but vary slowly as the synchronous velocity increases, (15.1). A reasonably good picture of the motion is obtained if we treat the phase as well as the other parameters as constant and examine the resulting trajectories. The important properties of these trajectories may be summarized in the form of a diagram such as Fig. 20. The ordinate is a parameter which characterizes the strength of the quadrupoles; E' or H' being, respectively, the gradients of the electric or magnetic fields. The diagram is drawn specifically for two focusing elements alternating with two defocusing elements, with the quadrupoles half as long as the unit cells[2].

[1] See preceding footnote 1 on p. 375 and J. S. BELL: AERE T/R 1114, 1953, for detailed expositions.

[2] Fig. 20 is specialized to the case of alternation in pairs; for grouping of one, three, etc., the shape of the bounding curves is approximately the same but the numerical values of ordinate and abscissa are proportional to N^{-2}, η is unchanged, but the g-values are proportional to N. Groups of two seem best suited for most present machines.

The most important feature of Fig. 20, is the delineation, by the heavy lines, of the range of parameters for which the transverse motion is bounded. That is, if the values of the parameters were strictly independent of axial distance, the particles would execute transverse oscillations of definite amplitude for values inside the area and would increase their transverse displacement exponentially with axial distance for values lying outside the bounding curves. The quantities g and η determine the amplitude of the oscillations in terms of the initial conditions. A trajectory parallel to the axis but displaced a distance x_f at the gap between focusing elements will have an amplitude X:

$$X = x_f. \tag{15.9}$$

If a particle crosses the axis at the same point with transverse velocity $\dfrac{d x_f}{d n}$ the amplitude is:

$$X = g \frac{d x_f}{d n}. \tag{15.10}$$

If it is parallel to the axis at the gap between defocusing elements, but displaced a distance x_d, its amplitude is:

$$X = \eta\, x_d. \tag{15.11}$$

If it crosses the axis with transverse velocity $d x_d / d n$ at that point, the amplitude is

$$X = \frac{g}{\eta} \frac{d x_d}{d n}. \tag{15.12}$$

g has the additional significance that if the parameters are varied slowly compared to the oscillation period, the amplitude of oscillation will vary as $g^{\frac{1}{2}}$.

Now a particular ion, with its instantaneous value of ψ and $\beta \sim \beta_s$ is represented in Fig. 20 by a point which oscillates horizontally about a value of the abscissa corresponding to ψ_s and β_s, and which moves vertically, as acceleration proceeds, in accordance with our choice of strengths for the successive quadrupole elements. If it were sufficient to consider only the synchronous ion, the specification of the focusing system would be quite simple; the quadrupole strengths need only be chosen so that the representative point stays well away from the stability boundaries as β_s increases. However, at injection, the ions are spread in phase from $+2\psi_s$ to $-\psi_s$, so that the points of interest cover an extended lateral region, and the question arises whether the quadrupole strengths can be chosen in such a way that all or most of the points fall within the area of stability. For instance, in the example of Sect. 14, the points range almost from one bounding curve to the other for the best quadrupole strength at injection.

Since the phase, ψ, can vary over so wide a range during a single transverse oscillation, it becomes necessary to integrate the trajectories numerically, using Fig. 20 as a guide in selecting the quadrupole strengths. Such computations indicate that the motion is, in reality, stable[1], and that the ion beam should be adequately confined for reasonably practical quadrupole strengths; however, there has not yet been a clear demonstration of the principle in a working machine.

Quadrupole focusing is extremely sensitive to imperfections in alignment of the elements. Since the focusing action depends on the near cancellation of alternate large focusing and defocusing impulses produced by fields with large gradients, an element which is displaced laterally will produce an undesirable

---

[1] If the phase oscillation frequency is twice the transverse oscillation frequency, a serious increase in amplitude can occur, and care should be taken to avoid this situation in specifying quadrupole strengths.

deflection that can easily be large enough to destroy the beam. To estimate the magnitude of the effect, consider the equation of motion in a quadrupole field:

$$\frac{d^2x}{dn^2} = \pm S x \tag{15.13}$$

where S is the quadrupole strength (the ordinate of Fig. 20), and ($\pm$) refers to defocusing and focusing elements. If an ion enters traveling along the nominal axis, but the element is displaced a distance δ, the ion will acquire a deflection:

$$\frac{dx}{dn} = \pm \frac{S}{2} \delta \tag{15.14}$$

if the quadrupole occupies half a cell length.

For $\eta \sim 1$, (15.10) or (15.12) gives the amplitude of the subsequent oscillation:

$$X = \frac{g\,S}{2} \delta.$$

Then, if N quadrupoles are misaligned at random, with r.m.s. displacement $\delta_{\text{r.m.s}}$, the r.m.s. amplitude after traversing the N drift tubes will be:

$$X_{\text{r.m.s}} = \frac{g\,S}{2} \delta_{\text{r.m.s}} \sqrt{N}. \tag{15.15}$$

If the beam is to remain within an area an inch or so in diameter through say, 100 drift tubes, (15.15) requires an alignment accuracy of the order of 10 mils. Such a lateral tolerance in 100 elements contained in a cavity 100 feet or so in length requires considerable care in design and assembly.

16. Injection. There are various considerations which determine the minimum initial energy for which an ion accelerator can be built. One is the transit time factor (12.8) for axial rays; for low velocities $K_1 a$ is approximately:

$$K_1 a = \frac{2\pi a}{\beta_s \lambda}. \tag{16.1}$$

The function $I_0(x)$ increases exponentially for large X, so that for given bore radius, a, free space wavelength, λ, and average gradient E_0, the amplitude of the accelerating wave component decreases rapidly as the injection velocity is reduced. Another is the difficulty of accurately constructing and installing many short drift tubes. A third is the problem of providing adequate focusing. If grids are used, a low velocity implies many grids per unit axial distance, with consequent loss of beam. If solenoids or magnetic quadrupoles are used, the fields required become larger as the velocity is decreased [Eq. (15.4) and Fig. 20], while at the same time the space available within the drift tube shell diminishes.

For all of these reasons it is advantageous to inject at as high an energy as possible, always provided that the injector does not become more complicated than the accelerator itself. For a 200 Mc accelerator with $1/_2''$ to $3/_4''$ bore diameter (or a 50 Mc accelerator with 2 to 3 inch bore diameter), an initial value of β_s of about 0.03 (500 kev protons) is about as low as one can go without major complications in the design of the accelerator. Since this energy can be obtained with a reasonably compact Cockcroft-Walton set operating at atmospheric pressure, most accelerators are designed to accept ions of about that velocity. The opposite extremes are the Brookhaven accelerator, designed for 750 kev protons to ease the problem of magnetic quadrupole focusing, and the heavy ion accelerator at Yale and the University of California, which use 500 kv injectors, but an initial β_s of 0.012, because of the low e/m of the ions produced in the source[1].

The Cockcroft-Walton units are of standard design, differing mainly in types of ion source. The cross-sectional area and divergence of the accelerated beams are small enough that the various focusing systems can keep them confined to an area considerably smaller than the area of the bore; however, if the output of ion sources is substantially increased from the present level (~ 5 ma of protons during the acceleration period) it may be necessary to modify the d.c. structure radically in order to overcome space-charge effects in the accelerating column and to take advantage of the larger available currents.

A valuable addition to the injection system is a buncher; that is a single unit cell run at the same frequency as the accelerator, but separated from it by a field-free section which serves as a drift space. As in a klystron, the buncher slightly increases or decreases the ion velocity, depending on the phase at which the ion passes through, the fast ones subsequently catching up to the earlier slow ones in the drift space. If the voltage on the buncher gap is small compared to the energy error the accelerator will accept [Eq. (14.7) and Fig. 16], then as many as half the injected ions can be caught and accelerated, even for a synchronous phase of 30° or less.

The presence of the buncher demands a rather strict control of the injector voltage during the beam pulse, not because of the energy acceptance of the accelerator, which is several percent, but because the drift velocity must be correct if the bunch is to arrive at the accelerator at the right time. That is, if the average bunch energy changes by an amount $\Delta w_i/w_i$, the time of flight, T, to the accelerator will change by an amount:

$$\frac{\Delta T}{T} = -\frac{1}{2}\frac{\Delta w_i}{w_i}, \tag{16.2}$$

in which time the field in the accelerator has accumulated an error in phase:

$$\Delta\psi = 2\pi\nu\,\Delta T = -\pi\frac{s}{\beta_i\lambda}\frac{\Delta w_i}{w_i} \tag{16.3}$$

where s is the drift distance. For a drift time of 20 cycles or so, which is necessary if the energy modulation is to be well within the energy acceptance of the accelerator, fluctuations of a few tenths of a percent in the d.c. voltage could decrease the efficiency of capture noticeably.

17. Tolerance requirements. In constructing a long cavity which is to resonate in its lowest mode, with average electrical gradient, E_0, independent of axial distance, it is necessary to understand the effects which arise from errors in dimensions of the cavity and drift tubes. An imperfect machine will differ from the ideal one in two ways; the frequency of the lowest mode will depart from the nominal value, and the average electrical gradient in this mode will, in general, be non-uniform. A frequency error is of no great consequence, for the frequency of the power source may be changed slightly to compensate for it, the net effect being a slight change in particle energy. That is, the final synchronous velocity

---

[1] Two exceptional cases are the original 32 Mev accelerator at Berkeley, which uses a 4 Mev Van de Graaff accelerator as injector because of the intended use of focusing foils, and the high current accelerator at Livermore, which uses a 100 kv injector specially designed for the purpose.

at any point along the machine is determined by the cell length, L, and wavelength, λ:

$$\beta_s = \frac{L}{\lambda} \qquad (17.1)$$

so that for small changes in λ,

$$\frac{\delta w_s}{w_s} = \frac{2\,\delta\beta_s}{\beta_s} = -\frac{2\,\delta\lambda}{\lambda} = \frac{2\,\delta\nu}{\nu}. \qquad (17.2)$$

On the other hand, errors in field distribution can be quite serious. The required change in β_s from one cell to the next is fixed by the change in cell lengths according to (13.2):

$$\frac{dL}{dn} = \lambda\,\frac{d\beta_s}{dn} = \frac{e\,E_0\,T\,\lambda^2}{m_0\,c^2}\cos\psi_s \qquad (17.3)$$

so that if E_0 varies with distance, ψ_s will also change. In order to maintain synchronism, ψ_s must be appreciably different from zero, so that for a downward variation of E_0, we require

$$-\frac{\delta E_0}{E_0} \ll 1 - \cos\psi_s. \qquad (17.4)$$

For $\psi_s \sim 25°$, then, a 10% decrease in E_0 will lose the ions completely, so that it would be desirable to keep the fluctuations in $E_0(z)$ to within a few percent[1].

An exact analysis of the effects of errors in such a complicated structure is not possible, but an estimate can be made in the following way:

Consider a cylindrical cavity of uniform diameter D, without drift tubes, but containing a material of dielectric constant ε and permeability μ which are functions of axial distance, thus simulating variations in the inductive and capacitive elements of the drift tube structure. MAXWELL's equations yield, for the axial component of the field:

$$\frac{1}{r}\frac{\partial}{\partial r}r\frac{\partial}{\partial r}E_z + \frac{\partial^2 E_z}{\partial z^2} + \frac{\omega_0^2}{c^2}\mu\,\varepsilon\,E_z = 0 \qquad (17.5)$$

where $\omega_0/2\pi$ is the resonant frequency of the loaded cavity.

The solution has the form:

$$E = J_0\left(j_1\frac{2\varrho}{D}\right)\left[E_0 + \delta E_0(z)\right] \qquad (17.6)$$

where j_1 is the first zero of J_0. Eq. (17.5) becomes, for $\delta E_0 \ll E_0$, μ and $\varepsilon \sim 1$

$$\frac{d^2}{dz^2}\frac{\delta E_0}{E_0} = -\frac{\omega^2(z) - \omega_0^2}{c^2} \sim -\frac{2\omega_0^2}{c^2}\frac{\delta\omega(z)}{\omega_0} \qquad (17.7)$$

where $\dfrac{\omega(z)}{2\pi}$ is the local frequency; that is, the resonant frequency of a cavity with the local values of μ and ε [†]. Eq. (17.7) is subject to the boundary conditions

$$\frac{d}{dz}\left(\frac{\delta E_0}{E_0}\right) = 0 \qquad (17.8)$$

at the ends of the cavity. Therefore:

$$-2\frac{\omega_0^2}{c^2}\int_0^L \frac{\delta\omega}{\omega_0}\,dz = 0 \qquad (17.9)$$

---

[1] An upward variation from the nominal value requires extra power and increases the danger of local sparking.

[†] Eq. (17.7) can also be derived by considering an empty cavity of diameter varying slightly with axial distance.

where L is the length of the cavity. Eq. (17.9) shows that the resonant frequency of the whole structure is the average of the local frequencies.

Returning to the real accelerator, $\delta\omega(z)$ represents the frequency errors of the individual unit cells. The sensitivity of the frequency of the unit cells to changes in the various dimensions can be determined from model studies so that the results following from solutions of (17.7) can be translated directly to dimensional tolerances on the cavity and required range of operation of possible tuning devices in the cavity walls.

As an example of the application of (17.7), consider a case in which the resonant frequencies of short sections, δL, at the two extreme ends are high and low by an amount $\delta\omega$. Then in the intervening portion,

$$\frac{d}{dz}\left(\frac{\delta E_0}{E_0}\right) = -2\frac{\omega_0^2}{c^2}\frac{\delta\omega}{\omega_0}\delta L. \tag{17.10}$$

Thus, if E_0 has the proper value at one end, it is wrong at the other end by an amount:

$$\frac{\delta E_0}{E_0} = -2\frac{\omega_0^2}{c^2}L\frac{\delta\omega}{\omega_0}\delta L. \tag{17.11}$$

If δL is equal to the last cell length, $\beta_F\lambda$, for example, then

$$\frac{\delta E_0}{E_0} = -8\pi^2\beta_F\frac{L}{\lambda}\frac{\delta\omega}{\omega_0}. \tag{17.12}$$

For a 50 Mev, 200 Mc accelerator 100 feet long, (17.12) requires that the end cells be correct in frequency to one part in 10^4. Conversely, the field distribution can be altered several percent by that amount of tuning at the ends of the cavity.

It is important to note that the sensitivity to frequency errors increases with cavity length. This is most apparent if the frequency error in (17.7) is Fourier-analyzed with the normal modes of the ideal cavity, $\cos\frac{n\pi z}{L}$, as a base. Then (17.7) yields:

$$\left(\frac{\delta E_0}{E_0}\right)_n = 8\left(\frac{L}{n\lambda}\right)^2\left(\frac{\delta\omega}{\omega_0}\right)_n \tag{17.13}$$

where the subscript n denotes the amplitude of the n-th Fourier component. Thus for errors varying slowly with axial distance, such as errors in cavity diameter or systematic errors in modeling of the unit cells, the effect on the field distribution increases as the square of the ratio of cavity length to free space wavelength. For this reason accelerator designers often elect to divide their machines into several short, electrically isolated, sections.

A frequency tolerance of 10^{-4} implies a fractional dimensional tolerance of the same order of magnitude. Whether or not one can obtain such precision in a large structure without exorbitant expense is a marginal question. Accordingly, single cavities more than a few wavelengths long generally make use of regularly spaced tuning devices, which may be rods, balls, or flat copper sheets to change the cavity volume near the outer wall, together with adjustable end walls or drift tubes to introduce a linear field variation (17.11).

18. Operating cycle. The operating cycle of a standing wave machine is somewhat different from that of a traveling wave machine (Sect. 9). In simplest terms, the r.f. power source is turned on and the cavity fields increase to the desired level in a time of the order of Q/ω; next, the ion source is turned on, and the acceleration period begins. This pattern is complicated by the phenomenon of multipactor action [3] which has the effect of absorbing large amounts of r.f. energy at very low fields and preventing further increase in stored energy. This effect can be avoided if the rate of increase of field strength at low fields is considerably greater than can be established by the oscillators. A separate set of low power oscillators, called pre-exciters, is used for this express purpose; if the high level oscillators are self-excited, the pre-exciters also serve to establish the correct mode and provide some excitation to start the main oscillators. The cycle proceeds as follows:

1. The pre-exciters are turned on and the field rises rapidly to a low level, about 10% of the final value.

2. The main power source, consisting of one or more tubes feeding the cavity through one or more coupling loops at the outer cavity wall, is turned on, and the field rises to the desired strength.

3. The ion source is turned on and the acceleration begins; it may last a few tens of micro-seconds or longer, all the way to continuous operation, depending on the purpose of the machine and the capabilities of the power tubes and their power supplies.

The focusing devices are usually run continuously except in the case of magnetic quadrupole focusing in 200 Mc machines, where the heat generated in the shortest drift tubes cannot be effectively removed. To insure a steady value for the field, the magnets are pulsed for a time somewhat longer than the acceleration period, the energy being supplied by a bank of condensers which are recharged during the off period.

19. Some existing machines. The original 32 Mev, 200 Mc accelerator (described at length in Ref. [3]) consists of a steel vacuum tank 40 feet long and 4 feet in diameter, inside of which is supported a 12-sided copper liner serving to define the volume of the electromagnetic fields (see Fig. 21). The drift tubes hang on silver plated brass stems supported from the top surface of the liner. Focusing grids are mounted in rings which can be screwed into the entering end of the bore of each drift tube. Power is supplied by nine self-excited oscillators, lined up along the outside of the vacuum tank, each delivering 250 kw to the cavity through its own transmission line and coupling loop.

The 10 Mev injector for the 6 Bev Berkeley proton synchrotron is a shorter (20 feet) version of the earlier machine, with changed dimensions appropriate to the different initial and final energies [3]. In addition to minor improvements two changes were made in the interest of reliability and to minimize down time; the design value of the electric gradient, E_0, was reduced from 0.9 Mv/foot to 0.6 Mv/foot, and the drift tubes were supported in such a way that alternate ones could be swung sideways for quick access to grids and drift tube surfaces. The superiority of the Cockcroft-Walton generator to the Van de Graaff in energy control during the beam pulse permitted the use of a buncher (Sect. 16), resulting in an increase in accelerated beam of a factor of three to four.

Successful attempts have been made to improve on this original design in both electrical and mechanical aspects. For higher energy machines requiring more radio-frequency power, it is advantageous to have more powerful tubes, thus avoiding a multiplicity of coupled systems. There has been a considerable effort in recent years to develop bigger tubes, largely in response to the demands of linear accelerator design. Today tubes capable of generating megawatts of power up to 100 Mc, at 200 Mc (resnatron) and 400 Mc (klystron) are available.

There is a current effort to extend conventional triode techniques at 200 Mc and to produce a multi-megawatt klystron at that frequency.

On the mechanical side, the original liner and single drift tube stem construction is not very sturdy, particularly if quadrupole focusing, with its strict mechanical tolerances, is to be used. Where copper-clad steel is available, design can be considerably improved by making the steel tank function as the cavity wall at the same time, thus providing a rigid frame for supporting drift tubes. With either type of tank, more recent designs involve two drift tube stems, approximately 90° apart, to support each drift tube.

The Minnesota 68 Mev, 200 Mc, accelerator is similar to the Berkeley machines in mechanical design, consisting of three separate tanks, 20, 40 and 40 feet long. For a power source, however, it uses one resnatron for each tank, with a fourth acting as a common driver for the others. Grid-focusing is used throughout, though conversion to quadrupole focusing is intended[1]. The frequency is established at low power level by a crystal-controlled oscillator driving the power tubes through intermediate stages of amplification; the resonant frequency of the three tanks is maintained at the proper value by moving copper vanes installed at regular intervals along the liner walls. In practice, there are only slow drifts in resonant frequency, probably associated with thermal effects, so that the positions of the vanes can be adequately controlled by the operator.

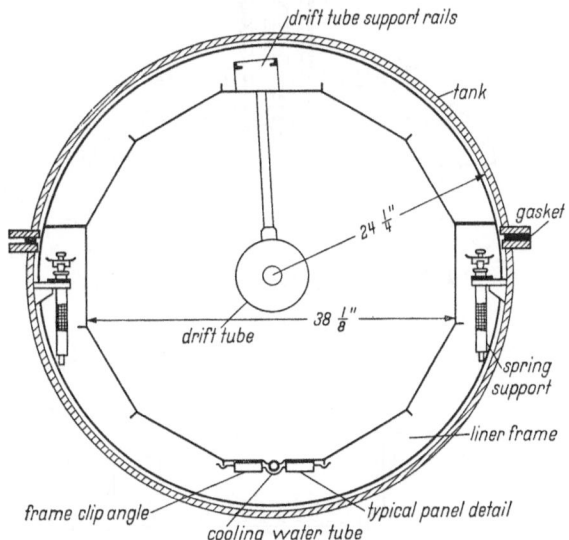

Fig. 21. Diagram showing arrangement of tank and liner [3].

The Harwell and CERN 50-Mev machines are essentially identical, the CERN group having adopted the Harwell design in order to save time and expense. They differ from the Minnesota machine primarily in that the drift tubes will each be supported from the bottom by two stems for increased stiffness; quadrupole focusing will be used from the beginning. The Brookhaven 50-Mev accelerator, also intended as an injector for a proton synchrotron, represents a more radical extension, in detail, of the original Berkeley design. Copper-clad construction will be used in a single cavity 110 feet long, split mechanically into sections approximately 10 feet long which may be removed for servicing without disturbing the rest of the machine. The drift tubes will be supported by two stems, one horizontal and one vertical, again to obtain the rigidity needed for quadrupole focusing. Fig. 22 is a sketch of a typical drift tube showing the focusing magnet and cooling system. The stainless steel bore tube and vacuum seal are to isolate the internal structure from the main vacuum system; the volume containing the magnet is pumped to rough vacuum from the outside through the heavier stem. Fig. 23 shows one of the 10-foot sections of the Brookhaven

---

[1] See L. C. Teng: Rev. Sci. Instrum. **25**, 264 (1954).

machine; the inner diameter of the cavity is about 3 feet and the drift tube diameter about 8 inches.

The heavy ion accelerators at Yale and the University of California were designed in a cooperative effort of the two laboratories and differ from each

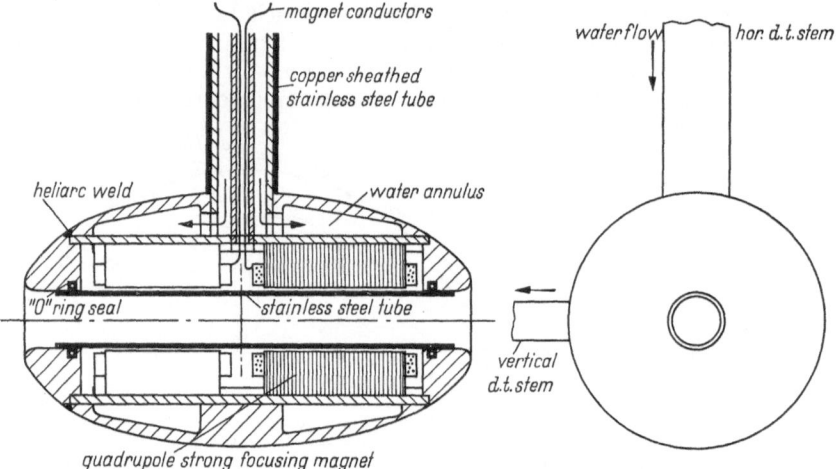

Fig. 22. Brookhaven linear accelerator drift tube. After BLEWETT, cf. footnote 2 on p. 386 below.

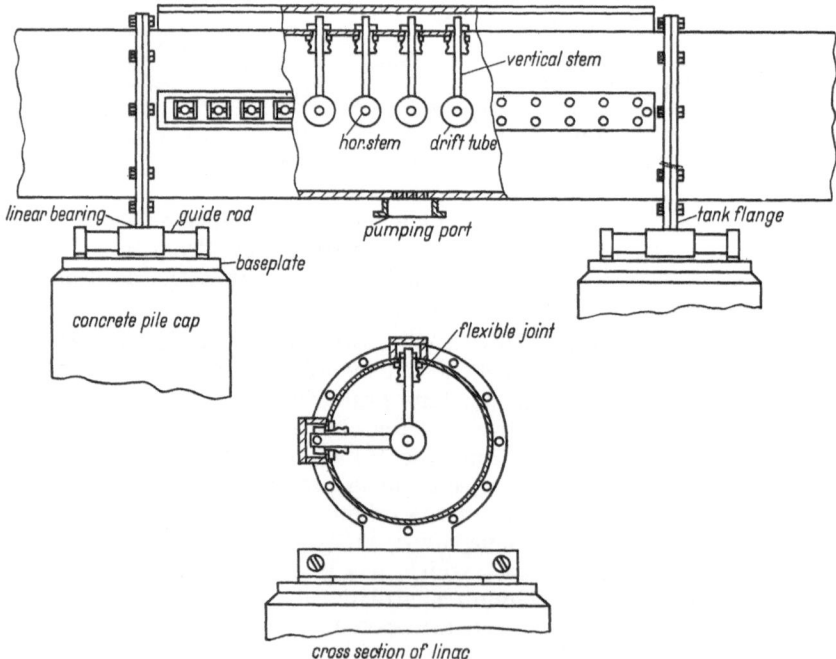

Fig. 23. Brookhaven linear accelerator injector for A.G.S. After BLEWETT, cf. footnote 2 on p. 386 below.

other only in minor details. Their frequency is 70 Mc, corresponding to a tank diameter (copper-clad) of about 9 feet. Because of the low specific charge of ions produced in the source it is necessary first to accelerate them to a velocity at which more electrons can be removed from the ions and to introduce them to the main accelerating cavity only after this has been done. For this reason the design calls for two tanks; the first, however, is so short that its power is supplied simply by a transmission line from the main tank. The small tank will use grids for focusing, and the main tank continuously excited magnetic quadrupoles in drift tubes supported on single heavy stems.

Fig. 24 is a photograph of a quadrupole magnet for one of the heavy ion accelerators, after assembly, but before mounting in the drift tube shell. Fig. 25

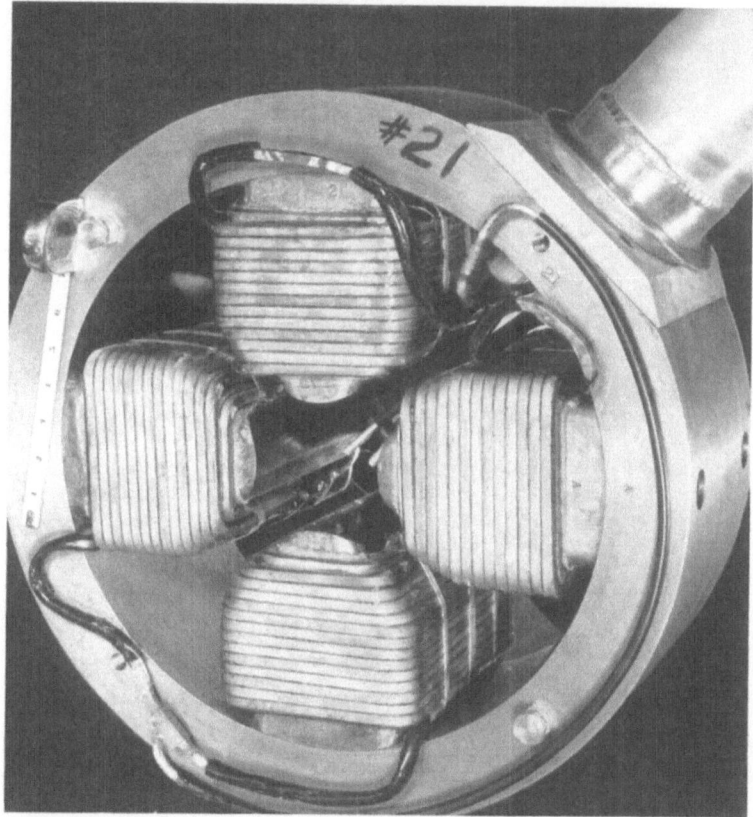

Fig. 24. Magnetic focusing quadrupole for the Berkeley-Yale heavy ion accelerators. The drift tube stem is seen at the upper right; the drift tube shell is only slightly larger in diameter than the iron flux return ring.

is a view of the interior of the short grid-focused tank; alternate stems are off-set to provide space on top of the tank for the clamps which hold the drift tubes in place. Fig. 26 is an over-all view of the University of California's heavy ion accelerator at a late stage of assembly.

20. Other types of ion accelerators. Varying amounts of developmental effort have been expended in studying other r.f. structures which compete favorably with the conventional form in effective shunt impedance and mechanical simplicity. In the extreme relativistic range, one would presumably borrow the techniques of electron acceleration, for the problems would be identical, but there is more immediate interest in extending the present energy range only moderately but in an efficient way.

A method which has received considerable attention is the traveling wave helix[1]. Use is made of the fact that a wave traveling along the axis defined by a conductor wound into a helix is slowed to a velocity less than that of light, the speed depending on the parameters of the helix. It has been shown that it is possible to make a 200 Mc helix for proton acceleration starting at 500 kev with excellent shunt impedance and helix radius of a few inches. Unfortunately, it turns out to be difficult to construct and support such an object adequately

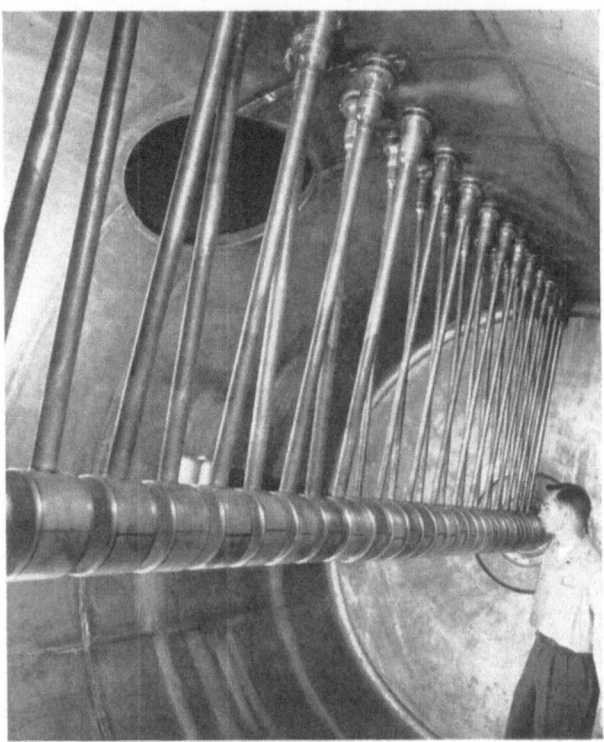

Fig. 25. Interior of the Berkeley heavy ion accelerator prestripper tank. The object on the far wall behind the drift tubes is a tuning loop.

and no satisfactory way has been found to incorporate the necessary focusing elements; as a result, the idea has not been developed further.

R.f. model studies at Brookhaven[2] have produced two variants which lead to a better shunt impedance in certain energy ranges. The so-called "interdigital" system consists of a succession of drift tubes attached alternately to heavy bars running along the top and bottom of a cylindrical cavity. The mode of interest is transverse electric away from the drift tubes, but strongly distorted at the gaps to provide an accelerating field. Such a structure has about twice the shunt impedance of the Alvarez design below 10 Mev, but falls behind rapidly with increasing particle energy. The "organ-pipe" system, has a cylindrical cavity which is loaded by a succession of stubs extending from one wall approximately to the axis of the cavity. Waves propagated through the cavity are

[1] H. Dahl and K. Johnsen: Chr. Michelsens Inst. Beretn. **14**, No. 4 (1951).

[2] J. P. Blewett: CERN Symposium, June 1956, Linear Accelerator Injectors for Proton Synchrotrons.

concentrated near the stub-ends, and phase velocity may be adjusted over a wide range by varying the length of the stubs. This system appears promising for energies above 50 Mev (protons), though little attention has as yet been given to details of design.

Theoretical studies at Harwell[1] made in preparing a design for 600 Mev protons[2] indicate that at about 100 Mev, it would become more economical of power to split the conventional accelerator into single cells operating alternately

Fig. 26. View of the Berkeley heavy ion accelerator. The tank coming from the right and ending just short of the men in the center is the pre-stripper. The rectifiers for the quadrupoles are seen mounted on top of the drift tube supports along the main tank.

180° out of phase (cell length $= \frac{1}{2}\beta_s \lambda$). A modification of this system, proposed by N. C. Christofilos, is to make the cells equal in size and couple each to the one preceding it by means of a hole of proper size and location in the common wall producing the required phase shift per cell in a wave traveling through the system. Such a design would tend with increasing energy toward the configuration of the electron accelerator in a natural way.

The only existing accelerator which differs substantially from the Alvarez design is the first section of the high current accelerator at the Livermore laboratory of the University of California. In this special case, the injector[3] delivers a d.c. deuteron beam of about one ampere at 100 kev in a cross-sectional area

---

[1] See footnote 1, p. 369; also R. T. Weir: UCRL-3150.
[2] Later abandoned in favor of a higher energy magnetic accelerator.
[3] See W. A. S. Lamb and E. J. Lofgren: Rev. Sci. Instrum. **27**, 907 (1956).

of two inches diameter. A bore diameter of three inches is necessary to contain this beam; at the chosen frequency of 48 Mc, the transit time factor (12.8) is very poor. Therefore an initial acceleration to 1 Mev is provided at 24 Mc by two electrically separated $\lambda/4$ lines with drift tubes hung on the ends; acceleration occurs as the deuterons enter and leave each drift tube. Since the energy increases by a factor of ten in four gaps, the analysis of Sect. 4 is quite unsuitable. The proper amplitudes and phases of the two resonant lines and the required strength of the focusing solenoids in the drift tubes were determined by numerical integration of orbits. The phase damping of the device is quite strong; in the short distance the deuterons can be bunched well within the limits demanded by Fig. 16 for the following 48 Mc accelerator.

Bibliography.

Until a very few years ago, ion accelerators have been meagerly described in the easily accessible literature, while electron accelerators have had a quite decent representation. This situation is improving rapidly, probably due in large part to the popularity of proton accelerators as injectors for the large magnetic machines. The following list is intended to cover the subject to the end of 1956, but the interested reader is warned to search the subsequent literature. The author regrets having to quote so often internal reports of various laboratories, but the most detailed expositions are still predominantly in that form.

[1] Fry, D. W., and W. Walkinshaw: Rep. Progr. Phys. 12, 102 (1949). — This is the earliest comprehensive review of the subject, including an extensive set of references. Unfortunately, there was almost no literature on ion accelerators at that time, so that the discussion is almost entirely devoted to electron machines.

[2] Chodorow, M., et al.: Rev. Sci. Instrum. 26, 134 (1955), and R. F. Post and N. S. Shiren, p. 205. — These two papers describe the important work of the Stanford group in considerable detail. The list of references, in conjunction with that of [1], covers the significant work on electron accelerators to the present time. Mention should be made, however, of the paper by G. Saxon, Proc. Phys. Soc. Lond. B 67, 705 (1954), with its references, on the subject of beam loading, an important effect when high intensity and high efficiency of converting electrical power to beam power are desired.

[3] Alvarez, L., et al.: Rev. Sci. Instrum. 26, 111 (1955), and Bruce Cork, p. 210. — These two papers constitute the first extensive descriptions of proton accelerators—the 32 Mev and the 10 Mev machines at Berkeley. The first paper also includes a history of the field and a list of the important references to the beginning of 1955.

[4] Chu, E. L.: Report ML 140 of the Stanford University Microwave Laboratory 1951. — This report is a very detailed presentation of the theoretical aspects of electron accelerators; i.e. particle dynamics, determination of electromagnetic field distributions, and discussion of effects of errors. Parts of the report appear in Ref. [2].

[5] Neal, R. B.: Report ML 185 of the Stanford University Microwave Laboratory 1953. — This report complements [4] in that it emphasizes the mechanical and electrical problems and presents some of the theory in less high-brow fashion. Parts of this report also appear in [2].

[6] Fry, D. W.: Philips Techn. Rev. 14, 1 (1952), and C. F. Bareford and M. G. Kelliher 15, 1 (1953) in the same journal. — These two papers describe in detail the latest electron accelerators of the United Kingdom. Fry's paper also constitutes an elegant, semi-popular, presentation of the problems involved, with excellent photographs and diagrams.

[7] Proc. Phys. Soc. Lond. 61 (1948). Walkinshaw, W., p. 246. — R.-S.-Harvie, R. B., p. 255. — Mullett, L. B., and B. G. Loach, p. 271. — This series of three papers presents the approach of the British group to electron accelerator design and describes some of the measurements of the electrical properties of disk-loaded wave guides.

[8] Slater, J. C.: Rev. Mod. Phys. 20, 473 (1948). — This paper is the earliest, and in many ways still the most extensive, theoretical treatise from the United States on the subject of long accelerators.

[9] Ginzton, E. L., W. W. Hansen and W. R. Kennedy: Rev. Sci. Instrum. 19, 89 (1948). — This paper is a much earlier version of [2], preceding most of the design and construction work, but expressing the point of view and method of attack of the Stanford group.

[*10*] Johnsen, K.: Chr. Michelsens Inst., Beretn. **16**, No. 3 (1954). — This excellent mono-graph, of almost book-length, develops the theory of the linear accelerator (both kinds) from the beginning and presents a discussion of various important special effects.

[*11*] Slater, J. C.: Ann. Rev. Nucl. Sci. **1**, 199 (1952). — This review article offers a con-cise description of machines in operation or under construction at that time, together with a comparative discussion of the reasons for differences in design.

In addition to the above articles specifically devoted to linear accelerators, the following references might also be of interest:

[*12*] Report of the 1956 CERN Symposium on high energy physics. — There are several noteworthy papers on linear accelerators and their application as injectors in the report beside the ones referred to in this article. In particular it is probably the most convenient source for Westerners who are interested in the developments in progress in the USSR.

[*13*] Pickavance, T. G.: Focusing in high energy accelerators, Progr. Nucl. Phys. **4**, 142 (1955). — This article for non-specialists includes a considerable discussion of linear accelerator focussing problems.

[*14*] Livingston, M. S.: High energy accelerators. Interscience, New York and London 1954. — This book includes an elementary chapter on linear accelerators.

Reactor Techniques.

By

D. J. HUGHES.

With 21 Figures.

In this article we are interested primarily in the techniques of research utilizing neutrons from reactors. We shall not be concerned with the fundamental theory of neutron interactions nor in a review of the results obtained with reactor research, for these subjects are treated elsewhere in the Handbuch. Thus the slowing down of neutrons, which is important in producing the particular neutron spectra available at reactors, is fully described in Part 2 of Vol. XXXVIII. Similarly, the *results* of reactor neutron research concerning neutron resonances and energy levels, as well as the general properties of the nuclear reactions involved, are considered in the articles by BURCHAM and by RAINWATER in Vol. XL. Fission, the particular reaction so important to nuclear reactors, is discussed in Part 2 of Vol. XLI. One of the most important *applications* of reactor neutrons is in the field of "neutron optics" and here the fundamental theory, as well as a survey of results, is given in the article of RINGO in Vol. XXXII; here we shall be careful to limit our treatment to the techniques themselves.

The neutron intensities available with reactors are so much larger than those produced by "natural" sources, such as Ra-α-Be, that many new techniques are thereby made possible. These techniques, which will occupy our major attention, are distinctly different from those used before the advent of reactors. As their details are not well known outside the large laboratories where high flux reactors are located, a description of these techniques is of importance. Many additional high flux research reactors are now being constructed or designed, hence the availability of intense neutron flux will be extended to a rapidly increasing number of scientists throughout the world. The flux available in reactors has been increasing rapidly with time ever since the first reactor went into operation in December of 1942, and as the flux is increased so does the complexity and sensitivity of the instruments used in reactor neutron research.

In order to present the techniques for reactor neutrons it is first necessary to consider some simple properties of nuclear reactors. We shall describe these properties in the next section, then go on to consider the research techniques that are used in the various energy regions for which reactor neutrons are available. Experiments done directly with reactors are limited almost completely to the use of neutrons and it is neutron research with which we shall be concerned. The radioisotopes made in reactors of course are often used to study the effects of large-scale gamma irradiation of materials, but it is not such secondary effects of reactors with which we are concerned. Although the field of use of radioisotopes is as wide as science itself, we are concerned only with the direct use of reactors in research, not with their products, radioisotopes and power.

I. The reactor as a neutron source.

1. The reproduction factor. In a typical nuclear reactor, neutrons are produced in the uranium fuel almost entirely by slow neutron fission of U^{235}. The usual research reactors contain *moderators* such as graphite or heavy water, in which the fast neutrons emitted at fission are slowed down to thermal energies, a few of them being absorbed in the various materials of the reactor during the process. Some of the slow neutrons produced by the moderation are absorbed in uranium and cause fission, thus beginning the neutron cycle anew. It is obvious that the neutrons produced in one fission must lead to at least one fission in the next generation in order to maintain the chain reaction. The number of fissions produced by a single fission in the preceding cycle is known as the *reproduction factor, k*.

The value of the reproduction factor for normal uranium (U^{238} to U^{235} ratio of 139) with no moderator is less than unity, and it is only by careful design, utilizing a moderator, that k is raised above unity for normal uranium. If we start our consideration of the neutron life cycle with the absorption of one thermal neutron in uranium, then k will be given by the number of thermal neutrons that are absorbed in uranium one generation later. When the thermal neutron is absorbed in uranium, a definite number of fission neutrons, η, is emitted from the uranium. Thermal neutron fission takes place only in U^{235}, of course, but the fission neutrons that are emitted from U^{235} have sufficiently high energy so that some fission of the abundant U^{238} occurs. The number of fission neutrons is thus augmented slightly by fast neutron fission in U^{238} and this increase (usually a few percent) is taken into account by multiplication by the *fast effect, ε*. The fast neutrons are then slowed down to thermal energy by elastic collision with the moderator atoms. In the moderating process some of the neutrons are absorbed in U^{238} when the neutrons have energies equal to the resonance absorption energies in this isotope. The number of neutrons that escape this resonance absorption and reach thermal energy is given by $\eta \varepsilon p$, where p is the *resonance escape probability*. Of the neutrons that reach thermal, some are absorbed in the moderator and some in the uranium. The fraction absorbed in uranium is the *thermal utilization, f*, and inclusion of the factor f brings us to the completion of the neutron cycle. The reproduction factor for an infinite medium, k_∞, is then

$$k_\infty = \eta \varepsilon p f, \tag{1.1}$$

a result sometimes alliteratively referred to as the "four factor formula".

2. Critical size. A k_∞ of 1.05, appropriate for a graphite-uranium reactor, implies that a lattice of infinite extent would have a neutron population that increases by 5% every neutron lifetime, which is about 10^{-3} sec. A chain reaction can be maintained in such a lattice if it is of finite size provided that no more than 5% of the neutrons are allowed to diffuse out of the boundaries of the lattice volume. If the percentage of neutrons that diffuse out of the pile is just equal to the *excess k* ($k_\infty - 1$), the *effective k* for the pile would be unity. The effective k, is given by k_∞ minus the loss of neutrons by diffusion out of the reactor, which is called the *leakage*.

In a "bare reactor" (one with no surrounding reflector) the neutron distribution must go to zero very close to (actually slightly beyond) the geometrical boundary of the reactor, because no neutrons are reflected back into it. The size appropriate for the steady state solution, i.e., the *critical size*, can be obtained in terms of the physical properties of the lattice. If we assume the pile is cubical with side a, then a possible form of the neutron distribution is

$$n(x, y, z) = n_0 \cos \frac{\pi x}{a} \cos \frac{\pi y}{a} \cos \frac{\pi z}{a}. \tag{2.1}$$

Eq. (2.1) will certainly go to zero at the surfaces $\pm a/2$, so the solution will satisfy the geometrical condition of a cube of side a. It is a solution of the fundamental neutron diffusion equation only for a particular value of a, however, which we obtain by differentiating the solution and substituting in the diffusion equation,

$$\frac{\nabla^2 n}{n} = \frac{3\pi^2}{a^2} = \frac{k_\infty - 1}{L^2}, \tag{2.2}$$

and

$$a = \sqrt{\frac{3\pi^2 L^2}{k_\infty - 1}}. \tag{2.3}$$

Thus a is evaluated in terms of the diffusion length, and the reproduction factor, which are properties of the lattice. The model used for the calculation (one-group model) was greatly oversimplified and the results can be markedly improved by a few simple changes. The neutrons, in slowing down, cover a region that is not negligible, but is given by τ, the neutron *age* from fission to thermal. The age has the dimensions cm² and is proportional to (actually is one-sixth of) the mean square distance traveled by the neutrons during moderation, just as L^2 corresponds to the distance traveled during diffusion. A quantity M^2, called the *migration area*,

$$M^2 \equiv L^2 + \tau, \tag{2.4}$$

if used instead of L^2 in Eq. (2.3), takes into account the neutron travel during moderation. Eq. (2.2) then becomes

$$-\frac{\nabla^2 n}{n} = \frac{k_\infty - 1}{M^2}, \tag{2.5}$$

and this result, which gives the Laplacian (sometimes referred to as the "buckling") in terms of the physical constants of the lattice, is fundamental to the description of reactor behavior. It is still an approximation to the exact "pile equation"[1], but the effect of the neutrons while still fast has been included. The equation gives the critical size of a "bare" reactor; actual reactors are always surrounded by *reflectors*, which by increasing the flux at the boundary of the lattice, lower the critical size by an amount of the order of the diffusion length. An actual reactor must be larger than the critical size of the ideal uranium-graphite lattice because of the inevitable presence of neutron-absorbing materials such as the aluminum jackets that cover the uranium rods, and because extra reactivity must be built into the reactor to allow for experimental holes and the absorbers destined to be in them.

3. Reactivity. We have been discussing the steady state solution, the one appropriate for an effective reproduction exactly unity, in which case the neutron density in the reactor is constant with time. If the Laplacian of the lattice is such that the neutron curvature is larger than that necessary to make the distribution go to zero at the boundary or, in other words, if the reactor is larger than the critical size, k will be larger than unity and the neutron density will increase with time. The reactor is then said to be *supercritical* and $k-1$ is

---

[1] An extensive treatment of the theory underlying the design of reactors is given by S. Glasstone and M. C. Edlund: The Elements of Nuclear Reactor Theory. New York: D. van Nostrand Co. 1952.

referred to as the *excess k* or the *reactivity*. Eq. (2.5), if rewritten as

$$k_\infty - \left(-\frac{\nabla^2 n}{n} M^2 \right) = 1,$$ (3.1)

shows that at the steady state the leakage is just sufficient to reduce k_∞ to unity. A larger k_∞ or a smaller M^2 means that the resulting k will be larger than unity, the reactor will be *supercritical*, and the steady state solution will no longer apply.

If the reactivity is positive, $k = (k_\infty - M^2 3 \pi^2/a^2)$ greater than unity, then the neutron flux will rise exponentially because it increases by the factor k each neutron lifetime, t_0. The time variation of the flux can then be written

$$n v = (n v)_0 e^{\frac{t}{T}},$$ (3.2)

where t is the time, $(n v)_0$ is the flux at zero time, and T is the reactor *period* or the time for the flux to rise by a factor e. If we assume that all the neutrons are emitted instantaneously in fission, then it is easy to calculate the period in terms of the excess reactivity, $k-1$. Differentiation of Eq. (3.2) gives the rate of rise of the neutron flux, at time zero, as

$$\frac{d(n v)}{(n v)_0 dt} = \frac{1}{T}.$$ (3.3)

The initial rate of rise is also given in terms of k and the neutron lifetime as

$$\frac{d(n v)}{(n v)_0 dt} = \frac{k-1}{t_0},$$ (3.4)

because the initial increase in flux in time t_0 is $(n v)_0 (k-1)$ by the definition of k. From the last two equations, we obtain the desired expression for T:

$$T = \frac{t_0}{k-1}.$$ (3.5)

The neutron lifetime is 1.4×10^{-3} sec in a typical thermal reactor, and we see that an excess reactivity of one percent ($k = 1.01$) leads to a period of $1.4 \times 10^{-3}/10^{-2}$ or 0.14 sec, corresponding to an extremely rapid rise of power level (which is proportional to neutron flux). In the present case, the power would rise by a factor of $e^{7.1} = 1200$ in one second. Such rapid changes in power would require correspondingly rapid control rod motions (which adjust k) and make control of the reactor extremely difficult. Fortunately, about one percent of the neutrons are emitted several seconds after fission has occurred and these *delayed neutrons* reduce the period greatly. For instance, the effect of delayed neutrons is to lengthen to seconds a period which is of the order of tenths of a second, that is, to a time scale in which mechanical motions of control rods are feasible.

4. Types of research reactors. The first nuclear reactor, then called a "pile", which was put in operation at the University of Chicago in 1942, was of extremely simple construction. As implied by its title, it was built by the simple process of "piling" blocks of graphite, which contained lumps of uranium, into a structure that became chain-reacting after exceeding its critical size. No shielding was present around this reactor and it could not be run above a power level of a few watts.

A wide variety of reactors now exists, made possible by the availability of large amounts of heavy water and enriched uranium. The various reactors can be classified according to the type of uranium used as fuel (normal or enriched in U^{235}), to the neutron energy (fast or thermal), or to the geometrical design of the lattice (homogeneous or heterogeneous). We have seen in the previous section that it requires careful design to attain a reproduction factor larger than unity for normal uranium mixed with graphite, and that the desired result is actually obtained by lumping the uranium fuel. The lumping of the uranium makes the reactor *heterogeneous*, as opposed to a *homogeneous* reactor in which the fuel and moderator are mixed, usually in liquid form. The graphite-normal uranium reactors that now exist all contain lumped uranium disposed in a lattice arrangement in the graphite moderator, together with some method for removal of heat from the lattice.

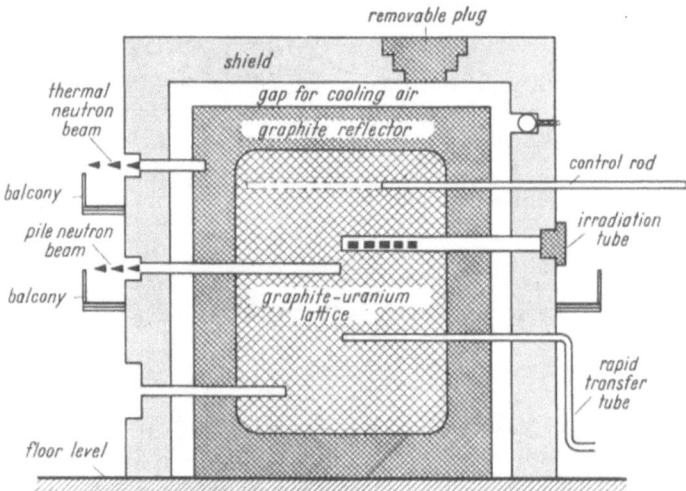

Fig. 1. Sketch of a typical graphite-uranium research reactor.

Examples of the graphite-moderated research reactors are those at Oak Ridge[1], Brookhaven[2], and Harwell, England[3]. These reactors are all of the same fundamental design, illustrated in Fig. 1, which actually is based on the Brookhaven reactor, but represents those common features of all the graphite reactors that are of interest for neutron research. The experimental holes shown are used to obtain neutrons of various energies and intensities, appropriate for particular experiments. A photograph of part of one face of the Brookhaven reactor, Fig. 2, gives the general appearance of the experimental holes and the equipment in operation for a few experiments.

Another type of thermal reactor is that in which the moderation is performed by heavy water. Examples of heavy water moderated normal uranium reactors are the CP 3 at Argonne[4] (Fig. 3) the NRX at Chalk River, Canada[5], the French reactor at Saclay[6], the Norwegian reactor at Oslo[7], and the Swedish reactor at

---

[1] Nucleonics **10**, No. 2, 24 (1952). — M. E. Ramsey and C. D. Cogle: Geneva Conf. Paper 486 (1955).

[2] L. B. Borst: Physics Today (Jan. 1951), p. 6. — M. Fox: Geneva Conf. Paper 860 (1955).

[3] Nucleonics **8**, No. 6, 36 (1951). — R. F. Jackson: Geneva Conf. Paper 762 (1955).

[4] H. S. Isben: Nucleonics **10**, No. 3, 10 (1952). — W. H. Zinn: Geneva Conf. Paper 861 (1955).

[5] F. W. Gilbert: Nucleonics **10**, No. 1, 6 (1952).

[6] L. Kowarski: J. Phys. Radium **12**, 751 (1951). — J. Yvon: Geneva Conf. Paper 387 (1955).

[7] O. Dahl and G. Randers: Nucleonics **9**, 11, 5 (1951). — Barendregt, Hidle, Lundby, Soeland and Trumpy: Geneva Conf. Paper 888 (1955).

Stockholm[1]. In these, the uranium is in the form of rods suspended in the heavy water, and cooling is usually obtained by circulation of the heavy water through an external heat exchanger. The same general principles already discussed in connection with graphite reactors apply to those containing heavy water, with minor numerical differences entering into the calculation of such constants as M and k_∞. Because of the lower absorption of heavy water, it is much easier to attain a k_∞ over unity than for graphite. A heavy water reactor is somewhat

Fig. 2. One section of the Brookhaven reactor, showing various experiments in place at the beam holes.

more inconvenient than a graphite reactor for those experiments that are performed inside the lattice, because of the necessary presence of a tight container for the water. For research involving beams of neutrons emerging from the reactor, however, there is no essential difference between the two reactor types.

If uranium enriched in the U^{235} isotope is used, a variety of reactor designs is possible, because the large available k_∞ permits the use of many materials that would absorb too many neutrons in a normal uranium reactor, and even makes possible the elimination of the moderator itself. The most extreme case of an enriched reactor is, of course, the atomic bomb, containing no moderator and as a result, essentially only fast neutrons. It is possible to design reactors at many intermediate stages between the bomb and the slow neutron reactor by varying the amount of moderator present. In general, the average neutron energy decreases with increasing amounts of moderator, from the energy of the fast neutrons in the bomb to the thermal energy in a fully moderated reactor. In general, an enriched reactor is small, even when fully moderated, so the total

[1] S. EKLUND: J. Nucl. Energy 1, 93 (1954).

power is small even when the flux is high, for the flux depends on the power per unit volume, or *specific power*.

An example of a highly enriched reactor is "Clementine"[1], the unmoderated mercury-cooled reactor at Los Alamos, which contains Pu[239] metal as fuel, rather than U[235]. A highly enriched, but moderated, reactor is exemplified by the "water boiler"[2] at Los Alamos, which is a thermal, enriched, homogeneous

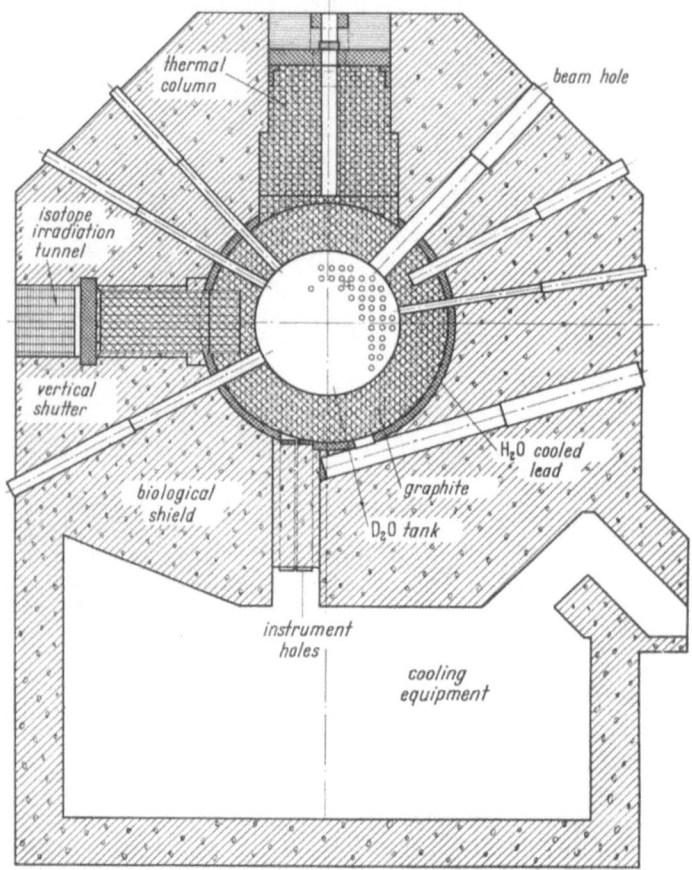

Fig. 3. Horizontal section of *CP3*, the heavy water research reactor at Argonne Laboratory.

reactor in which U[235] (about 800 gm) is present as a uranyl nitrate solution in ordinary water. The solution is contained in a stainless steel sphere 1″ in diameter, in which is mounted a helical coil carrying cooling water. A reactor similar to the Los Alamos water boiler has been built at North Carolina State College[3] as an unclassified research instrument. Another enriched, moderated reactor is the "swimming pool"[4], a large tank of water in which is mounted a small lattice of enriched uranium. All of these enriched reactors just mentioned are characteristically small in size because of the large k_∞, hence the ratio of the central flux to the total power output is much higher than it is for an unenriched reactor.

[1] D. B. Hall and J. Hall: AEC document MDDC 1080 (1947).
[2] L. D. P. King: Geneva Conf. Paper 488 (1955).
[3] C. Beck: Nucleonics 9, No. 5, 18 (1951). — Geneva Conf. Paper 487 (1955).
[4] Breazeale, Cochron and Donebon: Geneva Conf. Paper 489 (1955).

A disadvantage of the reactors of small size is that only a few experiments can be performed simultaneously, as contrasted to the large unenriched reactors. Several recent reactors have been constructed of intermediate size so as to provide high flux for a reasonable number of experiments without extremely high total

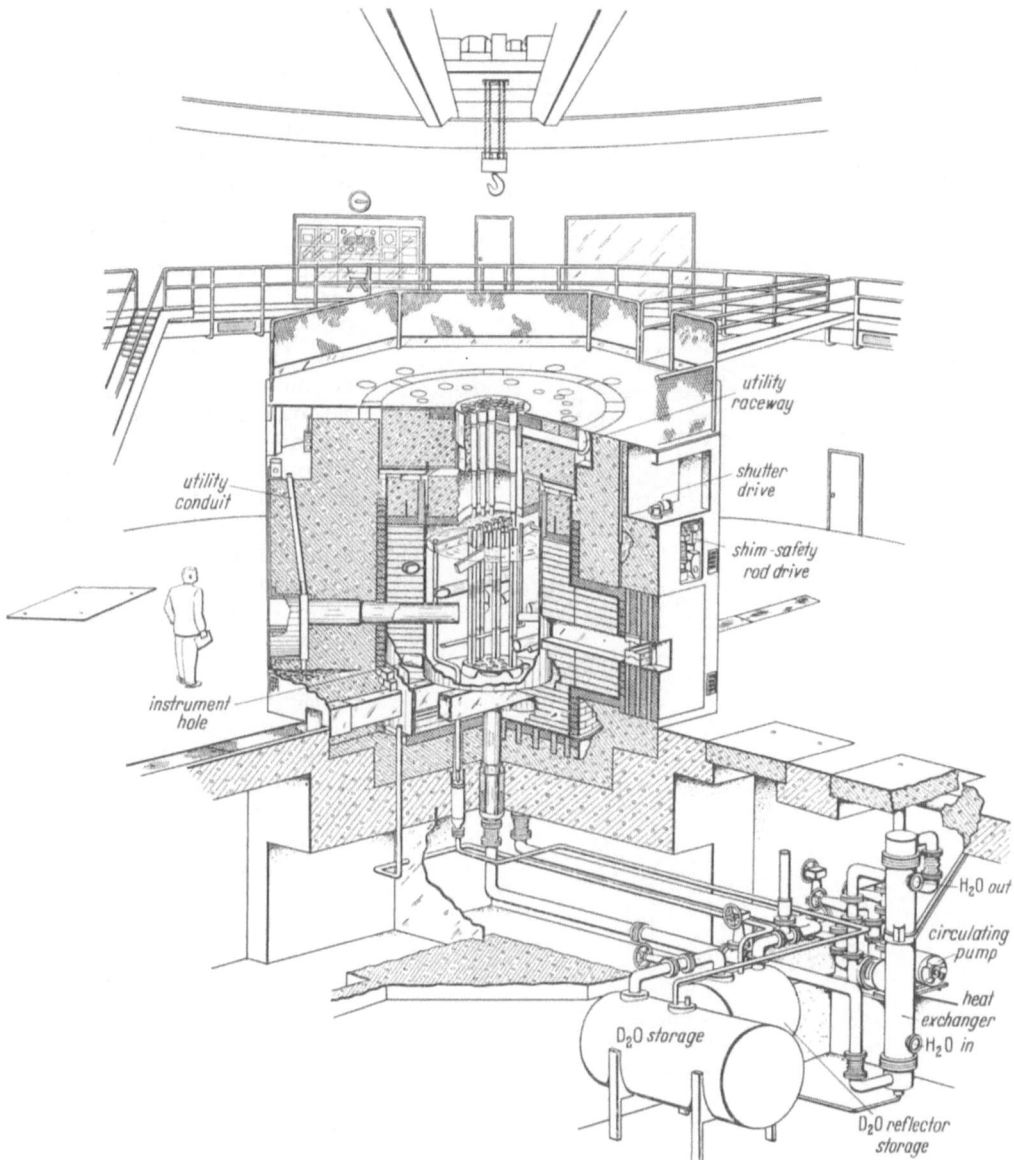

Fig. 4. *CP*5, the high flux research reactor at Argonne Laboratory.

power. Thus the Argonne heavy water research reactor CP5, Fig. 4, provides a reasonably large available space and a peak thermal flux of 10^{14} at 5 megawatts power. The Oak Ridge light water research reactor, Fig. 5, has about the same characteristics, as does the Harwell heavy water research reactor, DIDO.

Present research reactors have many features in common, and the general manner in which neutrons are obtained from them is illustrated in Fig. 1. This figure, as already mentioned, is based on a typical graphite reactor, but a heavy water reactor would differ only by the addition of a watertight enclosure of the lattice. The highest neutron fluxes are available inside the lattice itself, but because of space limitations and need for collimation, most experiments are actually performed with neutron beams that emanate from the reactor. If an experimental hole is opened through the shield and into the lattice itself, a beam of neutrons representative of the energy distribution present in the lattice is obtained, a beam of so-called "pile neutrons". On the other hand, if the hole is opened into a region where no uranium is present, such as the reflector, then thermal neutrons, more or less contaminated with fast neutrons, will be present in the beam. For some experiments, it is necessary to obtain extremely well thermalized neutrons, and an extension of the moderator, known as the *thermal column*, could be constructed for the reactor illustrated by replacing the removable shield plug with graphite, and the neutrons within this column, while of lower intensity than those in the reflector, would be less contaminated with fast neutrons.

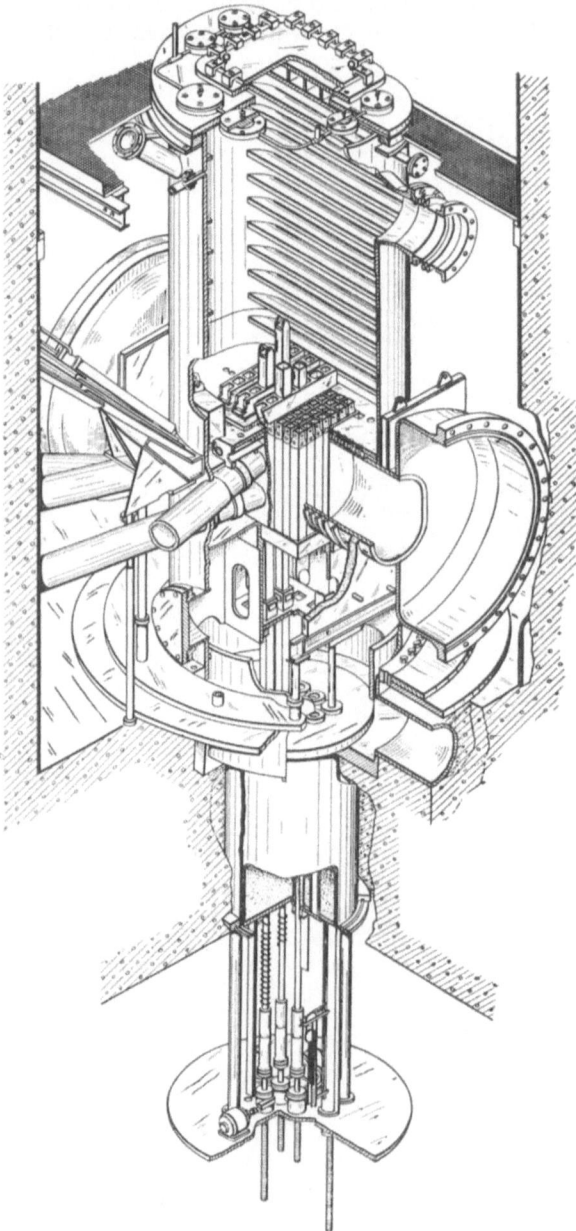

Fig. 5. The Oak Ridge research reactor, which is moderated and cooled with light water. Various experimental holes, the largest of which is 19 inches in diameter, are shown.

The most important characteristic relating to a reactor, as far as neutron research is concerned, is the value of the *neutron flux*. A secondary, although important, consideration is the *availability* of the flux, that is, the possibility of making openings into the lattice and reflector at desired locations. In comparing different reactors, it is customary to refer to the maximum flux, which is usually found in the moderator at the center of the lattice. Because the total power output is proportional to the integral of the flux over the volume of the reactor, those of small size have a relatively high flux at low power.

It is difficult to give a quantitative measurement of the availability of the flux, but in general it can be stated that the high flux of the water-cooled reactor (which is possible because of the superiority of liquid over gas cooling) is somewhat offset by the difficulty of opening experimental holes into the water tank. The graphite-moderated reactors have a lower central flux, but are larger in volume because of the lower k_∞, hence have more experimental space, which is also easily available, than the liquid-moderated reactors. The air-cooled reactors in general show more leakage of fast neutrons to the outer regions of the reflector because of the necessary openings into the center of the lattice for movement of cooling air. The control of the power level by neutron absorbing control rods in both types of reactors is simple, and the neutron flux can be easily held constant to a few tenths of a percent in either.

5. Distribution of fast neutrons. In considering neutrons produced in reactors with respect to their use in research, it is customary to divide them energywise into three groups; fast, resonance, and thermal. These energy classifications, in spite of their titles, refer rather to the manner in which the neutrons are produced in the reactor than to the nuclear reactions characteristic of them. The *fast* neutrons are those resulting from fission that have not been moderated by collisions. The *resonance* neutrons, comprising the dE/E spectrum produced by moderation, are usually considered to extend in energy from about 1 Mev down to 1 ev. The *thermal* neutrons have reached equilibrium with the atoms of the moderator and possess a Maxwellian distribution of velocities.

In the simple form of reactor theory discussed in Sect. 1, the neutrons were treated as if they were all of thermal energy. This one-group theory led to Eq. (2.2), and the solution of this equation for a rectangular lattice showed that the neutrons would have a cosine distribution. A more accurate form of theory must take into account the fact that fission neutrons move an appreciable distance before becoming thermal. In our discussion of the infinite lattice (evaluation of k_∞), the effect of fast neutrons was treated correctly by inclusion of the factors ε and p in Eq. (1.1), but in the case of the finite lattice this additional feature was included in an approximate way by changing L^2 to M^2 to allow for the migration of neutrons during moderation. Fortunately, the spatial distribution of neutrons of all energies in an actual pile is given reasonably well by the solutions of Eq. (2.5), which is just the one-group pile equation written so as to include the effect of fast neutrons.

The reason for the similarity in spatial distribution of fast and thermal neutrons is that the thermal neutron distribution determines the rate of fission in the uranium; as a result the emission rate of fast fission neutrons will have the same distributions as the thermal neutrons. As the fast neutrons are slowed down, they move over distances of the order of $(6\tau)^{\frac{1}{2}}$, the root mean square slowing-down distance, which for graphite is about 40 cm. Since this distance is small compared with the dimensions of the reactor, the moderated neutrons as well as the fast neutrons follow approximately the same distribution as the thermals. Outside the lattice itself, in the reflector of Fig. 1, the distributions of neutrons of various energies differ greatly. In the reflector there is no production of neutrons, hence the flux of neutrons of a particular energy is determined by the transmission through graphite for that energy.

Even before moderation, the fission neutrons exhibit an extremely wide energy range, from zero to at least 15 Mev. The energy distribution is represented within experimental error by the empirical formula,

$$N(E) = \sinh (2E)^{\frac{1}{2}} e^{-E}, \tag{5.1}$$

where E is the energy in Mev. This formula is consistent with a model in which the neutrons are emitted from or "boil off" the highly excited fission fragments after they have separated. According to this model, the observed neutron energy distribution is that of the evaporated neutrons relative to the fission fragment, modified by the motion of the fragment itself.

The neutrons of any particular energy in the fission spectrum attain a dE/E energy distribution after colliding with the moderator atoms. The actual energy distribution in the moderator will thus be a sum over many dE/E distributions, each one extending up to a particular energy in the fission spectrum. Because of the dE/E flux distribution, there are many more neutrons of low energy, per unit energy, than of the unmoderated fast neutrons in the moderator. It is difficult to observe the effects of the fast neutrons only in an experiment performed inside the lattice because of interference by the high flux of moderated neutrons.

It is easy to see that the flux of fast neutrons (of all energies in the fission spectrum) near a uranium fuel rod is of the same order of magnitude as the thermal neutron flux at the same point. Most of the thermal neutrons hitting the uranium rod will be absorbed in it, and for each one so absorbed $\eta \varepsilon$, or about 1.4 neutrons will reappear as fast neutrons. The number of these fast neutrons crossing a square centimeter at the surface of the rod, or the fast flux, will then be of the same order of magnitude as the thermal flux.

The fast neutrons will decrease with distance in the reflector in a much more complicated manner than do the thermal neutrons. The decrease of the fast neutrons is more rapid, hence the ratio of fast to thermal neutrons will decrease rapidly with distance in the reflector. The unmoderated fast neutrons (those that have suffered no collisions) will decrease in the moderator approximately as $e^{-x/\lambda}$, where λ is the mean free path for the fast neutrons. If we take a value of 2 barns for the scattering cross section in graphite for fast neutrons, and a density of 1.6, we obtain a mean free path of 6.2 cm, which is much shorter than the diffusion length in graphite, which is about 50 cm. The slightly moderated, but still fast, neutrons will decrease more slowly than the unmoderated neutrons, according to age theory, but they will still decrease much more rapidly than the diffusing thermals in the reflector.

6. Resonance neutrons. The neutrons of intermediate energy, those in the dE/E slowing-down spectrum, are of great value for study of cross-section resonances and are usually referred to as *resonance neutrons*. The number of neutrons in the resonance spectrum relative to the thermal neutrons is conveniently expressed in terms of the lifetime of the thermal neutrons. The number of neutrons becoming thermal per sec per cm³ is given by the *slowing-down density*, q, at an energy just above thermal and the *thermal flux* resulting from these neutrons is $q v t_0$, where v is the velocity, t_0 the lifetime of the thermal neutrons. The relationship of the resonance flux per unit energy interval to the slowing-down density is given by the nuclear properties of the moderator:

$$\text{resonance flux} \equiv (n v)_E \, dE = \frac{q_E}{\xi N \sigma_s E}, \tag{6.1}$$

where q_E is the slowing-down density at energy E, from which we see that the ratio of slowing down density to the resonance flux is determined by $\xi N \sigma_s$, or the *slowing-down power*.

The ratio of resonance flux to q, hence to fast flux, is about the same for a graphite as for a heavy water reactor, even though ξ is much higher for deuterium than for graphite, because the slowing down power is about the same for the two moderators. Because of the adjustment of the thermal absorption to attain a high k_∞, the ratio of the resonance to the thermal flux, determined by the thermal lifetime, is also about the same for a graphite and for a heavy water reactor. The resonance and thermal energy distributions are shown in Fig. 6 for a typical well-moderated reactor. There is the possibility, not exploited as yet, of obtaining an unusually high resonance flux for a given q (hence specific power) by use of a moderator of poor slowing-down power, for example of high atomic weight.

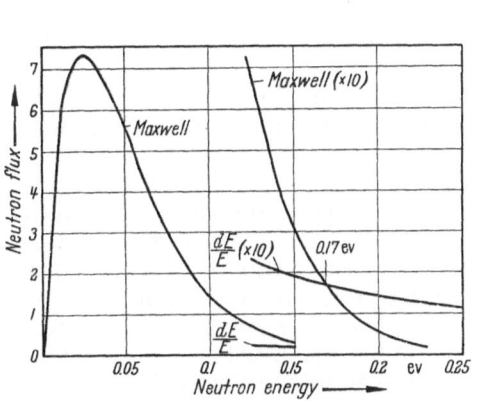

Fig. 6. The energy distribution in the region between thermal and resonance energies in a typical well-moderated reactor.

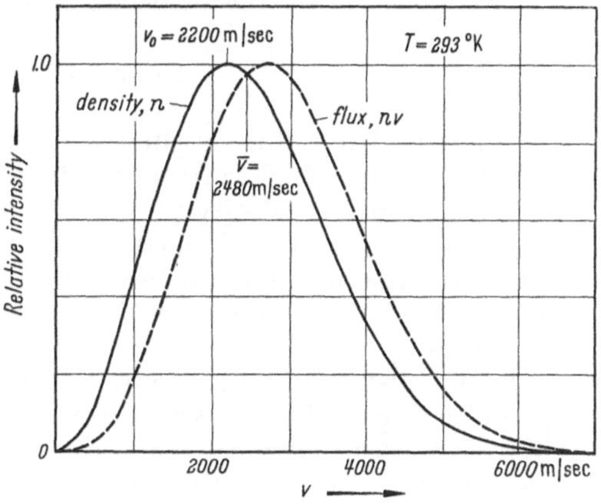

Fig. 7. The Maxwell density distribution and the flux distribution of thermal neutrons.

7. Thermal neutrons. We have just referred to the thermal flux, nv, as if all the thermal neutrons had the same velocity v. This procedure is a customary one in dealing with thermal neutrons, in spite of their relatively wide velocity spread, and in most cases no errors arise from this simplification. It is usually sufficient to interpret v in the formulas used as a particular velocity in the thermal distribution, at times the average velocity being correct, and at other times, the most probable velocity.

When neutrons are in thermal equilibrium with the moderator atoms they will have the well-known Maxwell distribution of velocity

$$dn = \frac{4n}{v_0^3 \sqrt{\pi}} v^2 \, e^{-\frac{v^2}{v_0^2}} \, dv, \tag{7.1}$$

where dn is the number of neutrons per unit volume in the velocity range dv at the velocity v, and n, as before, is the total number of neutrons per unit volume. The velocity v_0 is the most probable velocity, that is, the velocity for which $n(v)$ is a maximum, Fig. 7. The value of v_0 is determined from the fact that it corresponds to a kinetic energy of kT, where k is BOLTZMANN's constant (1.380×10^{-16} erg per degree) and T is the temperature of the distribution:

$$\tfrac{1}{2} m v_0^2 = kT,$$

$$v_0 = \left(\frac{2kT}{m}\right)^{\frac{1}{2}} = \left(\frac{2 \times 1.380 \times 10^{-16} \, T}{1.675 \times 10^{-24}}\right)^{\frac{1}{2}} = (1.648 \times 10^8 \, T)^{\frac{1}{2}}. \tag{7.2}$$

Thus the most probable velocity at room temperature, 20° C, will be 2200 m/sec (actually 2198 m/sec) and this velocity corresponds to an energy of 0.0253 ev.

The neutrons in a distribution of temperature T are sometimes referred to as "kT neutrons", although the average neutron kinetic energy is $\frac{3}{2}kT$ ($\frac{1}{2}kT$ per degree of freedom), and kT is the energy corresponding to the most probable velocity. Thermal cross sections, for instance those given in the Brookhaven compilation[1], are tabulated as cross sections for a velocity of 2200 m/sec, corresponding to $T = 293.6°$ K or a λ of 1.80 Å). The *average velocity* of the Maxwell distribution is greater than the most probable velocity, and integration of Eq. (7.1) shows that the average velocity is greater than the most probable by the factor $2/\sqrt{\pi}$ or 1.128. The root-mean-square velocity corresponds to the average energy and is obviously equal to $\sqrt{\frac{3}{2}}v_0$. The Maxwellian gives the density of neutrons, n, as a function of velocity, rather than the flux. The flux distribution, obtained by multiplication by the velocity, also shown in Fig. 7, is

$$d(nv) = \frac{4n}{v_0^3\sqrt{\pi}} v^3 e^{-\frac{v^2}{v_0^2}} dv. \tag{7.3}$$

The most probable flux, evaluated by differentiation of Eq. (7.3), is found at a velocity of $\sqrt{\frac{3}{2}}v_0$, which corresponds to a velocity greater by a factor of 1.22 than the most probable velocity.

If a neutron beam is formed by a hole that opens into the reflector of the pile or into the thermal column, the flux distribution in the beam will also be given by Eq. (7.3), because the beam flux has exactly the same distribution as that of the flux in the neutron-radiating medium. The effective flux distribution, that is, the flux distribution as seen by a particular detector, is obtained from Eq. (7.3) by multiplication by the sensitivity of the particular detector. For a detector having a $1/v$ sensitivity, such as a thin boron counter, the distribution of counting rate versus neutron velocity will again be just the Maxwell velocity distribution, Eq. (7.1). The actual velocity distribution of the thermal neutrons in the pile agrees well with the Maxwellian; except for some slight differences at very long wavelength, associated with diffraction effects.

The thermal flux in a beam is calculated from the geometry of the experimental hole in a straightforward manner. Thus a 10×10 cm hole, opened to a point in a reactor at which the flux is 5×10^{13} will produce a thermal flux in a beam just outside the shield, say 6 m from the center of the pile, given by

$$nv = \frac{5 \times 10^{13} \times 10^2}{4\pi(600)^2} = 1.1 \times 10^9 \text{ cm}^{-2} \text{ sec}^{-1}.$$

For the case of a beam from the thermal column, the flux at the radiating surface will be much less than that at the reactor center, and this decrease will be only partially compensated by the fact that the radiating surface is closer to the point just outside the shield where the neutrons are used. For example, for a 10×10 cm hole opened 2 m into the thermal column of the Argonne $CP3$ reactor, at which point the flux is 10^{10}, the flux at the outer end of the hole is

$$\frac{10^{10} \times 10^2}{4\pi(200)^2} = 2.0 \times 10^6 \text{ cm}^{-2}\text{-sec}^{-1}. \tag{7.4}$$

Although the thermal flux in this beam is very low, for some experiments the low intensity is far outweighed by the small number of resonance and fast neutrons.

---

[1] D. J. HUGHES and J. A. HARVEY: Neutron Cross Sections. Brookhaven National Laboratory Report BNL 325 (Superintendent of Documents, U. S. Government Printing Office, Washington, D. C. 1955).

II. Use of fast neutrons from reactors.

8. Fission neutrons inside lattice. Because of the great range of neutron energies present in the pile flux, irradiation inside the pile lattice is not useful for measurement of those fast neutron cross sections that have an appreciable value in the resonance energy range as well. The presence of resonance neutrons creates no difficulty, however, for such threshold reactions as (n, p), (n, α), and $(n, 2n)$ reactions, which usually have thresholds well over 1 Mev. These reactions, when produced in the lattice, are observed most simply by means of the radio-activity of the product nucleus, and the method is difficult to apply to those reactions for which the final nucleus is stable. Irradiation inside the pile lattice is particularly well suited for production of *threshold reactions* because their low cross sections necessitate much higher flux than is found in neutron beams.

As a result of the wide spread in energy of fission neutrons the interpretation of the observed rate of a threshold reaction produced by them, in terms of the cross section, which is a rapidly varying function of the energy, is not simple. For a light element, which usually has strong resonance structure in its cross section curve, the interpretation is even more complicated. In the case of a heavy element, however, the dominant feature in the cross section is the *barrier penetrability* of the outgoing proton or alpha-particle, which determines the width, or emission probability, of the particle. The fast neutron resonances are so closely

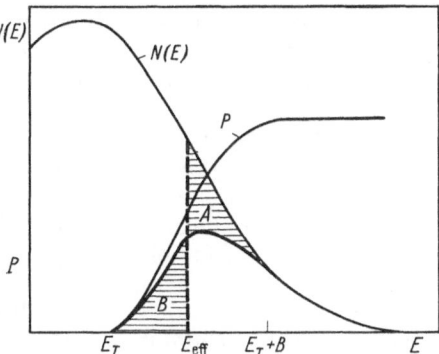

Fig. 8. The analysis of the production of a threshold reaction by fission neutrons in terms of the fission spectrum and the penetrability of the outgoing proton or α-particle.

spaced for all but the lightest nuclei that they overlap, and we may expect that as a result the cross section will rise smoothly, approximately as the penetrability factor, from zero at the threshold energy. This assumed shape of the cross section, together with the known energy distribution of fission neutrons, can be used to relate the measured reaction rate to the cross section.

When a sample of the material under study is irradiated with fission neutrons inside the reactor lattice, the activation rate produced is usually determined by "counting" the foil on a calibrated Geiger or scintillation detector. The activation, together with the known fission flux, will give the cross section averaged over the fission spectrum, $\bar{\sigma}$, for the particular reaction considered, by means of the usual *activation equation*,

$$\bar{\sigma} = I_0/n\,v\,N_t, \tag{8.1}$$

where I_0 is the *saturated* activity in disintegrations per sec, $n v$ is the neutron flux, and N_t the number of atoms in the sample. Of course, only those neutrons of energy greater than the threshold energy participate in the reaction, hence it is necessary to consider the reaction rate as a function of neutron energy in order to interpret the measured $\bar{\sigma}$.

The analysis of a threshold reaction is shown in Fig. 8, which includes the fission neutron distribution $N(E)$, as well as the penetrability P. Assuming that the cross section is proportional to the penetrability, we see that the reaction rate as a function of energy will be given by $N(E) \cdot P$, which product is also plotted in Fig. 8. In order to simplify the analysis, a vertical dotted line is drawn at such an energy that the area marked A is equal to the area marked B, and this energy

26*

is called the effective energy, E_{eff}. The meaning of the effective energy is that the correct reaction rate is obtained on the simple assumption that no neutrons below E_{eff} contribute to the reaction, but all above it contribute with a penetrability of unity.

It would be possible, from the known shape of the fission spectrum and from the value of $(E_{eff} - E_T)$, to predict the observed average cross section if the correct value were known for the cross section at unit penetrability, which we can call σ_0:

$$\bar{\sigma} = \sigma_0 \frac{\int_{E_{eff}}^{\infty} N(E)\,dE}{\int_0^{\infty} N(E)\,dE}. \qquad (8.2)$$

Little can be said theoretically about σ_0 beyond the fact that it should be proportional to the collision cross section πR^2. In general, σ_0 is only a fraction of πR^2 because of competition among various modes of decay of the compound nucleus. Observed cross sections, $\bar{\sigma}$, for a number of nuclei, show that σ_0 is, in fact, proportional to πR^2.

In order to obtain the highest possible intensity of unmoderated fission neutrons for a cross-section measurement, it is important to get the sample as close as possible to the uranium fuel rods in the lattice. A material can be irradiated with a high flux of fission neutrons by placing it inside one of the central lumps of uranium or, better still, inside a piece of U^{235}. A uranium (or U^{235}) fuel rod, constructed in the shape of a cylindrical shell to hold samples, is usually called a receptacle slug. The flux of fission neutrons produced will be of the same order of magnitude as the thermal flux because practically all the thermal neutrons will be converted to fission neutrons in the uranium. In principle, the fast flux could be calculated from the known constants of uranium, but the calculation is rather involved because of flux depression and self-protection. Actually, it is a simple matter to obtain the fission flux by comparing it with another fission flux produced in a simpler geometrical arrangement outside the reactor, such as the "converter" described in Sect. 10. The fission flux in the converter arrangement can be calculated quite easily from the thermal flux because of the simple geometry involved.

The method of irradiation inside a uranium slug is particularly valuable when it is desired to produce a radioactive isotope of high specific activity that can be made by a threshold reaction, because in that case chemical separation of the "carrier-free" radioactive product (that is, containing no inert atoms) is possible, as is never true for production by neutron capture (for which initial material and radioactive products are chemically identical). Several important radioisotopes, for example P^{32}, S^{35}, and Ca^{45}, are produced by (n, p) and (n, α) reactions at Oak Ridge for general distribution.

9. Radiation effects. Fast neutrons are very effective in producing changes in a variety of physical properties of materials, for example, resistivity, heat conductivity, hardness, and elasticity. These changes are referred to as "radiation damage", and their investigation now constitutes an important division of solid state physics. In addition to the obvious practical application of radiation damage investigations to the effects produced on materials (chiefly reactor components) by the high fluxes found in lattices, the results are useful in investigations of the nature of the structure of matter, particularly the relationship of physical properties to lattice imperfections.

The properties of solids that are particularly sensitive to radiation are those caused by the presence of imperfections in the crystal lattice. These imperfections may be lattice vacancies (the absence of atoms from their normal sites) or displacements (for which atoms are at positions that are not normal lattice points). The relationship between these imperfections and electrical resistivity, heat conductivity, hardness, elasticity, etc. is the subject of extensive theoretical work in solid state physics. The most efficacious radiations in the reactor for production of atomic displacements and vacancies are the fast neutrons, for which the cross section is mainly elastic scattering in which energetic recoil nuclei are produced, themselves capable of producing further recoil nuclei. As the average energy required to displace an atom from a lattice site is only about 25 ev, many of the recoil nuclei produce displaced atoms, and leave vacancies at the same time.

The dislocations produced by elastic recoil from fast neutrons result in changes in physical properties that resemble in many ways the effects of cold work, and much of the theory of cold work is applicable to radiation damage. Because crystalline imperfections result from cold work and radiation damage as well, it is not surprising that superficially similar physical results are produced from both. Cold work, however, disturbs a large number of atoms as a group, while individual atoms are displaced independently by radiation, hence it is expected that closer examination will reveal some differences. Impurity atoms are, of course, produced by nuclear disintegrations when materials receive neutron bombardment, and the effects of these impurities will have no analogue in cold work effects.

While irradiation with fast neutrons inside the reactor is an effective method of producing radiation damage, it is easily realized that the attendant difficulties are great. The work is made difficult by the intense radioactivity of the irradiated samples, which is a result of the high neutron flux necessary to produce the desired effects. Although the samples may contain several hundred curies of activity at removal, the desired physical properties may still be measured in the "hot laboratories" that are usually located at research reactors. At these laboratories, remote control equipment has been developed to the stage where almost any standard manipulation or measurement can be performed on these extremely active materials. Sample containers can be opened, the contents removed, weighed, machined, examined spectroscopically or by x-ray diffraction, and all the usual mechanical tests (hardness, elasticity, etc.) can be made without overexposure of personnel to radiation. The Argonne hot laboratory and its remote control equipment, described by HULL, GOERTZ, and FERGUSON[1] is typical of the facilities available for work with multicurie sources.

In order to obtain sufficient fast neutron fluxes, the irradiations are always performed inside the lattice, where the presence of low energy and resonance neutrons fortunately does not interfere with the radiation damage results, which are caused only by the displaced atoms. In addition, the precise energy of the fast neutrons is usually of no interest, so no special techniques are involved to determine the actual spectrum of the incident fast neutrons. The lattice is usually well above room temperature and this fact introduces definite difficulties in much radiation damage work. Many of the effects of displaced atoms disappear if the sample is not kept at a low temperature, that is, extensive "annealing" may take place. As a result, it is often desired to "freeze" the radiation effects by holding the sample at liquid air temperature during and after irradiation. Needless to say, the problem of maintaining a sample at liquid air temperature

---

[1] HULL, GOERTZ and FERGUSON: AEC report AECD 2990 (1949).

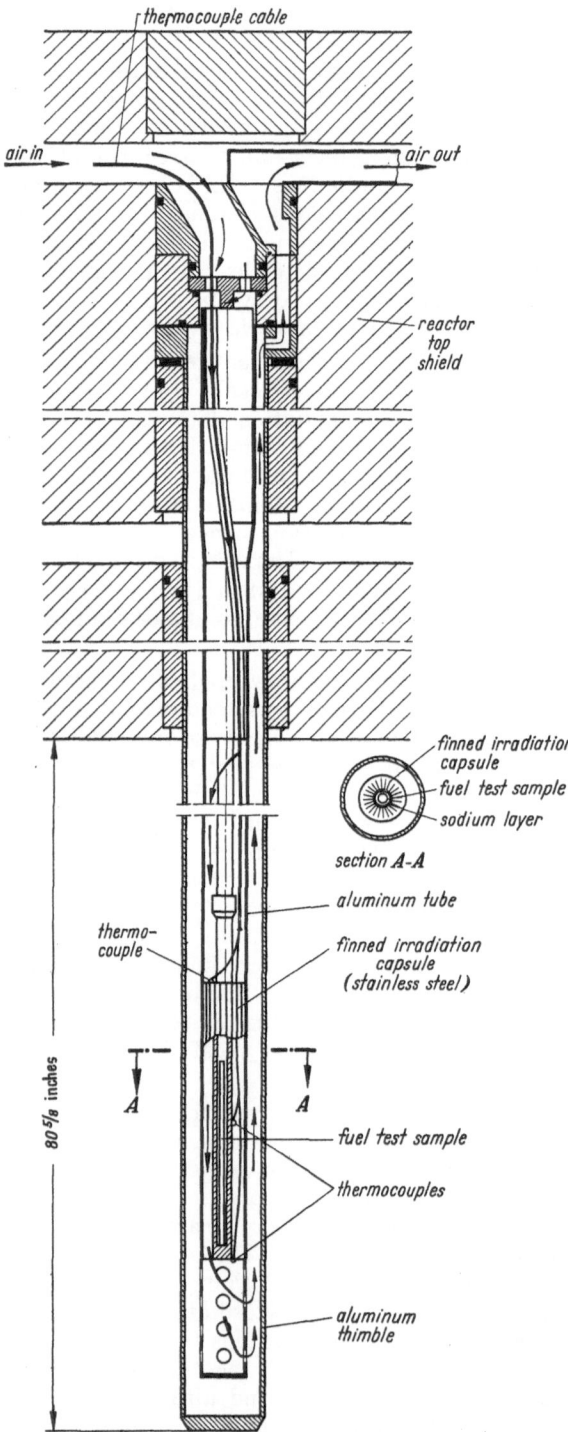

thermocouple cable

air in

air out

reactor top shield

finned irradiation capsule

fuel test sample

sodium layer

section A-A

aluminum tube

finned irradiation capsule (stainless steel)

thermo-couple

A A

fuel test sample

thermocouples

aluminum thimble

80 5/8 inches

Fig. 9. The arrangement in use at the Argonne reactor for irradiating fuel samples at high temperature.

for long periods at the center of a high flux reactor, and throughout subsequent study, is not simple. Even if the sample container is well insulated from its high temperature environment, neutron capture will produce heat; and if this source is lessened by use of low cross section material, heating by gammas is still a problem. Cryostats for low temperature irradiation, using circulating liquid air, are now in operation at Oak Ridge and Brookhaven Laboratories, although only a small amount of low temperature irradiation has been performed to date. In some investigations it is necessary to allow the sample to rise to a high temperature, fuel elements, for example. Fig. 9 shows a fuel element irradiation facility at the Argonne $CP5$ reactor in which the heat produced in the fuel sample (U^{235}) under test is removed by an air stream[1].

10. Fission source in thermal beam. It is sometimes desirable to obtain fission neutrons that are uncontaminated with resonance neutrons, for example in the measurement of (n, γ) cross sections at fission neutron energies. This selection is possible by irradiating uranium with well-thermalized neutrons outside the reactor. The fast neutron flux produced in this way will be much less intense than that inside the lattice, but the method is the only one that removes the resonance neutrons effectively. The thermal neutrons cause fission in the uranium with the emission of 2.5 fast neutrons per fission and the uranium thus acts as a neutron "converter",

[1] See Vol. 7 of the Geneva Conference Proceedings for many detailed reports on radiation damage by fast neutrons.

creating a fission flux that can be several times as large as the thermal flux. The sample to be irradiated with fast neutrons (usually in the form of a pressed pellet or metal foil) is wrapped in cadmium and held against the uranium plate, while a similar foil, not in cadmium, records the total flux (mainly fast).

The (n, γ) cross section measurement is carried out by activating a cadmium covered foil of the material of interest with the fission neutrons from the plate, while an identical bare foil is simultaneously irradiated with thermal (plus fission) neutrons. After the foils have been counted on a G-M counter, the fast activation cross section is obtained directly from the counting rate ratio of the cadmium-covered to the bare foil:

$$\sigma_{\text{fast}} = \frac{\sigma_{\text{th}}}{R} \frac{I_{\text{Cd}}}{I_{\text{bare}} - I_{\text{Cd}}}, \tag{10.1}$$

where σ_{th} is the known thermal cross section for the particular activity detected, and R is the calculated ratio of fast to slow flux at the plate (of the order of two). In this method no corrections are necessary for self-absorption of betas in the foil, complicated decay schemes, or efficiency of the counter. In some cases, however, where the thermal cross section is not well known, it must be determined from the bare foil activation, taking into consideration the thermal flux and the absolute counter efficiency.

In the method just described for measurement of (n, γ) cross sections, absolute knowledge of the fast flux is unnecessary. The ratio of the fast flux to the thermal flux is calculated from the geometrical arrangement. The flux at a point on the axis of the uranium plate, at a distance $(h - t)$ from the plate (the foil is not quite in contact with the plate), is obtained by integration over the volume of the plate:

$$(n \, v)_{\text{fast}} = (n \, v)_{\text{thermal}} \, N \, \sigma_F \, 2.5 \int\limits_0^t \int\limits_0^R \frac{2 \pi \, r \, dr \, e^{-N \sigma_T z} \, dz}{4 \pi \left[(h - z)^2 + r^2 \right]}, \tag{10.2}$$

where $N =$ density of nuclei, $\sigma_F =$ fission cross section, $\sigma_T =$ total cross section and t and R are the plate thickness and radius. For a plate of U^{235} 4″ in diameter and 3 mm thick, for example, the integration gives $n v_{\text{fast}} / n v_{\text{thermal}} = 2.2$ for a point 1.6 mm from the plate.

This result may seem surprising at first because each thermal neutron produces 2.5 fast neutrons and only half of these proceed in the direction of the foil, yet the fission flux at the foil is more than twice the incident thermal flux. It is seen from Eq. (10.2) that in principle there is no limit to the flux ratio as the radius R is increased. In the calculation the loss of fast neutrons by absorption in the plate is neglected, but this loss is partially compensated by the fact that additional neutrons are produced by fission resulting from absorption of the fast neutrons.

The fast neutron capture cross section obtained by the converter plate method is an average over the fission spectrum, but because the variation of $\sigma(n, \gamma)$ is about $1/E$ for all elements near 1 Mev, comparative cross sections are certainly meaningful. Calculation shows, furthermore, that the average cross section is just the cross section that would be observed with monoenergetic neutrons of 1 Mev energy, so the measured results can be considered the 1 Mev values, to a close approximation. The spread of energy present in the fission neutrons is of no disadvantage in the capture cross section work, for the cross section at 1 Mev would be an average over many resonances even with monoenergetic neutrons, for all but the lightest elements.

An apparatus has been developed[1] at Brookhaven Laboratory that is arranged in such a way that irradiations with fission neutrons can be accomplished without increasing the general neutron background around the reactor. The U^{235} plate is located in a space formed by removal of one of the large shielding blocks on the top of the reactor, and this cavity is made sufficiently large so that no difficulty is experienced with neutrons that are slowed down in the shield and return to the foil as resonance neutrons. The purpose of the design is to obtain a high flux of thermal neutrons (which do not need to be collimated) at the U^{235} plate with as few resonance neutrons as possible, and to shield the plate so that irradiations can be accomplished without pile shutdown.

It is possible to withdraw the plate, through an opening in the removable shield, to a shielded space outside the irradiation hole, and there remove foils from the plate without radiation hazard while the pile is operating at full power, 28 Mw. At 28 Mw, the thermal flux at the plate is 5×10^8, the fission flux about 10^9 and the total emission rate of the plate about 10^{11} fission neutrons per second.

11. Photo-neutron sources. It is possible to perform fast neutron research utilizing the reactor even though the fast neutrons are not produced in the reactor itself. Intense gamma sources may be made in the lattice by thermal neutron irradiation and the gamma sources used in turn to produce photoneutron sources. Unfortunately, only a few high intensity gamma sources can be made in the reactor that emit sufficiently energetic gammas to photodisintegrate beryllium and deuterium. The energies of the neutrons produced in these sources range from 30 kev to 1 Mev, about the same energy interval as that usually available from Van de Graaff generators, so the usefulness of the photoneutron sources is lessened by the availability of the Van de Graaffs. Nevertheless, the small size of the (γ, n) sources and the ease of their preparation have made them definitely useful in neutron research.

The gamma intensities of the reactor-produced radioisotopes and their efficiencies as photoneutron sources have been investigated by Wattenberg et al.[2-4]. Most of the sources emit single energy gamma-rays, and the neutrons produced in the (γ, n) reaction are therefore almost monoenergetic, possessing an energy spread of less than 1%. The neutron energy E_n, in terms of the gamma energy E_γ, and the threshold energy Q for the (γ, n) reaction, is[5]

$$E_n = \frac{A-1}{A} \left[E_\gamma - Q - \frac{E_\gamma^2}{1862 (A-1)} \right] + \delta, \tag{11.1}$$

where δ is a spread in energy depending on the angle of emission of the neutrons; this spread is usually negligible. However, in order to get useful Q values, the γ-emitters are surrounded by about one inch of D_2O or beryllium. These materials cause moderation of the initially monoenergetic neutrons, and the spread of the emergent neutrons from the source is about 30%.

III. Resonance neutrons.

12. Resonance integrals. Although resonance neutrons are used both in the lattice and in external beams, their use inside the lattice is limited by the difficulty of interpreting the results when such a wide range of neutron energy

---

[1] Garth, Hughes and Levin: Phys. Rev. **87**, 222 (1952).

[2] A. Wattenberg: Phys. Rev. **71**, 497 (1957); Preliminary Report No. 6 (1949), Nuclear Science Series, National Research Council, Washington, D. C.

[3] Russell, Sachs, Wattenberg and Fields: Phys. Rev. **73**, 545 (1948).

[4] V. Hummel and B. Hamermesh: Phys. Rev. **82**, 67 (1951).

[5] A. Wattenberg: Phys. Rev. **71**, 497 (1957); Preliminary Report No. 6 (1949), Nuclear Science Series, National Research Council, Washington, D. C.

is present. As a result, resonance measurements inside the lattice usually involve integral effects extending over large energy regions, while the selection of particular resonance energies is accomplished with neutron beams. The integral effects will be discussed first, then the use of resonance neutron beams.

When a material is subjected to the dE/E resonance spectrum, the rate of occurrence of a particular nuclear reaction is proportional to the integral of its cross section as a function of neutron energy multiplied by the flux density

$$\text{reaction rate} = \frac{q}{\xi N \sigma_s} \int\limits_{0.4\,\text{ev}}^{\infty} \sigma(E)\,\frac{dE}{E}, \tag{12.1}$$

as follows from Eq. (6.1). The integral of Eq. (12.1) is the *resonance integral* which is usually taken to extend from the *cadmium cut-off*, about 0.4 ev, to fission energies (denoted ∞ for convenience). The form of the integral is based on the assumption that the dE/E spectrum is not distorted by the absorber, an assumption that implies extreme dilution. If distortion of the spectrum is present, the integral over the distorted spectrum is called the *effective resonance integral*. The cross section in the integral is usually the absorption or the activation cross section, although the *resonance scattering integral* has a definite meaning as well.

Measurement of the resonance integral gives no information on the cross section at a particular energy, nor does it tell the location of the resonances. A high resonance integral may be caused by the simple fact that a resonance is near thermal energy, for in that case the cross section at resonance is large, as is the flux. In a few instances the resonance integral results almost completely from a single resonance, and under such conditions, some properties of the single resonance can be inferred from the resonance integral.

The simplest method of measuring the resonance integral inside the pile is by means of activation, for then simple counting of foils can be used for the determination. If the ratio of the resonance to the thermal flux is known, the resonance integral follows directly from the measured cadmium ratio for activation, R_{Cd}, and the thermal cross section, σ_{th}:

$$\int\limits_{0.4\,\text{ev}}^{\infty} \sigma(E)\,\frac{dE}{E} = \frac{\sigma_{\text{th}}}{R_{\text{Cd}} - 1}\,\frac{(n\,v)_{\text{th}}}{q/\xi N \sigma_s}. \tag{12.2}$$

The quantity $\dfrac{(n\,v)_{\text{th}}}{q/\xi N \sigma_s}$, which is the ratio of the thermal flux to the resonance flux in a logarithmic (base e) energy interval, can be determined from the R_{Cd} for a standard material, such as indium or boron. For indium, the resonance integral is known from the cross section as a function of energy, and for boron the integral can be calculated from the fact that the cross section is known to be $1/v$.

Indium is a very convenient material for determination of the resonance to thermal flux ratio because its resonance integral is determined almost completely from the large resonance at 1.44 ev, whose shape has been well determined by velocity selector[1] and crystal spectrometer[2] measurements. Indium also has a high thermal cross section, is easily made into thin foils, and has a convenient half-life (54 minutes) for activation. The cross-section measurements give the total cross section, but that for activation of the 54-minute period can

[1] HAVENS, WU, RAINWATER and MEAKER: Phys. Rev. **71**, 165 (1947).
[2] V. L. SAILOR and L. B. BORST: Phys. Rev. **87**, 161 (1952).

be obtained by correcting for the relatively small scattering and the contribution to the absorption of the activities other than the 54-minute. The value of the resonance absorption integral thus determined for indium with a lower limit of 0.4 ev, is 2800 barns, accurate to about 5%. The *standard* indium foils, 0.1 gm/cm^2 thick, often used for resonance and thermal flux measurements, have cadmium ratios about five times the values for thin indium foils because of self-protection of resonance neutrons. They are usually calibrated individually.

Although the flux ratio in theory will depend on the particular lattice, the value for the graphite-moderated reactor, for which the cadmium ratio is about 30 for a $1/v$ absorber, is nearly the same as it is for the heavy water reactor. Substances with appreciable resonance activation will have cadmium ratios much less than 30 in these lattices; thin indium, for example, has a cadmium ratio of the order of 2. The cadmium ratio is a sensitive and rapid method of measuring resonance activation integrals, the main difficulty arising when a low cadmium ratio, hence possible self-protection, occurs. The effective resonance integral, which applies to the case of appreciable self-protection, gives little information on nuclear structure but has important practical applications. It is a function not only of the thickness but of the specific shape of the absorber.

13. Resonance detectors. Because of the difficulty of interpretation, only a few attempts have been made to isolate discrete sections of the dE/E spectrum inside the lattice by filtration. For measurement of cross sections at discrete neutron energies, almost the entire effort has been spent on methods utilizing beams of resonance neutrons. The energy distribution in these beams is the same as that within the lattice itself, consisting of the dE/E resonance spectrum and the Maxwell distribution of thermal neutrons. Although neutron beams have fluxes a thousandfold smaller than those inside the lattice, the relatively low intensity is compensated by the manifold possibilities of varied experimental facilities that can be used with external beams. As the neutrons in a beam are approximately parallel, transmission cross-section measurements under conditions of "good geometry", simple to perform and capable of high accuracy, are possible. Many approaches have been used to isolate the effects of particular energies in the resonance spectrum; no doubt additional methods will be developed especially when higher fluxes of resonance neutrons became available. Actually, many of the resonance neutron techniques are handicapped because the presently available fluxes are not sufficiently high to use the instruments to their best advantage.

In the decade between the discovery of the neutron and the availability of research reactors, several ingenious techniques were developed for the study of resonance neutrons with low intensity sources. These techniques are indirect and usually do not give precise information on resonance properties. Some of these older methods were adapted to reactor research and, because of the high fluxes available, are still capable of giving useful results. Of the early methods we shall consider only resonance detectors, which are still useful in reactor research, omitting the ones replaced by superior techniques. Some of the earliest cross section measurements made with reactors utilized resonance detectors for isolating bands of neutron energies, just as they had been used in pre-pile days. The resonance detectors, however, can be used with reactor beams in a manner that is much more direct and easily interpretable than the experiments utilizing natural sources. The main reason for the difference is that a reactor beam is well collimated and transmission measurements can be measured under conditions of "good geometry". Indium is an obvious resonance detector and the *indium difference* measurement in a beam geometry is simply a matter of measuring the change in detector rate with insertion of the indium absorber. Contrasted to the lattice experiment, in which activation is used, however, the transmission or total cross section, of inherent high accuracy, is usually measured. The measured transmission of the sample for the neutrons absorbed by indium, gives the total cross section of the material at 1.44 ev. The intensity of the collimated beam is recorded with any convenient detector, usually a BF_3 proportional counter.

The use of resonance detectors has been extended by taking advantage of scattering resonances as well as absorption resonances, a technique that requires high flux. A resonance will be mainly scattering if the neutron width, Γ_n, is larger than the radiation width, Γ_γ. Most of the resonances observed for neutrons in the Mev region for light elements are scattering resonances, and these resonances are well known from work with Van de Graaff accelerators[1]. The existence of resonance scattering for slow neutrons was first demonstrated by Langsdorf and Arnold[2] with reactor neutron beams, by detection of the scattered neutrons and measurement of the high scattering cross section.

Resonance scatterers can be used for measurement of total cross sections in a manner very similar to the resonance activation detectors just discussed, with the single difference that the neutrons scattered from the detector foil, instead of the activation produced in it, are measured. The scattering resonances that are most useful are those at 373 volts in manganese and 132 volts in cobalt. The scattering foil must be thin, as for the activation method, so that only a negligible number of nonresonance neutrons will be scattered. The scattered neutrons are not truly monoenergetic but have an energy spread corresponding to the width of the resonances, a few ev for the resonances just mentioned. Many cross sections have been measured by Hibdon and Muehlhause[3] at Argonne National Laboratory using cobalt and manganese resonance scatterers. In their work the neutrons scattered from the detector foils are counted by a proportional counter in the shape of a cylindrical shell. The well-collimated beam of resonance neutrons passes down the axis of the cylinder, at the mid-point of which the thin resonance detector is placed.

14. Crystal monochromators. We shall now consider methods of resonance study that were impossible before the availability of high fluxes-but are now fundamental to the field of *neutron spectroscopy*. Below 10 kev there are at the present time two methods of neutron spectroscopy in general use, one based on Bragg reflection of neutrons by single crystals and the other on time-of-flight measurements. The instruments used in these methods, crystal monochromators and time-of-flight velocity selectors respectively, have in common a neutron source produced by moderation of fast neutrons, which results in a dE/E spectrum. In other respects there are important differences; these determine the specific energy regions and cross sections best determined by instruments of the two types.

The *crystal monochromator*, or "spectrometer", is an instrument that is simple in principle, quite direct in interpretation, but dependent on high neutron intensity. As soon as intense beams of resonance neutrons became available it was possible to make use of Bragg reflection at single crystals, analogous to

[1] D. J. Hughes and J. B. Harvey: Neutron Cross Sections, Brookhafen National Laboratory Report BNA 325 (Superintendent of Documents, U.S. Government Printing Office, Washington, D. C., 1955).

[2] A. Langsdorf and W. Arnold: Phys. Rev. **72**, 167 (1947).

[3] C. T. Hibdon and C. O. Muehlhause: Phys. Rev. **76**, 100 (1949).

x-ray diffraction, to select monoenergetic neutrons. Thus, shortly after the heavy water reactor commenced operation at Argonne Laboratory, the use of the crystal monochromator was begun by Zinn[1]; at about the same time a monochromator was put in operation by Borst[2] at Oak Ridge. The beam of neutrons for a crystal spectrometer, Fig. 10, is formed by opening a hole through the shield of the pile, usually to a point as near the center of the lattice as possible. The flux incident on the monochromating crystal, placed just outside the pile shield, is of course much less than at the lattice center, being reduced enormously (a factor of 10^5 to 10^6) by the small solid angle subtended at the crystal by the source area (the base of the hole). The crystal is met at a small, adjustable

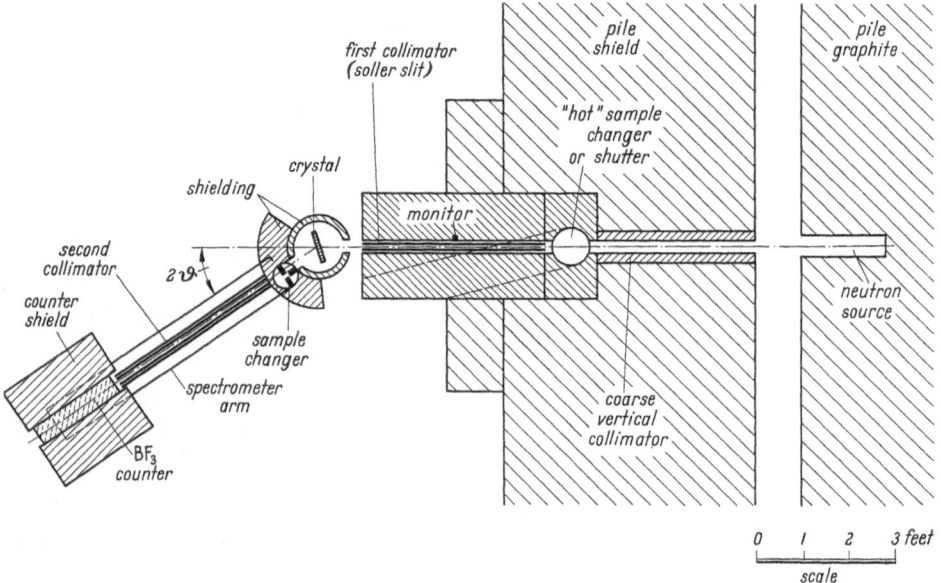

Fig. 10. The crystal monochromator or "spectrometer" for use in the resonance energy region.

angle to the incident collimated beam and the diffracted neutrons are detected by a counter, usually an enriched BF_3 proportional counter.

The neutron wavelength, λ, diffracted for a particular incident glancing angle, Θ, is given by the Bragg formula,

$$n\lambda = 2d\sin\Theta, \tag{14.1}$$

with n the "order" of reflection, usually unity. For a typical crystal, LiF, the value of d for the (111) planes is 2.32 Å hence for neutrons of 1 ev energy the glancing angle will be 3.5°. Whereas this value of Θ is a practicable one, although small, it becomes 0.35° for 100 ev neutrons, comparable to the angular collimation of the best monochromators in use today. For a collimation of 0.1° the conditions just described would imply an energy resolution $\Delta E/E$ ($\sim 2\Delta\lambda/\lambda$) of about 50% at 100 ev. The angular requirements just mentioned are important factors in reducing the range of utility of present crystal monochromators to energies well below 100 ev, in practice below about 20 ev. Another effect that restricts the instruments to low energy is the rapid decrease in reflected intensity with

[1] W. H. Zinn: Phys. Rev. **70**, 102 (1946).
[2] Borst, Ulrich, Osborne and Hasbrouck: Phys. Rev. **70**, 108 (1946).

increasing energy. This decrease is a result not only of the decreasing incident flux $(1/E)$ but of the reflectivity of the crystal itself, which varies as $1/E$. The resulting flux at the neutron detector drops off rapidly with energy, as $1/E^2$, and counting rate are prohibitively low for the high collimation associated with good resolution.

15. Time-of-flight spectrometers. The second general method in use for low energy neutron spectroscopy involves the production of neutrons in sharp bursts, of the highest possible intensity, together with the measurement of their time of arrival at a detector some 5 to 60 meter distant. The underlying principles of the time-of-flight techniques are the same whether the burst is produced by modulation of the output of a charged particle accelerator (cyclotron, betatron, or linear electron accelerator) or by mechanical interruption of a pile neutron beam. For all of the time-of-flight instruments the neutron energy (E, ev), velocity $(v, \text{m/sec})$ and flight time per meter $(t, \mu\text{sec/m})$ are related as follows:

$$v = 10^6/t, \tag{15.1}$$

$$E = 52.3 \times 10^2/t^2, \tag{15.2}$$

$$t = 72.3/(E)^{\frac{1}{2}}. \tag{15.3}$$

Thus for a typical path of 10 meters, the flight times to be measured for 1 kev and 1 ev neutrons are 22.9 and 723 μsec, respectively, easily accomplished electronically. The spread in energy actually existing for a "single" flight time, i.e. for a single time channel of the electronic recording circuits, is the result of a number of contributions to the uncertainty in timing. The most important of these are the length of the electronic time channel itself, the neutron path length in the detector, the time delay in the action of the detector, the length of the neutron burst, and the moderation time of the fast neutrons (the last not occurring for the pile choppers). The timing uncertainties just enumerated combine to give a "resolution function", whose full width at half-maximum, divided by the path length, is the "resolution" Δt in μsec/m. In the past few years resolutions have been improved about five-fold with the advent of the recent instruments, and at the present time the best available values are about 0.02 μsec/m, resulting in an energy spread of 0.02 ev at 10 ev (0.2%), rising to 18 ev at 1 kev (2%).

Although the variation of resolution with energy is the same as for the crystal monochromator, the counting rate will be constant with energy for the time-of-flight instruments. As a result, while the crystal has a very good intensity in the 0.1 to 10 ev region, it rapidly becomes inferior to the time-of-flight instruments as the energy is increased and is thus most useful in the lower energy range specified. For the time-of-flight methods in which the neutron source is pulsed, the accelerator itself (cyclotron, betatron, or linear accelerator) is not specifically designed for the purpose, and only the sharp pulsing of the ion source represents a special technique. At the present time the highest resolution of the pulsed source techniques is attained with the Columbia synchro-cyclotron, in which the intrinsic bunching of the accelerated protons produces a very short burst. The resolution of the order of magnitude 0.02 μsec/m already mentioned has been attained with this apparatus.

Whereas the charged particle accelerators used as time-of-flight sources are of conventional construction, the pile fast choppers are designed uniquely for neutron spectroscopy and hence merit a brief description. In order to produce sharp bursts of neutrons, a shutter, necessarily heavy to stop neutrons of all energies, must open and close quickly. The large size and high speed of the "fast" chopper (i.e. a chopper for fast neutrons) make it a much more complicated device than the "slow" chopper for which thin cadmium sheets suffice to stop the slow neutron beam. The requirement of short burst length for the fast chopper—about one μsec—leads to neutron beams of small area, hence low intensity; this latter tendency must be counteracted as much as possible to obtain a finite counting rate at the distant detector. These conflicting require-

Fig. 11. The fast chopper shown at the face of the Brookhaven reactor. The rotor, which operates at 10000 r.p.m. in the evacuated housing, contains plastic strips in which slits for passage of neutrons are machined.

ments result in a rather heavy, rapidly rotating shutter, containing shaped slits for the passage of neutrons, slits whose cross-sectional area is of the order of 0.01 in. by 1 in. As the slits in the moving shutter pass similar slits in a stationary collimator, neutron bursts of duration of about 1 to 5 μsec, depending on the particular chopper, are produced.

A typical high resolution chopper design is that of the Brookhaven machine, which has been in operation for about four years. The rotor is a horizontal disc, Fig. 11, 30 in. in diameter, suspended by a thin flexible shaft in an evacuated chamber. The narrow slits for the burst formation are machined in plastic pieces, which are held in the rotor between aluminium forgings. The present rotor has been in operation for about two years at 10000 rpm with a burst length slightly less than 1 μsec, the shortest burst yet attained in operating choppers.

With a scintillation neutron detector and a flight path of 20 meters, the measured resolution function (including all timing uncertainties) has a width of 1.0 μsec, hence the resolution is 0.05 μsec under these conditions.

Although the usual enriched BF_3 proportional counters or ion chambers have been used as detectors for time-of-flight work, their low efficiency (arising from the $1/v$ boron cross section), finite neutron path length, and long pulse collection time make it clear that better detectors are needed. The detectors mentioned can of course be improved to some extent. Scintillation counters, however, seem to hold the most promise. At the present time a boron-containing liquid scintillation counter is being used at Argonne in which neutrons are moderated somewhat to increase the efficiency although a time uncertainty of about 2 μsec. The Brookhaven chopper presents a more difficult detector problem than other designs, because of the larger beam area at the detector position and the higher gamma ray content of the beam.

Although fast choppers and the recent pulsed accelerators have approximately equal resolutions, there are several respects in which they differ, the principal differences pertaining to sample and counter sizes. The sample size is much less in the case of the fast chopper because of the inherently small slits. As a result samples as small as 10 mg have been used, and available supplies of separated isotopes are sufficient for investigation. On the other hand the counter area is usually smaller for the pulsed source machines and they are more suitable for measurements utilizing complex detection instruments, such as those appropriate for scattering and capture cross sections. However, we shall see that, as for the high energy region, by far the majority of the results pertain to total cross sections resulting from relatively simple transmission measurements. It is noteworthy that none of the time-of-flight methods can be used for activation cross sections, as can be done with the crystal monochromator; neutrons of all energies would activate the target, while in the latter instrument a monoenergetic beam is actually produced.

16. Determination of resonance parameters. Although the particular problems encountered in the conversion of a measured transmission curve to a neutron cross section resonance vary somewhat with the type of instrument, there are some principles, which we now consider, that are common to all methods. We shall be concerned primarily with total cross sections; the others have been measured to a much smaller extent and their analysis in terms of absolute cross sections, as well as the measurements themselves, is much more complex. A brief inspection of the Breit-Wigner formulae reveals that the total cross section alone, if accurately measured, gives much information concerning the parameters of the level involved. Considering $l = 0$ interactions only, the ones for which individual resonances are resolved in most nuclei, we see that measurement of σ_0, the peak height of the total cross section, and Γ, gives $g\Gamma_n$ as well. The peak height of the *resonance* in the *total* cross section, i.e. capture plus resonance scattering, but no potential scattering, is simply

$$\sigma_0 = 4\pi \lambda_0^2 g \frac{\Gamma_n}{\Gamma}. \tag{16.1}$$

If the statistical weight factor, g, is known, Γ_n and Γ_γ (given by $\Gamma - \Gamma_n$) are also obtained. Thus from a sufficiently accurate total cross section measurement alone all the level parameters can be obtained if g is known. For a zero-spin target nucleus g is unity and for high spin it is close to $\frac{1}{2}$, hence for these cases the total cross section alone gives all the parameters. The largest uncertainty in g occurs for a target spin of $\frac{1}{2}$ in which case it is either $\frac{1}{4}$ or $\frac{3}{4}$. Here a subsidiary measurement of the ratio of scattering or capture to total cross section (usually easier than an absolute determination) is useful as it gives Γ_n/Γ or Γ_γ/Γ and thus resolves the g uncertainty.

The difficulty of obtaining σ_0 and Γ from an experimental transmission curve varies enormously with experimental conditions, from the exceedingly few cases where essentially no corrections are necessary to those where only

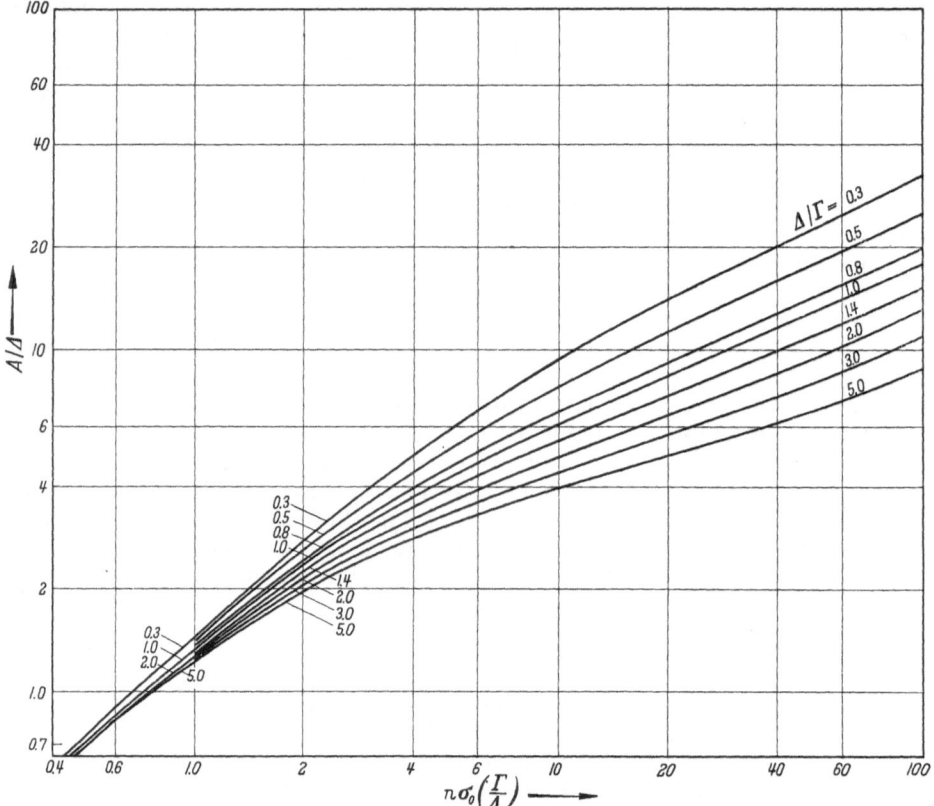

Fig. 12. Curves used for analysis of neutron resonances by the area method.

certain combinations of parameters, rather than σ_0 and Γ, can be determined accurately. As the resolution width becomes comparable to, then larger than Γ, increasingly serious complications ensue. We shall briefly describe the methods of analysis that are applicable in these various circumstances, starting with the good resolution region.

Because so few resonances are measured with negligible resolution width, we shall consider the "good" resolution to be that in which the resolution width is no larger than Γ, rather than negligible. Under these conditions the observed transmission can be corrected, before conversion to a cross section curve, for the distorting effect of the resolution function. The width and peak height of the cross-section resonance from the transmission curve corrected for resolution do not yet represent the true Γ and σ_0, however. Rather they refer to a modified resonance in which the cross section at each neutron energy actually corresponds to a spread of energies, caused by the temperature motion of the atoms of the sample. This spreading of the energies, the "Doppler broadening", lowers the peak height and increases the resonance width. The Doppler broadening is approximately a Gaussian with a width given by

$$\Delta = 2 (\mathrm{k} T E_0\, m/M)^{\frac{1}{2}} \tag{16.2}$$

where m/M is the mass ratio of neutron to nucleus. For a gas T would be the actual temperature, but for a solid the effective temperature is taken as a few degrees greater.

For heavy elements ($\Gamma_\gamma \gg \Gamma_n$) at energies where straightforward correction for resolution is almost impossible, measurements of combinations of parameters can still be made that give σ_0 and Γ_n indirectly. These measurements relate to the area of the "dip" in a transmission curve corresponding to the resonance, and if obtained for samples of widely different thicknesses give the desired parameters.

Various techniques of analysis have been developed in recent years for area analysis of resonances. A method that is rapid and sufficiently accurate is shown in Fig. 12, in which A/Δ (A is the dip area in ev) is plotted against $n\sigma_0(\Gamma/\Delta)$, for various values of Δ/Γ. All these quantities are dimensionless, and the input data, A/Δ, do not involve unknown parameters. The Doppler width, Δ, is given by Eq. (16.2). The use of the curves for the thick-thin sample case is very direct. Points are plotted, on a transparent sheet, for each sample at the correct ordinates, A/Δ, and a distance apart horizontally corresponding to the ratio of the sample thicknesses (n). If these two points are moved horizontally, they will fit only one of the curves, whose Δ/Γ value then gives Γ directly, and the abscissa gives σ_0, hence $g\Gamma_n$.[1] If the experimental errors in A are also plotted it is very simple to obtain the resultant errors in Γ and $g\Gamma_n$ by slight shifts of the sheet carrying the plotted points. Another type of analysis is involved when only one sample thickness is available. For this situation, which occurs frequently, the usual procedure is to assume a value of Γ_γ for the resonance and thus obtain $g\Gamma_n$ from the single sample area. The Γ_γ is usually known to about 20% accuracy from other considerations and this uncertainty produces an error in $g\Gamma_n$ that varies with sample thickness in a complicated manner. The curves of Fig. 12, however, give a simple solution to this problem also.

Table 1. *Resonance parameters of* Cd[113]*. The radiation width is taken as* (110 ± 2) mv *for levels where it is not measured and g is assumed to be* $\frac{1}{2}$ *except for the 0.178 ev level, for which it is taken as* $\frac{3}{4}$.

E_0 (ev)	Γ_γ (mv)	Γ_n (mv)	Γ_n^0 (mv)
0.178 ± 0.002	113 ± 5	0.65 ($J = 1$)	1.50 ± 0.05 ($J = 1$)
18.5 ± 0.2		0.28	0.064 ± 0.006
64.0 ± 0.7	80 ± 40	3.4	0.84 ± 0.07
85 ± 2		32	3.5 ± 0.5
109 ± 2		14	2.6 ± 0.3
192 ± 5		132	9.6 ± 1.0

As an example of the data now available we list in Table 1 the resonance parameters for the isotope Cd[113] as given in Supplement 1 of BNL 325, where the references to the data used in the evaluation may be found. The radiation width listed for the isotope, (110 ± 20) mv, is the value that can be assumed to apply to every level, its "error" being larger than the measured value for the

---

[1] The abscissa, $n\sigma_0\Gamma/\Delta$, is just $(4\pi \lambda_0^2 n/\Delta)g\Gamma_n$.

0.178 ev level to allow for some possible variation in Γ_γ from level to level. As the neutron widths can be obtained from total cross sections even when the total width Γ cannot, Γ_n's (and Γ_n^0, the reduced width, $\Gamma_n^0 = \Gamma_n/E_0^{\frac{1}{2}}$) appear for all levels, but Γ_γ's for two only. The radiation width is easily obtained when Γ is known by subtraction of Γ_n, which is usually small relative to Γ_γ. Actually $g\Gamma_n$, not Γ_n is obtained from the total cross section, but g is assumed to be $\frac{1}{2}$ (for Cd^{113} is actually $\frac{1}{4}$ or $\frac{3}{4}$ for $I = \frac{1}{2}$) unless J happens to be known as it is here for the first level. For this level, then, g is taken to be its true value of $\frac{3}{4}$ in getting Γ_n and Γ_γ, rather than $\frac{1}{2}$.

17. Measurement of strength function and nuclear radii.

The ratio $\overline{\Gamma_n^0}/D$, where $\overline{\Gamma_n^0}$ is the average reduced neutron width, is a quantity of great interest relative to nuclear theory. It can be obtained by averaging the parameters observed for individual resonances, but the effect of statistical fluctuations is a serious limitation. It is also possible to get the strength function by measurement of the *average* cross section, usually at higher energy, because it is proportional to the average of that part of the cross section curve corresponding to resonances. Even though individual resonances cannot be observed in U^{235} at 1 kev, for example, it is still possible to separate that part of the average cross section representing resonances from that representing potential scattering. The separation is possible because the resonance contribution is proportional to $1/v$ or to time of flight for a velocity selector measurement, while potential scattering is constant in the region of several kev.

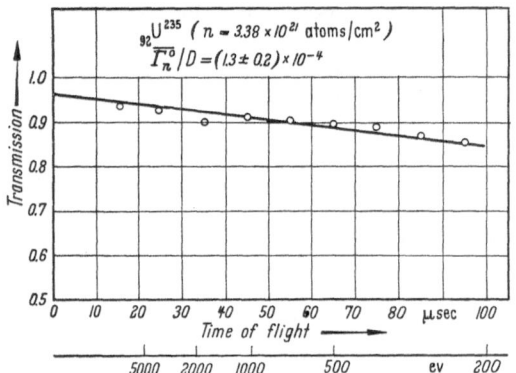

Fig. 13. The method of obtaining the strength function from the slope of the transmission curve in the kev region.

The method based on average cross sections is illustrated in Fig. 13 for U^{235}. The slope of the straight line on the transmission plot gives the strength function directly, and the intercept gives the value of the potential scattering, although the latter is not obtained with great accuracy. The method is extremely simple as illustrated by the fact that the cross section averaged over resonances at 1 kev is simply $13.0\,(\overline{\Gamma_n^0}/D) \times 10^4$ barns. We shall not consider the important bearing of the experimental values of the strength function on nuclear structure theory, particularly the "cloudy crystal ball" nuclear model; these matters are considered in detail by RAINWATER in Vol. XL.

The potential scattering, which can be determined accurately with slow neutrons, is of direct interest for nuclear models because it is intimately related to the size, shape and transparency of nuclei. Like the strength function, the potential scattering can be measured not only in the region of a few volts but at higher energies also, where resonances cannot be resolved. At low energies, as shown in Fig. 14 for U^{238}, the potential scattering is observed between resonances but is affected by the interference of the potential scattering with the resonance scattering, their amplitudes being added coherently. In the figure the cross section is shown as measured (open points) and as corrected by subtraction of the interference effects (solid points) by a computation involving the measured parameters of the nearby levels. The corrected curve is remarkably constant and represents accurately the potential scattering of 10.7 barns for U^{238}.

At higher energy, measurement of the cross section for a thick sample gives a result that is almost the same as potential scattering because neutrons of the resonance energies are removed in the first part of the sample and do not affect the results. Small corrections for the residual resonance effects can be made, however. The calculated correction for U^{238}, for example, involves consideration of an average level in U^{238} at 1 kev and the total cross section as affected by Doppler broadening. The *computed* transmission curve, based on a potential scattering cross section of 10.7 barns, agrees extremely well with the experimental results. The very close correspondence of calculated and measured values shows that the potential scattering in the energy region of several kev agrees within a few percent of that obtained at low energy.

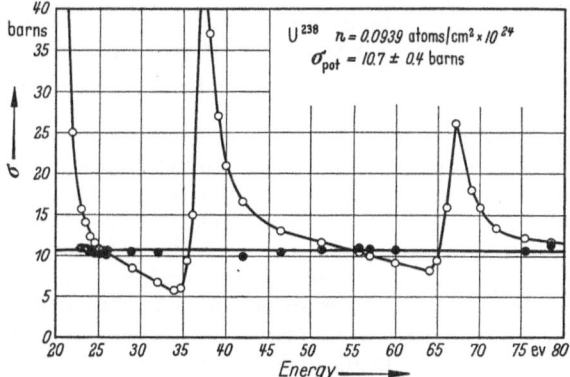

Fig. 14. Measurement of the potential scattering cross section by correction of the observed cross section for the interference effects of resonances.

The question of the relationship of the potential scattering thus determined to the nuclear radius, R, involves some recourse to nuclear theory. If the nucleus were a "hard sphere", the potential scattering would be simply $4\pi R^2$ but as nuclei are known to be partially transparent, some corrections to this simple relationship must be made. At the present time the details of the theory are not sufficiently well fixed to make the corrections accurately but theoretical developments are sufficiently rapid that shortly the corrections can probably be made with no uncertainty. The nuclear radii thus determined are of interest because they are definitely larger than the measured size of the nuclear charge distribution, the difference presumably related to the range of nuclear forces.

18. Resonances in fissionable nuclides. For the heavy nuclides, the possibility of fission adds complication to the investigation of the properties of the cross section resonances. In the low energy region where many resonances are observed, the analysis of the fissionable nuclides is very complex. This complexity is a result partly of the presence of fission, introducing another parameter, the fission width, in the analysis, and partly because the many closely spaced levels makes determination of the parameters of the individual levels difficult. The latter difficulty is accentuated by the possibility that the shape of a level may be appreciably distorted by neighboring levels because of the presence of *fission interference*.

Various special techniques have been devised for the study of fissionable nuclides in the resonance region and these, together with intensive effort by means of conventional techniques, have produced a large amount of information on the important fissionable nuclides. Although the state of knowledge of the parameters is still far from satisfactory, nevertheless great progress has been made recently and it now seems as if in a few years the parameters of the fissionable nuclides will be accurately known. Knowledge of these parameters is of great importance not only for the obvious applications to reactors but because of the close relationship to nuclear theory. Even though fission has been known for a long time the theory of fission is still in a very unsatisfactory state, and the fission resonance parameters are greatly needed in testing present theory and helping to point the way to improved theoretical models of fission.

As for any reaction cross section, the fission cross section is much harder to determine than the total cross section. Actually most fission cross sections are merely measured relative to some cross section standard, either to some other fissionable nuclide or to a material whose cross section as a function of energy is known. The principle of the measurement relative to another fissionable nuclide is very simple for it involves only a measurement of the rate of fission of a given sample relative to the rate of fission of the standard, in the same neutron beam, as the neutron energy is changed. The most widely used standard for the measurement relative to a known cross section is boron, whose cross section is strictly $1/v$ in the resonance neutron region. The use of boron is somewhat more complicated than a fissionable standard, for the rate of fission in the sample under investigation must be measured relative to the boron $(n-\alpha)$ disintegrations as a function of neutron energy, and care must be taken that the relative efficiency for detection of α's and fission fragments does not change with energy. The relative fission cross section measured as a function of energy by either method can be put on an absolute scale by referring the cross section to the value at 2200 m/sec, which is now known rather well for most fissionable nuclides.

In measuring fission cross sections, the detector of the fission event is either an ionization chamber or a gas scintillation counter. Until recently, practically all measurements of fission cross sections had been made with pulse-counting ionization chambers. In this type of chamber the fissionable material is present as a very thin foil and the ionization produced by the fission fragment in the gas of the chamber is detected by means of fast electronic circuits. Because of the short range of the fission fragments, a few mg/cm², only a small amount of fissionable isotopes can be used in the chamber. The result of this limitation in amount is that the absolute efficiency of the fission chamber is very low, hence for the usual neutron velocity selectors the counting rate in the fission chamber is extremely small.

Although it is possible to increase the efficiency of a chamber by building a multi-plate chamber, a limit in the amount of material that can be placed in an ionization chamber is fixed by the alpha activity for many of the fissionable isotopes. The intense emission rate of alpha particles from U^{233} and Pu^{239} is such that only a few milligrams of material can be placed in the chamber without providing inordinate background from the pile-up of alpha particles. At the Geneva meeting in 1955 it was reported by the Soviets[1] that they had been able to use as much as 22 milligrams of Pu^{239} in an ionization chamber by depositing the plutonium on thin nylon so that fission fragments would emerge into the gas on both sides of the film. The two fission fragment pulses were then added and better discrimination obtained against alpha pile-up in this manner.

A recent development that is of great value in the measurement of fission cross sections is the use of a gas scintillation chamber in which the ionization of the fission fragments causes a faint scintillation, which is then detected by a photomultiplier tube. Because of the great speed of the scintillation process and the photomultiplier tube, the alpha pile-up is much less serious than for an

---

[1] Adamchuk, Gerasimov, Yefimov, Zenkevich, Mostovoi, Pevzner, Chernyshov and Tsitovich: Proceedings of International Conference on Peaceful Uses of Atomic Energy (United Nations, New York 1956), Vol. 4, p. 216.

ionization chamber and a hundred or so milligrams of Pu^{239} can be used in the gas scintillation counter. This detector will certainly be a powerful aid in the analysis of the fissionable nuclides of high specific α activity.

A method for obtaining fission cross sections that has been developed only recently but holds great promise involves the detection of the neutrons produced in fission rather than the fission fragments themselves. If a thin foil of fissionable nuclide is used and the detector of fast neutrons is placed nearby, the counting rate in the detector will be a measure of the fission cross section. Actually the counting rate will be proportional to $\nu \sigma_F$ but as ν, the number of neutrons per fission, is constant with energy in the resonance region, σ_F is obtained. However, in order to get a higher neutron counting rate it is necessary to use a thick foil. In this case, the counting rate is not a measurement of the fission cross section, but rather a measurement of η, the number of neutrons observed per neutron absorbed. In the so-called "direct η experiment" as developed at Brookhaven[1], a foil of the fissionable isotope is used of sufficient thickness that all the neutrons stop in the foil. Under this condition, the flux incident on the foil, which can easily be measured with a thin BF_3 counter in the incident beam, gives the absorption rate of neutrons, and the counting rate in the fast neutron detector is then a measure of η:

$$\eta = k \frac{I_F}{v\,I_0}. \tag{18.1}$$

Here k is a constant, for the measurement is a relative one, I_F is the counting rate of the fission neutrons, I_0 of the incident neutrons and v the neutron velocity (because of the $1/v$ counter sensitivity).

The η experiment is particularly useful for very slow neutrons, but becomes more difficult to interpret as the neutron energy increases because a larger fraction of the incident neutrons are scattered from the foil. The simple equation given is based on the assumption that all neutrons hitting the foil are absorbed, an assumption that is quite accurate at low energy, where the scattering cross section is so small relative to absorption. Because it now seems certain that ν is constant with energy, the η experiment can be considered a measurement of the ratio of the total cross section to fission, that is a measurement of $1 + \alpha$, where α is the ratio of capture to fission,

$$\eta = \frac{\nu}{1 + \alpha}. \tag{18.2}$$

The neutron velocity selectors used for measurements of fissionable nuclides are the same as those described for the non-fissionable materials. However, because of the low counting rates of the fission detectors, a premium is placed on neutron flux and those instruments that give high neutron flux, even though not of the highest resolution, are particularly valuable for the fission measurements. Also, because it is difficult to build fission detectors of a very large area, the type of instrument such as the Brookhaven fast chopper, that uses a small beam at the chopper but a large detector area, is not very appropriate. This velocity selector is particularly good for total cross section measurements of rare materials, but an instrument of smaller target area and larger area at the chopper itself, such as the crystal monochromator, is a better instrument for fission cross section measurements.

In analysis of a fissionable nuclide, measurements are usually made of the total cross section as well as of the fission cross section or of the fission neutrons,

---

[1] PALEVSKY, HUGHES, ZIMMERMAN and EISBERG: J. Nucl. Energy 1, 177 (1956).

as in the η method. Actually, measurements of the three types are redundant, for only two of the three are necessary in order to give all of the parameters, in principle. However, because of the general difficulty of analysis of the fissionable isotopes, and some remaining uncertainties about the details of the methods, it is definitely advisable to measure as many types of cross sections as is possible for the important fissionable isotopes. As we have seen for the nonfissionable nuclides, the total cross section gives the total width and the peak height of the levels. Analysis of these total cross section results, as before, gives the values of $g\Gamma_n$ and Γ for the levels. From these total cross section measurements alone there is no knowledge of the fraction of the total width that is capture and the fraction that is fission. Knowledge of α then gives all the information necessary to obtain the remaining parameters. Thus, in addition to the total cross section all that is actually needed is a measurement of α at each resonance or the fission cross section. The necessary information thus needed, in addition to the total cross section, is supplied either by a measurement of the η type or by a fission cross section measurement.

IV. Thermal neutrons.

19. Activation cross sections. By far the largest amount of reactor research has been performed with thermal neutrons, which are present in much higher density than those of higher energy. Even when we restrict our attention to the use of thermal neutrons inside the reactor we find a great diversity of applications; cross-section measurements for activation and absorption; the use of activation for production of hundreds of radioisotopes, and for chemical "activation" analysis. Knowledge of the *activation cross sections* for the several hundred activities that can be produced by thermal neutrons has important application to the production of *radioisotopes* for use in medical treatment and extensive research in many fields.

The simple relationship between the neutron flux, nv, and the activation I, in disintegrations per second, produced in a sample is:

$$\left.\begin{aligned} I &= I_0(1 - e^{-\lambda t}), \\ I_0 &= n\,v\,\sigma_{\text{act}}\,N. \end{aligned}\right\} \tag{19.1}$$

Here λ (the reciprocal of the mean life, τ) is the disintegration constant, I_0 the saturation activation, t the irradiation time, and N the total number of nuclei present, of activation cross section σ_{act}. This equation can be used for a Maxwell distribution of neutrons if the cross section is taken as the value for 2200 m/sec.

The use of the 2200 m/sec cross section rests on the fact that the flux is practically always determined by a method that actually measures the neutron density n independently of v (because the response of a $1/v$ detector is proportional to n not to nv). The quoted result of a flux measurement is obtained by multiplying the measured neutron density by an arbitrary v, which is practically always taken as 2200 m/sec. The activation of the material, if $1/v$, also independent of v, but the cross section to be used in Eq. (19.1) *must* be the cross section at the velocity that was assumed in the quotation of the flux in the same equation (almost certainly 2200 m/sec) if the correct activation is to be obtained.

While this velocity independence of the activation holds strictly only for $1/v$ substances, most elements are $1/v$ within experimental error in the thermal region, and fortunately no further complications ensue for them. For those few materials that depart significantly from $1/v$, the calculation of activation in a Maxwell distribution must take this fact into account. The simplest way to correct for the non-1/v behavior is to compute the ratio of the actual activation to that for a 1/v material of the same 2200 m/sec cross-section value. This ratio, f, constitutes a correction factor for the non-1/v effect that is easily calculated from the actual cross section as function of velocity, for it is merely the ratio of the activation of the actual cross section to a 1/v cross section, both integrated over the Maxwell *flux* distribution:

$$f_x = \frac{\int n(v)\, v\, \sigma_x\, dv}{\int n(v)\, v\, \frac{k_x}{v}\, dv},\qquad (19.2)$$

$$f_x = \frac{1}{n} \int n(v)\, \frac{\sigma_x}{k_x/v}\, dv. \qquad (19.3)$$

In these equations, f_x is the correction factor for a material x whose cross section at 2200 m/sec is $k_x/2200$ (thus k_x/v is the 1/v cross section equal to that of x at 2200 m/sec). From Eq. (19.3), we see that f_x is just the cross-section ratio weighted according to the Maxwell density (not flux) distribution. The correction factor for a room temperature Maxwell distribution has been made for those isotopes that differ appreciably from 1/v in the thermal cross-section table in BNL 325. In use, the 2200 m/sec cross section in the table is simply multiplied by f before insertion into Eq. (19.1) to give the activation that will be obtained with a Maxwell distribution at room temperature. The f-factors for higher temperatures for the important fissionable nuclides are given in BNL 325.

Although most of the activation resulting from irradiation inside the pile lattice represents absorption of thermal neutrons, resonance neutrons do contribute a non-negligible activation. It is therefore necessary to take into account the resonance activation in a thermal cross-section measurement made inside the lattice. Fortunately, the resonance contribution can usually be ascertained experimentally by an additional irradiation inside cadmium.

Although the principle of activation measurements is quite simple, it is exceedingly difficult to obtain reproducible, accurate results in practice, with the result that activation cross sections in general are known only to about 10%. An activation cross section is made by irradiating a known amount of sample in a measured flux, counting the sample (usually with a calibrated end-window Geiger-Mueller counter, and calculating the cross section from Eq. (19.1). In each of these steps sizeable errors can, and often do, arise.

In addition to measurement of activation cross-sections, an important use of neutron activation is to produce short-lived activities for study. Short-lived activities are of very little use as radioisotopes, for instance in medical treatment or as tracers, because of the rapid disappearance of the activity. However, they are of great importance for basic nuclear research concerned with radioactivity itself. This research is largely a matter of the determination of the "decay schemes" of the radioisotopes, that is, of the nature, order, angular correlation, and energies of the radiations emitted during decay. While the decay schemes that are of interest to basic research are by no means confined to short-lived activities, it is mainly in connection with rapidly decaying activities that special techniques for rapid sample removal from reactors have been developed. For decay periods of the order of a few hours and less, it is usually necessary to insert and remove samples from reactors without a shutdown, to prevent interference with other pile research. For instance, the "bottling machine" in use at Brookhaven is very useful for activities of the order of several hours. The bottling machine is simply a device for pushing samples into the lattice in a long series of identical sample holders in conveyer belt fashion. The samples are pushed in a continuous line through the shield in such a way that radiation does not leak into the room, hence loading and unloading can be accomplished at full power. In the study of radioisotopes with decay periods of the order of minutes or seconds, samples must be inserted and removed from the reactor with extreme rapidity. Fast acting "rabbits", developed for these short irradiations, consist of pneumatic tubes extending into various parts of the reactor. Samples are conveyed in them usually by CO_2 pressure (air pressure can be used, although a minor radioactivity hazard caused by A^{41} develops).

The high thermal fluxes in present-day research reactors can give rise to appreciable activation even for extremely small amounts of material, and this fact can be made the basis of a method for analysis of trace elements. It is a method whose sensitivity varies greatly with the thermal cross sections of the elements of interest, but in a well-chosen situation the approach is powerful. Assuming a flux of 10^{12}, we obtain from Eq. (19.1) an activity at saturation of 600 disintegrations per sec for an impurity of 10 barns thermal cross section, present in a concentration of 10^{-4} atom percent in 10^{-3} mol of target material. Without question, this high counting rate from such a slight impurity represents great sensitivity. The general utility of this method of *activation analysis*, however involves consideration of the cross section of the elements under analysis, the ease of chemical isolation of the desired activities, and the amounts and cross sections of other constituents that would produce similar masking activities.

20. Absorption cross sections. The *activation* cross section may correspond to only part of the entire *absorption* cross section of an element, for some neutron absorption may result in the production of a stable isotope, hence no activity. For certain applications, for instance in connection with reactor behavior, the neutron absorption, whether resulting in activation or not, is the important quantity. The obviously desirable method for measurement of the neutron absorption cross section involves some process that is sensitive to the disappearance of the neutron, for example, the chain reaction itself. A technique of measuring the thermal absorption cross section by means of the effect on reactivity was developed[1] with the first nuclear reactor because of the importance of the method for the determination of the purity of materials such as graphite, uranium, and aluminum. As neutron-absorbing impurities in these components were particularly "dangerous" relative to the attainment of a chain reaction, the procedure came to be called the "danger coefficient" method of neutron absorption measurement.

The change in reactivity caused by a certain sample of material is obtained from the position of a calibrated *control rod*, with and without the sample in the pile. The sample is placed at the center of the lattice to secure the maximum sensitivity, for the effect of a neutron absorber on reactivity is proportional to the square of the neutron flux at the point of absorption[2]. The sample must be spread over a sufficient area so that the thickness does not produce appreciable self-absorption. Whereas a possible uncertainty in activation measurements is related to lack of understanding of the decay scheme, in the absorption measurements by danger coefficient, uncertainty may arise from chemical impurities, of high absorption cross section. The ultimate accuracy of the danger coefficient method depends on the size of the sample used, but considering the limits

---

[1] Anderson, Fermi, Wattenberg, Weil and Zinn: Phys. Rev. **72**, 16 (1947).

[2] The change in k by direct absorption is proportional to nv and, in addition, the absorber increases the pile leakage by an amount proportional to nv, so the net change in reactivity varies as $(nv)^2$.

on sample size set by self-absorption and by availability of pure materials, it is not surprising that the accuracy of danger coefficient methods attained in practice is about 5%.

The ultimate limit of accuracy in the danger coefficient method is related to the short-period irregular fluctuations of reactivity caused by changes in barometric pressure. The period of the barometric fluctuations is small compared with the time required for each reactivity measurement, and as a result it is difficult to avoid the disturbing effect of the barometric pressure. The *pile oscillator* was developed as a means of making reactivity measurement with sufficient rapidity to decrease the effect of random short-term fluctuations in reactivity to a negligible amount.

The pile oscillator follows the same principle as the danger coefficient method, although the reactor power does not reach equilibrium because of the rapid oscillation of the sample. Even though the change in power is a complicated function[1] of neutron absorption when short-period changes are considered, it is nevertheless true that the amplitude of the power variation is strictly proportional to the neutron absorption of the sample. If a sample is moved periodically in and out of the lattice, or even from one point of the reactor to another, and the neutron flux at a particular position recorded as a function of time, the oscillating component of the flux will be proportional to the neutron absorption of the sample. The fact that the position and extent of motion of the sample, as well as the position of the neutron detector can all be varied as desired is utilized to make the oscillator technique sensitive to various parts of the neutron spectrum.

One method of operation of the oscillator to attain sensitivity in a particular energy region is that of placing the oscillating sample inside a tube of cadmium. For this case, the effect on pile power is a result of resonance neutron absorption alone, and the oscillating component of the power gives the resonance absorption integral, $\int \sigma \, dE/E$, by comparison with a standard resonance absorber, such as indium or gold. On the other hand, the effect of resonance neutrons can practically be eliminated if the sample is oscillated in a part of the pile where no uranium is present, such as the reflector or thermal column. In these thermal neutron regions, the amplitude of flux oscillation, detected at some nearby point (the "local power"), will be caused almost entirely by the thermal absorption of the sample. Many absorption cross sections listed in the compilation BNL 325 have been measured with the pile oscillators at Argonne[2,3], Oak Ridge[4,5], and Harwell, England[6]. For those cases in which the absorption is at least of the same order of magnitude as the scattering, the absorption is known with an accuracy of about 5%.

The reaction rate of thermal neutrons for a material of high cross section is sufficient to cause a measurable change in the quantity of sample present. For example, if boron is irradiated in a flux of 10^{12} for one year ($\pi \times 10^7$ sec) the fraction of the B^{10} isotope ($\sigma = 4 \times 10^3$ b) that would disappear is given by

$$4 \times 10^3 \times 10^{-24} \times \pi \times 10^7 \times 10^{12} = 0.13 \, .$$

[1] A. M. WEINBERG and H. C. SCHWEINLER: Phys. Rev. **74**, 851 (1948).

[2] A. LANGSDORF: Phys. Rev. **74**, 1217 (1948).

[3] HARRIS, MUEHLHAUSE, RASMUSSEN, SCHROEDER and THOMAS: Phys. Rev. **80**, 342 (1950).

[4] HOOVER, JORDAN, MOAK, PARDUE, POMERANCE, STRONG and WOLLAN: Phys. Rev. **74**, 864 (1948).

[5] H. POMERANCE: Phys. Rev. **83**, 641 (1951).

[6] F. C. W. COLMER and D. J. LITTLER: Proc. Phys. Soc. Lond. A **63**, 1175 (1950).

The effect of the neutron irradiation would thus be a 13% change in the amount of B^{10}, that is, in the isotopic ratio of B^{10} to B^{11}, for the latter isotope would be essentially unchanged. The cross section for the reaction could be determined by a comparison of the isotopic ratio of boron after a long irradiation with the pre-irradiation ratio. As a result of the reaction, $B^{10}(n, \alpha)\ Li_7$, the Li^7 is formed in an amount sufficient to be measured with a mass spectrometer, hence the cross section could also be obtained from the Li^7 produced. We thus see that mass spectrometric methods can be used to measure σ_a by measuring the amount of the initial isotope that disappears, or to measure a cross section that is analogous to σ_{act} by measurement of the nucleus formed. Lapp, van Horn, and Dempster[1] first used this method in 1947 to show that the large absorption in samarium and gadolinium were caused by Sm^{149}, Gd^{155}, and Gd^{157}.

21. Diffusion length measurements. Experiments that best exploit the combination of high flux and large volumes typical of reactors are those involving the spatial distribution of neutrons on a large scale. These experiments are mainly of the diffusion type, in which the distribution of neutrons, usually thermal, in a medium of low absorption is studied.

The fundamental equation of neutron diffusion, or the *equation of continuity* for the case of absorption small compared with scattering is

$$\frac{\lambda v}{3} \nabla^2 n - \frac{v}{\Lambda} n + q = \frac{\partial n}{\partial t}, \tag{21.1}$$

where $\nabla^2 n$ is the Laplacian of the neutron density distribution, λ the scattering mean free path, Λ the absorption mean free path, v the neutron velocity, and q the production rate of thermal neutrons. In the use of neutron diffusion for the measurement of small absorption cross sections there is no need to invoke the complete and much more complex transport theory[2], which applies when the cross section is not small compared with capture.

A diffusion experiment consists essentially of a measurement of the Laplacian of the neutron distribution, $\nabla^2 n/n$, in a particular geometry, from which the diffusion length is obtained directly. The diffusion length, L, is defined as

$$L = \frac{1}{(3 N^2 \sigma_a \sigma_{tr})^{\frac{1}{2}}}, \tag{21.2}$$

where σ_{tr} is the *transport cross section* so Eq. (21.1) can be written, for $q=0$, as

$$\nabla^2 n - n/L^2 = 0, \tag{21.3}$$

the simple form that applies, for example, to diffusion in a material in which no neutrons are being produced. The diffusion length determination can be considered [Eq. (21.2)] as a measurement of the transport cross section if the absorption is well known or, more likely, as a measurement of σ_a in terms of the known σ_{tr}, related to the scattering cross section by $\sigma_{tr} = \sigma_s (1 - \overline{\cos \Theta})$ where Θ is the scattering angle.

An important point that arises in connection with diffusion experiments concerns the proper velocity average to be used for the cross sections. When the entire Maxwell distribution is considered, it is the *average mean free path* (that is, the average reciprocal cross section) that is effective in the diffusion,

[1] Lapp, van Horn and Dempster: Phys. Rev. **71**, 745 (1947).

[2] Glasstone and Edlund: Elements of Nuclear Reactor Theory. New-York: Van Nostrand 1952.

hence the correct absorption cross section to be inserted in Eq. (21.2), for a $1/v$ absorber, is

$$\overline{1/\sigma} = \overline{v/k} = 1/\sigma_{\bar{v}} \tag{21.4}$$

where $\sigma_{\bar{v}}$ is the cross section at the average velocity of the Maxwell distribution. This cross section, which is to be used in the diffusion equation, is $1/1.128$ of the usual cross section at the most probable velocity (2200 m/sec at room temperature), because the average velocity is 1.128 times the most probable velocity. In most thermal cross-section work it is correct to use the 2200 m/sec cross section uncritically, but neutron diffusion is one of the few instances where other values must be used.

The neutrons present in the *thermal column*, Fig. 3, provide an ideal source for diffusion experiments because of the high flux, large volume, and complete thermalization. The situation is much simpler than it was for use of point sources of fast neutrons (Ra—Be), because of the absence of resonance neutrons ($q = 0$ in the diffusion equation), and because the available space is sufficiently large so that the higher harmonics in the neutron distribution can be eliminated (assuming a sufficient quantity of the material under investigation is available). The circumstances are favorable enough so that diffusion lengths can be measured with high accuracy, but this accuracy in turn necessitates a consideration of effects that would be too small to observe in the previous experiments using natural sources.

The small effects that must be considered in an accurate diffusion length measurement are concerned mainly with the velocity distribution of the neutrons in the diffusing medium. Unlike the case of a reaction rate, which does not depend on neutron temperature but only on neutron density, the diffusion length is determined by the absorption cross section, which is a function of the neutron temperature. In a careful diffusion measurement the shape as well as the temperature of the neutron distribution must be considered in detail in obtaining the correct average cross sections. For example, extremely slow neutrons penetrate easily through solid materials, that is, they have a low transport cross section and, as a result, a large diffusion length. On the other hand, because the absorption cross section increases with decreasing velocity, the slow neutrons are preferentially absorbed and the distribution will "harden" as it diffuses. The net result of these two effects (filtration and hardening) is to shift the main part of the velocity distribution to higher velocities. Should the spectrum shape change appreciably, the diffusion length itself will not be constant and there will be no region of exponential behavior in the spatial distribution.

22. Thermal neutron beams. For experiments involving thermal neutrons outside the reactor lattice, that is, in the form of beams, Fig. 1, there are several desirable properties that are of more or less importance, depending on the particular experiment. These properties, which cannot be attained simultaneously, are high flux, complete thermalization, and good collimation. The thermal flux in a beam is increased if the hole is extended into the lattice rather than the thermal column, but such conditions permit a high contamination of resonance neutrons. Complete thermalization is attained in the thermal column at some distance from the lattice, but at a great loss in intensity. The conditions for good collimation reduce the intensity greatly, and if complete thermalization is called for at the same time the intensity will be still further lowered. The best way to illustrate the general characteristics of the available thermal beams is to describe several of the experimental arrangements in actual use.

A situation in which good (but not complete) thermalization is needed, is furnished by the U^{235} converter plate, Fig. 15. The source of neutrons is the graphite reflector of the reactor, where the intensity is much higher than in the thermal column, although the thermalization is not complete. The geometry is arranged so that the plate receives neutrons from a large area; by this method the intensity is increased together with a decrease in collimation, which is of no consequence for this purpose.

Another illustration of an experiment in which high neutron flux is the primary requisite, low gamma flux desirable, and collimation and complete thermalization unnecessary, is the study of capture gamma rays at the Argonne reactor[1].

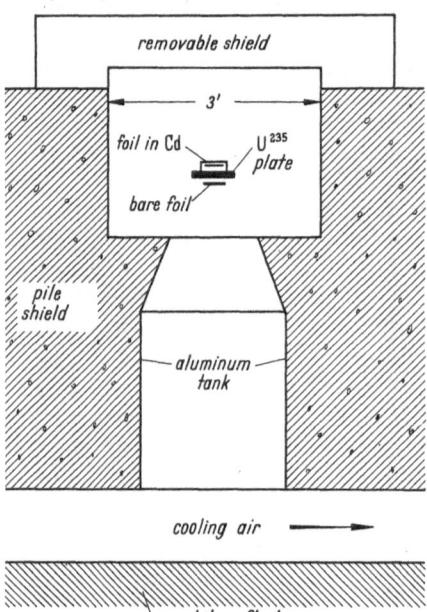

Fig. 15. The arrangement for obtaining a high flux of unmoderated fission neutrons without contamination by resonance or thermal neutrons.

Here, Fig. 16, the sample is placed inside the lattice itself to attain high flux and extraneous radiations are reduced because the sample is in a "through" hole. The capture gammas emitted from the sample are measured in the curved spectrometer outside the reactor. Lead collimators are located in such a way as to limit the "view" of the spectrometer to gammas emitted by the sample only. When fine collimation is the primary objective, a high-flux source point for the beam is necessarily implied, as for the optical experiments to be described later, neutron diffraction and reflection. The neutron source for optical experiments is usually the reflector of the reactor or even the lattice center (if fast neutrons can be tolerated), rather than the thermal column, in order to obtain high flux.

The actual methods used to delineate the thermal beam must sometimes be carefully considered. Diaphragms of cadmium define a beam of slow neutrons in a very satisfactory way, but they have no effect on the fast neutrons and, in addition, they produce capture gammas from the stopped neutrons. It is possible in principle to deflect the neutrons completely away from the gamma beam and thus obtain a gamma-free flux, although only at a great sacrifice in intensity. Deflection of the neutrons by crystal reflection has been used for this purpose in a study of capture gamma rays by Hamermesh[2].

The actual experiments in which a high-intensity, poorly collimated thermal neutron beam is used as a source range over a variety of subjects. In these experiments the collimation of the beam is of no concern and, in fact, an external beam is used only because of the size of, and necessity for access to, the equipment. It is not our purpose to describe in detail the kinds of research that utilize more or less standard techniques—scintillation counters, lens spectrometers, cloud chambers, photographic emulsions, etc.—although much work of fundamental importance has been performed in this way. However, we shall review

[1] W. H. McCorkle and W. H. Jenn: Geneva Conf. Paper 859 (1955).
[2] B. Hamermesh: Phys. Rev. **80**, 415 (1950); **81**, 487 (1951).

briefly the ways in which standard techniques have been applied to neutron beam sources, for it is the particular application rather than the instrument that is peculiar to reactor research. In these experiments, beams of neutrons from the lattice, reflector, or thermal column are collimated to a size convenient for the detecting instrument, which is usually operated in the conventional manner.

Some of the first cross-section measurements made with thermal neutrons involved transmissions of the complete Maxwell velocity distribution through samples. Total cross sections could easily be measured in this way because the well-collimated neutron beams ensured conditions of good geometry. The cross section obtained is a particular average over the Maxwell distribution and,

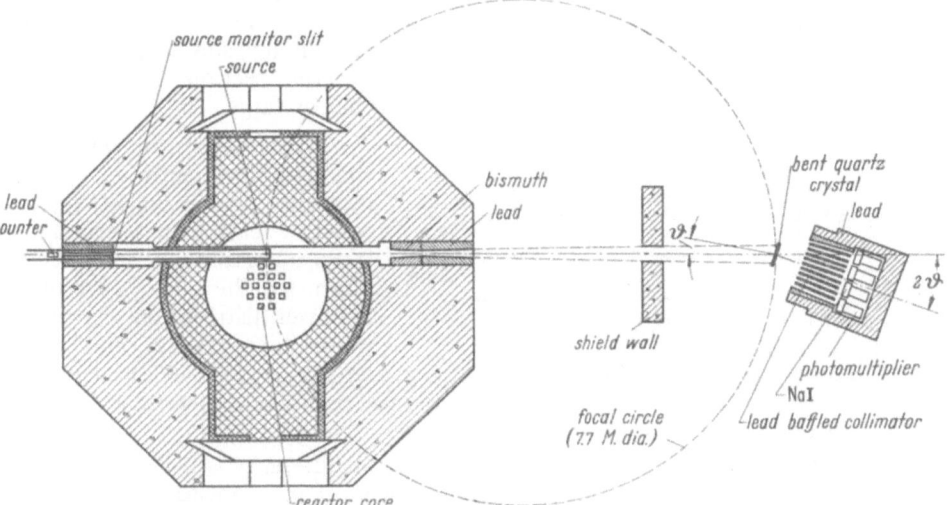

Fig. 16. The experimental arrangement used at Argonne Laboratory for study of capture gamma rays by a curved crystal spectrometer.

while it can be measured with great accuracy, its interpretation as a cross section at a particular velocity is difficult. The development of velocity selectors, which allow cross-section measurements at discrete velocities, made the average cross sections of less value, but for some materials the method is still of great value.

The relationship of the observed cross section to various averages involves a careful consideration of the shape of the neutron distribution and the type of counter used in he transmission measurement. The Maxwell velocity distribution contains a v^2 factor, and the flux a v^3 factor, thus the counting rate distribution for a $1/v$ detector contains the v^2 factor. The counting rate, I, for a $1/v$ detector is the integral over the Maxwell density distribution:

$$I = \text{const} \times \int_0^{0.4\,\text{ev}} v^2\, e^{-\left(\frac{v}{v_0}\right)^2}\, dv, \tag{22.1}$$

where v_0 is the most probable velocity, 2200 m/sec at room temperature.

The observed absorption cross section has a simple interpretation if the transmission sample is very thin. In that case (T almost unity), the transmission equation becomes simply

$$1 - T = N\, \sigma_T\, \chi, \tag{22.2}$$

and the $(1 - T)$ observed obviously gives the average total cross section, weighted according to the neutron distribution, and subtraction of scattering gives the average absorption. This average absorption cross section for a Maxwell distribution $\bar{\sigma}$, is seen to be, for a $1/v$ absorber $(\sigma = k/v)$,

$$\bar{\sigma} = \frac{\int \sigma v^2 e^{-(v/v_0)^2}\, dv}{\int v^2 e^{-(v/v_0)^2}\, dv} = \frac{2}{\sqrt{\pi}}\frac{k}{v_0},$$

$$\bar{\sigma} = \frac{2}{\sqrt{\pi}}\sigma_{v_0} = 1.1280\,\sigma_{v_0}. \tag{22.3}$$

Thus for a $1/v$ sample (and detector), the average absorption cross section obtained from a thin sample transmission measurement *is* 1.128 *times the commonly listed value at the most probable velocity*. It is interesting to note that, in contrast, the appropriate average cross section for a diffusion measurement, Eq. (21.4) $1/1.128$ times σ_{v_0}. We thus see that the average cross section is *not* the cross section at the average velocity but is $(1.128)^2$ times as large.

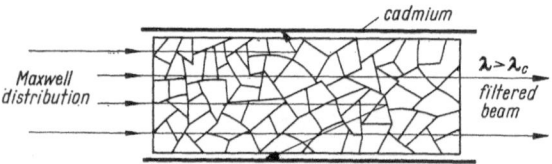

Fig. 17. The principle of a neutron filter.

23. Measurements with mono-energetic slow neutrons. The high intensity, which makes velocity selectors feasible for thermal neutrons, obviates the necessity for filters for the major part of the Maxwell distribution. At the extreme low velocity end of the distribution, that is for "cold neutrons", however, filters find an extensive use. These filters do not make use of neutron resonances to select certain velocities, but utilize instead the optical property of selective diffraction.

In order to explain the action of slow-neutron filters, it is necessary only to refer to the equation for Bragg reflection discussed in connection with crystal monochromators, Eq. (14.1). Let us consider a beam of thermal neutrons passing through a polycrystal containing many crystal grains oriented at random, Fig. 17. A neutron of a certain wavelength λ will pass through most of the crystal grains, scattering only when it meets one at such an angle that Eq. (14.1) is satisfied. As there are many planes of different spacings d, most of the neutrons will be reflected. We say "most" rather than "all" because there is a maximum d value, d_m, for each crystal lattice, which depends on the crystal structure. It is then clear that the longest wavelength that can be reflected is given by $\sin \Theta = 1$, that is, for the crystal plane perpendicular to the neutron motion, and this wavelength is

$$\lambda_m = 2 d_m. \tag{23.1}$$

Neutrons of wavelength greater than λ_m are practically not scattered at all, except for the small incoherent scattering, and are removed from the beam only by capture. As shown in Fig. 17, the neutrons with $\lambda < \lambda_m$ are scattered and then captured in the cadmium surrounding the filter, while those with $\lambda > \lambda_m$ pass through the filter if not removed by capture in the filter itself. Filters of low capture materials, such as beryllium and graphite, have been of great value in mirror reflection experiments, Sects. 26, 27.

The crystal monochromator, of great value in the resonance region, is not much used for thermal neutrons, because of the interfering effect of higher order reflections. From Eq. (14.1) we see that the n-th order reflection for a wavelength λ/n will coincide with the first order reflection for the wavelength λ. This interference by the shorter wavelength will become greater as λ increases, because the ratio of the neutron intensity at λ/n to that at λ increases with λ in the region of the Maxwell distribution. The effect of the higher order reflections is small in the resonance region where the crystal monochromator is principally used, but it is a serious drawback in the use of thermal neutrons, and as a result the mechanical velocity selector is preferable for use there. In some recent work with neutrons in the 5 to 10 Å range, McREYNOLDS[1] has shown that a filter can be combined with a crystal monochromator to decrease higher order reflections, and thus render the crystal useful for wavelengths beyond the filter cutoff.

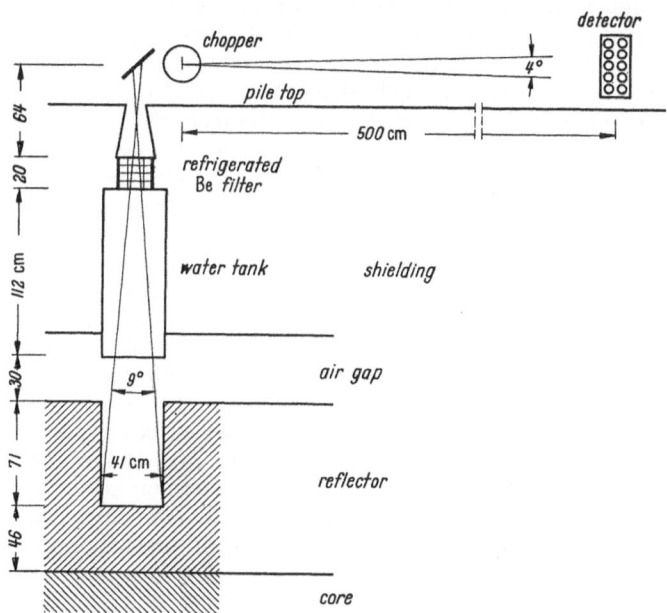

Fig. 18. The equipment in use at Brookhaven Laboratory for measurement of the energy gain of cold neutrons from lattice vibrations.

Of course, the monochromator crystal must have an unusually large lattice spacing in order to diffract neutrons in the 5 to 10 Å range. Magnetite (Fe_3O_4), for example[1], has a cutoff wavelength, λ_m of Eq. (23.1), of 9.66 Å and in addition a weak second order reflection.

A mechanical neutron "chopper" operating in the thermal region is the same in principle as the "fast chopper", but the "slow chopper" is of much simpler mechanical design. The slow chopper is of simple construction because the bursts produced are of the order of 20 instead of 1 microsecond and, in addition, the chopping of thermal neutrons can easily be accomplished with thin layers of cadmium instead of many inches of steel or plastic.

Slow choppers have been used for many cross-section measurements at thermal and sub-thermal energies. This energy range is much too small to include an appreciable number of resonances but the results are particularly useful for study of another phenomenon, *thermal inelastic scattering*. At sub-thermal energy, the cross section is markedly temperature dependent as exemplified by the results for beryllium in the compilation BNL 325. This cross section, which represents gain of energy by the neutron from the vibrations of the crystal lattice, gives

[1] A. W. McREYNOLDS: Phys. Rev. **88**, 958 (1952).

information on the properties of these vibrations, or *phonons*. The information is more meaningful if measurement of the amount of energy gained can be accomplished. This type of work requires high flux and is still in its beginning stages; encouraging results are now being obtained, however. A typical experimental arrangement is that at Brookhaven[1]. Fig. 18, designed to obtain a maximum flux of filtered neutrons at the sample. The energy of the scattered neutrons is measured by time of flight, using a conventional slow chopper.

The slow chopper is designed only for the production of neutron bursts with the actual velocity selection performed by electronic timing. A true *mechanical monochromator* is one in which neutrons of a narrow velocity band are produced by mechanical methods, with no time-of-flight measurement involved. A monochromator has certain advantages over a time-of-flight instrument, advantages that follow from the fact that actual separation of the different velocities is not accomplished in the latter. For example, activation with monoenergetic neutrons is possible with the monochromator and impossible with the time-of-flight method. Because no measured flight path is involved with the monochromator, the conditions for scattering cross-section measurements are much more favorable for it than for the time-of-flight apparatus. The

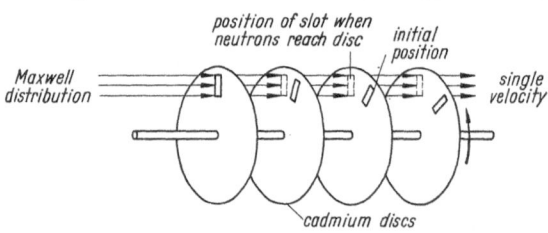

Fig. 19. The principle of the mechanical monochromator.

crystal monochromator has been used to some extent for both activation and scattering measurements in the energy region just above thermal, but is not useful for very slow neutrons because of the intereference of neutrons of higher order. The obvious advantages of a mechanical monochromator for activation and scattering have led to the construction of several models all of similar design; the principle of these is illustrated in Fig. 19.

24. Diffraction effects in transmission measurements. Thermal neutrons are extremely well adapted for two types of experiments that exhibit the optical properties of neutrons, *diffraction* and *reflection*. The former utilizes neutrons in the most intense part of the Maxwell distribution, where fortunately the wavelength is the correct magnitude for investigation of crystal lattice structure. Mirror reflection, on the other hand, is not dependent on crystal structure, and can be best studied with long wavelength neutrons, in the 5 to 10 Å range. The results of diffraction and reflection investigations are considered in detail by Ringo in Vol. XXXII of the Handbuch; we shall limit ourselves to a brief consideration of the techniques themselves.

The *powder diffraction* (Debye-Scherrer) method of crystal analysis can be used in the conventional x-ray manner, in which the angular distribution of scattered intensity is measured, or by study of the transmitted neutrons, which would be impossible with x-rays because of high absorption. Transmission as a method of studying neutron diffraction was first used by Fermi, Sturm, and Sachs[2] in 1947; they obtained the coherent scattering cross sections as a function of wavelength from the observed total cross section by subtracting capture and incoherent scattering. For wavelengths longer than the *cutoff wavelength*,

[1] I. Pelah, C. M. Eisenhauer, D. J. Hughes and H. Palevsky: Bull. Amer. Phys. Soc., Ser. II **2**, 43 (1957).

[2] Fermi, Sturm and Sachs: Phys. Rev. **71**, 589 (1947).

Eq. (23.1), the coherent scattering is zero for the Bragg condition cannot be satisfied. As λ decreases, scattering, from the planes of largest spacing suddenly appears; the scattered intensity then decreases with increasing energy, actually being proportional to λ^2.

Quantitatively, the cross section per unit cell from a set of planes (h, k, l) of spacing d and order of reflection n, will be [1,2], for $\lambda \leq 2d/n$:

$$\sigma_{h,k,l} = \frac{\lambda^2 N}{2} \left(|F|^2 M \frac{d}{n} \right)_{h,k,l} e^{-K\left(\frac{\sin\Theta}{\lambda}\right)^2}, \qquad (24.1)$$

where N is the number of unit cells per cm^3, M is the *multiplicity* [3] (number of possible orientations of the unit cell for the particular reflection h, k, l), the exponential is the Debye-Waller factor, and F is the *structure factor* for the unit cell. F is given by,

$$F_{h,k,l} = \sum_j a_j e^{2\pi(hx_j + ky_j + lz_j)}, \qquad (24.2)$$

where the summation is over the atoms of the unit cell, with coordinates x_j, y_j, z_j, and coherent amplitudes a_j. This structure factor is easily evaluated for different types of unit cells as described in any x-ray text [4,5]; it is simply an expression for the effect of the phase differences of the waves scattered by individual atoms on the net amplitude of the unit cell.

As the neutron wavelength decreases, more and more sets of planes can reflect neutrons (as λ becomes less than each value of $2d_{h,k,l}$), and the observed cross section at any wavelength is the sum of the contributions from all the reflecting planes. Precisely at the cut-off for any set of planes, the reflecting planes are those that are perpendicular to the neutron beam, and with decreasing wavelength the crystal grains with planes at other angles reflect the incident neutrons. At "high" energies (0.1 ev and higher) so many planes are active that the cross section becomes a smoothly varying function of energy. As the energy rises still higher, to several ev, the coherent scattering becomes very small because of the Debye-Waller factor, but the *total* scattering changes very little with energy, for the decrease in coherent scattering is compensated by the increase in incoherent inelastic scattering. The type of inelastic scattering in which energy is transferred to the lattice *increases* with neutron energy until the atoms act as if free at each collision. When this state of affairs is reached, at about 1 ev, the scattering cross section remains constant with energy (barring nuclear resonances) and the constant cross section is called the *free atom cross section*. It is this free atom cross section, after correction for center of gravity effects and for isotopic and spin-dependent incoherence that leads to the very useful coherent amplitude, appearing in Eq. (24.2), for example.

25. Powder method of neutron diffraction. This widely used method has been the subject of extensive development by WOLLAN and SHULL at Oak Ridge, and neutron powder diffraction work is now being carried out at almost all research reactors, although single crystal diffraction is now gaining in favor. Because the scattered neutrons are distributed throughout an entire cone in

[1] HALPERN, HAMERMESH and JOHNSON: Phys. Rev. **59**, 981 (1941).

[2] See footnote 1, page 432.

[3] C. S. BARRETT: Structure of Metals, p. 536. New York-Toronto-London: McGraw-Hill 1943.

[4] A. H. COMPTON and S. K. ALLISON: X-Rays in Theory and Experiment, pp. 357—364. New York: Van Nostrand 1935.

[5] C. S. BARRETT: Structure of Metals, pp. 79, 528—532. New York-Toronto-London: McGraw-Hill 1943.

the powder method, the intensity is low at the single point in the cone where the neutrons are detected by a counter; as a result, every effort must be made to obtain high incident neutron intensity and low background in order to observe the Bragg peaks with good resolution. In line with these requirements, the first or *monochromating* crystal is put as close to the neutron radiating surface

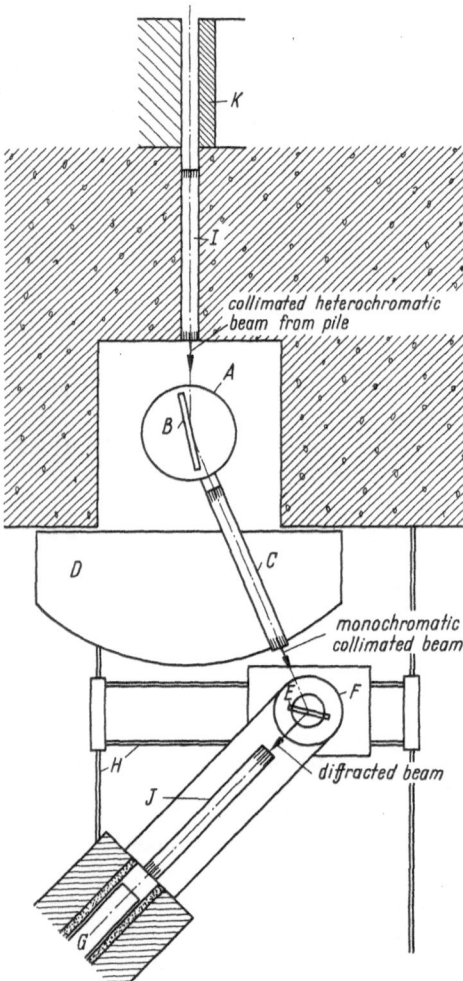

collimated heterochromatic beam from pile

monochromatic collimated beam

diffracted beam

Fig. 20. The neutron diffraction equipment used by CORLISS and HASTINGS at Brookhaven.

as possible and is surrounded by heavy shielding. The particular wavelength reflected by the first crystal (1.06 Å) is chosen for maximum intensity; this wavelength corresponds to the peak of the Maxwell distribution as "seen" by a crystal. The counter arm is geared to the crystal table so that the powder pattern can be traced automatically by rotation of the counter arm at twice the angle of rotation of the crystal table; the counting rate is recorded continuously during the process. Even though the neutron intensity at the monochromator is extremely high, the low reflectivity at the monochromator crystal (less than 1%), and again at the crystal under study, results in a reduction of counting rate at the detector to several hundred counts per minute.

A particular neutron diffraction apparatus, at Brookhaven[1], is shown in Fig. 20. The neutron beam incident on the lead monochromating crystal is collimated to 0.5 degrees. The intensity is kept high in spite of the fine collimation by use of multiple parallel beams formed by the long multiple slit sections shown at *I, C,* and *J* in Fig. 20. The movement of the crystal and counter, as well as the recording of data, is completely automatic. The intensity reflected from the monochromating crystal is much greater if, instead of consisting of a perfect crystal, it is made up of small regions misoriented at slight angles (mosaic structure). The optimum arrangement would be one for which the mosaic angular spread is equal to the collimation of the incident beam. The lead crystal used in the Brookhaven apparatus was chosen because of its large mosaic structure.

One important use of the powder method of neutron diffraction has been the determination of coherent scattering amplitudes. In the measurement of coherent amplitudes by diffraction, the intensities in various Bragg reflections

[1] L. M. CORLISS and J. M. HASTINGS: Phys. Rev. **90**, 1013 (1953).

are measured, and from these intensities and the calculated crystal structure factor, Eq. (24.2), the coherent amplitudes are calculated. The relationship between the intensity in a particular Bragg maximum and the coherent amplitude follows directly from the results already discussed for the cross section of mono-energetic neutrons, Eq. (24.1). The following equation, which is obtained from a standard x-ray diffraction formula[1], applies to the experimental arrangement of Fig. 20,

$$\frac{I_{h,k,l}}{I_0} = \frac{\lambda^2 N^2 l h}{4\pi r \sin 2\Theta \cos \Theta} \left(|F|^2 M \frac{d}{n} \right)_{h,k,l} e^{-K\left(\frac{\sin \Theta}{\lambda}\right)^2}. \tag{25.1}$$

Here $I_{h,k,l}$ is the intensity in the (h, k, l) reflection, measured with a counter slit of height 1; I_0 is the intensity incident on the sample of thickness h; r is the distance from sample to counter slit, and the other symbols have the same meaning as in Eq. (24.1). Additional obvious factors must be included if the transmission of the sample is appreciably less than unity or if the density of the powder is less than the crystalline density. The use of Eq. (24.3) would give the scattering amplitude in absolute units if a measurement of the diffracted intensity were made relative to the intensity of the incident beam I_0. However, because of difficulty in estimating the intensity of the incident beam, only the relative amplitudes are measured in practice. In other words, the amplitude of one nucleus is measured relative to another present in the same crystal. The coherent amplitudes are thus referred eventually to some standard nucleus, such as carbon, whose amplitude can be determined from a transmission measurement.

Coherent cross sections for a series of elements have been measured by means of powder diffraction by SHULL and WOLLAN[2], and their results furnish many of the coherent cross sections given in the cross section compilation BNL 325. The coherent cross sections were all measured relative to various substandards, which in turn were referred to carbon as the ultimate standard. In measuring the intensities in the Bragg reflections, the correction for thermal diffuse scattering was made by calculation of the Debye-Waller factor when the scatterer had a known Debye temperature, or if not, extrapolating to zero angle.

The coherent cross sections of the survey of SHULL and WOLLAN are accurate to about 7% for most cases. It is difficult to attain extreme accuracy in the powder diffraction method because of uncertainty in the Debye-Waller factor and the necessity of correcting the peak intensity for the diffuse scattering background, which arises from multiple scattering in the sample, as well as from isotopic, spin-dependent, and thermal incoherent scattering. In the case of the separated isotopes of nickel, however, extreme care resulted in 2% accuracy for the coherent cross sections[3].

26. Reflection of neutrons from mirrors. Measurement of the index of refraction of a particular medium for neutrons gives the coherent cross section in a very direct manner and, in addition, because the index is proportional to the amplitude rather than its square, the sign of the amplitude is obtained as well. The index of refraction, n, will be less than unity for a material of positive coherent scattering amplitude and greater than unity for a negative amplitude:

$$1 - n = \frac{\lambda^2 N a_{\text{coh}}}{2\pi}. \tag{26.1}$$

Here N is the number of nuclei per cm^3 and a_{coh} the coherent amplitude.

[1] C. S. BARRETT: Structure of Metals, pp. 534—541. New York-Toronto-London: McGraw-Hill 1943.
[2] C. G. SHULL and E. O. WOLLAN: Phys. Rev. **81**, 527 (1951).
[3] KOEHLER, WOLLAN and SHULL: Phys. Rev. **79**, 395 (1950).

Just as light is totally reflected from a medium of optical index of refraction less than unity when incident at a sufficiently small grazing angle, so neutrons will be totally reflected from a material of neutron index less than unity if the incident grazing angle is extremely small. The critical angle for beryllium, for example, will be 11.7' for a wavelength of 2 Å, so exceedingly well-collimated neutron beams are necessary in order to observe the total reflection. Mirror reflection was first demonstrated by Fermi and Zinn[1] and soon afterward use of critical angle measurements was made by Fermi and Marshall[2] to obtain coherent amplitudes. The purpose of their work was the determination of the sign of the coherent amplitude in connection with neutron diffraction studies, which, as we have seen, do not reveal absolute amplitudes. The experiments were used principally to obtain the sign of the amplitude, using the occurrence of total reflection to identify a positive amplitude and accurate numerical values of the coherent scattering cross section were not obtained in this work.

The index of refraction for neutrons, hence the critical angle, depends only on the average potential that the neutron experiences in a medium and is independent of molecular and crystalline structure. For a mirror of a single element with a small neutron absorption, the index of refraction and the critical glancing angle Θ_c are related as follows:

$$\left.\begin{aligned} \cos \Theta_c &= n = 1 - \frac{\lambda^2 N a_{\text{coh}}}{2\pi}, \\ \Theta_c &= \lambda (N a_{\text{coh}}/\pi)^{\frac{1}{2}}. \end{aligned}\right\} \qquad (26.2)$$

It is noteworthy that Θ_c depends on the coherent amplitude in a very direct way, there being no corrections, for form factors, temperature diffuse scattering or crystal effects. As pointed out by Hamermesh[3], this simplicity, which follows from the fact that the scattering is essentially forward, implies that the measurement of critical angles is in principle an accurate method of obtaining coherent amplitudes in such important cases as n-p scattering and the neutron-electron interaction.

In principle, the direct method that would be used to get the coherent amplitude would be to reflect monoenergetic neutrons from a mirror surface and look for the sharp drop in intensity at the critical angle. However, sufficient intensity is not available in the well-collimated beam that would be required. Because of the magnitude of the critical angle a resolution of one minute would be reasonable, but the collimation attainable with a crystal monochromator at good intensity is much poorer than one minute. In order to attain sufficiently high intensity at the requisite high collimation, a filter technique was developed by Hughes and Burgy[4] at Argonne Laboratory. Although a monoenergetic beam is not obtained by this method, a distribution with high intensity and an exceedingly sharp cut-off on the short wavelength side results. The combination of sharp cut-off wavelength and high intensity is especially desirable for the determination of critical angles. The incident angle Θ is measured in terms of the angle 2Θ between the reflected and the direct beam. In order to measure the critical angle, the intensity of the reflected beam is recorded as Θ is increased, the detector always being moved to keep it at an angle of 2Θ to the direct beam. A characteristic intensity pattern will be obtained that gives the critical angle for the cut-off wavelength of the filter directly. Another refinement in accurate

[1] E. Fermi and W. Zinn: Phys. Rev. **70**, 103 (1946).
[2] E. Fermi and L. Marshall: Phys. Rev. **71**, 666 (1947).
[3] M. Hamermesh: Phys. Rev. **77**, 140 (1950).
[4] D. J. Hughes and M. T. Burgy: Phys. Rev. **81**, 498 (1951).

work is necessitated by the fact that the reflectivity[1] above the critical angle does not go to zero immediately, but rather decreases as

$$R = \frac{1 - (1 - \Theta_c^2/\Theta^2)^{\frac{1}{2}}}{1 + (1 - \Theta_c^2/\Theta^2)^{\frac{1}{2}}} . \tag{26.3}$$

In theory[1], neutron absorption has an effect on the index of refraction, hence on the critical angle, but in practice the effect is quite negligible. Including absorption, Eq. (26.2) becomes

$$\Theta_c = \lambda \sqrt{\frac{N}{\pi} \left(\frac{\sigma_s}{4\pi} - \frac{\sigma_a^2}{4\lambda^2} \right)^{\frac{1}{2}}} , \tag{26.4}$$

from which we see that the effect on Θ_c is given by the ratio of $\pi (\sigma_a/\lambda)^2$ to σ_s. Even for a σ_a of 10^4 b, however, the first quantity is 0.1 b (for a λ of 5 Å), which would represent only a one percent effect on Θ_c for an average σ_s.

For a ferromagnetic material, the question of the correct coherent amplitude to use in the index of refraction introduces some complications. The neutrons in a domain [in which the magnetic induction always has the saturated value, (B_s)] experience a two-valued potential, $\pm \mu B_s$. It follows[2] that there are just two distinct indices of refraction for neutrons in iron, whether the iron is magnetized or not in the macroscopic sense:

$$n^2 = 1 - \lambda^2 N a/\pi \pm \mu B_s/E , \tag{26.5}$$

where E is the neutron energy in the same units as μB_s.

27. Mirror balancing techniques. The measurement with mirrors that we have mentioned has been a direct application of Eq. (26.2), that is, critical reflection of filtered neutrons from a single material. In this direct method, the resulting coherent scattering cross section is four times as inaccurate as the measured critical angle, and even under best conditions, gives an accuracy in coherent cross sections of 2%. While this accuracy in σ_{coh} is sufficient for general use in neutron diffraction, special problems arise from time to time for which accuracy better than 1% is required.

It is possible to increase the accuracy beyond that attainable in the direct method by balancing techniques in which several materials are combined in a single mirror, or in which the mirror surface is the interface between two media. In the former case the critical angle will be related to the average coherent scattering of the materials in the mirror, and in the latter to the difference between the two materials forming the interface. These two possibilities, together with the flexibility arising from the existence of both positive and negative amplitudes, mean that many unknown amplitudes can be measured by balancing them against a few known amplitudes. The critical angle is used to measure only the unbalanced amplitude, and a large error in the unbalanced amplitude will produce an arbitrarily small error in the final amplitude, provided the balance is sufficiently close. Of course, for a given wavelength, the critical angle decreases as the balance becomes closer and as a result becomes more difficult to measure. However, the critical angle varies as the square root of the unbalanced amplitude only, hence the amplitudes of two materials could be balanced to 1% and the critical angle would still be 10% of that for one of the materials used alone. The ultimate limit of the accuracy of the balancing method

[1] M. GOLDBERGER and F. SEITZ: Phys. Rev. **71**, 294 (1947).
[2] HUGHES, BURGY, HELLER and WALLACE: Phys. Rev. **75**, 565 (1949).

is the coherent amplitude of the standard material, which is obtained from the free atom cross section.

Coherent scattering cross sections, when obtained by diffraction, involve a correction for the temperature diffuse scattering (Debye-Waller factor). This correction makes accuracies better than 1% difficult to attain; especially for crystals containing several elements, which are inevitable in comparison work. With the neutron mirror balancing technique, however, no correction for temperature diffuse scattering enters. The essential reason for the disappearance of the Debye-Waller correction is that the index of refraction and the critical angle are determined by the scattering in the forward direction. The occurrence of $\sin \Theta/\lambda$ in the Debye-Waller correction factor makes it unity for all mirror work.

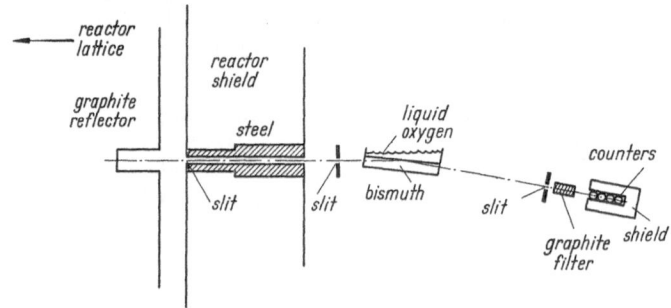

Fig. 21. Schematic diagram of an experiment in which neutrons are reflected from the interface of bismuth and liquid oxygen.

The fact that the critical angle is determined by the *average* coherent amplitude of the mirror material in an unambiguous way widens the possibilities of application of the technique. For instance, the critical angle for unmagnetized iron is unaffected by magnetic scattering even though the iron is composed of domains that are completely magnetized. The critical angle is determined by the average coherent scattering amplitude, and the magnetic amplitude, when averaged over the surface of an unmagnetized mirror, becomes zero, leaving only the nuclear amplitude to determine the critical angle. Because of the averaging property of a mirror, the critical angle can be measured for a mixture of substances (liquid, solid, or gas) and the coherent amplitude obtained will be an accurate average of the constituents, regardless of their molecular form or crystalline state.

A good example of the balancing technique involving the combination of several materials in a single mirror is given by the coherent hydrogen (neutron-proton) scattering measurement of Burgy, Ringo, and Hughes[1]. This experiment, the first in which the balancing technique was used, resulted in an accuracy of better than 1%, in spite of the extremely large incoherent scattering of hydrogen. The mirror was in the form of a liquid mixture of several hydro-carbons and the liquid surface proved to have excellent neutron-reflecting properties.

The balancing technique in which neutrons are reflected from the interface between two media has been used for a liquid-solid, and for a gas-liquid, interface. The liquid-solid combination was used for a measurement[2] of the neutron-

[1] Hughes, Burgy and Ringo: Phys. Rev. 77, 291 (1950). — Burgy, Ringo and Hughes: Phys. Rev. 84, 1160 (1951).

[2] D. J. Hughes, J. A. Harvey, M. D. Goldberg and Marilyn J. Stafne: Phys. Rev. 90, 497 (1953).

electron interaction, and the neutrons were reflected internally from a bismuth mirror, which was in contact with liquid oxygen, as shown in Fig. 21. The major difficulty proved to be the production of a good mirror surface on the bismuth block. The gas-liquid balance was used by McReynolds[1] for the coherent cross sections of the noble gases, which of course cannot be handled by ordinary crystal diffraction. The method consists of reflecting unfiltered neutrons from a liquid surface, above which the gas was held at high, adjustable pressure. The liquid was a hydrocarbon chosen to have an index of the same order of magnitude as the high pressure gas, the liquid index being known in terms of the hydrogen and carbon amplitudes. In principle, the measurement is made simply by adjusting the gas pressure until the indices of gas and liquid are equal, this state of affairs being identified by the disappearance of the reflected intensity. In practice, a method of extrapolation is used to obtain the balance point accurately.

28. Neutron polarization. The *polarization* of neutrons refers to the alignment of the neutron spins along a particular direction, compared with unpolarized neutrons in which the net spin along any axis is zero. The polarization P, or fraction of the spins in the $+$ direction not balanced by spins in the $-$ direction, is given by

$$P = \frac{n_+ - n_-}{n_+ + n_-}, \tag{28.1}$$

where n_+ and n_- are the numbers of neutrons with spins in the $+$ and $-$ directions. Some of the properties of polarized neutrons bear a slight resemblance to those of polarized light, but the resemblance is a matter of terminology rather than the intrinsic nature of the polarization. Polarized neutrons are particularly useful when used with polarized nuclei, as has been done recently at Oak Ridge[2].

Neutron polarization is an example of a phenomenon that is not exclusively a reactor technique, such as diffraction or mirror reflection. The latter have been developed with reactors because only in this way is sufficient intensity available. Polarized neutrons, on the other hand, have been studied with cyclotron-produced neutrons and even by means of Ra-Be sources. Because of this more extensive study of polarized neutrons, we shall not consider all its ramifications, but rather only the specific techniques developed in connection with reactors. The separation of the spin states, that is, production of polarized neutrons, can be obtained by several distinct methods, and we shall review these briefly, beginning with the *transmission* method, which for many years was the only method used.

In polarization by transmission through highly magnetized iron, the polarization of the beam is studied by measuring its transmission through a second piece of magnetized iron, the *analyzer,* as a function of magnetic field, the *double transmission effect.* The polarization produced by transmission is extremely sensitive to small deviations from magnetic saturation, and as a result the polarization can be used as a measure of the law of approach to magnetic saturation. The production of polarized neutrons by transmission involves a serious loss of intensity because of the large thicknesses of iron involved, and in addition necessitates high magnetizing fields.

It is fortunate, that other methods of production of highly polarized beams, not suffering from the disadvantages of the transmission method, have been developed. The reflectivities of the two spin states at a magnetized mirror differ,

---

[1] A. W. McReynolds and G. W. Johnson: Phys. Rev. **82**, 344 (1951).
[2] Bernstein, Roberts, Stanford and Dabbs: Phys. Rev. **94**, 1243 (1954).

hence reflection becomes a possible source of polarization. In order to produce polarized neutrons by reflection from magnetized iron, it would be necessary to use an angle of incidence that is greater than the critical angle for one spin state. Because the critical angles depend on wavelength, it would be necessary also to use monoenergetic neutrons to prevent overlapping of the critical angles. All wavelengths can be used, however, if the mirror is cobalt instead of iron. The magnetic term in Eq. (26.5) for cobalt is larger than the nuclear term, so one index will be less than unity and the other greater than unity for all wavelengths. As a result, the neutrons for which the index is greater than unity will reflect to a negligible amount, while the others will reflect totally and hence will be completely polarized.

The obvious method to demonstrate and measure the polarization of the reflected beam, is by means of a second cobalt mirror as an analyzer. For an analyzer magnetized in the same direction as the polarizer, it is expected that a completely polarized beam will reflect with no loss of intensity at the analyzer. For antiparallel magnetization of the analyzer, however, zero reflected intensity is expected because the polarized neutrons would then have an index of refraction greater than unity in the analyzer mirror. This analyzing mirror method is completely analogous to the transmission analysis already described, and hence can be considered the *double reflection effect*. Two distinct advantages of the reflection method of polarization are (1) no intensity loss occurs as in the transmission method, and (2) polarization of long wavelength neutrons is possible, unlike the transmission method, which is applicable only for $\lambda < 4.04$ Å (because iron becomes "transparent" to neutrons of longer wavelength).

Another method of production of highly polarized neutron beams, involving diffraction from magnetite (Fe_3O_4), has been demonstrated by Shull et al.[1]. The nuclear and magnetic amplitudes add algebraically in different ways for the various Bragg reflections in a crystal, according to the structure factor of the unit cell. In addition, the magnetic scattering varies from reflection to reflection because of the magnetic atomic form factor, whereas the nuclear amplitude is isotropic. Because of these variations, it is possible to find a reflection for which the form factor will be zero for one spin state (because the magnetic and nuclear amplitudes are equal but of opposite sign) and hence only the other will reflect.

Shull et al. found that the polarization was complete when monochromatic neutrons were reflected from the (220) planes of a single crystal of magnetite, the polarization being measured by transmission in an analyzing block of polycrystalline iron, in the manner we have already described. As the nuclear amplitude is 9.5×10^{-13} cm and the magnetic amplitude 9.7×10^{-13} cm for the (220) reflection, the scattering amplitude is expected to be extremely small for one spin state. The results of the transmission measurement in the analyzer showed that the polarization was 100% within an experimental uncertainty of 5%. This method of reflection from magnetite gives highly polarized, intense beams with an angular resolution of about 1° and a wavelength of about 1 Å. It would be very useful for experiments in which monochromatic polarized neutrons in this wavelength range are needed, while for higher degrees of collimation or for wavelengths greater than the crystal cutoff the mirror reflection method is superior. Single crystals of iron were also investigated by Shull et al. but the nuclear and magnetic amplitudes did not balance as closely as with magnetite, with the result that 41% polarization was the maximum obtained.

---

[1] C. G. Shull: Phys. Rev. **81**, 626 (1951). — Shull, Wollan and Koehler: Phys. Rev. **84**, 912 (1951).

V. Radiation protection at reactors.

29. Effects of radiation on tissue. Although radiations from reactors consist of neutrons, alphas, betas, and gammas, with each of these covering a wide energy range, the effects on living tissue are all related to the ionization resulting from the interactions[1] of all these radiations with matter. Electrons, for example, of the energy range usual for radioisotopes (about 1 Mev) lose energy mainly by direct ionization, as do the α-rays. Gamma-rays, however, lose energy to electrons by the processes of pair-production, Compton scattering, and production of photoelectrons, in proportions varying with the γ-energy. The electrons resulting from these processes then ionize directly, just as those from β-disintegration of radioisotopes. Neutrons, of course, produce no ionization directly, but do so indirectly by production of charged (hence ionizing) nuclei. In organic material, the production is mainly by elastic recoil of protons, although a significant contribution at low neutron energy is by means of the protons from the $N^{14}(n, p)\, C^{14}$ reaction.

For all the radiations mentioned, ionization is produced eventually, but the distribution of the ionization in tissue varies markedly with the type of radiation. Electronic ionization, whether produced by incident electrons or by those resulting from γ-rays, is distributed along the path of the electrons with a specific ionization (ion pairs per unit path length) of about 30 ion pairs per mg/cm^2 of tissue traversed. If the electrons are externally produced and incident on the body they will penetrate a distance of the order of only a few millimeters, whereas if resulting from gammas, for which the absorption is much smaller, the ionization produced by the secondary electrons is distributed more or less uniformly throughout large amounts of tissue. Because of the high specific ionization of heavy particles, such as protons, compared with electrons of the same energy, the specific ionization resulting from neutrons (i.e., from the recoil protons) in tissue is much greater than from gammas or electrons.

The biological results of the ionization vary with the density of ionization, the damage being greater when the ionization is concentrated, as for protons compared with electrons. As a result of the dependence on density, the *biological effectiveness* of a given number of ions produced by recoil protons may be as much as 10 times greater than if the same number of ions were produced more diffusely by electrons[2]. The relative biological effectiveness of the fast neutrons per ion produced or "RBE", compared with gammas, would then be as high as 10. The immediate effect of radiation of tissue is the destruction of some of the cells, hence a reduction of some bodily function; a more remote effect, however, can be an uncontrolled growth of tissue.

One of the most sensitive bodily reactions to radiation is a lowering of the white blood cell count (leucopenia), and it is for this reason that the blood count is used as a routine check of workers in a radiation environment. A result requiring higher radiation intensity, and one that was prevalent[3] after the atomic bomb explosions at Hiroshima and Nagasaki, is loss of hair. Many of those exposed to less than a fatal amount of radiation suffered loss of hair, with later

---

[1] Introduction to Pile Theory (R. D. EVANS), pp. 35—74. Cambridge, Mass.: Addison-Wesley 1952. — W. HEITLER: The Quantum Theory of Radiation, 2nd ed. Oxford 1944. — E. FERMI: Nuclear Physics, Chap. 2. University of Chicago 1950.

[2] Science and Engineering of Nuclear Power, Vol. II (R. D. EVANS), pp. 255—260. Cambridge, Mass.: Addison-Wesley 1949.

[3] The Effects of Atomic Weapons, p. 353. Washington, D. C.: Government Printing Office 1950.

complete recovery. Another effect of radiation produced by radiation below the fatal amount is sterility, which may be temporary or permanent[1].

The effects mentioned all appear soon after exposure, typically within a matter of days to weeks, but in addition to these short-term results other effects may not appear for years. The most striking of these long-term effects is the production of cancer, for example leukemia, which is an over-production of white blood cells by the cell-forming tissues in the bone narrow. Ingested radioactive materials are especially dangerous because they may cause long-term effects, such as the bone destruction resulting from ingestion of radium, which deposits in the bone narrow in a permanent manner. The deaths of radium watch-dial painters caused by radiation usually followed the intake of the radium by many years. A recent and most unfortunate effect of this type is the occurrence of cataracts in cyclotron workers several years after exposure to fast neutrons.

30. Units and limits of radiation exposure. One of the principal problems in radiation protection is the determination of "safe" limits of exposure[2]. Because of the complicated nature of the effects on living tissue, the lack of experimental data on the effects on man, and the delayed nature of some of the effects, it is exceedingly difficult to state even approximately this safe radiation limit. The question of the correct units in which to express radiation exposure is difficult in itself. This difficulty arises from the fact that the biological effect is of central interest, but the radiation intensity that produces it must be measured by physical effects, e.g., counting rates, ionization, darkening of photographic film, etc. It is not surprising therefore that biological units, expressed in terms of the effect of radiation on detecting instruments, acquire a certain complexity that requires care in their usage.

Considering the number of variables that effect radiation damage, we easily realize that the conversion of a physical unit of radiation dosage (amount of ionization in tissue) to biological effect will involve no single constant. The physically measured units always refer to some nonbiological effect in tissue, and, in fact, are usually units of energy absorption. The value of these units in terms of biological effect must therefore vary for different radiations according to the *relative biological efficiency* (RBE) for the tissue effect of interest.

The unit of radiation dosage is the *roentgen*, r, which represents the absorption of that amount of γ-radiation resulting in the production of one electrostatic unit of charge, of either sign, by ionization in 1 cm^3 (1.293 mg) of air[3]. The roentgen is thus a unit of ionization production or absorption of energy, but it is difficult to use it consistently in this manner. Usually we speak of 1 r of gammas producing 1 esu of charge in 1 cm^3 of air, which in energy units represents an absorption of 83 ergs per gram of air. This radiation, however, would produce 93 ergs per gram of water, the main constituent of tissue. The energy actually liberated in different kinds of tissues for a definite γ-intensity varies widely. For instance, a gamma dose that would be referred to as 1 r because it produces 83 ergs per gram in air would produce 42 ergs in fat and 883 in bone if the γ-energy is 12 kev. The gamma dose in each case would still be referred to as 1 r. For higher gamma energies, however, the energy liberated does not vary

[1] R. D. Evans: Nuclear Science Series, Preliminary Report No. 11, Problems Associated with the Transportation of Radioactive Substances, p. 27. National Research Council 1951.

[2] K. E. Morgan: AEC report TID 5031, pp. 176—210. Oak Ridge: Technical Information Service 1951.

[3] G. Failla: AEC report TID 388, pp. 54—58, 64—68. Oak Ridge: Technical Information Service 1951.

by such wide limits, and the reference to the roentgen as if it were a unit of gamma intensity does not lead to wide discrepancies in energy absorbed.

The r as originally defined refers only to gamma or x-ray absorption, but units equivalent to the r have been devised for other radiations. The equivalence may be in terms of a physical effect, in which case we have the roentgen-equivalent-physical (rep), defined as the absorption of 93 ergs per gram of tissue for the particular radiation considered. On the other hand the roentgen-equivalent-man (rem) is the absorption of an amount of radiation per gram of tissue that produces the same biological effect as the roentgen. Because of the great variation in biological effectiveness (RBE) for different types of tissue damage, it is obvious that no single rep-rem conversion factor exists. As a rough guide, however, it is customary to consider

$$0.1 \text{ rep} = 1 \text{ rem for ``fast'' neutrons } (\sim 2 \text{ Mev}),$$

$$\sim 0.2 \text{ rep} = 1 \text{ rem for thermal neutrons}.$$

It is useful to review some rough values for the effects of specific radiation doses, based on experience with human beings and animals. Because of the long use of x-rays and radium more of this practical information is here available than for neutrons. If the entire body receives a γ-radiation dosage of 1000 r, the actual energy absorbed is small[1], for it would raise the body temperature only 0.002° C, yet this dosage is almost certainly fatal. A dosage of 400 r to a small volume of tissue seems to do little permanent damage, but this same amount applied to the whole body, that is, as *whole-body radiation*, is about the LD 50 (60 days), which means that the radiation, if applied to a large number of human beings, would produce a 50% fatality within 60 days after the irradiation. A 50 r whole-body dosage, however, produces only a transient effect in the white blood cell count, with no permanent effect. A local dosage of 1000 r on the skin results only in a barely detectable reddening, and in tumor treatment local dosages of as much as 10000 r are used, spread over a few weeks.

It is seen that the matter of establishing a maximum permissible whole-body dosage rate, or tolerance rate, especially in view of such intangible matters as gene mutation, is practically impossible. The tolerance rate that has been picked, however, by the U.S. National Radiation Protection Committee in 1949, is 50 milliroentgens per day (or 0.3 r per week), and it seems certain that this rate, continued over many years, produces no detrimental effects. The tolerance dosage rate has been arrived at by consideration of x- and γ-ray experience, the results of accidental ingestion of radium in man, naturally occurring background radiation intensity, and animal experiments. The radiation received from cosmic rays at the surface of the earth is about $\frac{1}{500}$ of the maximum allowed rate, and other natural sources of radioactivity (soil, rocks, body constituents) increase this value to 1% of tolerance. A watch with a luminous dial (containing about a microgram of radium) produces a skin dosage rate of 100 mr per day, and the usual chest x-ray represents a dose of about 100 mr if x-ray photographic plates are used, and up to ten times this dosage with the fluorescent screen method. Calculations based on the known spontaneous rate of gene mutation shows that the allowable radiation limit will produce only a small effect of the natural rate of gene mutation. However, as is pointed out constantly by biologists and health physicists, the maximum rate is not a dosage to be received daily by all workers but a maximum rate, with the average being much less.

---

[1] R. E. LAPP and H. L. ANDREWS: Nuclear Radiation Physics, p. 435. Prentice-Hall 1948.

It is useful to express the tolerance dosage rate in terms of fluxes for the usual types of radiation. These values, taken from the results of Morgan, are given in Table 2; they are the fluxes that will produce 0.3 rem per week if the exposure lasts 24 hours per day, 5 days per week. For an eight-hour working day, the values will, of course, be three times larger. For gammas and electrons, an ionization chamber calibrated in mr/hr is more convenient than a measurement of flux in photons/cm²-sec, but for fast and thermal neutrons the flux measurement is very direct and can easily be converted to mrems/hr by Table 2.

Table 2. *Fluxes of various types of radiation corresponding to tolerance, 0.3* rem *per week* *(5 days, 24* hr/*day).*

Types fo radiation	Flux	Types of radiation	Flux
Gammas, 1 Mev . .	1300 photons/cm²-sec	Fast neutrons, 2 Mev	22 neutrons/cm²-sec
Electrons, 1 Mev . .	32 electrons/cm²-sec	Thermal neutrons. .	600 neutrons/cm²-sec

31. Methods of radiation protection. In many respects, the problem of radiation protection for reactor research is much simpler than for work with radioisotopes. The reason is simply that most experiments involve measurement of radiation, usually neutrons, at a level much below tolerance. Although the beam initially energing from the reactor may be many times tolerance, for example in the crystal diffraction apparatus, the final counting rate in the detector is only of the order of a few hundred counts per minute. Any background counting rate must be kept well below this signal counting rate or the experiment could not be performed. Thus the design of the experiment itself will tend automatically to reduce the background radiation below tolerance.

This automatic reduction of background that occurs with neutron beam experiments does not hold at all for work with radioisotopes. For example, in solid state studies a sample may be the equivalent of many grams of radium after removal from the reactor, and its physical properties must be measured while it is still of this strength. The so-called "hot labs" designed to use high-radiation samples contain various remote control equipment for measuring such intense samples. We are here not concerned with experiments involving radioisotopes, but only with the simpler radiation protection involved in reactor experiments. It is not always true, of course, that the interest of the experimenter will automatically reduce the intensity below tolerance and, in addition, there are times during the course of an experiment, for example when equipment must be installed or adjusted, when radiation hazards may be high.

The general methods of radiation protection consist of shielding, distance (simply remaining far enough away from the source of radiation), and time (waiting for the source to decay). As far as an operating experiment is concerned, the first method is the only practical one because the requirements of the experiment itself control the distance limit, and in the steady state there is no opportunity to wait for radioactive decay. The experiments we have described illustrate the manner in which collimators, designed to stop both fast and slow neutrons, are usually installed in the reactor shield itself. By this method most of the radiation is scattered out of the beam inside the shield so that the scattered radiation does not emerge into the room. If a large beam hole were open to the room and the collimation performed *outside* the shield, then the neutrons scattered in the collimator would raise the room background seriously. The smallest cross-sectional area of the beam should be located, if possible, at a position *inside* the shield, so that most of the beam is stopped before emerging into the room.

The best materials for construction of collimators are steel, which is reasonably effective for stopping fast neutrons and gammas and of course has excellent structural properties, combined with hydrogenous material for moderation of fast neutrons, and boron for capture of those so moderated. The surfaces of the collimator facing the lattice can be plated with cadmium to reduce the induced radioactivity in the collimator; this cadmium can be stripped off the collimator after removal in order to remove the activity in the cadmium itself. Paraffin containing uniformly distributed boron in some form is an excellent shield for fast neutrons because of the high slowing-down power of hydrogen and the large capture cross section of boron. Convenient shielding blocks are made of half-inch iron containing paraffin in which boric acid is uniformly mixed. These are $3 \times 6 \times 12''$ in size, so that they may be piled conveniently in various geometrical arrangements. They serve a very useful purpose as fast-neutron shielding as well as structural members of an experimental setup.

Because the stopping of fast neutrons is not a simple exponential it is impossible to give simple numerical values for the effectiveness of different materials in fast neutron shielding. For a thick shield the "tail" of the spatial distribution of moderated neutrons gives the effective decrease in neutron intensity, however. The intensity decrease in this "tail" region is expressed fairly well by $e^{-x/\lambda}$, where λ is the mean free path, given by the scattering cross section and x is distance in the shield.

In some situations it is undesirable to have shielding material in a beam because neutrons scattered by the shield may interfere seriously with the experiment. In this case the beam can be allowed to travel through the room (air scattering usually being negligible) to a so-called *beam catcher*. The general principle followed in the design of a beam catcher, as well as in any shielding of fast neutrons, is to incorporate hydrogenous material for slowing down with boron for neutron capture. A γ-ray of 0.5 Mev energy is produced when a neutron is absorbed by boron, and if this gamma production is disadvantageous lithium can be used instead, which results in no gamma production although the neutron cross section is somewhat lower than boron.

The gamma intensity in a neutron experiment is usually not very serious as far as the experiment itself is concerned, for BF_3 counters are extremely insensitive to gammas. Because of this fact, the "automatic" reduction mentioned for neutrons will not necessarily operate and the gamma background may be high even though the shielding is excellent from a neutron standpoint. For the purpose of gamma reduction the most useful material is lead, both because of

Table 3. *Shielding thicknesses for* 4.5-Mev *and* 0.7-Mev *Gamma radiation.*
Thickness of shield in inches.

Attenuation factor	4.5-Mev gamma				0.7-Mev gamma			
	water	concrete	iron	lead	water	concrete	iron	lead
0.2	30	11	3.8	1.6	12	5.1	1.8	0.7
0.1	40	15	5.2	2.3	16	6.8	2.4	1.0
0.02	70	25	8.3	3.9	25	11	3.9	1.7
0.01	80	30	9.5	4.6	29	13	4.5	2.0
0.001	110	40	14	6.7	41	19	6.4	3.1

its high cross section and its adaptability for structural use. In some cases other materials are used, usually because of particular structural reasons or cost, and it is convenient to estimate the gamma protection that will result from their use.

Because of the multiple scattering of gammas that takes place in a shield, the attenuation cannot be calculated simply. Approximate values of the shielding thickness necessary for a certain attenuation factor for common materials, at two typical gamma energies, are given[1] in Table 3. A simple exponential decrease cannot be assumed for γ-absorption because the interaction of a gamma with an atom sometimes removes it completely from the beam (photoelectric effect or pair production) and sometimes its energy is merely decreased (Compton scattering). These effects have been take into account approximately in Table 3.

General references.

[1] Bacon, G. E.: Neutron Diffraction. Oxford: Clarendon Press 1955. — Complete description of techniques and results.
[2] Blatt, J. M., and V. F. Weisskopf: Theoretical Nuclear Physics. New York: John Wiley & Sons 1952. — Thorough discussions of neutron cross-section and nuclear structure theory.
[3] Glasstone, S., and M. C. Edlund: The Elements of Nuclear Reactor Theory. New York: D. van Nostrand 1952. — General treatment of neutron behavior in reactors.
[4] Hughes, D. J.: Pile Neutron Research. Cambridge, Mass: Addison-Wesley 1953. — Most of the techniques reviewed in the present article are discussed in much greater detail in this book.
[5] Hughes, D. J., and J. A. Harvey: Neutron Cross Sections (Brookhaven National Laboratory Report, BNL 325, July 1, 1955), and D. J. Hughes and R. B. Schwartz BNL 325-Supplement No. 1 (Superintendent of Documents, Government Printing Office, Washington 25, D. C., January 1, 1957). — A compilation of many types of neutron cross sections, including essentially all those resulting from reactor research.
[6] Proceedings of the Geneva Conference, Vol. 2 Physics, Research Reactors. New York: United Nations Publications 1956. — Detailed papers on most of the world's research reactors, together with some discussion of specific research techniques.

---

[1] The Effects of Atomic Weapons, p. 353. Washington, D. C.: Government Printing Office 1950.

Robert R. Wilson:

Electron Synchrotons Additions and Corrections in Proof.

Additions:

p. 185. *At the end of Sect. 7 the following paragraph should be inserted:*

R. F. Christy (private communication (1958) has computed the *loss of particles by diffusion* out of the region of stable oscillation for the weak focusing case, $n = 0.6$, and for energies of about 1.5 Bev. He finds a 50% loss in the beam when the r-f voltage is a factor of two times that necessary to compensate for the radiation loss; a 10% loss requires a factor of roughly 2.5 in the r-f voltage; and a loss of 1% requires a factor of about 3 — all these factors being essentially independent of the particular machine energy in the range from 1.2 to 1.5 Bev and, presumably, over a much greater range of energies at large enough energies such that the characteristic damping time is small compared to the acceleration time.

p. 188. *The explanation given to the damping of vertical betatron oscillations is not correct. Replace the last paragraph of Sect. 8 by the following text:*

Vertical betatron oscillations are damped at a rate of $E_r/2E$ per turn—a result that was computed independently by Robinson and Ritson and by Kolomenski and Lebedev. The damping is not due to the extra curvature of the ion paths but comes about entirely because of the effect of the electric field of the radio-frequency accelerator, the effect being exactly the same as that just described above for the case of radial betatron oscillations, namely, a straightening out of the paths. Were the radio-frequency voltage to be turned off so that the electrons would coast, then there would be no dampening of the vertical oscillations. Furthermore, in that case, the radial oscillations would suffer the full antidamping factor of E_r/E and hence the beam size would grow rapidly in the radial direction as the beam spiralled in toward the inner wall of the vacuum chamber.

p. 190. *Between Sects. 9 and 10 the following additional Sect. 9a should be inserted:*

9a. *Intensity.* The usefulness of a synchrotron is largely dependent on the intensity of the beam of electrons accelerated to high energy. What are the limits to the number of electrons that can be accelerated? Clearly the initial source of electrons, i.e., a hot filament or a plasma, is not the limiting factor, for although space charge tends to keep the current density down, still the extracting field can be made high and the surface of the source can be large. Many amperes of electrons can be obtained from sources, especially in short pulses. Furthermore currents of the order of an ampere can be accelerated to high energy using a Van de Graaff or Linac. Thus one can expect to obtain about 10^{12} electrons in a pulse of 10^{-7} sec duration and at an energy of from 2 to 40 Mev to be used for injection into a synchrotron. Much higher pulse intensities would also be feasible.

The limiting factor in the intensity is not as yet well understood, but it does appear to occur immediately after injection—within the first few hundred µsec, say. Space charge forces, electric and magnetic, are the most obvious candidates for the role of beam destroyer. The electric field due to the beam tends to produce an expansion and the magnetic field tends to contract the beam. At highly relativistic energies the two forces would exactly cancel for an isolated beam, but other factors complicate this too-simple result—image charges and currents in the conducting walls or pole tips, neutralization or partial neutralization of the electric field by ionization of the residual gas, for example.

Actually the two fields do not exactly cancel—the fractional difference between the two decreases as mc^2/E—and the remaining force can be quite important at low injection energy. The effect of the force is to change the restoring force which holds the beam in the guide field. This change in restoring force will cause the frequency of the radial or vertical betatron oscillations to vary. When one of these frequencies becomes integral or half-integral, a resonance in the betatron oscillations will occur which will cause the beam size to increase until the space charge forces have decreased enough to take the beam off the resonance. Thus the size of the beam will depend on the current and there will be some critical or limiting current corresponding to the size of the beam reaching the size of the doughnut or other limiting aperture.

The MURA[1] workers have given a formula for this limiting current, i_{max}, which is given in amperes by

$$i_{max} = 10^4 \cdot f^{\frac{1}{2}} \nu \, \Delta \nu \, (a/R)^2 \, (E/mc^2)^3 \qquad (9a.1)$$

where f is a measure of the phase-bunching of the beam (unity for an unbunched beam, zero when the beam becomes completely bunched), ν is the number of betatron oscillations per turn, $\Delta \nu$ is the difference between ν and the nearest integral or half-integral number, a is the size of the beam when it fills the aperture, R is the ion radius, and E corresponds to the energy of the beam at the time of blow-up which will be close to injection.

As an example, for the Cornell machine ν is 2.25, $\Delta \nu$ may be about 0.2, a is about 1 cm, R is 384 cm, E at injection is 2 Mev, and let us assume f to be 0.1. The above formula gives 0.5 amp for i_{max}. In practice, the beam in the Cornell synchrotron seems to be limited by a mechanism close to injection, but at a current which is more like 10 ma. It seems unlikely that image charges and currents can explain the discrepancy of a factor of fifty.

It has been suggested that the positive ions produced in the residual gas by collisions with the beam will be trapped by the beam and will neutralize the electrical field of the beam, thereby leaving the full value of the magnetic force to act on the beam. This mechanism would change the exponent of the factor (E/mc^2) which appears in Eq. (9a.1) from three to unity and hence would account for a factor of 15 for our example of the Cornell machine. Now ionization will occur, except for unusual vacuum conditions, so the beam may be neutralized on the average. But the phase-bunching of the beam will cause the instantaneous values of the space charge field to be as much as ten times larger than the time averaged value. It is difficult to see how the positive ions can move rapidly enough to follow and neutralize an electric field having frequency components considerably higher than 80 Mc. Hence the neutralization hypothesis may not explain our discrepancy, although some curious plasma phenomenon might cause neutralization because of the motion of electrons.

T. GOLD[2] has suggested that the limitation in the intensity might arise because of the electromagnetic radiation by the electrons and in particular because of the

---

[1] MURA (Midwestern Universities Research Association): Proposal for a 15-Bev Clashing Beam Accelerator, 1958.

[2] T. GOLD: Private communication.

coherent part of the radiation. We can see by Eq. (6.9) that the energy radiated per turn in coherent radiation is independent of the energy of the electrons. Although unimportant at high energy, perhaps it could be serious at low energy.

Substituting numbers in Eq. (6.9) shows that the coherent energy radiated per turn per particle would be, say, one percent of the particle energy at injection (2 Mev) for a beam intensity of a few times 10^{12} electrons per pulse, i.e. about one ampere. If the antidamping of the radial oscillations is still given by $E_r/2E$, then the beam would blow up in a time of the order of one hundred turns, i.e. in 10 μsec for the Cornell machine. Perhaps the electric field of the coherent radiation can interact more directly with the beam to cause a loss at lower intensities. In any case, a complete theory of the effect of coherent radiation on the beam is yet to be developed.

A curious phenomenon has been observed at the California Institute of Technology with the operation of their synchrotron in its first phase of construction during which it gave a final energy of about 500 Mev. Under somewhat capricious conditions of pressure and beam current, the operators were dismayed to observe that their high energy beam fissioned into a vertical structure of two or more parts, each of which would eventually be lost to the walls. Evidently this was a manifestation of some sort of interaction between the beam and the plasma produced by the beam. G. BERNARDINI[1] has developed a theory of this interaction which may provide an explanation of these phenomena. The presently operating machines have not as yet been plagued by this disease, but it may turn up again to cause as a limitation when higher beam intensities are reached.

Finally we can mention an obvious practical limitation to the beam intensity that comes about when the beam loads down the r-f oscillator. This is a happy state of affairs in one sense, for at this point energy is being fed into the beam itself rather than to resistive losses in the copper. When the beam of the Cornell synchrotron reaches about 10^9 electrons per pulse, beam loading causes the resonator voltage to sag a few percent. For currents that are a few orders of magnitude larger, not only will the voltage sag become really significant but the reactance part of the beam loading will tend to detune seriously the cavity resonator. These effects will only raise problems of r-f technique, but their solution may become rather involved and expensive.

Misprints.

The following misprints have been overlooked in proof-reading.

p. 180, fourth line of figure caption (Fig. 5). Replace R_{Mev} by R_{m} ($=R$, measured in meters).

p. 183, Eq. (7.7). Delete comma and space before the symbol ψ. This is only *one* equation!

p. 184, Eq. (7.13). Replace the factor ($\hbar\,4R_0$) by ($\hbar c/R$).

Literature References.

p. 170, Ref. 1: Add the earlier paper of V.I. VEKSLER: C. R. Acad. Sci. USSR. 329 (1944).

p. 185, Ref. 2: The final report of KOLOMENSKY and LEBEDEV has been published in Proceedings of CERN Conference on High Energy Accelerators, Vol. 1, p. 447 (1956). — See also A. SOKOLOV and I. TERNOV: J. Exp. Theor. Phys. USSR. **28**, 431 (1955) and I.G. HENRY: Phys. Rev. **106**, 1057 (1957).

p. 186, Ref. 2: See also K.W. ROBINSON: Phys. Rev. **111**, 373 (1958).

---

[1] G. BERNARDINI, private communication.

Sachverzeichnis.

(Deutsch-Englisch.)

Bei gleicher Schreibweise in beiden Sprachen sind die Stichwörter nur einmal aufgeführt.

Ablenkplatte für den Zyklotronstrahl, *deflector for the cyclotron beam* 151.

Abschäler, *peeler* 153, 207—208.

Abschirmung zum Strahlenschutz, *shielding for radiation protection* 444—446.

— eines Zyklotrons, *of a cyclotron* 156 bis 158.

Abschirmungsringe in Bandgeneratoren, *shield rings in Van de Graaff generators* 73, 84—85.

Abschneidewellenlänge der kohärenten Neutronenstreuung, *cut-off wavelength for coherent neutron scattering* 430, 432.

Absorptionsquerschnitte für langsame Neutronen, *absorption cross sections for thermal neutrons* 424—426.

Abstand der Bänder von einander, *separation of belts* 76.

adiabatische Dämpfung, *adiabatic damping* 195—196, 203.

— — bei Betatronschwingungen, *of betatron-oscillations* 307, 315.

Admittanz, *admittance* 264, 270, 304, 307.

äquivalente Quanten (Röntgenintensität), *equivalent quanta (x-ray intensity)* 191.

AGS (alternating gradient synchrotron, Brookhaven) 326—335.

Aktivierungsgleichung, *activation equation* 403.

Aktivierungsquerschnitt für thermische Neutronen, *activation cross section for thermal neutrons* 422—424.

alternierende Feldgradienten, Theorie der Beschleuniger, *alternating gradient accelerators, theory* 300—319.

Alvarezscher Beschleunigertyp, *Alvarez-type accelerator* 366f., 386.

Analyse durch Aktivierung, *activation analysis* 424.

Anfachung durch Strahlung, *anti-damping due to radiation* 186.

Anregungsstärke der Resonanzen, *strength function of resonances* 418.

Arbeitszyklus eines Linearbeschleunigers, *operating cycle of a linear accelerator* 360, 381—382.

Aufladen des Bandes, *charging of a belt* 76 bis 78.

Aufladeband, *charging belt* 74—79.

Aufladestrom eines Generators, *charging current in a generator* 74.

Aufrechterhaltung des homogenen Feldes im Bandgenerator, *column gradient establishment* 73—74.

Ausbeute der eingeschossenen Teilchen, *efficiency of injection* 273.

AVF-Zyklotron s. azimutabhängiges Zyklotron, *AVF cyclotron see azimuthally varying field cyclotron*.

axiale Fokussierung im Betatron, *axial focussing in a betatron* 195, 202.

axiale Schwingungen s. vertikale Schwingungen, *axial oscillations see vertical oscillations*.

azimutabhängige Korrektionen, *azimuthal trimming* 243.

azimutabhängiges Zyklotron, *azimuthally varying field (AVF) cyclotron* 112.

azimutale Harmonische im Synchrotron mit konstantem Feldgradienten, *azimuthal harmonics in constant gradient synchrotron* 241.

Bänder, Abstand von einander, *belt separation* 76.

—, Auswechseln derselben, *replacement* 76.

Bahnstabilität im azimutabhängigen Zyklotron, *orbital stability in the AVF cyclotron* 112—115.

— im Zyklotron, *in cyclotrons* 107—121.

Band s. auch Aufladeband, *belt see also charging belt*.

—, Feld an der Oberfläche, *field at its surface* 75.

—, mechanische Kräfte darauf, *mechanical forces on it* 76.

Bandanordnung im Generator, *belt arrangement in a generator* 74—76.

Bandgeneratoren, Anordnungen, *Van de Graaff generators, design* 68—69, 70, 83, 96, 98—104.

—, aufrecht stehende, *vertical* 79—80.

—, früher offener Typ, *early open-air type* 66—67.

— mit Gasfüllung unter hohem Druck, *with high pressure gas* 67—71.

—, liegende, *horizontal* 80.

— als Synchrotron-Injektoren, *as synchrotron injectors* 176.

Bandstabilität, *belt stability* 76.

Bandtrommel als Induktion benutzt, *pulley used as inductor* 77, 79.

Subject Index.

(English-German.)

Where English and German spelling of a word is identical the German version is omitted.